Rules of Differentiation

Derivatives of Exponential Functions

$$\frac{d}{dx}(e^u) = e^u \frac{du}{dx}$$

$$\frac{d}{dx}(b^u) = (\ln b) b^u \frac{du}{dx}$$

Derivatives of Logarithmic Functions

$$\frac{d}{dx}(\ln u) = \frac{1}{u} \frac{du}{dx}$$

$$\frac{d}{dx}(\log_b u) = \frac{1}{\ln b} \frac{1}{u} \frac{du}{dx}$$

Derivatives of Trigonometric Functions

$$\frac{d}{dx}(\sin u) = \cos u \frac{du}{dx}$$

$$\frac{d}{dx}(\cos u) = -\sin u \frac{du}{dx}$$

$$\frac{d}{dx}(\tan u) = \sec^2 u \frac{du}{dx}$$

$$\frac{d}{dx}(\csc u) = -\csc u \cot u \frac{du}{dx}$$

$$\frac{d}{dx}(\sec u) = \sec u \tan u \frac{du}{dx}$$

$$\frac{d}{dx}(\cot u) = -\csc^2 u \frac{du}{dx}$$

CALCULUS
for
Management and the Life and Social Sciences

Index of Referenced Applications

The Wall Street Journal	Corporate Forecasts — 13
The Wall Street Journal	Costs — 14
American Scientist, March–April 1986	Memory Modules Kept Busy — 27
The Wall Street Journal	Stockbroker Fees — 31
Physician's Desk Reference, 1984	Erythromycin Dosage — 67
Higgins, *Analysis for Financial Management*, 1984	Return on Equity — 59
Scientific American, March 1986	Visual Reaction Time — 60
Scientific American, November 1986	Visual Reaction Time — 60
Newsweek	Greenhouse Effect — 61
The Wall Street Journal, September 29, 1988	Worldwide Platinum Use — 62
The Wall Street Journal, September 20, 1988	Stockbroker Salaries — 62
Higgins, *Analysis for Financial Management*, 1984	Determining Gross Profit — 63
Scientific American, January 1983	Growing Mammalian Cells — 64
Scientific American, October 1987	Cost in High Speed Computers — 65
Stevenson, *Production/Operations Management*, 1986	Plant Costs — 65
Scientific American, February 1987	Life Science Ph.D.s by Gender — 66
Scientific American, July 1987	Saving the U.S. Steel Industry — 67
American Scientist, July–August, 1988	Parallel Processing Speeds — 119
Scientific American, June 1987	Bond Duration — 140
Higgins, *Analysis for Financial Management*, 1984	Sustainable Growth Rates — 204
Budnick, McCleavey and Mojena, *Principles of Operations Research for Management*, 1988	Useful Life of Production Tools — 260
The Wall Street Journal	Yield Curves of Government Securities — 273
American Scientist, July–August 1988	Parallel Processing Speeds — 275
American Scientist, May–June 1988	Survival of Bald Eagle Colonies — 303
Coleman, *Introduction to Mathematical Sociology*, 1964	Consumer Purchases vs. Income — 311
Higgins, *Analysis for Financial Management*, 1984	Internal Rate of Return — 313
Stigum, *Money Market Calculations*, 1981	Average Daily Yield — 314
Scientific American, January 1987	Nuclear Testing — 321
Budnick, McCleavey and Mojena, *Principles of Operations Research for Management*, 1988	Length of Advertising Campaigns — 328
Budnick, McCleavey and Mojena, *Principles of Operations Research for Management*, 1988	Optimal Holding Period — 320
Physician's Desk Reference, 1984	Administering Anesthesia — 342
McGann and Russell, *Advertising Media*, 1988	Advertising Learning Curves — 343
Scientific American, December 1986	Cancer Drug Toxicity — 344
Scientific American, October 1987	Number of Components on Chips — 344
Scientific American, September 1988	Nuclear Strategy — 345
Chase and Aquilano, *Production and Operations Management*, 1985	Learning Curves — 345
Scientific American, January 1987	Nuclear Testing — 348
Scientific American, June 1986	Ideal Windmills — 352
Scientific American, April 1988	Energy-Efficient Buildings — 360
American Scientist, July–August 1988	A General Theory of Hurricanes — 386
American Scientist, July–August 1987	Computer Models of AIDS Epidemiology — 438
Niebel, *Motion and Time Study*, 1988	Machine Tending Times — 487
Niebel, *Motion and Time Study*, 1988	Object and Background Contrast — 488
McGann and Russell, *Advertising Media*, 1988	Duplication of Advertising Audience — 506
Scientific American, June 1988	Birth Rate vs. Stork Population — 553

Index of Referenced Applications

Scientific American, February 1987	
1989 World Almanac	
The Wall Street Journal, October 3, 1988	
The New York Times, July 3, 1988	
Scientific American, June 1988	
Scientific American, March 1986	
Scientific American, February 1987	
American Scientist, May–June, 1988	
American Scientist, July–August, 1988	
Coleman, *Introduction to Mathematical Sociology*, 1964	
Niebel, *Motion and Time Study*, 1988	
Scientific American, May 1986	
Niebel, *Motion and Time Study*, 1988	
Scientific American, May 1988	
Stevenson, *Production/Operations Management*, 1986	

Food Stamp Program Participation	553
World Gold Production	553
U.S. Oil Company Profits	553
Voter Participation	554
Ozone Levels over Antarctica	554
Nuclear Power Plant Construction Time	555
Distribution of Food to the Poor	556
Survival of Bald Eagle Colonies	557
Shoreham, NY Nuclear Power Plant Cost	557
Transition Rates	579
Object and Background Contrast	596
Males in a Family Who Marry	638
Machine Down-Time	647
Controlling Indoor Pollution	661
Inventory Control	676

The Irwin Series in Quantitative Analysis for Business
Consulting Editor   Robert B. Fetter   Yale University

Applied College Mathematics Series

This text is one of a collection intended for the study of applied finite mathematics and calculus at the introductory level. The only prerequisites assumed are $1\frac{1}{2}$ years of high school algebra or its equivalent. Each text contains a review of algebra which may be used selectively as needed.

Mathematics for Management and the Life and Social Sciences is a comprehensive introduction to applied finite mathematics and calculus appropriate for a two semester or three quarter course.

Calculus for Management and the Life and Social Sciences is intended for a two semester course in applied calculus. The topics in the later portion of the text are independent so that this text would also be appropriate for an ambitious one semester course with some topical omissions.

Brief Calculus for Management and the Life and Social Sciences is appropriate for a one semester course in applied calculus. It covers the first eight chapters of *Calculus*.

▲ ▼ ▲

CALCULUS
for
Management and the Life
and Social Sciences

Second Edition

Donald L. Stancl
Mildred L. Stancl
Both of St. Anselm College

Homewood, IL 60430
Boston, MA 02116

Photos: Journalism Services, Inc.

Cover photo (*I.M. Pei Pyramid at the Louvre, Paris*), © Imapress/N'Diaye

Chapter 1 (*New York*), © Matthew Rosenzweig, 2
Chapter 2 (*Chicago*), © Mike Kidulich, 88
Chapter 3 (*Munich*), © Fotex/Taubenberger, 176
Chapter 4 (*Calgary*), © Rinna Borgsteede, 226
Chapter 5 (*St. Louis*), © James Blank, 284
Chapter 6 (*Paris*), © Dave Brown, 364
Chapter 7 (*London*), © Gills J. Copeland, 426
Chapter 8 (*Paris*), © Imapress/N'Diaye, 476
Chapter 9 (*Batavia, IL*), © H. Rick Bamman, 558
Chapter 10 (*San Francisco*), © Snyder Photographic, 618
Chapter 11 (*Sydney, Australia*), © Dirk Gallian, 690
Chapter 12 (*Seattle*), © Dave Brown, 762

© RICHARD D. IRWIN, INC., 1988 and 1990

All rights reserved. No part of this publication may be reproduced, stored in a retrieval system, or transmitted, in any form or by any means, electronic, mechanical, photocopying, recording, or otherwise, without the prior written permission of the publisher.

Sponsoring editor: *Richard T. Hercher, Jr.*
Developmental editor: *Jim Minatel*
Project editor: *Waivah Clement*
Production manager: *Bette K. Ittersagen*
Interior designer: *Lucy Lesiak Design*
Cover designer: *Michael Finkelman*
Artist: *Benoit Design*
Compositor: *Arcata Graphics/Kingsport*
Typeface: *10/12 Times Roman*
Printer: *R. R. Donnelley & Sons Company*

Library of Congress Cataloging-in-Publication Data

Stancl, Donald L.
    Calculus for management and the life and social sciences / Donald L. Stancl, Mildred L. Stancl.—2nd ed.
      p.   cm.
    ISBN 0-256-08251-0
    1. Calculus. I. Stancl, Mildred L. II. Title
QA303.S833   1990
515—dc20   89–35298
                                                    CIP

Printed in the United States of America

1 2 3 4 5 6 7 8 9 0 DO 7 6 5 4 3 2 1 0

To each other

Preface

Calculus for Management and the Life and Social Sciences, Second Edition, is an applications-oriented text for an introductory calculus course addressed to students of business, management, economics, the life sciences, and the social sciences. The only prerequisite for the book is $1\frac{1}{2}$ to 2 years of high-school algebra or the equivalent.

Why We Wrote This Book. We believe that students who take the course for which this book is designed want to see and understand as many practical applications of mathematics as possible. They do not want to learn mathematics for its own sake, nor will they be satisfied with promises that the mathematics they are being taught will be useful at some indeterminate future time. On the other hand, we also believe that a "cookbook" approach to the subject—one that presents applications with little or no attempt to explain the concepts involved or to increase the mathematical sophistication of the reader—will not serve students well in the long run. Students must learn more than just the applications presented in the text: they must learn how to think mathematically so that they will possess the ability and confidence to use mathematics in problem solving. We wrote this book because we felt that other texts in the field did not strike the appropriate balance between the presentation of applications and the development of the student's ability to think mathematically. We have attempted to create such a balance by emphasizing mathematical modeling: each mathematical topic is introduced with and illustrated by as many practical applications as possible, but we pay as much attention to the construction and use of the relevant models as we do to the applications themselves. The result is an approach that not only presents a great variety of applications but does so in a manner that encourages the development of the student's ability to use mathematics in problem solving.

Features of the Book. Some noteworthy features of the book are:

Extensive Development Process. The book has undergone an extensive development process that began with prepublication reviews of the first edition. These were followed by postpublication reviews of the first edition, conversations with its users and other mathematicians, and surveys of opinions regarding what should be included in a book of this type. Using these sources as input, we prepared a draft of the second edition, which was itself extensively reviewed. The book was then rewritten to reflect the criticisms and suggestions of the reviewers. The revised version was reviewed again, and again revised in response to the reviewers, to produce the final version of the second edition. We believe that because of this careful development process this edition reflects the best current thinking as to what a text in this field should be.

Extensive Review Material. We feel that many books in the field do not contain enough review material for today's students with their diverse mathematical backgrounds. Accordingly we have provided in the first chapter of the book an unusually extensive review. For a discussion of the Chapter 1 review, see the section on Content below.

Examples. We have included substantially more examples than most other books in the field. There are over 500 examples, many of them applied. As soon as a new mathematical topic is introduced we illustrate it with mathematical examples and, when appropriate, with applied examples.

Exercises. Reviewers have commended the variety, choice, and number of exercises, especially the applied ones. There are over 3,900 exercises in this text, nearly 1,450 of them applied. Many of the applied exercises have several parts, and some 120 of them are taken from the literature and are referenced. At the end of each section, drill exercises appear first, followed by applied exercises organized by discipline (management, life science, and social science). Drill exercises and those organized by discipline are graded in difficulty and offer problems that will challenge students at every level. The most challenging exercises are marked by asterisks.

Exercises on New Topics. In order to make the exercise sets more interesting and keep the sections to a manageable length, we sometimes use the exercises to introduce topics and applications that are not treated in the text proper. Exercises that are concerned with such new topics are also marked by asterisks.

Figures. Whenever we felt that an explanation could be enhanced by a figure, we included the figure. The text contains nearly 500 figures used to illustrate concepts, substantially more than comparable texts.

Pedagogical Features. New terms are printed in **boldface.** Important definitions, facts, formulas, theorems, algorithms, and "keys to" are enclosed in boxes for easy reference. The rules of differentiation are listed in the front endpapers, and a table of integrals appears in the back endpapers.

Chapter Summaries. At the conclusion of each chapter we have included a summary that outlines the terms, notations, facts, and techniques that the student should know. These summaries can be used by the student as a study guide and checklist before attempting the review exercises for the chapter.

Review Exercises. Following each summary is a set of review exercises that covers all material discussed in the chapter. These exercises provide a thorough test of the student's mastery of the material covered in the chapter.

Additional Topics. Following the review exercises in each chapter is a list of additional topics. These topics are not treated in the text proper but might be of interest to some students. Each additional topic suggests a theorem, a technique, a concept, an application, or a historical subject for the student to investigate. The instructor can use the additional topics as a basis for short written or oral reports or for longer papers that might be appropriate in a writing across the curriculum program. They are particularly useful as a source of extra-credit assignments.

Chapter Supplements. Each chapter except the first is followed by a supplement containing additional material not usually covered in a textbook of this type. These supplements consider such diverse topics as consumption and the multiplier (Chapter 3), qualitative solutions of differential equations (Chapter 4), and the present value of a continuous stream of income (Chapter 7). Two of them are concerned with the historical development of the calculus. Each supplement is designed to build upon and enrich the material in its chapter. The supplements can be covered in class, but they are also appropriate for assignment as outside reading or as a basis for student presentations to the class. Each of the nonhistorical supplements has its own exercises, graded by difficulty.

Solutions to Exercises. Solutions to most odd-numbered section and supplementary exercises and to all review exercises may be found at the back of the book.

Applications Index. An applications index gives the location of all applications (whether they appear as examples or exercises), grouped by discipline.

Content. The first chapter of the book is a review chapter: its five sections review the real number system and its properties, functions and graphing, algebra, linear functions, and quadratic functions. Each of these sections is itself divided into subsections. The result is that the chapter is organized in such a manner that the instructor may cover all of it or selected portions of it in class, or omit it entirely, depending on the background of the students. Those portions not covered in class can be used by students on an individual basis for review as needed. Our goal with this chapter has been to provide all the review that students might possibly need while allowing the instructor the maximum amount of flexibility in his or her use of the material.

Following the review chapter, the text covers differential calculus (Chapters 2 through 4), exponential and logarithmic functions (Chapter 5), integral calculus (Chapters 6 and 7), multivariable calculus (Chapter 8), differential equations (Chapter 9), probability (Chapter 10), trigonometric functions (Chapter 11), and sequences and series (Chapter 12). A dependency diagram follows:

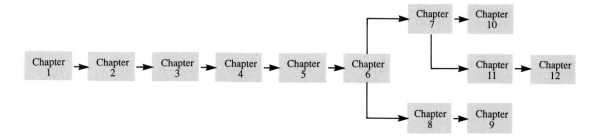

Characteristics of the Book. Many characteristics of this book have been praised by reviewers. Some of these include:

Emphasis on Modeling. As explained above, we emphasize the construction and use of mathematical models as an aid toward development of the student's ability to use mathematics. As part of this process, we strive whenever possible to show the reasoning that goes into the creation of the model so that the student may learn how to analyze practical situations mathematically.

Applications. Wherever possible, we illustrate the mathematical topic under discussion with one or more applications. We have made every effort to provide diverse and stimulating new applications as well as the standard ones that must be offered in a book of this type. Reviewers have praised both the variety and quantity of our applications.

Flexibility. We feel that this book is more flexible than other texts in the field. As noted above, the large amount of review material and its arrangement allow the instructor to tailor the review to the specific needs of the students. The additional topics and chapter supplements contribute to the flexibility of the book by allowing the instructor to individualize the course and vary it somewhat from year to year while still covering all required material.

Coverage. The book covers all the standard topics found in other books of this type, but offers more than others do in the additional topics and chapter supplements. It has more figures, examples, and exercises, and especially more applied exercises, than other texts.

Conversational Style. We have written the book using a straightforward conversational style. We have tried to avoid being overly chatty or cute. Many reviewers have remarked that our writing style is pleasant and appropriate to the material.

Changes in the Second Edition. In response to comments and suggestions from reviewers and users of the first edition, we have made the following changes in the second edition:

Figures, Examples, and Exercises. We have increased the number of figures, examples, and exercises in the new edition. In particular, we have substantially increased the number of applied examples and exercises and have included referenced exercises.

Content and Organization. We have rewritten many portions of the text (e.g., the section on relative optima) to tighten arguments, enhance precision, and clarify explanations. We have also added new material in several places and reorganized some chapters. Thus, Chapter 1 now contains more material on composite functions and new material on one-to-one and inverse functions, as well as an enhanced treatment of quadratic functions; we have dropped the treatment of rational functions from Chapter 1, preferring to treat these later with the methods of calculus. In Chapter 2 we have expanded the discussion of limits by creating a new Section 2.2 devoted to one-sided limits, infinite limits, and limits at infinity, and have dropped the section devoted to the Newton-Raphson method. Chapter 3 of the first edition has been split up into two chapters of more manageable length: a new Chapter 3, which covers the product rule, quotient rule, chain rule, differentials, implicit differentiation, and related rates, and a new Chapter 4, which covers relative and absolute optima and curve sketching. The supplement to Chapter 4 is new. The material on exponential and logarithmic functions (Chapter 5) has been extensively rewritten to provide more examples, a better discussion of bases other than e, and more emphasis on modeling. The supplement to Chapter 5 is new. The discussion of the definite integral as the limit of Riemann sums has been expanded and relocated at the end of the introductory chapter on integration as the new Section 6.5. The order of the topics covered in the chapter "More about Integration" has been changed. In the chapter on multivariable calculus (Chapter 8), the material on graphing has been simplified, and computer-generated graphs have been included. The explanation of the geometric significance of the first and second partial derivatives has also been expanded, as has the discussion of least-squares regression in the chapter supplement. In the chapter on differential equations (Chapter 9), the material on linear and exact equations has been enhanced and split into two new sections (Sections 9.3 and 9.4), one devoted solely to linear equations and the other to the total differential and exact equations. The chapter on trigonometric functions has been relocated to follow, rather than precede, the probability chapter, and a new section (Section 11.3) devoted to the trigonometric functions and their graphs has been added. The supplement to Chapter 11 is new.

Pedagogical Features. The chapter summaries and additional topics are features new to this edition.

Supplements. There are several supplements designed to accompany this text:

Student Solution Manual. The *Student Solution Manual* contains complete, annotated solutions to representative odd-numbered exercises from the text, as well as hints for solving other exercises. It is intended to help the student master the problem-solving techniques as well as to provide solutions to exercises.

Computer Manual. We have written an updated computer manual to accompany this edition of the book. The manual includes two IBM PC-formatted disks containing BASIC programs that illustrate the material in the text, as well as descriptions of these programs and directions and suggestions for their use. The programs allow students to solve exercises that require a great deal of computation, to check their answers to exercises from the text, and to experiment with many of the mathematical models covered in the text.

Instructor's Solution Manual. The *Instructor's Solution Manual* contains complete solutions to all exercises in the text and lecture notes for each section with examples and notes on subject matter that gives students difficulty.

Test Bank. A test bank is available in both a printed and computerized version. The printed version includes approximately 70 test questions per chapter. For details on use of the computerized test bank, which is known as COMPUTEST, please contact the publisher.

Complete Instructor's Package. The complete instructor's package consists of the text samples of all of the supplements, and the syllabi templates for your course, all packaged together in a handy three-ring binder format.

Companion Texts. We have prepared a shorter version of this text suitable for one-semester calculus courses; it is entitled *Brief Calculus for Management and the Life and Social Sciences, Second Edition,* and it consists of the first eight chapters of this book. We have also written a text entitled *Mathematics for Management and the Life and Social Sciences* designed for a one-year course in finite mathematics with calculus. This book covers matrices and linear programming, probability and statistics, the mathematics of finance, and calculus. Both of these texts are published by Richard D. Irwin, Inc.

Acknowledgments. Many people contributed to this book. We would particularly like to thank the reviewers. Reviewers of the first edition were: Ray Boersema, Front Range Community College; Genelle Beck, University of Texas, Austin; Larry Bowen, University of Alabama; Joseph Sukta, Moraine Valley Community College; Howard Frisinger, Colorado State University; John Hodges, University of Arkansas, Little Rock; John Hornsby, University of New Orleans; Gail Earles, St. Cloud State University; Donald Trahan, Naval Postgraduate School; Thalia Papageorgiou, Bentley College; Thomas Caplinger, Memphis State University; David Pentico, Duquesne University; Enrique Venta, Loyola University of Chicago; Robert L. Moore, University of Alabama; Jean Clarke, Virginia Commonwealth University.

The reviewers of the second edition were: Larry Bowen, University of Alabama; Chris Burditt, Napa Valley College; Angela Cobb, Sinclair Community College; Henry Decell, Jr., University of Houston; Joe S. Evans, Middle Tennessee State University; Betty L. Fein, Oregon State University; Alice M. Hughey, Brooklyn College/City University of New York; Paula Kemp, Southwest Missouri State University; John Krajewski, University of Wisconsin—Eau Claire; Giles Wilson Maloof, Boise State University; William Miller, Central Michigan University; Robert Moreland, Texas Tech University; Wayne B. Powell, Oklahoma State University; Winston A. Richards, The Pennsylvania State University at Harrisburg; Daniel Scanlon, Orange Coast College; Sally Sestini, Cerritos College; Paula R. Sloan, Vanderbilt University; Robert C. Smith, University of Wisconsin—Eau Claire; Thomas J. Woods, Central Connecticut State University; Earl J. Zwick, Indiana State University.

The patience of the reviewers in reading the various versions of the manuscript was admirable and their comments, criticisms, and suggestions were of the utmost value to us. Much of what is good about this book is due to them, and we gratefully acknowledge their contributions. We also thank those who took the time to complete one of our three surveys and otherwise provide insights for the development of this text.

We wish also to convey our thanks for their fine work to those who produced the supplements: to Daniel Scanlon and James Balch for the *Instructor's Solution Manual* and the *Student Solution Manual;* to Larry Bowen for the diagnostic pretests; to W. H. Howland and Giles Maloof for the Test Bank; and to Babak Ghayour for the lecture notes. Our sincere thanks also to Champak Panchal, Paul Patten, Winston Richards, Daniel Scanlon, and Mike Shirazi for solving the exercises; to Harold Bennett for verifying the accuracy of the examples; and to Mark Bridger of Bridge Software for the computer-generated artwork in Chapter 8.

Finally, we gratefully acknowledge the contributions of the superb professionals at Richard D. Irwin, Inc., who had so much to do with the making of this book. Sincere thanks to our sponsoring editor, Richard T. Hercher, Jr., for his unfailing support and enthusiasm. Special thanks also to our developmental editor, James Minatel, whose efforts on behalf of the book have been exceptional and whose patience with his authors has been monumental. We thank the copy editor, Tom Whipple, who did a fine job with a heavily rewritten manuscript. We are indebted also to our project editor, Waivah Clement, who patiently accepted our changes as she put the book together, as well as to the production manager, Bette Ittersagen. Together they have produced an exceptionally attractive book. Our heartfelt appreciation to these people and all the others at Irwin who worked so competently and to such good effect on this book.

Donald L. Stancl
Mildred L. Stancl

Contents

Chapter 1
Numbers and Functions, 3

1.1 Real Numbers, 4
1.2 Functions and Graphing, 16
1.3 Algebra, 33
1.4 Linear Functions, 48
1.5 Quadratic Functions, 68
Summary, 80
Review Exercises, 82
Additional Topics, 85

Chapter 2
The Derivative, 89

2.1 Limits, 90
2.2 One-Sided Limits, Infinite Limits, and Limits at Infinity, 103
2.3 The Derivative, 119
2.4 Rules of Differentiation, 132
2.5 Rates of Change, 140
2.6 The Theory of the Firm, 150
2.7 Differential Notation and Higher Derivatives, 163
Summary, 166
Review Exercises, 168
Additional Topics, 171
Supplement: Newton and Leibniz, 172

Chapter 3
More about Differentiation, 177

3.1 The Product Rule and the Quotient Rule, 178
3.2 The Chain Rule, 185
3.3 Differentials, 195
3.4 Implicit Differentiation, 205
3.5 Related Rates, 210
Summary, 218
Review Exercises, 219
Additional Topics, 220
Supplement: Consumption and the Multiplier, 221

Chapter 4
Applications of Differentiation, 227

4.1 Relative Maxima and Minima: The First Derivative Test, 228
4.2 Relative Maxima and Minima: The Second Derivative Test, 243
4.3 Absolute Maxima and Minima, 249
4.4 Curve-Sketching, 261
Summary, 275
Review Exercises, 276
Additional Topics, 279
Supplement: Qualitative Solutions of Differential Equations, 280

Chapter 5
Exponential and Logarithmic Functions, 285

5.1 Exponential Functions, 286
5.2 Logarithmic Functions, 303
5.3 Differentiation of Logarithmic Functions, 314
5.4 Differentiation of Exponential Functions, 322
5.5 Exponential Growth and Decline, 330
Summary, 348
Review Exercises, 350

Additional Topics, 353
Supplement: *S*-curves, 354

Chapter 6
Integration, 365

6.1 The Indefinite Integral, 366
6.2 Finding Indefinite Integrals, 373
6.3 The Definite Integral, 379
6.4 The Definite Integral and Area, 388
6.5 Riemann Sums, 403
Summary, 415
Review Exercises, 417
Additional Topics, 419
Supplement: The Development of the Definite Integral, 421

Chapter 7
More about Integration, 427

7.1 Consumers' Surplus, Producers' Surplus, and Streams of Income, 428
7.2 Integration by Parts and Tables of Integrals, 440
7.3 Improper Integrals, 448
7.4 Numerical Integration, 453
Summary, 466
Review Exercises, 467
Additional Topics, 470
Supplement: The Present Value of a Stream of Income, 470

Chapter 8
Functions of Several Variables, 477

8.1 Functions of More than One Variable, 478
8.2 Partial Derivatives, 489
8.3 Partial Derivatives as Rates of Change, 496
8.4 Optimization, 507
8.5 Lagrange Multipliers, 517
8.6 Double Integrals, 525
Summary, 539
Review Exercises, 541

Additional Topics, 543
Supplement: Least Squares, 544

Chapter 9
Differential Equations, 559

9.1 Solving Differential Equations, 560
9.2 Separable Equations, 567
9.3 Linear Equations, 580
9.4 Total Differentials and Exact Differential Equations, 588
9.5 Graphical Solutions, 598
9.6 Numerical Methods, 604
Summary, 610
Review Exercises, 611
Additional Topics, 612
Supplement: A Predator-Prey Model, 614

Chapter 10
Probability, 619

10.1 Probabilities, 620
10.2 Random Variables, 630
10.3 The Binomial and Poisson Distributions, 639
10.4 Continuous Random Variables, 649
10.5 The Normal and Exponential Distributions, 662
Summary, 678
Review Exercises, 680
Additional Topics, 683
Supplement: The Fundamental Theorem of Natural Selection, 683

Chapter 11
Trigonometric Functions, 691

11.1 Measuring Angles, 692
11.2 Trigonometry, 700
11.3 The Trigonometric Functions, 716
11.4 Differentiation of Trigonometric Functions, 727

11.5 Integration of Trigonometric Functions, 741
Summary, 752
Review Exercises, 753
Additional Topics, 755
Supplement: Periodicity and the Differential Equation $y'' + ky = 0$, 756

Chapter 12
Sequences and Series, 763

12.1 Sequences, 764
12.2 Infinite Series of Constants, 776
12.3 Power Series, 791
12.4 Taylor Series, 805
Summary, 815
Review Exercises, 817
Additional Topics, 819
Supplement: Discrete Income Streams, 820

Tables, 829

Solutions to Exercises, 833

Indexes, 881

CALCULUS
for
Management and the Life
and Social Sciences

1

Numbers and Functions

A function is a rule that describes a relationship. For example, a function might describe the relationship between

- The number of units of product a firm sells and its profits.
- The size of a city's population and the amount of pollution in its air.
- A society's intake of cholesterol and its incidence of heart disease.

As these examples suggest, functions are often used to model practical situations, so we must be able to analyze them. Calculus is a powerful tool for analyzing functions. As we proceed through this book, we will show how calculus can be used to study functions and to determine the rate of change of one quantity with respect to another, the maximum or minimum value that a quantity can attain, and many other interesting and important things. However, before we begin, we must develop some familiarity with functions and their properties. This is our goal in Chapter 1.

Since our functions will be described in terms of real numbers, we begin this chapter with an introduction to the system of real numbers and some of its properties. We then provide a general discussion of functions and their graphs, consider the algebra of functions, and conclude with brief studies of linear and quadratic functions.

1.1 REAL NUMBERS

We will be working with real numbers throughout this book, so we must know what they are and have an understanding of some of their properties. In this section we discuss the set of real numbers, the real line, inequalities among real numbers, intervals, the concept of the absolute value of a real number, and exponents and roots.

The Set of Real Numbers

A **set** is a collection of objects; the objects are called **elements** of the set. If object x is an element of set A, we write $x \in A$ and say that "x belongs to A." If object x is not an element of set A, we write $x \notin A$ and say that "x does not belong to A."

Often it is possible to specify a set by listing all its elements between braces. For example, if A is the set whose elements are the numbers -1, 0, and 1, then we may write $A = \{-1, 0, 1\}$. Note that $1 \in A$, but $2 \notin A$. If it is not convenient or possible to specify a set by listing its elements, then we may use the **set-builder notation** $\{x | x \text{ has property P}\}$. This is read as "the set of all x such that x has property P." For instance, the set $B = \{5\}$ may be written as $B = \{x | 2x - 10 = 0\}$, since the only object x that satisfies the equation $2x - 10 = 0$ is the number 5.

Two important sets of numbers are

$$\text{the } \textbf{natural numbers} \text{ (or } \textbf{positive integers)} \ \mathbf{N} = \{1, 2, 3, \ldots\}$$

and

$$\text{the } \textbf{integers} \ \mathbf{Z} = \{0, 1, -1, 2, -2, 3, -3, \ldots\}.$$

(The three dots . . . mean "and so on.") Another important set of numbers is the set $\mathbf{Q}$, which consists of all quotients of integers. Some elements of $\mathbf{Q}$ are

$$\frac{0}{1} = 0, \quad \frac{1}{2}, \quad -\frac{3}{5}, \quad -\frac{22}{33}, \quad \frac{319}{100}, \quad \frac{466}{13}, \quad \text{and} \quad \frac{14}{7} = 2.$$

Numbers that can be written as quotients of integers are called **rational numbers**. Numbers that cannot be written as quotients of integers, such as $\sqrt{2}$ and π, for example, are called **irrational numbers**. The set $\mathbf{Q}$ is thus referred to as the set of rational numbers, and we may write $\mathbf{Q}$ in set-builder notation as

$$\mathbf{Q} = \left\{ \frac{m}{n} \ \middle| \ m \in \mathbf{Z}, n \in \mathbf{Z}, n \neq 0 \right\};$$

note that $n \neq 0$, because division by 0 is not defined. Note also that every integer is an element of $\mathbf{Q}$, since every integer m may be written as $m/1$.

The set $\mathbf{R}$ of **real numbers** consists of all rational numbers together with all irrational numbers. Thus $\mathbf{R}$ contains all integers, all fractions of integers, and all irrational numbers such as $\sqrt{2}$, π, $-\sqrt[3]{5}$, and so on. Both rational and irrational numbers have decimal representations, such as

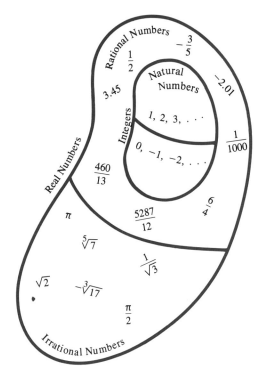

FIGURE 1–1
The Real Number System

$$\frac{1}{2} = 0.5, \quad \frac{2}{3} = 0.6666 \ldots, \quad \text{and} \quad \sqrt{2} = 1.414213 \ldots,$$

so every real number has a decimal representation. On the other hand, it can be shown that every decimal number represents either a rational number or an irrational number; hence we may think of the set **R** as consisting of all decimal numbers. Figure 1–1 shows the relationships among the number sets we have discussed here. From now on when we write "number," we shall mean "real number," and a set written in the form

$$\{x | x \text{ has property P}\}$$

will be understood to mean the set of all real numbers x such that x has property P.

The Real Line

The set of real numbers may be represented by a horizontal line called the **real line.** See Figure 1–2. Each number corresponds to a point on the line, and each point on the line corresponds to a number. For example, in Figure 1–3 we have

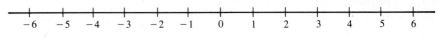

FIGURE 1–2
The Real Line

FIGURE 1–3

plotted the numbers $-\sqrt{5} = -2.23\ldots$, $-7/4$, -1, $-1/2$, 0, $2/3$, 1.75, 2, 2.8, and $\pi = 3.14\ldots$ on the real line.

Inequalities

The real line makes it possible for us to compare numbers using **inequalities.** If a and b are numbers and a is to the left of b on the real line, then we say that a **is less than b** (written $a < b$) or that b **is greater than a** (written $b > a$). See Figure 1–4. If $a < 0$, then a is a **negative** number; if $a > 0$, then a is a **positive** number. An expression of the form $a \leq b$ means that a is less than b or equal to b. Similarly, $b \geq a$ means that b is greater than a or equal to a.

EXAMPLE 1   The following are true:

$$-3 < -2, \quad -1 < -\frac{2}{3}, \quad 1 \leq 1, \quad 1 < 6.5 \quad 2 \geq 0, \quad \frac{5}{4} > \frac{9}{8}.$$

The following are false:

$$-3 < -4, \quad 1 < \frac{2}{3}, \quad 2 \leq -3.25, \quad 2 < 2, \quad 1.01 > 1.1, \quad \frac{5}{4} > \frac{11}{8}. \quad \square$$

Inequalities are often useful in describing practical situations.

EXAMPLE 2   Suppose a hospital uses at least 50 gallons of liquid soap and fewer than 85 gallons of disinfectant per week. If we let S denote the amount of soap and D the amount of disinfectant used per week, S and D in gallons, then we may express these facts by writing

$$S \geq 50 \quad \text{and} \quad D < 85. \quad \square$$

FIGURE 1–4
$a < b$ or $b > a$

NUMBERS AND FUNCTIONS

An inequality of the form $a < b < c$ is a **double inequality.** It is true if both $a < b$ and $b < c$ are true, that is, if b is between a and c on the real line.

EXAMPLE 3

The following are true:

$$-3 < -\frac{3}{2} < -1, \qquad -\frac{2}{3} < 0.02 \le 0.021, \qquad \frac{1}{2} < \frac{3}{4} < \frac{7}{8}.$$

The following are false:

$$-2 < -3 < -4, \qquad 2 < \frac{5}{2} < 2.25, \qquad 1 \le 2 < 2. \quad \square$$

Do not write statements such as $2 < 4 > 0$. All inequality symbols in a double inequality should point in the same direction, so the correct way to write the inequality relation among 0, 2, and 4 is either $0 < 2 < 4$ or $4 > 2 > 0$.

Double inequalities are also useful.

EXAMPLE 4

Suppose a hospital uses at least 50 but no more than 60 gallons of liquid soap per week and that each gallon of the soap costs \$0.80. If S again denotes the amount of soap used per week, in gallons, and if C denotes the weekly cost of the soap, then

$$50 \le S \le 60$$

and

$$\$0.80(50) \le C \le \$0.80(60),$$

or

$$\$40 \le C \le \$48. \quad \square$$

Intervals

It is often convenient to consider the set of all numbers that lie between two given numbers, or the set of all numbers that are greater than or less than a given number. Such sets of numbers are called **intervals.** For example,

$$\{x | 0 \le x < 1\}$$

is an interval because it consists of all the numbers between 0 and 1, including 0 but not including 1. Figure 1–5 shows a **line diagram** for this interval; the closed circle at 0 indicates that 0 belongs to the interval; the open circle at 1 indicates that 1 does not belong to the interval. We write

$$[0, 1) = \{x | 0 \le x < 1\};$$

FIGURE 1–5
The Interval $\{x | 0 \le x < 1\} = [0, 1)$

Interval	Interval notation	Line diagram
$a < x < b$	(a, b)	○——○ a b
$a \leq x < b$	$[a, b)$	●——○ a b
$a < x \leq b$	$(a, b]$	○——● a b
$a \leq x \leq b$	$[a, b]$	●——● a b
$x < a$	$(-\infty, a)$	←——○ a
$x \leq a$	$(-\infty, a]$	←——● a
$x > b$	$(b, +\infty)$	○——→ b
$x \geq b$	$[b, +\infty)$	●——→ b

Intervals

FIGURE 1–6

the expression $[0, 1)$ is called **interval notation.** Figure 1–6 depicts all possible forms of intervals and gives their interval notation. The symbols $+\infty$ and $-\infty$ are read as "plus infinity" and "minus infinity," respectively; they are not numbers, but are merely a way of indicating that the interval contains *all* the numbers greater than or less than a given number, as the case may be.

EXAMPLE 5 The interval $(1, 5)$ is $\{x | 1 < x < 5\}$; the interval $[1, +\infty)$ is $\{x | x \geq 1\}$. ☐

EXAMPLE 6 The line diagrams corresponding to the intervals $[1, 3]$, $(-1, 2)$, $(-\infty, 1)$, and $(3, +\infty)$ are shown in Figure 1–7. ☐

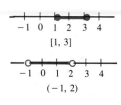

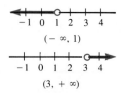

FIGURE 1–7

Note that an interval does *not* consist only of the integers that satisfy its definition, but rather of *all* real numbers, rational and irrational, that satisfy its definition. For example, the interval $(1, 5)$ does not consist only of the integers 2, 3, and 4, but rather of these numbers and all other real numbers that are greater than 1 and less than 5. For instance, the numbers

$$1.001, \sqrt{2}, 7/3, 2.95, \pi, 2\sqrt{3}, 23/5, \text{ and } 4.9257$$

are all in the interval $(1, 5)$ because all of them are greater than 1 and less than 5. (Check this.)

Intervals are often used to describe the ranges over which various alternatives hold.

EXAMPLE 7 Suppose that the recommended dosage of a certain drug is 30 milligrams (mg) if the patient weighs 200 pounds or less and 40 mg if the patient weighs more than

Numbers and Functions

200 pounds. If we let d denote the recommended dosage in milligrams and w the weight of the patient in pounds, we may express these facts using intervals by writing

$$d = \begin{cases} 30 & \text{if } w \text{ is in } (0, 200], \\ 40 & \text{if } w \text{ is in } (200, +\infty). \end{cases}$$

Absolute Value

The **absolute value** of a number is its distance from 0 on the real line, regardless of whether the number is greater than or less than 0. For example, the numbers 3 and -3 are both 3 units from 0, so both have an absolute value of 3. A more formal definition of the absolute value of a number is as follows:

Absolute Value

The **absolute value** of a number x, denoted $|x|$, is defined by

$$|x| = \begin{cases} x & \text{if } x \geq 0, \\ -x & \text{if } x < 0. \end{cases}$$

EXAMPLE 8

$|2| = 2, \quad |-2| = 2, \quad |0| = 0, \quad |\sqrt{2}| = \sqrt{2}, \quad \left|\dfrac{2}{3} - \dfrac{5}{3}\right| = 1.$

Absolute value is often used when it is the magnitude of a quantity (its distance from 0), rather than its sign, that is important.

EXAMPLE 9

An automatic pilot will correct a ship's heading if and only if the actual heading differs by at least 0.5 degrees from the programmed heading. Let x denote the actual heading and y the programmed heading, in degrees. Then $|x - y|$ is the difference between the headings, without regard to sign, and hence the automatic pilot will correct the heading if and only if $|x - y| \geq 0.5$.

Exponents and Roots

We conclude this section with a brief review of exponents and roots. In an expression of the form x^r, where x and r are numbers, x is called the **base** and r is called the **exponent** or **power**.

Nonnegative Integral Exponents

Let x be a number and n a positive integer. We define

$$x^n = \underbrace{x \cdot x \cdots x}_{n \ x\text{'s}}$$

Also, if $x \neq 0$, we define $x^0 = 1$.

EXAMPLE 10 $2^1 = 2$, $2^3 = 2 \cdot 2 \cdot 2 = 8$, $3^0 = 1$, $(-5)^3 = (-5)(-5)(-5) = -125$. □

If x is a number and n a positive integer, any number r such that $r^n = x$ is called an **nth root of** x. If x has just one nth root r, then r is the **principal nth root** of x; if x has two nth roots, its principal nth root is the positive one.

EXAMPLE 11   Since $(-3)^3 = -27$, -3 is a third root (or cube root) of -27; since $2^4 = 16$ and also $(-2)^4 = 16$, 2 and -2 are both fourth roots of 16. Since -3 is the only cube root of -27, it is the principal cube root of -27; since 2 is the positive fourth root of 16, it is the principal fourth root of 16. □

We denote the principal nth root of x by $\sqrt[n]{x}$. If $n = 2$, we write $\sqrt{x}$ instead of $\sqrt[2]{x}$ and call $\sqrt{x}$ the **square root** of x. Note that if n is even, then $\sqrt[n]{x}$ does not exist when x is negative and is positive when x is positive.

EXAMPLE 12   From Example 11,

$$\sqrt[3]{-27} = -3 \quad \text{and} \quad \sqrt[4]{16} = 2.$$

Note also that

$$\sqrt[2]{25} = \sqrt{25} = 5$$

is positive, but that $\sqrt{-25}$ does not exist. □

Now we can define rational exponents.

Rational Exponents

If x is a number, m and n are positive integers, and $\sqrt[n]{x}$ exists, then

$$x^{m/n} = (\sqrt[n]{x})^m.$$

EXAMPLE 13 $2^{1/2} = (\sqrt{2})^1 = \sqrt{2}$, $25^{1/2} = (\sqrt{25})^1 = 5$,

$$(-27)^{4/3} = (\sqrt[3]{-27})^4 = (-3)^4 = 81. \quad □$$

EXAMPLE 14   Suppose the cost of making x units of a product is $100x^{3/4}$ dollars. Then the cost of making 10,000 units will be

$$100(10{,}000)^{3/4} = 100(\sqrt[4]{10{,}000})^3 = 100(10)^3 = 100(1000) = 100{,}000$$

dollars. The cost of making 15,000 units will be $100(15{,}000)^{3/4}$ dollars. We can find this cost by using a calculator with an x^y or y^x key: from such a calculator,

$$15{,}000^{3/4} = 15{,}000^{0.75} \approx 1355.403.$$

(The symbol $\approx$ means "approximately equal to".) Hence the cost of making 15,000 units will be approximately

NUMBERS AND FUNCTIONS

$$\$100(1355.403) = \$135{,}540.30. \quad \square$$

Lastly, we define negative exponents.

Negative Exponents

If x is a nonzero number and r is a positive number, then

$$x^{-r} = \frac{1}{x^r}.$$

EXAMPLE 15 $\quad 2^{-1} = \frac{1}{2}, \quad (-3)^{-4} = \frac{1}{(-3)^4} = \frac{1}{81}, \quad 4^{-3/2} = \frac{1}{4^{3/2}} = \frac{1}{8}. \quad \square$

Now that we have defined x^r for rational numbers r, we state the rules of exponents.

Rules of Exponents

Let x, y, r, and s be numbers and assume that the following expressions are all defined. Then

a. $x^r x^s = x^{r+s}$

b. if $x \neq 0$, $\dfrac{x^r}{x^s} = x^{r-s}$

c. $(x^r)^s = x^{rs}$

d. $(xy)^r = x^r y^r$

e. if $y \neq 0$, $\left(\dfrac{x}{y}\right)^r = \dfrac{x^r}{y^r}.$

EXAMPLE 16

a. $3^2 3^{-4} = 3^{2+(-4)} = 3^{-2} = \dfrac{1}{9}$

b. $\dfrac{3^2}{3^{-4}} = 3^{2-(-4)} = 3^6 = 729$

c. $(3^2)^3 = 3^6 = 729$

d. $(3 \cdot 5)^2 = 3^2 \cdot 5^2 = 9 \cdot 25 = 225$

e. $\left(\dfrac{3}{5}\right)^2 = \dfrac{3^2}{5^2} = \dfrac{9}{25}. \quad \square$

EXAMPLE 17

$$(2x^3)(4x^2) = (2 \cdot 4)(x^3 \cdot x^2) = 8x^{3+2} = 8x^5$$

$$\frac{4x^3y^2z^4}{2xy^2z^{-3}} = 2x^{3-1}y^{2-2}z^{4-(-3)} = 2x^2y^0z^7 = 2x^2z^7$$

$$(x^2y^{-2}z)^3 = x^6y^{-6}z^3 = \frac{x^6z^3}{y^6}$$

$$\left(\frac{x^2}{y^3}\right)^2 = \frac{x^4}{y^6}$$

$$\sqrt[3]{\frac{x^3y^{-2}z^5x^{-6}}{y^4(z^2)^{-2}}} = (x^{3+(-6)}y^{-2-4}z^{5-(-4)})^{1/3}$$

$$= (x^{-3}y^{-6}z^9)^{1/3}$$

$$= x^{-1}y^{-2}z^3$$

$$= \frac{z^3}{xy^2}. \quad \square$$

■ *Computer Manual: Program INTERVAL*

1.1 EXERCISES

The Real Line

In Exercises 1 through 3, plot the given numbers on the real line.

1. $-5, -4.5, -7/4, -1, -2/5, 0, 1/3, 3/4, 1, 3/2, 17/8, 19/6, 4.2, 5, \sqrt{2}, 2\sqrt{3}, -\sqrt{5}, \pi/2, -3\pi/4$

2. $0, 1/12, 1/6, 1/4, 1/3, 5/12, 1/2, 7/12, 2/3, 3/4, 5/6, 11/12, 1$

3. $-5.2, -11/3, -2.1, -11/7, -9/7, -5/8, -1/8, 1/10, 0.3, 9/10, 1.25, 9/5, 2.125, 24/7$

In Exercises 4 through 6, what numbers are represented by the plotted points?

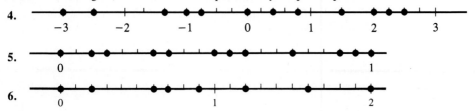

Inequalities

In Exercises 7 through 20, state whether the inequality is true or false.

7. $-3 < 2$
8. $-6 < -8$
9. $0 \leq 2$
10. $4 \geq 2$
11. $-3 > -4$
12. $2 < 2$
13. $\frac{2}{3} < \frac{3}{5}$
14. $-\frac{2}{3} < -\frac{3}{5}$
15. $-1 < 0 < 1$
16. $-3 < -4 < -5$
17. $-4 \leq 4 \leq 10$
18. $\frac{2}{3} < \frac{4}{7} \leq 1$
19. $\frac{4}{7} < \frac{2}{3} \leq 1$
20. $.5 < 5.1 < 5.01$

Management

The statements in Exercises 21 through 26 were suggested by recent articles in *The Wall Street Journal*. Write each statement using an inequality.

21. A corporation forecasts sales of at least $1,500,000 next quarter.
22. A corporation forecasts sales of no more than $1,500,000 next quarter.
23. A corporation forecasts sales of more than $1,500,000 next quarter.
24. A corporation forecasts sales of less than $1,500,000 next quarter.
25. The price of a corporation's stock ranged from $86.50 to $92.50 during the last six months.
26. A corporation's return on investment during the next quarter is forecast to be more than 10% but less than 12.5%.

Intervals

In Exercises 27 through 34, write the set of all real numbers x that satisfy the given inequality as an interval.

27. $-2 < x \leq 7$
28. $4 < x < 8$
29. $x > -3$
30. $x < \frac{2}{3}$
31. $\frac{1}{2} \leq x < \frac{3}{2}$
32. $-3 \leq x \leq \frac{3}{4}$
33. $x \leq -\frac{5}{3}$
34. $x \geq 4$

In Exercises 35 through 40, draw the line diagram for the given interval and then write it as a set of numbers that satisfy an inequality.

35. $(-4, 8)$
36. $(-6, 3]$
37. $[-5, +\infty)$
38. $[-\frac{1}{2}, 3]$
39. $(-\infty, -4)$
40. $(2, +\infty)$
41. Which of the following numbers are in the interval $[-2, 2)$? $-2.1, -2, -1.99, -\frac{32}{11}, -\frac{3}{4}, -1, \frac{2}{3}, 2, \pi, \pi/2, 3\sqrt{2}, \frac{9}{5}, \frac{9}{2}$

42. Which of the following numbers are in the interval [0, 1]? 0, 292/293, 293/292, 0.025, −0.025, 1/$\sqrt{2}$, 3/$\sqrt{5}$, π/3, π/4, 0.9999 . . . , 1.0001, 1

Management

The statements in Exercises 43 through 46 were suggested by recent articles in *The Wall Street Journal*. Write each statement using intervals.

43. A corporation's stock paid a dividend of $1.40 per share during the period 1980–1985 and a dividend of $1.60 per share during the period 1986–1989.
44. An industrial chemical costs $0.40 per gallon if less than 1000 gallons are purchased and $0.35 per gallon if 1000 or more gallons are purchased.
45. Sponsors of a TV show will pay $300,000 for each 30-second commercial if the show's rating is at least 21.0; otherwise they will pay $275,000 for each 30-second commercial.
46. A firm pays contract workers $4 per hour if they work fewer than 20 hours per week, $4.50 per hour if they work at least 20 but not more than 40 hours per week, and $5 per hour if they work more than 40 hours per week.

Life Science

In Exercises 47 and 48, use intervals to write the statement about the dosage.

47. The recommended dosage of a drug is 50 mg if the patient is less than 5 years old, 100 mg if the patient is between 5 and 12 years old, inclusive, and 200 mg if the patient is more than 12 years old.
48. The recommended dosage of a drug is 200 units per day for the first 14 days, 120 units per day for days 15 through 30, and 100 units per day thereafter.

Absolute Value

In Exercises 49 through 57, evaluate the expression.

49. $|4|$
50. $|-4|$
51. $-|-8|$
52. $\left|\dfrac{2}{3}\right|$
53. $\left|-\dfrac{4}{5}\right|$
54. $-|\sqrt{2}|$
55. $|7 - 12|$
56. $2|8 - 11|$
57. $2 - |-\sqrt{2}|$

58. The cruise control mechanism on a car will automatically change the speed of the car if and only if its actual speed is at least 1 mph faster or slower than the desired crusing speed. Express this fact using absolute values.
59. The pressure inside a natural gas tank is in the safe range if it is within ±2.5 pounds per square inch of the designed pressure of 20 pounds per square inch. Express this fact using absolute values.

Management

60. A production process is in control if the weight of each part it produces is no more than ±0.25 gram from the designed weight of 220 grams. Express this fact using absolute values.

Exponents and Roots

In Exercises 61 through 72, evaluate the expression.

61. 2^6 **62.** 3^{-4} **63.** π^0

64. $\left(\dfrac{2}{5}\right)^3$ **65.** $\left(-\dfrac{3}{4}\right)^2$ **66.** $36^{1/2}$

67. $\sqrt[3]{64}$ **68.** $81^{3/4}$ **69.** $\sqrt{-16}$

70. $100^{-1/2}$ **71.** $32^{-3/5}$ **72.** $(-8)^{1/3}$

In Exercises 73 through 81, simplify the expression. Use only positive exponents in your answer.

73. $\dfrac{5^2 y^3 z^8}{x^2 y^3 z^{-2}}$ **74.** $(3x^2 y^3 z^{-1})^3$ **75.** $(4x^2 y^{-2} z^5)^2 (2x^{-2} yz^4)^{-1}$

76. $\dfrac{x^2 y^{2/3} z^{-5/4}}{x^{1/2} y^{-1/3} z}$ **77.** $\dfrac{(x^4 y^2 z^6)^{-3/2}}{xyz}$ **78.** $\dfrac{(27x^3 y^{-3})^{1/3}}{(3x^{2/3} y^{4/3})^3}$

79. $\sqrt{x^2 y^3}$ **80.** $\sqrt[4]{32 x^4 y^9}$ **81.** $\sqrt[3]{\dfrac{27 x^5 y^3}{x^3 z^2}}$

Management

82. If a firm makes and sells x units of its product its net income will be $200x^{1/2}$ dollars. Find the firm's net income if it makes and sells 625 units.

83. Find the net income of the firm of Exercise 82 if it makes and sells 1,000,000 units.

84. If a corporation produces x units of its product, its average cost per unit is $40 + 120x^{-2/3}$ dollars. Find the corporation's average cost per unit if it makes 1000 units.

85. Find the average cost per unit for the corporation of Exercise 84 if it produces 125,000 units.

Life Science

86. Over a certain temperature range, the activity of a cold-blooded animal can be determined by means of the equation

$$A = kT^{3/2}.$$

Here A is a measure of the animal's activity, T is the temperature in degrees Celsius, and k is a numerical constant that depends on the animal. Suppose that for iguanas the constant k is 1. Find a measure for the activity of iguanas when the temperature is 20°C.

Social Science

*87. Suppose a political pollster asks N voters which candidate they will vote for. Let p_0 be the proportion of these voters who say they will vote for candidate A, and let p be the proportion of all voters who will vote for A. Statistical theory tells us that we can be 95% certain that the value of p lies in the interval

$$p_0 - 1.96\left[\frac{p_0(1-p_0)}{N}\right]^{1/2} < p < p_0 + 1.96\left[\frac{p_0(1-p_0)}{N}\right]^{1/2}.$$

Suppose that 500 of 1000 voters polled say they will vote for candidate A. We can be 95% certain that the proportion of all voters who will vote for A lies in what interval?

1.2 FUNCTIONS AND GRAPHING

As we noted at the beginning of this chapter, functions describe relationships. For instance, a function might describe the relationship between a company's sales and its profits, or the relationship between the average daily temperature and the amount of heating oil used to heat a building. Throughout this book we will be creating functions, describing them, graphing them, and examining their behavior. In this section we begin our study of functions.

Functions

A **function** is a rule that associates with each element of one set of numbers a single element of another set of numbers. Figure 1–8 depicts this situation symbolically: the rule shown there assigns to each number in the set X a single number in the set Y, and thus it is a function from X to Y. On the other hand, Figure 1–9 depicts a rule that assigns more than one number in the set Y to one of the numbers in the set X; hence it is not a function from X to Y.

Functions are often denoted by the letters f, g, or h and specified by a **defining equation** in **functional notation.** For example, the equation

$$f(x) = 3x - 2$$

defines a function f using functional notation. The letter x is the **independent variable** of f, and the notation $f(x)$ is read "f of x." If we substitute a particular real number for x in the defining equation, we can see what value the function assigns to that number. For instance, since

$$f(0) = 3(0) - 2 = -2,$$
$$f(1) = 3(1) - 2 = 1,$$

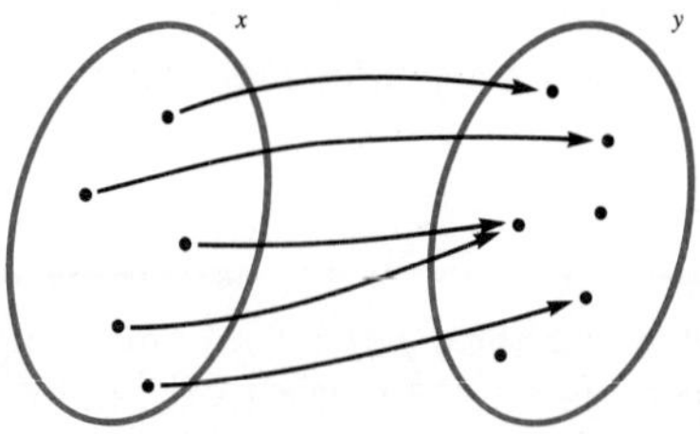

FIGURE 1–8 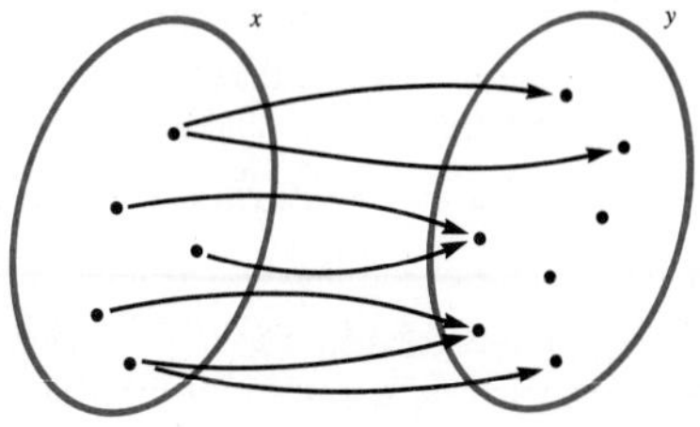   FIGURE 1–9

and
$$f(-2) = 3(-2) - 2 = -8,$$
we see that f assigns -2 to 0, 1 to 1, and -8 to -2. The numbers $f(0), f(1)$, and $f(-2)$ are called the **functional values** of f at 0, 1, and -2, respectively. Thus $f(x)$ denotes the functional value of f at x. Notice that a function f and a functional value $f(x)$ are not the same thing, since f is a *rule* but $f(x)$ is a *number*. Throughout this book we will specify functions by their defining equations, using phrases such as "let f be the function defined by the equation $f(x) = 3x - 2$," or "let f be the function given by $f(x) = 3x - 2$."

It is possible, and often useful, to substitute algebraic expressions for the independent variable of a function. For instance, consider the function g defined by the equation $g(x) = 5x + 2$: we have
$$g(t) = 5t + 2,$$
$$g(a^2) = 5a^2 + 2,$$
and
$$g(b - 1) = 5(b - 1) + 2 = 5b - 5 + 2 = 5b - 3.$$

EXAMPLE 1 Suppose it costs a firm $C(x) = x^2 + 5x + 100$ dollars to produce x units of its product. How much will it cost the firm to produce 20 units?

It will cost
$$C(20) = 20^2 + 5(20) + 100 = 600$$
dollars. If the firm is currently producing b units, its cost is
$$C(b) = b^2 + 5b + 100$$
dollars. ☐

The set of values that the independent variable of a function is allowed to assume is called the **domain** of the function. Sometimes the domain is stated explicitly along with the defining equation of the function. Thus, if a function f is defined by the statement
$$f(x) = 5x^2 + 2, \quad x > 1,$$
this indicates that the independent variable x is only allowed to take on values greater than 1; hence the domain of this function is $\{x | x > 1\} = (1, +\infty)$. Often the domain of the function is not stated explicitly. It is then understood that the domain consists of all numbers that may legitimately be substituted for the independent variable in the defining equation.

EXAMPLE 2 Consider the square-root function g defined by $g(x) = \sqrt{x}$. We cannot substitute negative numbers for x in this equation because the square root of a negative number is not defined as a real number. We can substitute any nonnegative number for x, however, and thus the domain of g is the interval $[0, +\infty)$.

Similarly, suppose the function h is defined by

$$h(x) = \frac{2}{x-1}.$$

Since division by 0 is not defined, we cannot substitute 1 for x in this equation, but we can substitute any other number for x. Hence the domain of h is $\{x | x \neq 1\}$. □

The **range** of a function is the set of all numbers produced by the function as its independent variable takes on all values in its domain. Thus, if f is a function,

range of $f = \{y | y = f(x)$ for some x in the domain of $f\}$.

For instance, the range of the square-root function g of Example 2 is the interval $[0, +\infty)$. This is so because $g(x) = \sqrt{x} \in [0, +\infty)$ for all x in the domain of g and every number $y \in [0, +\infty)$ can be obtained by applying g to the number y^2:

$$\text{if } y \in [0, +\infty), \text{ then } y = \sqrt{y^2} = g(y^2).$$

We will not always use functional notation to describe functions. Sometimes it is more convenient to define functions by equations involving the independent variable and a **dependent variable.** For example, the equation $y = 3x - 2$ specifies the dependent variable y as a function of the independent variable x. Substituting numerical values for x will yield numerical values for y. Note that the function defined by $y = 3x - 2$ is the same as the function defined by $f(x) = 3x - 2$. In fact, $s = 3t - 2$ and $g(u) = 3u - 2$ also define this same function. The letters used are not important; what is important is the relationship between the variables. It is common practice in the applications of mathematics to use letters that are suggestive of the things they represent. Thus, if a firm sells quantity q of its product at a price of \$20 per unit, its revenue will be given by the equation $R = 20q$. Here q is the independent variable, representing quantity sold; R is the dependent variable, representing revenue in dollars. If the firm sells 5000 units, its revenue is

$$R = 20(5000) = 100{,}000$$

dollars.

We have said that we will sometimes define functions by means of equations involving an independent variable and a dependent variable. However, not every such equation defines a function. For example, the equation $x^2 + y^2 = 1$ does not define y as a function of x, because it sometimes assigns more than one value of y to a single value of x. (For instance, if $x = 0$, then $y = 1$ and also $y = -1$.) A function must assign a *single* value of the dependent variable to each value of the independent variable.

Graphing

It is usually possible to depict a function geometrically by **graphing** it. A graph gives us information about the function in a visual way that is easy to understand.

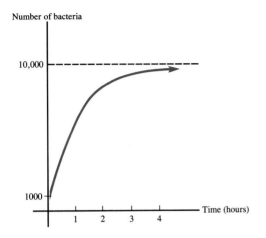

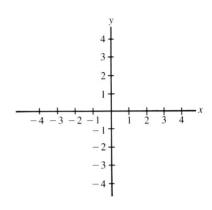

FIGURE 1–10

FIGURE 1–11
The Cartesian Coordinate System

For instance, the graph in Figure 1–10 shows the number of bacteria present in a culture as a function of time. The horizontal scale represents time in hours, the vertical scale the number of bacteria present. We immediately see from the graph that the initial number of bacteria present was 1000, that as time passed the bacteria population increased, but at a slower and slower rate, and that the number of bacteria was limited by some factor (the supply of nutrients, perhaps) to fewer than 10,000.

The process of graphing a function begins with two copies of the real line, one horizontal and one vertical, that intersect at their zero points, as in Figure 1–11. The horizontal real line and the vertical real line are called **axes:** the horizontal axis is often called the **x-axis,** the vertical one the **y-axis.** The two together form the **cartesian coordinate system.** If a and b are any real numbers, the point (a, b) is located on the cartesian coordinate system opposite a on the x-axis and opposite b on the y-axis, as shown in Figure 1–12. The number a is called the **x-coordinate** of the point (a, b), and the number b is its **y-coordinate.** The point $(0, 0)$ where the axes intersect is called the **origin.**

EXAMPLE 3 In Figure 1–13 we have plotted the points $(1, 0)$, $(2, 3)$, $(0, 2)$, $(-1, 1)$, $(-3, 0)$, $(-2, -4)$, $\left(0, -\frac{3}{2}\right)$, and $\left(\frac{5}{2}, -\frac{5}{2}\right)$. ☐

The **graph** of a function f is the set of all points $(x, f(x))$ for which x is in the domain of f. To draw the graph of a function, we proceed as follows:

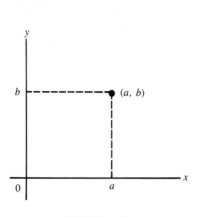

FIGURE 1–12   FIGURE 1–13

Graphing Procedure

To graph a function f:

1. Determine its domain.

2. Substitute representative numbers from the domain into the defining equation to obtain points $(x, f(x))$ that lie on the graph.

3. Plot the points obtained in step 2 on the cartesian coordinate system. When enough points have been plotted to make the shape of the graph clear, connect them with a curve.

It is also frequently useful to find the **intercepts** of a function. The **y-intercept** of a function f is the point where its graph crosses the y-axis; this is the point $(0, f(0))$. The **x-intercepts** of the function are the points where the graph crosses or touches the x-axis; we find them by solving the equation $f(x) = 0$ for x, if possible.

EXAMPLE 4 Let us graph the function f given by $f(x) = 2x - 1$.

1. Since any number can be substituted for x in the defining equation for f, the domain of f is the set **R** of real numbers.

2. We select some representative numbers from the domain of f and find the corresponding points on the graph:

NUMBERS AND FUNCTIONS

x	$f(x) = 2x - 1$	$(x, f(x))$
-2	$f(-2) = -5$	$(-2, -5)$
0	$f(0) = -1$	$(0, -1)$
1	$f(1) = 1$	$(1, 1)$
3	$f(3) = 5$	$(3, 5)$

The y-intercept of f is the point $(0, f(0)) = (0, -1)$. Since the equation $2x - 1 = 0$ has the solution $x = \frac{1}{2}$, there is one x-intercept, namely the point $(\frac{1}{2}, 0)$. The graph of f is shown in Figure 1–14.

Notice that because the domain of f is the set of all real numbers, there is a point $(a, f(a))$ on the graph for every number a on the x-axis; hence there are no "gaps" in the graph. Also, we see from the graph that for every number b on the y-axis there is some number a on the x-axis such that (a, b) is on the graph; this shows that the range of f is the entire y-axis, and hence is the set of all real numbers. ☐

EXAMPLE 5 Let us graph the **absolute value function** h defined by $h(x) = |x|$. The domain of this function is the set of real numbers (why?). We have

x	$h(x) = |x|$	$(x, h(x))$
-3	$y = 3$	$(-3, 3)$
-2	$y = 2$	$(-2, 2)$
-1	$y = 1$	$(-1, 1)$
0	$y = 0$	$(0, 0)$
1	$y = 1$	$(1, 1)$
2	$y = 2$	$(2, 2)$
3	$y = 3$	$(3, 3)$

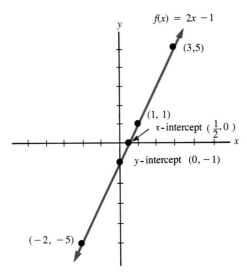

FIGURE 1–14

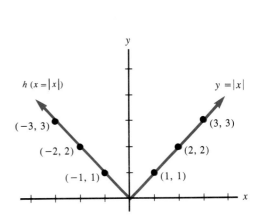

FIGURE 1–15

If $x = 0$, $h(0) = 0$, so the y-intercept is the point $(0, 0)$. Since the equation $|x| = 0$ has the single solution $x = 0$, $(0, 0)$ is also the only x-intercept. The graph of the absolute value function is shown in Figure 1–15.

Once again, because the domain of the function is the set of all real numbers there is a point $(a, h(a))$ on the graph for every number a on the x-axis, and thus the graph has no "gaps." Also, we see from the graph that if $b \geq 0$ is a number on the y-axis there is some number a on the x-axis such that (a, b) is on the graph, but that this is not so if $b < 0$. Hence the range of h is the nonnegative y-axis, that is, the interval $[0, +\infty)$. ∎

It is important to realize that a point (a, b) lies on the graph of a function f if and only if $f(a) = b$. Therefore:

- If (a, b) is on the graph of f, then a must be in the domain of f and b must be in the range of f.
- If a is *not* in the domain of f, then (a, b) cannot be on the graph of f for any number b, and hence a vertical straight line through a on the x-axis will not intersect the graph of f.
- If b is not in the range of f, then (a, b) cannot be on the graph of f for any number a, and hence a horizontal straight line through b on the y-axis will not intersect the graph of f.

A function may have different defining equations on different portions of its domain. When graphing such a function, we must be careful as we select numbers x from its domain to use the defining equation that is appropriate to the choice of x.

EXAMPLE 6

Let us graph the function defined by

$$y = \begin{cases} x/2, & -1 \leq x \leq 2, \\ 3 - x, & 2 < x \leq 4. \end{cases}$$

The domain of this function is the interval $[-1, 4]$. If we choose a number x from the domain and $-1 \leq x \leq 2$, then $y = x/2$, while if $2 < x \leq 4$, then $y = 3 - x$. Figure 1–16 shows the graph of this function. Note that the range of this function is $[-1, 1]$. ∎

EXAMPLE 7

Suppose a firm's profits in year t, where $t = 0$ represents 1974, are given by the function $P = t^2 - 16t + 48$, where P is in millions of dollars. The graph of this function is shown in Figure 1–17. From the graph we see that in 1974 the firm had a profit of $48 million. Profits declined after 1974 until in 1978 ($t = 4$) they were zero. Between 1978 and 1986 the firm had a negative profit (that is, a loss) each year; the largest loss was $16 million in 1982 ($t = 8$). The firm broke even again in 1986 ($t = 12$), and since then profits have been increasing. ∎

Our graphing procedure relies on finding points on the graph and then plotting them. Unless the function under consideration is a simple one, this is usually quite

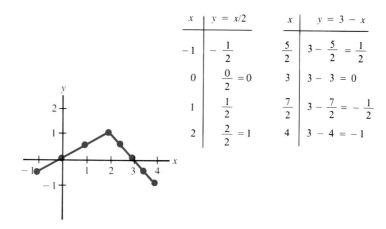

FIGURE 1–16

tedious. It is also potentially unreliable, for we cannot always be certain that we have plotted enough points to give us a true picture of the graph. One of the most important reasons for learning calculus is that it can be used to sketch graphs quickly and reliably. We will discuss curve sketching using calculus in Section 4.4.

It is possible to graph equations containing two variables even when they do not represent functions. For example, the graph of $x^2 + y^2 = 1$ is a circle of radius 1 (Figure 1–18). Note that it is possible for a vertical line to intersect the graph in more than one point: the y-axis, for instance, intersects the circle at the points $(0, 1)$ and $(0, -1)$. But this means that the single value $x = 0$ is assigned the two numbers 1 and -1 by the equation, and hence the equation does not define y as a function of x. Thus an examination of the graph of an equation can tell us whether or not the equation defines a function.

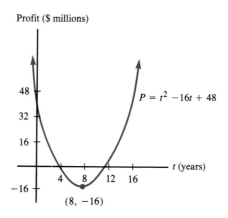

FIGURE 1–17

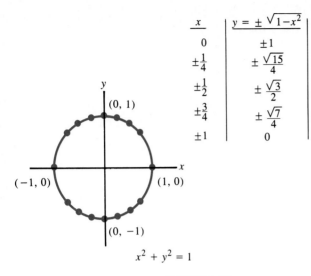

FIGURE 1–18

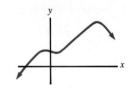

FIGURE 1–19
This Is the Graph of a Function

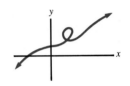

FIGURE 1–20
This Is Not the Graph of a Function

The Vertical Line Test

A graph is the graph of a function if and only if no vertical line intersects it in more than one point.

EXAMPLE 8   The graph of Figure 1–19 is the graph of a function, but that of Figure 1–20 is not. □

■ *Computer Manual: Programs FNCTNVAL, FNCTNTBL.*

1.2 EXERCISES

Functions

In Exercises 1 through 4, let f be the function defined by
$$f(x) = 4x - 2.$$

1. What is the domain of f?
2. Find $f(0)$, $f(1)$, and $f(-3)$.
3. Find $f(-\frac{1}{4})$, $f(\frac{5}{3})$, and $f(-2.3)$.
4. Find $f(b)$, $f(b+2)$, and $f(t^2-1)$.

In Exercises 5 through 8, let g be the function defined by
$$g(x) = \frac{1}{x}.$$

5. What is the domain of g?
6. Find $g(1)$, $g(4)$, $g(2)$, and $g(100)$.
7. Find $g(-3)$, $g(-9)$, $g(\frac{1}{2})$, and $g(\frac{1}{20})$.
8. Find $g(c)$, $g(c^3)$, and $g(1/c)$ if $c \neq 0$.

In Exercises 9 through 12, let h be the function defined by
$$h(x) = x^2 - 3x + 2.$$

9. What is the domain of h?
10. Find $h(0)$, $h(2)$, $h(1)$, and $h(-2)$.
11. Find $h(\frac{1}{2})$, $h(1.4)$, $h(\sqrt{2})$, and $h(\pi)$.
12. Find $h(a)$, $h(a+1)$, and $h(a^2)$.

In Exercises 13 through 16, consider the function defined by
$$y = \frac{x}{x-3}.$$

13. What is the domain of this function?
14. Find y when $x = -3$, $x = 5$, $x = 4$, and $x = 0$.
15. Find y when $x = \frac{3}{2}$, $x = -\frac{5}{3}$, $x = \frac{5}{2}$, and $x = \frac{7}{2}$.
16. Find y when $x = b$, $b \neq 3$, and when $x = 1/b$, $b \neq 0$.

In Exercises 17 through 20, consider the function defined by
$$y = 2\sqrt{x+1}.$$

17. What is the domain of this function?
18. Find y when $x = 0$, $x = 3$, $x = 8$, and $x = -1$.
19. Find y when $x = 1$, $x = 2$, $x = -\frac{1}{2}$, and $x = -\frac{3}{4}$.
20. Find y when $x = c - 1$, $c > 0$, and when $x = c^2 - 1$.

In Exercises 21 through 24, consider the function defined by the equation
$$p = \frac{2}{\sqrt{q}}.$$

21. What is the domain of this function?
22. Find p when $q = 1, 4, 9, \frac{1}{4}$, and $\frac{1}{100}$.
23. Find p when $q = 2, 8, 100$, and $10{,}000$.
24. Find p when $q = b$ and when $q = b^2$, $b > 0$.

*25. Which of the following equations define y as a function of x?

$$y = 2,\ 2x + y = 6,\ x + 2y = 6,\ y = x^2,\ y^2 = x,\ \frac{x^2}{4} + \frac{y^2}{16} = 1,\ y = |x|,\ |y| = x$$

Management

26. If a firm makes x units of its product, its cost is
$$C(x) = 20x + 30{,}000$$
dollars. Find the firm's cost if it makes the stated number of units.
(a) 0 units (b) 10 units (c) 100 units (d) a units

27. If a firm makes and sells x units of its product, its profit is
$$P(x) = -x^2 + 30x - 200$$
thousand dollars. Find the firm's profit if it makes and sells the stated number of units.
(a) 0 units (b) 10 units (c) 15 units (d) 25 units

28. If the price of a product is p dollars per unit, the quantity q demanded by consumers (that is, the number of units consumers will purchase) is given by
$$q = 200{,}000 - 8000p.$$
Find the quantity demanded at the given price.
(a) \$1 (b) \$5 (c) \$15 (d) \$25

29. If a firm makes x units of its product its average cost per unit is
$$A(x) = 32 + \frac{40{,}000}{x}$$
dollars. Find the firm's average cost per unit if it makes the given number of units.
(a) 100 units (b) 10,000 units (c) 1,000,000 units

Life Science

30. The size of an animal population in year t is given by
$$p(t) = \frac{210t + 400}{t + 2}.$$
Here $t = 0$ represents the year 1988 and $p(t)$ is the size of the population in thousands of individuals. Find the size of the population in the given year.
(a) 1988 ($t = 0$) (b) 1990 (c) 1994

Social Science

31. The average score of high-school seniors on a standardized mathematics test over a 10-year period is given by
$$M(t) = 480 - 12t + t^2, \quad 0 \le t \le 10.$$
Find the average score on the test at the stated time.
(a) The beginning of the 10-year period.
(b) The middle of the period.
(c) Two-thirds of the way through the period.
(d) The end of the period.

*32. Consumers are frequently asked to rank products or services using a **Likert scale** such as

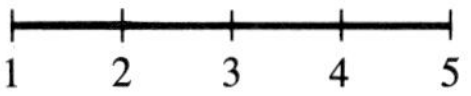

Here 1 is the lowest ranking and 5 the highest ranking. Suppose, for instance, that students are asked to evaluate a course using such a scale. Then the evaluation process can be thought of as a function f from the set of all possible judgments about the course to the interval $[1, 5]$. Thus f will assign to each judgment about the course a single number between 1 and 5, inclusive.

(a) Find f(very poor,) f(poor), f(average), f(above average), and f(excellent).
(b) Find f(halfway between average and above average).
(c) Find f(two-thirds of the way from poor toward average).

Computer Science

***33.** In an article on computing by Peter Denning in the March–April 1986 issue of *American Scientist,* it is stated that if a computer has n processors and m memory modules and $x = n/m$, then the fraction of modules kept busy per memory cycle is

$$f(x) = 1 + x - \sqrt{x^2 + 1}.$$

(a) What is the domain of this function? (*Hint:* n and m must be natural numbers.)
(b) Find the fraction of memory modules kept busy per memory cycle if a computer has 1 processor and 1 memory module; 1 processor and 2 memory modules.
(c) Suppose a computer has 2 processors and 3 memory modules, and then a fourth memory module is added to it. What proportion of memory modules is freed up by this addition?
(d) Suppose a computer has 2 processors and 10 memory modules and an eleventh memory module is added to it. What proportion of memory is freed up by this addition?
(e) Suppose a computer has 2 processors and 5 memory modules and another processor is added. How much additional memory is kept busy by this addition?

Graphing

34. Plot the following points on the cartesian coordinate system:
(a) $(3, 0)$, $(9/2, 0)$, $(-2, 0)$, $(-5.75, 0)$
(b) $(0, 4)$, $(0, -1)$, $(0, -2.3)$, $(0, 8/3)$
(c) $(1, 2)$, $(2, 1)$, $(7/3, 3)$, $(3/2, 5/2)$, $(-2, 4)$, $(-3, 3)$, $(-5, 7/2)$, $(-5/4, 12/5)$, $(-2, -2)$, $(-6, -2)$, $(-0.5, -1)$, $(-4.5, -7/3)$, $(3, -7)$, $(5, -2)$, $(1, -2.8)$, $(2/3, -1/3)$
(d) $(\sqrt{2}, 0)$, $(-\pi, 3)$, $(0, \sqrt{3})$, $(\sqrt{5}, -\sqrt{5})$, $(\pi/2, 3\pi/2)$, $(-2\sqrt{3}, -3\sqrt{2})$

In Exercises 35 through 44, graph the function f defined by the given equation. In each case find the intercepts and range of the function.

35. $f(x) = 3x$

36. $f(x) = 2x - 3$

37. $f(x) = -x^2$

38. $f(x) = x^2 + 2$

39. $f(x) = x^3 + 1$

40. $f(x) = \sqrt{x}$

41. $f(x) = \begin{cases} 1, & x < 2, \\ 3, & x \geq 2 \end{cases}$

42. $f(x) = \begin{cases} x, & 0 \leq x \leq 2, \\ 5, & x > 2 \end{cases}$

43. $f(x) = \begin{cases} -1, & x < 0, \\ 1, & 0 \le x \le 1, \\ x, & x > 1 \end{cases}$

44. $f(x) = \begin{cases} 2x, & 0 \le x \le 2, \\ 6 - x, & 2 < x \le 6 \end{cases}$

In Exercises 45 through 50, state whether the graph shown is the graph of a function.

45.

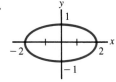

46.

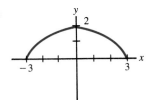

47.

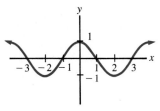

48.

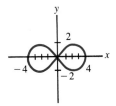

49.

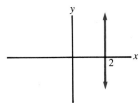

50.

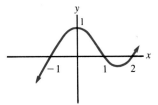

In Exercises 51 through 56, graph the given equation, and state whether the equation defines y as a function of x.

51. $y = -2x$

52. $y = |x - 1|$

53. $x^2 + y^2 = 4$

54. $4x^2 + y^2 = 4$

55. $y^2 = x + 1$

56. $x^2 = y + 1$

*57. For any real number x, define $[x]$ to be the largest integer less than or equal to x.

(a) Find $[0], [2], [8], [1000], [-3], [-7], [-99]$.

(b) Find $[½], [⅝], [-⅔], [-²²/₇], [510/509], [509/510], [-510/509], [-509/510]$.

(c) Find $[0.2], [6.99], [-4.26], [18.01], [-0.45], [-12.98], [-3.01]$.

(d) Define the **greatest integer function** f by $f(x) = [x]$. Graph the greatest integer function.

Management

58. The accompanying graph depicts the total cost to Alfa Company if it produces x units of its product.

(a) What is the total cost to Alfa if it produces 0 units of its product?

(b) What is the total cost to Alfa if it produces 100 units of its product?

(c) How many units can Alfa produce for a total cost of $150,000?

(d) As the number of units produced increases, what happens to Alfa's total cost?

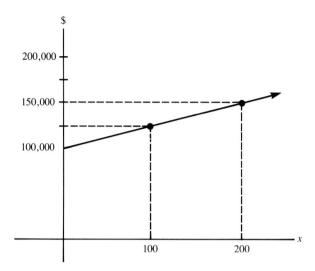

59. The accompanying graph depicts Bayta Corporation's average cost per unit if it makes x units of its product.

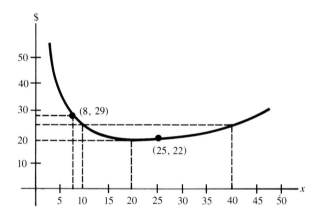

(a) What is Bayta's average cost per unit if it makes 8 units?
(b) What is Bayta's average cost per unit if it makes 25 units?
(c) At what number of units is Bayta's average cost per unit a minimum? What is Bayta's minimum average cost per unit?
(d) How many units must Bayta produce if its average cost per unit is to be $25.
(e) What can you say about Bayta's average cost per unit as the number of units made increases?

60. The accompanying graph depicts the profit of Ghamma Industries, in millions of dollars, if it makes and sells x thousand units of its product.

(a) What is Ghamma's profit if it makes and sells 0 units?
(b) What is Ghamma's profit if it makes and sells 20,000 units?

(c) What is Ghamma's profit if it makes and sells 50,000 units?
(d) How many units must Ghamma make and sell if its profit is to be $3 million?
(e) What is Ghamma's maximum profit? How many units must Ghamma make and sell in order to attain its maximum profit?
(f) What can you say about Ghamma's profit as the number of units produced and sold increases from 0 to 100,000?

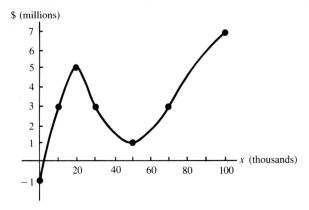

61. The accompanying graph depicts the average annual sales, in thousands of dollars, of Delda Stores salespeople who have been given t weeks of sales training.

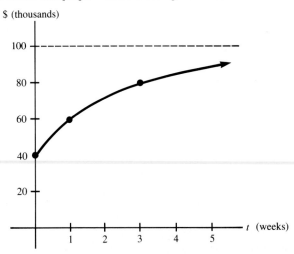

(a) What are average annual sales for salespeople who received no training?
(b) What are average annual sales for salespeople who received 1 week of training?
(c) How many weeks of training must Delda give its salespeople if it wants their average annual sales to be at least $80,000?
(d) What can you say about average annual sales as the number of weeks of training increases?

*62. A firm sells an industrial cleaner at a price of $2 per gallon for the first gallon or part of a gallon, and $1.50 per gallon for each additional gallon or part of a gallon. Let $C(x)$ denote the cost of purchasing x gallons, $x > 0$.

(a) Graph C.
(b) Can you find an equation (or a set of equations) that explicitly defines C as a function of x?

*63. According to information in a recent issue of *The Wall Street Journal*, a discount stockbroker charges a fee of

- $25 plus $0.10 per share traded on trades of fewer than 100 shares;
- $30 plus $0.05 per share traded on trades of at least 100 but fewer than 500 shares;
- $45 plus $0.02 per share traded on trades of 500 or more shares.

Let B(x) be the broker's total fee for executing a trade of x shares.
(a) Graph B.
(b) Can you find an equation (or a set of equations) that explicitly defines B as a function of x?

Life Science

64. The accompanying graph depicts the proportion of bacteria alive in a culture t minutes after it was treated with a bactericide.

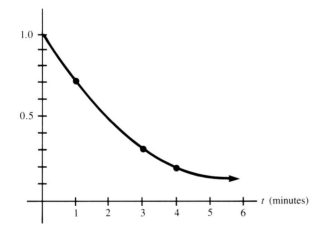

(a) What proportion of the bacteria were alive 1 minute after the application of the bactericide?
(b) What proportion of the bacteria were alive 3 minutes after the application of the bactericide?
(c) How long did it take for the bactericide to kill 80% of the bacteria?
(d) What can you say about the number of bacteria alive in the culture as the time since the bactericide was applied increased?

65. The accompanying graph depicts the planned cleanup of a polluted lake: it shows the concentration of pollutants in the lake, in parts per million (ppm), that are expected to be present t years after the cleanup begins.

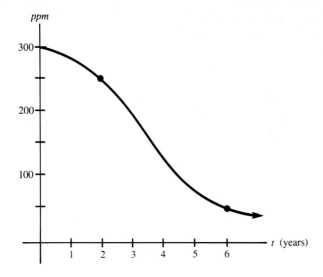

(a) What is the expected concentration of pollutants at the beginning of the cleanup?
(b) What is the expected concentration of pollutants two years after the beginning of the cleanup?
(c) How long will it take until the expected concentration of pollutants is 50 ppm?
(d) What can you say about the concentration of pollutants as the cleanup proceeds?

66. The accompanying graph depicts a patient's temperature t hours after the onset of a disease.

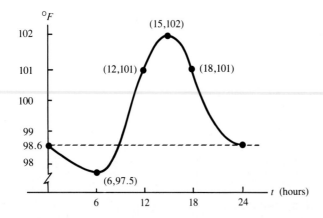

(a) What was the patient's temperature at the onset of the disease?
(b) What was the patient's temperature 6 hours after the onset of the disease?
(c) When was the patient's temperature 101°F?
(d) What was the patient's minimum temperature and when did it occur?
(e) What was the patient's maximum temperature and when did it occur?
(f) What can you say about the patient's temperature during the progress of the disease?

67. The *Physician's Desk Reference* gives the following dosage schedule for an oral suspension of erythromycin:

Body Weight (lb)	Dosage (mg/day)
(0, 10)	15 per lb
[10, 15)	200
[15, 25)	400
[25, 50)	800
[50, 100)	1200
[100, +∞)	1600

Let $d(x)$ be the dosage for a patient whose weight is x pounds.
(a) Graph the function d.
(b) Can you find an equation (or a set of equations) that explicitly defines d as a function of x?

Social Science

68. The accompanying graph depicts the percentage of children in a group who adopted a new fad within t weeks after its first appearance.

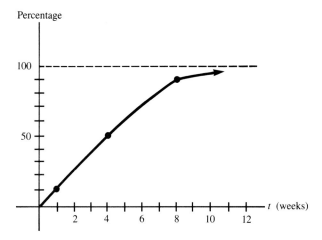

(a) What percentage of the group adopted the fad within one week of its first appearance?
(b) What percentage of the group adopted the fad within eight weeks of its first appearance?
(c) How long did it take until 50% of the group had adopted the fad?
(d) What can you say about the acceptance of the fad by the group?

1.3 ALGEBRA

We will frequently find it useful to combine functions to create new ones. For instance, we will soon see that if functions that describe a firm's cost and revenue are known, then the firm's profit function can be found by subtracting its cost function from its revenue function. In this section we show how new functions may be created from old ones, and we briefly review factorization of polynomial functions.

The Arithmetic of Functions

Functions can be added, subtracted, multiplied, and divided to form new functions. We illustrate with examples.

EXAMPLE 1   Let
$$f(x) = 2x^3 + 3x^2 + 5x - 3$$
and
$$g(x) = 4x^3 - 3x - 6.$$
We find the defining equation of the function $h = f + g$ by adding the corresponding terms of $f(x)$ and $g(x)$:
$$\begin{aligned} h(x) = f(x) + g(x) &= (2x^3 + 3x^2 + 5x - 3) + (4x^3 - 3x - 6) \\ &= (2x^3 + 4x^3) + 3x^2 + (5x - 3x) + (-3 - 6) \\ &= 6x^3 + 3x^2 + 2x - 9. \end{aligned}$$
Thus the defining equation for $h = f + g$ is
$$h(x) = 6x^3 + 3x^2 + 2x - 9.$$
Notice that both f and g have the set of real numbers as their domain and that $h = f + g$ does also.   □

If we can add functions, then we can also subtract one function from another. This is often useful when we are dealing with cost, revenue, and profit, since profit equals revenue minus cost.

Cost, Revenue, and Profit Functions

A **cost function** is a function that gives the cost of making x units of a product. A **revenue function** is a function that gives the revenue gained from selling x units of a product. A **profit function** is a function that gives the profit obtained from making and selling x units of a product. If C is the cost function, R the revenue function, and P the profit function for a product, then
$$P = R - C.$$

EXAMPLE 2   Suppose that if a firm makes and sells x units of its product, its cost function C is defined by the equation
$$C(x) = 20x + 1000$$
and its revenue function R by the equation

$$R(x) = 35x.$$

Here $x \geq 0$, and cost and revenue are in dollars. We find the firm's profit function P by subtracting C from R:

$$P(x) = R(x) - C(x) = 35x - (20x + 1000)$$
$$= (35x - 20x) - 1000 = 15x - 1000.$$

Thus P is given by

$$P(x) = 15x - 1000.$$

The profit earned by making and selling 100 units is therefore

$$P(100) = 15(100) - 1000 = 500$$

dollars. Note that the domain of the function P is the interval $[0, +\infty)$, which is also the domain of R and C. Figure 1–21 shows the graphs of the cost function C, the revenue function R, and the profit function P on the same set of axes. □

EXAMPLE 3 Suppose that

$$f(x) = 4x - 7 \quad \text{and} \quad g(x) = 2x + 6.$$

We find the defining equation of the function $h = fg$ by multiplying each term of $f(x)$ by each term of $g(x)$, using the rules of exponents, and then adding like terms:

$$h(x) = f(x)g(x) = (4x - 7)(2x + 6)$$
$$= (4x)(2x) + (4x)(6) + (-7)(2x) + (-7)(6)$$
$$= 8x^2 + 24x - 14x - 42$$
$$= 8x^2 + 10x - 42.$$

Therefore $h = fg$ has the defining equation

$$h(x) = 8x^2 + 10x - 42.$$

Notice that f, g, and $h = fg$ all have the set of real numbers as their domain. □

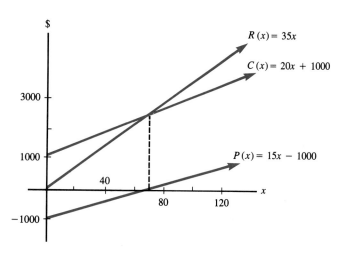

FIGURE 1–21

EXAMPLE 4

If $p(x) = 2x + 1$ and $q(x) = x - 3$, we may form the function $h = p/q$ by dividing $p(x)$ by $q(x)$:

$$h(x) = \frac{p(x)}{q(x)} = \frac{2x + 1}{x - 3}.$$

Notice that the domain of p is the set of real numbers, as is that of q, but that the domain of $h = p/q$ is $\{x | x \neq 3\}$. ☐

A **polynomial function** is a function whose defining equation is of the form

$$p(x) = a_n x^n + a_{n-1} x^{n-1} + \cdots + a_1 x + a_0,$$

where n is a nonnegative integer and the **coefficients** $a_n, a_{n-1}, \ldots, a_1, a_0$ are numbers. All the functions in Examples 1, 2, and 3 are polynomials, and, as these examples suggest, adding, subtracting, and multiplying polynomials always produces another polynomial. However, as Example 4 shows, dividing one polynomial by another does not necessarily produce another polynomial. A function that is a quotient of polynomials, such as the function $h = p/q$ in Example 4, is called a **rational function**. Examples 1 through 4 showed us how to perform arithmetic with polynomial functions. Now let us consider how to perform arithmetic with rational functions.

EXAMPLE 5

Suppose that

$$f(x) = \frac{x + 4}{x - 1} \quad \text{and} \quad g(x) = \frac{2x - 3}{x + 2}.$$

Let us find $f + g$, $f - g$, fg, and f/g.

To add f and g, we must find a common denominator for them: the product $(x - 1)(x + 2)$ of their denominators will serve as a common denominator. To write f and g as rational functions with the common denominator $(x - 1)(x + 2)$, we must multiply the numerator and denominator of $f(x)$ by $x + 2$ and the numerator and denominator of $g(x)$ by $x - 1$:

$$f(x) = \frac{x + 4}{x - 1} \cdot \frac{x + 2}{x + 2}, \qquad g(x) = \frac{2x - 3}{x + 2} \cdot \frac{x - 1}{x - 1}.$$

Therefore,

$$f(x) + g(x) = \frac{x + 4}{x - 1} \cdot \frac{x + 2}{x + 2} + \frac{2x - 3}{x + 2} \cdot \frac{x - 1}{x - 1}$$

$$= \frac{(x + 4)(x + 2) + (2x - 3)(x - 1)}{(x - 1)(x + 2)}$$

$$= \frac{x^2 + 6x + 8 + 2x^2 - 5x + 3}{x^2 + x - 2}$$

$$= \frac{3x^2 + x + 11}{x^2 + x - 2}.$$

Similarly,

$$f(x) - g(x) = \frac{x+4}{x-1} \cdot \frac{x+2}{x+2} - \frac{2x-3}{x+2} \cdot \frac{x-1}{x-1}$$

$$= \frac{(x+4)(x+2) - (2x-3)(x-1)}{(x-1)(x+2)}$$

$$= \frac{(x^2 + 6x + 8) - (2x^2 - 5x + 3)}{x^2 + x - 2}$$

$$= \frac{-x^2 + 11x + 5}{x^2 + x - 2}.$$

To multiply f by g, we multiply their numerators and their denominators:

$$f(x)g(x) = \frac{x+4}{x-1} \cdot \frac{2x-3}{x+2} = \frac{(x+4)(2x-3)}{(x-1)(x+2)} = \frac{2x^2 + 5x - 12}{x^2 + x - 2}.$$

To divide f by g, we invert the expression for $g(x)$ and multiply $f(x)$ by the result:

$$\frac{f(x)}{g(x)} = \frac{x+4}{x-1} \cdot \frac{x+2}{2x-3} = \frac{(x+4)(x+2)}{(x-1)(2x-3)} = \frac{x^2 + 6x + 8}{2x^2 - 5x + 3}. \quad \square$$

Let us summarize the rules of arithmetic for rational functions:

The Arithmetic of Rational Functions

Let f and g be rational functions, with

$$f(x) = \frac{p(x)}{q(x)} \quad \text{and} \quad g(x) = \frac{r(x)}{s(x)}.$$

Then

$$f(x) + g(x) = \frac{p(x)s(x) + r(x)q(x)}{q(x)s(x)},$$

$$f(x) - g(x) = \frac{p(x)s(x) - r(x)q(x)}{q(x)s(x)},$$

$$f(x)g(x) = \frac{p(x)r(x)}{q(x)s(x)},$$

$$\frac{f(x)}{g(x)} = \frac{p(x)s(x)}{q(x)r(x)}.$$

Composition of Functions

Besides addition, subtraction, multiplication, and division, there are other methods of creating new functions from old ones. One of the most important of these is called **composition of functions**. To illustrate composition of functions, let f and g be defined by

$$f(x) = x^2 \quad \text{and} \quad g(x) = x + 1.$$

We can substitute the expression $g(x)$ for x in the equation for $f(x)$ to obtain

$$f(g(x)) = g(x)^2 = (x + 1)^2.$$

The equation

$$f(g(x)) = (x + 1)^2$$

then defines a new function called the **composite of g with f**. This procedure can be carried out using any functions f and g as long as the substitution of $g(x)$ for x in $f(x)$ makes sense.

Composite Functions

If f and g are functions, then the function whose defining equation is given by $f(g(x))$ is called the **composite of g with f**.

EXAMPLE 6

Let

$$f(x) = 3x + 7 \quad \text{and} \quad g(x) = -2x.$$

The composite of g with f is defined by $f(g(x))$, where

$$f(g(x)) = 3(g(x)) + 7 = 3(-2x) + 7 = -6x + 7.$$

Thus,

$$f(g(0)) = -6(0) + 7 = 7,$$
$$f(g(1)) = -6(1) + 7 = 1,$$
$$f(g(2)) = -6(2) + 7 = -5,$$

and so on. □

Given two functions f and g, the composite of g with f is the function whose defining equation is given by $f(g(x))$. It is also possible to consider the composite of f with g: this is the function whose defining equation is given by $g(f(x))$.

EXAMPLE 7

Let

$$f(x) = 3x + 7 \quad \text{and} \quad g(x) = -2x.$$

NUMBERS AND FUNCTIONS

The composite of f with g is defined by $g(f(x))$, where
$$g(f(x)) = -2(f(x)) = -2(3x + 7) = -6x - 14.$$
Note that $g(f(x)) \neq f(g(x))$. □

EXAMPLE 8 Let
$$f(x) = x^2 - 2x + 2 \quad \text{and} \quad g(x) = 2x - 1.$$
Then
$$\begin{aligned}f(g(x)) &= (g(x))^2 - 2g(x) + 2 \\ &= (2x - 1)^2 - 2(2x - 1) + 2 \\ &= 4x^2 - 4x + 1 - 4x + 2 + 2 \\ &= 4x^2 - 8x + 5\end{aligned}$$
and
$$\begin{aligned}g(f(x)) &= 2f(x) - 1 \\ &= 2(x^2 - 2x + 2) - 1 \\ &= 2x^2 - 4x + 4 - 1 \\ &= 2x^2 - 4x + 3.\end{aligned}$$
Note that again $f(g(x)) \neq g(f(x))$. □

Composite functions are often useful when one variable is a function of a second and the second is a function of a third. We can then use composition of functions to write the first variable as a function of the third. Thus, if
$$y = x^2 + 1 \quad \text{and} \quad x = t + 5,$$
we may write y as a function of t by forming the composite function:
$$y = (t + 5)^2 + 1 = t^2 + 10t + 26.$$

EXAMPLE 9 Suppose the cats eat the rats and the rats eat the bats. Suppose also that if c is the number of cats, r the number of rats, and b the number of bats, then
$$c = 0.1r^{2/3} \quad \text{and} \quad r = \sqrt{b}.$$
How many cats can b bats support? The answer is
$$c = 0.1(\sqrt{b})^{2/3} = 0.1\sqrt[3]{b}.$$
For instance, 1000 bats can support 1 cat. □

Many functions can be written as composites of simpler ones. For example, the function k defined by $k(x) = (x + 1)^2$ may be written as the composite of $g(x) = x + 1$ with $f(x) = x^2$, because
$$f(g(x)) = f(x + 1) = (x + 1)^2 = k(x).$$

In Chapter 4 it will be important for us to be able to write functions as composites. We give an example to show how this is done.

EXAMPLE 10 Let $k(x) = (2x + 3)^{20}$. We seek to write the function k as a composite function by finding f and g such that $k(x) = f(g(x))$. The way to do this is to choose $g(x)$ to be the "inside" expression in $k(x)$ and $f(x)$ to be the "outside" expression in $k(x)$. Thus, if we take $g(x) = 2x + 3$ and $f(x) = x^{20}$, then

$$f(g(x)) = f(2x + 3) = (2x + 3)^{20} = k(x).$$

Similarly, if

$$h(x) = \sqrt{2x^3 + 4x},$$

and we take

$$g(x) = 2x^3 + 4x \quad \text{and} \quad f(x) = \sqrt{x},$$

then

$$f(g(x)) = f(2x^3 + 4x) = \sqrt{2x^3 + 4x} = h(x). \quad \square$$

Inverse Functions

The final method we will consider for creating new functions from old ones is that of finding the **inverse** of a function. Before we can define inverse functions, however, we must discuss one-to-one functions.

Some functions have the property of always assigning different numbers in their range to different numbers in their domain. Such functions are said to be **one-to-one**.

> **One-to-One Function**
>
> A function f is **one-to-one** if $f(x_1) \neq f(x_2)$ whenever $x_1 \neq x_2$.

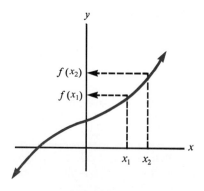

FIGURE 1–22
A One-to-One Function

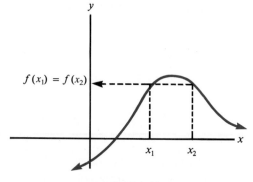

FIGURE 1–23
A Function that Is Not One-to-One

Numbers and Functions

Figure 1–22 shows the graph of a function that is one-to-one, while Figure 1–23 shows the graph of a function that is not one-to-one.

EXAMPLE 11 The function f defined by $f(x) = 2x$ is one-to-one, because if $x_1 \neq x_2$, then $f(x_1) = 2x_1 \neq 2x_2 = f(x_2)$. The function g defined by $g(x) = x^2$ is not one-to-one because $-2 \neq 2$, but $g(-2) = 4 = g(2)$. ☐

If a function f is one-to-one, we can define a function f^{-1} called the **inverse of f**:

The Inverse Function

Let f be a one-to-one function. The **inverse function of f** is the function f^{-1} defined as follows: for each y in the range of f,

$$f^{-1}(y) = x \quad \text{if and only if} \quad f(x) = y.$$

See Figure 1–24. Note that the domain of f^{-1} is the range of f, and the range of f^{-1} is the domain of f. Also, a point (x, y) is on the graph of f if and only if $y = f(x)$; but this is so if and only if $x = f^{-1}(y)$, and thus if and only if $(y, f^{-1}(y)) = (y, x)$ is on the graph of f^{-1}. Therefore the graphs of f and its inverse f^{-1} are symmetric about the line $y = x$, meaning that if either graph were rotated about this line, it would coincide with the other. See Figure 1–25.

If f is a one-to-one function, we can find the defining equation for its inverse function f^{-1} by setting $y = f(x)$, solving this equation for x in terms of y, and interchanging the variables x and y. We illustrate with an example.

EXAMPLE 12 Let $f(x) = 2x$. Since f is a one-to-one function, it has an inverse function f^{-1}. We set $y = 2x$ and solve this equation for x in terms of y, thus obtaining $x = y/2$. If we interchange x and y in this last equation, we have $y = x/2$. Thus the defining equation for f^{-1} is $f^{-1}(x) = x/2$. Figure 1–26 shows the graphs of f and f^{-1}: note the symmetry about the line $y = x$. ☐

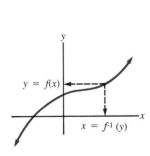

FIGURE 1–24
The Inverse Function

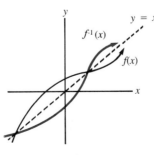

FIGURE 1–25

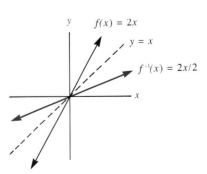

FIGURE 1–26

We must emphasize that the inverse function f^{-1} is defined *only* if f is one-to-one. This is so because if f is *not* one-to-one, then there are numbers x_1 and x_2, $x_1 \neq x_2$, such that $f(x_1) = f(x_2)$. Thus if $y = f(x_1) = f(x_2)$, then we would have to define $f^{-1}(y) = x_1$ *and* $f^{-1}(y) = x_2$, and hence f^{-1} would not be a function.

Factoring

We conclude this section with a brief review of factoring. Polynomial functions can sometimes be **factored** into a product of simpler polynomials. Factoring can be important if we need to solve an equation of the form $p(x) = 0$, where p is a polynomial. We illustrate with examples.

EXAMPLE 13 Let us factor the polynomial expression

$$p(x) = 3x^3 + 9x^2 - 30x.$$

Factorization begins with a search for the **common factors** of the terms of the polynomial. In this example each term has the common factor x, and the coefficients 3, 9, and -30 have the common factor 3. Such common factors can be **factored out** of the polynomial. Thus,

$$p(x) = 3x(x^2 + 3x - 10).$$

Now we seek to factor the polynomial expression in parentheses. If this expression does factor, it will do so as

$$x^2 + 3x - 10 = (x + a)(x + b),$$

where a and b are numbers. But since

$$(x + a)(x + b) = x^2 + (a + b)x + ab,$$

we see by comparing

$$x^2 + 3x - 10$$

with

$$x^2 + (a + b)x + ab$$

that we must have

$$ab = -10 \quad \text{and} \quad a + b = 3;$$

that is, we need to find a pair of numbers whose products is -10 and whose sum is 3. Such a pair is 5 and -2. Therefore

$$x^2 + 3x - 10 = (x + 5)(x - 2),$$

and

$$p(x) = 3x(x + 5)(x - 2).$$

We can check the factorization by multiplying out the right side to see that it is equivalent to the original expression for $p(x)$. □

Now suppose we wish to find the x-intercepts of the polynomial function p whose defining equation is $p(x) = 3x^3 + 9x^2 - 30x$. Then we must solve the equation $p(x) = 0$ for x. If we use the factored form of $p(x)$ obtained in Example 13, we must solve

$$3x(x + 5)(x - 2) = 0$$

for x. But

a product is zero if and only if at least one of its factors is zero.

Therefore the equation

$$3x(x + 5)(x - 2) = 0$$

tells us that either

$$x = 0, \; x + 5 = 0, \; \text{or} \; x - 2 = 0;$$

that is, that either

$$x = 0, \; x = -5, \; \text{or} \; x = 2.$$

Hence the x-intercepts of p are $(0, 0)$, $(-5, 0)$, and $(2, 0)$.

EXAMPLE 14   An ecologist studying a herd of minke whales has estimated that the population P of the herd in year t, where $t = 0$ represents 1988, is given by

$$P = -t^2 - 14t + 220.$$

Suppose the ecologist wants to know when the population will be 100 individuals. Then she must set $P = 100$ and solve the polynomial equation

$$100 = -t^2 - 14t + 220,$$

or

$$-t^2 - 14t + 120 = 0,$$

or

$$t^2 + 14t - 120 = 0.$$

To factor this polynomial, she must find two numbers whose product is -120 and whose sum is 14; 20 and -6 are such numbers. Therefore

$$t^2 + 14t - 120 = (t + 20)(t - 6),$$

so the equation in factored form is

$$(t + 20)(t - 6) = 0.$$

Setting each factor equal to zero, she finds that $t = -20$ and $t = 6$. Thus the population was 100 in 1968 ($t = -20$) and will be 100 again in 1994 ($t = 6$). ◻

■ *Computer Manual: Programs POLYRITH, RATLRITH, COMPOSIT*

1.3 EXERCISES

The Arithmetic of Functions

1. If $f(x) = 2x - 5$ and $g(x) = 3x + 9$, find the defining equations for $f + g$, $f - g$, fg, and f/g. Also find $f(2) + g(2)$, $f(0) - g(0)$, $f(3)g(3)$, and $f(-1)/g(-1)$. What is the domain of $f + g$? of f/g?

2. If $f(x) = 7x + 2$ and $g(x) = x^2 + x - 1$, find the defining equations for $f + g$, $f - g$, fg, and f/g. Also find $f(3) - g(3)$ and $f(2)/g(2)$.

3. If $f(x) = x^2 + 6x - 10$ and $g(x) = -3x^2 + 2x - 13$, find the defining equations for fg and f/g. Also find $f(1)g(1)$ and $f(0)/g(0)$.

4. If $f(x) = 5x^4 - 6x^2 + 2x$ and $g(x) = -3x^3 + 7x^2 - 8x + 11$, find the defining equations for $f + g$ and $f - g$. Also find $f(2) + g(2)$ and $f(-1) - g(-1)$.

5. If $f(x) = \dfrac{x - 2}{x + 3}$ and $g(x) = \dfrac{2x + 4}{x + 1}$, find the defining equations for $f + g$, $f - g$, fg, and f/g. Also find $f(1) + g(1)$ and $f(0)/g(0)$. What is the domain of $f + g$? of fg?

6. If $f(x) = \dfrac{x^2 + 9}{x + 4}$ and $g(x) = \dfrac{x^2 + x}{x^2 - 3}$, find the defining equations for $f + g$, $f - g$, fg, and f/g. Also find $f(2) - g(2)$ and $f(3)/g(3)$. What is the domain of fg?

7. If $f(x) = \dfrac{3x - 5}{x + 1}$ and $g(x) = \dfrac{(x - 3)(x + 2)}{4 - 2x}$, find the defining equations for $f + g$, fg, and f/g. What is the domain of $f + g$? of fg? of f/g?

8. If $f(x) = \dfrac{(x + 6)(2x - 8)}{(x - 1)(2x + 5)}$ and $g(x) = \dfrac{x(x + 6)}{x - 1}$, find the defining equations for $f - g$, fg, and f/g. What is the domain of $f - g$? of fg? of f/g?

Management

9. If a firm produces and sells x units of its product, its cost is $C(x)$ dollars and its revenue $R(x)$ dollars, where

$$C(x) = 200x + 10{,}000 \quad \text{and} \quad R(x) = 230x.$$

   (a) Find the firm's profit function P.
   (b) Find the firm's profit if it makes and sells 100 units. If the firm is currently producing and selling q units, what will be the additional profit if it makes and sells one more unit?

10. If a company produces and sells x units of its product, its cost is $C(x)$ dollars and its revenue $R(x)$ dollars, where

$$C(x) = 40x + 78{,}000 \quad \text{and} \quad R(x) = 45x.$$

    Find the company's profit function and then find its profit if it makes and sells 12,000 units; 15,600 units; 22,000 units.

11. If Puissant Flange Company makes and sells x flanges, its cost is $C(x) = 2x^2 + 2000$ dollars and its revenue $R(x) = x^2 + 50x + 1600$ dollars.

(a) Find Puissant's profit function and its profit when it makes and sells 5, 10, 30, 40, and 60 units.

(b) If Puissant is currently making and selling b units, find its additional cost, revenue, and profit if it makes and sells one more unit.

12. Each freezer unit is made by installing one cooling assembly in one insulated metal shell. If the cost of the raw materials used to make x cooling assemblies is given by
$$f(x) = 200x + 5000$$
and the cost of raw materials used to make x insulated metal shells is given by
$$g(x) = 400x + 2000,$$
find the function h that describes the total cost of raw materials used in making x freezer units. Graph h, f, and g on the same set of axes. Find the cost of raw materials if 20 freezer units are made.

Life Science

13. An ecologist studying wolves and their prey estimates that over an 8-year period the sizes of the wolf population and the rabbit population are given by
$$W(t) = 2000 + 400t - 50t^2$$
and
$$R(t) = 100{,}000 + 600t - 150t^2$$
respectively. Here $0 \leq t \leq 8$. Graph the functions W, R, and $W + R$ on the same set of axes and estimate when the peak population of wolves, the peak population of rabbits, and the peak population of wolves and rabbits combined occur.

Composition of Functions

In Exercises 14 through 21, find $f(g(x))$ and $g(f(x))$.

14. $f(x) = 4x$, $g(x) = 5 - 2x$.

15. $f(x) = 3x + 6$, $g(x) = x^2 + 4$.

16. $f(x) = x^2 - 3x + 12$, $g(x) = \sqrt{x + 1}$.

17. $f(x) = x^4 + 5x^2 + 1$, $g(x) = x + h$.

18. $f(x) = x^3 + x$, $g(x) = x^{2/3}$.

19. $f(x) = \dfrac{1}{x}$, $g(x) = x + h$.

20. $f(x) = \dfrac{x + 3}{x - 4}$, $g(x) = \dfrac{2}{x}$.

21. $f(x) = \dfrac{2x - 3}{7 + 2x}$, $g(x) = \dfrac{5x + 1}{x^2 - 1}$.

22. If $y = 5x + 3$ and $x = t^2 - 4$, express y as a function of t.

23. If $y = 4x - 16$ and $x = t + 4$, express y as a function of t.

24. If $u = 17v^2 + 2\sqrt{v} + 3$ and $v = w^{3/4}$, express u as a function of w.

25. If y and x are as in Exercise 23, express t as a function of y.

In Exercises 26 through 33, write k as a composite function.

26. $k(x) = (3x + 5)^{10}$

27. $k(x) = (5 - 7x)^{-4}$

28. $k(x) = (x^2 - 2x)^5$

29. $k(x) = \sqrt{4x - 1}$

30. $k(x) = (x^2 + 7)^{3/2}$

31. $k(x) = \sqrt[3]{x^2 + 1}$

32. $k(x) = \sqrt[5]{x^2 + 3}$

33. $k(x) = (1 - 2x^{-1})^{-3/2}$

Management

34. A firm's daily profit is P dollars, where $P(x) = 1.2x - 10{,}000$ and production x is a function of the number of employees n according to the equation $x = 125n$. Express the firm's daily profit as a function of n and find the daily profit when $n = 100$.

35. Suppose that if a commodity sells for $\$p$ per unit, consumers will purchase $q = 10{,}000/p$ units. Suppose also that the price has been increasing according to the equation $p = 0.02t + 5$, where t is in months, with $t = 0$ representing the current month. Find an expression for the quantity demanded by consumers as a function of time, and use it to forecast quantity demanded six months from now.

36. The quantity of a product demanded by consumers depends on its price: if the product sells for p dollars per unit, quantity demanded will be

$$q = 500p^2 + 3000p + 10{,}000$$

units. It is estimated that the price of the product will fall during the near future according to the equation

$$p = \frac{60}{t + 10},$$

where t is the time in days from the present. Find a function that expresses the quantity demanded as a function of time. What is the quantity demanded now? What will it be five days from now?

Life Science

37. It is estimated that the percentage of ozone remaining in the ozone layer when a total of x tons of chlorofluorocarbons (CFCs) have been released into the atmosphere is

$$f(x) = \frac{100}{0.16x + 1}.$$

It is also estimated that the total amount of CFCs released into the atmosphere is $x = 12t + 22$, where t is in years with $t = 0$ representing 1980. Find an expression for the percentage of ozone remaining in the ozone layer as a function of time.

38. The carbon monoxide concentration in the air of a large city depends on the number of cars driven in the city each day. The number of cars driven in the city is expected to increase with time. Suppose that if x cars are driven in the city each day, then the concentration of carbon monoxide in the atmosphere is

$$c(x) = 10^{-9}x^2 + 10^{-5}x + 1$$

parts per million (ppm); suppose also that the number of cars driven in the city per day is forecast to be

$$x = 10{,}000t + 100{,}000$$

in year t, where $t = 0$ represents 1987. Find a function that describes the carbon monoxide concentration as a function of time, and forecast this concentration in the year 1995.

Inverse Functions

In Exercises 39 through 44, state whether the graph is that of a one-to-one function; if it is, sketch the graph of its inverse function.

39.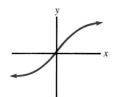

40.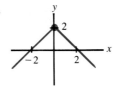

41.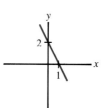

42.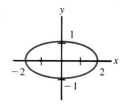

43.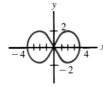

44.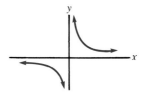

In Exercises 45 through 52, state whether the function f is one-to-one; if it is, find the defining equation for its inverse function f^{-1}.

45. $f(x) = 3x - 1$ **46.** $f(x) = 3 - 4x$ **47.** $f(x) = -x$ **48.** $f(x) = x^2 - 1$

49. $f(x) = x^3$ **50.** $f(x) = 2$ **51.** $f(x) = \sqrt{x}$ **52.** $f(x) = 1/x$

Factoring

In Exercises 53 through 64, solve the equation $f(x) = 0$ for x.

53. $f(x) = x^2 - x - 12$ **54.** $f(x) = x^2 + 8x + 15$ **55.** $f(x) = x^2 - 16$

56. $f(x) = 18x - 2x^3$ **57.** $f(x) = x^2 + 4x + 4$ **58.** $f(x) = 4x^4 - 24x^3 + 36x^2$

59. $f(x) = 3x^2 - 9x - 12$ **60.** $f(x) = -x^3 + 9x^2 - 14x$ **61.** $f(x) = x^3 + x^2 - 2x$

62. $f(x) = x^6 - 2x^5 + x^4$ **63.** $f(x) = 2x^2 + x - 3$ **64.** $f(x) = \frac{1}{2} + \frac{1}{2}x - x^2$

65. If a stone is thrown straight up from ground level its altitude t seconds later is $-16t^2 + 90t$ feet. Find the time it takes until the stone hits the ground.

66. Referring to Exercise 65: find the times at which the stone is at an altitude of 116 feet.

*67. An 8-inch by 12-inch block of text is to be printed on a rectangular poster board whose area is 396 square inches. If the top, bottom, and side margins around the block of text are to be of equal width, find this width.

*68. A rectangular sheet of tin is twice as long as it is wide. A small square 10 centimeters (cm) by 10 cm is cut from each corner, and the edges are then folded up to make a box. If the volume of the box is 5280 cm³, find the dimensions of the tin sheet.

Management

69. A firm is said to break even if its profit from making and selling its product is zero, since if this happens the firm is neither making nor losing money. Suppose Acme Company's profit is

$$P(x) = x^2 + 20x - 1500$$

dollars if it makes and sells x units. How many units must Acme make and sell in order to break even?

70. If a firm makes and sells x units of its product its profit will be $-x^2 + 170x - 6600$ dollars. How many units must it make and sell if its profit is to be $0?

71. How many units must the firm of Exercise 70 make and sell if its profit is to be $625?

72. Alef Company intends to depreciate its new headquarters building in the following way: t years after the construction of the building, its book value will be $-t^2 - 24t + 385$ million dollars. When will the book value of the building be zero?

*73. If a firm charges p for each unit of its product it can sell $5000 - 25p$ units. What price must it charge in order to have sales revenue of $246,400?

*74. Suppose a firm can sell 1000 units of its product at $50 each, but that if it raises the price it will sell five units fewer for each $1 rise in price. What price must the firm charge in order to have its revenue from sales amount to $58,280?

Life Science

75. It is estimated that the concentration of sulfur dioxide in the air over a large city will average

$$y = 100 + 6t - t^2$$

parts per million in year t, where $t = 0$ represents 1990. In what year will the concentration average 28 ppm?

Social Science

76. The number of riders on a city's transit system is given by

$$y = t^2 - 40t + 800,$$

where t is in years, with $t = 0$ representing 1970, and y is thousands of riders. When will ridership reach one half of the level of 1970?

1.4 LINEAR FUNCTIONS

A **linear function** is a polynomial function whose defining equation can be written in the form

$$y = mx + b,$$

where m and b are numbers. A knowledge of linear functions will be required when we begin our study of calculus in the next chapter. In addition, linear functions are

NUMBERS AND FUNCTIONS

important because they can be used to model many practical situations. In this section we discuss graphing linear functions, finding the equation of a line, and finding the point where two lines intersect. We also illustrate linear models and their uses with descriptions of linear break-even analysis and supply–demand analysis.

Graphing Linear Functions

The graph of a linear function is a line. (Throughout this book "line" means "straight line.") Since two points determine a line, when graphing a linear function we need only find two points on its graph.

EXAMPLE 1 The graph of the linear function defined by

$$y = 2x + 4$$

is shown in Figure 1–27. Check that the points $(-1, 2)$ and $(1, 6)$ are on the line. ☐

EXAMPLE 2 The equation

$$6x + 2y = 4$$

may be solved for y to obtain the equation

$$y = -3x + 2;$$

thus the graph of the equation is a line. The line is depicted in Figure 1–28: here we used the x-intercept $(\frac{2}{3}, 0)$ and the y-intercept $(0, 2)$ as the two points we needed to draw the graph. ☐

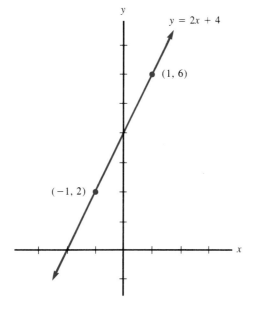

FIGURE 1–27

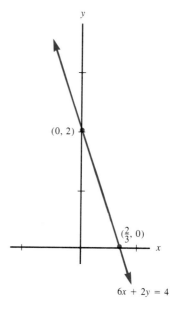

FIGURE 1–28

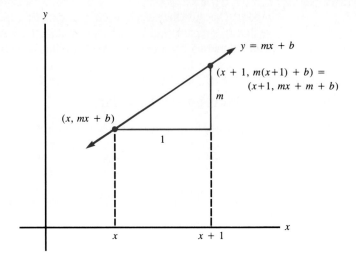

FIGURE 1–29

For a linear function with defining equation $y = mx + b$, every increase of 1 in the value of x causes a change of m in the value of y. See Figure 1–29. Therefore $y = mx + b$ increases (if m is positive) or decreases (if m is negative) at the constant rate m. The number m is called the **slope** of the line. For example, the line of Figure 1–27 has slope $m = 2$: thus every increase of 1 in the value of x causes the value of y to increase by 2. Similarly, the line of Figure 1–28 has slope $m = -3$, so every increase of 1 in the value of x causes the value of y to decrease by 3. Note also that the function defined by $y = mx + b$ has y-intercept $(0, b)$.

A vertical line such as that pictured in Figure 1–30 is *not* the graph of a function. (Why not?) A vertical line has no slope.

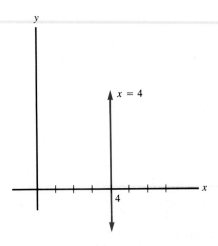

FIGURE 1–30

Finding the Equation of a Line

Every line is the graph of some equation, called the **equation of the line.** If a line is vertical, its equation is of the form $x = c$, where $(c, 0)$ is the x-intercept of the line. Thus the line of Figure 1–30 is the graph of the equation $x = 4$. If a nonvertical line has slope m and y-intercept $(0, b)$, then its equation is $y = mx + b$.

EXAMPLE 3 Let us find the equation of the line that has slope -4 and passes through the point $(0, 3)$. Since we are told the slope and y-intercept of the line, we can immediately write its equation $y = -4x + 3$. □

EXAMPLE 4 Let us find the equation of the line that has slope 2 and passes through the point $(3, 5)$. Here we know the slope of the line but not its y-intercept. However, since $m = 2$, the equation must be of the form

$$y = 2x + b.$$

Since the point $(3, 5)$ is on the line, it must satisfy this equation. Hence

$$5 = 2(3) + b = 6 + b,$$

so $b = -1$. Therefore the equation is

$$y = 2x - 1. \quad □$$

We have noted that if a function is linear, then it changes at a constant rate equal to the slope of the line. The converse of this statement is also true: if a function changes at a constant rate, it is linear and the rate is the slope of the line.

EXAMPLE 5 Suppose Alpha Company's sales were $100,000 in 1980 and have increased by the constant amount of $20,000 each year. Let us find Alpha's annual sales S as a function of time t.

Because the change in Alpha's sales is constant at $20,000 each year, the function we seek will be linear with slope $m = 20,000$. If we measure time in years, with $t = 0$ representing 1980, then we have $b = 100,000$. Hence the function is given by

$$S = 20,000t + 100,000. \quad □$$

We often want to find the equation of a line when its slope is not given explicitly. However, we can find the slope of a line if we know two points on the line. Thus, if the equation of the line is $y = mx + b$ and (x_1, y_1), (x_2, y_2) are points on the line, then we must have

$$y_1 = mx_1 + b \quad \text{and} \quad y_2 = mx_2 + b.$$

Subtracting the first equation from the second, we obtain

$$y_2 - y_1 = mx_2 - mx_1 = m(x_2 - x_1).$$

Therefore

$$m = \frac{y_2 - y_1}{x_2 - x_1}.$$

> **The Slope of a Line through Two Points**
>
> A nonvertical line passing through the points (x_1, y_1) and (x_2, y_2) has slope
>
> $$m = \frac{y_2 - y_1}{x_2 - x_1}.$$

Geometrically, this result says that we can find the slope of a nonvertical line by choosing any two points (x_1, y_1), (x_2, y_2) on the line, using them to draw a right triangle as in Figure 1–31, and setting

$$m = \frac{\text{Vertical ``rise''}}{\text{Horizontal ``run''}} = \frac{y_2 - y_1}{x_2 - x_1}.$$

EXAMPLE 6 Let us find the equation of the line passing through the points $(2, 5)$ and $(-2, 13)$. Our strategy will be to find the slope m first, then substitute one of the points into the resulting equation to find b. Setting

$$(x_1, y_1) = (2, 5) \quad \text{and} \quad (x_2, y_2) = (-2, 13),$$

we have

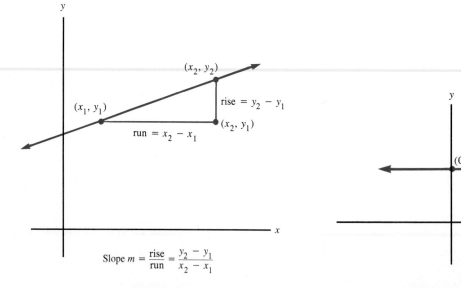

FIGURE 1–31

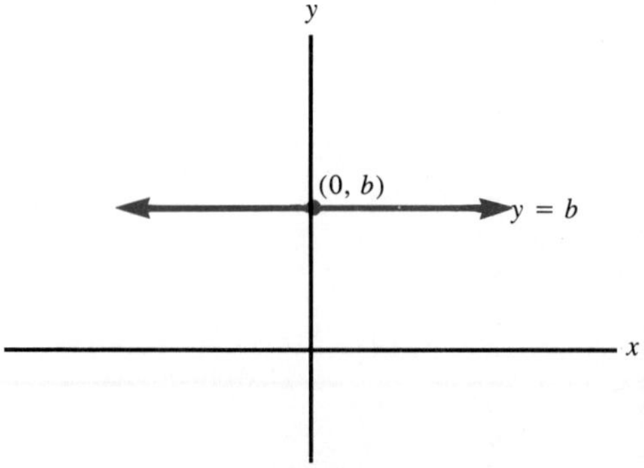

FIGURE 1–32
Graph of the Constant Function $y = b$

Numbers and Functions

$$m = \frac{y_2 - y_1}{x_2 - x_1} = \frac{13 - 5}{-2 - 2} = -2.$$

Therefore $y = -2x + b$. But since the point (2, 5) is on the line, we must have $5 = -2(2) + b$, from which we find that $b = 9$. Hence the equation of the line is

$$y = -2x + 9. \quad \square$$

If the slope of a line is zero, the line is horizontal (why?) and its equation reduces to $y = b$. See Figure 1–32. A function whose defining equation is of the form $y = b$ is called a **constant function.** Also, if a line has no slope then it is vertical and will have an equation of the form $x = c$.

Sometimes it is convenient to write the equation of a line in the form

$$Ax + By = C.$$

This is called the **general form** of the equation of a line.

EXAMPLE 7 Beta Company produces large and small boilers. Each large boiler uses 20 kilograms (kg) of steel, and each small one uses 10 kg of steel. If Beta has 1000 kg of steel on hand and wants to use it all up, it can make any combination of x large boilers and y small boilers as long as

$$20x + 10y = 1000, \qquad x \geq 0, \quad y \geq 0.$$

The graph of this equation is the line segment shown in Figure 1–33. $\square$

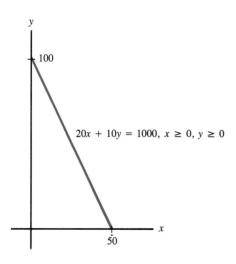

FIGURE 1–33

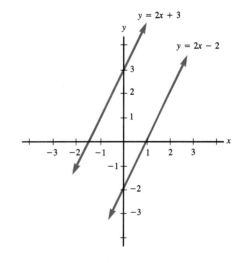

FIGURE 1–34
Parallel Lines

Intersection of Lines

Lines with the same slope are **parallel;** parallel lines do not intersect. The lines whose equations are

$$y = 2x - 2 \quad \text{and} \quad y = 2x + 3$$

are parallel and do not intersect. See Figure 1–34.

Lines whose slopes are not equal are not parallel, and they intersect in a point. Since the y-coordinate of the point of intersection must be the same for both lines, we can find the point of intersection by setting the right sides of their equations equal, solving for x, and then back-substituting to find y. We illustrate with an example.

EXAMPLE 8   Find the point of intersection of the lines whose equations are

$$y = 2x - 9 \quad \text{and} \quad y = x + 3.$$

We set

$$2x - 9 = x + 3$$

and solve for x to obtain $x = 12$. Substituting 12 for x in either $y = 2x - 9$ or $y = x + 3$ yields $y = 15$. Therefore the point of intersection of the lines is (12, 15). See Figure 1–35. □

Finding the point of intersection of two lines can often be of practical importance.

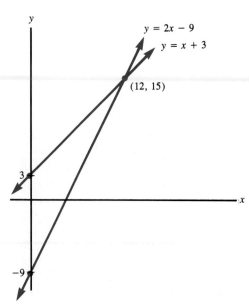

FIGURE 1–35

EXAMPLE 9 Suppose each Brand X vitamin pill contains 10 milligrams (mg) of vitamin A and 20 mg of vitamin C, and each Brand Y pill contains 10 mg of A and 30 mg of C. If you need exactly 50 mg of A and 120 mg of C each day, how many pills of each brand should you take?

Let x denote the number of pills of Brand X and y the number of pills of Brand Y that you should take. Then your vitamin requirements imply that

$$\text{(vitamin A)} \qquad 10x + 10y = 50,$$
$$\text{(vitamin C)} \qquad 20x + 30y = 120.$$

Solving these equations for y, we have

$$y = -x + 5 \quad \text{and} \quad y = \frac{2}{3}x + 4.$$

Therefore

$$-x + 5 = -\frac{2}{3}x + 4,$$

or

$$-\frac{1}{3}x = -1,$$

so $x = 3$. Hence $y = -3 + 5 = 2$. □

Break-Even Analysis

We often use linear functions to model a firm's cost and revenue when performing **break-even analysis.** Break-even analysis begins with a **cost function** C, which gives the total cost of producing x units of a product, and a **revenue function** R, which gives the total revenue obtained from selling x units of the product. If the cost and revenue functions are linear, they will have defining equations of the form

$$C(x) = vx + F \quad \text{and} \quad R(x) = px,$$

respectively. Here F is the **fixed cost,** which is the cost of producing zero units of the product, v is the **variable cost per unit,** which is the cost of producing each unit of product, and p is the selling price per unit. Note that since x is the number of units produced and sold, $x \geq 0$.

Figure 1–36 shows the graphs of typical linear cost and revenue functions on the same set of axes. The point of intersection of the cost line and the revenue line is called the firm's **break-even point,** and its x-coordinate is the **break-even quantity.** As Figure 1–36 shows, if the firm makes and sells its break-even quantity, then its cost will equal its revenue and it will break even. If the firm makes and sells fewer units than its break-even quantity, it will suffer a loss; if it makes and sells more, it will enjoy a gain.

EXAMPLE 10 Suppose that Gamma Company's cost and revenue functions are given by

$$C(x) = 5x + 100{,}000 \quad \text{and} \quad R(x) = 7x,$$

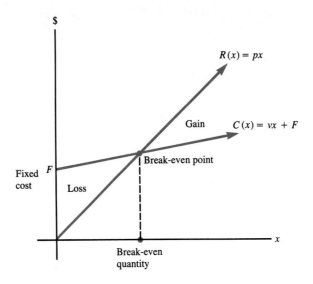

FIGURE 1–36

respectively. Then Gamma's fixed cost is $100,000, its variable cost per unit is $5, and it sells each unit for $7. We find Gamma's break-even quantity by setting cost equal to revenue and solving for x:

$$5x + 100{,}000 = 7x$$

yields $x = 50{,}000$. Therefore the break-even point is (50,000, 350,000). Thus if Gamma makes and sells 50,000 units, its cost and its revenue will both equal $350,000 and the company will break even.

Note also that Gamma's profit function P is given by

$$P(x) = R(x) - C(x) = 2x - 100{,}000.$$

From this linear function we see that each unit produced and sold contributes $2 toward profit. Note that we can also obtain the break-even quantity of 50,000 units by solving the equation $P(x) = 0$. Gamma Company's cost, revenue, and profit functions are graphed on the same set of axes in Figure 1–37. □

Supply and Demand

Linear functions are also often used to model situations of supply and demand. A **supply function** is a function that gives the number of units of a commodity that producers will supply to the market at any given price per unit. A **demand function** gives the number of units of the commodity that consumers will purchase at any given price per unit. Figure 1–38 shows typical linear supply and demand functions graphed on the same set of axes. (The graph in Figure 1–38 indicates that some units are supplied at every price greater than zero; this need not always be the case.) Note that as the unit price of the commodity increases, the quantity supplied to the

NUMBERS AND FUNCTIONS 57

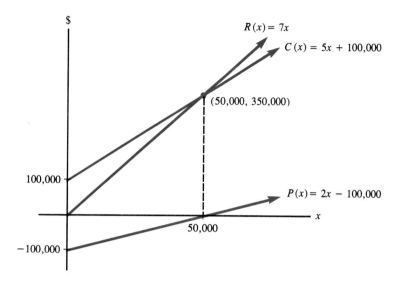

FIGURE 1-37

market by producers increases while the quantity demanded by consumers decreases. The point of intersection of the supply and demand functions is called the **market equilibrium point,** and its price coordinate is the **market equilibrium price.** At the market equilibrium price, quantity supplied and quantity demanded are equal, so there will be neither a shortage nor a surplus of the product at this price.

EXAMPLE 11   Suppose the supply and demand functions for microcomputer stands are given by

$$q = 500p \quad \text{and} \quad q = -1000p + 120{,}000,$$

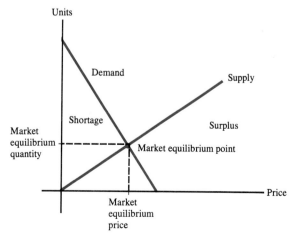

FIGURE 1-38

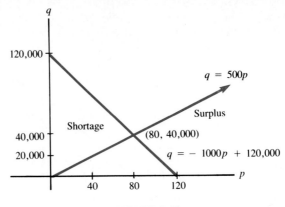

FIGURE 1–39

respectively. Here p is the unit price of the stands, in dollars, and q is the quantity of stands supplied by producers or demanded by consumers, as the case may be. If we set quantity supplied equal to quantity demanded, we have

$$500p = -1000p + 120,000,$$

or

$$1500p = 120,000,$$

which yields $p = 80$. It then follows that the market equilibrium point is $(80, 40,000)$. See Figure 1–39. If the stands sell for \$80 each, both quantity supplied and quantity demanded will be 40,000 units. If they sell for less than \$80, quantity demanded will exceed quantity supplied, and there will be a shortage of the stands; if they sell for more than \$80 each, quantity supplied will exceed quantity demanded and there will be a surplus. ☐

■ *Computer Manual: Programs SLPNRCPT, INTRSECT, BREAKEVN*

1.4 EXERCISES

Graphing Linear Functions

In Exercises 1 through 8, graph the equation.

1. $y = 4x - 8$
2. $y = -x + 12$
3. $y = 2x$
4. $2x + 3y = 12$
5. $2x - 5y = 10$
6. $2x + 4y = 0$
7. $y = 4$
8. $x = -2$

Management

9. Acme Company's sales are described by the equation
$$S = 200t + 500,$$

where t is time in years with $t = 0$ representing 1985 and S is annual sales in thousands of dollars. Graph this function. What were Acme's sales in 1985? Are sales increasing or decreasing each year? by how much?

10. Baker Corporation makes regular and deluxe model radios. The numbers of regular and deluxe radios the firm can produce are related by the equation
$$20x + 30y = 60{,}000, \qquad x \geq 0, \quad y \geq 0,$$
where x is the number of regular radios and y is the number of deluxe radios. Graph this equation. Can Baker produce 1200 regular and 1200 deluxe radios? 1800 regular and 800 deluxe? 1600 regular and 1000 deluxe? What is the maximum number of each kind of radio that the firm can produce?

*11. This Exercise was suggested by material in Higgins, *Analysis for Financial Management*, Irwin, 1984. It concerns the effect of leverage on a corporation's return on equity. Return on equity, or *ROE*, is a financial ratio defined by
$$ROE = \frac{\text{Profit after taxes}}{\text{Owners' equity}}$$
Owners' equity is the amount invested in the company by the shareholders; it is the total ownership interest of the shareholders and is equal to the value of the corporation's assets minus its debts. Return on equity is important because it estimates the return to owners on their investment in the company. Corporations strive to make their *ROE* as large as possible. Suppose we let

D = outstanding debt,

OE = owners' equity,

L = leverage = D/OE,

A = asset value = $D + OE$,

E = earnings before interest and taxes,

i = interest rate paid on debt, in decimal form,

t = tax rate, in decimal form.

Define the return on assets *(ROA)* by
$$ROA = \frac{\text{Earnings after interest and taxes}}{\text{Value of assets}} = \frac{E - t(E - iD)}{A}.$$
Then *ROE* can be written in the form
$$ROE = ROA + (ROA - i)L.$$

(a) Find a corporation's *ROE* if its asset value is $600 million, its debt is $400 million, its earnings before interest and taxes are $200 million, the interest rate it pays on debt is 15%, and its tax rate is 40%.

(b) What is a corporation's *ROE* if it has no debt?

(c) Graph *ROE* as a function of L if $ROA > i$. What happens to *ROE* when L increases by 1?

(d) Repeat part (c) when $ROA = i$.

(e) Repeat part (c) when $ROA < i$.

(f) A corporation will adopt different financial strategies with regard to its debt and owners' equity depending on whether $ROA > i$, $ROA = i$, or $ROA < i$. To see this, determine how a corporation can increase its *ROE* if its *ROA* is greater than *i*; if its *ROA* is equal to *i*; if its *ROA* is less than *i*.

Life Science

12. If the concentration of pollutants in a lake is x mg/L, then the proportion of fish eggs that will hatch is

$$y = 0.25 - 0.005x.$$

    Graph this equation. For every increase of 1 mg/L in pollution, what additional proportion of fish eggs fail to hatch? At what level of pollution will all the eggs fail to hatch?

13. Weber's Law states that if a subject is exposed to a stimulus of level s, the smallest change c in the stimulus that will be noticeable to the subject is given by $c = ms$, for some number m. Suppose that a subject is exposed to a sound level of s decibels (dB) and that $m = 0.02$. Find the smallest noticeable change in the sound level when the sound level is 5000 dB; when it is 50,000 dB. Graph $c = 0.02s$. For every increase of 100 dB in the sound level, what happens to the smallest noticeable change in the sound level?

14. In an article entitled "Mental Imagery and the Visual System" (*Scientific American,* March 1986), Ronald Finke reported on an experiment in which subjects were shown a drawing containing a pattern of dots and an arrow and were asked if the arrow was pointing at any of the dots. It was found that if x was the distance of the arrow from the dots, in centimeters, and y the time it took the subject to make a decision, in milliseconds, then y was given by

$$y = 850 + 26x.$$

    Find the reaction time y when the arrow was 2.5 cm from the dots; when it was 25 cm from the dots. Graph the function. What is the significance of its y-intercept? What is the significance of its slope?

*15. In an article entitled "Features and Objects in Visual Processing" (*Scientific American,* November 1986), Anne Treisman reported on an experiment in which a "target" symbol either was or was not placed among a group of "distractor" symbols and subjects were asked to locate the target symbol.

  (a) If there were x distractor symbols and no target symbol present, the average search time until the subject decided that the target was not present was y milliseconds, where

$$y = 40x + 440.$$

  Find the average search time when there were 10 distractor symbols present. Graph the function. What is the significance of its slope?

  (b) If there were x distractor symbols and one target symbol present, the average search time until the subject located the target was y milliseconds, where

$$y = 20x + 430.$$

  Find the average search time when there were 10 distractor symbols present. Graph the function. What is the significance of its slope?

  (c) If subjects examine each symbol individually until they either find the target symbol or decide that it is not present, then on average it should take about half as long to locate a target symbol as it should to decide that one is not present.

Do the graphs of parts (a) and (b) support the hypothesis that subjects examine each symbol individually?

Social Science

16. In a certain state the proportion of voters who are Independents is given by
$$-0.2t + 20y = 9,$$
where t is time in years, coded so that $t = 0$ represents 1988, and y is the proportion of Independents in the state in year t. Graph this equation. What proportion of the voters in the state were Independents in 1978? in 1988? in 1990? By how much is the proportion of Independents changing each year?

Finding the Equation of a Line

In Exercises 17 through 24, find the equation of the line.

17. The line with slope -3 and y-intercept (0, 2).
18. The line with slope 5 and y-intercept (0, -3).
19. The line passing through the point (4, 8) with slope 6.
20. The line passing through the point (-5, -9) with slope -2.
21. The line passing through the points (2, 6) and (3, 8).
22. The line passing through the points (5, -2) and (0, -7).
23. The line passing through the points (6, 2) and (4, 2).
24. The line passing through the points (5, -2) and (5, 8).
25. Let F be temperature in degrees Fahrenheit, and let C be temperature in degrees Celsius. When $F = 32$, $C = 0$; when $F = 212$, $C = 100$. The relation between the two temperature scales is linear. Find an equation that relates F and C.
26. If the there were 20 million tons of ozone in the ozone layer in 1987 and the amount of ozone is decreasing at a rate of 1.25 million tons per year, find an equation that describes the amount of ozone present as a function of time.
    (a) How much ozone will be present in 1996?
    (b) How long will it take until 75% of the ozone is gone?
27. According to a recent article in *Newsweek*, the greenhouse effect may increase the average global temperature by approximately 0.6°F per decade. If the average global temperature is now 59.5°F, find an equation that describes average global temperature as a function of time.
    (a) When will the average global temperature reach 61°F?
    (b) How long will it be until the average global temperature is 64°F?
28. The concentration of a chemical accidentally spilled into a river is decreasing at a constant rate and was 600 ppm one month after the spill and 510 ppm three months after the spill.
    (a) Find a function that describes the concentration of the chemical t months after the spill.
    (b) What was the initial concentration of the chemical?
    (c) How long will it take until all traces of the chemical are gone?

Management

29. Epsilon, Inc. has been increasing the size of its sales force by 10 persons per year. If the sales forces consisted of 150 salespersons in 1988, find an equation that describes the size of the sales force as a function of time, and forecast
    (a) the size of the sales force in 1995;
    (b) the year in which the sales force will consist of 210 salespeople.

30. Zeta Corporation has been losing market share at the rate of 1.5% per year. If Zeta's market share in 1985 was 45.7%, find an equation that describes Zeta's market share as a function of time, and find
    (a) the corporation's market share in 1990;
    (b) the year in which its market share will be 29.2%.

31. According to an article in the September 29, 1988 issue of *The Wall Street Journal*, worldwide usage of platinum was expected to be 3,320,000 tons in 1988 and 3,860,000 tons in 1993. Find a linear equation that describes worldwide usage of platinum as a function of time.
    (a) Forecast usage in 1991.
    (b) When will usage reach the level of 4,000,000 tons per year?

32. When Atlantic Hardware, Inc. has 10 stores, its total annual sales were $125 million; when it had 17 stores, they were $212.5 million. Find a linear equation that describes annual sales as a function of the number of stores. What is the significance of the slope of this equation? How many stores would Atlantic need in order to have annual sales of $300 million?

33. A camera store orders 200 cameras at a time from its supplier and sells them at the rate of 4 per day. Find an equation that expresses the number of cameras the store has on hand as a function of the time since the arrival of the last order.
    (a) How many cameras will the store have on hand 22 days after an order arrives?
    (b) If it takes 10 days from the time an order is sent in until the cameras arrive at the store, what is the latest the store can order cameras if it does not want to run out of them?

34. According to an article in the September 20, 1988 issue of *The Wall Street Journal*, most stockbrokers earn no salary, but instead receive a commission equal to 40% of their sales. Find an equation that describes a stockbroker's commissions as a function of sales.
    (a) What is the commission of a stockbroker who sells $140,000?
    (b) Find the sales that will result in a commission of $100,000.

35. Farson Corporation produces small and large clocks. Their production cost is $25 for each small clock and $35 for each large one. Farson's production budget for the clocks is $105,000, and management intends to use the entire amount. Find and graph an equation describing the number of each size clock the company can produce. How many small clocks can the firm produce if it wants to make 1250 large clocks? What is the maximum number of each size clock the firm can produce?

36. The advertising director of Theta Corporation knows that when the firm spent $100,000 on broadcasting TV ads the ads were seen by 2,500,000 people and when the firm spent $300,000 broadcasting the ads, they were seen by 3,450,000

people. Find a function that expresses audience size as a function of amount spent, assuming that these quantities are linearly related. What is the significance of the slope of this linear function? If the ads must be seen by at least 5,000,000 people, how much must be budgeted for them? What size audience can be reached if the budget for the ads is $1,000,000?

37. A haberdasher carries two lines of overcoats: a high-priced line, on which the profit is $80 per overcoat, and a low-priced line, on which the profit is $45 per overcoat. If the haberdasher's profit from sales of overcoats is $18,000, find an equation that describes the number of each type of overcoat sold. Could the haberdasher have sold 150 of the high-priced coats and 200 of the low-priced ones? 135 of the high-priced ones and 160 of the low-priced ones? What is the maximum number of high-priced coats the haberdasher could have sold? the maximum number of low-priced coats?

*38. This exercise was suggested by material in Higgins, *Analysis for Financial Management,* Irwin, 1984. It concerns the **percentage-of-sales method** of determining gross profit.

(a) Suppose a firm's sales are given by a linear equation

$$S = mt + b,$$

where S is sales in dollars and $t = 0$ represents the present. Suppose also that the cost to the firm of the goods it sells (its **cost of goods sold**) is a certain proportion P of its sales. Find an equation that describes the firm's cost of goods sold as a function of time. Assuming that m and b are positive and that $0 < P < 1$, graph the sales equation and the cost of goods sold equation on the same set of axes.

(b) A firm's **gross profit** is equal to its sales minus its cost of goods sold. For the firm of part (a), find an equation that describes gross profit as a function of time. Then graph the gross profit equation on the same set of axes with the sales and cost of goods sold equations when

(i) $P > 0.5$ (ii) $P = 0.5$ (iii) $P < 0.5$.

(c) Suppose a firm's sales are given by $S = 8.5t + 22$, where S is in millions of dollars and t is in years with $t = 0$ representing the present. If the firm's cost of goods sold is 82% of sales, use the result of part (b) to find the firm's gross profit two years from now.

(d) Repeat part (c) if the cost of goods sold is 45% of sales.

Life Science

39. A pediatrician studying babies finds that, on average, newborns are 53 cm tall and grow at the rate of 2.1 cm per month until the age of 9 months. Find

(a) a function that describes the relationship between average height and age in months;

(b) the average height of babies at age 4.5 months;

(c) the number of months required for the average height to reach 64 cm.

40. It is estimated that the coyote population of New England is increasing by 30 individuals per year, and that there were 250 coyotes in New England in 1988. Find an equation that describes the size of the coyote population in New England as a function of time.

(a) How many coyotes will there be in New England in 1995?
(b) When will the New England coyote population reach 500?

41. A patient needs to receive exactly 2000 units of a certain drug each day. Brand A pills each contain 80 units of the drug, while Brand B pills each contain 125 units. Find an equation that describes the relationship between the number of pills of each brand that the patient can take. How many Brand A pills can he take if he takes no Brand B pills? if he takes eight Brand B pills?

*42. In an article entitled "The Large-Scale Cultivation of Mammalian Cells" (*Scientific American,* January 1983), Joseph Feder and William Tolbert reported on an experiment that involved growing cells in a glucose solution: the cells took up the glucose and produced lactic acid. Eighteen hours after the beginning of the experiment the concentration of glucose in the solution was 3.92 milligrams per milliliter (mg/mL) and that of lactic acid was 0.90 mg/mL; 32 hours later the corresponding figures were 2.80 mg/mL of glucose and 2.50 mg/mL of lactic acid. Assuming linearity, find equations that describe both the concentration of glucose and the concentration of lactic acid in the solution as a function of time. Graph these equations on the same set of axes and use the graphs to estimate the time when the concentrations were equal.

Social Science

43. Newly hired salespersons were given an aptitude test, and one year later their test scores were correlated with the total dollar value of the sales they made during their first year of employment. It was found that those who scored 52 on the test averaged $119,000 in sales their first year, and those who scored 74 averaged $163,000 in first-year sales. Assuming a linear relationship between score on the test and average first-year sales, find a function describing this relationship. What is the significance of the slope of this linear function? What is the significance of its y-intercept?

44. A town has local elections every two years. In 1986, 62% of the eligible voters actually voted, and since then the percentage of eligible voters who have voted has decreased at the rate of 4.2% per election. Find an equation that relates the percentage of eligible voters who actually vote to time in years. Also find:
(a) the percentage of eligible voters who voted in 1990;
(b) the first election year in which the percentage who vote will fall below 50%.

45. In a certain city the number of high-school dropouts has been increasing at a rate of 160 individuals per year. If there were 2240 high-school dropouts in the city last year, find an equation that describes the number of dropouts as a function of time.
(a) How many dropouts will there be six years from now?
(b) When will the droput rate reach 3000 per year?

46. If 63.5% of the population of a developing nation was literate in 1980 and 65.8% was literate in 1985, and if literacy has been increasing in a linear fashion, find an equation that describes literacy as a function of time.
(a) What percentage of the population will be literate in 1990?
(b) When will 90% of the population be literate?

Intersection of Lines

In Exercises 47 through 52, find the point of intersection of the given lines.

47. $y = 2x - 5, y = 5x + 7$
48. $y = -3x + 1, y = -2x + 2$
49. $-2x + 4y = 8, 2x - 3y = 12$
50. $3x + y = 4, -5x + 4y = 16$
51. $-3x + 6y = 2, y = 0.5x - 5$
52. $5x + 4y = 10, 15x + 12y = 21$

53. An investor has $10,000 and wants to invest part of the money in stock A and part in stock B. Stock A pays a dividend of 6% per year, stock B a dividend of 8% per year. How much should be invested in each stock if the annual dividends must amount to $750?

Management

54. Triangle Company manufactures small and large wheelbarrows. Two worker-hours are needed to produce each small wheelbarrow, and three worker-hours are required to make each large one. Each wheelbarrow has one rubber wheel. There are 126 worker-hours and 52 rubber wheels available today. How many of each size wheelbarrow will Triangle produce today?

55. (This exercise draws on material in Peled, "The Next Computer Revolution," *Scientific American*, October 1987.) The speed of computers is measured by how many million instructions per second (MIPS) they can execute. Suppose a business must decide whether to purchase a computer with a single high-speed processor or one with several slower processors. The computer with the single processor will cost $250,000 plus $12 per MIPS, while the one with the multiprocessor will cost $200,000 plus $22 per MIPS.
  - **(a)** Find and graph equations that express the total cost of each computer as a function of the number of MIPS required.
  - **(b)** Over what range of MIPS will the computer with the single processor be the cheaper one?

*56. Circle Company intends to purchase a machine for producing a new product. They can purchase any one of three machines, A, B, or C. Machine A costs $20,000 and can make up to 50,000 units of the product per year at a cost of $5.00 per unit. Thus, the cost equation for machine A is

$$C_A = 5.00x + 20{,}000,$$

where C_A is in dollars, x is the number of units of the product made, and

$$0 \le x \le 50{,}000.$$

Machine B costs $35,000 and can make up to 40,000 units of the product per year at a cost of $3.50 per unit. Machine C costs $50,000 and can make up to 100,000 units per year at a cost of $3.00 per unit.
  - **(a)** Write the cost equations for machine B and machine C.
  - **(b)** Graph the cost equations for the three machines on the same set of axes.
  - **(c)** For what annual production levels will machine A be the least costly? machine B? machine C?

*57. (This exercise is based on material in Stevenson, *Production/Operations Management*, 2nd edition, Irwin, 1986.) A manufacturer can build a plant at one of four locations. The fixed costs and variable costs per unit for production at each location are as follows:

Location	Fixed Cost	Variable Cost/Unit
A	$100,000	$2.00
B	80,000	2.50
C	86,000	2.20
D	120,000	1.80

For each location, find the range of production for which it will be the best alternative.

Life Science

58. According to an article by Cole and Zuckerman ("Marriage, Motherhood and Research Performance in Science," *Scientific American,* February 1987), during the period 1970–1985 the number of males earning Ph.D.'s in the life sciences remained approximately constant at 4000 per year, while the number of females earning such Ph.D.'s increased from approximately 600 in 1970 to 1700 in 1985. If the increase in female Ph.D.'s was linear, when will the annual number of Ph.D.'s in the life sciences earned by females equal those earned by males?

59. A certain habitat currently supports 1618 individuals of animal species A and none of species B. Biologists plan to introduce two individuals of species B into the area. They estimate that if this is done the population of species A will thereafter decrease by 50 individuals per year while that of species B will increase by 30 individuals per year. How long will it take after the introduction of species B until the populations of the two species in the habitat are the same?

60. Each Brand X vitamin pill contains 10 mg of vitamin A and 20 mg of vitamin C. Each brand Y vitamin pill contains 10 mg of vitamin A and 30 mg of vitamin C. If you require 50 mg of vitamin A and 120 mg of vitamin C each day, how many of each pill must you take?

Social Science

61. National mathematics test scores for first-year college students have been decreasing linearly at a rate of 4.5 points per year. The average score in 1985 was 478. A certain college knows that its students' mathematics test scores have been increasing linearly, with the average score being 385 in 1983 and 397 in 1986. When will the average test score for this college reach the national average?

62. In 1980, 48.6% of registered voters in a certain state were Democrats, 36.4% were Republicans, and the rest were independents. The percentage of registered Democrats has been declining at a constant rate of 2% per year, the percentage of registered Republicans has been declining at a constant rate of 0.8% per year, and the percentage of registered independents has been increasing at a constant rate of 1.9% per year. When will the number of independents equal the number of
    (a) Republicans?
    (b) Democrats?
    (c) Republicans and Democrats combined?

Break-Even Analysis

63. Ajax Company's fixed cost is $100,000, its variable cost per unit is $6, and it sells each unit it makes for $10. Find Ajax's cost and revenue functions and its break-even point.

64. Boodle, Inc. has cost and revenue functions given by

$$C(x) = 15x + 35,000 \quad \text{and} \quad R(x) = 20x,$$

respectively. Graph these functions on the same set of axes. What are the significances of the slopes and intercepts of these functions? Find Boodle's break-even point. Find Boodle's profit function. What are the significance of its slope and intercepts?

65. Delta Corporation's cost function is given by $C = 42.50x + 38,500$ and it sells each unit it makes for $55.40. Graph the cost and revenue functions on the same set of axes and find the break-even point.

66. Epsilon, Ltd., has fixed cost of $23,000 and variable cost per unit of $2.40. It sells each unit it makes for $3.20. Find its break-even point.

67. Chaco Corporation's profit function is given by $P(x) = 7x - 280,000$. Graph this function and find Chaco's break-even quantity.

68. Zeta Company's profit function is given by $P = 2.2x - 319,600$. Graph this function and find the firm's break-even quantity.

69. Repeat Exercise 68 if Zeta's profit function is given by $P = 55.25x - 1,025,000$.

*70. An article by Julian Szekely ("Can Advanced Technology Save the U.S. Steel Industry?", *Scientific American,* July 1987) contains the following cost information for three types of steel mills:

Type of Mill	Fixed Cost ($ per ton of capacity)	Variable Cost ($ per ton)
Old integrated	$230–$280	$ 95–$120
Modern integrated	$140–$190	$ 85–$115
Minimill	$ 30–$105	$110–$205

(a) Suppose a mill has 100,000 tons of capacity. Assuming linearity, find its cost function in the best possible case (lowest fixed cost, lowest variable cost per ton) if it is an old integrated mill; if it is a modern integrated mill; if it is a minimill.

(b) Graph the cost functions of part (a) on the same set of axes. If steel sells for $250 per ton, find the break-even quantity for each type of mill and compare it with the mill's capacity.

(c) Suppose steel sells for p per ton. For the cost functions of part (a), over what range of prices p will each type of mill reach break-even at or before its capacity of 100,000 tons?

(d) Repeat part (a) in the worst possible case (highest fixed cost, highest variable cost per ton).

(e) Graph the cost functions of part (d) on the same set of axes. If steel sells for $250 per ton, find the break-even quantity for each type of mill.

(f) Suppose steel sells for p per ton. Using the cost functions of part (d), over what range of prices p will each type of mill reach break-even at or before its capacity of 100,000 tons?

Supply and Demand

71. The supply and demand functions for a certain product are given by
$$q = 8000p \quad \text{and} \quad q = -12{,}000p + 200{,}000,$$
respectively. Graph these functions on the same set of axes. Find the market equilibrium point for the product. What are the significances of the slopes and intercepts of these functions?

72. Repeat Exercise 71 if $q = 8000p - 6000$.

73. A commodity has demand function given by $q = -6250p + 112{,}000$ and supply function given by $q = 8300p - 4400$. Find the market equilibrium point.

74. Repeat Exercise 73 if $q = -300p + 9348$ and $q = 460p$.

75. When book bags sold for $10 each, the quantity supplied to the market was 15,000 units and the quantity demanded by consumers was 90,000 units. When they sold for $20 each, the quantity supplied was 40,000 units and the quantity demanded was 60,000 units. Assuming linearity of supply and demand, find the supply and demand functions for book bags and their market equilibrium point.

76. A commodity has equilibrium point (9.50, 20,000) and its supply and demand functions are linear. At a price of $4.50 per unit, no units will be supplied and quantity demanded will be 35,000 units. Find the supply and demand functions for the commodity.

77. A product's equilibrium points is (125, 16,250). Its supply and demand functions are linear and when the unit price is $100, quantity demanded is 29,000 units while quantity supplied is 10,750 units. Find the supply and demand functions.

78. Suppose the supply and demand functions for a commodity are given by $p = 0.02q$ and $p = -0.03q + 50$, respectively. Here p is the unit price of the commodity and q is quantity supplied or demanded. Graph these equations on the same set of axes and find the equilibrium point for the commodity.

1.5 QUADRATIC FUNCTIONS

It will be helpful for us to have some knowledge of the characteristics of quadratic functions and models based on them. In this section we discuss quadratic functions and quadratic models.

Quadratic Functions

A polynomial function whose defining equation is of the form
$$y = ax^2 + bx + c,$$
where a, b, c are numbers and $a \neq 0$, is called a **quadratic function.** The graph of a quadratic function is called a **parabola;** if $a > 0$, the parabola opens upward, as in Figure 1–40; if $a < 0$, it opens downward, as in Figure 1–41. The point labeled V in each figure is the **vertex** of the parabola. Every parabola that is the graph of

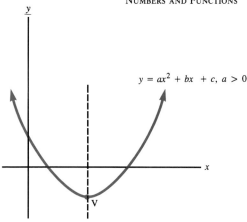

FIGURE 1–40

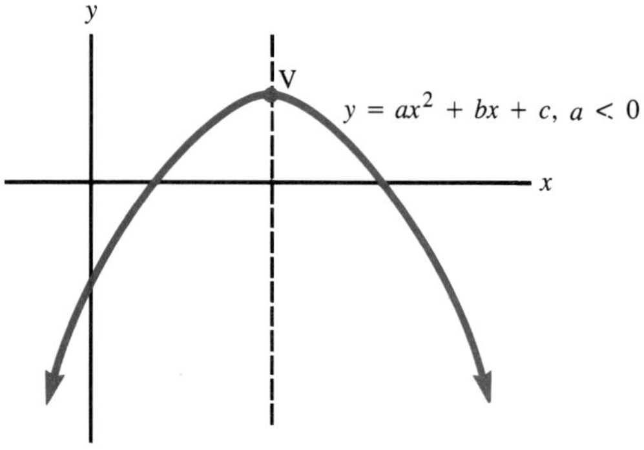

FIGURE 1–41

a function is symmetric with respect to a vertical line through its vertex; in other words, the two halves of the graph on either side of such a vertical line are mirror images of one another. When drawing the graph of a quadratic function, we must find its vertex and its intercepts. Let us determine how to do this.

The standard form of the defining equation of a quadratic function can be written as

$$y = a\left(x + \frac{b}{2a}\right)^2 + c - \frac{b^2}{4a}.$$

(You should check this by simplifying the right-hand side of the equation to obtain $y = ax^2 + bx + c$.) If a is positive, the parabola that is the graph of the equation opens upward and hence has its vertex at its lowest point. Since

$$\left(x + \frac{b}{2a}\right)^2$$

is nonnegative and since a is positive, the smallest value that y can assume occurs when

$$x + \frac{b}{2a} = 0,$$

that is, when

$$x = -\frac{b}{2a}.$$

Hence, the lowest point of the parabola, its vertex, is at the point with coordinates

$$x = -\frac{b}{2a} \quad \text{and} \quad y = c - \frac{b^2}{4a}.$$

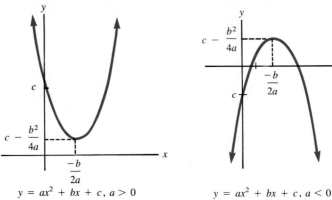

FIGURE 1–42

Similar reasoning shows that the vertex has these same coordinates if a is negative. Consequently, the vertex of the parabola that is the graph of $y = ax^2 + bx + c$ is the point

$$\left(-\frac{b}{2a},\, c - \frac{b^2}{4a}\right).$$

See Figure 1–42.

Next we consider the intercepts of a quadratic. The y-intercept of the quadratic function defined by $y = ax^2 + bx + c$ is the point $(0, c)$ (why?), and its x-intercepts may be found by solving the equation

$$ax^2 + bx + c = 0$$

for x. We can attempt to solve this equation by factoring, as discussed in Section 1.3, or by means of the **quadratic formula.**

The Quadratic Formula

Given a quadratic equation $ax^2 + bx + c = 0$:

1. If $b^2 - 4ac \geq 0$, the solutions of the equation are

$$x = \frac{-b \pm \sqrt{b^2 - 4ac}}{2a}.$$

2. If $b^2 - 4ac < 0$, the equation has no real solutions.

The quantity $b^2 - 4ac$ is called the **discriminant** of the quadratic.

(A derivation of the quadratic formula is outlined in Exercise 17 at the end of this section.)

NUMBERS AND FUNCTIONS

If we now put all the information we have developed together, we obtain the following procedure for graphing a quadratic function.

Graphing a Quadratic Function

The graph of the quadratic function defined by
$$y = ax^2 + bx + c$$
is a parabola that:

1. opens upward if a is positive and opens downward if a is negative;
2. has its vertex at $\left(-\dfrac{b}{2a},\ c - \dfrac{b^2}{4a}\right)$;
3. has y-intercept $(0, c)$; and
4. has x-intercepts, if the parabola intersects the x-axis, at the solutions of the quadratic equation $ax^2 + bx + c = 0$.

We illustrate the process of graphing quadratic functions with examples.

EXAMPLE 1   Let us graph the quadratic function whose defining equation is
$$y = x^2 - 2x - 3.$$
Comparing $y = x^2 - 2x - 3$ with the standard form $y = ax^2 + bx + c$, we see that $a = 1$, $b = -2$, and $c = -3$. Hence $a > 0$, so the parabola opens upward, and its vertex is at
$$(-b/2a,\ c - b^2/4a) = (-(-2)/2(1)),\ -3 - (-2)^2/4(1)) = (1, -4).$$
The y-intercept is the point $(0, -3)$. To find the x-intercepts (if any), we must solve the equation
$$x^2 - 2x - 3 = 0$$
for x. Since this can be factored as
$$(x - 3)(x + 1) = 0,$$
its solutions are $x = 3$ and $x = -1$. Hence the x-intercepts are $(3, 0)$ and $(-1, 0)$. The graph is shown in Figure 1–43. ☐

EXAMPLE 2   Let us graph the quadratic function defined by
$$y = -x^2 + 2x + 4.$$
Here $a = -1$, $b = 2$, and $c = 4$. Since $a < 0$, the parabola opens downward. Its vertex is at
$$(-b/2a,\ c - b^2/4a) = (-2^2/2(-1),\ 4 - (2)^2/4(-1)) = (1, 5).$$

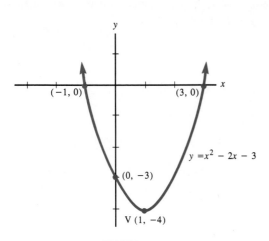

FIGURE 1–43   FIGURE 1–44

The y-intercept is $(0, 4)$. We find the x-intercepts by using the quadratic formula to solve

$$-x^2 + 2x + 4 = 0.$$

We have

$$x = \frac{-b \pm \sqrt{b^2 - 4ac}}{2a} = \frac{-2 \pm \sqrt{2^2 - 4(-1)(4)}}{2(-1)} = \frac{-2 \pm \sqrt{20}}{-2}$$

$$= \frac{-2 \pm 2\sqrt{5}}{-2} = 1 \pm \sqrt{5}.$$

Thus the x-intercepts are $(1 + \sqrt{5}, 0)$ and $(1 - \sqrt{5}, 0)$. [Since $\sqrt{5} \approx 2.24$, these points are approximately $(3.24, 0)$ and $(-1.24, 0)$.] The graph is shown in Figure 1–44. ◻

EXAMPLE 3   The graph of the quadratic function given by

$$y = 2x^2 + 8x + 9$$

is depicted in Figure 1–45. The parabola opens upward because $a = 2 > 0$, the vertex is at

$$(-8/2(2), 9 - 8^2/4(2)) = (-2, 1),$$

the y-intercept is $(0, 9)$, and there are no x-intercepts because

$$b^2 - 4ac = 8^2 - 4(2)(9) = -8 < 0. \quad ◻$$

Quadratic Models

Quadratic functions are often useful in creating models for situations that cannot be handled with linear functions. We illustrate with examples.

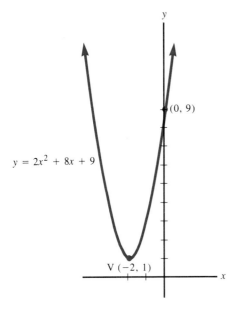

FIGURE 1–45

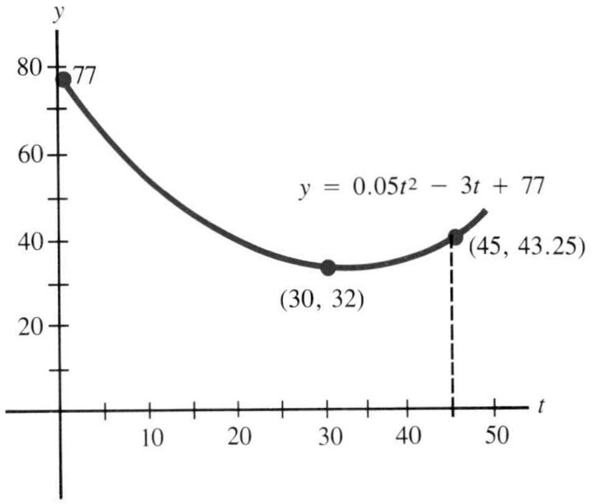

FIGURE 1–46

EXAMPLE 4 Suppose that t months after the inauguration of a president of the United States, $y\%$ of the public approves of the job he is doing, where

$$y = 0.05t^2 - 3t + 77.$$

The graph of this quadratic is the parabola of Figure 1–46. Note that public approval of the president is high (77%) at the time of his inauguration, that it declines as his term proceeds, reaching a low point of 32% at 30 months after inauguration, and that it then increases somewhat as the next election approaches. If the next election will take place 45 months after inauguration, we can use the model to predict that at that time the president will have the approval of

$$y = 0.05(45)^2 - 3(45) + 77 = 43.25,$$

or 43.25% of the population. ☐

EXAMPLE 5 Cost, revenue, and profit functions can be quadratics. For instance, suppose that Acme Company's cost and revenue functions are given by

$$C = x^2 + 30x + 500 \quad \text{and} \quad R = 90x.$$

The graphs of these functions are shown in Figure 1–47. We see from the figure that Acme has two break-even points, which may be found by setting cost equal to revenue and solving for x:

$$C = R,$$
$$x^2 + 30x + 500 = 90x,$$
$$x^2 - 60x + 500 = 0,$$
$$(x - 10)(x - 50) = 0.$$

Thus, $x = 10$ and $x = 50$. The break-even points are $(10, 900)$ and $(50, 4500)$, as shown in Figure 1–47.

Notice that Acme's profit function is given by $P = -x^2 + 60x - 500$. The graph of this quadratic is shown in Figure 1–48. We see from the graph that Acme's maximum profit occurs when it makes and sells $x = 30$ units and that the firm's maximum profit is $400. □

EXAMPLE 6 Supply and demand functions can also be quadratics. To illustrate, suppose that the supply and demand functions for gadgets are given by

$$q = p^2 + 30p \quad \text{and} \quad q = -p^2 - 100p + 2400,$$

respectively. See Figure 1–49. If we set supply equal to demand and solve for price p, we obtain $p = 15$ and $p = -80$. (Check this.) Discarding the negative solution as meaningless, we see that the market equilibrium point is $(15, 675)$ as indicated in Figure 1–49. □

Just as any two points determine a line, any three points not on a line determine a parabola. Thus we can fit a quadratic model to any three observations that do not lie on a line.

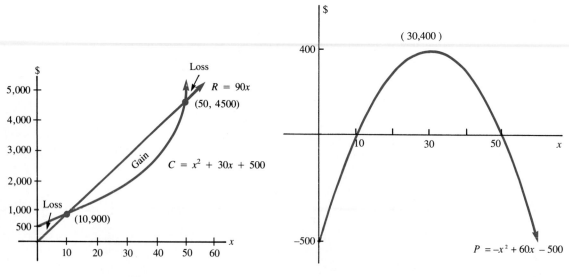

FIGURE 1–47

FIGURE 1–48

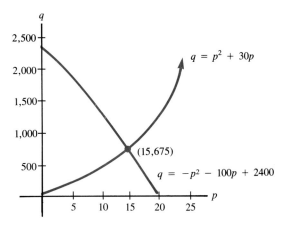

FIGURE 1–49

EXAMPLE 7 One hour after being admitted to a hospital a patient's temperature was 2.1° above normal, 2 hours after being admitted it was 3.0° above normal, and 3 hours after being admitted it was 3.7° above normal. Let us fit a quadratic model to these observations.

We let y denote the patient's temperature t hours after being admitted to the hospital, y measured in degrees above normal. Then $y = 2.1$ when $t = 1$, $y = 3.0$ when $t = 2$, and $y = 3.7$ when $t = 3$. If y is a quadratic function, so that $y = at^2 + bt + c$ for some a, b, and c, we have

$$2.1 = a \cdot 1^2 + b \cdot 1 + c,$$
$$3.0 = a \cdot 2^2 + b \cdot 2 + c,$$

and

$$3.7 = a \cdot 3^2 + b \cdot 3 + c,$$

or

$$a + b + c = 2.1,$$
$$4a + 2b + c = 3.0,$$
$$9a + 3b + c = 3.7.$$

Now we must solve this system of equations for a, b, and c. One way to do this is to choose one of the variables, c say, and use the first equation to solve for c in terms of a and b:

$$a + b + c = 2.1 \quad \text{implies that} \quad c = 2.1 - a - b.$$

If we substitute this expression for c into the other two equations, we will obtain two equations in the two variables a and b: we have

$$4a + 2b + (2.1 - a - b) = 3.0,$$
$$9a + 3b + (2.1 - a - b) = 3.7,$$

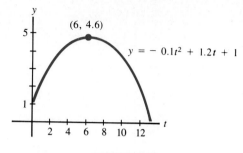

FIGURE 1–50

or

$$3a + b = 0.9,$$
$$8a + 2b = 1.6.$$

Now we use the first of these equations to solve for b in terms of a and then substitute the result into the second equation:

$$b = 0.9 - 3a$$

yields

$$8a + 2(0.9 - 3a) = 1.6,$$

or

$$2a = -0.2.$$

Hence

$$a = -0.1,$$
$$b = 0.9 - 3(-0.1) = 1.2,$$

and

$$c = 2.1 - (-0.1) - 1.2 = 1.0.$$

Therefore

$$y = -0.1t^2 + 1.2t + 1.$$

See Figure 1–50. Note that the patient's maximum temperature is 4.6° above normal and that this occurs 6 hours after admittance. ☐

■ *Computer Manual: Programs QUADFORM, PARABOLA*

1.5 EXERCISES

Quadratic Functions

In Exercises 1 through 16, graph the given quadratic, labeling the vertex and intercepts.

1. $y = 2x^2$
2. $y = -x^2$
3. $y = x^2 + 2$
4. $y = -x^2 + 9$
5. $y = -x^2 + 4x$
6. $y = x^2 + 5x$

7. $y = 3x^2 - 6x$
8. $y = x^2 + 2x - 15$
9. $y = -x^2 + 2x - 3$
10. $y = 2x^2 - x - 6$
11. $y = 2x^2 - 16x + 32$
12. $y = x^2 + x + 1$
13. $y = 2x^2 - 6x + 1$
14. $y = -2x^2 + 8x + 9$
15. $y = x^2 + 5x + 10$
16. $y = -9x^2 + 6x - 1$

*17. Derive the quadratic formula. (*Hint:* In

$$y = a\left(x + \frac{b}{2a}\right)^2 + \left(c - \frac{b^2}{4a}\right),$$

let $y = 0$, solve for $(x + b/2a)^2$, and take square roots.)

Quadratic Models

18. The number of sunspots on the sun varies over an 11-year cycle. Suppose that the average number of sunspots visible t years into the cycle is y, where

$$y = 0.5t^2 - 5.5t + 19,$$

for $0 \le t \le 11$. Graph this quadratic, and find the minimum average number of sunspots visible and the time in the cycle when this minimum occurs.

19. A rocket's altitude, in miles, t minutes after launching is given by the equation

$$y = -0.5t^2 + 18t.$$

Graph this quadratic.
   (a) What is the rocket's maximum altitude? When will it be attained?
   (b) How long will it take after launching for the rocket to return to earth?

20. A mortar is fired from a hillside into a valley. The height of the mortar shell above the valley floor t seconds after firing is

$$y = -12t^2 + 72t + 100$$

feet. Graph this quadratic.
   (a) How far above the valley floor is the mortar?
   (b) What is the maximum altitude reached by the mortar shell, and when does this maximum occur?
   (c) When does the mortar shell reach the ground?

Management

21. Alpha Company's cost function is given by the equation

$$C(x) = x^2 + 5000,$$

where x is the number of units produced and $C(x)$ is in dollars. Alpha Company sells each unit of its product for $150.
   (a) What is the defining equation of Alpha Company's revenue function?
   (b) Draw the graphs of C and R on the same set of axes.
   (c) Find Alpha Company's break-even quantity or quantities.
   (d) Find Alpha Company's profit function and graph it.
   (e) What is Alpha Company's maximum profit, and how many units must Alpha Company produce and sell in order to realize it?

22. Beta Company's cost function is given by the equation

$$C(x) = x^2 + 40x + 4200,$$

and its revenue function is given by the equation
$$R(x) = x^2 + 100x,$$
where x is the number of units produced and sold and both $C(x)$ and $R(x)$ are in dollars.

(a) Draw the graphs of C and R on the same set of axes.
(b) Find Beta Company's break-even quantity or quantities.
(c) Find Beta Company's profit function and graph it.
(d) What is Beta Company's maximum profit, and how many units must Beta Company produce and sell in order to realize it?

23. Answer the questions of Exercise 22 if Beta Company's cost function is given by the equation
$$C(x) = 80 + 3x,$$
and its revenue function is given by the equation
$$R(x) = 0.5x^2.$$

24. Gamma Corporation's profit function is given by
$$P = -x^2 + 100x - 1600.$$
Graph this function.

(a) Over what range of x will Gamma make money?
(b) What is Gamma's maximum profit and where does it occur?

25. Repeat Exercise 24 if $P = -0.02x^2 + 1.5x - 27$, where x is in thousands of units and P is in millions of dollars.

26. The demand function for a certain product is given by the equation
$$q = -p^2 + 10{,}000,$$
and the supply function for the product is given by the equation
$$q = 150p.$$
Graph the demand and supply functions on the same set of axes. Find the market equilibrium price.

27. Repeat the work of Exercise 26, but assume that
$$q = -p^2 - 20p + 8000$$
and
$$q = p^2 + 100p.$$

28. Suppose the supply and demand equations for a commodity are
$$q = p^2 + 20p - 50 \quad \text{and} \quad q = -p^2 - 10p + 150,$$
respectively, where q is in thousands of units. Find the market equilibrium price and quantity for the commodity.

29. Repeat Exercise 28 if
$$q = 8p^2 + 2p - 10 \quad \text{and} \quad q = -4p^2 - 3p + 15,$$
where q is in millions of units.

30. A firm's fixed cost is $50,000, the cost of making 10 thousand units of its product is $100,000, and the cost of making 20,000 units is $250,000. Find the firm's cost function if it is known to be a quadratic.

31. When a firm sold five units of its product, its revenue was $15; when it sold 10 units, its revenue was $40. Assuming that revenue from sales of 0 units is $0 and that the revenue function is a quadratic, find the revenue function.

32. A firm's fixed cost is $60,000, it lost $32,000 when it made and sold 4000 units, and its profit when it made and sold 16,000 units was $4,000. Find a quadratic model for the firm's profit function and find the firm's break-even quantities and its maximum profit.

Life Science

33. In certain cases in which patients have rapidly increasing body temperatures, powerful drugs are employed to decrease temperature. Suppose that t minutes after the injection of such a drug a patient's temperature is
$$y = -0.006t^2 + 0.36t + 4.2$$
degrees above normal.
   (a) What will be the patient's maximum temperature above normal and how long after the injection of the drug will it occur?
   (b) How long will it take for the patient's temperature to return to normal?

34. The concentration of sulfur dioxide in the air of a city from 7 A.M. to 7 P.M. is given by
$$y = -0.3t^2 + 5t + 3,$$
where y is the concentration (measured in parts per million) at time t, $t = 0$ being 7 A.M. Graph this function. At what time of day is the concentration at its greatest? What is the maximum concentration?

35. A drug is administered to a patient and t hours later there are y milligrams of the drug present in the patient's liver, where
$$y = -0.025t^2 + 1.8t.$$
What is the maximum amount of the drug present in the liver, and when does this occur? When will all the drug be gone from the liver?

36. The furbish lousewort plant was placed on the endangered species list and t years later there were
$$y = 2t^2 - 32t + 800$$
living furbish louseworts. How many were there when the plant was placed on the endangered species list? What was the minimum number of plants, and when did this minimum occur?

37. A drug is ingested and t days later $y\%$ of it remains in the body. If y is a quadratic function of t, and it is known that one day after ingestion 87% of the drug remains in the body, two days after ingestion 72% remains, and three days after ingestion 55% remains, find the function. Approximately how long will it take for all of the drug to be eliminated from the body?

38. A census of the bald eagles along an Alaskan river counted 200 eagles; two years later another census counted 276 of them, and two years after the second census a third one counted 384 of them. If the size of the eagle population can be modeled by a quadratic function, find the function and then find the minimum size of the population and when it occurred.

Social Science

39. Altruistic Charities has found that if it sends out x mailings per year soliciting donations, it nets y dollars, where
$$y = -30{,}000x^2 + 240{,}000x.$$
How many mailings should Altruistic Charities send out each year in order to maximize net contributions? What will the maximum net contributions be?

40. Voter interest, as expressed by the percentage of eligible voters who state that they intend to vote, usually peaks before election day. Suppose a political scientist finds that t days before an election ($0 \le t \le 30$), $y\%$ of the eligible voters intend to vote, where
$$y = -0.07t^2 + 1.26t + 60.$$
Graph this function. How many days before the election is voter interest at its peak? What percentage of eligible voters actually vote?

41. When a governor came into office her voter approval rating was 60%; six months later it was 78%, and after one year in office it was 74.4%. If voter approval can be modeled by a quadratic function, find the function. Then find the governor's maximum approval rating and when it occurred.

SUMMARY

This summary of Chapter 1 consists of terms and symbols whose meaning you should know, facts you should know, and techniques and methods you should be able to employ.

Section 1.1

Set, element, $\in$, $\notin$. Set-builder notation: $\{x \mid x \text{ has property } P\}$. Natural number, integer, positive integer, rational number, irrational number, real number. Number systems **N, Z, Q, R** and the relationships between them. The real line, plotting points on the real line. Inequality: $<, \le, >, \ge$. Positive number, negative number. Double inequality. Interval, line diagram. Interval notation: (a, b), $[a, b]$, $(a, +\infty)$, $(-\infty, a]$, etc. Absolute value of a real number: $|a|$. Describing practical situations in terms of inequalities, intervals, and absolute values. Base, exponent, power. Nonnegative integral exponent:
$$x^n = \underbrace{x \cdot x \cdots x}_{n \ x\text{'s}}, \quad x^0 = 1.$$

nth root, principal nth root $\sqrt[n]{x}$. Square root $\sqrt{x}$. Rational exponent: $x^{m/n} = (\sqrt[n]{x})^m$. Negative exponent: $x^{-r} = 1/x^r$. Rules of exponents: when defined,

$$x^r x^s = x^{r+s} \qquad x^r/x^s = x^{r-s} \qquad (x^r)^s = x^{rs} \qquad (xy)^r = x^r y^r \qquad (x/y)^r = x^r/y^r.$$

Evaluating and simplifying exponential expressions.

Numbers and Functions

Section 1.2

Function. Defining equation. Functional notation. Independent variable. Functional values. Finding functional values. The domain of a function. Finding the domain. The range of a function. Dependent variable. Functions defined by equations involving a dependent variable. Not every equation defines a function. The graph of a function. The x-axis, the y-axis. The cartesian coordinate system. The x-coordinate and y-coordinate of a point, the origin. Plotting points. The graphing procedure. Intercepts: the y-intercept, x-intercepts. Finding intercepts. Graphing functions using the graphing procedure. A point (a, b) is on the graph of f if and only if $f(a) = b$. If a is not in the domain of f or b is not in the range of f, then (a, b) cannot be on the graph of f. Graphing functions that have more than one defining equation. Graphing equations that do not define functions. The vertical line test.

Section 1.3

Adding functions by adding like terms. Cost, revenue, and profit functions, $P = R - C$. Finding the profit function from the revenue and cost functions. Polynomial and rational functions. Adding, subtracting, multiplying and dividing polynomial functions. The arithmetic of rational functions. Adding, subtracting, multiplying, and dividing rational functions. Composition of functions. The composite of g with f: $f(g(x))$. The composite of f with g: $g(f(x))$. Finding the defining equations of composite functions. Writing a function as a composite of other functions. One-to-one function. Determining whether a function is one-to-one. The inverse f^{-1} of a one-to-one function f: $f^{-1}(y) = x$ if and only if $f(x) = y$. Determining the defining equation for the inverse of a one-to-one function. Common factors. Factoring a polynomial. A product is zero if and only if at least one of its factors is zero. Solving polynomial equations by factoring.

Section 1.4

Linear function. Standard form $y = mx + b$. Graphing linear functions. The slope of a line. An increase of 1 in x causes $y = mx + b$ to change by m. A vertical line has no slope. The equation of a line. Finding the equation of a line when its slope and a point on it are known. The slope of a line through two points:

$$m = \frac{y_2 - y_1}{x_2 - x_1}.$$

Finding the equation of a line through two points. Constant function. The general form of the equation of a line. Parallel lines. Finding the point of intersection of two lines. Linear cost model: fixed cost, variable cost per unit, $C = vx + F$. Linear revenue model: $R = px$. Break-even point, break-even quantity. Finding the break-even point. Supply function, demand function. Linear supply and demand functions. The market equilibrium point, the market equilibrium price. Finding the market equilibrium point.

Section 1.5

Quadratic function. Standard form $y = ax^2 + bx + c$. Parabola. Parabola opens upward when $a > 0$, opens downward when $a < 0$. Vertex of a parabola. Finding the vertex of a parabola: $V = (-b/2a, c - b^2/4a)$. Intercepts of a parabola: $(0, c)$, solutions of $ax^2 + bx + c = 0$. The quadratic formula:

$$x = \frac{-b \pm \sqrt{b^2 - 4ac}}{2a}.$$

The discriminant $b^2 - 4ac$. Graphing quadratic functions. Quadratic models: the vertex as maximum or minimum. Quadratic cost, revenue, profit models. Quadratic supply and demand models. Fitting a quadratic model to any three points not on a line.

REVIEW EXERCISES

In Exercises 1 through 12, indicate whether the given statement is true or false.

1. $2 < 3$
2. $|-5| = -5$
3. $-3 > -2$
4. $4 < 6 \leq 5$
5. $|-2| = 2$
6. $-|-3| = 3$
7. $-3 < -3$
8. $-4 < -2 < 1$
9. $|8 - 2| = 6$
10. $-4 \geq -5$
11. $3 \leq 6 < 8$
12. $-2|5 - 9| = 8$

In Exercises 13 through 16, draw the line diagram corresponding to the given interval.

13. $(0, 1)$
14. $[0, 1]$
15. $(-\infty, 1]$
16. $(0, +\infty)$

17. Write the given statement using an inequality.
    (a) The pressure in a boiler is in the safe range if it is no more than 25 pounds per square inch.
    (b) A chemical reaction will take place if a solution is at a temperature of at least 20°C.
    (c) Profits are forecast to be less than $2 per share during the next quarter.
    (d) Profits are forecast to be at least $2 per share, but not as much as $2.25 per share, during the next quarter.

18. Evaluate:
    (a) $\dfrac{4^2 4^3 4^{-1} 3^3}{3^2 3^{-4} 4^5}$
    (b) $3(2^{-3})^2$
    (c) $2(5^4)^{-3/4}$
    (d) $121^{3/2}$
    (e) $(216 \cdot 1000)^{-2/3}$

Numbers and Functions

19. Simplify the following expressions. Write your answers using only positive exponents.

(a) $(x^3 y^{-3} z^{3/2})^{2/3}$

(b) $\sqrt[5]{\dfrac{x^4 y^{3/4} z^{-2} y^{1/4} z^{-3}}{x^{-1} y^7 z^5}}$

In Exercises 20 through 31, let

$$f(x) = 5 - 3x, \quad g(x) = 2 - x^2, \quad \text{and} \quad h(x) = \dfrac{x+1}{x-2}.$$

Find the indicated functional value if it exists.

20. $f(0)$ **21.** $f(2)$ **22.** $f(-1)$ **23.** $f(-a)$

24. $g(4)$ **25.** $g(\tfrac{1}{2})$ **26.** $g(\sqrt{2})$ **27.** $g(\sqrt{b}), b \geq 0$

28. $h(3)$ **29.** $h(0)$ **30.** $h(-1)$ **31.** $h(2)$

32. If a company makes x units of its product, its average cost per unit is $A(x) = 35 + 1200x^{-1/2}$ dollars. Find the average cost per unit when the company makes 100, 2500, and 10,000 units.

33. Graph the given function and state its domain and range:

(a) $f(x) = (x-1)^3$

(b) $y = \begin{cases} 4+x, & -2 \leq x \leq 1, \\ 3, & 1 < x < 2, \\ 6 - 2x, & 2 \leq x \leq 3 \end{cases}$

34. Let $f(x) = -2x^3 + 5x^2 + 3x - 6$ and $g(x) = 4x^3 - 6x + 4$. Find defining equations for $f + g$, $f - g$, fg, and f/g.

35. Let $f(x) = \dfrac{x^2 - 2}{2x - 6}$ and $g(x) = \dfrac{x+5}{3-x}$. Find defining equations for $f + g$, $f - g$, fg, and f/g.

36. Let $f(x) = 2x + 3$, $g(x) = 5x + 2$, and $h(x) = \dfrac{2}{x-1}$. Find $f(g(x))$, $g(f(x))$, $f(h(x))$, and $h(g(x))$.

37. It costs a firm $C(x) = 0.5(x+2)^2 + 100$ dollars to produce x units of one of its products, and the number of units produced depends on the number of worker-hours w devoted to production according to the equation $x = 3\sqrt{w+1}$. Express the cost of making the product in terms of the worker-hours devoted to it.

38. Write $k(x) = \sqrt[3]{3x^2 + 5x + 1}$ as a composite function.

39. State whether the given function is one-to-one; if it is, find a defining equation for the inverse function and sketch its graph.

(a) $f(x) = x^{-2}, x > 0$

(b) $f(x) = |x|$

(c) $y = x^4$

(d) $y = ax + b$, a,b real, $a \neq 0$

In Exercises 40 through 45, solve the equation $f(x) = 0$ for x.

40. $f(x) = 64 - x^2$ **41.** $f(x) = x^2 + 4x - 32$ **42.** $f(x) = 3x^2 - 24x + 48$

43. $f(x) = x^3 - x$ **44.** $f(x) = 2x^4 - 2x^3 - 24x^2$ **45.** $f(x) = 4x^2 + 4x - 3$

46. A retailer can sell y sweaters at p dollars each, where $y = 2280 - 40p$. What price must the retailer charge if the revenue from the sale of the sweaters is to be $31,280?

In Exercises 47 through 50, graph the equation and find the slope of the line if possible.

47. $y = 4x - 3$ **48.** $2x + 5y = 20$ **49.** $y = 6$ **50.** $x = 6$

In Exercises 51 through 54, find the equation of the given line.

51. The line with slope $\tfrac{1}{3}$ passing through the point (2, 6).

52. The line passing through the points (2, −7) and (−9, 26).

53. The horizontal line passing through the point (5, 12).

54. The vertical line passing through the point (5, 12).

55. The number of bacteria alive in a culture that has been treated with a bactericide declines at a constant rate of 6000 bacteria per hour. If 120,000 bacteria were alive in the culture at the time the bactericide was applied, find an equation describing the number of living bacteria present as a function of time. How long will it take until all the bacteria are dead?

56. A firm's sales were $368,000 in 1983 and $740,000 in 1988. Assuming linearity, find an equation describing the firm's sales as a function of time, predict the firm's sales in 1993, and predict the year in which sales will reach $1,484,000.

57. Find the point of intersection of the lines
(a) $y = 3x + 6$ and $y = 5x - 12$;
(b) $4x - 3y = 9$ and $2x + 5y = 6$.

58. A farmer wants to feed his cattle a diet supplement containing both calcium and an antibiotic. One brand of diet supplement contains 100 units of calcium and 50 units of antibiotic per pound, and another contains 140 units of calcium and 25 units of antibiotic per pound. If the cattle require 446 units of calcium and 160 units of antibiotic per week, how much of each brand should be fed to them?

59. A firm's fixed cost is $92,000, its variable cost per unit is $46, and it sells each unit it makes for $62. Find its cost, revenue, and profit functions and its break-even point.

60. The supply and demand functions for a commodity are given by
$$q = 6250p - 15{,}625 \quad \text{and} \quad q = -8500p + 183{,}500,$$
respectively. What are the significances of the slopes and intercepts of these functions? Find the market equilibrium price for the commodity.

61. A firm's cost and revenue functions are given by
$$C(x) = x^2 + 140x + 1050 \quad \text{and} \quad R(x) = 0.5x^2 + 190x.$$
(a) Find the firm's break-even point or points.
(b) Find and graph the firm's profit function.
(c) What is the firm's maximum profit, and how many units must be made and sold in order for the maximum profit to be attained?

62. A manufacturer who intends to put out a line of electronic desk calendars estimates that the calendars can be produced for $10 each, that a selling price of $20 per calendar would result in annual sales of 30,000 calendars, and that each rise of $5 in the selling price would result in the sale of 5000 fewer calendars.
(a) Find the demand function for the calendars.
(b) Find the profit function for the calendars and sketch its graph. (*Hint:* Total profit equals profit per unit times demand.)

(c) What is the maximum profit the manufacturer could make on the calendars? At what selling price would the maximum profit occur?

63. The supply curve for a commodity is a parabola. When the unit price of the commodity was $10, 38,000 units were supplied to the market; similarly, when the price was $15, 118,000 units were supplied, and when the price was $20, 238,000 units were supplied. Find the equation of the supply curve and find the approximate price at which quantity supplied will be zero.

ADDITIONAL TOPICS

Here are some suggestions for topics not covered in the text that you might want to investigate on your own.

1. A **terminating decimal** is one that stops, such as 2.675. Every terminating decimal is the result of dividing out some rational number p/q. (For instance, $2.675 = 107/40$.) Explain why this is so. For any given terminating decimal, show how to find the rational number that it is equal to.

2. A **repeating decimal** is one in which some block of digits repeats forever, such as 1.2754545454.... Every repeating decimal is the result of dividing out some rational number p/q. (For instance, $1.27545454\ldots = 12{,}627/9900$.) Explain why this is so. For any given repeating decimal, show how to find the rational number that it is equal to.

3. Every rational number, when divided out, becomes either a terminating or a repeating decimal. (For instance, $9/16 = 0.5625$ and $13/6 = 2.166666\ldots$) Explain why this is so.

4. A more formal definition of the concept of a function than the one we used in Section 1.2 makes use of the **cartesian product of two sets.** Find out what the cartesian product of two sets is, and how it can be used to define the notion of a function. How is this definition using the cartesian product related to our definition of a function?

5. Rene Descartes (1596–1650) was a French mathematician and philosopher who discovered the coordinate system named after him (the cartesian coordinate system), invented the mathematical subject of analytic geometry, and founded the philosophical school known as cartesianism. Find out about Descartes' life and his contributions to mathematics and philosophy.

6. Points can be plotted by using **polar coordinates** as well as cartesian coordinates; polar coordinates locate points using angles and distances from the origin. Find out how this is done.

7. In Section 1.3 we discussed addition, subtraction, and multiplication of polynomials, but we did not consider the long division of one polynomial by another. When one polynomial is divided by another, the result is a polynomial quotient plus a remainder that is a rational expression. (The rational expression may be zero.) For instance,

$$(x^3 + 2x + 3) \div (x - 1) = x^2 + x + 3 + \frac{6}{x - 1}.$$

Here $x^2 + x + 3$ is the quotient and $\dfrac{6}{x-1}$ is the remainder. Find out how long division of one polynomial by another is carried out.

8. In Section 1.4 we stated that lines having the same slope are parallel. In fact, two nonvertical lines are parallel(that is, they do not intersect) if and only if they have the same slope. Explain why this is so. (You must explain two things: why nonvertical lines that are parallel have the same slope, and why nonvertical lines that have the same slope are parallel.)

9. Lines are **perpendicular** if they meet at a right angle. Thus any horizontal line and any vertical line are perpendicular. Two nonvertical lines with equations

$$y = m_1 x + b_1 \quad \text{and} \quad y = m_2 x + b_2$$

are perpendicular if and only if $m_1 m_2 = -1$. Explain why this is so. (You must explain two things: why nonvertical lines that are perpendicular have slopes whose product is -1, and why nonvertical lines whose slopes have product equal to -1 are perpendicular.)

10. Complex numbers are numbers of the form $a + bi$, where a and b are real numbers and $i^2 = -1$. Find out how complex numbers are added, subtracted, multiplied, and divided. Show that the real number system is a subsystem of the complex number system.

11. Show that if we allow equations to have complex numbers as solutions, then every quadratic equation has as its solutions either two distinct real numbers, one real number, or a pair of complex numbers of the form $a + bi$ and $a - bi$.

12. The quadratic formula gives us the solutions of any quadratic equation $ax^2 + bx + c = 0$. There is an analogous cubic formula which gives us the solutions of any cubic equation $ax^3 + bx^2 + cx + d = 0$. The cubic formula was discovered ca. 1535 by Nicolo Fontana and stolen from him and published in 1545 by Girolano Cardano. Find out about this cubic formula and its colorful history. (A good place to start would be with the book *Great Moments in Mathematics before 1650* by Howard Eves, published by The Mathematical Association of America in 1980.) You might also find out about the **quartic formula,** which gives us the solutions of any fourth-degree equation $ax^4 + bx^3 + cx^2 + dx + e = 0$. It has been proved that for $n > 4$ there is no formula for finding the solutions of the nth-degree equation $a_n x^n + a_{n-1} x^{n-1} + \cdots + a_1 x + a_0 = 0$.

2

The Derivative

Now we begin our study of calculus. Calculus is a powerful method for studying functions. For instance, suppose a demographer concludes that a certain function will describe a nation's population size as a function of time over the next few years. The techniques of calculus can be applied to this function to answer questions such as: At what rate will the population be changing at any given time? When will the population be increasing and when will it be decreasing? What will be the maximum and minimum sizes of the population over the next several years? The particular concept of calculus needed to answer questions like these is called the **derivative.**

In this chapter we study the derivative. The derivative is defined in terms of another concept of calculus called the **limit,** so we will begin with a brief intuitive introduction to limits. We will then define the derivative, develop some rules for finding derivatives of functions, show how to use the derivative to find rates of change, exhibit some other useful applications of the derivative, and conclude with a discussion of the various notations for the derivative and for derivatives of the derivative.

2.1 LIMITS

The **limit** of a function f as its independent variable x approaches a number a is a number L with the property that, as x gets closer and closer to a (without ever being equal to a), the functional values $f(x)$ get closer and closer to L. For a given function f and limiting value a such a number L may or may not exist. The existence or nonexistence of the limit, and its value if it does exist, thus tells us something about the behavior of the function for values of x near a. Hence limits are important in analyzing functions. In this section we illustrate the notion of the limit of a function geometrically, state the properties of limits, and show how to use these properties to find limits of polynomial and rational functions.

The Limit of a Function

Consider the linear function f whose defining equation is $f(x) = 2x + 1$. We ask the following question: if we let x approach the number 1 on the x-axis, from either side of 1 and without ever letting $x = 1$, what can we say about the behavior of the corresponding functional values $f(x)$? For instance, Table 2–1 shows some functional values $f(x) = 2x + 1$ as x approaches 1 from the left and from the right:

TABLE 2–1

	\multicolumn{5}{c}{x approaches 1 from the left}
x	0.8	0.9	0.99	0.999	0.9999
$f(x) = 2x + 1$	2.6	2.8	2.98	2.998	2.9998
	\multicolumn{5}{c}{x approaches 1 from the right}
x	1.2	1.1	1.01	1.001	1.0001
$f(x) = 2x + 1$	3.4	3.2	3.02	3.002	3.0002

Notice that as x gets closer and closer to 1, from either side of 1, the values $f(x)$ get closer and closer to the single number 3. Hence we say that the **limit of $f(x)$ as x approaches 1 is 3,** written

$$\mathop{\mathrm{Lim}}_{x \to 1} f(x) = 3,$$

or

$$\mathop{\mathrm{Lim}}_{x \to 1} (2x + 1) = 3.$$

We can also tell that $\mathop{\mathrm{Lim}}_{x \to 1} (2x + 1) = 3$ from the graph of the function. Figure 2–1 shows the graph of f. As the figure indicates, as x approaches 1 from either side on the x-axis the functional values $f(x)$ approach 3 on the y-axis.

Having illustrated the idea of the limit of a function, we now present an informal definition of the concept.

The Derivative

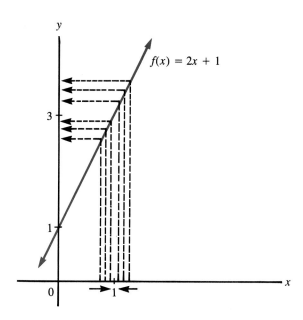

FIGURE 2–1
$\text{Lim}_{x \to 1} (2x + 1) = 3$

The Limit of a Function

If f is a function and a is a number, the **limit of f as x approaches a is L,** written

$$\text{Lim}_{x \to a} f(x) = L,$$

provided that as x comes arbitrarily close to a from either side, without ever being equal to a, the corresponding functional values $f(x)$ come arbitrarily close to the single number L. If there is no such single number L, the limit does not exist.

EXAMPLE 1   If $f(x) = x^2 + 1$ and $a = 0$, we claim that $\text{Lim}_{x \to 0} f(x) = 1$. Figure 2–2 verifies this: it shows that as x approaches 0 on the x-axis, from either side and without ever being equal to 0, the corresponding functional values approach (that is, come arbitrarily close to) the single number 1 on the y-axis.   □

EXAMPLE 2   If f is defined by

$$f(x) = \begin{cases} x - 1 & \text{if } 0 \leq x \leq 3, \\ x + 1 & \text{if } x > 3, \end{cases}$$

then $\text{Lim}_{x \to 3} f(x)$ does not exist. To see this, look at Figure 2–3, which depicts the graph of this function. Notice that as x approaches 3 on the x-axis from the left,

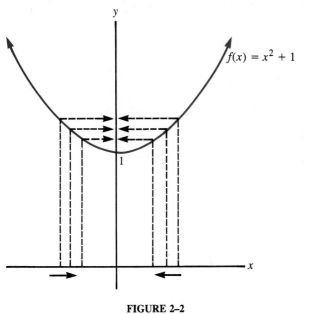
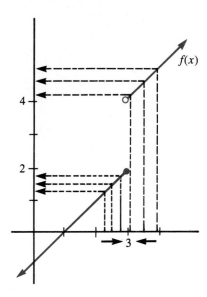

FIGURE 2–2
$\operatorname*{Lim}_{x \to 0} (x^2 + 1) = 1$

FIGURE 2–3
$\operatorname*{Lim}_{x \to 3} f(x)$ Does Not Exist

the corresponding functional values approach 2 on the y-axis, but as x approaches 3 from the right, the corresponding functional values approach 4 on the y-axis. Therefore there is no single number L that the functional values approach as x approaches 3, so the limit does not exist. ☐

EXAMPLE 3 Let h be defined by

$$h(x) = \frac{x+1}{x-1}.$$

Figure 2–4 shows the graph of h. From the figure, we see that as x approaches -1 on the x-axis, from either the right or the left, the functional values $h(x)$ approach 0. Hence

$$\operatorname*{Lim}_{x \to -1} h(x) = \operatorname*{Lim}_{x \to -1} \frac{x+1}{x-1} = 0.$$

On the other hand, as x approaches 1 from the right, the functional values $h(x)$ become larger and larger without bound, and thus cannot approach any number L. Hence

$$\operatorname*{Lim}_{x \to 1} h(x) = \operatorname*{Lim}_{x \to 1} \frac{x+1}{x-1} \quad \text{does not exist.} \quad \square$$

EXAMPLE 4 Figure 2–5 shows the graph of a function g. From the graph, we see that

  a. $\operatorname*{Lim}_{x \to -1} g(x) = 0$, because as x approaches -1 from either side, without ever being equal to -1, the functional values $g(x)$ approach 0;

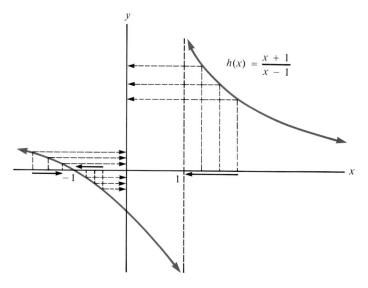

FIGURE 2–4

b. $\text{Lim}_{x \to 0} g(x) = 3$, because as x approaches 0 from either side, without ever being equal to 0, the functional values actually become equal to 3 (and hence are certainly arbitrarily close to 3);

c. $\text{Lim}_{x \to 1} g(x) = 5$, because as x approaches 1 from either side, without ever being equal to 1, the functional values approach 5; note that this is so even though the functional value $g(1) \neq 5$;

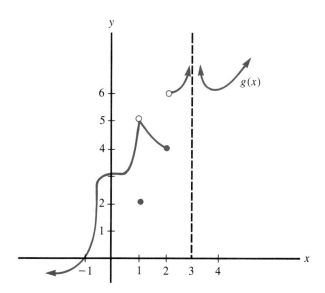

FIGURE 2–5

d. $\text{Lim}_{x\to 2}\, g(x)$ does not exist, because as x approaches 2 from the left the functional values approach 4, but as x approaches 2 from the right they approach 6;
e. $\text{Lim}_{x\to 3}\, g(x)$ does not exist, because as x approaches 3 from the left (and also from the right), the functional values grow larger and larger without bound and hence do not approach a single number L. ☐

We should remark that our definition of limit is highly intuitive and not mathematically precise. A more rigorous definition of the limit of a function is presented in suggestion 1 of the additional topics at the end of this chapter.

Properties of Limits

The following properties are useful when it is necessary to evaluate limits. They may be proved using the rigorous definition of limit referred to in the preceding paragraph.

Properties of Limits

Let a be any number and assume that both $\text{Lim}_{x\to a} f(x)$ and $\text{Lim}_{x\to a} g(x)$ exist.

1. If $f(x) = g(x)$ whenever $x \neq a$, then $\text{Lim}_{x\to a} f(x) = \text{Lim}_{x\to a} g(x)$.
2. For any number c, $\text{Lim}_{x\to a} c = c$.
3. For any number r, $\text{Lim}_{x\to a} x^r = a^r$, provided that a^r is defined; if a^r is not defined as a real number, the limit does not exist.
4. For any real number c, $\text{Lim}_{x\to a} cf(x) = c \cdot \text{Lim}_{x\to a} f(x)$.
5. $\text{Lim}_{x\to a} [f(x) \pm g(x)] = \text{Lim}_{x\to a} f(x) \pm \text{Lim}_{x\to a} g(x)$.
6. $\text{Lim}_{x\to a} [f(x)g(x)] = [\text{Lim}_{x\to a} f(x)][\text{Lim}_{x\to a} g(x)]$.
7. If $\text{Lim}_{x\to a} g(x) \neq 0$, then
$$\text{Lim}_{x\to a} \frac{f(x)}{g(x)} = \frac{\text{Lim}_{x\to a} f(x)}{\text{Lim}_{x\to a} g(x)}.$$
8. If $\text{Lim}_{x\to a} g(x) = 0$ and $\text{Lim}_{x\to a} f(x) \neq 0$, then $\text{Lim}_{x\to a} \frac{f(x)}{g(x)}$ does not exist.

EXAMPLE 5

The functions f and g defined by

$$f(x) = x^2 + 1 \quad \text{and} \quad g(x) = \frac{x^3 + x}{x}$$

THE DERIVATIVE

are equal except when $x = 0$, and we know from Example 1 that

$$\lim_{x \to 0} f(x) = 1.$$

Therefore by Property 1 we have

$$\lim_{x \to 0} g(x) = \lim_{x \to 0} f(x) = 1. \quad \square$$

EXAMPLE 6 By Property 2, we have $\lim_{x \to a} 4 = 4$ for any number a. $\quad \square$

EXAMPLE 7 Property 3 shows that $\lim_{x \to -2} x = -2$, $\lim_{x \to 3} x^2 = 3^2 = 9$, and

$$\lim_{x \to 4} \sqrt{x} = \lim_{x \to 4} x^{1/2} = 4^{1/2} = 2.$$

But also by Property 3,

$$\lim_{x \to 0} x^{-1} \text{ does not exist, since } 0^{-1} = 1/0 \text{ is not defined,}$$

and

$$\lim_{x \to -1} \sqrt[4]{x} \text{ does not exist, since } \sqrt[4]{-1} \text{ is not defined as a real number.} \quad \square$$

EXAMPLE 8 By Properties 4 and 3,

$$\lim_{x \to 2} 3x^2 = 3 \cdot \lim_{x \to 2} x^2 = 3(2^2) = 12. \quad \square$$

EXAMPLE 9 By Properties 2 through 5, we have

$$\lim_{x \to 3} (2x^3 + 4x^2 - x + 5) = \lim_{x \to 3} (2x^3) + \lim_{x \to 3} (4x^2) - \lim_{x \to 3} x + \lim_{x \to 3} 5$$

$$= 2 \cdot \lim_{x \to 3} x^3 + 4 \cdot \lim_{x \to 3} x^2 - \lim_{x \to 3} x + 5$$

$$= 2(3^3) + 4(3^2) - 3 + 5$$

$$= 92. \quad \square$$

EXAMPLE 10 Properties 2 through 6 imply that

$$\lim_{x \to 3} [(2x^{1/2})(3x - 1)] = \left[\lim_{x \to 3} 2\sqrt{x}\right]\left[\lim_{x \to 3} (3x - 1)\right]$$

$$= [2\sqrt{3}][3(3) - 1]$$

$$= 16\sqrt{3}. \quad \square$$

EXAMPLE 11 Properties 2 through 7 show that

$$\operatorname*{Lim}_{x\to 0} \frac{x^2 + 2}{2x + 3} = \frac{\operatorname*{Lim}_{x\to 0}(x^2 + 2)}{\operatorname*{Lim}_{x\to 0}(2x + 3)}$$

$$= \frac{0^2 + 2}{2(0) + 3}$$

$$= \frac{2}{3}. \quad \square$$

EXAMPLE 12 By Property 8,

$$\operatorname*{Lim}_{x\to 1} \frac{x^2 + 1}{x - 1}$$

does not exist because

$$\operatorname*{Lim}_{x\to 1}(x^2 + 1) = 2$$

while

$$\operatorname*{Lim}_{x\to 1}(x - 1) = 0. \quad \square$$

Recall that a **polynomial function** is one whose defining equation is of the form

$$p(x) = a_n x^n + a_{n-1} x^{n-1} + \cdots + a_1 x + a_0.$$

As Example 9 suggests, Properties 2 through 5 imply that we can find the limit of any polynomial function as x approaches a simply by substituting a for x in the defining equation of the function.

Limit of a Polynomial Function

If p is a polynomial function, then $\operatorname*{Lim}_{x\to a} p(x) = p(a)$ for any a.

EXAMPLE 13 We have

$$\operatorname*{Lim}_{x\to 2}(x^3 - 5x^2 + 3x - 12) = 2^3 - 5(2^2) + 3(2) - 12 = -18. \quad \square$$

Now let us consider limits of rational functions. (Recall that a rational function is one that is a quotient of polynomial functions; for instance, the function h of Example 4 of this section is a rational function.) If $f = p/q$ is a rational function, where the polynomials p and q have a common factor, Property 1 implies that we can cancel this common factor from p/q without affecting the limit of f. Therefore to find the limit of a rational function, we first reduce the function to its lowest

THE DERIVATIVE

terms by canceling common factors from its numerator and denominator. We then use our result for the limit of polynomials and apply either Property 7 or Property 8. We illustrate with examples.

EXAMPLE 14 Consider
$$\lim_{x \to 3} \frac{x^2 + x - 12}{x^2 - x - 6} = \lim_{x \to 3} \frac{(x + 4)(x - 3)}{(x + 2)(x - 3)}.$$

We cancel the common factor $x - 3$ from the rational expression. Then by Property 7 we have

$$\lim_{x \to 3} \frac{x^2 + x - 12}{x^2 - x - 6} = \lim_{x \to 3} \frac{x + 4}{x + 2} = \frac{\lim_{x \to 3} (x + 4)}{\lim_{x \to 3} (x + 2)}$$

$$= \frac{3 + 4}{3 + 2}$$

$$= \frac{7}{5}. \quad \square$$

EXAMPLE 15 We use the same rational function as in Example 14, but this time take its limit as x approaches -2. Now,

$$\lim_{x \to -2} \frac{x^2 + x - 12}{x^2 - x - 6} = \frac{\lim_{x \to -2} (x + 4)}{\lim_{x \to -2} (x + 2)},$$

and
$$\lim_{x \to -2} (x + 4) = -2 + 4 = 2,$$

while
$$\lim_{x \to -2} (x + 2) = 0.$$

Hence by Property 8,

$$\lim_{x \to -2} \frac{x^2 + x - 12}{x^2 - x - 6} = \frac{\lim_{x \to -2} (x + 4)}{\lim_{x \to -2} (x + 2)} = \frac{2}{0}$$

does not exist. $\square$

EXAMPLE 16 Since

$$\lim_{x \to 3} \frac{x^3 - 9x}{x^3 - 6x^2 + 9x} = \lim_{x \to 3} \frac{x(x - 3)(x + 3)}{x(x - 3)^2}$$

$$= \lim_{x \to 3} \frac{x + 3}{x - 3} = \frac{\lim_{x \to 3} (x + 3)}{\lim_{x \to 3} (x - 3)} = \frac{6}{0},$$

this limit does not exist. However,

$$\operatorname*{Lim}_{x\to 3}\frac{x^3 - 6x^2 + 9x}{x^3 - 9x} = \operatorname*{Lim}_{x\to 3}\frac{x-3}{x+3} = \frac{\operatorname*{Lim}_{x\to 3}(x-3)}{\operatorname*{Lim}_{x\to 3}(x+3)} = \frac{0}{6} = 0. \quad \square$$

As we saw in Section 1.2, sometimes functions are defined in "pieces," with each piece being a polynomial or rational expression. (The function of Example 2 is such a function.) Our techniques for finding limits of polynomial and rational functions can be applied to each piece of such a function.

EXAMPLE 17   Let f be defined by

$$f(x) = \begin{cases} 2 & \text{if } 0 \le x < 3, \\ x - 1 & \text{if } 3 \le x \le 5, \\ 16 - 2x & \text{if } x > 5. \end{cases}$$

Figure 2–6 shows the graph of f. Clearly,

$$\operatorname*{Lim}_{x\to 3} f(x) = 2 \quad \text{and} \quad \operatorname*{Lim}_{x\to 5} f(x) \text{ does not exist.}$$

If $0 < a < 3$, then $f(x) = 2$ when x is near a, so

$$\operatorname*{Lim}_{x\to a} f(x) = \operatorname*{Lim}_{x\to a} 2 = 2;$$

if $3 < a < 5$, then $f(x) = x - 1$ when x is near a, so

$$\operatorname*{Lim}_{x\to a} f(x) = \operatorname*{Lim}_{x\to a} (x - 1) = a - 1;$$

if $a > 5$, then $f(x) = 16 - 2x$ when x is near a, so

$$\operatorname*{Lim}_{x\to a} f(x) = \operatorname*{Lim}_{x\to a} (16 - 2x) = 16 - 2a. \quad \square$$

EXAMPLE 18   The cost of cleaning up an oil spill of x gallons in a harbor is $C(x)$ dollars, where

$$C(x) = \begin{cases} 250{,}000 + 2x, & 0 < x \le 10{,}000, \\ 210{,}000 + 6x, & x > 10{,}000. \end{cases}$$

See Figure 2–7. Thus if $0 < a < 10{,}000$,

$$\operatorname*{Lim}_{x\to a} C(x) = \operatorname*{Lim}_{x\to a} (250{,}000 + 2x) = 250{,}000 + 2a.$$

Similarly, if $a > 10{,}000$, then

$$\operatorname*{Lim}_{x\to a} C(x) = \operatorname*{Lim}_{x\to a} (210{,}000 + 6x) = 210{,}000 + 6a.$$

Note also from Figure 2–7 that

$$\operatorname*{Lim}_{x\to 10{,}000} C(x) = 270{,}000. \quad \square$$

■ *Computer Manual: Program LIMIT*

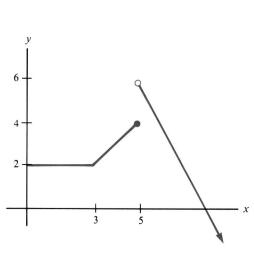

FIGURE 2–6

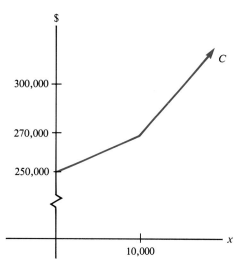

FIGURE 2–7

2.1 EXERCISES

The Limit of a Function

In Exercises 1 through 4, complete the table and find the limit if it exists; if the limit does not exist, so state.

1. $\lim_{x \to 3} f(x)$, where $f(x) = x^2 - 9$.

x	x approaches 3 from the left				
	2.8	2.9	2.99	2.999	2.9999
f(x)					

x	x approaches 3 from the right				
	3.2	3.1	3.01	3.001	3.0001
f(x)					

2. $\lim_{x \to 3} f(x)$, where $f(x) = \dfrac{x^2 - 9}{x - 3}$.

x	x approaches 3 from the left				
	2.8	2.9	2.99	2.999	2.9999
f(x)					

x	x approaches 3 from the right				
	3.2	3.1	3.01	3.001	3.0001
f(x)					

CHAPTER 2

3. $\lim\limits_{x \to 0} f(x)$, where $f(x) = \dfrac{1}{x}$.

	x approaches 0 from the left				
x	−0.1	−0.01	−0.001	−0.0001	−0.00001
f(x)					

	x approaches 0 from the right				
x	0.1	0.01	0.001	0.0001	0.00001
f(x)					

4. $\lim\limits_{x \to 1} f(x)$, where $f(x) = \begin{cases} x & \text{if } 0 \leq x < 1, \\ 2x & \text{if } 1 \leq x < 2. \end{cases}$

	x approaches 1 from the left				
x	$1/2$	$3/4$	$7/8$	$15/16$	$31/32$	$63/64$
f(x)						

	x approaches 1 from the right				
x	$3/2$	$5/4$	$9/8$	$17/16$	$33/32$	$65/64$
f(x)						

The following is the graph of a function f:

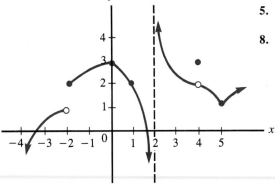

In Exercises 5 through 10, find the indicated limit if it exists; if it does not exist, so state.

5. $\lim\limits_{x \to -2} f(x)$ 6. $\lim\limits_{x \to 0} f(x)$ 7. $\lim\limits_{x \to 1} f(x)$

8. $\lim\limits_{x \to 2} f(x)$ 9. $\lim\limits_{x \to 4} f(x)$ 10. $\lim\limits_{x \to 5} f(x)$

The following is the graph of a function g:

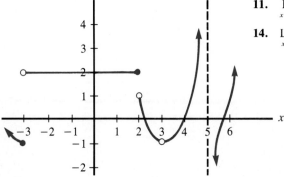

In Exercises 11 through 16, find the indicated limit if it exists; if it does not exist, so state.

11. $\lim\limits_{x \to -3} g(x)$ 12. $\lim\limits_{x \to -1} g(x)$ 13. $\lim\limits_{x \to 0} g(x)$

14. $\lim\limits_{x \to 2} g(x)$ 15. $\lim\limits_{x \to 3} g(x)$ 16. $\lim\limits_{x \to 5} g(x)$

The following is the graph of a function h:

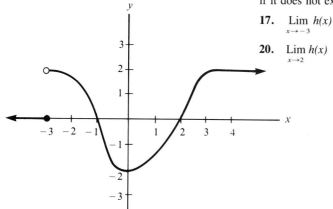

In Exercises 17 through 22, find the indicated limit if it exists; if it does not exist, so state.

17. $\lim\limits_{x \to -3} h(x)$
18. $\lim\limits_{x \to -1} h(x)$
19. $\lim\limits_{x \to 0} h(x)$
20. $\lim\limits_{x \to 2} h(x)$
21. $\lim\limits_{x \to 3} h(x)$
22. $\lim\limits_{x \to 4} h(x)$

Properties of Limits

In Exercises 23 through 64, find the indicated limit if it exists; if it does not exist, so state.

23. $\lim\limits_{x \to 2} 5$
24. $\lim\limits_{x \to -2} 3$
25. $\lim\limits_{x \to -4} \sqrt{3}$
26. $\lim\limits_{x \to a} \pi$
27. $\lim\limits_{x \to 2} x^3$
28. $\lim\limits_{x \to 9} \sqrt{x}$
29. $\lim\limits_{x \to 3} 1/x$
30. $\lim\limits_{x \to 4} x^{-2}$
31. $\lim\limits_{x \to 16} x^{3/4}$
32. $\lim\limits_{x \to 2} 5x$
33. $\lim\limits_{x \to -3} 3x^4$
34. $\lim\limits_{x \to 4} -5x^{3/2}$
35. $\lim\limits_{x \to 0} 4x^{-1/2}$
36. $\lim\limits_{x \to 8} 3x^{-1/3}$
37. $\lim\limits_{x \to 4} (3x + 2)$
38. $\lim\limits_{x \to 0} (5x - 7)$
39. $\lim\limits_{x \to 5} (5 - x)$
40. $\lim\limits_{x \to 2} (x^2 + x)$
41. $\lim\limits_{x \to 3} (x^2 + 3x)$
42. $\lim\limits_{x \to 1} (x^2 - 1)$
43. $\lim\limits_{x \to 0} (2x^3 - 5x^2 + 3x - 17)$
44. $\lim\limits_{x \to 2} (x^3 + 2x^2 - 6x + 1)$
45. $\lim\limits_{x \to -2} (-3x^3 + 2x^2 + 5x + 3)$
46. $\lim\limits_{x \to 0} (x^{99} + 2x^{98} + 3x^{97} + \cdots + 99x + c)$
47. $\lim\limits_{x \to 4} (x^{-2} + 5x)$
48. $\lim\limits_{x \to 2} \left(\frac{2}{x} - \frac{3}{x^2} + 1 \right)$
49. $\lim\limits_{x \to 0} (2x + 4)(3x - 2)$
50. $\lim\limits_{x \to 2} (5x - 3)(x^2 + 5)$
51. $\lim\limits_{x \to 1} (x^2 - 4x - 6)(x^3 + x + 2)$
52. $\lim\limits_{x \to 1} \dfrac{2}{x + 1}$
53. $\lim\limits_{x \to 1} \dfrac{2}{x - 1}$
54. $\lim\limits_{x \to 0} \dfrac{x + 1}{2x + 3}$
55. $\lim\limits_{x \to 1} \dfrac{4x + 7}{5x - 2}$
56. $\lim\limits_{x \to 3} \dfrac{x^2 - 9}{x + 2}$
57. $\lim\limits_{x \to 1} \dfrac{x^2 + 4x + 3}{x - 2}$
58. $\lim\limits_{x \to 2} \dfrac{x^2 + 4x + 3}{x - 2}$
59. $\lim\limits_{x \to 3} \dfrac{x^2 - 5x + 6}{x^2 + x - 12}$
60. $\lim\limits_{x \to 3} \dfrac{x^2 - 5x + 6}{x^2 + 10x + 24}$
61. $\lim\limits_{x \to 2} \dfrac{x^2 - 5x - 14}{x^2 - 4}$
62. $\lim\limits_{x \to 7} \dfrac{x^2 - 5x - 14}{x^2 - 4}$
63. $\lim\limits_{x \to 1} \dfrac{x^3 - 2x^2 - 15x}{x^4 - 4x^3 - 12x^2}$
64. $\lim\limits_{x \to 0} \dfrac{x^3 - 2x^2 - 15x}{x^4 - 4x^3 - 12x^2}$

65. Let f be defined by
$$f(x) = \begin{cases} 0, & 0 \le x < 2, \\ 1, & x \ge 2. \end{cases}$$

Graph f and find
(a) $\lim\limits_{x \to 2} f(x)$ (b) $\lim\limits_{x \to a} f(x)$ if $0 < a < 2$ (c) $\lim\limits_{x \to a} f(x)$ if $a > 2$.

66. Let f be defined by
$$f(x) = \begin{cases} -1, & 0 \le x \le 2, \\ 1, & 2 < x \le 3, \\ 2, & x > 3. \end{cases}$$

Graph f and find
(a) $\lim\limits_{x \to 2} f(x)$, $\lim\limits_{x \to 3} f(x)$ (b) $\lim\limits_{x \to a} f(x)$ if $0 < a < 2$, $2 < a < 3$, and $a > 3$

67. Let f be defined by
$$f(x) = \begin{cases} x, & 0 \le x \le 3, \\ 6 - x, & 3 < x \le 6. \end{cases}$$

Graph f and find
(a) $\lim\limits_{x \to 3} f(x)$ (b) $\lim\limits_{x \to a} f(x)$ if $0 < a < 3$ and $3 < a < 6$

68. Let f be defined by
$$f(x) = \begin{cases} 1, & -1 \le x \le 1, \\ 2x, & 1 < x < 2, \\ 3, & x \ge 2. \end{cases}$$

Graph f and find
(a) $\lim\limits_{x \to 1} f(x)$, $\lim\limits_{x \to 2} f(x)$ (b) $\lim\limits_{x \to a} f(x)$ if $-1 < a < 1$, $1 < a < 2$, and $a > 2$

69. Let f be defined by
$$f(x) = \begin{cases} 1 + x, & 0 \le x \le 3, \\ 4, & 3 < x \le 4, \\ 12 - 2x, & 4 < x < 5, \\ 0, & x \ge 5. \end{cases}$$

Graph f and find
(a) $\lim\limits_{x \to 3} f(x)$, $\lim\limits_{x \to 4} f(x)$, $\lim\limits_{x \to 5} f(x)$
(b) $\lim\limits_{x \to a} f(x)$ if $0 < a < 3$, $3 < a < 4$, $4 < a < 5$, and $a > 5$

70. Let f be defined by
$$f(x) = \begin{cases} x^2, & 0 \le x \le 1, \\ 2x, & 1 < x \le 2, \\ 10 - 3x, & 2 < x < 4, \\ 16 - x^2, & x \ge 4. \end{cases}$$

Graph f and find
(a) $\lim\limits_{x \to 1} f(x)$, $\lim\limits_{x \to 2} f(x)$, $\lim\limits_{x \to 4} f(x)$
(b) $\lim\limits_{x \to a} f(x)$ if $0 < a < 1$, $1 < a < 2$, $2 < a < 4$, and $a > 4$

Management

71. If a biochemical company produces x grams of interferon, its variable cost per gram is $v(x)$ dollars, where

$$v(x) = \begin{cases} 50, & 0 < x \le 1000, \\ 40, & 1000 < x \le 5000, \\ 35, & x > 5000. \end{cases}$$

Graph the function v and find
   (a) $\lim_{x \to 1000} v(x)$, $\lim_{x \to 5000} v(x)$.
   (b) $\lim_{x \to a} v$ if $0 < a < 1000$, $1000 < a < 5000$, and $a > 5000$.

72. A software company sold its word-processing package for $s(t)$ dollars t years after its introduction, where

$$s(t) = \begin{cases} 120, & 0 < t < 2, \\ 99, & 2 \le t < 4, \\ 79, & t \ge 4. \end{cases}$$

Graph the function s and find
   (a) $\lim_{t \to 2} s(t)$, $\lim_{t \to 4} s(t)$.
   (b) $\lim_{t \to a} s(t)$ if $0 < a < 2$, $2 < a < 4$, and $a > 4$.

73. Suppose the cost of recovering x barrels of oil from a stripper well is C dollars, where

$$C(x) = \begin{cases} 40{,}000 + 18x, & 0 < x < 5000, \\ 50{,}000 + 20x, & x \ge 5000. \end{cases}$$

Graph this function and find $\lim_{x \to a} C(x)$ if $0 < a < 5000$, $a = 5000$, and $a > 5000$.

74. If the price of a gallon of gasoline is p dollars, gasoline consumption will be y billion gallons per year, where

$$y = \begin{cases} 500 - 10p, & 0 \le p \le 1.50, \\ 560 - 50p, & p > 1.50. \end{cases}$$

Graph this function, and find $\lim_{p \to a} y$ if $0 < a < 1.50$, $a = 1.50$, and $a > 1.50$.

75. Repeat Exercise 74 if

$$y = \begin{cases} 500 - 10p, & 0 \le p \le 1.50, \\ 400 - 40p, & p > 1.50. \end{cases}$$

2.2 ONE-SIDED LIMITS, INFINITE LIMITS, AND LIMITS AT INFINITY

In the previous section we introduced the notion of the limit of a function f as its independent variable x approaches a limiting value a, and we saw how the existence or nonexistence of $\lim_{x \to a} f(x)$ describes the behavior of f when x is near a. There is more to be said about limits, however. There are several refinements of the notion of limit that can be used to describe the behavior of functions with greater precision.

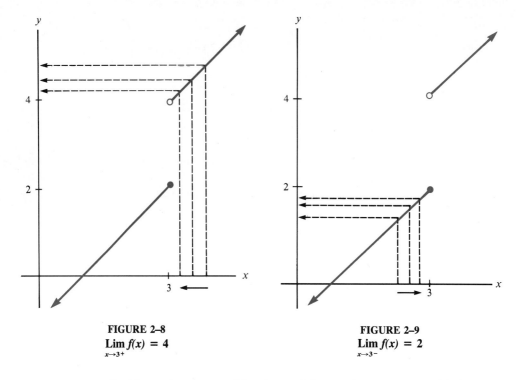

FIGURE 2–8
$\text{Lim}_{x \to 3^+} f(x) = 4$

FIGURE 2–9
$\text{Lim}_{x \to 3^-} f(x) = 2$

In this section we consider three of these refinements: one-sided limits, infinite limits, and limits at infinity.

One-Sided Limits

Consider the function f defined by

$$f(x) = \begin{cases} x - 1, & 0 \le x \le 3, \\ x + 1, & x > 3. \end{cases}$$

See Figure 2–8. Note that if we let x approach 3 *only* from the right, then the corresponding functional values $f(x)$ approach the number 4. We express this fact by means of the **one-sided limit**

$$\text{Lim}_{x \to 3^+} f(x) = 4.$$

Notice the notation $x \to 3^+$, which indicates that x approaches 3 only from the right. Similarly, letting $x \to 3^-$ indicate that x approaches 3 only from the left, we have the one-sided limit

$$\text{Lim}_{x \to 3^-} f(x) = 2.$$

This is so because, as Figure 2–9 shows, as x approaches 3 from the left, the functional values $f(x)$ approach 2. One-sided limits such as these are often convenient in describing the behavior of functions.

The Derivative

EXAMPLE 1 Let f be defined by

$$f(x) = \begin{cases} -1, & x < 0, \\ 1, & x \geq 0. \end{cases}$$

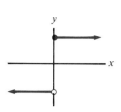

FIGURE 2–10

See Figure 2–10. We have

$$\lim_{x \to 0^-} f(x) = -1 \quad \text{and} \quad \lim_{x \to 0^+} f(x) = 1.$$

Note that $\lim_{x \to 0^-} f(x) \neq \lim_{x \to 0^+} f(x)$ and that $\lim_{x \to 0} f(x)$ does not exist. □

EXAMPLE 2 Let f be defined by

$$f(x) = \begin{cases} x, & x < 1, \\ 2, & x = 1, \\ x, & x > 1. \end{cases}$$

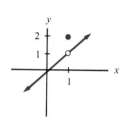

FIGURE 2–11

See Figure 2–11. Then

$$\lim_{x \to 1^-} f(x) = 1 \quad \text{and} \quad \lim_{x \to 1^+} f(x) = 1.$$

Note that $\lim_{x \to 1^-} f(x) = \lim_{x \to 1^+} f(x) = 1$ and that $\lim_{x \to 1} f(x) = 1$ also. □

EXAMPLE 3 Let g be given by $g(x) = \sqrt{x}$. Because $g(x)$ is not defined for $x < 0$, we can only consider the one-sided limit as $x \to 0^+$. From Figure 2–12 we see that

$$\lim_{x \to 0^+} g(x) = \lim_{x \to 0^+} \sqrt{x} = 0. \quad \square$$

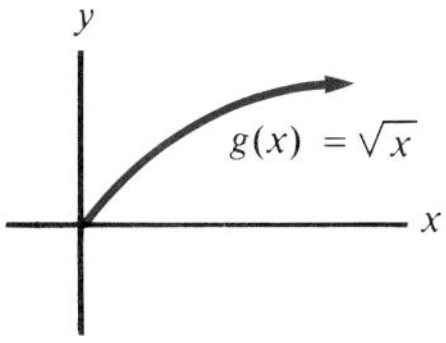

FIGURE 2–12

It should be clear that if the one-sided limits $\lim_{x \to a^-} f(x)$ and $\lim_{x \to a^+} f(x)$ exist and are equal (as in Example 2), then $\lim_{x \to a} f(x)$ exists and is equal to them. On the other hand, if either of the one-sided limits does not exist, or if they both exist but are not equal (as in Example 1), then $\lim_{x \to a} f(x)$ does not exist.

EXAMPLE 4 Let f be the function whose graph is shown in Figure 2–13. Then

$$\lim_{x \to 1^-} f(x) = -2$$

and $\lim_{x \to 1^+} f(x)$ does not exist, so $\lim_{x \to 1} f(x)$ does not exist. □

EXAMPLE 5 If a firm makes and sells 6000 or fewer units, its profit is $P(x) = 250x - 400{,}000$ dollars. However, in order to make more than 6000 units, it must put on a second shift, which causes its profit to be $P(x) = 200x - 300{,}000$ dollars. Thus

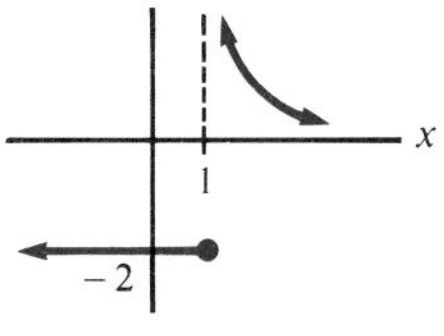

FIGURE 2–13

$$P(x) = \begin{cases} 250x - 400{,}000 & 0 \leq x \leq 6000, \\ 200x - 300{,}000, & x > 6000. \end{cases}$$

Thus $\lim_{x \to 6000^-} P(x) = 1{,}100{,}000$ and $\lim_{x \to 6000^+} P(x) = 900{,}000$. (See Figure 2–14.) Hence as the firm's production approaches 6000 units from the left, its profit

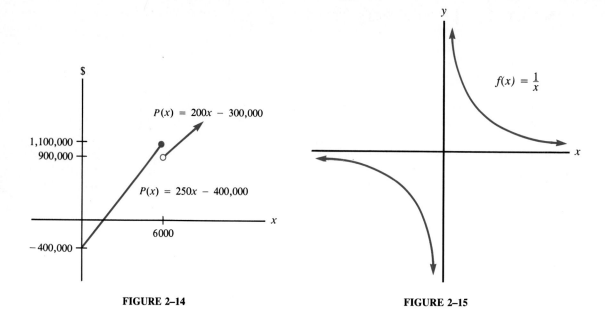

FIGURE 2–14

FIGURE 2–15

approaches $1,100,000, and as its production approaches 6000 units from the right, its profit approaches $900,000. ☐

Infinite Limits

Consider the function f defined by $f(x) = 1/x$. See Figure 2–15. The one-sided limit $\text{Lim}_{x \to 0^+} f(x)$ does not exist because as $x \to 0^+$ the corresponding functional values $f(x)$ increase without bound. When this happens, we say that the limit is **positive infinity,** written $+\infty$. Thus we write

$$\text{Lim}_{x \to 0^+} f(x) = \text{Lim}_{x \to 0^+} \frac{1}{x} = +\infty.$$

This does *not* mean that the one-sided limit exists: $+\infty$ is not a number but a symbol for a process of increasing without bound. The statement $\text{Lim}_{x \to 0^+} f(x) = +\infty$ is thus a way of saying that the limit does not exist *because* the functional values increase without bound.

Returning to Figure 2–15, we see also that $\text{Lim}_{x \to 0^-} 1/x$ does not exist because as $x \to 0^-$ the corresponding functional values decrease without bound. In such a case we say that the limit is **negative infinity,** written $-\infty$. Hence we have

$$\text{Lim}_{x \to 0^-} \frac{1}{x} = -\infty.$$

Again, this does not mean that the limit exists but that it does *not* exist because the functional values decrease without bound.

If $\text{Lim}_{x \to a^+} f(x) = \pm\infty$ or $\text{Lim}_{x \to a^-} f(x) = \pm\infty$, the line $x = a$ is called a

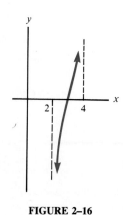

FIGURE 2–16

vertical asymptote for f. For instance, the line $x = 0$ (the y-axis) is a vertical asymptote for $f(x) = 1/x$ as $x \to 0^+$ and also as $x \to 0^-$. Notice in Figure 2–15 how the graph of f approaches the y-axis more and more closely as x approaches 0 from either side. Finding asymptotes will be important to us when we discuss curve sketching in Chapter 4.

EXAMPLE 6 The function g whose graph is shown in Figure 2–16 has $\text{Lim}_{x \to 2^+} g(x) = -\infty$ and $\text{Lim}_{x \to 4^-} g(x) = +\infty$. Therefore $x = 2$ is a vertical asymptote for g as $x \to 2^+$, and $x = 4$ is a vertical asymptote for g as $x \to 4^-$. Notice how the graph of g approaches these lines more and more closely as $x \to 2^+$ and as $x \to 4^-$. ☐

Polynomial functions cannot have vertical asymptotes, because if p is such a function and a is any real number, then $\text{Lim}_{x \to a} p(x) = p(a)$, and $p(a) \neq \pm\infty$. However, rational functions can have vertical asymptotes: these will occur at those values of x for which the denominator of the function is zero while its numerator is nonzero.

EXAMPLE 7 Consider the rational function f defined by

$$f(x) = \frac{x}{x-1}.$$

Because $x = 1$ makes the denominator of f zero and its numerator nonzero, the line $x = 1$ is a vertical asymptote for the function, and hence $\text{Lim}_{x \to 1^+} f(x) = \pm\infty$. We can decide between the alternatives $+\infty$ and $-\infty$ for the limit by examining the sign of the fraction $x/(x-1)$ as $x \to 1^+$. Note that as x approaches 1 from the right, x is greater than 1, so both x and $x - 1$ are positive. Therefore as $x \to 1^+$, the fraction $x/(x-1)$ is always positive, and thus

$$\text{Lim}_{x \to 1^+} \frac{x}{x-1} = +\infty.$$

Similarly, since $x/(x-1)$ is negative when x is near 1 but less than 1, it follows that

$$\text{Lim}_{x \to 1^-} \frac{x}{x-1} = -\infty.$$

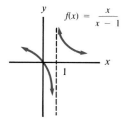

FIGURE 2–17

See Figure 2–17. ☐

EXAMPLE 8 It is estimated that the cost of removing $x\%$ of the pollutants from the air in the Los Angeles basin would be

$$C(x) = \frac{60x}{100 - x}$$

billion dollars. (This model implies that it would be impossible to remove all pollutants from the air. Why?) Since $C(x)$ is positive when x is near 100 and less than 100, we have $\text{Lim}_{x \to 100^-} C(x) = +\infty$. Thus the cost of removing $x\%$ of the pollutants would become prohibitively large as x approached 100. ☐

Limits at Infinity

The infinity symbols $\pm\infty$ are also useful when considering how a function behaves as its independent variable is allowed to increase or decrease without bound. We write

$x \to +\infty$ to indicate that x is allowed to increase without bound (move to the right along the x-axis without bound).

$x \to -\infty$ to indicate that x is allowed to decrease without bound (move to the left along the x-axis without bound).

Now consider the function f whose graph is shown in Figure 2–18. Notice that as $x \to +\infty$, the corresponding functional values $f(x)$ approach the number L; as $x \to -\infty$, the functional values approach the number K. We express these facts by writing the **limits at infinity**

$$\mathop{\text{Lim}}_{x \to +\infty} f(x) = L \quad \text{and} \quad \mathop{\text{Lim}}_{x \to -\infty} f(x) = K,$$

respectively, and referring to the lines $y = L$ and $y = K$ as **horizontal asymptotes for f**.

EXAMPLE 9 The function f whose graph is depicted in Figure 2–19 has $\text{Lim}_{x \to +\infty} f(x) = 3$ and $\text{Lim}_{x \to -\infty} f(x) = 0$. Hence the line $y = 3$ is a horizontal asymptote for f as x approaches $+\infty$ and the line $y = 0$ (the x-axis) is a horizontal asymptote for f as x approaches $-\infty$. □

If the functional values $f(x)$ increase without bound as $x \to +\infty$, we write $\text{Lim}_{x \to +\infty} f(x) = +\infty$; if they decrease without bound, we write $\text{Lim}_{x \to +\infty} f(x) = -\infty$. Similar remarks apply to $\text{Lim}_{x \to -\infty} f(x)$.

EXAMPLE 10 The function f of Figure 2–20 has $\mathop{\text{Lim}}_{x \to +\infty} f(x) = +\infty$ and $\mathop{\text{Lim}}_{x \to -\infty} f(x) = -\infty$. □

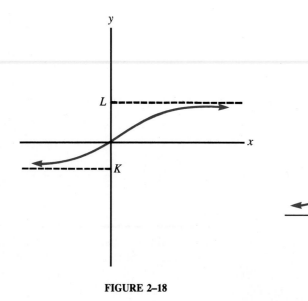

FIGURE 2–18

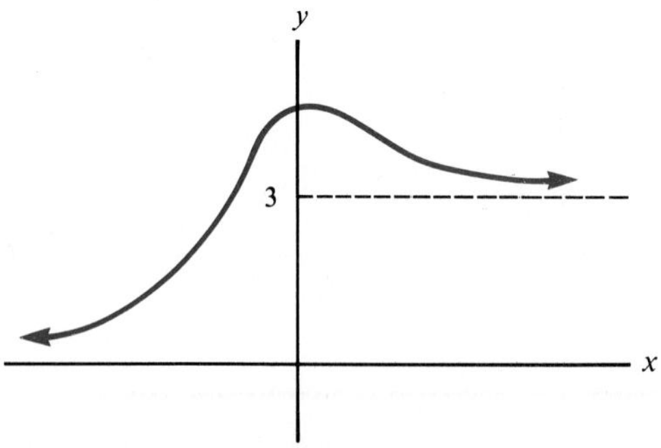

FIGURE 2–19
$\mathop{\text{Lim}}_{x \to +\infty} f(x) = 3, \quad \mathop{\text{Lim}}_{x \to -\infty} f(x) = 0$

THE DERIVATIVE

EXAMPLE 11   Let us find the limits at infinity of the function f defined by $f(x) = 1/x$. When x is a large positive number, $1/x$ is a small positive fraction; we can make $1/x$ as close to 0 as we want by taking x sufficiently large. Therefore

$$\operatorname*{Lim}_{x \to +\infty} \frac{1}{x} = 0.$$

This shows that the x-axis is a horizontal asymptote for the function f as $x \to +\infty$. Similar reasoning as $x \to -\infty$ shows that the x-axis is also a horizontal asymptote for f as $x \to -\infty$. The graph of f is shown in Figure 2–21. □

In Example 11 we showed that $\operatorname{Lim}_{x \to \pm\infty} (1/x) = 0$. The reasoning used there holds also for $1/x^n$, where n is any positive integer. On the other hand, as x increases without bound, so does x^n for any positive integer n, so $\operatorname{Lim}_{x \to +\infty} x^n = +\infty$. Similarly, as $x \to -\infty$, x^n approaches $+\infty$ if n is even and approaches $-\infty$ if n is odd.

Limits at Infinity for Powers of x

For any positive integer n:

1. $\operatorname*{Lim}_{x \to +\infty} \dfrac{1}{x^n} = 0$ and $\operatorname*{Lim}_{x \to -\infty} \dfrac{1}{x^n} = 0.$

2. $\operatorname*{Lim}_{x \to +\infty} x^n = +\infty$, $\operatorname*{Lim}_{x \to -\infty} x^n = +\infty$ if n is even, and
   $\operatorname*{Lim}_{x \to -\infty} x^n = -\infty$ if n is odd.

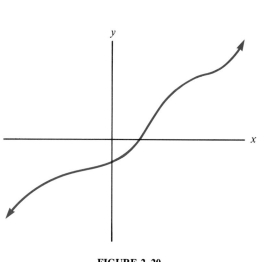

FIGURE 2–20
$\operatorname*{Lim}_{x \to +\infty} f(x) = +\infty$, $\operatorname*{Lim}_{x \to -\infty} f(x) = -\infty$

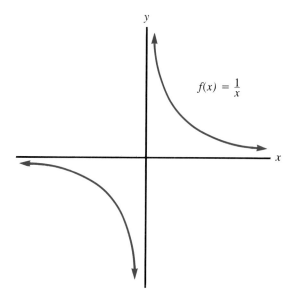

$f(x) = \frac{1}{x}$

FIGURE 2–21

EXAMPLE 12

$$\text{Lim}_{x \to +\infty} -\frac{2}{x^2} = -2 \cdot \text{Lim}_{x \to +\infty} \frac{1}{x^2} = -2(0) = 0,$$

$$\text{Lim}_{x \to +\infty} 4x^3 = 4 \cdot \text{Lim}_{x \to +\infty} x^3 = 4(+\infty) = +\infty,$$

$$\text{Lim}_{x \to -\infty} 4x^3 = 4 \cdot \text{Lim}_{x \to -\infty} x^3 = 4(-\infty) = -\infty. \quad \square$$

Because $\text{Lim}_{x \to \pm\infty} x^n = \pm\infty$, polynomial functions do not have horizontal asymptotes. Rational functions can have horizontal asymptotes, however; to find them, we must be able to evaluate the limits at infinity of such functions. This can be accomplished by dividing each term of the numerator and denominator by the largest power of x that appears in the denominator and using our results about limits at infinity for powers of x. We illustrate with examples.

EXAMPLE 13

Let us evaluate

$$\text{Lim}_{x \to +\infty} \frac{2x^2 + 1}{x^2 - 5x + 4}.$$

The largest power of x in the denominator is x^2, so we rewrite the limit as

$$\text{Lim}_{x \to +\infty} \frac{2x^2/x^2 + 1/x^2}{x^2/x^2 - 5x/x^2 + 4/x^2} = \text{Lim}_{x \to +\infty} \frac{2 + 1/x^2}{1 - 5(1/x) + 4(1/x^2)}.$$

But $\text{Lim}_{x \to +\infty} (1/x) = 0$ and $\text{Lim}_{x \to +\infty} (1/x^2) = 0$ by our results for limits at infinity for powers of x. Hence

$$\text{Lim}_{x \to +\infty} \frac{2x^2 + 1}{x^2 - 5x + 4} = \frac{2 + 0}{1 - 5(0) + 4(0)} = \frac{2}{1} = 2.$$

Thus $y = 2$ is a horizontal asymptote for f as $x \to +\infty$. $\quad \square$

EXAMPLE 14

Let us evaluate

$$\text{Lim}_{x \to -\infty} \frac{x^3 + 8}{x^2 - 5}.$$

The highest power of x present in the denominator is x^2, so

$$\text{Lim}_{x \to -\infty} \frac{x^3/x^2 + 8/x^2}{x^2/x^2 - 5/x^2} = \text{Lim}_{x \to -\infty} \frac{x + 8/x^2}{1 - 5/x^2}.$$

But $\text{Lim}_{x \to -\infty} x = -\infty$ and $\text{Lim}_{x \to -\infty} (1/x^2) = 0$, so

$$\text{Lim}_{x \to -\infty} \frac{x^3 + 8}{x^2 - 5} = \text{Lim}_{x \to -\infty} \frac{x + 8(1/x^2)}{1 - 5(1/x^2)} = \frac{-\infty + 8(0)}{1 - 5(0)} = -\infty.$$

Therefore this function has no horizontal asymptotes. $\quad \square$

EXAMPLE 15

If it makes x units of its product, a firm's average cost per unit is $A(x)$ dollars, where

$$A(x) = \frac{12x + 200}{x}.$$

THE DERIVATIVE

Since

$$\operatorname*{Lim}_{x\to +\infty} A(x) = \operatorname*{Lim}_{x\to +\infty} \frac{12x + 200}{x} = \operatorname*{Lim}_{x\to +\infty} \frac{12 + 200/x}{1} = \frac{12 + 0}{1} = 12,$$

we see that as the firm makes more and more units, its average cost per unit approaches $12. ☐

In addition to vertical and horizontal asymptotes, functions can also possess **oblique asymptotes.** An oblique asymptote is a line $y = mx + b$, $m \neq 0$, such that as $x \to \pm\infty$, the functional values $f(x)$ approach $mx + b$. This means that as $x \to \pm\infty$, the graph of f will approach the graph of $y = mx + b$ more and more closely. We illustrate with an example.

EXAMPLE 16 Consider the rational function g defined by

$$g(x) = \frac{x^2 + 1}{x}.$$

Notice that we can write g in the form

$$g(x) = x + \frac{1}{x}.$$

Then

$$\operatorname*{Lim}_{x\to 0^+} g(x) = \operatorname*{Lim}_{x\to 0^+} \left(x + \frac{1}{x}\right) = 0 + (+\infty) = +\infty$$

and

$$\operatorname*{Lim}_{x\to 0^-} g(x) = \operatorname*{Lim}_{x\to 0^-} \left(x + \frac{1}{x}\right) = 0 + (-\infty) = -\infty.$$

Therefore the y-axis is a vertical asymptote for g as $x \to 0^+$ and as $x \to 0^-$. Also,

$$\operatorname*{Lim}_{x\to +\infty} g(x) = \operatorname*{Lim}_{x\to +\infty} \frac{x^2 + 1}{x} = \operatorname*{Lim}_{x\to +\infty} \frac{x + 1/x}{1} = \frac{+\infty + 0}{1} = +\infty$$

and

$$\operatorname*{Lim}_{x\to -\infty} g(x) = \operatorname*{Lim}_{x\to -\infty} \frac{x^2 + 1}{x} = \operatorname*{Lim}_{x\to -\infty} \frac{x + 1/x}{1} = \frac{-\infty + 0}{1} = -\infty,$$

so g has no horizontal asymptotes. However, closer examination of the behavior of $g(x)$ as $x \to \pm\infty$ shows that the line $y = x$ is an oblique asymptote for g. This is so because as $x \to \pm\infty$, $1/x$ approaches 0, and hence

$$g(x) = x + \frac{1}{x} \approx x + 0 = x.$$

Therefore as $x \to \pm\infty$, the functional values $g(x)$ approach x. Figure 2–22 shows the graph of g: notice how the graph approaches the line $y = x$ as an asymptote. ☐

■ *Computer Manual: Programs LIMIT, INFLIMIT*

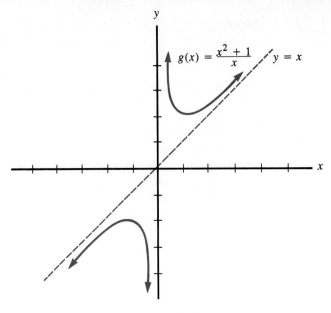

FIGURE 2–22

2.2 EXERCISES

One-Sided Limits

1. The following is the graph of a function f. Find $\lim\limits_{x \to 0^+} f(x)$, $\lim\limits_{x \to 0^-} f(x)$, and $\lim\limits_{x \to 0} f(x)$.

   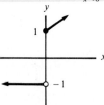

2. The following is the graph of a function f. Find $\lim\limits_{x \to 2^+} f(x)$, $\lim\limits_{x \to 2^-} f(x)$, and $\lim\limits_{x \to 2} f(x)$.

   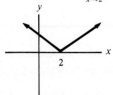

THE DERIVATIVE

3. The following is the graph of a function f. Find $\lim_{x \to 1^+} f(x)$, $\lim_{x \to 1^-} f(x)$, and $\lim_{x \to 1} f(x)$.

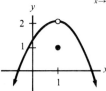

4. The following is the graph of a function f. Find $\lim_{x \to -3^+} f(x)$, $\lim_{x \to -3^-} f(x)$, $\lim_{x \to -3} f(x)$, $\lim_{x \to 2^+} f(x)$, $\lim_{x \to 2^-} f(x)$, and $\lim_{x \to 2} f(x)$.

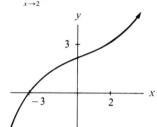

5. The following is the graph of a function f. Find $\lim_{x \to -1^+} f(x)$, $\lim_{x \to -1^-} f(x)$, $\lim_{x \to -1} f(x)$, $\lim_{x \to 1^+} f(x)$, $\lim_{x \to 1^-} f(x)$, and $\lim_{x \to 1} f(x)$.

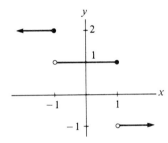

In Exercises 6 through 21, find the indicated limit.

6. $\lim_{x \to 5^+} 2$

7. $\lim_{x \to 2^-} x$

8. $\lim_{x \to 2^+} 3x$

9. $\lim_{x \to 3^+} (5x + 4)$

10. $\lim_{x \to 4^-} (x - 2)$

11. $\lim_{x \to 0^+} |x|$

12. $\lim_{x \to 0^-} |x|$

13. $\lim_{x \to 1^+} \sqrt{x - 1}$

14. $\lim_{x \to -5^+} \sqrt{x + 5}$

15. $\lim_{x \to (1/2)^-} \sqrt{1 - 2x}$

16. $\lim_{x \to 3^-} 2\sqrt{3 - x}$

17. $\lim_{x \to 0^+} f(x)$, $\lim_{x \to 0^-} f(x)$, $\lim_{x \to 0} f(x)$ if $f(x) = \begin{cases} -1, & x < 0, \\ x, & x > 0. \end{cases}$

18. $\lim_{x \to 2^+} f(x)$, $\lim_{x \to 2^-} f(x)$, $\lim_{x \to 2} f(x)$ if $f(x) = \begin{cases} x + 1, & x \leq 2, \\ 5 - x, & x > 2. \end{cases}$

19. $\lim_{x \to 0^+} f(x)$, $\lim_{x \to 0^-} f(x)$, $\lim_{x \to 0} f(x)$ if $f(x) = \begin{cases} x, & x < 0, \\ 2, & x = 0, \\ x, & x > 0. \end{cases}$

20. $\lim_{x \to -2^+} f(x)$, $\lim_{x \to -2^-} f(x)$, $\lim_{x \to -2} f(x)$, $\lim_{x \to 1^+} f(x)$, $\lim_{x \to 1^-} f(x)$, and $\lim_{x \to 1} f(x)$ if

$$f(x) = \begin{cases} 2x + 1, & x < -2, \\ x - 1, & -2 \le x \le 1, \\ x + 2, & x > 1. \end{cases}$$

*21. $\lim_{x \to 0^+} f(x)$, $\lim_{x \to 0^-} f(x)$, $\lim_{x \to 0} f(x)$ if $f(x) = \begin{cases} |x|/x, & x \ne 0, \\ 0, & x = 0. \end{cases}$

Management

22. A company's cost function is defined by

$$C(x) = \begin{cases} 20x + 15{,}000, & 0 \le x \le 1000, \\ 24x + 16{,}000, & x > 1000. \end{cases}$$

Find and interpret $\lim_{x \to 1000^+} C(x)$ and $\lim_{x \to 1000^-} C(x)$.

23. A firm's profit function is given by

$$P(x) = \begin{cases} 10x - 5000, & 0 \le x \le 2500, \\ 6x + 5000, & 2500 < x \le 5000, \\ 5x + 6000, & 5000 < x \le 7500. \end{cases}$$

Find and interpret $\lim_{x \to 2500^+} P(x)$, $\lim_{x \to 2500^-} P(x)$, $\lim_{x \to 5000^+} P(x)$, and $\lim_{x \to 5000^-} P(x)$.

24. If a printing company purchases x gallons of ink it costs them $C(x)$ dollars, where

$$C(x) = \begin{cases} 5.50x, & 0 \le x < 25, \\ 5.00x, & 25 \le x < 100, \\ 4.75x, & x \ge 100. \end{cases}$$

Find and interpret $\lim_{x \to 25^+} C(x)$, $\lim_{x \to 25^-} C(x)$, $\lim_{x \to 100^+} C(x)$, and $\lim_{x \to 100^-} C(x)$.

25. It costs a mail-order company y dollars to ship a package that weighs x pounds, where

$$y = \begin{cases} 0.5 + 5.0x, & 0 < x < 2, \\ 1.5 + 4.5x, & 2 \le x < 10, \\ 6.5 + 4.0x, & x \ge 10. \end{cases}$$

Find and interpret $\lim_{x \to 2^+} y$, $\lim_{x \to 2^-} y$, $\lim_{x \to 10^+} y$, and $\lim_{x \to 10^-} y$.

Infinite Limits

In Exercises 26 through 30 use the symbols $\pm \infty$.

26. The following is the graph of a function f. Find $\lim_{x \to 2^+} f(x)$, $\lim_{x \to 2^-} f(x)$, and the vertical asymptotes of f.

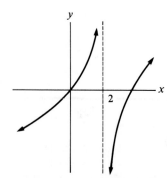

THE DERIVATIVE   115

27. The following is the graph of a function f.
Find $\lim_{x \to 1^+} f(x)$, $\lim_{x \to 1^-} f(x)$,
and the vertical asymptotes of f.

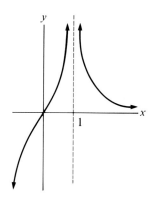

28. The following is the graph of a function f.
Find $\lim_{x \to 2^+} f(x)$, $\lim_{x \to 2^-} f(x)$, $\lim_{x \to -2^+} f(x)$,
$\lim_{x \to -2^-} f(x)$, and the vertical asymptotes of f.

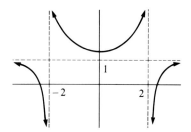

29. The following is the graph of a function f.
Find $\lim_{x \to 3^+} f(x)$, $\lim_{x \to 3^-} f(x)$,
and the vertical asymptotes of f.

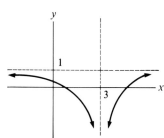

30. The following is the graph of a function f.
Find $\lim_{x \to -2^+} f(x)$, $\lim_{x \to 2^+} f(x)$, $\lim_{x \to 2^-} f(x)$, $\lim_{x \to 4^-} f(x)$,
and the vertical asymptotes of f.

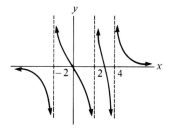

In Exercises 31 through 42, find the indicated limits. Use the symbols $\pm \infty$ as appropriate.

31. $\lim_{x \to 0^+} \dfrac{1}{x^3}$, $\lim_{x \to 0^-} \dfrac{1}{x^3}$

32. $\lim_{x \to 0^+} \dfrac{2}{x^4}$, $\lim_{x \to 0^-} \dfrac{2}{x^4}$

33. $\lim_{x \to 0^+} \dfrac{-2}{x}$, $\lim_{x \to 0^-} \dfrac{-2}{x}$

34. $\lim_{x \to 2^+} \dfrac{1}{x-2}$, $\lim_{x \to 2^-} \dfrac{1}{x-2}$

35. $\lim\limits_{x \to -1^+} \dfrac{1}{1-x}$, $\lim\limits_{x \to -1^-} \dfrac{1}{1-x}$

36. $\lim\limits_{x \to 1^+} \dfrac{x}{x-1}$, $\lim\limits_{x \to 1^-} \dfrac{x}{x-1}$

37. $\lim\limits_{x \to -3^+} \dfrac{x}{x+3}$, $\lim\limits_{x \to -3^-} \dfrac{x}{x+3}$

38. $\lim\limits_{x \to 5^+} \dfrac{|x|}{x-5}$, $\lim\limits_{x \to 5^-} \dfrac{|x|}{x-5}$

39. $\lim\limits_{x \to -1^+} \dfrac{-2x}{x+1}$, $\lim\limits_{x \to -1^-} \dfrac{-2x}{x+1}$

40. $\lim\limits_{x \to 1^+} \dfrac{x^2}{x-1}$, $\lim\limits_{x \to 1^-} \dfrac{x^2}{x-1}$

41. $\lim\limits_{x \to 2^+} \dfrac{x}{x^2-4}$, $\lim\limits_{x \to 2^-} \dfrac{x}{x^2-4}$

42. $\lim\limits_{x \to -1^+} \dfrac{x^2}{x^2-1}$, $\lim\limits_{x \to -1^-} \dfrac{x^2}{x^2-1}$

43. Suppose the cost of eliminating all air pollution in the Los Angeles basin is $C(x)$ billion dollars, where
$$C(x) = \dfrac{60x}{120-x}.$$
Find and interpret $\lim\limits_{x \to 100^-} C(x)$ (cf. Example 8).

Management

44. The demand function for a commodity is given by
$$q = 20\sqrt{100-p}.$$
Find and interpret $\lim\limits_{x \to 100^-} q$.

45. Reducing the pollution caused by auto emissions by $x\%$ over the next five years would cost the auto manufacturers y billion dollars, where
$$y = \dfrac{10x}{20 - 0.4x}.$$
Find and interpret $\lim\limits_{x \to 50^-} y$.

46. The cost of decommissioning a nuclear power plant that has operated for t years is estimated to be y million dollars, where
$$y = \dfrac{40t + 30}{30 - t}.$$
Find and interpret $\lim\limits_{t \to 30^-} y$.

47. An oil field is estimated to contain 100 million barrels of oil. The cost of obtaining $x\%$ of this oil will be
$$C(x) = \dfrac{400x - 3.5x^2}{95 - x}$$
million dollars. Find and interpret $\lim\limits_{x \to 95^-} C(x)$.

48. The supply function for a commodity is given by
$$q = \dfrac{p^2 + 100p - 50{,}000}{500 - p}.$$
Find and interpret $\lim\limits_{p \to 500^-} q$.

Limits at Infinity

In Exercises 49 through 53, find $\lim_{x \to \pm\infty} f(x)$ and the horizontal asymptotes for the function f whose graph is shown. Use the symbols $\pm\infty$ as appropriate.

49.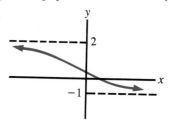

50.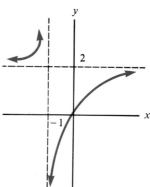

51.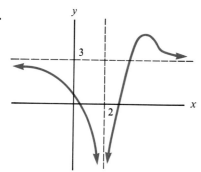

52.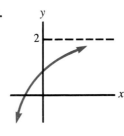

53.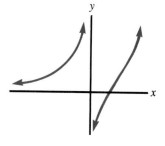

In Exercises 54 through 71, find the indicated limit at infinity. Use the symbols $\pm\infty$ as appropriate.

54. $\lim\limits_{x \to +\infty} (2x + 3)$

55. $\lim\limits_{x \to -\infty} (x^2 - 2)$

56. $\lim\limits_{x \to -\infty} (x^3 + 4)$

57. $\lim\limits_{x \to +\infty} \dfrac{1}{x^2}$

58. $\lim\limits_{x \to -\infty} \dfrac{2}{x^3}$

59. $\lim\limits_{x \to -\infty} \dfrac{3}{x^4}$

60. $\lim\limits_{x \to +\infty} -\dfrac{2}{x^2}$

61. $\lim\limits_{x \to +\infty} \dfrac{x}{x + 5}$

62. $\lim\limits_{x \to -\infty} \dfrac{3x}{2x + 4}$

63. $\lim\limits_{x \to +\infty} \dfrac{2x + 1}{2 - 3x}$

64. $\lim\limits_{x \to -\infty} \dfrac{3x}{x^2 + 1}$

65. $\lim\limits_{x \to +\infty} \dfrac{2x^2 + 1}{3x - 2}$

66. $\lim\limits_{x \to +\infty} \dfrac{3x^2}{5x^2 + 3}$

67. $\lim\limits_{x \to +\infty} \dfrac{x^2 + 5x - 7}{x^3 - 2x + 1}$

68. $\lim\limits_{x \to +\infty} \dfrac{2x^2 + 5x - 7}{4x^2 + 8x + 2}$

69. $\lim\limits_{x \to -\infty} \dfrac{5x^3 + 6x + 12}{11x^3 - 2x + 3}$

70. $\lim\limits_{x \to +\infty} \dfrac{4x^2 + 6x + 9}{2x^4 + 3x^3 + 2}$

71. $\lim\limits_{x \to +\infty} \dfrac{6x^4 - 2x^3 + 8x + 1}{x^2 - 3x + 2}$

In Exercises 72 through 81, find all asymptotes, vertical, horizontal, and oblique, of the given function f.

72. $f(x) = \dfrac{x}{x-3}$

73. $f(x) = \dfrac{6x}{3x-2}$

74. $f(x) = \dfrac{x+1}{x-3}$

75. $f(x) = \dfrac{5x+2}{2x-6}$

76. $f(x) = \dfrac{2-2x}{1+x}$

77. $f(x) = \dfrac{2+3x}{x}$

78. $f(x) = \dfrac{3x}{x^2-1}$

79. $f(x) = \dfrac{x-1}{x^2-4}$

80. $f(x) = \dfrac{x^2+1}{x}$

81. $f(x) = \dfrac{2x^2-2}{x}$

Management

82. If x units of a product are made, its average cost per unit is $A(x)$ dollars, where
$$A(x) = 22 + \dfrac{50{,}000}{x}.$$
Find and interpret $\lim\limits_{x \to +\infty} A(x)$.

83. Acme Company's average cost per unit produced is given by
$$A(x) = \dfrac{5x+80}{x+1},$$
where x is the number of units produced and $A(x)$ is the average cost per unit in dollars. Find and interpret $\lim\limits_{x \to +\infty} A(x)$.

84. A firm's average cost per unit if it makes x units is $A(x)$ dollars, where
$$A(x) = \dfrac{2x^2+50}{x}.$$
Find and interpret $\lim\limits_{x \to +\infty} A(x)$.

85. A computer company is introducing a new model of computer. The firm estimates that it will take $T(x)$ worker-hours to build the xth unit of the new computer, where
$$T(x) = \dfrac{300x^3 + 50x + 100}{x^3}.$$
Find and interpret $\lim\limits_{x \to +\infty} T(x)$.

Life Science

86. If a person who suffers a bite by a poisonous snake receives an immediate shot of antivenin, then t seconds after the shot is given there will be
$$y = \dfrac{0.5t + 2000}{2t + 3}$$
ppm of poison in the victim's blood. Find and interpret $\lim\limits_{t \to +\infty} y$.

Social Science

87. Subjects of a memory experiment are asked to memorize a list of numbers, and it is found that t days later the typical subject can recall

$$y = 2 + \frac{20}{t+2}$$

of the numbers. Find and interpret $\lim\limits_{t \to +\infty} y$.

Computer Science

*88. According to an article by Peter Denning ("The Science of Computing: Speeding Up Parallel Processing," *American Scientist*, July–August 1988), if the size of a problem is held constant while the number of processors in a computer increases, then the speedup in processing is

$$F = \frac{aN^2 p}{aN^2 + pb}.$$

Here p is the number of processors, $p \geq 1$, N is a measure of the size of the problem, aN^2 is the total computation time, and b is the communication time between processors. Find and interpret $\lim\limits_{p \to +\infty} F$.

2.3 THE DERIVATIVE

The concept of the derivative is the basis of calculus. The derivative of a function tells us many things about the function. For instance, as we shall see later in this chapter, we can use the derivative to examine the rate of change of a function. Thus, suppose a certain function describes the concentration of an antibiotic in a patient's blood as a function of time; then the derivative of the function will tell us the rate at which the concentration of the antibiotic is changing with respect to time. Derivatives can also help us graph functions, find their maximum and minimum values, and in general analyze them in many useful ways. Before we can do any of these things, however, we must first develop a thorough understanding of what the derivative is. In this section we introduce the derivative, discuss its geometric significance, and briefly consider the concepts of differentiability and continuity of a function.

Definition of the Derivative

In order to make our initial discussion of the derivative specific, we choose to consider the function f defined by $f(x) = x^2$ and the point $(1, 1)$ on its graph. We may construct a unique line that just touches the graph at the point $(1, 1)$. See Figure 2–23. This line is called the **line tangent to f at $(1, 1)$**. If we knew the slope of this tangent line, we could write its equation; we will now try to find its slope.

Suppose we add a number h to 1. The point

$$(1 + h, f(1 + h)) = (1 + h, (1 + h)^2)$$

will lie on the graph of f, and we can draw a line through the points $(1, 1)$ and $(1 + h, (1 + h)^2)$. This line is called a **secant line**. Figure 2–24 depicts the situation if h is positive. The secant line has

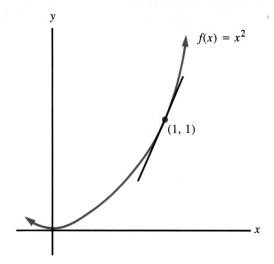

FIGURE 2–23　　　　　　　　　　　　FIGURE 2–24

$$\text{Slope} = \frac{(1+h)^2 - 1}{1 + h - 1} = \frac{1 + 2h + h^2 - 1}{h} = \frac{h(2+h)}{h} = 2 + h.$$

Now let h approach 0. As h gets smaller and smaller, $1 + h$ approaches 1 on the x-axis, the point $(1 + h, (1 + h)^2)$ "slides" along the graph toward the point $(1, 1)$, and the secant line approaches the tangent line at $(1, 1)$. Figure 2–25 shows how as h approaches 0 the secant line "pivots" at $(1, 1)$ and approaches the tangent line. Since the secant line approaches the tangent line, it follows that the slope of the secant line must approach the slope of the tangent line. In other words,

$$\lim_{h \to 0} (\text{slope of secant line}) = \text{Slope of tangent line},$$

or

$$\lim_{h \to 0} (2 + h) = 2 + 0 = 2.$$

Thus the slope of the line tangent to f at $(1, 1)$ is equal to 2. This slope is called the **derivative of f at $x = 1$.**

The preceding analysis can be applied to any function f at any number in its domain as long as the function has a unique nonvertical tangent line at the point $(x, f(x))$ on its graph. Figure 2–26 depicts the process. We begin by adding h to x to obtain the point $(x + h, f(x + h))$ on the graph. We then form the secant line through $(x, f(x))$ and $(x + h, f(x + h))$. The secant line will have

$$\text{Slope} = \frac{f(x + h) - f(x)}{x + h - x} = \frac{f(x + h) - f(x)}{h},$$

and thus

$$\text{Slope of tangent line} = \lim_{h \to 0} \frac{f(x + h) - f(x)}{h}.$$

The limit on the right side of this equation is called the **derivative of f at x.**

The Derivative

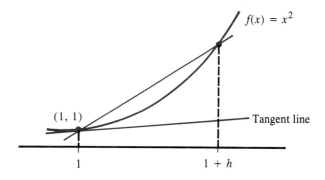

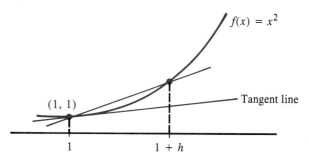

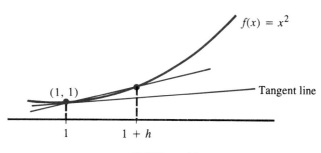

FIGURE 2–25

The Derivative of a Function

Let f be a function. The **derivative of f at x,** denoted by $f'(x)$, is defined to be

$$f'(x) = \lim_{h \to 0} \frac{f(x + h) - f(x)}{h},$$

provided this limit exists. The notation $f'(x)$ is read "f prime of x."

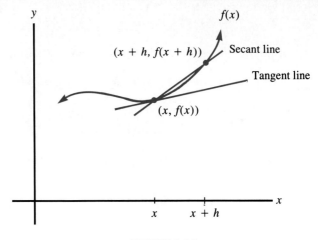

FIGURE 2–26

As we have seen, the derivative $f'(x)$ is equal to the slope of the line tangent to f at the point $(x, f(x))$.

EXAMPLE 1 Let $f(x) = x^2$. Let us find the derivative $f'(x)$. We begin by forming the quotient

$$\frac{f(x+h) - f(x)}{h},$$

which is called the **difference quotient of f at x**. Since $f(x) = x^2$, we have

$$f(x+h) = (x+h)^2,$$

so

$$\frac{f(x+h) - f(x)}{h} = \frac{(x+h)^2 - x^2}{h}.$$

Next we simplify the difference quotient, our goal being the removal of the factor h from the denominator:

$$\frac{f(x+h) - f(x)}{h} = \frac{(x+h)^2 - x^2}{h}$$

$$= \frac{x^2 + 2xh + h^2 - x^2}{h}$$

$$= \frac{h(2x+h)}{h}$$

$$= 2x + h.$$

Therefore the derivative of $f(x) = x^2$ at x is

$$f'(x) = \lim_{h \to 0} \frac{f(x+h) - f(x)}{h} = \lim_{h \to 0} (2x + h) = 2x + 0 = 2x. \quad \square$$

The Derivative

Notice that the derivative of the function f defined by $f(x) = x^2$ is another function f' defined by $f'(x) = 2x$. If we let $x = a$ in the equation $f'(x) = 2x$, we obtain the slope of the line tangent to f at the point $(a, f(a))$. For instance, if we take $x = 1$, then $f'(1) = 2(1) = 2$, so the slope of the line tangent to f at $(1, f(1)) = (1, 1)$ is equal to 2. This is the result we obtained at the beginning of this section.

The procedure we used in Example 1 to find the derivative is known as the **Three-Step Rule.**

The Three-Step Rule

To find the derivative $f'(x)$ of $f(x)$:

1. Form the difference quotient $\dfrac{f(x + h) - f(x)}{h}$.

2. Simplify the difference quotient to eliminate the factor h from the denominator.

3. Then
$$f'(x) = \lim_{h \to 0} \frac{f(x + h) - f(x)}{h}$$
if this limit exists.

EXAMPLE 2 Let f be given by $f(x) = x^2 + 2x + 4$. We will use the Three-Step Rule to find the defining equation for the derivative f'. We have

$$\frac{f(x + h) - f(x)}{h} = \frac{[(x + h)^2 + 2(x + h) + 4] - [x^2 + 2x + 4]}{h}$$

$$= \frac{x^2 + 2xh + h^2 + 2x + 2h + 4 - x^2 - 2x - 4}{h}$$

$$= \frac{h(2x + h + 2)}{h}$$

$$= 2x + h + 2.$$

Therefore

$$f'(x) = \lim_{h \to 0} \frac{f(x + h) - f(x)}{h} = \lim_{h \to 0} (2x + h + 2) = 2x + 2. \quad \square$$

EXAMPLE 3 If

$$g(x) = \frac{1}{x},$$

find $g'(x)$ and then find the equation of the line tangent to g at $(2, g(2))$.
We have

$$\frac{g(x+h) - g(x)}{h} = \frac{1/(x+h) - 1/x}{h}$$

$$= \frac{\frac{x - (x+h)}{x(x+h)}}{h}$$

$$= \frac{-h}{hx(x+h)}$$

$$= \frac{-1}{x(x+h)}.$$

Thus

$$g'(x) = \lim_{h \to 0} \frac{g(x+h) - g(x)}{h} = \lim_{h \to 0} \frac{-1}{x(x+h)} = \frac{-1}{x(x+0)} = \frac{-1}{x^2}.$$

Hence if $g(x) = 1/x$, then $g'(x) = -1/x^2$. Thus the line tangent to g at $(2, g(2)) = (2, \frac{1}{2})$ has slope equal to

$$g'(2) = -\frac{1}{2^2} = -\frac{1}{4},$$

and the equation of the tangent line is

$$y = -\frac{1}{4}x + b.$$

But the point $(2, \frac{1}{2})$ is on the tangent line, so

$$\frac{1}{2} = -\frac{1}{4}(2) + b,$$

which implies that $b = 1$. Hence the equation of the tangent line is

$$y = -\frac{1}{4}x + 1.$$

See Figure 2–27. □

Differentiability and Continuity

If a function f has a derivative $f'(a)$ at $x = a$, we say that the function is **differentiable at a**; if it does not have a derivative at a, we say it is **not differentiable at a**.

EXAMPLE 4

The function f defined by $f(x) = \sqrt{x}$ is differentiable at a if $a > 0$, but it is not differentiable at 0. To see this, suppose $a \geq 0$ and note that

$$\lim_{h \to 0} \frac{\sqrt{a+h} - \sqrt{a}}{h} = \lim_{h \to 0} \frac{(\sqrt{a+h} - \sqrt{a})(\sqrt{a+h} + \sqrt{a})}{h(\sqrt{a+h} + \sqrt{a})}$$

$$= \lim_{h \to 0} \frac{a + h - a}{h(\sqrt{a+h} + \sqrt{a})}$$

$$= \lim_{h \to 0} \frac{1}{\sqrt{a+h} + \sqrt{a}}.$$

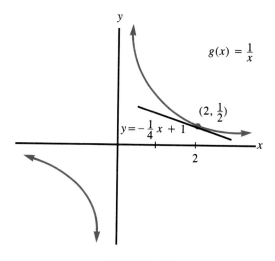

FIGURE 2–27

If $a > 0$, this limit is equal to

$$\frac{1}{\sqrt{a + 0} + \sqrt{a}} = \frac{1}{2\sqrt{a}}.$$

Thus $f'(a) = \dfrac{1}{2\sqrt{a}}$ exists if $a > 0$. However, if $a = 0$, the limit becomes

$$\operatorname*{Lim}_{h \to 0} \frac{1}{\sqrt{0 + h} + \sqrt{0}} = \operatorname*{Lim}_{h \to 0} \frac{1}{\sqrt{h}},$$

which does not exist, and therefore f is not differentiable at 0. ☐

If the derivative $f'(a)$ exists, it is the slope of the unique nonvertical line tangent to f at $(a, f(a))$. Therefore if f does not have a unique tangent line at $(a, f(a))$, as in Figures 2–28 and 2–29, or if the tangent line is vertical, as in Figure 2–30, then the derivative $f'(a)$ cannot exist. For instance, the square-root function of Example 4 has a vertical tangent line at $(0, 0)$ (see Figure 2–31), so it is not differentiable at $x = 0$.

Closely allied to the notion of differentiability is the notion of continuity. A function f is **continuous at $x = a$** if we can trace its graph through the point $(a, f(a))$ without lifting our pencil from the paper. This means that f is continuous at $x = a$ if there is neither a jump nor a gap in its graph at $x = a$. Figure 2–32 shows the graph of a function that is continuous at every real number; such a function is said to be **continuous everywhere.** If f is not continuous at $x = a$, we say it is **discontinuous at $x = a$.** Figures 2–33, 2–34, and 2–35 show different ways in which a function can be discontinuous at $x = a$: in Figure 2–33 the function has a jump in its graph at $x = a$; in Figures 2–34 and 2–35 it has a gap in its graph at $x = a$.

We can state the definition of continuity of a function at $x = a$ in terms of limits.

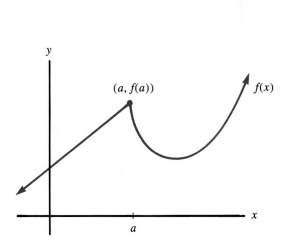

FIGURE 2–28

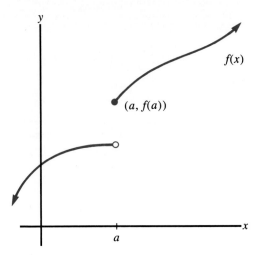

FIGURE 2–29

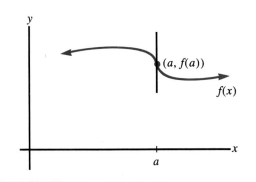

FIGURE 2–30

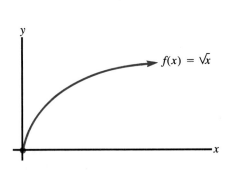

FIGURE 2–31

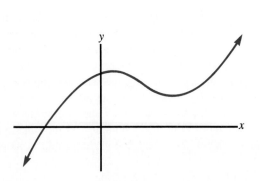

FIGURE 2–32

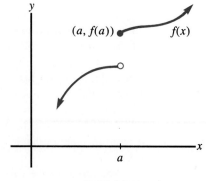

FIGURE 2–33

The Derivative

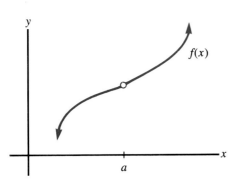

FIGURE 2–34

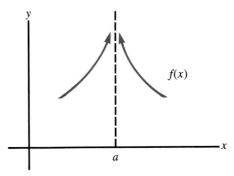

FIGURE 2–35

The First Limit Definition of Continuity

A function f is continuous at $x = a$ if $f(a)$ exists and

$$\lim_{h \to 0} f(x) = f(a).$$

This definition of continuity agrees with our previous geometric one: if $f(a)$ exists, then the point $(a, f(a))$ is on the graph, so there can be no gap in the graph of f at $x = a$; if $\lim_{x \to a} f(x) = f(a)$, then there can be no jump in the graph at $x = a$. (Why?)

Note that if we set $x = a + h$, then we may write

$$\lim_{x \to a} f(x) \quad \text{as} \quad \lim_{h \to 0} f(a + h).$$

Hence we may rewrite the first limit definition of continuity as

The Second Limit Definition of Continuity

A function f is continuous at $x = a$ if $f(a)$ exists and

$$\lim_{h \to 0} f(a + h) = f(a).$$

This second version of the limit definition of continuity will be useful to us later. We saw in Section 2.1 that if p is a polynomial function, then

$$\lim_{x \to a} p(x) = p(a).$$

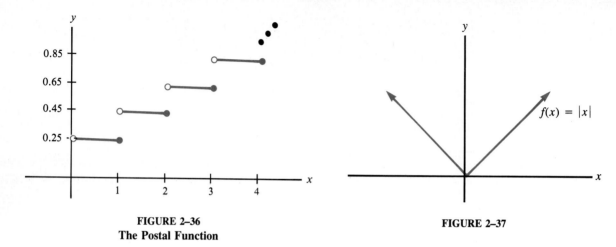

FIGURE 2–36
The Postal Function

FIGURE 2–37

Thus polynomial functions are continuous at every real number. Rational functions that have been reduced to lowest terms are continuous at every real number for which they are defined. However, there are functions that are *not* continuous at every real number for which they are defined.

EXAMPLE 5   Let P be the postal function defined by

$$P(x) = \text{Cost of mailing a letter that weighs } x \text{ ounces.}$$

Currently postage is $0.25 for a letter weighing 1 ounce or less, and an additional $0.20 for each additional ounce or part of an ounce. Therefore

$$P(x) = \begin{cases} \$0.25, & 0 < x \le 1, \\ \$0.45, & 1 < x \le 2, \\ \$0.65, & 2 < x \le 3, \\ \vdots & \end{cases}$$

The graph of P is shown in Figure 2–36. Note that P is discontinuous at $x = 1, 2, 3, \ldots$ and continuous in the intervals $(0, 1), (1, 2), (2, 3), \ldots$. □

The relationship between continuity and differentiability is this: a function differentiable at $x = a$ is continuous at $x = a$. To see this, suppose the function f has a derivative at $x = a$. If f were not continuous at $x = a$, it would have either a jump or a gap in its graph at $x = a$. But if this were the case, it could not have a unique tangent line at $(a, f(a))$, and hence it could not have a derivative at $x = a$. Thus f must be continuous at $x = a$.

Differentiability Implies Continuity

If the function f is differentiable at $x = a$, then it is continuous at $x = a$.

The converse of the preceding statement is *not* true; continuity does *not* imply differentiability. In other words, it is possible for a function to be continuous at

$x = a$ but not differentiable at $x = a$. For example, the absolute value function defined by $f(x) = |x|$ has the graph shown in Figure 2–37; from the graph we see that this function is continuous at $x = 0$, but it is not differentiable at $x = 0$ because it does not have a unique tangent line there.

■ **Computer Manual: Program DFQUOTNT**

2.3 EXERCISES

Definition of the Derivative

In Exercises 1 through 16, use the Three-Step Rule to find $f'(x)$.

1. $f(x) = x$
2. $f(x) = -2x$
3. $f(x) = 3x + 2$
4. $f(x) = 3x^2$
5. $f(x) = 1 - x^2$
6. $f(x) = x - x^2$
7. $f(x) = x^3$
8. $f(x) = 2x^3 + x^2$
9. $f(x) = 3/x$
10. $f(x) = x^{-2}$
11. $f(x) = x + \dfrac{1}{x}$
12. $f(x) = \dfrac{x}{x-1}$
13. $f(x) = \dfrac{x-1}{x+1}$
14. $f(x) = \sqrt{2x}$
15. $f(x) = \sqrt{x+1}$
16. $f(x) = x^{-1/2}$

In Exercises 17 through 32, find the equation of the line tangent to f at $(a, f(a))$. You may wish to use the results of Exercises 1 through 16.

17. $f(x) = x$, $a = 0$
18. $f(x) = -2x$, $a = 4$
19. $f(x) = 3x + 2$, $a = -\tfrac{1}{2}$
20. $f(x) = 3x^2$, $a = 1$
21. $f(x) = 1 - x^2$, $a = 3$
22. $f(x) = x - x^2$, $a = -2$
23. $f(x) = x^3$, $a = -1$
24. $f(x) = 2x^3 + x^2$, $a = 0$
25. $f(x) = 3/x$, $a = 2$
26. $f(x) = x^{-2}$, $a = \tfrac{1}{2}$
27. $f(x) = x + \dfrac{1}{x}$, $a = 1$
28. $f(x) = \dfrac{x}{x-1}$, $a = -1$
29. $f(x) = \dfrac{x-1}{x+1}$, $a = -1$
30. $f(x) = \sqrt{2x}$, $a = 2$
31. $f(x) = \sqrt{x+1}$, $a = -1$
32. $f(x) = x^{-1/2}$, $a = 9$

*33. If $f(x) = x^{1/3}$, use the Three-Step Rule to find $f'(x)$. (*Hint:* Multiply numerator and denominator of the difference quotient by $(x + h)^{2/3} + x^{1/3}(x + h)^{1/3} + x^{2/3}$.)

*34. If $f(x) = |x|$, use the Three-Step Rule to find $f'(x)$. (*Hint:* Consider the cases $x > 0$, $x = 0$, and $x < 0$ separately.)

Differentiability and Continuity

35. Find the values of x at which the functions of Exercises 1 through 8 are defined but not differentiable.

36. Find the values at x at which the functions of Exercises 9 through 16 are defined but not differentiable.

In Exercises 37 through 46, use the given graph to find
(a) the intervals on which the function is continuous;
(b) the values of x at which the function is defined but discontinuous;
(c) the values of x at which the function is defined but not differentiable.

37.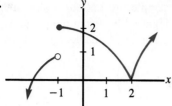

38.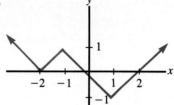

39.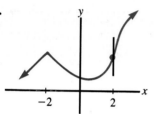

40.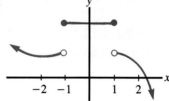

41.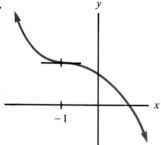

42.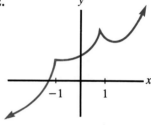

43.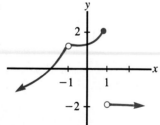

44.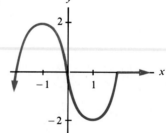

45.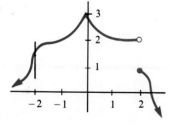

46.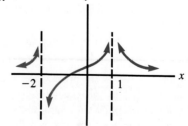

The Derivative

In Exercises 47 through 58, find the intervals on which the function f is continuous and the values of x for which it is defined but discontinuous.

47. $f(x) = x^2 - 3x + 2$

48. $f(x) = x^{15} - 19x^{13} + 27x^3 + 1$

49. $f(x) = \dfrac{x}{x-2}$

50. $f(x) = \dfrac{2x-4}{x^2+1}$

51. $f(x) = \dfrac{x^2 - 6x + 8}{x^2 - 4}$

52. $f(x) = \dfrac{x^3 - 1}{x^2 + 2x - 3}$

53. $f(x) = \begin{cases} 2, & 0 \le x \le 1, \\ 4, & 1 < x \le 2, \\ 6, & 2 < x \le 3 \end{cases}$

54. $f(x) = \begin{cases} 1, & 0 \le x \le 1, \\ 2 - x, & 1 < x < 2, \\ 0, & x > 2 \end{cases}$

55. $f(x) = \begin{cases} 2x + 1, & 0 \le x < 2, \\ 5, & 2 \le x \le 3, \\ 5 - x, & x > 3 \end{cases}$

56. $f(x) = \begin{cases} x^2, & -1 \le x < 0, \\ x^3, & 0 \le x \le 1, \\ 4 - 2x, & 1 < x \le 2 \end{cases}$

57. $f(x) = \begin{cases} 2/x, & x < 2, \\ x/2, & x \ge 2 \end{cases}$

*58. $f(x) = [x]$, where $[x]$ denotes the greatest integer less than or equal to x.

59. The cost of renting a VCR is \$12 per week or any part of a week. Find and graph a function that describes the cost of renting a VCR for x weeks, $0 \le x \le 4$. Where is this function continuous? Where is it discontinuous? Where is it not differentiable?

60. The cost of an international telephone call is \$3 per minute for the first minute and another \$2 per minute for each additional minute or part of a minute. Find and graph a function that describes the cost of making a call that lasts for t minutes, $0 \le t \le 5$. Where is this function continuous? Where is it discontinuous? Where is it not differentiable?

Management

61. (See Exercise 71, Exercises 2.1.) If a biochemical company makes x grams of interferon, its variable cost per gram is $v(x)$ dollars, where

$$v(x) = \begin{cases} 50, & 0 < x \le 1000, \\ 40, & 1000 < x \le 5000, \\ 35, & x > 5000. \end{cases}$$

Where is this function continuous? Where is it discontinuous? Where is it not differentiable?

62. (See Exercise 72, Exercises 2.1.) A software company sold its word-processing package for $s(t)$ dollars t years after its introduction, where

$$s(t) = \begin{cases} 120, & 0 < t < 2, \\ 99, & 2 \le t < 4, \\ 79, & t \ge 4. \end{cases}$$

Where is this function continuous? Where is it discontinuous? Where is it not differentiable?

63. (See Exercise 74, Exercises 2.1.) If the price of a gallon of gasoline is p dollars, gasoline consumption will be y billion gallons per year, where

$$y = \begin{cases} 500 - 10p, & 0 \le p \le 1.50, \\ 560 - 50p, & p > 1.50. \end{cases}$$

Where is this function continuous? Where is it discontinuous? Where is it not differentiable?

64. (See Exercise 75, Exercises 2.1.) Repeat Exercise 63 if
$$y = \begin{cases} 500 - 10p, & 0 \le p \le 1.50, \\ 400 - 40p, & p > 1.50. \end{cases}$$

65. In order for a firm to produce more than 10,000 units of its product, it must put on a second shift, and this increases its variable cost per unit. Thus its cost function is given by
$$C(x) = \begin{cases} 100,000 + 8x, & 0 \le x \le 10,000, \\ 80,000 + 10x, & x > 10,000. \end{cases}$$
Graph this function. Where is it continuous? Where is it discontinuous? Where is it not differentiable?

66. Repeat Exercise 65 if
$$C(x) = \begin{cases} 100,000 + 8x, & 0 \le x \le 10,000, \\ 80,000 + 12x, & x > 10,000. \end{cases}$$

67. A firm that sells computer supplies charges $9.00 per ribbon for printer ribbons on orders of fewer than 10 ribbons, $8.60 per ribbon on orders of 10 through 24 ribbons, and $8.20 per ribbon on orders of 25 or more ribbons. Find and graph a function that expresses the cost of an order of x ribbons. Where is the function continuous? Where is it discontinuous? Where is it not differentiable?

68. A candy company has three machines that can make chocolate bonbons. Machine A, which is new, can make 2000 bonbons per hour at a cost of 1.8 cents each. Machine B, which is older, can make 1500 bonbons per hour at a cost of 2.2 cents each. Machine C, which is ancient, can make 800 bonbons per hour at a cost of 3.5 cents each. Find and graph a function that expresses the cost of making x bonbons per hour. (*Hint:* If $x \le 2000$, the company will use only machine A; if $2000 < x \le 3500$, they will use A and B, etc.) Where is this function continuous? Where is it discontinuous? Where is it not differentiable?

Life Science

69. The number of bacteria present t hours after the start of an experiment is y thousand, where
$$y = \begin{cases} 2t^2 + 100, & 0 \le t < 2, \\ 160 - 26t, & 2 \le t \le 6. \end{cases}$$
Graph this function. Where is it continuous? Where is it discontinuous? Where is it not differentiable?

2.4 RULES OF DIFFERENTIATION

The process of finding the derivative of a function is called **differentiation.** As of now, we can only differentiate functions by means of the Three-Step Rule, which is often both time-consuming and tedious to apply. Fortunately, it is possible to use the Three-Step Rule to prove some very general rules of differentiation. Once we have these rules of differentiation available and understand how to use them, we will be able to differentiate entire classes of functions (such as the polynomial functions) almost automatically. In this section we prove some of the most important rules of differentiation and illustrate their usage.

Let c be any number and consider the **constant function** f defined by

$$f(x) = c \quad \text{for all numbers } x.$$

By the Three-Step Rule,

$$f'(x) = \lim_{h \to 0} \frac{f(x+h) - f(x)}{h} = \lim_{h \to 0} \frac{c - c}{h} = \lim_{h \to 0} 0 = 0.$$

Thus we have proved the **constant rule.**

The Constant Rule

If c is a number and $f(x) = c$ for all x, then

$$f'(x) = 0.$$

That is, the derivative of a constant function is zero.

EXAMPLE 1 Let $f(x) = 2$ and $g(x) = \sqrt{2}$ for all x. Then $f'(x) = 0$ and $g'(x) = 0$ for all x; thus $f'(5) = 0$, $f'(\pi) = 0$, $g'(0) = 0$, and so on. □

Now let n be a positive integer and consider the **power function** defined by

$$f(x) = x^n \quad \text{for all numbers } x.$$

We have

$$f'(x) = \lim_{h \to 0} \frac{f(x+h) - f(x)}{h} = \lim_{h \to 0} \frac{(x+h)^n - x^n}{h}.$$

But by the Binomial Theorem,

$$(x+h)^n = x^n + nx^{n-1}h + \frac{n(n-1)}{2} x^{n-2}h^2 + \cdots + h^n,$$

where all the terms after the first two on the right side of the equality have h^2 as a factor. Therefore if we factor h out of all terms after the first one, we will obtain

$$(x+h)^n = x^n + h\left[nx^{n-1} + \frac{n(n-1)}{2} x^{n-2}h + \cdots + h^{n-1}\right],$$

where now all the terms inside the brackets after the first one still have h as a factor, and hence will approach 0 as $h \to 0$. Thus

$$f'(x) = \lim_{h \to 0} \frac{x^n + h[nx^{n-1} + (n(n-1)/2)x^{n-2}h + \cdots + h^{n-1}] - x^n}{h}$$

$$= \lim_{h \to 0} \left[nx^{n-1} + \frac{n(n-1)}{2} x^{n-2} h + \cdots + h^{n-1} \right]$$

$$= nx^{n-1} + 0 + \cdots + 0$$

$$= nx^{n-1}.$$

We have shown that if $f(x) = x^n$, then $f'(x) = nx^{n-1}$. This fact, which is known as the **power rule,** remains true if we replace n by any nonzero number (though we shall not prove this).

> **The Power Rule**
>
> If r is any number except 0 and $f(x) = x^r$, then
>
> $$f'(x) = rx^{r-1}.$$
>
> That is, the derivative of x raised to a nonzero power is the power times x raised to the power minus 1.

EXAMPLE 2 If $f(x) = x$, then $f(x) = x^1$, and $f'(x) = 1x^{1-1} = 1x^0 = 1 \cdot 1 = 1$. Thus if $f(x) = x$, then $f'(x) = 1$. ☐

EXAMPLE 3 If $g(x) = x^2$, then $g'(x) = 2x^{2-1} = 2x^1 = 2x$. ☐

EXAMPLE 4 If $h(x) = x^5$, then $h'(x) = 5x^{5-1} = 5x^4$. ☐

EXAMPLE 5 If $k(x) = \sqrt{x}$, then $k(x) = x^{1/2}$, so

$$k'(x) = \frac{1}{2} x^{(1/2)-1} = \frac{1}{2x^{1/2}} = \frac{1}{2\sqrt{x}}. \quad \square$$

EXAMPLE 6 If $s(x) = 1/x = x^{-1}$, then $s'(x) = (-1)x^{-1-1} = -x^{-2} = -1/x^2$. ☐

The power rule shows how to differentiate $f(x) = x^r$ when r is any nonzero number. We would like to be able to differentiate functions whose defining equations are of the form $f(x) = cx^r$, where c is a number. But if g is any function for which $g'(x)$ exists, and $f(x) = c \cdot g(x)$, then

$$f'(x) = \lim_{h \to 0} \frac{f(x+h) - f(x)}{h} = \lim_{h \to 0} \frac{c \cdot g(x+h) - c \cdot g(x)}{h}$$

$$= c \cdot \lim_{h \to 0} \frac{g(x+h) - g(x)}{h} = c \cdot g'(x).$$

and we have proved the **constant-multiple rule.**

The Constant-Multiple Rule

If c is a number and g is a function such that $f(x) = c \cdot g(x)$ and $g'(x)$ exists, then

$$f'(x) = c \cdot g'(x).$$

That is, the derivative of a constant times a function is the constant times the derivative of the function.

EXAMPLE 7   If $f(x) = 2x$, then $f'(x) = (2x)' = 2(x)' = 2 \cdot 1 = 2.$ □

EXAMPLE 8   If $h(x) = 3x^2$, then $h'(x) = (3x^2)' = 3(x^2)' = 3(2x) = 6x.$ □

EXAMPLE 9   If $k(x) = \dfrac{1}{3}x^{-3}$, then $k'(x) = -x^{-4}.$ □

Now we can differentiate functions whose defining equations are of the form $f(x) = cx^r$. Polynomials are formed by taking sums and differences of such functions, and we would like to be able to differentiate polynomials. But if $k(x) = f(x) + g(x)$, where $f'(x)$ and $g'(x)$ exist, then

$$k'(x) = \lim_{h \to 0} \frac{k(x+h) - k(x)}{h}$$

$$= \lim_{h \to 0} \frac{[f(x+h) + g(x+h)] - [f(x) + g(x)]}{h}$$

$$= \lim_{h \to 0} \frac{f(x+h) - f(x)}{h} + \lim_{h \to 0} \frac{g(x+h) - g(x)}{h}$$

$$= f'(x) + g'(x).$$

Similarly, if $k(x) = f(x) - g(x)$, then $k'(x) = f'(x) - g'(x)$. Thus we have the **sum rule.**

The Sum Rule

If $f'(x)$ and $g'(x)$ both exist and $k(x) = f(x) \pm g(x)$, then

$$k'(x) = f'(x) \pm g'(x).$$

That is, the derivative of a sum (difference) is the sum (difference) of the derivatives.

EXAMPLE 10   If $k(x) = -2x^2 + 3$, then

$$k'(x) = (-2x^2 + 3)' = -2(x^2)' + (3)' = -2(2x) + 0 = -4x. \quad \square$$

EXAMPLE 11 If $s(x) = x^4 - 5x^3$, then
$$s'(x) = 4x^3 - 5(3x^2) = 4x^3 - 15x^2. \quad \square$$

We note that the sum rule also holds if there are more than two summands. Thus if
$$k(x) = f_1(x) \pm f_2(x) \pm \cdots \pm f_n(x),$$
then
$$k'(x) = f_1'(x) \pm f_2'(x) \pm \cdots \pm f_n'(x).$$

The four rules of differentiation we have proved in this section allow us to differentiate polynomials and also functions that are formed by taking sums and differences of nonintegral powers of the variable.

EXAMPLE 12 If $f(x) = 2x^3 + 5x^2 - 12x + 6$, then
$$f'(x) = 2(3x^2) + 5(2x) - 12(1) + 0 = 6x^2 + 10x - 12. \quad \square$$

EXAMPLE 13 If $g(x) = 2\sqrt{x} - 4x^{6/5} + 6x - 10/x^2 = 2x^{1/2} - 4x^{6/5} + 6x - 10x^{-2}$, then
$$g'(x) = 2\left(\frac{1}{2}x^{-1/2}\right) - 4\left(\frac{6}{5}x^{1/5}\right) + 6(1) - 10(-2x^{-3})$$
$$= x^{-1/2} - \frac{24}{5}x^{1/5} + 6 + 20x^{-3}$$
$$= \frac{1}{\sqrt{x}} - \frac{24}{5}\sqrt[5]{x} + \frac{20}{x^3} + 6. \quad \square$$

EXAMPLE 14 Let $f(x) = x^5 - 2x^3 + 7x + 2$. Find the equation of the line tangent to f at the point $(1, f(1)) = (1, 8)$.

We find the derivative
$$f'(x) = 5x^4 - 6x^2 + 7$$
and evaluate it at $x = 1$ to obtain
$$f'(1) = 5(1^4) - 6(1^2) + 7 = 6.$$
Since $f'(1) = 6$, the tangent line has slope equal to 6. Hence its equation is
$$y = 6x + b.$$
But since the point $(1, 8)$ is on the line, we have
$$8 = 6(1) + b = 6 + b.$$
Thus $b = 2$. Therefore the equation of the tangent line is
$$y = 6x + 2. \quad \square$$

EXAMPLE 15 A patient's white blood cell count t hours after being admitted to a hospital is
$$f(t) = 5t^2 - 80t + 500$$

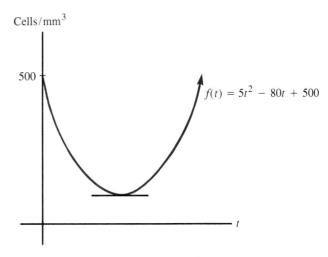

FIGURE 2–38

cells per cubic millimeter. The graph of this quadratic function is the parabola of Figure 2–38. Note that the line tangent to the graph at the vertex of the parabola is horizontal and therefore has slope zero. But the slope of the line tangent to f at $(t, f(t))$ is $f'(t)$. And if

$$f'(t) = 10t - 80 = 0,$$

then $t = 8$. Thus the vertex is at $(8, f(8))$; since

$$f(8) = 5(8)^2 - 80(8) + 500 = 180,$$

the vertex is $(8, 180)$. Therefore the patient's minimum white blood cell count occurs 8 hours after admission and is equal to 180 cells per cubic millimeter. ☐

■ *Computer Manual: Program DIFFPOLY*

2.4 EXERCISES

In Exercises 1 through 50, use the rules of differentiation to find $f'(x)$.

1. $f(x) = 2$
2. $f(x) = \sqrt{3}$
3. $f(x) = -4$
4. $f(x) = x^3$
5. $f(x) = x^7$
6. $f(x) = x^{100}$
7. $f(x) = x^{-2}$
8. $f(x) = 1/x^3$
9. $f(x) = 1/x^5$
10. $f(x) = x^{3/2}$
11. $f(x) = x^{7/5}$
12. $f(x) = x^{-2/3}$
13. $f(x) = x^{-2.7}$
14. $f(x) = \sqrt[3]{x}$
15. $f(x) = \sqrt[5]{x}$
16. $f(x) = 1/\sqrt{x}$
17. $f(x) = 3x$
18. $f(x) = 5x^2$
19. $f(x) = -2x^3$
20. $f(x) = 3x^9$
21. $f(x) = 2x^{3/2}$

22. $f(x) = -3x^{-2/3}$
23. $f(x) = -4\sqrt{x}$
24. $f(x) = 5/x$
25. $f(x) = -3/x^2$
26. $f(x) = 4/x^5$
27. $f(x) = 3/\sqrt[3]{x}$
28. $f(x) = 6x + 2$
29. $f(x) = 3 - 8x$
30. $f(x) = 15x - 9$
31. $f(x) = x^2 - 5x$
32. $f(x) = 2x^2 - 8x + 12$
33. $f(x) = 8x^2 - 12x + 3$
34. $f(x) = -1.5x^2 + 2.2x + 11.3$
35. $f(x) = x^3 - 7x^2 + 9x$
36. $f(x) = -x^3 + 8x + 2$
37. $f(x) = x^4 - 3x^3 - 3x^2 + 5x + 1$
38. $f(x) = -\frac{3}{4}x^4 + \frac{7}{3}x^3 - \frac{7}{2}x^2 + 1$
39. $f(x) = -2x^{100} + 3x^{99} + 10$
40. $f(x) = 4\sqrt{x} + 3x^{2/3}$
41. $f(x) = 3\sqrt{x} + 6x^{-3/2} + 1$
42. $f(x) = \frac{1}{x} - \frac{2}{x^2} + \frac{3}{x^3} - \frac{4}{x^4}$
43. $f(x) = \frac{4}{x} + \frac{1}{x^{2/5}}$
44. $f(x) = 2x^{-2} + 5x^{0.6} + 10x^{-0.4}$
45. $f(x) = 6\sqrt[3]{x} + 5\sqrt{x} - \frac{2}{\sqrt{x}}$
46. $f(x) = (x + 1)(x - 4)$
47. $f(x) = (2x - 5)(3x + 6)$
48. $f(x) = (x^2 - 2x)(3x + 9)$
49. $f(x) = \frac{x + 2x^2}{x^2}$
50. $f(x) = \frac{2x^{3/4} - 4}{x^{5/4}}$

In Exercises 51 through 60, find the equation of the line tangent to f at $(a, f(a))$.

51. $f(x) = 5x - 3$, $a = 2$
52. $f(x) = 2x^2 - 3$, $a = 1$
53. $f(x) = 1 - x - x^2$, $a = 3$
54. $f(x) = x^2 - 3x + 1$, $a = 0$
55. $f(x) = x^3 - 1$, $a = -2$
56. $f(x) = -x^3 + 2x^2 - x + 1$, $a = -1$
57. $f(x) = 4\sqrt{x} - 5x$, $a = 9$
58. $f(x) = 3x^{4/3} - 2x^{2/3}$, $a = 27$
59. $f(x) = \frac{2}{x} - \frac{4}{x^2}$, $a = 2$
60. $f(x) = 6x^{-2} + x^{-3/4}$, $a = 16$

Management

61. A firm's profit function is given by $P(x) = -x^2 + 40x - 350$, where x is in units and profit is in thousands of dollars.
    (a) Graph this function.
    (b) The firm's maximum profit occurs at the point on the graph where the tangent line is horizontal. Use this fact to find the firm's maximum profit and the value of x at which it occurs.

62. Repeat Exercise 61 if $P(x) = -x^2 + 800x - 42{,}000$, where x is in units and profit is in dollars.

63. Over the past five years the average productivity of a company's plant has been $y = 0.4t^2 - 30t + 1400$ units per day in month t, where $t = 0$ represents the beginning of the five-year period.
    (a) Graph this function.
    (b) The plant's minimum average productivity occurred at the point on the graph where the tangent line is horizontal. Find the plant's minimum average productivity over the five-year period and the time when it occurred.

64. Repeat Exercise 63 if $y = -0.02t^2 + 1.6t + 12$. (In this case you will be finding the plant's *maximum* average productivity.)

65. A firm's average cost per unit is given by $A(x) = 10x + 1000/x$. As x increases, the average cost per unit decreases, then increases again; the firm's minimum average cost per unit occurs at the value of x for which the slope of the tangent line is 0. Find this value of x and the firm's minimum average cost per unit.

66. A firm's monthly sales over the past year are given by
$$S(t) = 2t^3 - 39t^2 + 240t + 100,$$
where t is in months, with $t = 0$ representing the beginning of the year, and sales are in thousands of dollars. During the year monthly sales increased for a time, then decreased, and finally increased again. The points on the graph where sales changed direction have horizontal tangent lines. Find the time when sales stopped increasing and began to decrease and the time when they stopped decreasing and began to increase.

67. Over the past decade, the productivity of a firm's employees decreased, then increased, and finally decreased again. Suppose employee productivity in year t was y units produced per week, where $y = -t^3 + 18t^2 - 96t + 300$. Here $t = 0$ represents the beginning of the decade. The points on the graph where productivity changed direction have horizontal tangent lines. Find the time when productivity stopped decreasing and began to increase and the time when it stopped increasing and began to decrease.

Life Science

68. The number of black bear cubs born in New England in year t was estimated to be y, where $y = -0.5t^2 + 5t + 100$ and $t = 0$ represents 1980.
    (a) Graph this function.
    (b) Find the maximum number of cubs born in New England and the year in which this occurred by finding the value of t for which the tangent line is horizontal.

69. Repeat Exercise 68 if $y = t^2 - 8t + 100$. (In this case you will be finding the *minimum* number of cubs born.)

70. Biologists estimate that over the next several years the number of elephants in a certain region of East Africa will decrease somewhat, then increase temporarily, and finally begin a pronounced decline. Suppose the number of elephants alive t years from now will be y, where $y = -2t^3 + 45t^2 - 300t + 200$. The points on the graph where the temporary increase and then the final decline occur have horizontal tangent lines. Find the time when the temporary increase and the final decline will occur.

71. A new antibiotic is tested by applying it to a culture containing bacteria, with the result that t days later there are y thousand bacteria alive in the culture, where $y = t^5/5 - 29t^3/3 + 100t + 140$. The graph of this function shows that after the application of the antibiotic the number of bacteria grows for a time, then declines, and then begins to grow again. The points on the graph where growth turns into decline and decline into growth have horizontal tangent lines. Find these points.

Social Science

72. A politician's approval rating is given by $y = 0.0002t^2 - 0.01t + 0.6$, where t is the number of months since the last election and y is the proportion of voters who approve of the politician's performance in office.

(a) Graph this function.
(b) Find the politician's minimum approval rating and the time that it occurred by finding the value of t for which the slope of the tangent line is 0.

73. Over the past four years the unemployment rate in a state has been given by $u(t) = 0.0002t^3 - 0.0147t^2 + 0.216t + 5.9$. Here $u(t)$ is the unemployment rate, as a percentage of the work force, in month t, with $t = 0$ being the beginning of the four-year period. The graph of this function shows that the unemployment rate rose, then fell, then rose again, and the points on the graph where the rate changed from rising to falling and from falling to rising occurred where the tangent lines are horizontal. Find these points.

Bond Duration

The yield-to-maturity of a bond is an interest rate that tells investors how much an investment in the bond would return to them. Thus, a bond that has a yield-to-maturity of 10% gives a greater return on investment to its purchaser than one that has a yield-to-maturity of 9%. The price p at which a bond sells and its yield-to-maturity $y\%$ are related: if y rises, then p must fall; if y falls, then p must rise. In fact, for each possible value of y, there is a single price p at which the bond must sell in order to have y as its yield-to-maturity. Thus we may consider the price of a bond to be a function of its yield-to-maturity. According to an article by Martin Leibowitz ("Analytical Measures in Today's Bond Market," an advertisement in *Scientific American*, June 1987), if $p = f(y)$ describes bond price as a function of yield-to-maturity, then the **duration** D of the bond is defined by

$$D = \frac{f'(y)}{f(y)} 100.$$

Duration measures the short-term risk to the bondholder of a change in interest rates: D is the approximate percentage change in price that will result from a 1% rise in interest rates. (We will see why this is so in Section 3.3.) Exercises 74 and 75 refer to bond duration.

*74. Suppose a bond's price as a function of its yield to maturity is given by $p = 8000/y$, $0 < y \le 100$. Find and interpret the bond's duration if its current yield-to-maturity is 8%; 10%; 16%.

*75. Repeat Exercise 74 if $p = y^2 - 120y + 2000$.

2.5 RATES OF CHANGE

We have several times referred to the fact that the derivative of a function describes the rate of change of the function with respect to its independent variable. In this section we see why this is true, and we demonstrate how to use this property of the derivative in some practical applications.

Suppose we make a journey in a small airplane. We fly in a straight line, and after t hours in the air we have traveled a distance of

$$s(t) = 4t^2 + 150t$$

miles from our starting point, where $0 \le t \le 4$. What is the average velocity of our plane during the second hour of the trip?

The Derivative

Average velocity is distance traveled divided by elapsed time. During the second hour of the trip the time changes from $t = 1$ to $t = 2$, so the distance traveled during the second hour is $s(2) - s(1)$ miles, and therefore the average velocity during the second hour is

$$\frac{s(2) - s(1)}{2 - 1} = \frac{316 - 154}{2 - 1} = 162$$

miles per hour. The same reasoning holds for any time interval: the average velocity of the plane from time t to time $t + h$ is

$$\frac{s(t + h) - s(t)}{t + h - t} = \frac{s(t + h) - s(t)}{h}$$

miles per hour. Notice that the right side of the equation is the difference quotient of the **distance function** $s = s(t)$. If we take the limit of this difference quotient as h approaches 0, we obtain the derivative $s'(t)$. On the other hand, as h approaches 0 the difference quotient gives the average velocity over shorter and shorter intervals of time. Hence if we take the limit as h approaches 0, we obtain the instantaneous velocity at time t. Thus the derivative

$$s'(t) = 8t + 150$$

of $s(t)$ gives the instantaneous velocity of the airplane at time t. For example, when $t = 2$ we have

$$s'(2) = 8(2) + 150 = 16 + 150 = 166.$$

Therefore at the instant when $t = 2$ the velocity of the airplane is 166 miles per hour.

A distance function such as the one we have just considered is simply a means of specifying the position of an object at time t relative to its starting point. The analysis we performed for $s(t) = 4t^2 + 150t$ can be repeated for any function $s = s(t)$ that gives the position of an object relative to some reference point, as long as the motion is in a straight line.

Velocity

If an object moves in a straight line so that its position at time t relative to some reference point is given by the function $s = s(t)$, then its instantaneous velocity at time t is

$$v(t) = s'(t).$$

If we take the reference point to be the zero point on the line of motion, then one direction along the line is positive and the other is negative. Thus if the velocity is positive at time $t = a$, the object is moving along the line in the positive direction; if the velocity is negative, it is moving in the negative direction.

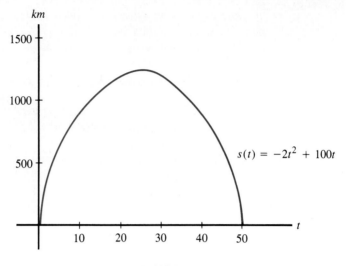

FIGURE 2–39

EXAMPLE 1 A rocket is launched straight up into the atmosphere. The altitude of the rocket t minutes after it is launched is

$$s(t) = -2t^2 + 100t$$

kilometers. We have chosen the rocket's zero point to be its launch pad and its upward motion to be positive motion. Figure 2–39 shows the graph of the position function $s = s(t)$. Note that this is the graph of the rocket's distance from its launch pad (that is, its altitude) at time t, and not a picture of its motion; the rocket moves in a straight line.

The instantaneous velocity of the rocket at time t is

$$v(t) = s'(t) = -4t + 100$$

kilometers per minute. For instance, exactly 10 minutes after launching, the rocket is at an altitude of

$$s(10) = 800$$

kilometers and its velocity at the instant when $t = 10$ is

$$v(10) = 60$$

kilometers per minute. Therefore at the instant when $t = 10$, the rocket's altitude is increasing at a rate of 60 kilometers per minute; we can tell that the altitude is increasing because $v(10)$ is positive. See Figure 2–40.

Similarly, 45 minutes after launching, the rocket is at an altitude of

$$s(45) = 450$$

kilometers and has a velocity of

$$v(45) = -80$$

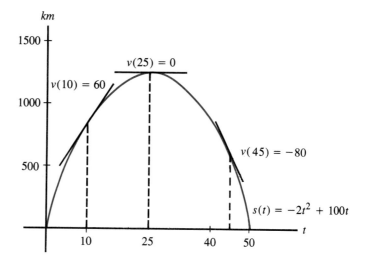

FIGURE 2–40

kilometers per minute. Hence when $t = 45$ the rocket's altitude is decreasing at the rate of 80 kilometers per minute; the fact that the velocity is negative tells us that the altitude is decreasing. See Figure 2–40.

Notice that $v(25) = 0$. Thus at the instant when $t = 25$, the rocket's altitude is neither increasing nor decreasing. Therefore the rocket reaches its peak altitude 25 minutes after launching. Again, see Figure 2–40. □

We have seen that the derivative of a position function is a velocity function. But velocity is just the name we give to the rate of change of position with respect to time. Thus the derivative of a position function is the rate of change of the position function with respect to time. We can apply the same reasoning to any function to conclude that its derivative represents the rate of change of the function with respect to the independent variable.

The Derivative as Rate of Change

For any function f, the derivative f' gives the instantaneous rate of change of f with respect to its independent variable.

EXAMPLE 2 A stone dropped from a height falls

$$s(t) = 16t^2$$

feet in t seconds. The velocity of the stone at time t is thus

$$v(t) = s'(t) = 32t$$

feet per second. For instance, the velocity of the stone exactly 2 seconds after it is dropped is

$$v(2) = 64$$

feet per second. The derivative of $v(t)$ will give the stone's rate of change of velocity with respect to time; this is known as its **acceleration.** If we let $a(t)$ denote the stone's acceleration at time t, then

$$a(t) = v'(t) = 32$$

feet per second per second. Note that the acceleration of the stone is constant. ◻

In the preceding example we said that the acceleration of the stone was the derivative of its velocity. This is true in general.

Acceleration

If $v(t)$ is the velocity of an object at time t and $a(t)$ its acceleration at time t, then

$$a(t) = v'(t).$$

The fact that the derivative of a function can be viewed as its rate of change with respect to its independent variable has many applications in addition to finding velocity and acceleration.

EXAMPLE 3 Alpha Company has studied the productivity of its employees and has found that after t years of experience an employee's monthly productivity can be expected to be

$$f(t) = -2t^2 + 120t + 100$$

units. The derivative

$$f'(t) = -4t + 120$$

gives the rate of change of monthly productivity with respect to years of experience. For instance, when $t = 10$ we have $f'(10) = 80$. Thus employees with 10 years of experience are increasing their monthly productivity at a rate of 80 units per year. Similarly, since $f'(20) = 40$, employees with exactly 20 years of experience are also increasing their monthly productivity, but only at the rate of 40 units per year.

Note that $f'(30) = 0$, and hence employees with exactly 30 years of experience will be neither increasing nor decreasing their productivity. Since

$$f'(t) = -4t + 120$$

is negative for $t > 30$, employee productivity decreases when $t > 30$. It follows that greatest productivity is reached when $t = 30$, and hence that maximum monthly productivity is $f(30) = 1900$ units. See Figure 2–41 for a graph of the productivity function. ◻

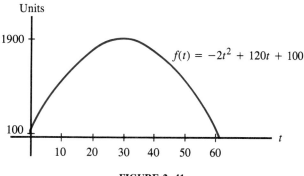

FIGURE 2-41

EXAMPLE 4 If Beta Corporation makes x units of product, its average cost per unit is

$$A(x) = 0.10x + \frac{40}{x}$$

dollars. Thus, if Beta produces 10 units, its average cost per unit is

$$A(10) = 0.10(10) + \frac{40}{10} = 5$$

dollars. The derivative

$$A'(x) = 0.10 - \frac{40}{x^2}$$

gives the rate of change of Beta's average cost per unit. For instance, since

$$A'(10) = 0.10 - \frac{40}{10^2} = -0.30,$$

when Beta produces 10 units its average cost per unit is decreasing by $0.30 per unit. ☐

■ *Computer Manual: Program DIFFPOLY*

2.5 EXERCISES

1. In t hours an automobile travels $s(t) = 40t$ miles in a straight line from its starting point. Find its velocity and acceleration t hours after it started.
2. In t minutes a rocket travels $s(t) = 250t$ kilometers in a straight line. Find its velocity and acceleration t minutes after it started.
3. A particle moves in a straight line in such a manner that t seconds after its start it has traveled $s(t) = 0.2t^{5/2}$ meters.
   (a) Find the distance it has traveled 1, 4, and 9 seconds after its start.
   (b) Find its velocity 1, 4, and 9 seconds after its start.
   (c) Find its acceleration 1, 4, and 9 seconds after its start.

4. A vehicle travels $s(t) = 40t^2 - 3t^3$ kilometers in a straight line in t hours, where $0 \leq t \leq 4$. Find the vehicle's
   (a) velocity 1, 2, and 3.5 hours after it started;
   (b) acceleration 1, 2, and 3.5 hours after it started.

5. A drag racer's car has traveled $s(t) = 12t^2 + 110t$ feet from the starting line t seconds after the start of a race.
   (a) What is the car's velocity 1 second after the start of the race? 4 seconds after the start?
   (b) What is the car's acceleration 1 second after the start of the race? 4 seconds after the start?

You throw a ball straight up. The altitude of the ball t seconds after you release it is $s(t) = -2t^2 + 20t$ feet. Exercises 6 through 10 refer to this situation.

6. What is the altitude of the ball 2 seconds after you release it? What is its velocity at this time? Is it going up or coming down? What is its acceleration?

7. Repeat Exercise 6 for $t = 5$.

8. Repeat Exercise 6 for $t = 7$.

9. Repeat Exercise 6 for $t = 10$.

10. What is the maximum altitude of the ball? When will it attain its maximum altitude?

A mortar is fired, and t seconds later the shell is at an altitude of
$$s(t) = -t^2 + 8t + 220$$
feet. Exercises 11 through 15 refer to this situation.

11. What is the altitude of the mortar shell 2.5 seconds after firing? What is its velocity at this time? Is it going up or coming down? What is its acceleration?

12. Answer the questions of Exercise 11 for $t = 4$.

13. Answer the questions of Exercise 11 for $t = 5.5$.

14. Answer the questions of Exercise 11 for $t = 8$.

15. What is the maximum altitude of the shell? When does it attain its maximum altitude?

A stone dropped from rest falls in a straight line. Suppose that its velocity t seconds after it was dropped is $v(t) = 32t$ feet per second. Exercises 16 through 20 refer to this situation.

16. Find the stone's acceleration t seconds after it was dropped.

*17. Find a function that describes the distance the stone has fallen as a function of the time since it was dropped. (*Hint:* Find a function $s(t)$ whose derivative is $v(t)$, using the fact that $s(0) = 0$.)

18. Find the distance the stone fell, its velocity, and its acceleration 1 second after it was dropped.

19. Repeat Exercise 18 for $t = 2$.

20. Repeat Exercise 18 for $t = 4$.

On the moon the velocity of a stone dropped from a height is $v(t) = 1.6t$ meters per second. Exercises 21 through 25 refer to this situation.

The Derivative

21. Find the stone's acceleration at time t.
*22. Find a function that describes the distance the stone falls in the first t seconds after it is dropped.
23. Find the distance the stone falls, its velocity, and its acceleration 1.1 seconds after it is dropped.
24. Repeat Exercise 23 for $t = 2.2$.
25. Repeat Exercise 23 for $t = 4.4$.
26. Stefan's Law states that the amount of energy radiated by a body per unit of surface area is directly proportional to the fourth power of its temperature in degrees Kelvin: $y = kT^4$, where T is the temperature of the body in $°K$, y is the amount of energy radiated by the body per unit area of its surface, and k is a constant that depends on the body. The surface temperature of the sun is approximately 6000° K. Find the rate of change of the energy radiated by the sun per unit of surface area.

Management

27. A firm's cost function is given by $C(x) = 0.025x^2 + 25,000$. Find the rate at which its cost is changing when production is 10 units; 100 units; 1000 units.
28. Delta Company's cost function is given by $C(x) = 5x^2 + 2000$, where x is in thousands of units produced. Delta's average cost per thousand units is therefore given by $A(x) = C(x)/x$. At what rate is Delta's average cost per thousand units changing when production is 5000 units? 10,000 units? 20,000 units? 30,000 units? At each of these production levels, is average cost per thousand units increasing or decreasing?
29. Total monthly sales of Gamma Corporation t months after the founding of the company are
$$S(t) = t^3 - 7t^2 + 10t + 100$$
thousand dollars.
   (a) What were total monthly sales one month after the founding of the corporation? At what rate were sales changing then? Were sales increasing or decreasing?
   (b) Answer the questions of part (a) for $t =$ three months.
30. A firm's profit function is given by
$$P(x) = -0.02x^3 + 4.5x^2 - 300x - 5000.$$
At what rate is profit changing when 40 units are being produced? when 80 units are being produced? when 150 units are being produced? At each of these production levels, is profit per unit increasing or decreasing?
31. If the supply function for a product is given by
$$q = p^2 + 2p,$$
where q is in thousands of units, find the rate at which the quantity of the product supplied to the market is changing when the price is $5 per unit; $100 per unit.
32. If the demand function for a product is given by $q = -p^2 - 30p + 4000$, where q is in thousands of units, find the rate at which the quantity of the product demanded by consumers is changing when the price is $10 per unit; when the price is $50 per unit.

33. The supply function for a commodity is $q = 12p - 2p^{1/2}$, where q is in millions of units and p is in dollars. Find the rate of change of quantity supplied with respect to price when the price per unit is
    (a) $1   (b) $4   (c) $9   (d) $16   (e) $25

34. The demand function for a product is given by $q = 10 - 0.5\sqrt{p}$, where q is in thousands of units.
    (a) Find the rate of change of quantity demanded with respect to price when the price is $100 per unit; $256 per unit.
    (b) Find the rate of change of price with respect to quantity demanded when quantity demanded is 5000 units; when quantity demanded is 2000 units. (*Hint:* Solve the demand equation for p in terms of q.)

35. A firm's revenue function is $R(x) = x^3/3 - 15x^2 + 216x$. Find the rate of change of revenue with respect to sales when the number of units sold is
    (a) 0   (b) 6   (c) 12   (d) 14   (e) 16   (f) 18   (g) 21   (h) 24

36. A firm's monthly sales over the past year are given by
    $$S(t) = 2t^3 - 39t^2 + 240t + 100,$$
    where t is in months, with $t = 0$ representing the beginning of the year, and sales are in thousands of dollars. Find the rate at which sales are changing when $t = 2$, 5, 6, 8, and 10. In each case state whether sales are increasing or decreasing at the given time. (See Exercise 66 of Exercises 2.4.)

37. Employee productivity in a firm in year t was y units produced per week, where
    $$y = -t^3 + 18t^2 - 96t + 300.$$
    Here $t = 0$ represents the beginning of the decade. Find the rate at which employee productivity was changing when $t = 1$, 4, 6, 8, and 9. In each case state whether productivity was increasing or decreasing at the given time. (See Exercise 67 of Exercises 2.4.)

Life Science

38. The volume of an artery of length 10 centimeters (cm) is $V = 10\pi r^2$ cm³, where r is its radius. If a drug increases the radius of the artery, find the rate at which its volume is changing when
    (a) $r = 0.2$ cm   (b) $r = 0.3$ cm   (c) $r = 0.4$ cm

39. A reagent is added to a chemical solution containing an enzyme and the enzyme precipitates out of solution. The amount of enzyme precipitated out t seconds after the reagent has been added is $y = -0.04t^2 + 0.20t$ grams, where $0 \le t \le 2.5$. Find the rate at which the enzyme is precipitating out of solution 0.5 second, 1 second, 1.5 seconds, and 2 seconds after the reagent is added.

40. The number of bacteria present in a culture t hours after the start of an experiment is given by $f(t) = t^3 + 6t^2 + 100$. Find the rate at which the size of the bacteria population in the culture is changing when $t = 2$, $t = 5$, and $t = 10$.

41. If there are x million cars on the road in a city, there will be y parts per million of carbon monoxide in its air, where
    $$y = 0.2x^3 + 4.$$
    Find the rate at which the carbon monoxide concentration is changing when the number of cars on the road, in millions, is
    (a) 1.0   (b) 2.0   (c) 3.0   (d) 3.25   (e) 3.5

42. A field is sprayed with weed killer and t days later, $0 \leq t \leq 15$, it contains an average of
$$W = t^3 - 3t^2 - 144t + 900$$
weeds per acre. Find the rate at which the average number of weeds per acre is changing 1 day after spraying; 5 days after spraying; 10 days after spraying; 12 days after spraying. Is the average number of weeds per acre increasing or decreasing at these times?

43. The number of bacteria alive in a culture t days after an antibiotic is applied to it is y thousand, where
$$y = \frac{t^5}{5} - \frac{29t^3}{3} + 100t + 140.$$
Find the rate at which the bacteria population is changing when $t = 1, 2, 3, 4, 5$, and 6. In each case state whether the bacteria population is increasing or decreasing. (See Exercise 71 of Exercises 2.4.)

44. The proportion of a scar that remains visible t days after an operation is y, where
$$y = 0.2 + 0.8t^{-1/2}, \qquad t \geq 1.$$
Find the rate at which the visibility of the scar is changing 16 days after the operation; 30 days after the operation.

Social Science

45. In a memory experiment, subjects memorize a list of 20 numbers and then watch rock videos for various lengths of time. After t minutes of watching the videos, the typical subject can remember approximately
$$y = -0.02t^2 + 20$$
of the numbers. Find the rate at which the typical subject forgets the numbers when $t = 5$, $t = 10$, $t = 15$, and $t = 25$ minutes.

46. Students who take an intensive foreign language course in which only the language being taught is spoken to them typically understand
$$P = 0.005t^3 - 0.05t^2 + 2t$$
percent of what is being said to them by the end of day t of the course, $0 \leq t \leq 25$. Find the rate at which the comprehension of a typical student is changing 5 days, 10 days, and 15 days into the course.

47. The unemployment rate in a state over a four-year period has been $u(t)$, where
$$u(t) = 0.0002t^3 - 0.0147t^2 + 0.216t + 5.9.$$
Here $u(t)$ is the unemployment rate as a percentage of the work force and t is in months, with $t = 0$ representing the beginning of the four-year period. Find the rate at which unemployment was changing when $t = 5, 9, 30$, and 40. In each case state whether the unemployment rate was increasing or decreasing at the given time. (See Exercise 73, Exercises 2.4.)

48. In a certain state expenditures on welfare y depend on the unemployment rate x according to the equation
$$y = 0.005x^4 + 0.1x^2 + 15,$$
where x is the percentage unemployment rate and y is in millions of dollars. Find the rate of change of welfare expenditures when the unemployment rate is
   (a) 4%   (b) 6%   (c) 8%   (d) 10%

49. A politician's campaign manager estimates that the proportion of voters who support the politician t days before an election is y, where

$$y = 0.45 + \frac{0.10}{t}$$

and $1 \le t \le 10$. Find the rate at which the politician's support is changing 10 days, 5 days, and 1 day before the election.

2.6 THE THEORY OF THE FIRM

The **theory of the firm** is concerned with questions such as

- How many units of its product must a firm make and sell in order to maximize its profit?
- What price should the firm charge for its product in order to attain its maximum profit?
- How will changing the product's price affect the firm's revenue?

In this section we show how to use calculus to answer these questions. We do this by exploiting both viewpoints of the derivative: that it is the slope of the tangent line, and that it gives the rate of change with respect to the independent variable.

Marginality

If a firm produces and sells x units of its product, then, as we have seen, it has a **cost function C**, a **revenue function R**, and a **profit function**

$$P = R - C.$$

These functions need not be linear, but as in the linear case we can find the firm's **break-even quantities** by setting $R(x) = C(x)$ and solving for x, or, alternatively, by setting $P(x) = 0$ and solving for x.

EXAMPLE 1 Suppose Acme Company's cost and revenue functions are given by

$$C(x) = x^2 + 500{,}000 \quad \text{and} \quad R(x) = 1500x,$$

respectively. Then Acme's profit function is given by

$$P(x) = R(x) - C(x) = -x^2 + 1500x - 500{,}000.$$

Since

$$P(x) = -(x - 500)(x - 1000),$$

setting $P(x) = 0$ and solving for x yields $x = 500$ and $x = 1000$. Therefore the firm's break-even quantities are 500 units and 1000 units. Figures 2–42 and 2–43 show the graphs of C, R, and P. Note that Acme will make a profit if it makes and sells between 500 and 1000 units, but will suffer a loss if it makes and sells fewer than 500 or more than 1000 units. ☐

THE DERIVATIVE

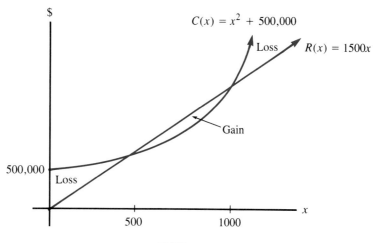

FIGURE 2–42

The derivatives of the cost, revenue, and profit functions are called **marginal functions.** Thus C' is the **marginal cost function,** R' the **marginal revenue function,** and P' the **marginal profit function.** The economic interpretation of the marginal cost function is as follows: since the derivative $C'(a)$ is the instantaneous rate of change of the cost when $x = a$, $C'(a)$ represents the approximate additional cost of producing one more unit. In other words, $C'(a)$ is the approximate additional cost of producing the $(a + 1)$st unit. Similarly, $R'(a)$ is the approximate additional revenue obtained from selling the $(a + 1)$st unit, and $P'(a)$ is the approximate additional contribution to profit made by the $(a + 1)$st unit.

EXAMPLE 2 Let Acme company's cost, revenue, and profit functions be as in Example 1. Then Acme's marginal cost, revenue, and profit functions are given by

$$C'(x) = 2x, \quad R'(x) = 1500, \quad \text{and} \quad P'(x) = -2x + 1500,$$

respectively. Since $C'(600) = 2(600) = 1200$, it will cost Acme approximately $1200 to make the 601st unit. Similarly, since $R'(600) = 1500$, Acme's revenue from selling the 601st unit will be approximately $1500. (Actually, in this example Acme's revenue from selling the 601st unit will be *exactly* $1500, since every unit

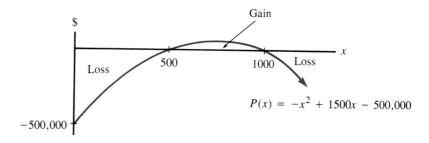

FIGURE 2–43

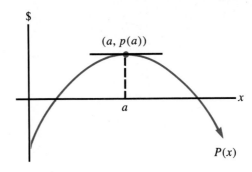

FIGURE 2-44

sells for $1500.) Also, the approximate additional contribution to profit made by the 601st unit will be $P'(600) = \$300$. ◻

Figure 2-44 shows the graph of a typical profit function P. Notice that the maximum profit occurs when $x = a$ units are made and sold. Since the tangent line to P at $(a, P(a))$ is horizontal, its slope will be zero; hence we must have $P'(a) = 0$. (Why?) But since $P(x) = R(x) - C(x)$,

$$0 = P'(a) = R'(a) - C'(a),$$

and thus $R'(a) = C'(a)$. Therefore profit is maximized at the value of x for which

$$R'(x) = C'(x),$$

or, in the language of economics, at the quantity where marginal revenue equals marginal cost.

> **Profit Maximization**
>
> Profit is maximized at the quantity for which marginal revenue equals marginal cost.

EXAMPLE 3 Consider Acme Company of Examples 1 and 2 again. If we set marginal revenue equal to marginal cost and solve for x, we have

$$R'(x) = C'(x)$$
$$1500 = 2x$$
$$750 = x.$$

Therefore, as Figure 2-45 indicates, Acme's profit is maximized when it makes and sells 750 units, and its maximum profit is $P(750) = \$62{,}500$. ◻

What if a firm has not yet determined the price it should charge for its product? For a single firm, the price that must be charged in order to obtain a given revenue

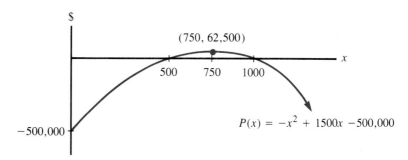

FIGURE 2–45

depends on the quantity of the product demanded: if the quantity demanded is large, the firm can sell many units at a low price; if the quantity demanded is small, it cannot sell very many units, and therefore must charge a higher price per unit in order to obtain the same revenue. If we let p denote the unit price of the product and x the quantity demanded, then p depends on x by means of a demand function $p = d(x)$. Since revenue is quantity sold times unit price, it follows that the firm's revenue is given by

$$R(x) = xp = xd(x).$$

EXAMPLE 4   As in the previous examples, suppose Acme Company's cost function is given by

$$C(x) = x^2 + 500{,}000,$$

but now suppose that the firm has not yet established a price for its product. Let the demand function for the product be defined by

$$p = 2100 - 0.5x.$$

Then Acme's revenue function is given by

$$R(x) = xp = x(2100 - 0.5x) = 2100x - 0.5x^2.$$

As before, we maximize profit by setting $R'(x) = C'(x)$ and solving for x: we have

$$R'(x) = C'(x)$$
$$2100 - x = 2x$$
$$700 = x.$$

Therefore Acme will maximize its profit by making and selling 700 units. In order to sell 700 units, the firm must charge

$$p = 2100 - 0.5(700) = \$1750$$

per unit. We have thus determined that if Acme prices its product at \$1750 per unit, it will sell 700 units for a maximum profit of $P(700) = \$235{,}000$. ◻

Compare Example 4 with Example 3. In Example 3 we assumed that Acme had set a price of \$1500 per unit and found that under this condition the firm should

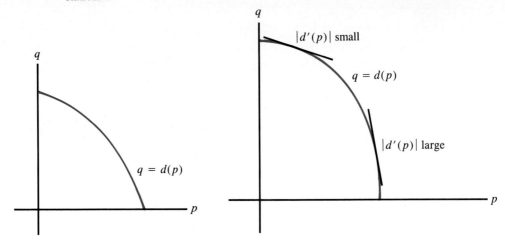

FIGURE 2–46   FIGURE 2–47

make and sell 750 units for a maximum profit of $62,500. In Example 4 it turned out that if Acme's demand function is defined by $p = 2100 - 0.5x$, then the firm should make fewer units, sell them at a higher price, and make a larger profit. A different choice of demand function could have led to the opposite conclusion: that Acme would be better off selling more units at a lower price. The property of the demand function that leads to such results is called **elasticity of demand.**

Elasticity of Demand

A demand function for a product relates quantity demanded q to unit price p. Here we consider the price p to be the independent variable so that the demand function is defined by an equation of the form $q = d(p)$. Figure 2–46 shows the graph of a typical demand function. Note that as price p increases, quantity demanded q decreases.

The derivative $q' = d'(p)$ gives the rate of change of quantity demanded with respect to price. Since the line tangent to the demand curve at any point will have negative slope, as indicated in Figure 2–47, $d'(p)$ will always be negative. If $|d'(p)|$ is large, then quantity demanded is decreasing quite rapidly as price increases; if $|d'(p)|$ is small, then quantity demanded is decreasing quite slowly as price increases. In other words, if $|d'(p)|$ is large, a small change in price results in a large change in quantity demanded; if $|d'(p)|$ is small, a small change in price results in a small change in quantity demanded.

EXAMPLE 5   Suppose

$$d(p) = \frac{10{,}000}{p}.$$

Then

$$d'(p) = -\frac{10{,}000}{p^2},$$

so $d'(2) = -2500$ and $d'(100) = -1$. Thus when the price is $2, a small increase in price will cause a large decrease in quantity demanded (in fact, we can say that

The Derivative

when $p = \$2$, an increase of \$1 in the price will cause quantity demanded to decrease by approximately 2500 units). On the other hand, when the price is \$100, a small increase in price will cause only a small decrease in quantity demanded. See Figure 2–48. ☐

When a small change in price results in a large change in quantity demanded, we say that demand is **elastic.** When a small change in price results in only a small change in quantity demanded, we say that demand is **inelastic.** In order to make these notions precise, we define the **elasticity of demand** for a commodity.

> ### The Elasticity of Demand Formula
> If the demand function for a commodity is defined by an equation of the form $q = d(p)$, the commodity's **elasticity of demand at price p** is $E(p)$, where
> $$E(p) = \frac{p[d'(p)]}{d(p)}.$$

Exercise 74 at the end of this section outlines the derivation of the elasticity formula.

The elasticity of demand $E(p)$ measures the proportional change in quantity demanded relative to price. If $|E(p)| > 1$, demand is elastic; if $|E(p)| < 1$, demand is inelastic. A product can have elastic demand at one price and inelastic demand at another.

EXAMPLE 6 If the demand function for a product is defined by

$$q = -1000p + 10{,}000,$$

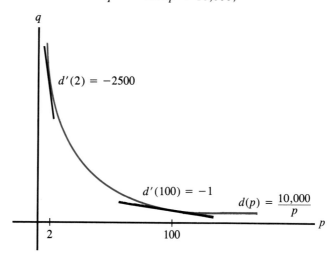

FIGURE 2–48

then the elasticity of demand for the product is

$$E(p) = \frac{p[d'(p)]}{d(p)} = \frac{-1000p}{-1000p + 10{,}000}.$$

Suppose that $p = \$2$. Then

$$|E(p)| = |E(2)| = |-0.25| = 0.25 < 1,$$

so demand for the product is inelastic at a price of $2. Note that when $p = \$2$, quantity demanded $q = 8000$ units and revenue is thus $\$2(8000) = \$16{,}000$. If the price is increased by 10% to $2.20, the quantity demanded decreases to 7800 and the revenue increases to $\$2.20(7800) = \$17{,}160$. Hence, when $p = \$2$, a 10% increase in price results in a small decrease in quantity demanded and an increase in revenue.

Now suppose that $p = \$8$. Then

$$|E(p)| = |E(8)| = |-4| = 4 > 1,$$

so demand for the product is elastic at a price of $8. When $p = \$8$, quantity demanded $q = 2000$ units and revenue is $\$8(2000) = \$16{,}000$ again. But now a 10% increase in the price, from $8 to $8.80, results in a decrease in quantity demanded to 1200 units, which in turn yields revenue of $10,560. Hence when $p = \$8$, a 10% increase in price leads to a large decrease in quantity demanded and a large decrease in revenue.

When $|E(p)| = 1$, demand is said to be of **unit elasticity.** For instance, in this example demand for the product is of unit elasticity when $p = \$5$. Notice also that $|E(p)| < 1$ when $p < \$5$ and $|E(p)| > 1$ when $p > \$5$. Hence demand is inelastic for $p < \$5$, of unit elasticity when $p = \$5$, and elastic when $p > \$5$. □

Note that in Example 6, when demand was inelastic at a price of $2, increasing the price by a small amount led to an increase in revenue; when demand was elastic at a price of $8, increasing the price by a small amount led to a decrease in revenue. These facts are examples of general truths:

Elasticity and Revenue

If demand is inelastic at a price,

- increasing the price by a small amount will result in an increase in revenue;
- decreasing the price by a small amount will result in a decrease in revenue.

If demand is elastic at a price,

- increasing the price by a small amount will result in a decrease in revenue;
- decreasing the price by a small amount will result in an increase in revenue.

If demand is of unit elasticity at a price, increasing or decreasing the price by a small amount will leave revenue approximately constant.

We will ask for a proof of these facts in Exercises 4.1.

EXAMPLE 7 Let $q = -1000p + 10{,}000$ be a demand function. Then, as we saw in Example 6, demand is inelastic for $p < \$5$; hence when $p = \$3$, for instance, increasing the price by a small amount will result in an increase in revenue and decreasing it by a small amount will result in a decrease in revenue. Similarly, when $p = \$7.50$ demand is elastic, so increasing the price by a small amount will result in a decrease in revenue and decreasing it by a small amount will lead to an increase in revenue. □

We conclude this section by remarking that instead of considering quantity demanded to be a function of price, as we have done here, economists usually consider price to be a function of quantity demanded. Therefore they define demand functions by equations of the form $p = d(q)$. When price is thus considered to be a function of quantity demanded, the formula for elasticity of demand has the alternative form

$$E(q) = \frac{d(q)}{q[d'(q)]}.$$

■ *Computer Manual: Program DIFFPOLY*

2.6 EXERCISES

Marginality

Zeta Company's cost function is given by $C(x) = 0.5x^2 + 100{,}000$ and Zeta sells each unit it makes for $600. Exercises 1 through 5 refer to this situation.

1. Find Zeta's total cost, revenue, and profit if it makes and sells 300 units.
2. Find Zeta's approximate additional cost from making the 301st unit.
3. Find Zeta's approximate additional revenue from selling the 301st unit.
4. Find Zeta's approximate additional profit from making and selling the 301st unit.
5. Find Zeta's maximum profit and the number of units it must make and sell in order to attain its maximum profit.

Eta Corporation's cost and revenue functions are given by
$$C(x) = 5x^2 + 500{,}000 \quad \text{and} \quad R(x) = -x^3 - 25x^2 + 14{,}400x,$$
respectively. Exercises 6 through 9 refer to this situation.

6. Find Eta's cost if it makes 25 units, and find the approximate additional cost of making the twenty-sixth unit.
7. Find Eta's revenue if it sells 50 units, and find the approximate additional revenue from selling the fifty-first unit.
8. Find Eta's profit if it makes and sells 75 units, and find the approximate additional profit from making and selling the seventy-sixth unit.
9. Find Eta's maximum profit and the number of units it must make and sell in order to attain its maximum profit.

Zeta Company's cost function is given by $C(x) = 0.5x^2 + 100{,}000$ and the demand function for its product by $p = 1200 - 2x$. Exercises 10 through 14 refer to this situation.

10. Find Zeta's total cost, revenue, and profit if it makes and sells 100 units.
11. Find Zeta's approximate additional cost from making the 101st unit.
12. Find Zeta's approximate additional revenue from selling the 301st unit.
13. Find Zeta's approximate additional profit from making and selling the 301st unit.
14. Find Zeta's maximum profit, the number of units it must make and sell, and the price per unit it must charge in order to attain its maximum profit.

Zeta Company's cost function is given by $C(x) = 0.5x^2 + 100{,}000$, and the demand function for its product by $p = 700 - 0.5x$. Exercises 15 through 19 refer to this situation.

15. Find Zeta's total cost, revenue, and profit if it makes and sells 200 units.
16. Find Zeta's approximate additional cost from making the 201st unit.
17. Find Zeta's approximate additional revenue from selling the 301st unit.
18. Find Zeta's approximate additional profit from making and selling the 301st unit.
19. Find Zeta's maximum profit, the number of units it must make and sell, and the price per unit it must charge in order to attain its maximum profit.

Eta Corporation's cost function is given by $C(x) = 5x^2 + 500{,}000$, and the demand function for its product by $p = -x^2 - 130x + 66{,}000$. Exercises 20 through 23 refer to this situation.

20. Find Eta's cost if it makes 50 units, and find the approximate additional cost of making the 51st unit.
21. Find Eta's revenue if it sells 50 units, and find the approximate additional revenue from selling the 51st unit.
22. Find Eta's profit if it makes and sells 75 units, and find the approximate additional profit from making and selling the 76th unit.
23. Find Eta's maximum profit, the number of units it must make and sell, and the price per unit it must charge in order to attain its maximum profit.

Eta Corporation's cost function is given by $C(x) = 5x^2 + 500{,}000$, and the demand function for its product by $p = -x^2 - 10x + 9000$. Exercises 24 through 27 refer to this situation.

24. Find Eta's cost if it makes 75 units, and find the approximate additional cost of making the 76th unit.
25. Find Eta's revenue if it sells 50 units, and find the approximate additional revenue from selling the 51st unit.
26. Find Eta's profit if it makes and sells 75 units, and find the approximate additional profit from making and selling the 76th unit.
27. Find Eta's maximum profit, the number of units it must make and sell, and the price per unit it must charge in order to attain its maximum profit.
28. A Corporation's cost function is given by $C(x) = 20x^2 + 40{,}000$, and the demand function for its product by $p = 6000 - 100x$. Find its maximum profit, the num-

ber of units the company must make and sell, and the price per unit it must charge in order to attain its maximum profit.

29. Repeat Exercise 28 if $C(x) = 2.3x^2 + 4000x$, $p = 10,000 - 0.10x$, and x is in thousands of units.

30. A firm's fixed cost is $100,000, its marginal cost function is given by $C'(x) = 500x$, and its demand function is linear, with quantity demanded being 0 units when price is equal to $214,500 and price being $0 when quantity demanded is equal to 8580 units. Find the firm's maximum profit and the number of units and price per unit required to attain it.

31. Theta, Inc. knows the following:

   —Its fixed cost is $20,000.
   —Its marginal cost function is given by $C'(x) = 10x$.
   —Its demand function is linear.
   —At a price of $2650 per unit, it sold 90 units.
   —At a price of $2800 per unit, it sold 80 units.

   Find Theta's maximum profit, the number of units the company must make and sell, and the price per unit it must charge in order to attain its maximum profit.

The average cost function is the cost function divided by the number of units produced; the **marginal average cost function** is the derivative of the average cost function. Similarly, marginal average revenue is the derivative of average revenue, and marginal average profit is the derivative of average profit. Exercises 32 through 35 are concerned with marginal average cost, revenue, and profit.

*32. Suppose Zeta Company's cost function is given by $C(x) = 0.5x^2 + 100,000$ and that Zeta sells each unit it produces for $600. Find and interpret Zeta's marginal average cost, marginal average revenue, and marginal average profit when it makes and sells 300 units.

*33. Repeat Exercise 32 if Zeta makes and sells 200 units.

*34. Suppose Eta Corporation's cost function is given by $C(x) = 5x^2 + 500,000$ and its revenue function by $R(x) = -x^3 - 25x^2 + 14,400x$, respectively. Find and interpret Eta's marginal average cost, marginal average revenue, and marginal average profit when it makes and sells 50 units.

*35. Repeat Exercise 34 if Eta's cost function is given by $C(x) = 5x^2 + 500,000$ and the demand function for its product by $p = -x^2 - 130x + 66,000$.

Elasticity of Demand

The demand function for a product is given by $q = -3000p + 120,000$. Exercises 36 through 38 refer to this product.

36. Find the elasticity of demand for the product when $p = \$15$. Find quantity demanded and revenue when $p = \$15$, $p = \$13.50$, and $p = \$16.50$.

37. Find the elasticity of demand for the product when $p = \$20$. Find quantity demanded and revenue when $p = \$20$, $p = \$19$, and $p = \$21$.

38. Find the elasticity of demand for the product when $p = \$30$. Find quantity demanded and revenue when $p = \$30$, $p = \$27$, and $p = \$33$.

The demand function for a commodity is given by $q = -100p + 25{,}000$. Exercises 39 through 41 refer to this commodity.

39. Find the elasticity of demand for the commodity when $p = \$150$. Find quantity demanded and revenue when $p = \$150$, $p = \$153$, and $p = \$147$.

40. Find the elasticity of demand for the commodity when $p = \$125$. Find quantity demanded and revenue when $p = \$125$, $p = \$127.50$, and $p = \$122.50$.

41. Find the elasticity of demand for the commodity when $p = \$100$. Find quantity demanded and revenue when $p = \$100$, $p = \$102$, and $p = \$98$.

The demand function for a commodity is given by $q = -50p^2 + 60{,}000$. Exercises 42 through 44 refer to this commodity.

42. Find the elasticity of demand for the commodity when $p = \$25$. Find quantity demanded and revenue when $p = \$25$, $p = \$25.25$, and $p = \$24.75$.

43. Find the elasticity of demand for the commodity when $p = \$20$. Find quantity demanded and revenue when $p = \$20$, $p = \$20.20$, and $p = \$19.80$.

44. Find the elasticity of demand for the commodity when $p = \$15$. Find quantity demanded and revenue when $p = \$15$, $p = \$15.15$, and $p = \$14.85$.

The demand function for a product is defined by $q = -5p + 3000$. Exercises 45 through 47 refer to this product.

45. Find the elasticity of demand for the product when $p = \$100$ and when $p = \$500$.

46. Find quantity demanded and revenue when $p = \$100$ and when $p = \$500$, and also when the price is increased by 10% in each case.

47. Find the price at which the product has unit elasticity of demand. Find quantity demanded and revenue when the price is increased or decreased by 5% from this value.

The demand function for a firm's product is given by $q = -2250p + 65{,}250$. Exercises 48 through 50 refer to this product.

48. Find an expression for the elasticity of demand for the product.

49. Over what range of prices is demand for the product elastic? inelastic? At what price is demand of unit elasticity?

50. Suppose the firm is currently selling the product for $10 per unit. Will increasing the price by a small amount increase or decrease revenue? Answer the same question if the firm is currently selling the product for $20 per unit.

The demand function for a commodity is given by $q = -200p + 180{,}000$. Exercises 51 through 54 refer to this commodity.

51. Find an expression for the elasticity of demand of the commodity as a function of p.

52. Find the price at which demand for the commodity is of unit elasticity. Find the range of prices over which demand for the commodity is elastic and the range over which it is inelastic.

53. If the commodity is currently selling for $500 per unit, what will happen to revenue if the price is increased by a small amount? if it is decreased by a small amount?

54. Answer the questions of Exercise 53 if the commodity is currently selling for $450; for $400.

The demand function for a commodity is given by $q = -160p^2 + 1{,}200{,}000$. Exercises 55 through 58 refer to this commodity.

55. Find an expression for the elasticity of demand of the commodity as a function of p.

56. Find the price at which demand for the commodity is of unit elasticity. Find the range of prices over which demand for the commodity is elastic and the range over which it is inelastic.

57. If the commodity is currently selling for $45 per unit, what will happen to revenue if the price is increased by a small amount? If it is decreased by a small amount?

58. Answer the questions of Exercise 57 if the commodity is currently selling for $50; for $55.

59. The daily demand for ice cream bars at the beach is given by
$$q = 10{,}000 - 2500p.$$
Find the price at which demand is of unit elasticity, the price range where it is elastic, and the price range where it is inelastic. If ice cream bars currently sell for $1.50 each, what will happen to sales revenue if the price is increased to $1.75?

60. The monthly demand for a certain brand of automobile is given by
$$q = 12{,}000 - 0.75p.$$
Find the price at which demand is of unit elasticity, the price range where it is elastic, and the price range where it is inelastic. If the automobile currently sells for $10,000, will raising the price by $500 increase or decrease sales revenue?

61. Cutter Cotter Pin Company needs to increase its revenue. The marketing department has recommended a 5% decrease in the price the company charges for its cotter pins, arguing that this would increase the quantity demanded so much that revenue would actually increase. The accounting department has recommended a 5% increase in the price, arguing that this would increase revenue because the resulting decrease in quantity demanded would be more than offset by the higher price. Suppose the demand function for cotter pins is given by $q = -1250p + 25{,}000$, where p is price per dozen cotter pins and q is in dozens of cotter pins. Whose recommendation should be taken if the current price per dozen of cotter pins is
(a) $6? (b) $10? (c) $12?

62. The demand for gasoline in a certain country is given by
$$q = \begin{cases} -10p + 50, & 0 \leq p < 1.20, \\ -35p + 80, & 1.20 \leq p \leq 2.00, \end{cases}$$
where p is the price in dollars per gallon and q is in millions of gallons per day.
(a) Find the elasticity of demand for gasoline when $p = 0.80$, $p = 1.15$, $p = 1.25$, and $p = 1.50$.
(b) Graph the demand function. Can you explain why the price per gallon must rise to at least $1.20 before significant conservation of gasoline occurs?

The demand function for a commodity is given by $p = -0.01q + 1600$. Exercises 63 through 67 refer to this commodity.

63. Use the elasticity of demand formula in the form $E(q) = d(q)/q[d'(q)]$ to find an expression for the elasticity of demand for the commodity as a function of q.

64. Use the expression obtained in Exercise 63 to find the elasticity of demand for the commodity when $q = 20{,}000$ units.

65. Use the expression obtained in Exercise 63 to find the elasticity of demand for the commodity when $p = \$400$.

66. Solve the demand equation for p in terms of q and then use the elasticity of demand formula in the form $E(p) = p[d'(p)]/d(p)$ to find an expression for the elasticity of demand for the commodity as a function of p.

67. Use the expression obtained in Exercise 66 to find the elasticity of demand for the commodity when $p = \$1400$ and when $q = 120{,}000$.

The demand function for a commodity is given by $p = -0.005q^2 + 80$. Exercises 68 through 70 refer to this commodity.

68. Use the elasticity of demand formula in the form $E(q) = d(q)/q[d'(q)]$ to find an expression for the elasticity of demand for the commodity as a function of q.

69. Use the expression obtained in Exercise 68 to find the elasticity of demand for the commodity when $q = 100$ units.

70. Use the expression obtained in Exercise 68 to find the elasticity of demand for the commodity when $q = 50$ units.

*71. Show that a commodity whose demand function is given by an equation of the form $q = a/p$, where a is a positive number, will have unit elasticity of demand at every price and that revenue from sales of the commodity will therefore remain constant no matter what the price.

*72. If a commodity has a demand function whose equation is of the form $q = c$, where c is a constant, show that its elasticity of demand is 0. This is called **perfect inelasticity;** when it occurs, revenue depends only on price. (Why?)

*73. If the demand function for a commodity has an equation of the form $p = c$, where c is a constant, show that its elasticity of demand does not exist. This is known as **perfect elasticity;** when it occurs, revenue depends only on quantity demanded. (Why?)

*74. If the price of a commodity changes by some percentage, then in response the quantity of the commodity demanded by consumers will change by some percentage. Elasticity of demand is defined to be the ratio of the percentage change in quantity demanded to the percentage change in price. If $d(p)$ is a demand function and the price changes from p to $p + h$, then the quantity demanded changes from $d(p)$ to $d(p + h)$, and hence

$$\text{Percentage change in price} = \frac{(p+h)-p}{p} \cdot 100 = \frac{h}{p} \cdot 100$$

and

$$\text{Percentage change in quantity demanded} = \frac{d(p+h)-d(p)}{d(p)} \cdot 100.$$

Take the limit as h approaches 0 of the ratio

$$\frac{\text{Percentage change in quantity demanded}}{\text{Percentage change in price}}$$

and interpret the result.

2.7 DIFFERENTIAL NOTATION AND HIGHER DERIVATIVES

Thus far we have written the derivative of a function f as f'. This is known as the **prime notation** for derivatives. Another common way of writing derivatives, known as **differential notation**, is sometimes more convenient and useful than prime notation. In this section we discuss differential notation for derivatives. We also consider the higher derivatives of a function: higher derivatives are derivatives of derivatives. For instance, as we saw in Section 2.5, velocity is the derivative of position, and acceleration is the derivative of velocity; thus acceleration is a higher derivative of position.

Differential Notation

Consider the function f defined by the equation
$$f(x) = x^3 + 4x + 1.$$
Using prime notation, we write the derivative of this function as
$$f'(x) = 3x^2 + 4.$$
If the function is given in the form $y = x^3 + 4x + 1$, we can still use prime notation and write the derivative as $y' = 3x^2 + 4$.

Another notation often used for the derivative of a function is **differential notation**: if f is a function of x, its derivative is written in differential notation as
$$\frac{df}{dx}$$
and referred to as the **derivative of f with respect to x**. The notation
$$\left.\frac{df}{dx}\right|_{x=a}$$
indicates that the derivative is to be evaluated at $x = a$.

EXAMPLE 1 Let $f(x) = x^3 + 2x^2 + 3x$, $y = \dfrac{2}{x} + 3x$, and $g(t) = 3\sqrt{t}$. Then the derivative of f with respect to x is
$$\frac{df}{dx} = 3x^2 + 4x + 3$$
and
$$\left.\frac{df}{dx}\right|_{x=1} = 3(1)^2 + 4(1) + 3 = 10.$$
Similarly, the derivative of y with respect to x is
$$\frac{dy}{dx} = -\frac{2}{x^2} + 3$$

and
$$\left.\frac{dy}{dx}\right|_{x=2} = \frac{5}{2}.$$

Finally,
$$\frac{dg}{dt} = \frac{3}{2\sqrt{t}}$$

and
$$\left.\frac{dg}{dt}\right|_{t=4} = \frac{3}{2\sqrt{4}} = \frac{3}{4}. \quad \square$$

At this time we will not consider df/dx and dy/dx to be ratios, but merely symbols for the derivative. Later we will show how these symbols can indeed be interpreted as ratios.

Differential notation is especially convenient when we must make explicit which variable is to be regarded as the independent one.

EXAMPLE 2 Let $y = x^3$. Then dy/dx is the derivative of y with respect to x. As usual, x is the independent variable and y is the dependent one. Thus

$$\frac{dy}{dx} = \frac{d}{dx}(x^3) = 3x^2.$$

On the other hand, dx/dy is the derivative of x with respect to y, so here y is to be regarded as the independent variable and x as the dependent one. If we solve the equation $y = x^3$ for x in terms of y to obtain $x = y^{1/3}$, we find that

$$\frac{dx}{dy} = \frac{d}{dy}(y^{1/3}) = \frac{1}{3}y^{-2/3}. \quad \square$$

Higher Derivatives

If f' is the derivative of f, then f' may itself have a derivative $(f')'$, which we write as f''. Similarly, f'' may have a derivative $(f'')' = f'''$, and so on. The derivative f' is the **first derivative of f**, f'' is the **second derivative of f**, and f''' is the **third derivative of f**. If $n > 3$, we denote the nth derivative of f by $f^{(n)}$.

EXAMPLE 3 Let $f(x) = x^4 + 5x^3 + 3x^2 + 4x - 1$. Then we have

first derivative: $f'(x) = 4x^3 + 15x^2 + 6x + 4$
second derivative: $f''(x) = 12x^2 + 30x + 6$
third derivative: $f'''(x) = 24x + 30$
fourth derivative: $f^{(4)}(x) = 24$
fifth derivative: $f^{(5)}(x) = 0$

Note that $f^{(n)}(x) = 0$ for $n \geq 5$. $\quad \square$

THE DERIVATIVE

When using differential notation, we write the second derivative as

$$\frac{d^2f}{dx^2},$$

with the obvious extension to higher derivatives.

EXAMPLE 4   Let $y = 1/x$. Then

$$\frac{dy}{dx} = -\frac{1}{x^2}, \quad \frac{d^2y}{dx^2} = \frac{2}{x^3},$$

$$\frac{d^3y}{dx^3} = -\frac{6}{x^4}, \quad \frac{d^4y}{dx^4} = \frac{24}{x^5}. \quad \square$$

■ *Computer Manual: Program HIGHPOLY*

2.7 EXERCISES

Differential Notation

In Exercises 1 through 22, find the indicated derivatives.

1. $\dfrac{dy}{dx}$, $y = x^2 - 5x + 2$

2. $\dfrac{dy}{dx}$, $y = 2x^3 - 3x + 12$

3. $\dfrac{df}{dx}$, $f(x) = x^3 - 5x^2 + 6x - 2$

4. $\dfrac{df}{dt}$, $f(t) = t^4 - 3t^2 + 9$

5. $\dfrac{dy}{dx}$, $y = 5\sqrt{x} - 2x^{2/3}$

6. $\dfrac{df}{dx}$, $f(x) = 7x^{-2} + 3x + 1$

7. $\dfrac{df}{dt}$, $f(t) = 3t^7 + t^{-2} + 12$

8. $\dfrac{dW}{dq}$, $W = 14q^2 + 10q - q^{-1}$

9. $\dfrac{dg}{dx}$, $g(x) = \dfrac{2}{x} - \dfrac{4}{x^3}$

10. $\dfrac{ds}{dt}$, $s = \sqrt{t} + 3t^2 + 2$

11. $\dfrac{df}{dx}\bigg|_{x=2}$, $f(x) = x^2 + 5x - 2$

12. $\dfrac{dy}{dx}\bigg|_{x=1}$, $y = 5x^4 - 20x + 2$

13. $\dfrac{dp}{dq}\bigg|_{q=9}$, $p = q^{3/2} + 5q$

14. $\dfrac{dy}{dx}\bigg|_{x=4}$, $y = x^5 - 5x^{3/2} + 2$

15. $\dfrac{dh}{dv}\bigg|_{v=8}$, $h(v) = 2v^{-3} + 2v^{-1/3}$

16. $\dfrac{df}{dy}\bigg|_{y=-1}$, $f(y) = 2y^{2/3} + 4$

17. $\dfrac{df}{dx}\bigg|_{x=a}$, $f(x) = x^3 + 2x^2$

18. $\dfrac{dq}{dp}\bigg|_{p=b}$, $q = 400p^2 - 1200p + 100$

19. $\dfrac{dy}{dx}, \dfrac{dx}{dy}$, $3x + 2y = 5$

20. $\dfrac{dp}{dq}, \dfrac{dq}{dp}$, $40p - 35q = 1250$

21. $\dfrac{dy}{dt}, \dfrac{dt}{dy}$, $4ty = 1$

22. $\dfrac{du}{dv}, \dfrac{dv}{du}$, $u^3 = 27v$

Higher Derivatives

In Exercises 23 through 44, find the indicated higher derivatives.

23. $f''(x)$, $f'''(x)$, $f(x) = x^3 + x^2 + x + 1$
24. $f''(x)$, $f^{(4)}(x)$, $f(x) = x^5 - 2x^3 + 4x$
25. $f''(x)$, $f(x) = 3x^5 - 12x^3 + 1$
26. $g'''(x)$, $g(x) = x^4 + 1$
27. $f^{(5)}(x)$, $f(x) = 6x^5 + 20x^3 + 3$
28. $f^{(100)}(x)$, $f(x) = 2x^{76} + 36x^{48} + 99x^{25} - 455x^7 + 123$
29. $\dfrac{d^2f}{dx^2}$, $\dfrac{d^3f}{dx^3}$, $f(x) = x^4 - 4x^3 + 3x + 12$
30. $\dfrac{d^2f}{dx^2}$, $f(x) = x^3 + 3x^2 - 9x + 10$
31. $\dfrac{d^2y}{dx^2}$, $\dfrac{d^4y}{dx^4}$, $y = \dfrac{2}{x^2} - \dfrac{1}{x}$
32. $\dfrac{d^2g}{dt^2}$, $g(t) = \dfrac{5}{t^2} + \dfrac{1}{t} + 2$
33. $\dfrac{d^2y}{dx^2}$, $\dfrac{d^3y}{dx^3}$, $y = x^{-3}$
34. $\dfrac{d^3y}{dx^3}$, $\dfrac{d^4y}{dx^4}$, $y = 2\sqrt{x}$
35. $\dfrac{d^5y}{dx^5}$, $y = x^3 + x^{-3}$
36. $f''(3)$, $f(u) = u^4 + 3u + 2$
37. $f'''(0)$, $f(t) = 2t^5 + t^4$
38. $g''(4)$, $g^{(4)}(4)$, $g(p) = 2p^{5/2}$
39. $\left.\dfrac{d^2y}{dx^2}\right|_{x=0}$, $y = 3x^3 + 2x - 7$
40. $\left.\dfrac{d^3y}{dx^3}\right|_{x=1}$, $y = x^2 + 2\sqrt{x}$
41. $\left.\dfrac{d^2p}{dq^2}\right|_{q=1}$, $\left.\dfrac{d^3p}{dq^3}\right|_{q=1}$, $p = 3q^{-1} + q^2$
42. $\left.\dfrac{d^2f}{dx^2}\right|_{x=4}$, $\left.\dfrac{d^3f}{dx^3}\right|_{x=-4}$, $f(x) = \dfrac{5}{x} - \dfrac{1}{x^2}$
43. $f^{(4)}(a)$, $f(x) = x^8 + x^4 + x$
44. $g^{(4)}(b)$, $g(x) = x^3 + 2x^2 + 3x + 15$

SUMMARY

This summary consists of terms and symbols whose meaning you should know, facts you should know, and techniques and methods you should be able to employ.

Section 2.1

Meaning of "x approaches a": $x \to a$. Limit of a function f as x approaches a: $\mathrm{Lim}_{x \to a} f(x)$. Finding limits using tables of functional values. Finding limits using graphs. Properties of limits. Finding limits using the properties of limits. Limits of polynomial functions. Finding limits of polynomial functions. Rational function. Finding limits of rational functions. Finding limits of functions having different defining equations on different intervals.

Section 2.2

Meaning of "x approaches a from the right," "x approaches a from the left": $x \to a^+$, $x \to a^-$. One-sided limits: $\mathrm{Lim}_{x \to a^+} f(x)$, $\mathrm{Lim}_{x \to a^-} f(x)$. Finding one-sided limits from graphs. $\mathrm{Lim}_{x \to a} f(x) = L$ if and only if $\mathrm{Lim}_{x \to a^+} f(x) = \mathrm{Lim}_{x \to a^-} f(x) = L$. Infinite limits: $\mathrm{Lim}_{x \to a^+} f(x) = \pm\infty$, $\mathrm{Lim}_{x \to a^-} f(x) = \pm\infty$.

The Derivative

Meaning of $\pm\infty$ as limits. Vertical asymptote. Polynomials have no vertical asymptotes. A rational function has vertical asymptotes at those values that make its denominator zero and its numerator nonzero. When $x = a$ is an asymptote of a rational function, determining whether a one-sided limit as x approaches a is $+\infty$ or $-\infty$. Meaning of "x approaches $+\infty$," "x approaches $-\infty$": $x \to +\infty$, $x \to -\infty$. Limits at infinity: $\lim_{x \to +\infty} f(x)$, $\lim_{x \to -\infty} f(x)$. Finding limits at infinity from graphs. Horizontal asymptote. Limits at infinity for powers of x. Polynomial functions do not have horizontal asymptotes. Finding limits at infinity for rational functions. Finding horizontal asymptotes for rational functions. Oblique asymptote. Finding oblique asymptotes for rational functions.

Section 2.3

Line tangent to f at $(x, f(x))$. Secant line. The derivative of f at x:

$$f'(x) = \lim_{h \to 0} \frac{f(x + h) - f(x)}{h}$$

The derivative $f'(x)$ as the slope of the line tangent to f at the point $(x, f(x))$. The difference quotient. The Three-Step Rule. Using the Three-Step Rule to find derivatives. Using the derivative to find the equation of the line tangent to f at $(x, f(x))$. Function differentiable at $x = a$. Function not differentiable at $x = a$. Determining whether a function is differentiable or not differentiable at $x = a$ from its graph. Function continuous at $x = a$. Function discontinuous at $x = a$. Function continuous everywhere. Determining whether a function is continuous or discontinuous at $x = a$ from its graph. First limit definition of continuity. Second limit definition of continuity. Polynomial functions are continuous everywhere. Rational functions are continuous everywhere they are defined. Differentiability implies continuity. Continuity does not imply differentiability.

Section 2.4

Constant function. The constant rule. Using the constant rule. Power function. The power rule. Using the power rule. The constant-multiple rule. Using the constant-multiple rule. The sum rule. The sum rule for more than two summands. Using the sum rule.

Section 2.5

Distance function. Average velocity. Instantaneous velocity. Velocity as the derivative of position. Finding velocities. The derivative as rate of change. Acceleration as the derivative of velocity. Finding accelerations. Finding rates of change.

Section 2.6

The theory of the firm. Cost, revenue, profit functions. Break-even quantities. Finding break-even quantities. Marginal cost, marginal revenue, marginal profit functions. The meaning of marginality: the meaning of $C'(a)$, $R'(a)$, $P'(a)$. Finding marginal cost, revenue, profit. Maximum profit occurs when marginal revenue equals marginal cost. Finding maximum profit and when it occurs. Revenue equals price per unit times quantity demanded. Finding the price that must be charged in order to attain maximum profit. Quantity demanded q as a function of price p. The rate of change of quantity demanded with respect to price. Elastic demand, inelastic demand. The elasticity of demand formula. Demand is elastic when $|E(p)| > 1$, inelastic when $|E(p)| < 1$, of unit elasticity when $|E(p)| = 1$. Finding elasticities. Finding changes in revenue when price is changed. Relationships between elasticity, price changes, and revenue changes. Elasticity formula $E(q)$ when price p is considered to be a function of quantity demanded.

Section 2.7

Prime notation f'. Differential notation df/dx. The derivative evaluated at $x = a$: $df/dx|_{x=a}$. Using differential notation. First derivative f', second derivative f'', third derivative f''', nth derivative $f^{(n)}$. Finding higher derivatives. Differential notation for higher derivatives: $d^n f/dx^n$, $d^n f/dx^n|_{x=a}$. Using differential notation for higher derivatives.

REVIEW EXERCISES

Let the function f have the following graph:

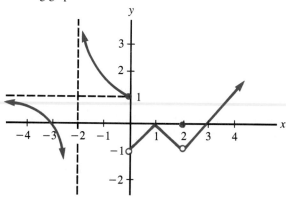

In Exercises 1 through 20, find the indicated limit. Use the symbols $\pm\infty$ as appropriate.

1. $\lim\limits_{x \to -2} f(x)$
2. $\lim\limits_{x \to -2^+} f(x)$
3. $\lim\limits_{x \to -2^-} f(x)$
4. $\lim\limits_{x \to -1} f(x)$
5. $\lim\limits_{x \to -1^+} f(x)$
6. $\lim\limits_{x \to -1^-} f(x)$
7. $\lim\limits_{x \to 0^+} f(x)$
8. $\lim\limits_{x \to 0^-} f(x)$
9. $\lim\limits_{x \to 0} f(x)$
10. $\lim\limits_{x \to 1^+} f(x)$
11. $\lim\limits_{x \to 1^-} f(x)$
12. $\lim\limits_{x \to 1} f(x)$
13. $\lim\limits_{x \to 2^+} f(x)$
14. $\lim\limits_{x \to 2^-} f(x)$
15. $\lim\limits_{x \to 2} f(x)$
16. $\lim\limits_{x \to 3} f(x)$
17. $\lim\limits_{x \to 3^+} f(x)$
18. $\lim\limits_{x \to 3^-} f(x)$
19. $\lim\limits_{x \to +\infty} f(x)$
20. $\lim\limits_{x \to -\infty} f(x)$

The Derivative

In Exercises 21 through 42, find the indicated limit. Use the symbols $\pm\infty$ as appropriate.

21. $\lim\limits_{x \to 2} (3x^2 - 5x - 2)$

22. $\lim\limits_{x \to 1} (x^4 + x^3 - 2x^2 + 5x - 1)$

23. $\lim\limits_{x \to 2^+} (5x^2 + 3x - 2)$

24. $\lim\limits_{x \to 0^-} (3x^3 + 5x^2 - 7x + 13)$

25. $\lim\limits_{x \to 9} \left(4\sqrt{x} - \dfrac{2}{x}\right)$

26. $\lim\limits_{x \to 1} \left(\dfrac{1}{x} - \dfrac{2}{x^2}\right)$

27. $\lim\limits_{x \to -1} (x^4 + x + 1)(x^2 - x - 1)$

28. $\lim\limits_{x \to 1} \dfrac{2}{x - 1}$

29. $\lim\limits_{x \to 2} \dfrac{x + 3}{x + 2}$

30. $\lim\limits_{x \to 4^+} \dfrac{x + 3}{x - 4}$

31. $\lim\limits_{x \to -2^-} \dfrac{12x + 5}{3x + 6}$

32. $\lim\limits_{x \to 2} \dfrac{x + 3}{x - 2}$

33. $\lim\limits_{x \to 0^+} \dfrac{2x + 5}{x^2 + 2}$

34. $\lim\limits_{x \to 0^+} \dfrac{2x + 5}{x^2 + 2x}$

35. $\lim\limits_{x \to 0} \dfrac{3x + 1}{2x^2 + 5}$

36. $\lim\limits_{x \to 5} \dfrac{x^2 - 7x + 10}{x^2 + 2x - 15}$

37. $\lim\limits_{x \to 5} \dfrac{x^2 + 7x + 10}{x^2 + 2x - 15}$

38. $\lim\limits_{x \to +\infty} \dfrac{4x^2 - 3x}{x + 2}$

39. $\lim\limits_{x \to 2^-} \dfrac{x^2 + x - 6}{x^2 + 3x - 10}$

40. $\lim\limits_{x \to 3^+} \dfrac{x^3 + x}{x^2 - 2x - 3}$

41. $\lim\limits_{x \to -\infty} \dfrac{x^2 + 2x}{4x^2 - 3x}$

42. $\lim\limits_{x \to +\infty} \dfrac{x + 2}{4x^2 - 3x}$

In Exercises 43 through 45, find all asymptotes of the given function.

43. $f(x) = \dfrac{3x - 7}{4 - 2x}$

44. $y = \dfrac{x^2 + 1}{x - 4}$

45. $g(x) = \dfrac{x^2 - 3x + 2}{x}$

46. Use the Three-Step Rule to find $k'(x)$ if $k(x) = 5x^3$. Find the equation of the line tangent to k at $(-3, k(-3))$.

47. Use the Three-Step Rule to find $g'(x)$ if $g(x) = \sqrt[3]{3x}$. For what values of x does $g'(x)$ not exist?

48. The following is the graph of a function h:

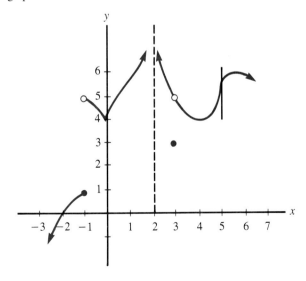

(a) At what values of x is h defined but not differentiable?
(b) At what values of x is h defined and discontinuous?

In Exercises 49 through 51, find the intervals over which the function f is continuous, the values of x at which f is defined but discontinuous, and the values of x at which f is defined but $f'(x)$ does not exist.

49. $f(x) = 2x^2 - 8x + 12$

50. $f(x) = \dfrac{x^2 - 2x - 8}{x^2 + 9x + 20}$

51. $f(x) = \begin{cases} x + 4, & x < -1, \\ -2x + 1, & -1 \le x \le 0, \\ 2, & 0 < x < 1, \\ x^2 + 1, & 1 \le x < 2, \\ 2, & x = 2, \\ x^2 - 8x + 17, & x > 2 \end{cases}$

In Exercises 52 through 59, find $f'(x)$.

52. $f(x) = -402$

53. $f(x) = \sqrt{3}x$

54. $f(x) = 3x^2 + 6x + 12$

55. $f(x) = -x^6 + 4x^5 - 12x^3 + 16x^2 - 8x - 3$

56. $f(x) = \dfrac{2}{x^3} - \dfrac{1}{x} + 3$

57. $f(x) = 4x^{-2} + 2x^{-3}$

58. $f(x) = 4\sqrt{x}$

59. $f(x) = 5x^{4/5} - 2x^{-1/2} + \dfrac{3}{x^{2/3}}$

In Exercises 60 through 65, find the equation of the line tangent to f at the point $(a, f(a))$.

60. $f(x) = 2/\sqrt{3}, \quad a = -5$

61. $f(x) = 4x^2 - 6x + 2, \quad a = 3$

62. $f(x) = x^5 + 2x^4 - 9x^2 + 3x + 12, \quad a = 0$

63. $f(x) = 3\sqrt{x} + \dfrac{2}{x} - 2, \quad a = 1$

64. $f(x) = 2x^{6/5}, \quad a = 32$

65. $f(x) = 4x^{-3/4} + 1, \quad a = 81$

66. A mortar shell is fired straight up into the air. The altitude of the shell t seconds after firing is

$$s(t) = -5t^2 + 150t$$

meters. Find the shell's velocity and acceleration 3 seconds and 6 seconds after firing. Find the shell's maximum altitude. How long is the shell in the air?

67. A firm estimates that the value of a computer t years after it is purchased new will be

$$V(t) = 576 - 192\sqrt{t}$$

thousand dollars. Here $0 \le t \le 9$. Find the rate at which the value of the computer is changing at the end of one year; at the end of four years; at the end of eight years.

68. A firm's cost function is given by $C(x) = x^2 + 320$, and the demand function for its product by $p = 56 - x$. Find the firm's maximum profit, the number of units that must be made and sold in order to attain the maximum profit, and the price per unit the firm must charge in order to attain the maximum profit.

69. The demand function for a commodity is given by $q = -6000p + 240{,}000$. Find
   (a) the elasticity of demand for the commodity when $p = \$10$, $p = \$20$, and $p = \$30$;
   (b) the revenue when $p = \$10$, $p = \$20$, and $p = \$30$;
   (c) the revenue when $p = \$11$, $p = \$22$, and $p = \$33$.

70. For the commodity of Exercise 69, find the price at which unit elasticity occurs, the range of prices where demand for the commodity is elastic, and the range where it is inelastic. If the commodity sells for $25 per unit, what will happen to revenue if the price is increased to $26? if it is decreased to $24?

In Exercises 71 through 75, find the indicated derivatives.

71. $f(x) = 2x^4 - 5x^2 + 12$, $\dfrac{df}{dx}, \dfrac{d^2f}{dx^2}, \dfrac{d^3f}{dx^3}\bigg|_{x=2}$

72. $y = 2\sqrt{x} + \dfrac{2}{\sqrt{x}}$, $\dfrac{dy}{dx}, \dfrac{d^2y}{dx^2}$

73. $g(x) = 3x^3 - 4x^{4/5} + 2$, $g'(x), g''(x), g'''(1)$

74. $w = 2u^2 + 3bu + 2\sqrt{u} + 1$, b a constant, $\dfrac{dw}{du}, \dfrac{d^2w}{du^2}$

75. $\dfrac{8x}{y} = 11$, $\dfrac{dy}{dx}, \dfrac{dx}{dy}$

ADDITIONAL TOPICS

Here are some additional topics not covered in the text that you might wish to investigate on your own.

1. A more rigorous definition of the limit of a function than the one presented in Section 2.1 is the following:

   $\text{Lim}_{x \to a} f(x) = L$ if for every positive real number ϵ there exists a positive real number δ such that $|f(x) - L| < \epsilon$ whenever $0 < |x - a| < \delta$.

   Explain how this definition of limit "translates" into the definition we gave in Section 2.1.

2. Show how the rigorous definition of limit given in suggestion 1 is used to prove the following properties of limits:

   $\text{Lim}_{x \to a} c = c$, $\quad \text{Lim}_{x \to a} cf(x) = c \cdot \text{Lim}_{x \to a} f(x)$,

   $\text{Lim}_{x \to a} [f(x) + g(x)] = \text{Lim}_{x \to a} f(x) + \text{Lim}_{x \to a} g(x)$,

   $\text{Lim}_{x \to a} f(x)g(x) = [\text{Lim}_{x \to a} f(x)][\text{Lim}_{x \to a} g(x)]$.

3. In Section 2.3 we gave an intuitive argument to show that if a function is differentiable at a point then it must be continuous there. Use the Second Limit Definition of Continuity to prove this.

4. Given a function f, it is often important to be able to find numbers x_r such that $f(x_r) = 0$. Such numbers are called **roots** of f, and are often difficult to find. **Newton's method** (also called the **Newton-Raphson method**) uses the derivative of f to find approximate roots of f. Find out how Newton's method works, why it works, why it sometimes fails to work, and some of its applications.

5. The derivative was introduced to mathematics in the seventeenth century by Isaac Newton and Gottfried Wilhelm Leibniz. Find out about the early development and use of the derivative: why the concept of the derivative was needed by Newton and Leibniz, the different ways in which each defined and used his discovery, and how

each of them introduced his version of the derivative to the mathematical world. A good place to begin research on this topic would be with the supplement to this chapter.

SUPPLEMENT: NEWTON AND LEIBNIZ

No one person "invented" calculus. The development of calculus was a lengthy process contributed to by many mathematicians over many centuries. Nevertheless, there was a short period of time during which the partial results and incomplete methods previously discovered were greatly extended and then organized into a coherent mathematical theory that was thereafter known as "the calculus." This rapid development of the subject occurred in the latter part of the seventeenth century and was the independent work of Isaac Newton (1643–1727) of England and Gottfried Wilhelm Leibniz (1646–1716) of Germany.

Isaac Newton was one of the greatest of human intellects. He made monumental contributions to physics by enunciating the law of universal gravitation and the three laws of motion that bear his name, working out many of the consequences of these laws as they apply to astronomy and mechanics, experimenting in optics, and proposing a particle theory of light. His development of calculus went hand-in-hand with his work in physics as he created new mathematical tools to solve problems in celestial and earthly mechanics. In addition to these activities, Newton also experimented in chemistry and alchemy, studied and wrote on theology, and served as Master of the Royal Mint.

Newton apparently developed what he called his "fluxional calculus" during the period 1665–1666. Quantities that changed with time he referred to as **fluents;** he called their rates of change with respect to time **fluxions.** For instance, if a particle moves along a curve its x- and y-coordinates change with time and hence are fluents, and therefore their rates of change with respect to time, which Newton denoted by $\dot{x}$ and $\dot{y}$, respectively, are fluxions. Thus the fluxions $\dot{x}$ and $\dot{y}$ are what we would now write as dx/dt and dy/dt, the derivatives of x and y with respect to time. The notation $\dot{x}$ for the derivative with respect to time survives today in physics and engineering.

The method by which Newton found derivatives was rather cumbersome, proceeding much in the manner of our Three-Step Rule but using the idea of infinitesimal quantities rather than limits. His reasoning was as follows: if h is the duration of an infinitely short time period, then $\dot{x}h$ and $\dot{y}h$ are infinitesimals, that is, infinitely small nonzero amounts by which x and y change during the time period. If we substitute $x + \dot{x}h$ for x and $y + \dot{y}h$ for y in the relationship between x and y, subtract the original equation from the result, and then drop all terms containing powers of h that are 2 or greater because they are infinitesimally small compared with the remaining terms, we can obtain an expression for $\dot{y}$ in terms of $\dot{x}$. To illustrate, suppose

$$y = x^2.$$

Then

$$y + \dot{y}h = (x + \dot{x}h)^2,$$

or

$$y + \dot{y}h = x^2 + 2x\dot{x}h + (\dot{x})^2 h^2.$$

Subtracting $y = x^2$ from this last equation, and dropping the term containing h^2 because it is infinitesimal compared with the rest of the terms, we have

$$\dot{y}h = 2x\dot{x}h,$$

or

$$\dot{y} = 2x\dot{x}.$$

This last equation can be rewritten in the form

$$\frac{\dot{y}}{\dot{x}} = 2x,$$

which expresses what we today call the derivative of y with respect to x as a ratio of the time derivatives $\dot{y}$ and $\dot{x}$.

Newton seems to have been a secretive man who was not anxious to share his discoveries with others. For many years he did not communicate his discoveries in calculus and physics to the world at large, but instead revealed them in piecemeal fashion to a few selected colleagues. Finally, at the instigation of his friend Edmund Halley, he published in 1687 a book entitled *Philosophiae Naturalis Principia Mathematica*, known today as the *Principia*. In the *Principia* Newton presented for the first time his work on gravitation and celestial mechanics, as well as the fluxional calculus that made it possible. Unfortunately, Newton recast all the calculus into geometrical form, in approved classical style, and this served to obscure the power and utility of his method of fluxions.

During the years 1673–1676, after Newton had developed his version of the calculus but before he had published it, Leibniz independently created a slightly different version. He, too, utilized infinitesimals, but he called his infinitesimals **differentials:** the differentials for x and y he denoted by dx and dy, respectively. These were defined geometrically as shown in Figure 2–49: dx is an arbitrary infinitesimal increment to x, and dy is determined by dx and the tangent line at (x, y). Leibniz defined the derivative of y with respect to x to be the ratio dy/dx of the differentials dy and dx.

Leibniz's terminology and notation have survived to the present day. This is why we use the differential notation dy/dx for the derivative. We will discuss differentials in more detail in Section 3.3.

The method by which Leibniz found derivatives was much like Newton's. In the relationship between x and y, x was replaced by $x + dx$, y was replaced by $y + dy$, and the original equation was subtracted from the result. Then all terms containing powers of dx or dy that were 2 or greater, or all terms with a factor of the form $dx\, dy$, were eliminated on the grounds that they were infinitesimally small

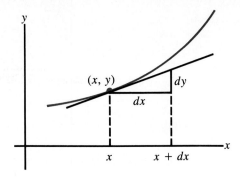

FIGURE 2–49

compared with the remaining terms. The end result was an expression relating dy and dx. However, Leibniz also worked out and used many of the rules of differentiation, such as (in his notation)

$$d(cx) = c \cdot dx$$

and

$$d(x^n) = nx^{n-1}\, dx.$$

Unlike Newton, Leibniz was eager to let others know of his discoveries. Even before the formal publication of his results in 1684, knowledge of his methods was quickly spread through the European mathematical community by personal correspondence and word of mouth.

In 1676 Newton learned that Leibniz was working on a version of calculus; in an attempt to assert his priority, he sent Leibniz a letter in which he mentioned his earlier work in the field. Leibniz replied with a letter outlining his own methods. When Leibniz published his results in 1684, he did not mention Newton's claim of priority; in response Newton, in the *Principia,* told of this exchange of letters. This started a quarrel, not so much between Newton and Leibniz but between their followers, over who had priority of discovery. Accusations of plagiarism were made by both sides, and the situation became, particularly in England, one of national pride. As a result, English mathematicians for many years insisted on Newton as the ultimate mathematical authority and adopted his terminology and notation exclusively. Since the terminology and notation used by Leibniz were generally superior to those of Newton, this had an unfortunate stultifying effect on English mathematics.

Newton and Leibniz both used infinitesimals in their development of calculus. Such "infinitely small" but nonzero quantities seemed illogical to mathematicians and made them uneasy. Newton himself was not satisfied with his reliance on infinitesimals, and in his later work he tried to get by without them; in so doing, he nearly arrived at the notion of the limit of a function. However, it was not until 1821, when the French mathematician Augustin Louis Cauchy introduced the idea of limit in its modern form, that it was possible for calculus to be put on a firm logical foundation.

Calculus is one of the greatest tools ever discovered for the investigation of nature. It was developed as a method for solving problems about the movement of objects, particularly astronomical bodies, but its applications soon began to expand, first throughout the physical sciences and engineering, and then to the biological and social sciences, including management and economics. The usefulness of calculus continues to grow today as it is used to investigate new problems as they arise in these and other fields.

Suggestion

Many books describe the development of calculus in much greater detail than we have been able to do here. If you are interested in this topic and would like to read more about it, here are some places to start:

Andrade, E. N. Da C. "Isaac Newton." In *The World of Mathematics,* vol. 1, ed. James R. Newman. New York: Simon and Schuster, 1956.

Bell, E. T. *Men of Mathematics.* New York: Simon and Schuster, 1937.

Boyer, Carl B. *History of the Calculus and Its Conceptual Development.* New York: Dover, 1959.

Boyer, Carl B. *A History of Mathematics.* New York: John Wiley & Sons, 1968.

Eves, Howard. *An Introduction to the History of Mathematics.* New York: Holt, Rinehart, and Winston, 1961.

Eves, Howard. *Great Moments in Mathematics After 1650.* Dolciani Mathematical Expositions No. 7, The Mathematical Association of America, 1983.

3

More about Differentiation

In Chapter 2 we introduced the derivative and presented some of the basic rules of differentiation: the constant rule, the power rule, the constant-multiple rule, and the sum rule. These rules are not sufficient to handle all functions of interest, however. For instance, they do not tell us how to differentiate functions that are products or quotients, such as those defined by

$$f(x) = (x^2 + 2x + 3)(x^5 + 4x^3 - 12)$$

and

$$g(x) = \frac{3x^2 + 2x - 5}{5x + 3},$$

or those that are defined as generalized powers, such as

$$h(x) = (2x^3 - 5x + 7)^{12}.$$

In this chapter we continue our study of differentiation by developing the product rule, quotient rule, and chain rule for derivatives. These rules will allow us to differentiate functions such as those referred to above. We also discuss the topic of differentials, showing how the differential notation dy/dx can indeed be regarded as a ratio, and demonstrate the useful technique of implicit differentiation. We conclude the chapter with a section devoted to the solution of related rates problems.

3.1 THE PRODUCT RULE AND THE QUOTIENT RULE

In this section we develop and illustrate rules of differentiation that allow us to find the derivatives of functions that are products and quotients of other functions.

The Product Rule

Let f and g be functions and suppose that $f'(x)$ and $g'(x)$ both exist. We would like to find the derivative of the product function $k = fg$ at x. Using the Three-Step Rule, we form the difference quotient of k at x, then rewrite it in terms of the difference quotients of f and g by adding and subtracting $f(x)g(x + h)$ in its numerator:

$$\frac{k(x + h) - k(x)}{h} = \frac{f(x + h)g(x + h) - f(x)g(x)}{h}$$

$$= \frac{f(x + h)g(x + h) - f(x)g(x + h) + f(x)g(x + h) - f(x)g(x)}{h}$$

$$= \frac{[f(x + h) - f(x)]g(x + h) + f(x)[g(x + h) - g(x)]}{h}$$

$$= \frac{f(x + h) - f(x)}{h} g(x + h) + f(x) \frac{g(x + h) - g(x)}{h}.$$

Because $f'(x)$ and $g'(x)$ exist, we have

$$\operatorname*{Lim}_{h \to 0} \frac{f(x + h) - f(x)}{h} = f'(x) \quad \text{and} \quad \operatorname*{Lim}_{h \to 0} \frac{g(x + h) - g(x)}{h} = g'(x).$$

Also, since $g'(x)$ exists $g(x)$ is differentiable and hence continuous at x; therefore by the second limit definition of continuity (Section 2.3),

$$\operatorname*{Lim}_{h \to 0} g(x + h) = g(x).$$

Thus

$$k'(x) = \operatorname*{Lim}_{h \to 0} \frac{f(x + h) - f(x)}{h} \operatorname*{Lim}_{h \to 0} g(x + h)$$

$$+ \operatorname*{Lim}_{h \to 0} f(x) \operatorname*{Lim}_{h \to 0} \frac{g(x + h) - g(x)}{h}$$

$$= f'(x)g(x) + f(x)g'(x).$$

This is the **product rule** for derivatives.

The Product Rule

If $k = fg$ and $f'(x)$ and $g'(x)$ both exist, then

$$k'(x) = f(x)g'(x) + f'(x)g(x).$$

In words: the derivative of a product is the first factor times the derivative of the second factor plus the derivative of the first factor times the second factor. Notice that the derivative of a product is *not* the product of the derivatives.

EXAMPLE 1 If $k(x) = (6x + 7)(3x - 2)$, then by the product rule,

$$k'(x) = (6x + 7)(3x - 2)' + (6x + 7)'(3x - 2)$$
$$= (6x + 7)3 + 6(3x - 2)$$
$$= 18x + 21 + 18x - 12$$
$$= 36x + 9.$$

(Of course, we could have found $k'(x)$ without using the product rule by multiplying out its right-hand side and then taking the derivative:

$$k(x) = 18x^2 + 9x - 14, \quad \text{so} \quad k'(x) = 36x + 9.) \quad \square$$

EXAMPLE 2 If $k(x) = (x^2 - 3x + 4)(x^3 + x)$, then

$$k'(x) = (x^2 - 3x + 4)(x^3 + x)' + (x^2 - 3x + 4)'(x^3 + x)$$
$$= (x^2 - 3x + 4)(3x^2 + 1) + (2x - 3)(x^3 + x)$$
$$= 5x^4 - 12x^3 + 15x^2 - 6x + 4. \quad \square$$

EXAMPLE 3 Let $f(x) = (x^2 + 1)(x^3 + 2)(x^4 + 3)$. We can find $f'(x)$ by applying the product rule twice. Since $f(x)$ is the product of $x^2 + 1$ and $(x^3 + 2)(x^4 + 3)$, a first application of the product rule gives

$$f'(x) = (x^2 + 1)[(x^3 + 2)(x^4 + 3)]' + (x^2 + 1)'(x^3 + 2)(x^4 + 3).$$

A second application of the product rule now allows us to find the derivative of $(x^3 + 2)(x^4 + 3)$:

$$f'(x) = (x^2 + 1)[(x^3 + 2)(x^4 + 3)' + (x^3 + 2)'(x^4 + 3)]$$
$$\quad + (x^2 + 1)'(x^3 + 2)(x^4 + 3)$$
$$= (x^2 + 1)[(x^3 + 2)4x^3 + 3x^2(x^4 + 3)] + 2x(x^3 + 2)(x^4 + 3),$$

so

$$f'(x) = (x^2 + 1)[7x^6 + 8x^3 + 9x^2] + 2x(x^7 + 2x^4 + 3x^3 + 6)$$
$$= 9x^8 + 7x^6 + 12x^5 + 15x^4 + 8x^3 + 9x^2 + 12x. \quad \square$$

The Quotient Rule

For functions f and g, suppose that $f'(x)$ and $g'(x)$ both exist and that $g(x) \neq 0$. We would like to find the derivative of the quotient function $k = f/g$ at x. Using a trick similar to that employed in deriving the product rule, we write

$$\frac{k(x+h)-k(x)}{h} = \frac{f(x+h)/g(x+h) - f(x)/g(x)}{h}$$

$$= \frac{f(x+h)g(x) - f(x)g(x+h)}{hg(x)g(x+h)}$$

$$= \frac{f(x+h)g(x) - f(x)g(x) + f(x)g(x) - f(x)g(x+h)}{hg(x)g(x+h)}$$

$$= \frac{f(x+h) - f(x)}{h} \cdot \frac{g(x)}{g(x)g(x+h)}$$

$$- \frac{f(x)}{g(x)g(x+h)} \cdot \frac{g(x+h) - g(x)}{h}.$$

As was the case for the product rule, the existence of $f'(x)$ and $g'(x)$ implies that

$$\lim_{h \to 0} \frac{f(x+h) - f(x)}{h} = f'(x) \quad \text{and} \quad \lim_{h \to 0} \frac{g(x+h) - g(x)}{h} = g'(x).$$

Also, the existence of $g'(x)$ implies that g is differentiable and hence continuous at x; therefore by the second limit definition of continuity (Section 10.3),

$$\lim_{h \to 0} g(x+h) = g(x).$$

Thus

$$k'(x) = \lim_{h \to 0} \frac{f(x+h) - f(x)}{h} \lim_{h \to 0} \frac{g(x)}{g(x)g(x+h)}$$

$$- \lim_{h \to 0} \frac{f(x)}{g(x)g(x+h)} \lim_{h \to 0} \frac{g(x+h) - g(x)}{h}$$

$$= f'(x) \frac{g(x)}{g(x)g(x)} - \frac{f(x)}{g(x)g(x)} g'(x)$$

$$= \frac{g(x)f'(x) - f(x)g'(x)}{[g(x)]^2}.$$

This is the **quotient rule** for derivatives.

The Quotient Rule

If $k = f/g$, $f'(x)$ and $g'(x)$ both exist, and $g(x) \neq 0$, then

$$k'(x) = \frac{g(x)f'(x) - f(x)g'(x)}{[g(x)]^2}.$$

In words: the derivative of a quotient is the denominator times the derivative of the numerator minus the numerator times the derivative of the denominator, all divided

More about Differentiation

by the square of the denominator. Notice that the derivative of a quotient is *not* the quotient of the derivatives.

EXAMPLE 4 If

$$k(x) = \frac{2x + 3}{5x + 4},$$

then by the quotient rule,

$$k'(x) = \frac{(5x + 4)(2x + 3)' - (2x + 3)(5x + 4)'}{[5x + 4]^2}$$

$$= \frac{(5x + 4)2 - (2x + 3)5}{[5x + 4]^2}$$

$$= \frac{10x + 8 - 10x - 15}{[5x + 4]^2}$$

$$= -\frac{7}{25x^2 + 40x + 16}. \quad \square$$

EXAMPLE 5 If

$$f(x) = \frac{x^2 + x + 1}{2x^3 + 1},$$

then

$$f'(x) = \frac{(2x^3 + 1)(x^2 + x + 1)' - (x^2 + x + 1)(2x^3 + 1)'}{[2x^3 + 1]^2}$$

$$= \frac{(2x^3 + 1)(2x + 1) - (x^2 + x + 1)(6x^2)}{[2x^3 + 1]^2}$$

$$= \frac{4x^4 + 2x^3 + 2x + 1 - 6x^4 - 6x^3 - 6x^2}{[2x^3 + 1]^2}$$

$$= \frac{-2x^4 - 4x^3 - 6x^2 + 2x + 1}{4x^6 + 4x^3 + 1}. \quad \square$$

Sometimes it is necessary to use the product and quotient rules together.

EXAMPLE 6 Let

$$f(x) = (x^2 + 1)\left(\frac{2x}{4x + 9}\right).$$

To find $f'(x)$, we apply the product rule followed by the quotient rule:

$$f'(x) = (x^2 + 1)\left(\frac{2x}{4x + 9}\right)' + (x^2 + 1)'\left(\frac{2x}{4x + 9}\right)$$

$$= (x^2 + 1)\left(\frac{(4x + 9)(2x)' - 2x(4x + 9)'}{(4x + 9)^2}\right) + 2x\left(\frac{2x}{4x + 9}\right)$$

$$= (x^2 + 1)\frac{(4x + 9)2 - 2x(4)}{[4x + 9]^2} + 2x\frac{2x}{4x + 9}$$

$$= (x^2 + 1)\frac{18}{[4x + 9]^2} + \frac{4x^2}{4x + 9}$$

$$= \frac{16x^3 + 54x^2 + 18}{[4x + 9]^2}. \quad \square$$

EXAMPLE 7 A patient is placed on a regimen that includes taking a daily dose of a certain drug. The amount of the drug concentrated in the patient's liver t days after the start of the regimen is y grams, where

$$y = \frac{20t}{t + 15}.$$

At what rate is the amount concentrated in the liver changing after 30 days on the regimen?

By the quotient rule,

$$\frac{dy}{dt} = \frac{(t + 15)20 - 20t(1)}{[t + 15]^2},$$

or

$$\frac{dy}{dt} = \frac{300}{[t + 15]^2}.$$

Therefore at the end of 30 days the amount of the drug concentrated in the liver is increasing at the rate of

$$\left.\frac{dy}{dt}\right|_{t=30} = \frac{300}{[30 + 15]^2} = \frac{4}{27}$$

gram per day. $\square$

■ *Computer Manual: Programs PRODPOLY, QUOPOLY*

3.1 EXERCISES

In Exercises 1 through 40, use the product rule and/or the quotient rule to find the derivative of the given function.

1. $k(x) = (3x + 5)(2x - 1)$
2. $k(x) = (2x + 6)(4x - 15)$
3. $f(x) = (5x + 1)(x^2 - 2)$
4. $f(x) = (x^2 + 3)(2x + 5)$

5. $f(x) = (x^2 + 5)(x^2 - 3)$

6. $k(x) = (3x^5 + 2x)(x^2 + 4x - 1)$

7. $k(x) = \dfrac{4}{x + 2}$

8. $k(x) = -\dfrac{2}{3x - 1}$

9. $f(x) = \dfrac{3}{\sqrt{x} - 2}$

10. $f(x) = -\dfrac{2x}{3x + 1}$

11. $f(x) = \dfrac{x}{2x - 1}$

12. $f(x) = \dfrac{x + 1}{2x + 3}$

13. $k(x) = \dfrac{2x + 4}{5x - 1}$

14. $k(x) = \dfrac{4 - 3x}{7 - x}$

15. $f(x) = \dfrac{x^2 + 3x}{x + 2}$

16. $f(x) = \dfrac{x^2 + 3x}{3x + 4}$

17. $k(x) = \dfrac{x^2 + 3x - 2}{9 - 3x}$

18. $k(x) = (5x + 3)^2$

19. $k(x) = (x^2 + 4x)^2$

20. $k(x) = (x^4 + 2)x^{-3}$

21. $k(x) = 2x^{-3}(x^3 + 4)$

22. $f(x) = (2 + \sqrt{x})(x^3 + x^2)$

23. $f(x) = 2(x^{-3} + x^{-2})(x^2 + 4x)$

24. $f(x) = \dfrac{5x^2 + 3}{5x - 2}$

25. $f(x) = \dfrac{x^2 + 3x}{x^2 - 2}$

26. $k(x) = (3x^{1/3} - 2\sqrt{x})(x^2 - 3x + 1)$

27. $k(x) = \dfrac{2x^{-1} + 3x^{-2}}{x^{-2} - 1}$

28. $k(x) = x(2x + 1)(3x + 2)$

29. $f(x) = (x + 1)(x^2 + 1)(x^3 + 1)$

30. $k(x) = \dfrac{(3x + 5)(2x - 7)}{6x + 1}$

31. $k(x) = \dfrac{5x + 6}{(2x - 1)(3x + 9)}$

32. $k(x) = \dfrac{8x + 3}{(x^2 + 1)(7 - 2x)}$

33. $k(x) = \dfrac{(x^2 - 3)(4 - 3x)}{5x + 11}$

34. $k(x) = \dfrac{(x^2 + 1)(x^3 + 2)}{x^2 - 2x}$

35. $k(x) = \dfrac{x^2 - 2x}{(x^2 + 1)(x^3 + 2)}$

36. $k(x) = (1 - 2x)(x^2 - 1)(3 + 4x)(x^3 + 2x)$

37. $k(x) = x(x^2 + 1)(x^3 + x^2 + 1)$

38. $k(x) = (x^2 + 3)(x^2 + 8)(x^3 + 2x^2 + 1)(x^4 - 1)$

39. $k(x) = \dfrac{5x + 3}{7x + 2} \cdot \dfrac{6x + 1}{x - 1}$

40. $k(x) = \dfrac{4x + 3}{5 - 2x} \cdot \dfrac{x^2 + 1}{x^2 - 1}$

Management

41. Weekly revenue from box office receipts of a movie t weeks after the movie is released is y million dollars, where

$$y = \dfrac{20t}{0.5t + 5}.$$

Find the rate of change of revenue when $t = 1, 2,$ and 5 weeks. Also find and interpret $\lim\limits_{t \to +\infty} y$.

42. A firm's cost function is given by

$$C(x) = \frac{x^2}{x+1} - 1, \qquad x \geq 1.$$

Find the firm's marginal cost when $x = 1$, $x = 10$, and $x = 20$, and interpret your results.

43. If Acme Company makes $x \geq 1$ units of its product, its average cost per unit is given by

$$A(x) = \frac{20x - 4}{5x - 0.5}.$$

Find the rate at which Acme's average cost per unit is changing when $x = 1$, $x = 10$, and $x = 100$. Also find $\lim\limits_{x \to +\infty} A(x)$ and interpret your result.

44. If a firm produces x units of its product, its average cost per unit is $A(x)$ dollars, where

$$A(x) = \frac{6x + 53}{2x + 1}, \qquad x \geq 1.$$

(a) Find and interpret $A'(1)$, $A'(2)$, and $A'(10)$.
(b) Find and interpret $\lim\limits_{x \to +\infty} A(x)$.

45. Repeat Exercise 44 if

$$A(x) = \frac{0.2x^2 + 100}{3x + 2}.$$

46. The demand function for a commodity is given by

$$p = \frac{1000}{q + 1},$$

where q is the quantity demanded in thousands of units and p is the unit price in dollars. Find the marginal revenue when the quantity demanded is 9000, 99,000 and 999,000 units. Also find $\lim\limits_{q \to +\infty} p$ and interpret your result.

47. The demand function for a product is given by

$$p = \frac{2q + 200}{8q + 10}.$$

Here q is in millions of units and p is in dollars. Find the marginal revenue when $q = 5$, 25, and 100 million units. Also find $\lim\limits_{q \to +\infty} p$ and interpret your result.

Life Science

48. Suppose that t hours after the introduction of a single bacterium into a culture there are

$$y = \frac{200t}{t + 10}$$

bacteria in the culture. Find the rate at which the number of bacteria is growing when $t = 1$, $t = 5$, and $t = 10$. Also find $\lim\limits_{t \to +\infty} y$ and interpret your result.

49. The number of signs a chimpanzee knows after t months of being trained in sign language is y, where

$$y = \frac{540t}{2t + 7}.$$

Find the rate at which the chimpanzee in learning signs when $t = 0$, 6, and 12 months. Also find and interpret $\lim_{t \to +\infty} y$.

50. A drug is given to reduce a benign tumor. A dosage of x units will shrink the tumor by y cm, where

$$y = \frac{x^2 + 1}{2 + 3x}.$$

Find the rate at which the tumor will shrink if the dosage is 5 units; if the dosage is 10 units.

Social Science

51. A psychologist asks the subjects of an experiment to perform a series of complex tasks, and finds that after x repetitions of the series they can, on average, complete

$$y = \frac{100x + 30}{x + 1}$$

percent of the tasks without mistakes. Find the rate at which learning takes place as a function of the number of repetitions if the number of repetitions is 0, 2, and 5.

52. A municipality estimates that t years from now the proportion of its budget that will be devoted to debt service is y, where

$$y = (2t + 1)(7t + 8)^{-1}.$$

(a) Find and interpret $\left.\dfrac{dy}{dt}\right|_{t=0}$, $\left.\dfrac{dy}{dt}\right|_{t=1}$, and $\left.\dfrac{dy}{dt}\right|_{t=2}$.

(b) Find and interpret $\lim_{t \to +\infty} y$.

3.2 THE CHAIN RULE

We can create new functions from old ones by addition, subtraction, multiplication, and division, and the sum rule, product rule, and quotient rule for derivatives show us how to differentiate such new functions in terms of the old ones. But in Section 1.3 we introduced another method of creating new functions from old ones, a method known as **composition of functions**. (Recall that if f and g are functions, the **composite of g with f** is the function k defined by $k(x) = f(g(x))$; for instance, if $f(x) = \sqrt{x}$ and $g(x) = 2x + 1$, then $k(x) = f(g(x)) = \sqrt{2x + 1}$.) In order to complete our rules of differentiation, we need a result that will tell us how to differentiate composite functions; this result is known as the **chain rule**. In this section we introduce the chain rule and illustrate some of its uses.

Let us develop the chain rule. Suppose $k(x) = f(g(x))$, where $f'(g(x))$ and $g'(x)$ both exist. We form the difference quotient

$$\frac{k(x+h) - k(x)}{h} = \frac{f(g(x+h)) - f(g(x))}{h}.$$

If $g(x+h) - g(x) \neq 0$, we may rewrite this difference quotient as

$$\frac{k(x+h) - k(x)}{h} = \frac{f(g(x+h)) - f(g(x))}{g(x+h) - g(x)} \cdot \frac{g(x+h) - g(x)}{h}.$$

Since $g'(x)$ exists, g is continuous at x. Therefore

$$\lim_{h \to 0} g(x+h) = g(x),$$

or

$$\lim_{h \to 0} [g(x+h) - g(x)] = 0.$$

Let us set $r = g(x+h) - g(x)$. As h approaches 0, r approaches 0, and

$$g(x+h) = g(x) + r.$$

Thus

$$k'(x) = \lim_{h \to 0} \frac{f(g(x+h)) - f(g(x))}{g(x+h) - g(x)} \lim_{h \to 0} \frac{g(x+h) - g(x)}{h}$$

$$= \lim_{r \to 0} \frac{f(g(x) + r) - f(g(x))}{r} g'(x)$$

$$= f'(g(x))g'(x).$$

We have thus shown that if $k(x) = f(g(x))$, then $k'(x) = f'(g(x))g'(x)$; this is the **chain rule.** We have proved the chain rule under the assumption that $g(x+h) - g(x) \neq 0$, but it is true in general.

The Chain Rule: First Formulation

If $k(x) = f(g(x))$ and $f'(g(x))$ and $g'(x)$ both exist, then

$$k'(x) = f'(g(x))g'(x).$$

EXAMPLE 1 Suppose $k(x) = (2x + 3)^{20}$. We will use the chain rule to find $k'(x)$. In section 1.3 we showed how to write k as a composite function by choosing f to be the "outside function" and g the "inside function":

$$k(x) = f(g(x))$$

where

$$f(x) = x^{20} \quad \text{and} \quad g(x) = 2x + 3.$$

Then $k(x) = f(g(x))$, and since

$$f'(x) = 20x^{19} \quad \text{and} \quad g'(x) = 2,$$

the chain rule tells us that

$$\begin{aligned} k'(x) &= f'(g(x))g'(x) \\ &= f'(2x + 3)g'(x) \\ &= 20(2x + 3)^{19}(2) \\ &= 40(2x + 3)^{19}. \quad \square \end{aligned}$$

EXAMPLE 2   Let $h(x) = \sqrt{2x^3 + 4x}$. We write h as a composite function

$$h(x) = f(g(x)),$$

where

$$f(x) = \sqrt{x} \quad \text{and} \quad g(x) = 2x^3 + 4x.$$

Since

$$f'(x) = \frac{1}{2\sqrt{x}} \quad \text{and} \quad g'(x) = 6x^2 + 4,$$

by the chain rule we have

$$\begin{aligned} h'(x) &= f'(g(x))g'(x) \\ &= f'(2x^3 + 4x)g'(x) \\ &= \frac{1}{2\sqrt{2x^3 + 4x}}(6x^2 + 4) \\ &= \frac{3x^2 + 2}{\sqrt{2x^3 + 4x}}. \quad \square \end{aligned}$$

We do not really need to write out $f(x)$, $g(x)$, $f'(x)$, and $g'(x)$ every time we differentiate a composite function. Roughly speaking, what the chain rule says is that to differentiate a composite expression $f(g(x))$ we must differentiate from the "outside in." In other words, we first differentiate the "outside" expression $f(x)$, then multiply the result by the derivative of the "inside" expression $g(x)$. With a little practice this becomes easy to do.

EXAMPLE 3   Let us differentiate $k(x) = (2x + 3)^{20}$ again. This time we will not write out $f(x)$, $g(x)$, $f'(x)$, and $g'(x)$, but instead will differentiate from the outside in as follows:

$$\begin{aligned} k'(x) &= [(2x + 3)^{20}]' = 20(2x + 3)^{19}(2x + 3)' \\ &= 20(2x + 3)^{19}(2) \\ &= 40(2x + 3)^{19}. \quad \square \end{aligned}$$

EXAMPLE 4 Let us differentiate $h(x) = \sqrt{2x^3 + 4x}$ again. This time we will differentiate from the outside in. We have

$$h'(x) = (\sqrt{2x^3 + 4x})'$$
$$= [(2x^3 + 4x)^{1/2}]'$$
$$= \frac{1}{2}(2x^3 + 4x)^{-1/2}(2x^3 + 4x)'$$
$$= \frac{1}{2\sqrt{2x^3 + 4x}}(6x^2 + 4)$$
$$= \frac{3x^2 + 2}{\sqrt{2x^3 + 4x}}.$$

The same differentiation using differential notation would be written as follows:

$$\frac{dh}{dx} = \frac{d}{dx}(\sqrt{2x^3 + 4x}) = \frac{1}{2\sqrt{2x^3 + 4x}}\frac{d}{dx}(2x^3 + 4x)$$
$$= \frac{1}{2\sqrt{2x^3 + 4x}}(6x^2 + 4)$$
$$= \frac{3x^2 + 2}{\sqrt{2x^3 + 4x}}. \quad \square$$

The chain rule may be used with the other rules of differentiation.

EXAMPLE 5 Let

$$p(x) = (x^2 + 1)^3(x^4 - 1)^5.$$

Then by the product rule we have

$$p'(x) = (x^2 + 1)^3[(x^4 - 1)^5]' + [(x^2 + 1)^3]'(x^4 - 1)^5.$$

Now we apply the chain rule to find

$$[(x^4 - 1)^5]' = 5(x^4 - 1)^4(x^4 - 1)' = 5(x^4 - 1)^4(4x^3)$$
$$= 20x^3(x^4 - 1)^4$$

and

$$[(x^2 + 1)^3]' = 3(x^2 + 1)^2(x^2 + 1)' = 3(x^2 + 1)^2(2x)$$
$$= 6x(x^2 + 1)^2.$$

Therefore

$$p'(x) = (x^2 + 1)^3[20x^3(x^4 - 1)^4] + [6x(x^2 + 1)^2](x^4 - 1)^5$$
$$= 2x(x^2 + 1)^2(x^4 - 1)^4[10x^2(x^2 + 1) + 3(x^4 - 1)]$$
$$= 2x(x^2 + 1)^2(x^4 - 1)^4[13x^4 + 10x^2 - 3]. \quad \square$$

EXAMPLE 6 Let $q(x) = [(x + 1)/(x - 1)]^{2/3}$. To differentiate $q(x)$, we use the chain rule followed by the quotient rule:

$$\begin{aligned}
q'(x) &= \left[\left(\frac{x+1}{x-1}\right)^{2/3}\right]' = \frac{2}{3}\left(\frac{x+1}{x-1}\right)^{-1/3}\left(\frac{x+1}{x-1}\right)' \\
&= \frac{2}{3}\left(\frac{x+1}{x-1}\right)^{-1/3} \frac{(x-1)(x+1)' - (x+1)(x-1)'}{(x-1)^2} \\
&= \frac{2}{3}\left(\frac{x+1}{x-1}\right)^{-1/3} \frac{(x-1)(1) - (x+1)(1)}{(x-1)^2} \\
&= -\frac{4}{3(x-1)^2}\left(\frac{x+1}{x-1}\right)^{-1/3}. \quad \square
\end{aligned}$$

The functions to which we have thus far applied the chain rule have all been defined by equations of the form

$$k(x) = [g(x)]^r.$$

But the chain rule tells us that the derivative of such a function is

$$k'(x) = r[g(x)]^{r-1}g'(x).$$

(Why?) This special case of the chain rule is known as the **general power rule**.

The General Power Rule

If

$$k(x) = [g(x)]^r, \qquad r \neq 0,$$

then

$$k'(x) = r[g(x)]^{r-1}g'(x).$$

EXAMPLE 7 If

$$k(x) = (x^3 - 2x^2 + 17)^{12},$$

then

$$\begin{aligned}
k'(x) &= 12(x^3 - 2x^2 + 17)^{11}(x^3 - 2x^2 + 17)' \\
&= 12(3x^2 - 4x)(x^3 - 2x^2 + 17)^{11}. \quad \square
\end{aligned}$$

There is another formulation of the chain rule that is often useful. If $y = f(x)$ defines y as a function of x and $x = g(t)$ defines x as a function of t, then

$$y = f(g(t))$$

defines y as a function of t. Differentiating with respect to t by means of the chain rule, we have

$$\frac{dy}{dt} = \frac{df}{dx}\frac{dg}{dt},$$

which we may write as

$$\frac{dy}{dt} = \frac{dy}{dx}\frac{dx}{dt}.$$

This is the second formulation of the chain rule.

The Chain Rule: Second Formulation

If $y = f(x)$ and $x = g(t)$ and dy/dx and dx/dt both exist, then

$$\frac{dy}{dt} = \frac{dy}{dx}\frac{dx}{dt}.$$

EXAMPLE 8   Suppose $y = x^2 + 1$ and $x = 3t - 2$. Then by the second formulation of the chain rule,

$$\frac{dy}{dt} = \frac{dy}{dx}\frac{dx}{dt} = \frac{d}{dx}(x^2 + 1)\frac{d}{dt}(3t - 2) = 2x \cdot 3 = 6x.$$

If we wish, we may write dy/dt as a function of t by substituting $3t - 2$ for x:

$$\frac{dy}{dt} = 6x = 6(3t - 2) = 18t - 12. \quad \square$$

In the previous example it would have been easy to substitute $3t - 2$ for x in the equation that defines y and then find dy/dt directly:

$$y = x^2 + 1 = (3t - 2)^2 + 1 = 9t^2 - 12t + 5,$$

so

$$\frac{dy}{dt} = \frac{d}{dt}(9t^2 - 12t + 5) = 18t - 12.$$

Nevertheless, as we shall see later in this chapter, we often need to find dy/dt without writing y explicitly as a function of t. We then must use the second formulation of the chain rule.

The second formulation of the chain rule may be used with variables other than x, y, and t.

EXAMPLE 9   Let $u = 3v^2 - 2v + 2$ and $v = 2\sqrt{w}$. Then

$$\frac{du}{dw} = \frac{du}{dv}\frac{dv}{dw}$$

$$= \frac{d}{dv}(3v^2 - 2v + 2)\frac{d}{dw}(2\sqrt{w})$$

$$= \frac{6v - 2}{\sqrt{w}}. \quad \square$$

It is important to realize that the chain rule is a statement about rates of change: it says that if y depends on x and x depends on t, then the rate of change of y with respect to t is the product of the rate of change of y with respect to x and the rate of change of x with respect to t.

EXAMPLE 10 Suppose that the demand function for a product is given by

$$q = 100{,}000 - 200p^2$$

and that the price p depends on the time t according to the equation

$$p = 2\sqrt{t} + 100,$$

where t is time in years with $t = 0$ representing the current year. Then the rate of change of quantity demanded with respect to time is dq/dt, and by the chain rule,

$$\frac{dq}{dt} = \frac{dq}{dp}\frac{dp}{dt} = (-400p)\left(\frac{1}{\sqrt{t}}\right) = -\frac{400p}{\sqrt{t}}.$$

Four years from now the price will be

$$p = 2\sqrt{4} + 100 = 104$$

dollars, and thus

$$\left.\frac{dq}{dt}\right|_{t=4} = -\frac{400(104)}{\sqrt{4}} = -20{,}800.$$

Hence four years from now quantity demanded will be decreasing at a rate of 20,800 units per year. $\square$

■ *Computer Manual: PROGRAM POWRPOLY*

3.2 EXERCISES

In Exercises 1 through 32, use the chain rule to find $k'(x)$.

1. $k(x) = (3x + 5)^{10}$
2. $k(x) = (2 - 3x)^{12}$
3. $k(x) = (5 - 7x)^{-4}$
4. $k(x) = (x^2 + 4)^{11}$
5. $k(x) = (x^2 - 2x)^5$
6. $k(x) = (2x^2 + 3x - 1)^{-5/2}$
7. $k(x) = (5x^2 - 3x + 1)^{100}$
8. $k(x) = \sqrt{4x - 1}$
9. $k(x) = \sqrt{1 - 3x}$
10. $k(x) = (x^2 + 7)^{3/2}$
11. $k(x) = \sqrt[3]{x^2 + 1}$
12. $k(x) = \sqrt[5]{x^2 + 3}$
13. $k(x) = (1 - 2x^{-1})^{-3/2}$
14. $k(x) = \sqrt[3]{x^5 - 3x^2}$
15. $k(x) = x^3(5x^2 - 1)^2$

16. $k(x) = (2x + 1)^4(3x - 2)^5$
17. $k(x) = (2x^2 + 1)^{11}(x^3 - 1)^9$
18. $k(x) = 2x^2\sqrt{2x + 3}$
19. $k(x) = x^3\sqrt[3]{4x + 3}$
20. $k(x) = 2x\sqrt[3]{x^2 + 8}$
21. $k(x) = \sqrt{x^2 - 1}\sqrt[3]{x^3 - 1}$
22. $k(x) = [(3 - 2x - x^4)(x^2 - 4)]^7$
23. $k(x) = [(4\sqrt{x} - 3x^{-2})(4x^{4/3} + 10)]^{-3}$
24. $k(x) = \sqrt{(5x + 1)(x^2 + 8)}$
25. $k(x) = \sqrt{(x^2 + 1)(x^2 - 3)}$
26. $k(x) = \left(\dfrac{2x}{x - 3}\right)^5$
27. $k(x) = \left(\dfrac{x + 2}{x - 2}\right)^4$
28. $k(x) = \left(\dfrac{5x - 1}{x + 2}\right)^{-5}$
29. $k(x) = \dfrac{3x + 1}{\sqrt{x^2 + 3}}$
30. $k(x) = \sqrt{\dfrac{x^2 + 2}{x^2 - 1}}$
31. $k(x) = \left(\dfrac{x^{-1} + \sqrt{x}}{x^{-1} - \sqrt{x}}\right)^3$
32. $k(x) = \dfrac{(x^2 + 2x)^5}{(x^3 - 5x^2)^{12}}$

33. Use the second formulation of the chain rule to find $\dfrac{dy}{dt}$ if $y = x^2$ and $x = 4 - 2t$.

   Check your answer by expressing y as a function of t and finding $\dfrac{dy}{dt}$ directly.

34. Repeat Exercise 33 if $y = 5x + 3$ and $x = t^2 + 4$.
35. Repeat Exercise 33 if $y = 2x^2 - 3x + 1$ and $x = 5t + 1$.
36. Use the second formulation of the chain rule to find $\dfrac{du}{dv}$ if $u = \sqrt{x + 1}$ and $x = v^2 - 1$. Check your answer by expressing u as a function of v and finding $\dfrac{du}{dv}$ directly.
37. Repeat Exercise 36 if $u = 3x^3 + 1$ and $x = \sqrt{1 - v}$.
38. Use the second formulation of the chain rule to find $\dfrac{dy}{dt}$ if

   (a) $y = \sqrt{x^2 + 1}$ and $x = t^2$
   (b) $y = x^2$ and $x = t^2 + 1$

39. Use the second formulation of the chain rule to find $\dfrac{dy}{dt}$ if

   (a) $y = \dfrac{1}{x}$ and $x = 4t - 5$
   (b) $y = (x^3 + 3x - 1)^5$, $x = u^2$, and $u = 5t$

40. Use the second formulation of the chain rule to find $\dfrac{dz}{dw}$ if

   (a) $z = 3\sqrt{v}$, $v = \dfrac{1}{q}$, and $q = w^2 + 1$
   (b) $z = \dfrac{p - 1}{2 - p}$, $p = 5s + 1$, and $s = w^{-3}$

41. Find $\dfrac{dy}{dt}$ if $y = 3x^2 + 5x - 7$ and $x = 2t + 11$.

42. Find $\dfrac{dx}{ds}$ if $x = \dfrac{t}{8} + 4$ and $t = 4s^2 - 12$.

43. Find $\dfrac{dp}{dr}$ if $p = \sqrt{5q^2 + 2}$ and $q = \dfrac{8}{r}$.

44. Find $\dfrac{dy}{dv}$ if $y = 2(x - 1)^{-1}$, $x = 3t + 5$, and $t = v^2 + 8$.

Management

45. A firm's revenue function is given by
$$R(x) = x\left(1 - \frac{x^2}{100}\right)^2$$
and its cost function by
$$C(x) = \frac{x^2}{x+1} + 1,$$
both valid for $x \geq 1$. Find the firm's marginal revenue and marginal profit when $x = 1$, $x = 10$, and $x = 20$, and interpret your results.

46. Suppose a firm's cost and revenue functions are given by
$$C(x) = \sqrt{x + 100} \quad \text{and} \quad R(x) = \sqrt{x^2 + 2},$$
respectively, for $x \geq 1$. Find the firm's marginal cost, marginal revenue, and marginal profit when $x = 10$, and interpret your results.

47. A firm's cost and revenue functions are given by
$$C(x) = (x + 0.25)^{1/2} \quad \text{and} \quad R(x) = (x^2 + 32)^{1/2},$$
respectively, where x is in thousands of units and cost and revenue are in millions of dollars. Find the firm's maximum profit and the number of units required to attain it.

48. A firm's daily profit is P dollars, where $P(x) = 1.2x - 10{,}000$ and production x is a function of the number of employees n according to the equation $x = 125n$. Find the rate of change of profit with respect to the number of employees when the number of employees is 100.

49. The supply function for a commodity is given by $q = \sqrt{100p - 1000}$, $p \geq 10$. Find the rate of change of quantity supplied with respect to price when $p = \$20$, \$30, and \$50.

50. Suppose the demand function for a product is given by $q = 10{,}000/p$ and that the price has been increasing according to the equation $p = 0.02t + 5$, where t is in months, with $t = 0$ representing the current month. Find the rate of change of quantity demanded with respect to time six months from now.

Life Science

51. A rare species of seal was placed on the endangered species list in 1984. It is estimated that the size of the population of this species in year t is given by
$$y = \sqrt{\frac{8t + 2}{2t + 9}},$$
where $t = 0$ represents 1984 and y is in thousands. Find the rate at which the population was changing in 1986 and 1990. Also find and interpret $\lim_{t \to +\infty} y$.

52. The number of bacteria in a culture t hours after the start of an experiment is y thousand, where
$$y = \left(\frac{t+2}{5t+3}\right)^{3/4}.$$
Find the rate of change of the bacteria population when $t = 0$ and when $t = 1$.

53. The number of bacteria in a culture t hours after the start of an experiment is y, where $y = 1000\sqrt{300t^2 + 10{,}000}$. Find the rate of change of the bacteria population when $t = 1, 2,$ and 10.

54. It is estimated that t years from now y percent of the population of a country will be afflicted with the disease bilharzia, where $y = 100(0.3t + 4)^{-1/2}$. Find both the incidence of the disease and the rate at which it will be declining (a) now (b) 40 years from now.

55. It is estimated that the percentage of ozone remaining in the ozone layer when a total of x tons of chlorofluorocarbons (CFCs) have been released into the atmosphere is

$$f(x) = \frac{100}{0.16x + 1}.$$

It is also estimated that the total amount of CFCs released into the atmosphere is $x = 12t + 22$, where t is in years with $t = 0$ representing 1980. Find the rate at which the ozone concentration was changing in 1988.

56. The carbon monoxide concentration in the air of a large city depends on the number of cars driven in the city each day. The number of cars driven in the city is expected to increase with time. Suppose that if x cars are driven in the city each day, then the concentration of carbon monoxide in the atmosphere is

$$c(x) = 10^{-9}x^2 + 10^{-5}x + 1$$

parts per million (ppm); suppose also that the number of cars driven in the city per day is forecast to be

$$x = 10{,}000t + 100{,}000$$

in year t, where $t = 0$ represents 1987. Forecast the rate of change of the CO concentration in 1995.

Social Science

57. Suppose the proportion of eligible voters who participated in presidential elections in the United States since World War II is y, where

$$y = \frac{1}{\sqrt{0.5t + 2}}$$

and $t = 0$ represents 1948, $t = 1$ represents 1952, and so on. Find the rate at which voter participation was changing in the election of 1980 and in the election of 1984.

Marginal Revenue Product

If $p = f(q)$ is the demand function for a firm's product, and output q is a function of the number of employees n, then revenue R is a function of n. The derivative of R with respect to n is called the **marginal revenue product**. Exercises 58 through 60 refer to the marginal revenue product.

*58. Show that the marginal revenue product $\dfrac{dR}{dn}$ is given by

$$\frac{dR}{dn} = \left(p + q\frac{dp}{dq}\right)\frac{dq}{dn}.$$

*59. Find and interpret the marginal revenue product when $p = 100 - 0.1q$, $q = 10n - 0.02n^2$, and
 (a) $n = 10$ (b) $n = 90$.

*60. If p and q are as in Exercise 59, set $\dfrac{dR}{dn} = 0$ and solve for n, and then examine the sign of the marginal revenue product on either side of your solution. What does this tell you about the relationship between the firm's number of employees and its revenue?

3.3 DIFFERENTIALS

In Section 2.7 we discussed the use of differential notation dy/dx for the derivative and stated that for the time being we would regard this expression not as a ratio but merely as another symbol for the derivative. Now it is time to demonstrate how the symbol dy/dx can indeed be regarded as the ratio of the **differentials** dy and dx. This view of the derivative as a ratio will be extremely useful to us in later chapters when we discuss the process of integration. We also show in this section how differentials can be used to approximate the change in a function's value brought about by a small change in its independent variable.

The Differentials dy and dx

Let f be a function and let $(x, f(x))$ be a point on its graph. Let Δx be a number added to x. (The symbol Δ is the Greek letter delta, and Δx is read "delta x.") The number Δx is called an **increment** and is referred to as the **change in x**; Δx may be positive or negative. Figure 3–1 depicts the situation if Δx is positive.

Now suppose we set $dx = \Delta x$ and dy equal to the vertical distance from the

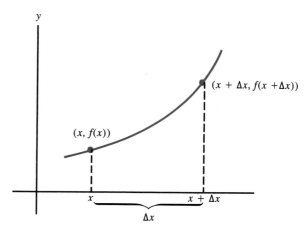

FIGURE 3–1

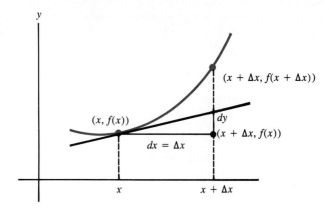

FIGURE 3–2

point $(x + \Delta x, f(x))$ to the tangent line to f at $(x, f(x))$. The number dx is the **differential of x,** and dy is the **differential of y.** See Figure 3–2. Since dy/dx is the slope of the tangent line, and since this slope is the derivative $f'(x)$, we have

$$\frac{dy}{dx} = f'(x).$$

This shows how the derivative may be regarded as a ratio of the differentials dy and dx.

Now that we can regard dy and dx as individual quantities, we can solve the equation

$$\frac{dy}{dx} = f'(x)$$

for dy to obtain the differential formula:

The Differential Formula

If f is differentiable at x, then

$$dy = f'(x)\,dx.$$

EXAMPLE 1   If $f(x) = x^2 - 5x + 3$, then $dy = (2x - 5)\,dx$. □

Approximation by Differentials

As before, we let Δx be an increment added to x. The point $(x + \Delta x, f(x + \Delta x))$ is on the graph of f, and the quantity

$$\Delta f = f(x + \Delta x) - f(x)$$

is the **change in f** brought about by the change Δx in x. See Figure 3–3.

More about Differentiation

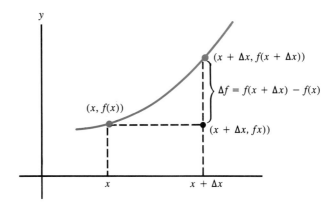

FIGURE 3-3

EXAMPLE 2 Let $f(x) = x^2$, $x = 1$, and take $\Delta x = 0.1$. Then

$$x + \Delta x = 1.1,$$

and thus

$$\Delta f = f(1.1) - f(1) = (1.1)^2 - 1^2 = 0.21.$$

See Figure 3-4. □

What happens if the increment Δx is allowed to approach zero? Figure 3-5 shows that as Δx approaches 0, the point

$$(x + \Delta x, f(x + \Delta x)) = (x + \Delta x, f(x) + \Delta f)$$

"slides" down the graph toward the point (x, y) and hence Δf approaches dy. Therefore for values of Δx that are close to (but not equal to) zero, we have

$$\Delta f \approx dy.$$

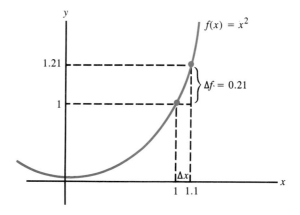

FIGURE 3-4

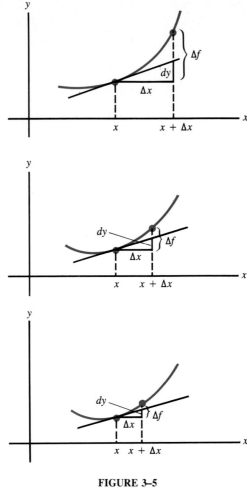

FIGURE 3-5
As $\Delta x \to 0$, $\Delta f \to dy$

Approximation by Differentials

If f is differentiable at x and Δx is close to but not equal to zero, then

$$\Delta f \approx f'(x)\,\Delta x.$$

Approximation by differentials is often used to produce formulas for the approximate amount of change in the dependent variable as a function of the amount of change in the independent variable. We illustrate with examples.

EXAMPLE 3   If $f(x) = x^2$, then

$$\Delta f \approx 2x\,\Delta x$$

when Δx is close to zero. Thus, for instance, if x is changed from $x = 1$ to $x = 1.1$, so that $\Delta x = 0.1$, then the change Δf in f will be

$$\Delta f \approx 2(1)(0.1) = 0.2.$$

Compare this result with that of Example 2, where we found the *exact* change in f as x went from 1 to 1.1 to be 0.21. □

EXAMPLE 4   Suppose the supply function for a commodity is $p = 3q^{2/3}$. The change in price Δp brought about by a small change Δq in quantity supplied is given by

$$\Delta p \approx 2q^{-1/3} \Delta q$$

or

$$\Delta p \approx \frac{2}{\sqrt[3]{q}} \Delta q.$$

Thus, for instance, if quantity supplied q changes from 1000 units to 990 units, Δq will be -10, and the corresponding change in price will be

$$\Delta p \approx \frac{2}{\sqrt[3]{1000}}(-10) = -\$2. \quad \square$$

EXAMPLE 5   Suppose the cost of making x units of a product is $C(x)$ dollars. If production is increased from x units to $x + 1$ units, then $\Delta x = 1$ and hence the change ΔC in cost is given by

$$\Delta C \approx C'(x) \Delta x = C'(x) \cdot 1 = C'(x).$$

In other words, the additional cost of making one more unit is approximately equal to the marginal cost, a fact we stated without proof in Section 2.6. □

As Figure 3–3 shows, when x changes by an amount Δx, the functional value $f(x)$ changes by an amount Δf. The **proportional change in $f(x)$** is thus defined to be the ratio

$$\frac{\Delta f}{f(x)}.$$

Since $\Delta f \approx f'(x) \Delta x$ when Δx is close to zero, we have

> **The Proportional Change in $f(x)$**
>
> If f is differentiable at x and Δx is close to but not equal to zero, then
>
> $$\frac{\Delta f}{f(x)} \approx \frac{f'(x) \Delta x}{f(x)}.$$

If we multiply the proportional change in $f(x)$ by 100, we obtain the percentage change in $f(x)$.

EXAMPLE 6 If one side of a square plot of land is measured with a possible measurement error of Δx units, estimate the proportional and percentage errors that might occur in the value for the area of the plot.

If the side of the plot is x units, its area is $A(x) = x^2$ square units. Hence, the proportional change in area that could be introduced by a change of Δx in the length of a side is

$$\frac{\Delta A}{A(x)} \approx \frac{2x\,\Delta x}{x^2},$$

or

$$\frac{\Delta A}{A(x)} \approx \frac{2\,\Delta x}{x}.$$

The percentage error is 100 times the proportional error. For instance, if the side were measured to be 10 meters long, with a possible measurement error of ± 0.1 meter, then the proportional error in the area would be

$$\frac{\Delta A}{A(10)} \approx \frac{2(\pm 0.1)}{10} = \pm 0.02.$$

The percentage error would thus be approximately $\pm 2\%$. □

3.3 EXERCISES

The Differentials dy and dx

In Exercises 1 through 8, find dy.

1. $f(x) = 3x^2 - 2x + 5$
2. $y = -5x^3 + 4x^2 - 2x + 11$
3. $y = (x^3 - 5x + 2)^8$
4. $f(x) = \sqrt{x^2 - 2x + 5}$
5. $y = (x^2 + 2x)(x^3 + 4)$
6. $y = (3x^2 - 2x + 1)(x^2 + 4)^4$
7. $f(x) = \dfrac{x^2 - 1}{x^3 + 2x}$
8. $y = \sqrt{\dfrac{3x + 2}{5x - 1}}$

Approximation by Differentials

In Exercises 9 through 16, find an expression for the approximate change in f brought about by a change of Δx in x. Then use this expression to find the approximate change in f brought about by the stated change in x.

9. $f(x) = x^3$, x changes from 1 to 1.2
10. $f(x) = x^3$, x changes from 1 to 1.02
11. $f(x) = \dfrac{1}{x}$, x changes from 10 to 10.5
12. $f(x) = \dfrac{1}{x}$, x changes from 100 to 100.5
13. $f(x) = \sqrt{x}$, x changes from 4 to 3.9
14. $f(x) = \sqrt{x}$, x changes from 100 to 99
15. $f(x) = \sqrt[3]{x}$, x changes from 64 to 63.7
16. $f(x) = x^{2/3}$, x changes from 27 to 27.1

*17. Show that when t is near 0, $\sqrt{1+t} \approx 1 + \frac{1}{2}t$. (*Hint:* Let $f(x) = \sqrt{x}$, $x = 1$, $\Delta x = t$.)

*18. Show that when t is near 0, $(1 + t)^r \approx 1 + rt$.

In Exercises 19 through 24, find an expression for the approximate proportional change in f brought about by a change of Δx in x. Then use this expression to find the approximate proportional and percentage changes in f brought about by the stated change in x.

19. $f(x) = x$, x changes from 2 to 2.1

20. $f(x) = x^2$, x changes from 10 to 9.5

21. $f(x) = x^3$, x changes from 4.2 to 4.1

22. $f(x) = x^4$, x changes from 5 to 5.02

23. $f(x) = \sqrt{x}$, x changes from 4 to 4.1

24. $f(x) = \frac{1}{x}$, x changes from 100 to 98.5

25. The side of a square is measured with a possible measurement error of Δx units.
  (a) Find an expression for the approximate proportional error in the perimeter of the square due to measurement error.
  (b) Use the expression of part (a) to find the approximate proportional and percentage errors in the perimeter of the square due to measurement error if its side measures 10 cm with a possible error of ± 0.1 cm.

26. The radius of a circle is measured with a possible measurement error of Δx units.
  (a) Find an expression for the approximate proportional error in the circumference of the circle due to measurement error.
  (b) Use the expression of part (a) to find the approximate proportional and percentage errors in the circumference of the circle due to measurement error if its radius measures 10 cm with a possible error of ± 0.1 cm.

27. Repeat Exercise 26, substituting the word "area" for "circumference."

28. The edge of a cube is measured to be 4 feet with a possible measurement error of ± 0.5 inch. Find the approximate proportional and percentage errors in
  (a) the surface area   (b) the volume
of the cube due to measurement error.

29. The radius of a sphere is measured to be 12 meters with a possible measurement error of ± 2 millimeters. Find the approximate proportional and percentage errors in
  (a) the surface area $(4\pi r^2)$ (b) the volume $\left(\frac{4\pi r^3}{3}\right)$
of the sphere due to measurement error.

30. Find the approximate proportional and percentage changes in the area of a square if the length of each of its sides is increased by 1%.

31. Find the approximate proportional and percentage changes in the volume of a cube if the length of each of its sides is increased by 1%.

32. Find the approximate proportional and percentage changes in the area of a circle if the length of its radius is decreased by 2%.

33. Find the approximate proportional and percentage changes in the volume of a sphere if the length of its radius is decreased by 1%.

*34. How accurately must the side of a square be measured in order that
  (a) its area   (b) its perimeter
be determined to within $\pm 5\%$ of the true value? (*Hint:* For (a) you want $-0.05\, A(x) \leq \Delta A(x) \leq 0.05\, A(x)$.)

*35. How accurately must the radius of a circle be measured in order that
(a) its area   (b) its circumference
be determined to within ±2% of the true value?

*36. How accurately must the edge of a cube be measured in order that
(a) its volume   (b) its surface area
be determined to within ±1% of the true value?

*37. If the earth is a sphere, how accurately must its radius be measured if
(a) its surface area   (b) its volume
is to be determined to within ±0.1%?

Management

38. The supply function for a product is given by
$$p = 0.02\sqrt{q}$$
   (a) Find an expression for the approximate change in price brought about by a small change in quantity supplied.
   (b) Use the expression of part (a) to find the approximate change in price when quantity supplied changes from 400 units to 405 units.

39. The demand function for a product is given by $p = 5000q^{-1/2}$.
   (a) Find an expression for the approximate change in price brought about by a small change in quantity demanded.
   (b) Use the expression of part (a) to find the approximate change in price when quantity demanded changes from 3600 units to 3580 units.

40. The supply function for a commodity is given by $q = 200(p - 1)^{4/3}$.
   (a) Find an expression for the approximate change in quantity supplied brought about by a small change in price.
   (b) Use the expression of part (a) to find the approximate change in quantity supplied when the price changes from $9 per unit to $9.50 per unit.

41. A firm's weekly output of goods is worth $y = 20x^{1/5}$ thousand dollars, where x is the amount (in thousands of dollars) that the firm spends on labor during the week. If the firm is currently spending $32,000 per week on labor, find the approximate change in the weekly output if new hires increase labor costs to $35,000 per week.

42. With an input of x thousand dollars of capital, a firm can produce y units of product, where $y = 200(x + 10)^{4/5}$. If the firm's capital input is currently $22 thousand per week, find the approximate change in output if capital input is decreased to $20 thousand per week.

43. A firm's profit function is given by $P(x) = -x^2 + 80x - 200$.
   (a) Find an expression for the approximate proportional change in profit brought about by a small change in the number of units made and sold.
   (b) Use the expression of part (a) to find the approximate proportional and percentage change in profit if the number of units made and sold changes from 20 to 21.

44. A company's cost function is given by $C(x) = 0.5x^2 + 5x + 2000$.
   (a) Find an expression for the approximate proportional change in the company's cost brought about by a change in the number of units made.

(b) Use the expression of part (a) to find the approximate proportional and percentage change in cost if the number of units sold changes from 25 to 27.

45. A firm's revenue function is given by $R(x) = x^2 + 4x$. Find the approximate proportional and percentage change in revenue if sales fall from 100 units to 96 units.

Life Science

46. The number of bacteria in a culture t hours after an experiment has begun is
$f(t) = 200t^{3/2} + 4000$.
  (a) Find an expression for the approximate change in the number of bacteria during a small period of time.
  (b) Use the expression of part (a) to find the approximate change in the number of bacteria during the period from $t = 4$ hours to $t = 4$ hours and 5 minutes.

47. The number of fish remaining in a lake t months after it was stocked is
$y = 20(4000 - 120t)^{1/2}$.
  (a) Find an expression for the approximate change in the number of fish in the lake during a small period of time.
  (b) Use the expression of part (a) to find the approximate change in the number of fish during the period from $t = 30$ to $t = 32$.

48. The concentration of pollutants in a reservoir as of year t is given by
$$y = 10 + 2t - 0.5\sqrt{t + 3}.$$
Here y is in parts per million and $t = 0$ corresponds to January 1, 1989. Find the approximate change in the concentration of pollutants during the first three months of 1990.

49. A patient is injected with a drug, and t minutes later the concentration of the drug in the patient's bloodstream is y milligrams per cubic centimeter, where $y = 10(1 + 2t)^{-1/2}$. Find the approximate change in the concentration of the drug from time $t = 0$ to time $t = 1$.

50. An artery 100 millimeters (mm) long has a circular cross section. As plaques build up in the artery, its radius decreases. Suppose that the radius decreases from its original value of 2.0 mm to 1.75 mm. Find the proportional and percentage change in
  (a) the cross-sectional area of the artery;
  (b) the volume of the artery. (*Hint:* The volume of a circular cylinder of radius r and height h is $\pi r^2 h$.)

Social Science

51. A rumor is started and t days later, $t \geq 0$, $y = 200\sqrt{t} + 1$ people have heard it.
  (a) Find an expression for the approximate change in the number of people who hear the rumor during a short period of time.
  (b) Use the expression of part (a) to find the approximate number of people who hear the rumor during the tenth day after it was started.

52. If a person learns a foreign language but does not use it, it is forgotten. Suppose that t years after learning a language, a person who has never used it recalls $f(t)\%$ of it, where $f(t) = 100(t + 1)^{-3/4}$. Find the approximate change in the percentage of the language retained in the first year after it is learned.

Sustainable Growth Rates

Exercises 53 through 61 were suggested by material in Higgins, *Analysis for Financial Management,* Irwin, 1984. They are concerned with the sustainable growth rate of a corporation: if S is the annual sales of a corporation and ΔS is its change in annual sales, then its **sustainable growth rate** is the relative change in sales $\Delta S/S$. Sales growth requires new assets, which a corporation can finance by increasing its debt (i.e., by borrowing), by retaining its profits, or by selling new stock. Suppose we let

P = earnings/sales (= profit margin).
D = dividends paid/earnings (= dividend payout ratio).
T = asset value/sales (= asset-to-sales ratio).
L = debt/equity (= debt-to-equity ratio).

(Equity is the total ownership value of the corporation; it is equal to the corporation's assets minus its liabilities and can be increased by selling stock or by retaining profits.) In what follows we will assume that the corporation wants its debt-to-equity ratio to remain constant and that it cannot or will not sell more stock.

*53. Show that over the next year the corporation's retained profits will be equal to $P(S + \Delta S)(1 - D)$. (*Hint:* Over the next year sales will be $S + \Delta S$ and retained profits = earnings minus dividends paid.)

Each dollar of retained profit increases equity by \$1, and since L is to remain constant, this means that the firm's debt must also increase by \$1. Thus over the next year,

$$\text{New debt} = P(S + \Delta S)(1 - D)L.$$

The retained profits and new debt will be used to finance new assets. Hence the change in the value of the corporation's assets over the next year will be

$$\Delta T = \text{Retained profits plus new debt}.$$

*54. Show that $\Delta T = P(S + \Delta S)(1 - D)(1 + L)$.

But a change in sales ΔS requires a proportional change in assets ΔT. Therefore the asset-to-sales ratio will remain constant, so $\Delta T/\Delta S = T$, or $(\Delta S)T = \Delta T$.

*55. Show that

$$\frac{\Delta S}{S} = \frac{P(1 - D)(1 + L)}{T - P(1 - D)(1 + L)}.$$

(*Hint:* Write out $(\Delta S)T = \Delta T$ and solve for ΔS.)

The equation of Exercise 55 is called the **sustainable growth equation.** It gives the relative rate of change in sales that will occur for given values of P, D, T, and L.

*56. Suppose $P = 0.1$, $D = 0.4$, $T = 0.8$, and $L = 0.5$. Find the sustainable growth rate $\Delta S/S$.

*57. Find the sustainable growth rate of a corporation if its sales this year were \$50 million, its earnings were \$10 million, its dividend payout was \$500,000, its debt was \$30 million, its equity was \$50 million, and its assets were worth \$80 million.

*58. How can a firm increase its sustainable growth rate? (Remember: L is to remain fixed.)

*59. Many new firms pay no dividends, find they cannot use their assets more efficiently, and refuse to dilute their ownership by issuing new stock. How must such firms finance their growth? Are there dangers to such firms in trying to grow too fast?

If sales S are a function of time t, so that $S = S(t)$, then

$$\frac{\Delta S}{S} \approx \frac{S'(t)\,\Delta t}{S(t)}.$$

(Why?) Hence we can estimate the sustainable growth rate of a corporation from its sales function.

*60. Suppose a corporation's sales function is given by $S = 20t + 200$, where sales S is in millions of dollars and t is time in years, with $t = 0$ representing the present. Estimate the firm's sustainable growth rate over the next year.

*61. Suppose a corporation's sales function is given by $S = 0.02t^2 + 1.5t + 10$, where S is in sales in millions of dollars and t is time in years with $t = 0$ representing the present. Estimate the firm's sustainable growth rate over the next six months.

3.4 IMPLICIT DIFFERENTIATION

A defining equation of the form $y = f(x)$ is said to define y **explicitly** as a function of x. For instance, the equation $y = x^2 + 3x - 5$ defines y explicitly as a function of x. Sometimes, however, an equation involving x and y will not be solved for y in terms of x; examples are

$$4x + 2y = 12 \qquad \text{and} \qquad x^2 + y^2 = 1.$$

Such equations are said to define y **implicitly** as a function of x. It is sometimes necessary to be able to find the derivative of y with respect to x when y is defined implicitly as a function of x. In this section we show how to use **implicit differentiation** to accomplish this.

Equations that define y implicitly as a function of x can sometimes be solved to yield one or more explicitly defined functions. For example, the equation

$$4x + 2y = 12$$

can be solved for y to obtain the explicitly defined function

$$y = -2x + 6.$$

Similarly, the equation

$$x^2 + y^2 = 1$$

can be solved for y to yield the two explicitly defined functions

$$y = +\sqrt{1 - x^2} \qquad \text{and} \qquad y = -\sqrt{1 - x^2}.$$

However, not all implicitly defined functions can be converted to explicit ones, and even when such conversion is possible, it is often far too tedious to carry out. Therefore we determine the derivatives of implicitly defined functions by the method of **implicit differentiation,** which requires that we apply the rules of differentiation to all terms of the defining equation while keeping in mind the fact that y is actually a function of x. We illustrate with examples.

EXAMPLE 1 Let us find the derivative dy/dx if y is defined implicitly as a function of x by the equation

$$xy + y^2 = 1.$$

To remind ourselves that y is indeed a function of x, we will replace y by $f(x)$ in the equation:

$$xf(x) + [f(x)]^2 = 1.$$

Now we differentiate each term of this equation, using the product rule to differentiate the term $xf(x)$ and the chain rule to differentiate the term $[f(x)]^2$. We have

$$\frac{d}{dx}(xf(x)) + \frac{d}{dx}([f(x)]^2) = \frac{d}{dx}(1).$$

or

$$x\frac{d}{dx}(f(x)) + \frac{d}{dx}(x) \cdot f(x) + 2[f(x)]\frac{d}{dx}(f(x)) = 0,$$

or

$$x\frac{d}{dx}(f(x)) + 1 \cdot f(x) + 2f(x)\frac{d}{dx}(f(x)) = 0,$$

or

$$x\frac{dy}{dx} + y + 2y\frac{dy}{dx} = 0,$$

or

$$(x + 2y)\frac{dy}{dx} = -y.$$

Solving the last equation for dy/dx, we find that

$$\frac{dy}{dx} = -\frac{y}{x + 2y}. \quad \square$$

EXAMPLE 2 Let us find dy/dx if

$$\frac{2x^2}{y} = x + y^3.$$

This time we will not replace y by $f(x)$, but will simply remember as we differentiate that y is defined implicitly as a function of x. It will also be convenient to use prime notation in this example. We have

More about Differentiation

$$\left(\frac{2x^2}{y}\right)' = (x + y^3)'.$$

Now we differentiate the left side of the equation by the quotient rule and the right side by the sum rule to obtain

$$\frac{y(2x^2)' - 2x^2 y'}{y^2} = (x)' + (y^3)',$$

or

$$\frac{4xy - 2x^2 y'}{y^2} = 1 + 3y^2 y',$$

or

$$4xy - 2x^2 y' = y^2 + 3y^4 y',$$

so

$$(3y^4 + 2x^2)y' = 4xy - y^2,$$

and thus

$$\frac{dy}{dx} = y' = \frac{4xy - y^2}{3y^4 + 2x^2}. \quad \square$$

EXAMPLE 3 We find the equation of the line tangent to the circle

$$x^2 + y^2 = 5$$

at the point (1, 2). We have

$$\frac{d}{dx}(x^2) + \frac{d}{dx}(y^2) = \frac{d}{dx}(5),$$

or

$$2x + 2y\frac{dy}{dx} = 0.$$

Hence

$$\frac{dy}{dx} = -\frac{x}{y},$$

so at (1, 2) we have

$$\frac{dy}{dx} = -\frac{1}{2}.$$

Thus the slope of the line tangent to the circle at (1, 2) is $-\frac{1}{2}$, and the equation of the tangent line is therefore

$$y = -\frac{1}{2}x + b.$$

Since $(1, 2)$ is on the line, $b = 5/2$, and the equation is

$$y = -\frac{1}{2}x + \frac{5}{2}. \quad \square$$

If y is defined implicitly as a function of x, and x is a function of a third variable t, then y is a function of t, and we can use implicit differentiation and the chain rule to find dy/dt.

EXAMPLE 4 Suppose

$$xy + y^2 + x = 1$$

and that x is a function of t. Let us find dy/dt.
Differentiating with respect to t, we have

$$\frac{d}{dt}(xy) + \frac{d}{dt}(y^2) + \frac{d}{dt}(x) = \frac{d}{dt}(1)$$

or

$$x\frac{dy}{dt} + y\frac{dx}{dt} + 2y\frac{dy}{dt} + \frac{dx}{dt} = 0$$

or

$$(x + 2y)\frac{dy}{dt} + (1 + y)\frac{dx}{dt} = 0.$$

Thus

$$\frac{dy}{dt} = -\frac{1 + y}{x + 2y} \cdot \frac{dx}{dt}. \quad \square$$

3.4 EXERCISES

In Exercises 1 through 16, use implicit differentiation to find $\dfrac{dy}{dx}$.

1. $y^2 = x$
2. $xy^2 - 2x = 4$
3. $x^2 - y^3 = 2$
4. $x^2 - 2y^2 = 1$
5. $y^2 + 4x^3 - 2y = 0$
6. $x^2y^2 + x = 6$
7. $y^3 = x^2$
8. $x^3y^2 = 4$
9. $\dfrac{5x}{y^2} = 1$

10. $\dfrac{x}{y^3} - x^2 = 0$

11. $\dfrac{x}{y} - \dfrac{y^2}{x} = 1$

12. $(x^2 + y^2)^5 = 100$

13. $(x^2 + 2y)^3 = 1$

14. $\sqrt{xy} + y^2 = 0$

15. $\dfrac{y}{x} - \sqrt{xy} + 2x - 3y = 8$

16. $\sqrt{\dfrac{x}{y}} + 2x + y^{-2} = 6$

In Exercises 17 through 26, find the equation of the line tangent to the graph of the function at the given point.

17. $2x - 5y^2 = 0$, $(10, 2)$
18. $x^2 + y^2 = 10$, $(1, 3)$
19. $x^3 - y^3 = 0$, $(1, 1)$
20. $3xy^2 = 1$, $(3, ⅓)$
21. $2x^3y^2 - 8x = 0$, $(1, -2)$
22. $xy^2 - yx^2 + 6 = 0$, $(-1, -3)$
23. $\dfrac{x}{y^2} + 2y = 3$, $(1, 1)$
24. $\dfrac{y}{x^2} + \dfrac{y^2}{x} = 0$, $(1, -1)$
25. $(5x - 1)^2(y^2 - 1)^2 = 25$, $(1, \tfrac{3}{2})$
26. $(x + 2)^2 + (y - 2)^2 = 4$, $(0, 2)$

In Exercises 27 through 32, x and y are functions of t. Find dy/dt.

27. $x^2 + y^2 = 4$
28. $x^3 + y^3 = 1$
29. $\dfrac{x^2}{2} - \dfrac{y^2}{4} = 1$
30. $xy^2 + yx^2 = 1$
31. $x^2 - y^2x + 2x - 3y = 0$
32. $(xy + 1)^{1/2} = 9$

In Exercises 33 through 38, find $\dfrac{d^2y}{dx^2}$.

*33. $2x + y^2 = 1$
*34. $x^2 - 3y^2 = 0$
*35. $xy^3 - 3x = 5$
*36. $2xy^2 + y^3 = 8$
*37. $2x^2 - \dfrac{y^2}{x} = 2$
*38. $\dfrac{x}{y} - \sqrt{y} = 1$

Management

39. A demand function is given by the equation

$$p = 10{,}000 - 0.05q.$$

(a) Solve the equation for q in terms of p and find $\dfrac{dq}{dp}$.

(b) Use implicit differentiation to find $\dfrac{dq}{dp}$ without solving the equation for q.

40. A demand function is given in the form

$$p = 200 - 0.002q^2.$$

Use implicit differentiation to find $\dfrac{dq}{dp}$ and then find the rate at which demand is changing when
   (a) $q = 200$
   (b) $p = \$195$

41. A demand function is given in the form

$$p = -0.25q^2 - 0.5q + 100.$$

Use implicit differentiation to find $\dfrac{dq}{dp}$ and then find the rate at which demand is changing when
(a) $q = 9$ (b) $p = \$10$

*42. (a) Suppose quantity demanded q is written as a function of price p, say $q = f(p)$. Show that the elasticity of demand formula of Section 2.6 may be written in the form

$$E(p) = \frac{p(dq/dp)}{q}.$$

(b) Suppose a demand function is written in the form $p = g(q)$. Show that the elasticity of demand formula may be written in the form

$$E(q) = \frac{p}{q(dp/dq)}.$$

3.5 RELATED RATES

There is a common type of problem in which information about the rate of change of one variable with respect to another is known and it is required to find the rate of change of the second variable with respect to a third one. Such a problem is known as a **related rates** problem. Some examples of related rates problems are

- Finding the rate at which the quantity demanded of a product is changing if the rate at which its price is changing is known.
- Finding the rate at which the area of a circle is changing if the rate at which its radius is changing is known.
- Finding the rate at which the top of a ladder is sliding down a wall if the rate at which the bottom of the ladder is being pulled away from the base of the wall is known.

Related rates problems are solved with the aid of the chain rule. The key to solving related rates problems is to find an equation that relates the variable whose rate of change we want to the one whose rate of change we know; we can then differentiate this equation with the aid of the chain rule and solve for the unknown rate of change. Sometimes implicit differentiation is helpful. We illustrate with examples.

EXAMPLE 1 A stone is dropped into a pool and the ripples created move outward in a circular pattern from the point where the stone entered the water. The radius of the outermost ripple increases at a rate of 0.2 meter per second. Find the rate at which the area covered by the ripples is increasing when the radius of the outermost ripple is 1 meter.

The radius r of the outermost ripple and hence the area A of the circle covered by the ripples are both functions of time t. See Figure 3–6 for a picture of the situation. We wish to find the rate at which A is changing with respect to time when $r = 1$; that is, we wish to find dA/dt when $r = 1$. Since the area A is the area of a circle of radius r, we have

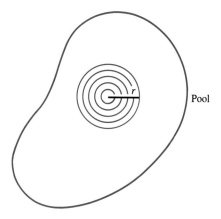

FIGURE 3–6

$$A = \pi r^2.$$

Because we want to find dA/dt, we use the chain rule to differentiate this expression with respect to t and thus obtain an equation involving dA/dt:

$$\frac{dA}{dt} = 2\pi r \frac{dr}{dt}.$$

But the radius r of the outermost ripple is changing at the rate of 0.2 meter per second (mps); that is, $dr/dt = 0.2$ mps. Also, we want dA/dt when $r = 1$. Therefore when $r = 1$,

$$\frac{dA}{dt} = 2\pi(1)(0.2) = 0.4\pi$$

m²ps. ☐

The method we used to solve the related rates problem of Example 1 can be applied to any related rates problem. We summarize it here:

> **Keys to Solving Related Rates Problems**
>
> 1. Assign variables to the quantities that vary with time and draw a picture of the situation if possible.
>
> 2. Identify the desired rate of change as a derivative.
>
> 3. Find a mathematical relationship between the variables (the picture may be helpful here) and differentiate it with respect to time to obtain an equation involving the desired derivative.
>
> 4. The equation of step 3 will contain derivatives and/or variables. Find the values of the derivatives and the values of the variables at the desired time, substitute these values into the equation of step 3, and solve for the desired derivative.

In the examples that follow, notice how we use this method to solve related rates problems.

EXAMPLE 2 Suppose the demand function for a product is given by

$$q = 10{,}000 - 400p.$$

where q is quantity demanded and p is the unit price of the product. Suppose also that the price is given as a function of time by the equation

$$p = 5 + 2\sqrt{t},$$

where t is time in months, with $t = 0$ representing the present. Let us find the rate at which quantity demanded will be changing with respect to time four months from now.

The variables p and q have already been assigned to the quantities that vary with time. No picture is possible here. We want to find the rate of change of quantity demanded q with respect to time t when $t = 4$; that is, we want to find dq/dt when $t = 4$. The relationship between p and q is given as

$$q = 10{,}000 - 400p,$$

so we differentiate it with respect to t to obtain

$$\frac{dq}{dt} = -400\frac{dp}{dt}.$$

Now,

$$p = 5 + 2\sqrt{t},$$

so

$$\frac{dp}{dt} = \frac{1}{\sqrt{t}}.$$

Hence

$$\frac{dq}{dt} = -\frac{400}{\sqrt{t}},$$

and therefore

$$\left.\frac{dq}{dt}\right|_{t=4} = -\frac{400}{\sqrt{4}} = -200.$$

Thus four months from now quantity demanded will be decreasing at the rate of 200 units per month. ☐

EXAMPLE 3 A 10-foot ladder is leaning against a wall. The bottom of the ladder is pulled away from the wall at a rate of 2 feet per second. Find the rate at which the top of the ladder is sliding down the wall when the bottom is 6 feet from the base of the wall.

We begin by sketching a picture of the situation. See Figure 3–7. Here x is

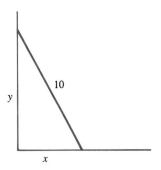

FIGURE 3–7

the distance from the base of the wall to the bottom of the ladder, and y is the distance from the base of the wall to the top of the ladder. Both x and y change with time, so they are functions of time t. We must find dy/dt when x equals 6 feet. We need an equation relating x and y that we can differentiate to find dy/dt. But the triangle in Figure 3–7 is a right triangle, so by the Theorem of Pythagoras we have

$$x^2 + y^2 = 10^2,$$

or

$$x^2 + y^2 = 100.$$

Differentiating this last equation implicitly with respect to t, we find that

$$2x\frac{dx}{dt} + 2y\frac{dy}{dt} = 0,$$

or

$$\frac{dy}{dt} = -\frac{x}{y}\frac{dx}{dt}.$$

We know that x is increasing at the rate of 2 feet per second (fps). That is,

$$\frac{dx}{dt} = 2$$

fps. Substituting 6 for x in the equation $x^2 + y^2 = 100$, we obtain

$$6^2 + y^2 = 100,$$

which yields

$$y = \sqrt{64} = 8.$$

Hence when $x = 6$,

$$\frac{dy}{dt} = -\frac{6}{8}(2) = -1.5$$

fps. ☐

EXAMPLE 4 A person who is 5 feet tall is walking toward a building at a rate of 3 feet per second. The building is 100 feet high and has a light on its top. At what rate is the length of the person's shadow changing?

Figure 3–8 depicts the situation: x is the distance from the building to the person, and s is the length of the person's shadow. Both x and s are functions of time t. We must find ds/dt.

We need to find an equation relating s and x that we can differentiate to obtain ds/dt. From Figure 3–8 we see that the triangles ABE and CDE are similar, so their corresponding sides are in proportion. Thus

$$\frac{100}{x+s} = \frac{5}{s},$$

which yields

$$s = \frac{x}{19}.$$

Therefore

$$\frac{ds}{dt} = \frac{1}{19}\frac{dx}{dt}.$$

But the person is approaching the building at a rate of 3 feet per second; that is, $dx/dt = -3$ fps. (Note that dx/dt is *negative:* this is so because the distance x is *decreasing* as the person walks toward the building.) Hence

$$\frac{ds}{dt} = \frac{1}{19}(-3) = -\frac{3}{19}$$

fps. ☐

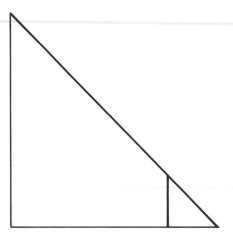

FIGURE 3–8

3.5 EXERCISES

1. If you blow up a balloon so that its radius increases at a rate of 2 centimeters per second (cps), find the rate at which its surface area is changing 4 seconds after you begin to blow it up. (The surface area of a sphere of radius r is $4\pi r^2$.)

2. Referring to Exercise 1, find the rate at which the volume of the balloon is changing 4 seconds after you begin to blow it up. (The volume of a sphere of radius r is $\frac{4}{3}\pi r^3$.)

3. The length of the edge of a cube is increasing at a rate of 2 cps. How fast is the volume of the cube changing at the instant when the edge is 20 cm long?

4. Boyle's Gas Law states that $V = T/P$, where V denotes the volume of a gas in cubic inches, P the pressure in pounds per square inch, and T the temperature in degrees Kelvin. Suppose the temperature is constant at 100°K and the pressure is increasing at 2 pounds per square inch (psi) per second. Find the rate at which the volume of the gas is changing with respect to time when the pressure is 20 psi.

5. Refer to Exercise 4: Suppose P is fixed at atmospheric pressure (14.7 psi) and the temperature is increasing at 2°K per minute. Find the rate of change of volume with respect to time.

6. A person who is 6 feet tall walks toward an 18-foot high street light at a rate of 2 fps. How fast is the length of the person's shadow changing?

7. A person who is 6 feet tall walks away from a 30-foot-high street light at a rate of 5 fps. How fast is the length of the person's shadow changing?

8. Two airplanes take off from the same airport at the same time. One flies due south at 300 mph, the other due west at 400 mph. At what rate is the distance between the planes changing 3 hours after takeoff?

9. Two cars are approaching an intersection, one from due north and the other from due east. The car from the north has a constant speed of 30 mph. When the cars are each 1 mile from the intersection, the distance between them is decreasing at a rate of 40 mph. How fast is the car coming from the east traveling at the instant when both cars are 1 mile from the intersection?

10. A baseball player runs from first base to second base at a rate of 20 fps. How fast is the player's distance from third base changing when the player is 10 feet from second base? (*Hint:* A baseball diamond is a square with a side of 90 feet.)

11. A side street is perpendicular to a straight main highway. A police cruiser is situated on the side street 40 feet from the highway, and its radar is locked onto a car that is traveling along the highway toward the side street. The radar indicates that at the instant when the straight-line distance from the cruiser to the car is 200 feet, the distance between them is decreasing at a rate of 100 fps. How fast is the car traveling along the highway at this instant?

12. A helicopter is ascending straight up at a rate of 10 fps. An observer is 100 feet from the point where the helicopter took off. Find the rate at which the distance between the helicopter and the observer is changing when the helicopter is 50 feet above the ground.

13. A water tank is cylindrical with a circular base of radius 2 feet. If water is drawn from the tank at a rate of 1 gallon per minute, how fast is the water level in the tank changing? (*Hint:* The volume of a circular cylinder is $V = \pi r^2 h$, where r is the radius of its base and h is its height.)

14. A water tank has the shape of a cone 30 meters tall with a circular base whose radius is 3 meters. Water is being drawn out of the tank at a rate of 5 cubic meters per minute. How fast is the water level falling when the water is 20 meters deep? (The volume of a cone of height h whose base is a circle of radius r is $\frac{1}{3}\pi r^2 h$.)

15. An observer on the shore is watching a ship steam through a canal. The ship heads down the middle of the canal, which is one-quarter mile wide, at a rate of 20 mph. Find the rate at which the distance between the ship and the observer is changing when the ship is 1 mile from the observer.

Management

16. A firm's weekly cost function is given by $C(x) = 40x + 50{,}000$, where x is units produced per week and the number of units produced is increasing at the rate of 100 per week. Find the rate of change of cost with respect to time measured in weeks.

17. A company's daily revenue function is given by
$$R(x) = 0.03x^2 + 0.9x,$$
where x is the number of units sold per day. If the number of units sold is decreasing at a rate of 15 per day, find the rate of change of the company's revenue when $x = 100$.

18. The supply function for a product is given by
$$q = p^2 + 10p,$$
where q is quantity supplied and p is unit price. The price as of month t is given by
$$p = 25 - 0.5t,$$
where $t = 0$ represents the present. Find the rate at which quantity supplied will be changing three months from now.

19. The demand function for a commodity is given by
$$q = 100 - 0.02p^2,$$
with q in thousands of units and p in dollars. The price t months from now is forecast to be
$$p = 1.1t^{1/2} + 5$$
dollars. Find the rate at which quantity demanded will be changing with respect to time four months from now.

20. A firm's monthly cost function is given by
$$C(x) = 0.02x^2 + 100x + 20{,}000.$$
If production t months from now is expected to be
$$x = 1000 - 40t$$
units, find the rate at which cost will be changing six months from now.

21. The demand function for a company's product is given by
$$q = 100{,}000 - 5p$$

and the price t years from now is expected to be
$$p = 12 + 0.2t^{3/2}.$$
Find the rate at which the company's revenue will be changing one year from now.

22. A firm plans to build a new manufacturing plant. It is estimated that if the plant produces x units per month, its contribution to the firm's profits will be P dollars, where
$$P = 0.5x^2 - 3x - 50{,}000.$$
It is also estimated that t months after it opens, the plant will produce
$$x = 200\sqrt{t} + 50$$
units per month. Find the rate at which the plant's contribution to profits will be changing 16 months after it opens.

23. A store's monthly sales depend on the number of newspaper ads it runs, according to the equation
$$S = 20\sqrt{x + 100}.$$
Here S is monthly sales in thousands of dollars, and x is the number of ads run during a month. If the store is increasing the number of ads it runs at the rate of five per month, find the rate at which sales are increasing when 96 ads are run during a month.

24. If a firm has access to x thousand dollars of capital it can produce $y = 30(x + 10)^{2/3}$ units of its product. If the firm will have access to $x = 1.5t^{3/4} + 5$ units of capital t weeks from now, find the rate at which its production will be changing with respect to time 16 weeks from now.

Life Science

25. A circular oil slick is created as the result of the blowout of an offshore oil well. The radius of the slick increases at the rate of 400 feet per hour. Find the rate at which the area polluted by the slick is increasing 12 hours after the blowout.

26. The daily amount of smog produced by cars coming into the city is given by
$$y = 2 + 0.01x^{3/2},$$
where y is smog concentration in parts per million and x is the number of cars in thousands. The number of cars coming into the city is increasing at the rate of 250 per day. Find the rate at which the smog concentration will be changing when 225,000 cars are coming into the city.

27. Cases of a new disease are increasing at a rate of 10,000 per year, and the number of deaths y depends on the number of cases x according to the equation $y = 0.1\sqrt{x}$. Find an expression that gives the annual death rate due to the disease.

28. A patient's systolic blood pressure S depends on the concentration C of medication in the patient's blood stream according to the equation $S = 180 - 12(C + 10)^{1/2}$, and the concentration of the medication depends on the time t since the patient began taking the medication according to the equation $C = 20t^{1/2}$. Find the rate at which the patient's blood pressure is changing with respect to time nine days after medication was begun.

29. If a blood vessel has a circular cross section of radius r mm, Poiseuille's Law states that the velocity V of blood that is a mm from the central axis of the blood vessel is

given by $V = k(r^2 - a^2)$, where k is a constant. Suppose a drug that dilates blood vessels causes an artery whose radius is 2 mm to expand at a rate of 0.1 mm per minute. Find the rate at which the velocity of the blood in the artery is changing. Treat a as a constant.

Social Science

30. If the population of a state is x million people, the state will need y thousand social workers in its department of social services, where
$$y = 3\sqrt{x} + 20.$$
Suppose the population of the state is increasing at a rate of 50,000 people per year. Find the annual rate at which the number of social workers needed is changing when the state's population is 5 million people.

SUMMARY

This summary consists of terms and symbols whose meaning you should know, facts you should know, and techniques and methods you should be able to employ.

Section 3.1

The product rule. Using the product rule to find derivatives. The quotient rule. Using the quotient rule to find derivatives.

Section 3.2

The first formulation of the chain rule: if $k(x) = f(g(x))$, then $k'(x) = f'(g(x))g'(x)$ Using the first formulation of the chain rule to find derivatives: differentiating "from the outside in." The general power rule: if $k(x) = [g(x)]^r$, $r \neq 0$, then $k'(x) = r[g(x)]^{r-1}g'(x)$ Using the general power rule to find derivatives. The second formulation of the chain rule: if y is a function of x and x is a function of t, then $\dfrac{dy}{dt} = \dfrac{dy}{dx}\dfrac{dx}{dt}$ Using the second formulation of the chain rule to find derivatives.

Section 3.3

Increment Δx, change in x. The differentials dx, dy. The derivative dy/dx as a ratio of the differentials dy, dx. The differential formula. Using the differential formula to find dy. The change in f: Δf. Approximation by differentials. Using the approximation by differentials formula to find expressions for the approximate change in f. The proportional change in f. The percentage change in f. Approximating the proportional and percentage change in f.

Section 3.4

Explicitly defined function. Implicitly defined function. The technique of implicit differentiation. Using implicit differentiation to find derivatives.

Section 3.5

Related rates problem. Method for solving related rates problems. Solving related rates problems.

REVIEW EXERCISES

In Exercises 1 through 16, find $f'(x)$.

1. $f(x) = (6x^2 - 2x)(x^3 + 4)$
2. $f(x) = \dfrac{3x + 7}{5x - 3}$
3. $f(x) = \dfrac{2x^2 - x}{4x + 6}$
4. $f(x) = (2\sqrt{x} + 3)(x^3 - 1)$
5. $f(x) = \dfrac{(5x + 2)(x^2 + 3)}{2x^2 + 1}$
6. $f(x) = \dfrac{x^2 - 4}{(5x + 2)(x^2 + 1)}$
7. $f(x) = (2x^2 + x)\dfrac{x^3 - 1}{x^2 + 1}$
8. $f(x) = \left(x^2 + \dfrac{x}{x - 1}\right)\left(x^2 - \dfrac{x}{x + 1}\right)$
9. $f(x) = (7x - 9)^{10}$
10. $f(x) = (3x^2 + 5x + 1)^{12}$
11. $f(x) = \sqrt{6x^3 - 2x + 3}$
12. $f(x) = \sqrt{\dfrac{5x + 2}{x^2 - 2}}$
13. $f(x) = (2x + 3)^6(x^2 - 5x)^8$
14. $f(x) = [(2x - 2)^4(x^2 + 1)]^5$
15. $f(x) = \left(\dfrac{3x - 1}{x^2 + 2}\right)^7$
16. $f(x) = \dfrac{(3x + 4)^5}{(5x^2 - 1)^4}$

17. If $y = 6x^2 - 2x + 1$ and $x = 4t^2 - 9$, find $\dfrac{dy}{dt}$.

18. If $y = x^2 - 5x + 2$ and $x = 8t - 3$, find $\dfrac{dy}{dt}$.

19. If $v = 7w - 3\sqrt{w}$ and $u = (v^4 - 2v + 11)^{11}$, find $\dfrac{du}{dw}$.

20. The supply function for a commodity is given by
$$q = 5000(0.5p^2 - 2\sqrt{p})^{3/4}.$$
Find the rate of change of quantity supplied q when $p = \$16$.

21. A demand function is given by $p = 80{,}000(10q)^{-1/2}$.
    (a) Find an expression for the approximate change in price brought about by a change of Δq in quantity demanded.
    (b) Find the approximate change in price brought about by a change in quantity demanded from 1000 units to 1005 units.

22. For the demand function of Exercise 21,
    (a) Find an expression for the approximate proportional change in price brought about by a change of Δq in quantity demanded.
    (b) Find the approximate proportional and percentage change in price if quantity demanded changes from 250 units to 248 units.

23. A conical water tank whose base is a circle with a radius of 20 feet has its height measured as 50 feet with an error of ± 3 inches. Find the approximate proportional and percentage errors in its volume $\left(\dfrac{\pi r^2 h}{3}\right)$ due to measurement error.

24. How accurately must the radius of the base of a conical water tank whose height is 50 feet be measured if its volume is to be determined to within $\pm 1\%$?

In Exercises 25 through 28, find $\dfrac{dy}{dx}$.

25. $8y^2 - 2x = 5$
26. $xy^3 - x^2y = 10$
27. $x^2 + y^2 + \dfrac{x}{y} = 1$
28. $x^2 + y^2 - \dfrac{y}{x} = 1$

In Exercises 29 and 30, find the equation of the line tangent to the graph of the function at the given point.

29. $-x^2 + y^2 + 3x = 9$, $(0, 3)$
30. $x^2 + 3\dfrac{x}{y^2} = 8$, $(-3, 3)$

31. Find $\dfrac{d^2y}{dx^2}$ if $2x^2y^2 = 4$.
32. Find $\dfrac{d^2y}{dx^2}$ if $\dfrac{2x}{\sqrt{y}} - x^2 = 0$.

33. If x and y functions of u and $3xy^2 + \dfrac{2y}{x} = 1$, find $\dfrac{dy}{du}$.

34. If price p and quantity demanded q for a certain commodity are related by the equation
$$q = -p^2 - 50p + 5000$$
and the price is increasing at a rate of $0.50 per month, find the rate at which quantity demanded is changing with respect to time when the price is $25.

35. If one person drives due north from point A at 55 mph and another person leaves point A two hours after the first and drives due east at 50 mph, find the rate at which the distance separating them is changing 4 hours after the second person left.

ADDITIONAL TOPICS

Here are some additional topics not covered in the text that you might wish to investigate on your own.

1. In Section 3.2 we proved the chain rule under the assumption that $g(x + h) - g(x) \neq 0$. Show how the chain rule is proved when this is not assumed to be the case.

2. There is one general rule of differentiation that we have not presented. It involves inverse functions: Recall that if f is a one-to-one function, then f^{-1} is the function defined by $f^{-1}(y) = x$ if and only if $f(x) = y$. (See Section 1.3) There is a relationship between the derivatives df/dx and df^{-1}/dx; this relationship is known as the **inverse function rule**. Find out about the inverse function rule, how it is proved, and how it is used.

3. In Section 2.4 we introduced the power rule
$$\dfrac{d}{dx}(x^r) = rx^{r-1}, \quad r \neq 0,$$
and outlined a proof of the rule in the case where r is a positive integer. Now prove the power rule when r is a negative integer.

4. Prove the power rule when r is a rational number p/q.
5. A pair of equations of the form $x = f(t)$, $y = g(t)$ are called **parametric equations** for x and y with **parameter** t. Find out how parametric equations are used to describe curves. Given parametric equations for x and y, can you find expressions for dy/dx and dx/dy?

SUPPLEMENT: CONSUMPTION AND THE MULTIPLIER

It is not uncommon to find news items such as this: "Today XYZ Corporation announced that it will build a new plant in Smallville. The plant will provide employment for 250 workers at an operating cost of some $1 million per year. It is estimated that the presence of the plant will add $4 million per year to the Smallville economy."

How was the figure of $4 million per year determined? It must have taken into account the fact that some of the income generated by an investment (such as the $1 million per year that XYZ is investing in Smallville) will be saved and invested by those who receive it; this new round of investment will in turn produce more income, some of which will again be saved and invested to produce still more income, and so on. This "chain reaction" of income generation is called the **multiplier effect**. Let us see if we can quantify the multiplier effect.

Evidently the strength of the multiplier effect depends on the proportion of income saved and invested at each stage of the process. Therefore we need an expression that relates income to savings and investment. For simplicity we assume that all income that is saved is invested, so savings equals investment. Thus we need a relationship between income and savings.

Economists divide income into the portion that is saved and the portion that is not saved; the latter is called **consumption**. Thus,

$$\text{Income} = \text{Savings} + \text{Consumption},$$

or

$$\text{Consumption} = \text{Income} - \text{Savings}.$$

If we denote income by Y, consumption by C, and savings by S, then consumption is given as a function of income by a **consumption function**

$$C = Y - S.$$

EXAMPLE 1   Let

$$C = 5 + 8\sqrt{Y}$$

be a consumption function, with Y and C in millions of dollars. Thus when $Y = 100$ million dollars, consumption is

$$C(100) = 5 + 8\sqrt{100} = 85$$

million dollars. The remaining $100 - 85 = 15$ million dollars of income is saved. □

CHAPTER 3

The derivative of C with respect to Y is the rate of change of consumption with respect to income, which is known as the **marginal propensity to consume.** The marginal propensity to consume is the proportion of an additional $1 of income that would be spent rather than saved.

EXAMPLE 2 Let
$$C = 5 + 8\sqrt{Y}$$
as in Example 1. Then the marginal propensity to consume is
$$\frac{dC}{dY} = \frac{4}{\sqrt{Y}}$$

For instance, when $Y = 100$, we have
$$\text{Marginal propensity to consume} = \frac{4}{\sqrt{100}} = 0.40.$$

Hence at an income level of $100 million, $0.40 of each additional dollar of income would be spent, and the remaining $0.60 would be saved. Note that the marginal propensity to consume $dC/dY = 4/\sqrt{Y}$ is a decreasing function of Y. Therefore in this example, as income increases a smaller proportion of it is spent and a larger proportion is saved. ☐

The equation $C = Y - S$ shows that savings S is a function of income Y also. As we remarked previously, the strength of the multiplier effect depends on the proportion of income invested at each stage of the income-generation process. Thus, since we are assuming that savings equals investment, we can say that the multiplier effect is measured by the rate at which *income* changes with respect to *savings*, or dY/dS. The derivative dY/dS is called the **multiplier.**

From $C = Y - S$ we have
$$\frac{dC}{dS} = \frac{dY}{dS} - \frac{dS}{dS}.$$

Applying the chain rule to the left side gives
$$\frac{dC}{dY}\frac{dY}{dS} = \frac{dY}{dS} - 1.$$

Solving this last equation for dY/dS, we obtain
$$\frac{dY}{dS} = \frac{1}{1 - dC/dY} = \frac{1}{1 - \text{marginal propensity to consume}}.$$

EXAMPLE 3 Again let $C = 5 + 8\sqrt{Y}$. Then the multiplier is
$$\frac{dY}{dS} = \frac{1}{1 - 4/\sqrt{Y}} = \frac{\sqrt{Y}}{\sqrt{Y} - 4}.$$

Thus, at an income level of $100 million,
$$\frac{dY}{dS} = \frac{\sqrt{100}}{\sqrt{100} - 4} = \frac{5}{3}.$$
Hence each additional $1 of savings generates an additional $1.67 in income. □

EXAMPLE 4 Suppose the consumption function for Smallville is
$$C = 20 + 0.75Y,$$
where Y and C are in millions of dollars. The multiplier is then
$$\frac{1}{1 - dC/dY} = \frac{1}{1 - 0.75} = 4,$$
so if XYZ Corporation invests $1 million per year in Smallville it will generate approximately $4 million in additional income for the town. □

The multiplier effect is what is behind the idea of "pump priming" in the economy: if there are many opportunities for investment, the multiplier is large; therefore a relatively small investment (the "priming of the pump") can lead to a large increase in income. Unfortunately, the reverse is also true: a relatively small decrease in investment can lead to a large decrease in income.

SUPPLEMENTARY EXERCISES

1. Let $C = 20 + 0.6Y$ be a consumption function, with Y and C in millions of dollars, $Y \geq 0$. Find and interpret the marginal propensity to consume when the income level is $10 million; when it is $100 million.

2. Let $C = 350 + 0.32Y$ be a consumption function, with Y and C in billions of dollars, $Y \geq 0$. Find and interpret the marginal propensity to consume at every income level.

3. Let $C = 200 + 20\sqrt{Y}$ be a consumption function, with Y and C in billions of dollars, $Y \geq 100$. Find and interpret the marginal propensity to consume when the income level is
   (a) $400 billion   (b) $900 billion

4. Let $C = 50 + 10\sqrt{Y - 10}$ be a consumption function, with Y and C in billions of dollars, $Y \geq 35$. Find and interpret the marginal propensity to consume when the income level is
   (a) $410 billion   (b) $2510 billion

5. Let $C = 245 + 75\sqrt[3]{Y}$ be a consumption function, with Y and C in millions of dollars, $Y \geq 125$. Find and interpret the marginal propensity to consume when the income level is
   (a) $125 million   (b) $1000 million   (c) $8000 million

6. Find and interpret the multiplier for the consumption function of Exercise 1.

7. Find and interpret the multiplier for the consumption function of Exercise 2.
8. Find and interpret the multiplier for the consumption function of Exercise 3 when the income level is
   (a) $400 billion
   (b) $900 billion
9. Find and interpret the multiplier for the consumption function of Exercise 4 when the income level is
   (a) $410 billion
   (b) $2510 billion
10. Find and interpret the multiplier for the consumption function of Exercise 5 when the income level is
    (a) $8000 million
    (b) $1000 million
    (c) $125 million

Since $C = Y - S$, we have $S = Y - C$. The derivative dS/dY is called the **marginal propensity to save.** In Exercises 11 through 15 find and interpret the marginal propensity to save for the given consumption function at the given level of income.

11. The consumption function of Exercise 1, at income levels of $10 million and $100 million.
12. The consumption function of Exercise 2, at every income level.
13. The consumption function of Exercise 3, at income levels of $400 billion and $900 billion.
14. The consumption function of Exercise 4, at income levels of $410 billion and $2510 billion.
15. The consumption function of Exercise 5, at income levels of $125 million, $1000 million, and $8000 million.

4

Applications of Differentiation

The graph of a function is usually of considerable help in analyzing the behavior of the function. For instance, Figure 4–1 shows the graph of a function that predicts high-school enrollment in a large city for the years 1990 through 2020. From the graph we see the following:

- From 1990 to 1996 enrollment is predicted to decline, with the rate of decline decreasing as 1996 approaches.
- The minimum enrollment will be 200,000 students in 1996.
- From 1996 to 2004 enrollment will grow at an increasing rate.
- From 2004 to 2010 enrollment will continue to grow, but at a decreasing rate.
- The maximum enrollment will be 250,000 students in 2010.
- From 2010 to 2020 enrollment will decline at an increasing rate.

To analyze the behavior of a function as we have just done, we must be able to draw the graph accurately enough to show

- Where the function is increasing and where it is decreasing.
- What its maximum and minimum values are and where they occur.
- The general shape of the curve.

In this chapter we show how to use the first and second derivatives to determine these things. The first two sections of the chapter are devoted to finding relative maxima and minima of functions, the third to finding absolute maxima and minima, and the last demonstrates how calculus may be used to sketch the graph of a function.

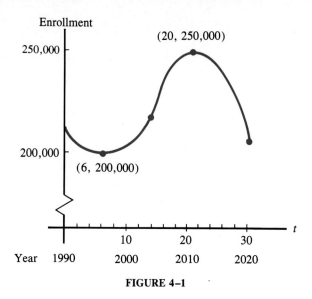

FIGURE 4-1

4.1 RELATIVE MAXIMA AND MINIMA: THE FIRST DERIVATIVE TEST

Suppose a function has the graph shown in Figure 4-2. Notice that the graph comes to a peak at the point $(-1, 10)$ and forms a trough at the point $(2, 1)$. The point $(-1, 10)$ is an example of a **relative maximum point;** the point $(2, 1)$ is an example of a **relative minimum point.** When graphing functions we must find their relative maximum and relative minimum points. In this section we show how to use the first derivative to do this.

Figure 4-3 shows the graph of a function f which has a relative maximum point at $(a, f(a))$ and a relative minimum point at $(b, f(b))$. Consider the point $(a, f(a))$: from the figure it is clear that for all values of x near a, on either side of a, $f(x)$ exists and $f(x) \leq f(a)$. Similarly, for all values of x near b, on either side of b, $f(x)$ exists and $f(x) \geq f(b)$. These properties define the concepts of relative maximum and relative minimum:

Relative Maximum and Relative Minimum

Let f be a function.

1. The point $(a, f(a))$ is a **relative maximum point** for f if for all x near a, on either side of a, $f(x)$ exists and $f(x) \leq f(a)$. In such a case we say that $f(a)$ is a **relative maximum for f.**

2. The point $(b, f(b))$ is a **relative minimum point** for f if for all x near b, on either side of b, $f(x)$ exists and $f(x) \geq f(b)$. In such a case we say that $f(b)$ is a **relative minimum for f.**

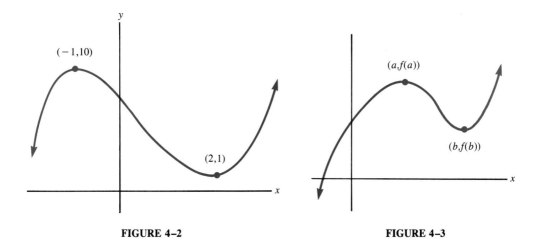

FIGURE 4–2 FIGURE 4–3

EXAMPLE 1 The function whose graph is shown in Figure 4–4 has relative maxima at the points (4, 8) and (9, 7) and relative minima at the points (3, 2) and (7, 4). □

How can we find the relative maxima and relative minima for a function f? As Figure 4–5 shows, if $(a, f(a))$ is a relative maximum point *and* there is a unique tangent line at $(a, f(a))$, then the tangent line must be horizontal; in other words, if $f(a)$ is a relative maximum and $f'(a)$ exists, then $f'(a) = 0$. The same reasoning holds for relative minima. Therefore we can find at least *some* of the relative maxima and relative minima of f by finding the values of x for which $f'(x) = 0$. We might not find *all* relative maxima and relative minima for f by doing this, because f might have relative maxima or relative minima at points where the derivative $f'(x)$ does not exist. As an example of such behavior, the function of Figure 4–4 has a relative maximum point at (9, 7) and a relative minimum point at (3, 2), but $f'(9)$ and $f'(3)$ do not exist. (Why not?)

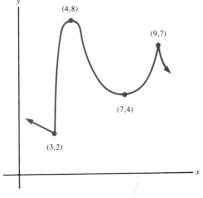

FIGURE 4–4

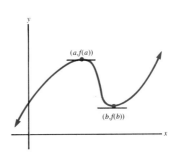

FIGURE 4–5

Our conclusion, then, is that if we are looking for the relative maxima and relative minima of a function f, we must find all values of x in the domain of f such that $f'(x) = 0$ or $f'(x)$ does not exist. Such values are called **critical values** of the function.

> **Critical Values**
>
> Let f be a function. The value $x = a$ is a **critical value** for f if a is in the domain of f and either $f'(a) = 0$ or $f'(a)$ does not exist.

EXAMPLE 2 Let us find the critical values for the functions f and g defined by

$$f(x) = 2x^3 - 12x^2 + 18x + 1 \quad \text{and} \quad g(x) = x^{2/3}.$$

We begin by finding the derivative of f:

$$f'(x) = 6x^2 - 24x + 18 = 6(x - 1)(x - 3).$$

Now we set $f'(x) = 0$ and solve for x to obtain $x = 1$ and $x = 3$. Since $f'(1) = 0$ and $f'(3) = 0$ and both 1 and 3 are in the domain of f, $x = 1$ and $x = 3$ are critical values for f. Because $f'(x)$ exists for all x, these are the only critical values for f.

Now we find the critical values for g. We have

$$g'(x) = \frac{2}{3x^{1/3}},$$

so $g'(x)$ is never zero, but $g'(x)$ does not exist at $x = 0$. Since $g(0) = 0$, 0 is in the domain of g, and hence $x = 0$ is the only critical value for g. □

Having found the critical values for a function, we must somehow test them to find out if they yield relative maxima or relative minima. In this connection, it is important to realize that if $x = a$ is a critical value for f, it is not the case that $f(a)$ must be either a relative maximum or a relative minimum: it might be neither. For instance, Figure 4–6 shows the graph of a function f that has a critical value

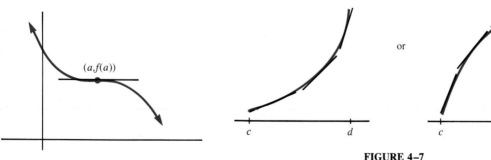

FIGURE 4–6

FIGURE 4–7
$f'(x) > 0$ for x in (c, d)

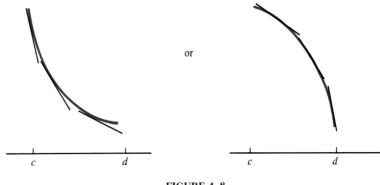

FIGURE 4–8
$f'(x) < 0$ for x in (c, d)

at $x = a$ (because $f'(a)$ is the slope of tangent line at $(a, f(a))$ and this slope is 0), but the point $(a, f(a))$ is neither a relative maximum point nor a relative minimum point for f.

How can we test a critical value to determine whether it yields a relative maximum, a relative minimum, or neither? The answer lies in the sign of the derivative. Since $f'(x)$ is the slope of the line tangent to f at $(x, f(x))$, it follows that if $f'(x) > 0$ for all x in an interval (c, d), all the tangent lines to f over the interval have positive slope, as in Figure 4–7. However, as Figure 4–7 shows, this means that the functional values $f(x)$ increase as x increases from c to d; in such a case, we say that the function f is **increasing on the interval.** Similarly, if $f'(x) < 0$ for all x in an interval, then as Figure 4–8 demonstrates, the functional values $f(x)$ decrease as x increases over the interval, and we say that f is **decreasing on the interval.**

The Sign of the Derivative

1. If $f'(x) > 0$ for all x in an interval (c, d), then f is increasing on the interval.

2. If $f'(x) < 0$ for all x in an interval (c, d), then f is decreasing on the interval.

EXAMPLE 3   Let f be defined by $f(x) = x^2 - 4x + 2$. Then
$$f'(x) = 2x - 4 = 2(x - 2).$$
Thus:

if $x < 2$, $x - 2 < 0$, so $f'(x) = 2(x - 2) < 0$ and therefore f is decreasing on the interval $(-\infty, 2)$;

if $x > 2$, $x - 2 > 0$, so $f'(x) = 2(x - 2) > 0$ and therefore f is increasing on the interval $(2, +\infty)$.

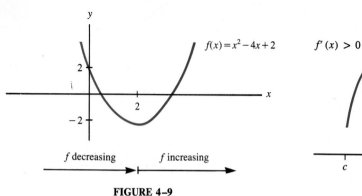
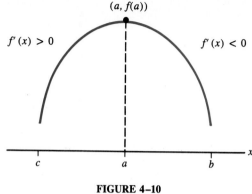

FIGURE 4–9

FIGURE 4–10

The graph of f is shown in Figure 4–9. Note that the graph of f is indeed decreasing over $(-\infty, 2)$ and increasing over $(2, +\infty)$. ☐

Now suppose that $x = a$ is a critical value for f. If $f'(x) > 0$ for all x in an interval (c, a) and $f'(x) < 0$ for all x in an interval (a, b) then f is increasing to the left of the point $(a, f(a))$ and decreasing to the right of $(a, f(a))$, as in Figure 4–10; hence $(a, f(a))$ must be a relative maximum point for f. Similarly, if $f'(x) < 0$ on an interval (c, a) and $f'(x) > 0$ on an interval (a, b), then $(a, f(a))$ must be a relative minimum point for f. See Figure 4–11. Thus we can test a critical value $x = a$ to see whether it yields a relative maximum or a relative minimum (or neither) by examining the sign of $f'(x)$ for x near a. We illustrate with an example.

EXAMPLE 4   Let f be defined by $f(x) = 2x^3 - 12x^2 + 18x + 1$. Then

$$f'(x) = 6x^2 - 24x + 18 = 6(x - 1)(x - 3),$$

and as we have seen, the critical values for f are $x = 1$ and $x = 3$. Note that

if $x < 1$, $x - 1 < 0$ and $x - 3 < 0$, so

$$f'(x) = 6(x - 1)(x - 3) > 0;$$

if $1 < x < 3$, $x - 1 > 0$ and $x - 3 < 0$, so

$$f'(x) = 6(x - 1)(x - 3) < 0;$$

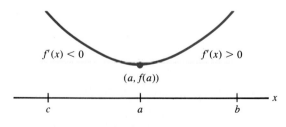

FIGURE 4–11

if $x > 3$, $x - 1 > 0$ and $x - 3 > 0$, so

$$f'(x) = 6(x - 1)(x - 3) > 0.$$

We record this information on a diagram of the x-axis:

Here we have used the critical values to subdivide the x-axis into intervals; a $+$ sign over an interval means that f' is positive on the interval, and a $-$ sign over an interval means that f' is negative on the interval. Therefore f is increasing on $(-\infty, 1)$ and $(3, +\infty)$, and decreasing on $(1, 3)$. We add this information to the diagram:

Here an upward-slanting arrow ↗ indicates that f is increasing and a downward-slanting arrow ↘ indicates that f is decreasing. This last diagram shows which critical values yield relative maxima and which yield relative minima: the pattern of arrows ↗↘ at $x = 1$ shows that f has a relative maximum at $(1, f(1)) = (1, 9)$; the pattern of arrows ↘↗ at $x = 3$ shows that f has a relative minimum at $(3, f(3)) = (3, 1)$. Figure 4–12 shows the graph of f: note the relative maximum at $(1, 9)$ and the relative minimum at $(3, 1)$, and note also that f is increasing over the intervals $(-\infty, 1)$ and $(3, +\infty)$ and decreasing over the interval $(1, 3)$. ☐

The procedure we used in Example 4 to find relative maxima and minima is called the **first derivative test.**

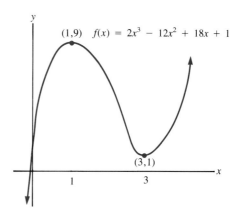

FIGURE 4–12

> **The First Derivative Test**
>
> To find the relative maxima and relative minima of a function f, proceed as follows:
>
> 1. Find all values of x such that $f'(x) = 0$ or $f'(x)$ does not exist.
> 2. Plot the values obtained in step 1 on the x-axis, thereby dividing the x-axis into intervals.
> 3. **a.** On each interval of the diagram of step 2, determine whether the derivative f' is positive or negative; record this information on the diagram using $+$ and $-$ signs.
>
>   **b.** f is increasing on those intervals where f' is positive and decreasing on those intervals where f' is negative; record this information on the diagram using arrows.
> 4. The diagram of step 3b will show which critical values yield relative maxima, which yield relative minima, and which yield neither.

(Notice that in step 1 of the first derivative test we will certainly obtain all the critical values of f, but we may also obtain numbers x that are not critical values because neither $f(x)$ nor $f'(x)$ exists. Since such numbers are not critical values, relative maxima and minima cannot occur at them; however, the derivative can have different signs on either side of such values, and when we discuss curve sketching in Section 4.4 it will be convenient to have the behavior of the function around them analyzed as part of the first derivative test.)

We illustrate the use of the first derivative test with examples.

EXAMPLE 5   Let us use the first derivative test to find the relative maxima and relative minima of the function f defined by $f(x) = -x^2 + 8x - 12$.

1. We have
$$f'(x) = -2x + 8 = -2(x - 4).$$
Setting $f'(x) = 0$ and solving for x yields $x = 4$, so $x = 4$ is a critical value for f. There are no values of x for which $f'(x)$ does not exist.

2. We plot the critical value $x = 4$ on the x-axis:

$$\xrightarrow{\qquad\qquad\qquad\underset{4}{|}\qquad\qquad\qquad} x$$

3. **a.** We determine the sign of $f'(x) = -2(x - 4)$ on the intervals $(-\infty, 4)$ and $(4, +\infty)$:

   if $x < 4$, $x - 4$ is negative, so $f'(x)$ is positive;

   if $x > 4$, $x - 4$ is positive, so $f'(x)$ is negative.

APPLICATIONS OF DIFFERENTIATION

We record this information on the diagram:

b. Since f' is positive on $(-\infty, 4)$ and negative on $(4, +\infty)$, f is increasing on $(-\infty, 4)$ and decreasing on $(4, +\infty)$. We add this information to the diagram:

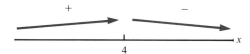

4. The diagram of step 3b shows that f has a relative maximum at $(4, f(4)) = (4, 4)$. There are no relative minima. The graph of f is shown in Figure 4–13. □

EXAMPLE 6 We find the relative maxima and relative minima for the function g defined by $g(x) = x^{2/3}$.

1. We have

$$g'(x) = \frac{2}{3x^{1/3}}.$$

Thus $g'(x)$ is never 0, but it does not exist at $x = 0$. (We saw in Example 2 that since 0 is in the domain of g, $x = 0$ is a critical value for g.)

2. We plot $x = 0$ on the x-axis:

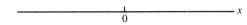

3. a. We determine the sign of

$$g'(x) = \frac{2}{3x^{1/3}}$$

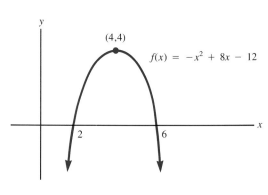

FIGURE 4–13

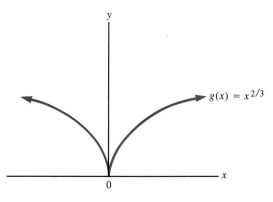

FIGURE 4–14

on the intervals $(-\infty, 0)$ and $(0, +\infty)$. Since $x^{1/3} = \sqrt[3]{x}$ is negative when $x < 0$ and positive when $x > 0$, $g'(x)$ is negative over $(-\infty, 0)$ and positive over $(0, +\infty)$. Hence we have the diagram

b. Since g' is negative over $(-\infty, 0)$ and positive over $(0, +\infty)$, g is decreasing over $(-\infty, 0)$ and increasing over $(0, +\infty)$. Hence the diagram becomes

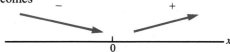

4. The diagram of step 3b shows that g has a relative minimum at $(0, g(0)) = (0, 0)$. There are no relative maxima. The graph of g is shown in Figure 4–14. □

Step 3a of the first derivative test requires that we determine the sign of f' on each of the intervals between critical points of f. If f' is continuous on such an interval, we can determine its sign by choosing a number from the interval and substituting it for x in $f'(x)$. This fact can often be used to simplify the work in step 3a.

EXAMPLE 7   Let us find the relative maxima and relative minima of the function h given by
$$h(x) = -x^3 + 6x^2 + 15x.$$

1. Since
$$h'(x) = -3x^2 + 12x + 15 = -3(x + 1)(x - 5),$$
setting $h'(x) = 0$ and solving for x yields $x = -1$ and $x = 5$. There are no values of x for which $h'(x)$ does not exist.

2. We have

3. a. Since h' is a polynomial, it is continuous on the intervals $(-\infty, -1)$, $(-1, 5)$, and $(5, +\infty)$. Therefore we can determine its sign on each of these intervals by selecting a number from the interval and substituting it into $h'(x)$. For instance, if we select -2 from $(-\infty, -1)$, 0 from $(-1, 5)$, and 6 from $(5, +\infty)$, we have
$$h'(-2) = -3(-2 + 1)(-2 - 5) < 0,$$
$$h'(0) = -3(0 + 1)(0 - 5) > 0,$$
$$h'(6) = -3(6 + 1)(6 - 5) < 0.$$

Thus $h'(x)$ is negative on $(-\infty, -1)$ and $(5, +\infty)$, and positive on $(-1, 5)$. Hence we have

b. Thus h is increasing over $(-1, 5)$, and decreasing over $(-\infty, -1)$ and $(5, +\infty)$. Therefore the diagram becomes

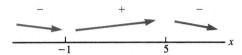

4. The diagram of step 3b shows that $(-1, h(-1)) = (-1, -8)$ is a relative minimum point for h and that $(5, h(5)) = (5, 100)$ is a relative maximum point for h. Figure 4–15 shows the graph of h. □

EXAMPLE 8 We find the relative maxima and relative minima of the function k defined by

$$k(x) = -x^4 + 8x^3 - 18x^2.$$

1. Since

$$k'(x) = -4x^3 + 24x^2 - 36x = -4x(x - 3)^2,$$

setting $k'(x) = 0$ and solving for x yields $x = 0$ and $x = 3$. There are no values for which $k'(x)$ does not exist.

2. We have

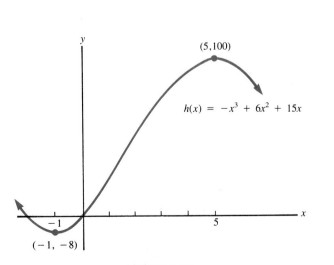

FIGURE 4–15

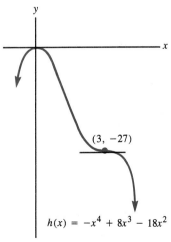

FIGURE 4–16

3. Note that k' is continuous on $(-\infty, 0)$, $(0, 3)$, and $(3, +\infty)$. (Why?) If we choose $x = -1$ from $(-\infty, 0)$, $x = 1$ from $(0, 3)$, and $x = 4$ from $(3, +\infty)$, we find that $k'(-1) = 64$ and $k'(1) = k'(4) = -16$.

   a. Hence k' is positive on $(-\infty, 0)$, and negative on $(0, 3)$ and on $(3, +\infty)$:

   b. Thus k is increasing on $(-\infty, 0)$, and decreasing on $(0, 3)$ and on $(3, +\infty)$:

   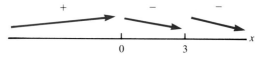

4. The diagram of step 3b shows that there is a relative maximum at $(0, k(0)) = (0, 0)$. Note that since k is decreasing on both sides of $x = 3$, there is neither a relative maximum nor a relative minimum for k at $x = 3$. In fact, the graph of k "flattens out" at $(3, k(3)) = (3, -27)$, because $k'(3) = 0$ implies that the graph has a horizontal tangent line at $(3, -27)$. See Figure 4-16. ☐

EXAMPLE 9   Let us find the relative maxima and minima of the rational function f defined by

$$f(x) = \frac{x^2}{x^2 - 1}.$$

Note that $f(1)$ and $f(-1)$ do not exist.

Since

$$f'(x) = -\frac{2x}{(x^2 - 1)^2},$$

setting $f'(x) = 0$ and solving for x yields $x = 0$. Thus 0 is a critical value for f. The derivative f' does not exist at $x = \pm 1$, but ± 1 are not critical values because they are not in the domain of f. Thus our step 1 diagram is as follows:

Here we have employed dashed lines at $x = \pm 1$ to remind us that f cannot have relative maxima or minima at these values because they are not critical values.

Now,

if $x < -1$, $f'(x) > 0$;
if $-1 < x < 0$, $f'(x) > 0$;

APPLICATIONS OF DIFFERENTIATION

if $0 < x < 1$, $f'(x) < 0$;
if $x > 1$, $f'(x) < 0$.

Therefore we obtain the diagram

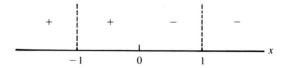

which leads in turn to

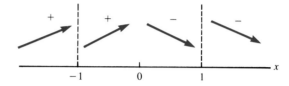

This last diagram shows that f has a relative maximum at $(0, f(0)) = (0, 0)$. The graph of f is shown in Figure 4–17. ☐

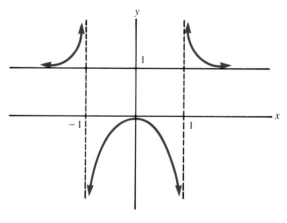

FIGURE 4–17

■ *Computer Manual: Program DERTEST1*

4.1 EXERCISES

In Exercises 1 through 6, find all critical values, relative maxima, and relative minima of the function whose graph is shown.

1.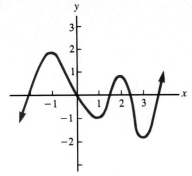

2.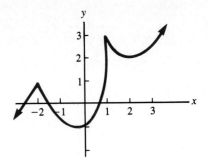

3.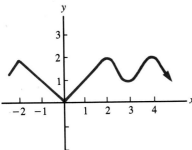

4.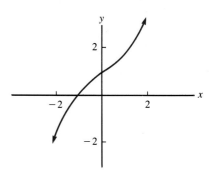

5.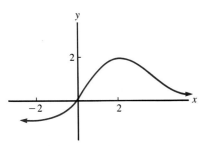

6.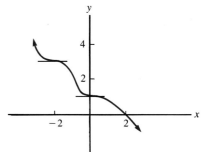

In Exercises 7 through 22, find the intervals on which f is increasing and those on which it is decreasing.

7. $f(x) = x^2 - 10x + 24$
8. $f(x) = -x^2 + 2x + 3$
9. $f(x) = x^3$
10. $f(x) = -x^3 + 3x^2 - 3x$
11. $f(x) = -2x^3 - 9x^2 + 60x + 10$
12. $f(x) = 2x^3 - 27x^2 - 120x$
13. $f(x) = \dfrac{x^3}{3} - \dfrac{3x^2}{2} + \dfrac{5x}{4} + 1$
14. $f(x) = x^4 - 1$
15. $f(x) = x^4 - 8x^2 + 1$
16. $f(x) = 3x^4 + 4x^3 - 12x^2$
17. $f(x) = \sqrt{x}$
18. $f(x) = x^{1/3}$
19. $f(x) = \dfrac{1}{x}$
20. $f(x) = \dfrac{x}{x+1}$
21. $f(x) = \dfrac{x+1}{2x-4}$
22. $f(x) = \dfrac{x^2}{x+1}$

In Exercises 23 through 53, use the first derivative test to find all relative maxima and relative minima for f.

23. $f(x) = x^2 - 4x + 3$
24. $f(x) = -2x^2 - 8x + 14$
25. $f(x) = 0.25x^2 + 3x - 9$
26. $f(x) = x^3 + 4$
27. $f(x) = x^3 - 2x^2$
28. $f(x) = 3x^3 - 9x$

29. $f(x) = -x^3 + 12x$
30. $f(x) = x^3 + 6x^2 + 12x + 8$
31. $f(x) = x^3 + 3x^2 - 9x$
32. $f(x) = -x^3 + 21x^2 - 120x + 20$
33. $f(x) = x^3 - 6x^2 + 12x + 2$
34. $f(x) = -2x^3 + 15x^2 - 24x + 8$
35. $f(x) = 2x^3 + 9x^2 - 108x + 20$
36. $f(x) = x^4$
37. $f(x) = x^4 - 2x^2$
38. $f(x) = 3x^4 + 4x^3$
39. $f(x) = 3x^4 + 4x^3 - 12x^2$
40. $f(x) = -3x^4 + 4x^3 + 6x^2 - 12x$ [Hint: $f'(x) = -12(x-1)(x^2-1)$.]
41. $f(x) = 3x^5 - 20x^3 + 10$
42. $f(x) = 3x^5 - 10x^3 + 15x + 1$ [Hint: $f'(x) = 15(x^2-1)^2$.]
43. $f(x) = 2x^6 - 27x^4$
44. $f(x) = x^{4/3}$
45. $f(x) = \dfrac{2x-1}{x+2}$
46. $f(x) = \dfrac{3x+2}{x-4}$
47. $f(x) = \dfrac{x+2}{x^2-3}$
48. $f(x) = \dfrac{x+1}{-3-x^2}$
49. $f(x) = \dfrac{x^2}{x+1}$
50. $f(x) = \dfrac{10x^2+4}{20x+3}$
51. $f(x) = x + \dfrac{1}{x}$
52. $f(x) = x^2 + \dfrac{2}{x}$
53. $f(x) = \dfrac{1}{3}x^3 + \dfrac{16}{x}$

54. During a superconductivity experiment, a solution is cooled rapidly and then heated more slowly. If the temperature of the solution t minutes after the start of the experiment is T degrees Kelvin, where $T = 2t + 4/t$, find the intervals on which T is increasing and those on which it is decreasing, and find the minimum temperature of the solution and the time when it occurs.

Management

55. A worker's productivity is given by
$$f(t) = -0.05t^2 + 1.8t + 150,$$
where t is the number of years the worker has been on the job and $f(t)$ is the dollar value of the worker's average daily production in year t. Find the intervals on which the worker's productivity is increasing and those on which it is decreasing. Find the worker's maximum productivity and the year in which it occurs.

56. A firm's profit function is given by
$$P(x) = -x^3 + 135x^2 - 2400x - 75{,}000.$$
Find the intervals on which profit is increasing and those on which it is decreasing, and find the maximum profit and the number of units required to attain it.

57. If Beatific Corporation spends x thousand dollars on advertising, its sales will be $f(x)$ dollars, where
$$f(x) = 2x^3 - 210x^2 + 6000x + 100{,}000, \qquad 0 \le x \le 40.$$
Find the intervals on which sales are increasing and those on which they are decreasing. Find the firm's maximum sales and the amount that must be spent on advertising to attain maximum sales.

58. Amity Company's average cost function is given by
$$A(x) = 10x + \dfrac{4000}{x},$$

where x is the number of units produced, $x \geq 1$. Find the intervals on which the average cost is increasing and those on which it is decreasing, and find the minimum average cost per unit.

59. A radio station's share of the audience during 1989 was given by
$$s = \frac{t-2}{t^2+12} + \frac{3}{10},$$
where t is in months, $1 \leq t \leq 12$, with $t = 1$ representing January 1989. Find the intervals on which the audience share was increasing and those on which it was decreasing, and find the maximum audience share and the time during the year when it occurred.

*60. Show that
   (a) if marginal revenue is less than marginal cost on an interval, then profit is decreasing on the interval;
   (b) if marginal revenue is greater than marginal cost on an interval, then profit is increasing on the interval.

61. Suppose the demand function for a product is given by
$$q = -1000p + 10{,}000.$$
   (a) Find the intervals on which revenue from sales of the product is increasing and those on which it is decreasing.
   (b) Find the maximum revenue and the price at which it occurs.
   (c) Find the intervals on which demand for the product is elastic and those on which it is inelastic, and find the price at which demand is of unit elasticity. Compare your answers to those of parts (a) and (b).

62. Repeat Exercise 61 if the demand function is $q = 216{,}000 - 20p^2$.

*63. Show the following:
   (a) if demand for a product is elastic on a price interval, then revenue from sales of the product decreases as the price increases within the interval;
   (b) if demand for a product is inelastic on a price interval, then revenue from sales of the product increases as the price increases within the interval;
   (c) maximum revenue occurs at the price for which demand is of unit elasticity.

Life Science

64. The number of birds in a wildlife sanctuary in month t of the year is estimated to be
$$y = -t^3 + 5.25t^2 + 33t + 1000.$$
The maximum bird population occurs at a relative maximum of this function. Find the intervals on which the size of the bird population is increasing and those on which it is decreasing, and find the maximum bird population and the time during the year when it occurs.

65. The number of gypsy moths at time t is given by
$$y = \frac{10t}{t^2+4}, \quad t \geq 0,$$
where t is in months, with $t = 0$ representing the present, and y is in millions. Find the intervals on which the gypsy moth population is increasing and those on which it is decreasing, and find the time at which the population will peak.

66. During an epidemic the number of people inoculated during month t was given by
$$y = -24t^5 + 165t^4 + 800t^3, \quad 0 \le t \le 10.$$
Find the intervals on which the number inoculated was increasing and those on which it was decreasing, and find the maximum number inoculated during a month and when it occurred.

Social Science

67. Enrollment in the higher education system of a nation for the period 1990–2010 is forecast to be
$$y = 0.20t^5 - 7.5t^4 + 99t^3 - 594t^2 + 1620t + 5000$$
thousand students in year t, where $t = 0$ represents 1990. Find the intervals on which enrollment is increasing and those on which it is decreasing. Find the maximum and minimum enrollments and the years in which they will occur. [Hint: $y' = (t - 3)(t - 6)^2(t - 15)$.]

68. Political advisors for a certain candidate estimate that if x thousand people vote in the next election, their candidate will receive y thousand votes, where
$$y = 10\sqrt{x} - 0.5x - 5.$$
Find the intervals on which the candidate's vote is increasing and those on which it is decreasing, and find the candidate's maximum number of votes.

4.2 RELATIVE MAXIMA AND MINIMA: THE SECOND DERIVATIVE TEST

In this section we present another method for testing the critical values of a function to find out whether they yield relative maxima or relative minima. The test we will develop, which is called the **second derivative test,** involves substituting critical values into the second derivative of the function.

In the previous section we noted that if f' is positive on an interval, then f is increasing over the interval. Since this is true for any function and its derivative, let us apply it to the function f' and its derivative f'': if $f''(x) > 0$ for all x in an interval, then f' must be increasing over the interval. But if f' is increasing over the interval, then the slopes of the tangent lines to f are increasing over the interval, as shown in Figure 4–18, for instance. Therefore in order for the graph of f to fit the tangent lines, it must lie above them and bend upward when $f''(x) > 0$.

Similarly, if the derivative of a function is negative over an interval, then the function is decreasing over the interval. Thus if $f''(x) < 0$ for all x in an interval, then f' is decreasing over the interval. But then the slopes of the tangent lines to f are decreasing, as in Figure 4–19. Thus, in order to fit the tangent lines, the graph of f must lie below them and bend downward over the interval.

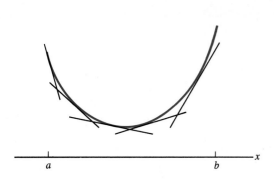
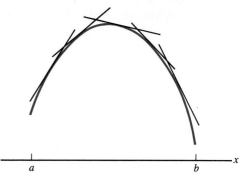

FIGURE 4-18
Slopes of Tangent Lines Increasing on *(a, b)*

FIGURE 4-19
Slopes of Tangent Lines Decreasing on *(a, b)*

Concavity

1. If $f''(x) > 0$ for all x in an interval, the graph of f bends upward over the interval, as shown in Figure 4-20, and f is said to be **concave upward** over the interval.

2. If $f''(x) < 0$ over an interval, the graph of f bends downward over the interval, as shown in Figure 4-21, and f is said to be **concave downward** over the interval.

EXAMPLE 1   If $f(x) = x^2$, then $f''(x) = 2$. Hence $f''(x)$ is positive for all x, and therefore f is concave upward for all x. See Figure 4-22. ☐

EXAMPLE 2   If $g(x) = \dfrac{1}{3}x^3 - x^2 - 3x + 2$, then $g''(x) = 2(x - 1)$. Note that if $x < 1$, then $g''(x) < 0$, while if $x > 1$, then $g''(x) > 0$. Thus $g''(x) < 0$ over the interval $(-\infty, 1)$, and $g''(x) > 0$ over the interval $(1, +\infty)$. Therefore g is concave downward over $(-\infty, 1)$ and concave upward over $(1, +\infty)$. See Figure 4-23. ☐

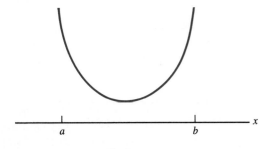
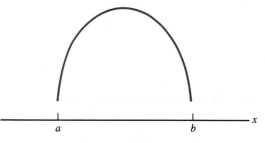

FIGURE 4-20
f Concave Upward over *(a, b)*

FIGURE 4-21
f Concave Downward over *(a, b)*

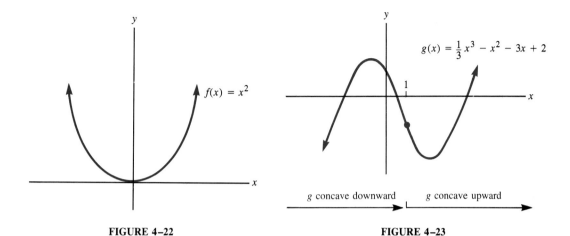

FIGURE 4–22 **FIGURE 4–23**

Now suppose that $f'(a) = 0$ and $f''(a) < 0$. Then f is concave downward at $(a, f(a))$ and also has a horizontal tangent there (see Figure 4–24), and thus $(a, f(a))$ must be a relative maximum point for f. Similarly, if $f'(a) = 0$ and $f''(a) > 0$, f is concave upward and has a horizontal tangent at $(a, f(a))$ (see Figure 4–25), so $(a, f(a))$ must be a relative minimum point for f. This shows that we can sometimes test critical values by substituting them into the second derivative.

The Second Derivative Test

1. If $f'(a) = 0$ and $f''(a) > 0$, then $(a, f(a))$ is a relative minimum point for f.

2. If $f'(a) = 0$ and $f''(a) < 0$, then $(a, f(a))$ is a relative maximum point for f.

3. If $f'(a) = 0$ and $f''(a) = 0$ or $f''(a)$ does not exist, the test fails.

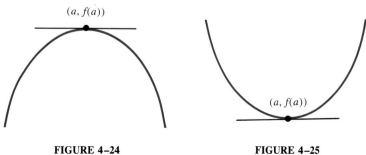

FIGURE 4–24
Relative Maximum

FIGURE 4–25
Relative Minimum

EXAMPLE 3   Let $f(x) = -2x^3 - 3x^2 + 36x + 5$. We have
$$f'(x) = -6x^2 - 6x + 36 = -6(x+3)(x-2)$$
and
$$f''(x) = -12x - 6 = -6(2x+1).$$
Since
$$f'(-3) = 0 \quad \text{and} \quad f''(-3) = 30 > 0,$$
f has a relative minimum at $(-3, f(-3)) = (-3, -76)$. Since
$$f'(2) = 0 \quad \text{and} \quad f''(2) = -30 < 0,$$
f has a relative maximum at $(2, f(2)) = (2, 49)$. ☐

The second derivative test is often easier to apply than the first derivative test, but it is not as general as the latter. First of all, if $x = a$ is a critical value and $f'(a)$ does not exist, then the second derivative test does not apply. Second, even if $f'(a) = 0$, the test tells us nothing about whether $(a, f(a))$ is a relative maximum or a relative minimum if $f''(a) = 0$ or if $f''(a)$ does not exist. In such cases we must use some other means (for example, the first derivative test) to analyze the behavior of f at $(a, f(a))$.

EXAMPLE 4   Let g be defined by $g(x) = x^{2/3}$. Then $x = 0$ is a critical value for g (see Example 2, Section 4.1), but we cannot use the second derivative test to determine whether $(0, g(0)) = (0, 0)$ is a relative maximum or relative minimum point for g because $g'(0)$ does not exist. The first derivative test (see Example 6, Section 4.1) shows that in fact g has a relative minimum at $(0, 0)$. ☐

EXAMPLE 5   Let $f(x) = -x^3 + 6x^2 - 12x + 8$. Then
$$f'(x) = -3(x-2)^2 \quad \text{and} \quad f''(x) = -6(x-2),$$
so $f'(2) = 0$ and $f''(2) = 0$. Thus the second derivative test cannot be used to classify the point $(2, f(2)) = (2, 0)$ as a relative maximum or a relative minimum point for f. The first derivative test leads to the diagram

(Check this.) Hence $(2, 0)$ is neither a relative maximum nor a relative minimum point for f, but rather a point where the graph "flattens out." ☐

EXAMPLE 6   Let $g(x) = x^{4/3}$. We have
$$g'(x) = \frac{4}{3}x^{1/3} \quad \text{and} \quad g''(x) = \frac{4}{9}x^{-2/3}.$$

Therefore $g'(0) = 0$, but $g''(0)$ does not exist, and again the second derivative test cannot be used. The first derivative test leads to the diagram

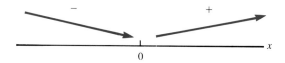

Hence $(0, g(0)) = (0, 0)$ is a relative minimum for g. ∎

■ **Computer Manual: Program DERTEST2**

4.2 EXERCISES

In Exercises 1 through 12, find the intervals over which the given function is concave upward and those over which it is concave downward.

1. $f(x) = x^2 - 4x$
2. $f(x) = -x^2 + 8x - 9$
3. $f(x) = x^3$
4. $f(x) = -x^3 + 2x$
5. $f(x) = 2x^3 - 12x^2 + 6x - 9$
6. $f(x) = -2x^3 + 15x^2 - 12x + 4$
7. $f(x) = x^4 - 16$
8. $f(x) = x^4 - 24x^2$
9. $f(x) = -x^4 + 2x^3 + 36x^2 + 4x$
10. $f(x) = x^5 - 10x$
11. $f(x) = -x^5 + 80x^2$
12. $f(x) = 3x^5 - 5x^4$

In Exercises 13 through 32, use the second derivative test to find all relative maxima and minima for the given function. If the second derivative test cannot be applied or if it fails, use the first derivative test.

13. $f(x) = -x^2 + 6x + 5$
14. $f(x) = 3x^2 - 30x + 1$
15. $f(x) = x^3 + 9x^2$
16. $f(x) = x^3 + 12x^2 + 48x + 3$
17. $f(x) = 2x^3 - 15x^2 - 216x + 25$
18. $f(x) = -4x^3 + 6x^2 - 3x + 5$
19. $f(x) = -x^3 + 6x^2 + 63x$
20. $f(x) = -x^4 + 4x^3 + 30$
21. $f(x) = x^4 - 4x^2 + 10$
22. $f(x) = x^4/4 + 2x^3 + 5x^2/2 - 10$
23. $f(x) = -x^4 + 8x^3 - 22x^2 + 24x$ [Hint: $f'(x) = -4(x - 1)(x^2 - 5x + 6)$]
24. $f(x) = 3x^4 - 40x^3 + 150x^2 - 10$
25. $f(x) = 3x^5 + 5x^3$
26. $f(x) = 6x^5 - 15x^4 + 10x^3 + 25$
27. $f(x) = -6x^5 + 45x^4 - 110x^3 + 90x^2 - 50$ [Hint: $f'(x) = -30x(x - 1)(x^2 - 5x + 6)$]
28. $f(x) = 4x^5 + 15x^4 - 40x^2$ [Hint: $f'(x) = 20x(x - 1)(x^2 + 4x + 4)$]
29. $f(x) = x^{8/5}$
30. $f(x) = x^{4/5}$
31. $f(x) = (x - 1)^{11/5}$
32. $f(x) = -2(x - 1)^{7/5}$

*33. The second derivative test is not always easier to apply than the first derivative test. To see this, use the first derivative test and then the second derivative test to find the relative maxima and relative minima of f if

$$f(x) = \frac{x + 1}{x^2 + 3}.$$

Management

34. Caritas, Inc., knows that the demand function for its product is given by
$$p = 10{,}000 - 8x.$$
The company's maximum revenue occurs at a relative maximum of its revenue function. Use the second derivative test to find the maximum revenue.

35. Dismal Company's cost function is given by $C(x) = x^2 + 12x + 900$. The firm's minimum average cost per unit occurs at a relative minimum of its average cost function. Use the second derivative test to find the minimum average cost per unit.

36. If a firm spends x thousand dollars on advertising, its net profit will be $P(x)$ dollars, where
$$P(x) = x^3 - 240x^2 + 18{,}900x - 100{,}000, \qquad 0 \le x \le 100.$$
The firm's maximum profit occurs at a relative maximum of this function. Use the second derivative test to find the firm's maximum profit and the amount that must be spent on advertising in order to attain it.

37. A firm's profit function is given by
$$P(x) = -0.25x^4 + 20x^3 - 800{,}000.$$
The firm's maximum profit occurs at a relative maximum of P. Use the second derivative test to find the maximum profit.

*38. Show that average cost per unit is minimized at the value of x for which average cost per unit equals marginal cost. You may assume that the cost function is concave upward for all $x \ge 0$.

*39. Exercise 38 shows that the average cost per unit is minimized at the value $x = a$ for which average cost equals marginal cost. Show also that
  (a) average cost is greater than marginal cost on the interval $(0, a)$;
  (b) average cost is less than marginal cost on the interval $(a, +\infty)$.

Life Science

40. Suppose that t minutes after the injection of a temperature-reducing drug a patient's temperature is $y°F$, where
$$y = -0.006t^2 + 0.36t + 102.8.$$
The maximum temperature occurs at a relative maximum of this function. How long after the injection of the drug does the maximum temperature occur? What is the maximum temperature? Use the second derivative test to answer these questions.

41. The number of bacteria in a culture at time t was y thousand, where
$$y = 2\sqrt{t+1} + \frac{8}{\sqrt{t+1}}.$$
Here t is in hours since the start of an experiment. The minimum number of bacteria present occurred at a relative minimum of this function. Use the second derivative test to find the minimum number of bacteria present and when this minimum occurred.

Social Science

42. Demographics suggest that over the next 20 years, the cohort of 18- to 25-year-olds in a city will contain approximately
$$y = 0.2t^5 - 6.625t^4 + 51t^3 + 40{,}000$$
individuals. Use the second derivative test to find the maximum and minimum number in the cohort, assuming that these occur at a relative maximum and relative minimum of the function.

43. Police records for a large city suggest that the city's crime rate may be described by the equation
$$y = 12.5\sqrt{t} + \frac{400}{t}.$$
Here y is the number of crimes per thousand population in year t, and $t = 1$ represents 1980. The minimum crime rate occurs at a relative minimum of this function. Use the second derivative test to find the minimum crime rate and the year in which it will occur.

4.3 ABSOLUTE MAXIMA AND MINIMA

The largest value a function can assume on an interval is called its **absolute maximum** on the interval; the smallest value it can assume is its **absolute minimum** on the interval. For instance, Figure 4–26 shows the graph of a function which describes

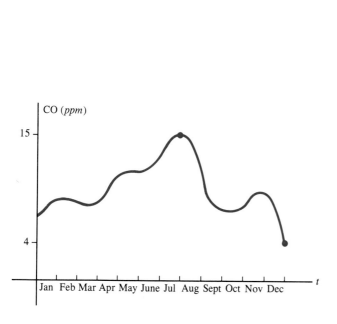

FIGURE 4–26

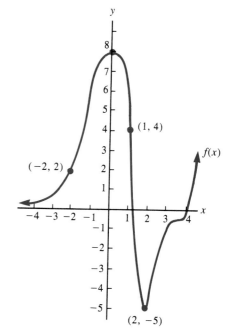

FIGURE 4–27

the concentration in parts per million of carbon monoxide in the air of a large city during 1989. The absolute maximum of this function is 15 ppm; it occurred on August 1. The absolute minimum of the function is 4 ppm, which occurred on December 31. It is often necessary to find the absolute maximum or absolute minimum of a function. In this section we show how to use the first derivative test to do this.

The existence and value of the absolute maximum and absolute minimum of a function depend not only on the function itself but also on the interval over which it is considered. For instance, consider the function f whose graph is pictured in Figure 4–27:

- On the entire x-axis f has an absolute minimum of -5 at $x = 2$; however, f has no absolute maximum on the entire x-axis because as x approaches $+\infty$, the functional values $f(x)$ increase without bound.

- On the interval $[-1, 4]$, f has an absolute minimum of -5 at $x = 2$ and an absolute maximum of 8 at $x = 0$.

- On the interval $[-2, 1]$, f has an absolute minimum of 2 at $x = -2$ and an absolute maximum of 8 at $x = 0$.

- On the interval $(-\infty, -2]$, f has an absolute maximum of 2 at $x = -2$; however, f has no absolute minimum on $(-\infty, -2]$ because as x approaches $-\infty$, the functional values $f(x)$ approach but never equal zero.

It should now be clear that the absolute maximum or absolute minimum of a function on an interval may or may not exist; if it does exist, it may occur at either a relative maximum or relative minimum within the interval or at an endpoint of the interval. Therefore in order to find the absolute maximum or absolute minimum of a function on an interval, we must examine its relative maxima or minima and the values of the function at the endpoints of the interval.

Finding the Absolute Maximum and Absolute Minimum

Let f be a continuous function defined on an interval $[a, b]$.

1. The absolute maximum of f on $[a, b]$ is the largest of $f(a)$, $f(b)$, and all the relative maxima that lie within $[a, b]$.

2. The absolute minimum of f on $[a, b]$ is the smallest of $f(a)$, $f(b)$, and all the relative minima that lie within $[a, b]$.

The first derivative test is very useful when we must find absolute maxima and absolute minima, because the diagram of step 3b of the test tells us where to look for them. We illustrate with examples.

EXAMPLE 1 Let

$$f(x) = x^3 - 3x^2 - 45x + 1.$$

Then
$$f'(x) = 3x^2 - 6x - 45 = 3(x + 3)(x - 5),$$
and the diagram of step 3b of the first derivative test looks like this:

(Check this.) Suppose we want to find the absolute maximum and absolute minimum of f on the interval $[0, 8]$. Then we can only consider the function over this interval, so our diagram becomes

The diagram shows that the absolute minimum must occur at $x = 5$; therefore the absolute minimum is equal to $f(5) = -174$. It also shows that the absolute maximum must occur at either $x = 0$ or $x = 8$. Since $f(0) = 1$ and $f(8) = -39$, the absolute maximum occurs at $x = 0$ and is equal to 1. ☐

EXAMPLE 2 Let f be as in Example 1, but this time let us find its absolute maximum and absolute minimum on the interval $[-5, 6]$. The diagram of step 3b of the first derivative test then becomes

Therefore the absolute maximum of f on the interval $[-5, 6]$ must occur at either $x = -3$ or $x = 6$. Since $f(-3) = 82$ and $f(6) = -161$, the absolute maximum is 82 at $x = 3$. Similarly, the absolute minimum must occur at either $x = -5$ or $x = 5$. Since $f(-5) = 26$ and $f(5) = -174$, the absolute minimum is -174 at $x = 5$. ☐

EXAMPLE 3 Again let f be as in Example 1.

a. On the interval $[0, +\infty)$ the diagram becomes

Thus on $[0, +\infty)$ the function has an absolute minimum of -174 at $x = 5$. It has no absolute maximum, since its functional values increase without bound as $x \to +\infty$.

b. On the entire x-axis the diagram becomes

so on the x-axis the function has neither an absolute maximum nor an absolute minimum. (Why not?) ☐

Now let us turn to some practical problems involving the absolute maximum and absolute minimum. Our first problem concerns profit maximization.

EXAMPLE 4 The profit function of Altruism, Inc., is given by

$$P(x) = -0.01x^3 + 300x - 10{,}000.$$

Let us find Altruism's maximum daily profit if the firm can make at most 400 units of product per day.

We seek to maximize

$$P(x) = -0.01x^3 + 300x - 10{,}000$$

for $0 \le x \le 400$, that is, on the interval $[0, 400]$.

Setting $P'(x) = 0$, we have

$$P'(x) = -0.03x^2 + 300 = 0,$$

which has solutions $x = 100$ and $x = -100$. Since -100 is not in our interval, we discard it, and our diagram is

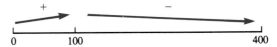

Therefore the maximum profit is $P(100) = \$10{,}000$. Note that if we had said that Altruism could make at most 50 units per day, then the maximum profit would have been $P(50) = \$3750$. (Why?) ☐

EXAMPLE 5 Suppose we wish to minimize the amount of fencing that will be required to enclose a rectangular field of 900 square meters. The field will look like this:

Here x and y are in meters. We must minimize the amount of fencing

$$F = 2x + 2y$$

subject to the constraint that the area of the field is 900 square meters, that is, that

$$xy = 900$$

square meters. In order to apply our methods of minimization to F, we must write it as a function of a single variable. The constraint equation $xy = 900$ allows us to do this, for we may solve it for y to obtain

$$y = \frac{900}{x}.$$

Hence

$$F = 2x + 2\left(\frac{900}{x}\right) = 2x + \frac{1800}{x}.$$

Thus we must minimize

$$F = 2x + \frac{1800}{x}, \qquad x > 0.$$

But

$$\frac{dF}{dx} = 2 - \frac{1800}{x^2} = 0$$

implies that

$$2x^2 - 1800 = 0,$$

or

$$x = \pm 30.$$

Since we require $x > 0$, we discard the negative critical value. To test the value $x = 30$, we use the first derivative test: the diagram is

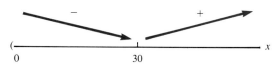

(Check this, and note that we have placed a parenthesis "(" at $x = 0$ to show that 0 is not included in our interval.) Hence $x = 30$ yields a minimum. Therefore $y = 900/30 = 30$, and the minimum amount of fencing is $F = 2(30) + 2(30) = 120$ meters. ☐

EXAMPLE 6   We wish to construct a box with a square bottom and top and rectangular sides. The material for the sides costs $2.00 per square foot and that for the top and bottom costs $0.25 per square foot. We intend to spend $24.00 on materials. Let us find the dimensions that will give us the box of maximum volume.

The box will look like this:

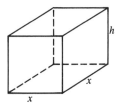

Here x and h are in feet. The volume V of the box is its length times its width times its height, or
$$V = x^2 h$$
cubic feet. The cost of the material used to make the box is
$$\$0.25 \text{ (area of top + area of bottom)} + \$2.00 \text{(area of four sides)}$$
$$= \$0.25(2x^2) + \$2.00(4xh)$$
$$= \$0.50x^2 + \$8.00xh,$$
and this must equal $\$24.00$. Thus we seek to maximize
$$V = x^2 h$$
subject to the constraint
$$0.5x^2 + 8xh = 24.$$
Now we use the constraint equation to express V in terms of a single variable we solve
$$0.5x^2 + 8xh = 24$$
for h to obtain
$$h = \frac{24 - 0.5x^2}{8x} = \frac{3}{x} - \frac{x}{16}$$
and then substitute this into the expression for V to get
$$V = x^2 \left(\frac{3}{x} - \frac{x}{16} \right) = 3x - \frac{x^3}{16}.$$
Thus we seek to maximize
$$V = 3x - \frac{x^3}{16}, \qquad x > 0.$$
But
$$\frac{dV}{dx} = 3 - \frac{3x^2}{16} = 0$$
implies that
$$x^2 = 16,$$
so $x = \pm 4$. We may discard the negative critical value. Since
$$\frac{dV}{dx} > 0 \quad \text{if} \quad 0 < x < 4 \quad \text{and} \quad \frac{dV}{dx} < 0 \quad \text{if} \quad x > 4,$$
We have

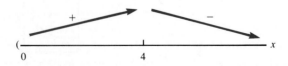

APPLICATIONS OF DIFFERENTIATION

Hence $x = 4$ yields a maximum. Therefore the dimensions that give the box of maximum volume are $x = 4$ feet and

$$h = \frac{3}{4} - \frac{4}{16} = \frac{1}{2}$$

foot, and the maximum volume is

$$V = 4^2 \left(\frac{1}{2}\right) = 8$$

cubic feet. □

An **inventory problem** is one concerned with minimizing the total cost of purchasing, ordering, and carrying a product in inventory. Total inventory cost is thus the sum of purchasing cost, ordering cost, and carrying cost, where

Purchasing cost = (Number of units purchased) (Price per unit),
Ordering cost = (Number of orders placed) (Cost of placing an order),

and

Carrying cost = (Average number of units in inventory) × (Cost of carrying a unit in inventory).

We solve an inventory problem by finding the **optimal reorder quantity,** which is the number of units that should be ordered each time to minimize total inventory cost.

EXAMPLE 7 Suppose the Shoe Shop sells 2000 pairs of Lazy Loafers each year. Each pair of Lazy Loafers is purchased for $10, and it costs $100 to place an order for them. Also, it costs $1.60 to carry a pair of shoes in inventory for one year. Let us find the optimal reorder quantity for this problem.

Let x denote the number of pairs of Lazy Loafers that should be ordered each time an order is made. Since 2000 pairs of Lazy Loafers are required during the course of a year, they must be ordered $2000/x$ times per year. Suppose we assume that pairs of Lazy Loafers are sold at a constant rate and that they sell out just as a new order arrives; then since each order consists of x pairs, the average number of pairs in inventory at any time will be $x/2$. Therefore

Annual purchasing cost = (2000 pairs)($10 per pair) = $20,000,

Annual ordering cost = $\left(\dfrac{2000}{x} \text{ orders}\right)$ ($100 per order) = $\dfrac{\$200{,}000}{x}$,

Annual carrying cost = $\left(\dfrac{x}{2} \text{ pairs}\right)$ ($1.60 per pair) = $0.80x$.

Thus if $I(x)$ is the total annual inventory cost that results from ordering x pairs of Lazy Loafers per order, then

$$I(x) = 20{,}000 + \frac{200{,}000}{x} + 0.80x,$$

where $x > 0$. We must minimize the inventory cost function I on the interval $(0, +\infty)$.

Since
$$I'(x) = -\frac{200{,}000}{x^2} + 0.80,$$

setting $I'(x) = 0$ and solving for x yields $x = \pm 500$. Because x must be positive, we may discard the critical value -500, and the diagram of the first derivative test looks like this:

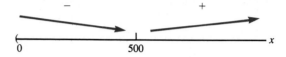

(Check this.) It is clear that the minimum inventory cost occurs at $x = 500$. Therefore the optimal reorder quantity is 500 pairs of Lazy Loafers per order. This reorder quantity will result in $2000/500 = 4$ orders per year, and the store's minimum annual inventory cost will be

$$I(500) = \$20{,}000 + \frac{\$200{,}000}{500} + \$0.80(500) = \$20{,}800. \quad \square$$

■ **Computer Manual: Program ABMAXMIN**

4.3 EXERCISES

1. Let f have the following graph:

   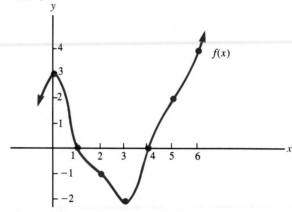

   Find the absolute maximum and absolute minimum of f, if they exist, on
   (a) the interval $[0, 4]$
   (b) the interval $[2, 6]$
   (c) the interval $(-\infty, 0]$
   (d) the interval $[3, +\infty)$

2. Let g have the following graph:

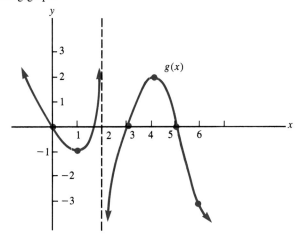

Find the absolute maximum and absolute minimum of g, if they exist, on
(a) the interval [0, 2)   (b) the interval (2, 6]
(c) the interval [3, 6]   (d) the x-axis

In Exercises 3 through 14, find the absolute maximum and absolute minimum of the given function on the given intervals.

3. $f(x) = 3x^2 - 18x - 5$, [0, 5], [6, 8], [0, +∞), the x-axis.
4. $f(x) = -2x^2 + 8x + 14$, [-1, 1], [1, +∞), (-∞, -1], the x-axis.
5. $f(x) = x^3 - 12x$, [0, 6], [-1, 1], [-6, 6], the x-axis.
6. $f(x) = -x^3 + x^2$, (-∞, 1], [1, +∞), [0, 1], the x-axis.
7. $f(x) = -x^3 + 21x^2 - 120x + 200$, [0, 12], [0, 5], [5, 12], the x-axis.
8. $f(x) = 2x^3 + 9x^2 - 60x + 1$, [-6, 6], [-6, 3], [1, +∞), (-∞, 0].
9. $f(x) = x^4 - 32x$, (-∞, 0], [-3, 3], [0, 4], [0, +∞).
10. $f(x) = -x^4 + 2x^2 + 6$, [-2, 2], [-1, 0], [0, 1], [1, 2].
11. $f(x) = \dfrac{x - 3}{x - 1}$, (-∞, 1), (1, +∞), [0, 2], [2, 4].
12. $f(x) = \dfrac{x + 1}{2x - 3}$, [-1, 3/2), (3/2, 2], [-1, 2], [-1, 1].
13. $f(x) = \dfrac{x^2}{x^2 - 9}$, (-3, 3), (3, +∞), [-2, 2], the x-axis.
14. $f(x) = \dfrac{x^2 - 1}{2x - 4}$, [0, +∞), (-∞, 0], [0, 2), (2, 4].

15. A boat that travels at a speed of x miles per hour uses $y = 6x^{-1} + x/150$ gallons of fuel per mile. What speed gives the minimum fuel cost?

16. What are the dimensions of the rectangular field of maximum area that can be fenced in with exactly 800 feet of fencing?

*17. Prove that the rectangle of maximum area having a fixed perimeter P is actually a square with side $P/4$.

18. What are the dimensions of the rectangular field whose area is 10,000 square meters if the fencing around the field is to be a minimum?

*19. Prove that the rectangle of minimum perimeter having a fixed area A is actually a square with side $\sqrt{A}$.

20. You wish to fence in a rectangular field. One side of the field will be bordered by a stone wall and therefore will not require any fencing. The area of the field must be 12,800 square feet, and fencing costs $2 per foot. What should the dimensions of the field be in order to minimize the cost of the fencing if the stone wall is 200 feet long?

21. Do Exercise 20 if the stone wall is 100 feet long.

22. A square plot of land has sides 30 meters long. You fence off a corner of the plot with a straight fence 20 meters long. Find the dimensions that will maximize the area of the fenced-off triangle.

23. A rectangular plot of land with an area of 600 square meters is fenced in on all four sides and then divided into two plots of equal area by another fence parallel to one of the outside fences. Find the minimum fencing required.

24. A window is in the shape of a rectangle surmounted by a semicircle. If the perimeter of the window is 16 feet, find the dimensions that maximize its area.

25. A wire 100 centimeters (cm) long is cut into two pieces and one of the pieces is formed into a square while the other is bent into an equilateral triangle. Where should the wire be cut and which part of it should be bent into the square if the total area of the two figures is to be maximized? (*Hint:* The area of an equilateral triangle of side s is $s^2\sqrt{3}/4$.)

26. Repeat Exercise 25 if the total area of the two figures is to be minimized.

27. A road is built from point A to point B in the figure:

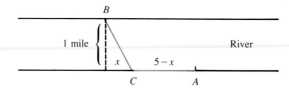

Suppose it costs $10,000 per mile to build the portion of the road along the river from A to C and $20,000 per mile to build the bridge from C to B. Where should C be located in order to minimize the cost of the road?

28. Repeat Exercise 27 if it costs $9000 per mile to build the bridge from C to B.

29. The base of a toy chest is to be a rectangle twice as long as it is wide. The material for the top and bottom of the chest costs $5 per square foot, and that for the sides costs $3 per square foot. The materials budget is $240. Find the dimensions that maximize the volume of the toy chest.

30. A box with an open top is to have a square base and a volume of 500 cubic centimeters. If the material for the box costs $0.50 per square centimeter, find the dimensions of the box that minimize its cost.

APPLICATIONS OF DIFFERENTIATION

31. A cylindrical container with a circular base and lid is to have a volume of 54 cubic inches. Find the dimensions that minimize the amount of material from which the container is made.

Management

32. A firm's profit function is given by
$$P(x) = -2x^3 + 270x^2 - 4800x - 6000,$$
where x is in thousands of units and profit is in dollars. Find the firm's maximum profit if it can produce at most 100,000 units; if it can produce at most 50,000 units.

33. A firm's profit function is given by
$$P(x) = -2x^3 + 216x^2 - 4320x - 20{,}000.$$
Find the firm's maximum profit if it can make and sell at most
   (a) 100 units
   (b) 50 units

34. Derma Corporation's cost function is given by
$$C(x) = 0.01x^2 + 10x + 40{,}000.$$
Find Derma's minimum average cost per unit if it can produce at most 5000 units; if it can produce at most 1500 units.

35. The demand function for a company's product is defined by
$$p = 1000 - \frac{q}{200}.$$
Find the company's maximum revenue and the price per unit that the company must charge to attain it if the firm can produce at most 200,000 units; if it can produce at most 80,000 units.

36. A chain of furniture stores sells 10,000 sofas each year. Each sofa costs the chain $500, and it costs $200 to order sofas from the supplier. It also costs the chain $100 to carry a sofa in inventory for a year. Assuming an average inventory equal to one half the optimal reorder quantity, find the optimal reorder quantity, the number of orders the chain must make each year, and the minimum annual inventory cost.

37. The Drive N' Shop supermarket sells 100 boxes of soap each week. Each box costs the store $0.50, each order of soap costs $2.00, and it costs $0.01 to carry a box of soap in inventory for one week. Assuming an average inventory equal to one half the optimal reorder quantity, find the optimal reorder quantity, the number of orders required each week, and the minimum weekly inventory cost.

*38. Octopus Industries makes radios. The annual demand for its radios is 50,000 units. It costs Octopus Industries $70 to make a radio, $2000 to set up a production run of radios, and $2 to carry a radio in inventory for a year. Assuming an average inventory equal to one half the number of radios produced per production run, find the number of radios that Octopus should produce in each production run, the number of production runs required each year, and the minimum inventory cost for the radios. (*Hint:* Let x denote the number of production runs the company should make each year.)

*39. Suppose that the quantity of a product demanded is Q units per time period, that it costs K dollars to place each order for the product, and that it costs A dollars to carry 1 unit of the product in inventory for the time period. Assuming that the average inventory is one half the optimal reorder quantity, show that the optimal reorder quantity is

$$x = \sqrt{\frac{2QK}{A}}.$$

40. A health club offers to give employees of a certain corporation a discount on its membership fee. If 20 employees (the minimum number the club will accept) join, the fee will be $100 each. For every additional employee who joins, the club will reduce the fee for each employee by $1. Find the number of employees that will maximize the club's revenue if the club can accept at most
    (a) 80 new members   (b) 50 new members

41. A promoter can sell 20,000 tickets to a rock concert at $10 each. For every dollar the price of a ticket is raised, it is estimated that 200 fewer tickets will be sold. Find the promoter's maximum revenue.

42. Slicendice Company has $120,000 to spend on advertising. Each radio ad will cost $3000, and each newspaper ad $6000. If Slicendice runs x radio ads and y newspaper ads, it will obtain

    $$E = 10{,}000x - 30{,}000xy + 20{,}000y + 6{,}000{,}000$$

    effective exposures to its ads. How many of each type of ad should the company run in order to maximize effective exposures?

43. Gruel World Corporation makes regular and deluxe model globes. Each regular globe costs $4 to make, and each deluxe one costs $8 to make. The corporation has budgeted $800 for daily production costs. If it makes x regular and y deluxe globes per day, its daily profit is $P = -x^2 + 3xy - 400$ dollars. How many of each type of globe must the company produce in order to maximize daily profit?

44. An advertising handbill is to have 24 square inches of printed material, with margins of 1 inch along each of the two sides, 2 inches at the top, and 1 inch at the bottom. Find the dimensions of the handbill that will minimize the cost of the paper used.

45. A florist sells 160 bouquets a day at $12 each. For each $1 rise in the price of a bouquet, sales will decrease by 10 bouquets per day. Find the florist's maximum revenue from sales of bouquets and the price that must be charged in order to attain it.

46. Repeat Exercise 45 if each $1 rise in price causes sales to decrease by 15 bouquets per day.

*47. (This exercise was suggested by material in Budnick, McLeavey, and Mojena, *Principles of Operations Research for Management*, second edition, Irwin, 1988.)
    A drill press costs $500,000 new and depreciates until its scrap value is $50,000. Its average operating cost t years after purchase is given by

    $$\overline{C}(t) = 40{,}000 + 4500t.$$

    Find the time when the drill press should be replaced. (*Hint:* Minimize the average cost per time period, which is given by

    $$\frac{500{,}000 - 50{,}000}{t} + \overline{C}(t).)$$

*48. (This exercise was suggested by material in Budnick, McLeavey, and Mojena, *Principles of Operations Research for Management*, second edition, Irwin, 1988.)

Let p denote the purchase price of an asset and s its salvage value. Assume that s is independent of the asset's age. Then the total capital cost of the asset is $p - s$, and its average cost at time t is therefore

$$A(t) = \frac{p - s}{t}.$$

Let the average operating cost of the asset at time t be linear,

$$\overline{C}(t) = mt + b.$$

Show that the asset should be replaced t_0 years after purchase, where

$$t_0 = \sqrt{\frac{p - s}{m}}.$$

Life Science

49. An egg ranch has 120 chickens, each of which produces 250 eggs per year. If more chickens are squeezed into the chicken coop, the resulting overcrowding will reduce egg production by one egg per chicken per year for each additional chicken squeezed in. Find the maximum egg production and the number of chickens that will yield this maximum production.

50. Do Exercise 49 if each additional chicken squeezed in reduces egg production by three eggs per chicken per year.

51. A lake was stocked with trout in 1965, and t years later there were

$$y = -\frac{1}{3}t^3 + 12.5t^2 - 100t + 1000$$

trout in the lake. Find the maximum and minimum numbers of trout in the lake from 1965 through 1985.

52. If a farmer sows x seeds per square foot, the resulting yield will be y bushels of grain, where

$$y = -2x^3 - 39x^2 + 4620x.$$

Find the farmer's maximum yield if there are enough seeds on hand to sow 30 seeds per square foot; if there are enough to sow 20 seeds per square foot.

53. Poiseuille's law states that (if the units of measurement are properly chosen) the velocity of blood x units from the central axis of an artery of radius r is

$$V(x) = r^2 - x^2.$$

Find the position within the artery where blood flows the fastest.

54. If a blood vessel contracts, the velocity at which blood flows through it is affected. If the units of measurement are properly chosen, the average velocity of the blood flowing through an artery of radius x, where x is less than or equal to the normal radius r, is

$$V(x) = x^2(r - x).$$

Find the amount of contraction that maximizes the average velocity of the blood.

4.4 CURVE SKETCHING

Sometimes a knowledge of the maxima and minima of a function, relative or absolute, is all we need to know in order to answer a question. However, if we want to analyze a function in the manner described earlier, we need to know

- The intervals on which the function is increasing and those on which it is decreasing.
- The location of its relative maxima and minima.
- The intervals on which the function is concave upward and those on which it is concave downward.

It is also useful to know the intercepts of the function and its asymptotes, if any. Given all this information, we can then make an accurate analysis of the behavior of the function. In this section we collect our results of Sections 4.1 and 4.2 in a procedure that will allow us to analyze functions with relative ease.

Recall that the first derivative test locates relative maxima and minima. The diagram used in the test also tells us the intervals on which the function is increasing and those on which it is decreasing. The second derivative tells us about the concavity of the function: if the second derivative is positive on an interval, the function is concave upward there; if the second derivative is negative, the function is concave downward on the interval. A point on its graph where the function changes its concavity is called **a point of inflection.** The function f whose graph is shown in Figure 4–28 has points of inflection at $(a, f(a))$ and $(b, f(b))$. (Note that in Figure 4–28, $x = b$ is a critical value for f, because the tangent line at $(b, f(b))$ is horizontal, so that $f'(b) = 0$; hence critical values can yield points of inflection.)

If $(x, f(x))$ is a point of inflection for a function f, the function changes concavity at $(x, f(x))$, and therefore f'' has different signs on either side of x. But this can happen only if $f''(x) = 0$ or $f''(x)$ does not exist. Thus we can find the points of inflection for f by finding those values of x in its domain such that $f''(x) = 0$ or $f''(x)$ does not exist.

Suppose we use the first and second derivatives of a function to find its relative maxima, its relative minima, its points of inflection, and to determine where the function is increasing, where it is decreasing, where it is concave upward, and where it is concave downward. Suppose we also find the intercepts and asymptotes of the function. All this information will enable us to sketch an accurate graph of the function. Let us outline a procedure for this.

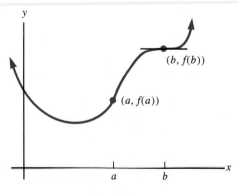

FIGURE 4–28

Applications of Differentiation

The Curve-Sketching Procedure

To sketch the graph of a function f, proceed as follows:

1. Use the first derivative test to locate all relative maxima and relative minima of the function, and to find the intervals on which it is increasing and those on which it is decreasing.

2. a. Find all solutions of the equation $f''(x) = 0$ and also all values of x for which $f''(x)$ does not exist, and plot these values on the x-axis.
   b. On each interval of the diagram of step 2a, determine whether f'' is positive or negative, and add this information to the diagram.
   c. On intervals where f'' is positive, f is concave upward; on intervals where f'' is negative, f is concave downward. Add this information to the diagram. The diagram now shows the intervals on which f is concave upward and those on which it is concave downward and indicates the values of x that yield points of inflection.

3. Find the intercepts of f.
   a. If $f(0)$ is defined, the y-intercept is $(0, f(0))$.
   b. To find the x-intercepts, set $f(x) = 0$ and solve for x. (In practice, this step is often omitted when the equation $f(x) = 0$ is not easily solved.)

4. Find the asymptotes for f.
   a. If the denominator of $f(x)$ is zero and the numerator is nonzero when $x = c$, then the vertical line $x = c$ is an asymptote for f. To determine how the graph approaches the vertical asymptote as x approaches c, find
   $$\lim_{x \to c^+} f(x) \quad \text{and} \quad \lim_{x \to c^-} f(x).$$
   b. If
   $$\lim_{x \to +\infty} f(x) = K \quad \text{or} \quad \lim_{x \to -\infty} f(x) = K,$$
   then $y = K$ is an asymptote for f as $x \to +\infty$ or $x \to -\infty$, respectively. (Here K can be either a number or an expression involving x.)

5. Sketch the graph of f using the information obtained in steps 1 through 4.

EXAMPLE 1 Let us use the curve-sketching procedure to draw the graph of the quadratic function defined by
$$f(x) = x^2 - 8x + 15.$$

1. Since $f'(x) = 2x - 8$, the first derivative test yields the following diagram:

Hence the function has a relative minimum at $(4, f(4)) = (4, -1)$ and is decreasing on the interval $(-\infty, 4)$ and increasing on the interval $(4, +\infty)$.

2. **a.** Since $f''(x) = 2$, the equation $f''(x) = 0$ has no solution. Also, $f''(x)$ exists for all x. Therefore the diagram is

———————————————————————— x

b. Since $f''(x) = 2$ is positive for all x, we have

$+$
———————————————————————— x

c. Since $f''(x)$ is positive for all x, f is concave upward for all x, so we have

Since the function never changes concavity, it has no points of inflection.

3. **a.** The y-intercept is $(0, f(0)) = (0, 15)$.
   b. The x-intercepts occur at the solutions of
   $$x^2 - 8x + 15 = 0,$$
   or
   $$(x - 3)(x - 5) = 0,$$
   and thus are $(3, 0)$ and $(5, 0)$.

4. The function is a polynomial, and polynomials have no asymptotes.

5. The graph of the function is shown in Figure 4–29. Notice how each piece of information determined in steps 1 through 4 has been used to sketch the graph. □

EXAMPLE 2 Let us sketch the graph of the polynomial function defined by
$$f(x) = x^3 - 3x^2 - 24x + 3.$$

1. Since
$$f'(x) = 3x^2 - 6x - 24 = 3(x + 2)(x - 4),$$

Applications of Differentiation

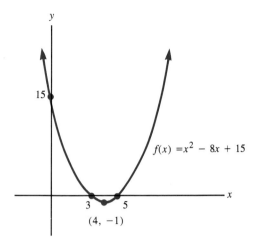

FIGURE 4-29

the diagram of the first derivative test is

Hence the function

- has a relative maximum at $(-2, f(-2)) = (-2, 31)$;
- has a relative minimum at $(4, f(4)) = (4, -77)$;
- is increasing on the intervals $(-\infty, -2)$ and $(4, +\infty)$;
- is decreasing on the interval $(-2, 4)$.

2. a. Since

$$f''(x) = 6x - 6 = 6(x - 1),$$

setting $f''(x) = 0$ and solving for x yields $x = 1$. The second derivative exists for all x. Hence we have the diagram

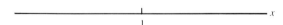

b. Since $f''(x) < 0$ if $x < 1$ and $f''(x) > 0$ if $x > 1$, we have

c. Therefore f is concave downward on $(-\infty, 1)$ and concave upward on $(1, +\infty)$, and we have the diagram

Since f changes concavity at $(1, f(1)) = (1, -23)$, this is a point of inflection for the function.

3. **a.** The y-intercept is $(0, f(0)) = (0, 3)$.
   b. The x-intercepts occur at the values of x for which
   $$x^3 - 3x^2 - 24x + 3 = 0.$$
   Since this equation is not easily solved, we omit this step.

4. The function is a polynomial and hence has no asymptotes.

5. The function is graphed in Figure 4–30. ∎

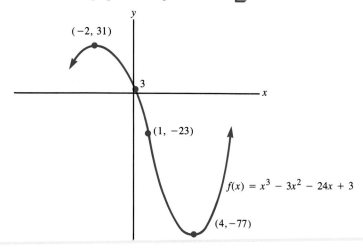

FIGURE 4–30

EXAMPLE 3   We sketch the graph of the rational function whose defining equation is
$$f(x) = \frac{4x - 3}{x - 2}.$$

Note that $f(2)$ is not defined.

1. Since
$$f'(x) = -\frac{5}{(x-2)^2},$$
the equation $f'(x) = 0$ has no solution. Since $f'(2)$ does not exist and $x = 2$ is not a critical value for f, we have the diagram

APPLICATIONS OF DIFFERENTIATION

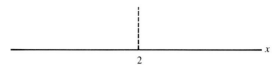

Note the vertical dashed line at $x = 2$: this reminds us that $x = 2$ is not a critical value and thus cannot yield a relative maximum or minimum. Since $f'(x) < 0$ for all values of x, the function is decreasing on the intervals $(-\infty, 2)$ and $(2, +\infty)$:

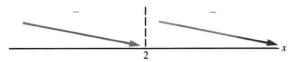

2. a. Since

$$f''(x) = \frac{10}{(x-2)^3},$$

the equation $f''(x) = 0$ has no solution. The second derivative does not exist at $x = 2$. Our diagram is

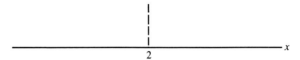

Again, note the dashed line at $x = 2$ to remind us that $f(2)$ is not defined, and hence that $x = 2$ cannot yield a point of inflection.

b. Since $f''(x) < 0$ if $x < 2$ and $f''(x) > 0$ if $x > 2$, we have

c. The diagram of step 2b shows that f is concave downward on $(-\infty, 2)$ and concave upward on $(2, +\infty)$. Hence we have

The function has no points of inflection.

3. a. The y-intercept is $(0, f(0)) = (0, \tfrac{3}{2})$.
b. The x-intercepts occur at the values of x for which

$$\frac{4x - 3}{x - 2} = 0.$$

The only solution of this equation is $x = 3/4$. Thus the function's single x-intercept is $(3/4, 0)$.

4. **a.** When $x = 2$, the denominator of $f(x)$ is zero but the numerator is not zero. Therefore the line $x = 2$ is a vertical asymptote for f. As $x \to 2^+$, $4x - 3$ and $x - 2$ are both positive, so

$$\lim_{x \to 2^+} f(x) = \lim_{x \to 2^+} \frac{4x-3}{x-2} = +\infty.$$

Similarly, as $x \to 2^-$,

$$\lim_{x \to 2^-} f(x) = -\infty.$$

b. Since

$$\lim_{x \to \pm\infty} \frac{4x-3}{x-2} = \lim_{x \to \pm\infty} \frac{4 - 3(1/x)}{1 - 2(1/x)} = \frac{4 - 3(0)}{1 + 2(0)} = 4,$$

the line $y = 4$ is a horizontal asymptote for f as $x \to +\infty$ and as $x \to -\infty$.

5. The graph of the function is shown in Figure 4–31. ∎

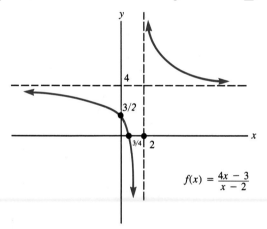

FIGURE 4–31

EXAMPLE 4 Let us sketch the graph of the rational function f defined by

$$f(x) = x + \frac{1}{x}.$$

Note that $f(0)$ is not defined.

1. Since

$$f'(x) = 1 - \frac{1}{x^2},$$

$f'(0)$ does not exist and the equation $f'(x) = 0$ has solutions $x = \pm 1$. Therefore we have

Since $f'(x) > 0$ for $x < -1$ and $x > 1$, and $f'(x) < 0$ for $-1 < x < 0$ and $0 < x < 1$, the diagram becomes

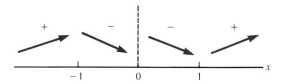

Therefore f is increasing on $(-\infty, -1)$ and $(1, +\infty)$, decreasing on $(-1, 0)$ and $(0, 1)$, and has relative maxima at $(-1, -2)$ and $(1, 2)$.

2. Since

$$f''(x) = \frac{2}{x^3},$$

$f''(0)$ does not exist and the equation $f''(x) = 0$ has no solution. Therefore we have

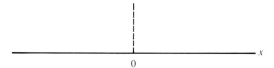

Since $f''(x) < 0$ for $x < 0$ and $f''(x) > 0$ for $x > 0$, the diagram becomes

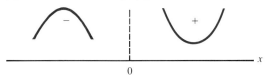

Hence f is concave downward on $(-\infty, 0)$, concave upward on $(0, +\infty)$, and has no points of inflection.

3. Since f is not defined at 0, the function has no y-intercept. It has no x-intercepts because the equation $f(x) = 0$ has no solution.

4. The line $x = 0$ (the y-axis) is a vertical asymptote for f. Note that

$$\operatorname*{Lim}_{x \to 0^+} f(x) = \operatorname*{Lim}_{x \to 0^+} \left(x + \frac{1}{x}\right) = 0 + (+\infty) = +\infty$$

and

$$\operatorname*{Lim}_{x \to 0^-} f(x) = \operatorname*{Lim}_{x \to 0^-} \left(x + \frac{1}{x} \right) = 0 + (-\infty) = -\infty.$$

Also,

$$\operatorname*{Lim}_{x \to +\infty} f(x) = \operatorname*{Lim}_{x \to +\infty} \left(x + \frac{1}{x} \right) = +\infty + 0 = +\infty$$

and

$$\operatorname*{Lim}_{x \to -\infty} f(x) = \operatorname*{Lim}_{x \to -\infty} \left(x + \frac{1}{x} \right) = -\infty + 0 = -\infty.$$

Therefore f has no horizontal asymptotes. However, as $x \to \pm\infty$,

$$f(x) = x + \frac{1}{x} \to x + 0 = x.$$

Hence the line $y = x$ is an oblique asymptote for f.

5. The graph of f is shown in Figure 4–32. ∎

EXAMPLE 5 Suppose that Acme Company's profit function is given by

$$P(x) = -x^3 + 36x^2 - 240x - 100,$$

where x is in thousands of units and profit is in thousands of dollars. Let us analyze Acme's profits. We begin by sketching the graph of the profit function. Since x

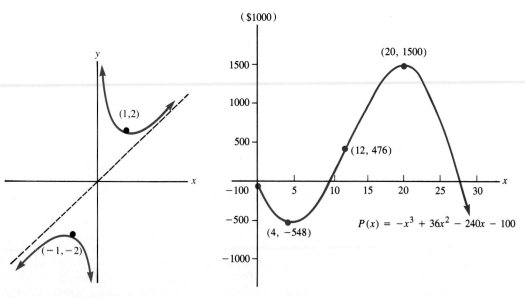

FIGURE 4–32

FIGURE 4–33

must be nonnegative, we can restrict our diagrams to the interval $[0, +\infty)$. We will leave out some of the details of the curve-sketching procedure: you should check these details.

1. The diagram of the first derivative test is

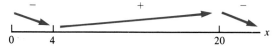

The relative maximum occurs at $(20, 1500)$, and the relative minimum occurs at $(4, -548)$.

2. The concavity diagram is

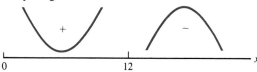

The point of inflection is $(12, 476)$.

3. The y-intercept is $(0, -100)$.
4. There are no asymptotes.
5. The graph of Acme's profit function is shown in Figure 4–33.

Now let us use the graph to analyze Acme's profit. Starting at $x = 0$ and moving to the right along the x-axis, we observe the following:

a. Since $P(0) = -100$, when Acme produces 0 units of its product it suffers a loss of \$100,000. Thus Acme's fixed cost is \$100,000.
b. As production increases from 0 units to 4000 units, Acme's profit decreases, but at a slower and slower rate, until it reaches a minimum of $-\$548,000$ at 4000 units.
c. As production increases from 4000 units to 12,000 units, profit increases, and it does so at a faster and faster rate as the point of inflection is approached.
d. At a production level of 12,000 units, Acme's profit is \$476,000. As production increases from 12,000 units to 20,000 units, profit continues to increase, but at a slower and slower rate, until it reaches a maximum of \$1.5 million at 20,000 units.
e. If production increases beyond 20,000 units, profit will decrease, and it will do so at a faster and faster rate as production goes up; eventually losses will begin to accrue at an ever-increasing rate.

Note also that careful drawing of the graph shows us that Acme will break even somewhere between 9000 and 10,000 units, and again somewhere near 27,000 units. (The exact break-even quantities are the solutions of the equation

$$-x^3 + 36x^2 - 240x - 100 = 0.)$$

■ *Computer Manual: Program CURSKECH*

4.4 EXERCISES

In Exercises 1 through 38, use the curve-sketching procedure to sketch the graph of the given function.

1. $f(x) = x^2 - 2x - 15$
2. $f(x) = -2x^2 + 18x - 28$
3. $f(x) = 2x^3 + 3x^2 - 12x + 2$
4. $f(x) = -2x^3 - 12x^2 - 24x + 30$
5. $f(x) = -x^3 + 21x^2 - 120x + 20$
6. $f(x) = -3x^3 + 6x^2$
7. $f(x) = x^3 + 3x^2 - 9x + 10$
8. $f(x) = -2x^3 + 27x^2 - 108x + 12$
9. $f(x) = x^4 - 32x$
10. $f(x) = x^4 - x^2$
11. $f(x) = 3x^4 + 4x^3$
12. $f(x) = 6x^5 + 15x^4 + 10x^3$
13. $f(x) = x^4 - 8x^2 + 1$
14. $f(x) = 3x^5 + 15x^4 - 10x^3$
15. $f(x) = \frac{1}{5}x^5 - \frac{8}{3}x^3 + 7x$
16. $f(x) = x^5 - 20x^2 + 1$
17. $f(x) = 2x^6 - 27x^4$
18. $f(x) = 5x^6 - 18x^5 + 30x^4 - 40x^3 + 45x^2 - 30x$
19. $f(x) = x^{2/3}$
20. $f(x) = x^{4/3}$
21. $f(x) = (x - 3)^{11/5}$
22. $f(x) = (x - 3)^{6/5}$
23. $f(x) = (x - 3)^{1/5}$
24. $f(x) = \dfrac{2}{x^2}$
25. $f(x) = \dfrac{3}{x^3}$
26. $f(x) = \dfrac{1}{x - 1}$
27. $f(x) = \dfrac{2}{(x + 1)^2}$
28. $f(x) = \dfrac{x - 3}{x + 2}$
29. $f(x) = \dfrac{2x - 1}{3x + 2}$
30. $f(x) = \dfrac{3x + 2}{x - 4}$
31. $f(x) = \dfrac{x}{x^2 - 1}$
32. $f(x) = \dfrac{x^2}{x + 1}$
33. $f(x) = \dfrac{x^2 - 1}{x}$
34. $f(x) = \dfrac{x^2}{x^2 - 3}$
35. $f(x) = x + \dfrac{4}{x}$
36. $f(x) = 0.5x^2 - 1/x$
37. $f(x) = \dfrac{x^2 - 3}{x^2}$
38. $f(x) = \dfrac{x^3 + 1}{x^2}$

*39. Use the curve-sketching procedure to graph
   (a) $f(x) = ax^2 + bx + c$, $a > 0$
   (b) $f(x) = ax^2 + bx + c$, $a < 0$

40. An economist estimates that if inflation is q percent per year, then families will save approximately s percent of their income, where
$$s = -0.001q^3 + 0.06q^2 - 1.2q + 12.$$
Draw the graph and analyze the effect of inflation on savings.

Management

41. Brawny Corporation's profit function is given by
$$P(x) = -x^3 + 90x^2 - 1500x - 9,$$
where x is in thousands of units and profit is in thousands of dollars. Sketch the graph of the profit function and analyze Brawny's profit in the manner of Example 5.

42. Since it was founded 15 years ago, a firm's annual sales have been S million dollars per year, where
$$S = 0.025t^3 - 1.2t^2 + 18t + 12$$
and $t = 0$ represents the year the firm was founded. Sketch the graph of S and analyze annual sales since the firm's founding.

43. Suppose the cost of removing $x\%$ of the pollutants in a chemical plant's waste water is
$$y = \frac{8x}{100 - x}$$
million dollars, $0 \leq x < 100$. Sketch the graph and analyze the cost.

44. Repeat Exercise 43 if
$$y = \frac{6x}{110 - x}, \quad 0 \leq x \leq 100.$$

45. The cost of decommissioning a nuclear power plant that produces electricity for t years, $0 \leq t \leq 25$, is estimated to be y million dollars, where
$$y = \frac{60t + 10}{30 - t}.$$
Graph this function, and analyze the cost.

46. Each weekday *The Wall Street Journal* publishes a **yield curve** for U.S. government securities. The yield curve depicts the percentage yields y on government securities as a function of their maturities, which run from 3 months to 30 years. Suppose a yield curve is given by
$$y = \frac{5t + 0.5}{0.5t + 0.15}, \quad 0.25 \leq t \leq 30,$$
t in years. Graph this function and analyze yield as a function of maturity.

47. The average cost function of Gormless, Inc., is given by
$$A(x) = 2x + \frac{50}{x}, \quad x > 0.$$
Here x is in hundreds of units and $A(x)$ is in dollars. Draw the graph of A and discuss Gormless's average cost per hundred units produced.

48. Delta Company's average cost function is given by
$$A(x) = \frac{300}{x} + 0.6x, \quad x > 0.$$
Graph the function A and analyze Delta's average cost per unit.

49. A firm's average cost per thousand units is given by
$$A(x) = 0.02x + 600/x^2, \quad x > 0.$$
Sketch the graph and analyze average cost per thousand.

Life Science

50. The number of deer in a region is
$$y = -t^3 + 18t^2 - 96t + 2000,$$

where $t = 0$ represents 1976, t in years. Sketch the graph and discuss the changes in the deer population from 1976 through 1986. At one time deer hunting was banned in the region for a four-year period. Can you tell when this occurred?

51. The cost of removing $x\%$ of the pollutants from a lake is estimated to be

$$C(x) = \frac{3x}{100 - x}, \quad 0 \le x < 100.$$

Draw the graph of C and analyze the cost.

52. An immunization campaign is described by the equation

$$y = (9t + 40)/(10t + 125),$$

where t is time in months since the campaign began and y is the proportion of the population immunized at time t. Sketch the graph and analyze the campaign.

53. Suppose that in the t years since they have been declared an endangered species, the population of whooping cranes has grown to y birds, where

$$y = \frac{20t^2 + 25}{t + 1}, \quad t \ge 0.$$

Graph this function and analyze the recovery of the whooping crane.

54. It is estimated that t years after the start of an international effort to save the whales the number of blue whales alive was y hundred, where

$$y = \frac{(t + 1)^3 + 128}{t + 1}, \quad t \ge 0.$$

Graph this function and discuss the changes in the blue whale population.

Social Science

55. Housing starts in a city are described by the equation

$$y = -3t^4 + 80t^3 - 750t^2 + 3000t + 500,$$

where t is in years, with $t = 0$ representing 1978, and y is the number of new houses started in the city in year t. Draw the graph and analyze housing starts in the city from 1978 through 1990. (*Hint:* $t - 10$ and $(t - 5)^2$ are factors of the derivative of y with respect to t.)

56. The number of children attending public school in a large city in year t is

$$y = -\frac{t^4}{4} + \frac{14t^3}{3} - 20t^2 + 200,$$

where $t = 0$ represents 1989 and y is in thousands. Draw the graph and analyze enrollment from 1989 through 2001. Do you think it would make sense to sell school buildings in 1992 because of declining enrollment?

57. The cost of providing x million elementary school children with a free glass of milk at lunch is estimated to be

$$y = \frac{25x}{10 + x}$$

million dollars, $x \ge 0$. Graph this function and analyze the cost.

58. It is estimated that if a state's economy continues to grow at its current pace, the unemployment rate in the state t months from now will be $y\%$, where

$$y = \frac{2.4t + 1.3}{t + 0.25}, \quad 0 \le t \le 12.$$

Graph this function and discuss the estimated unemployment rate over the next year.

Computer Science

An article by Peter Deming ("The Science of Computing: Speeding Up Parallel Processing," *American Scientist*, July–August, 1988) states that the best possible speed-up in running a program on a computer with $p > 1$ processors rather than 1 processor is given by

$$f(p) = \frac{np}{n + s(p - 1)}.$$

Here n is the number of operations required by the program and s is the number of operations in the longest sequential path in the program. Exercises 59 through 61 refer to the function f. (Note that p is not a continuous variable, since $p = 2, 3, 4, \ldots$; however, for the purposes of these exercises, you may treat p as if it were continuous.)

*59. Graph f for $p > 1$ when $n = 1000$ and
 (a) $s = 10$ (b) $s = 100$ (c) $s = 1000$

*60. Graph f for $p > 1$ and any n and s if
 (a) $s < n$ (b) $s = n$

*61. Show that for all p, n, and s, $f(p) \le n/s$. This inequality is known as Amdahl's Law, and it was long taken as showing that there is a fundamental limit to how much a program could be speeded up by the use of multiprocessors. Can you explain why the inequality implies such a limit? (In fact, it has recently been demonstrated that Amdahl's Law does not hold under all circumstances.)

SUMMARY

This summary consists of terms and symbols whose meaning you should know, facts you should know, and techniques and methods you should be able to employ.

Section 4.1

Relative maximum, relative minimum. If $(a, f(a))$ is a relative maximum or relative minimum point and $f'(a)$ exists, then $f'(a) = 0$. Critical values. Relative maxima and relative minima occur at critical values. Not all critical values yield relative maxima or relative minima. Finding critical values. Function increasing on an interval, decreasing on an interval. Increasing and decreasing functions and the first derivative: if $f'(x) > 0$ on an interval, f is increasing on the interval; if $f'(x) < 0$ on an interval, f is decreasing on the interval. Determining the intervals over which a function is increasing and those over which it is decreasing. The first derivative test. Using the first derivative test to find relative maxima and relative minima.

Section 4.2

Concavity: concave upward, concave downward. Concavity and the second derivative: if $f''(x) > 0$ on an interval, f is concave upward over the interval; if $f''(x) < 0$ on an interval, f is concave downward over the interval. Determining the intervals over which a function is concave upward and those over which it is concave downward. The second derivative test. Using the second derivative test to find relative maxima and relative minima. The second derivative test fails when the second derivative is zero or does not exist.

Section 4.3

Absolute maximum, absolute minimum. Finding the absolute maximum and absolute minimum of a continuous function on an interval. Setting up and solving maximum/minimum problems. Inventory problems.

Section 4.4

Point of inflection. The curve-sketching procedure. Using the curve-sketching procedure to sketch the graphs of functions. Analyzing the graphs of models.

REVIEW EXERCISES

Exercises 1 through 5 refer to the function f whose graph is as follows:

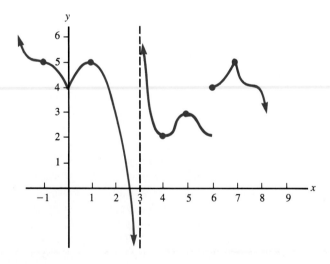

1. Find all relative maxima and relative minima for f and where they occur.
2. Find all critical values for f.
3. Find all intervals where f is increasing and all those where it is decreasing.

4. Find all intervals where f is concave upward and all those where it is concave downward.

5. Find all points of inflection of f.

In Exercises 6 through 11, use the first derivative test to find all relative maxima and relative minima for the given function.

6. $f(x) = -x^2 + 7x - 4$

7. $f(x) = x^3 + 6x^2 - 180x - 1200$

8. $f(x) = x^4 + 4x^3$

9. $f(x) = 3x^5 - 20x^3 - 480x + 200$

10. $f(x) = \dfrac{x^2 - 8}{2x + 1}$

11. $f(x) = \dfrac{2x + 1}{x^2 + 8}$

12. A firm's average cost per unit is given by
$$A(x) = 0.5x^2 - 54\sqrt{x} + 200.$$
Find the firm's minimum average cost per unit.

13. A new disease that affects birch trees is introduced into a locality, and t years later the proportion of birches that have the disease is
$$y = \dfrac{t^2 + 8t + 2}{4t^2 + 24}.$$
What was the maximum proportion of birches that contracted the disease?

In Exercises 14 through 19, use the second derivative test to find all relative maxima and relative minima for the given function. If the second derivative test fails, use the first derivative test.

14. $f(x) = -2x^2 + 9x - 3$

15. $f(x) = x^3 + 6x^2 - 15x$

16. $f(x) = x^4 - 4.5x^2 + 1$

17. $f(x) = \dfrac{1}{5}x^5 - 4x^2 + 1$

18. $f(x) = x^{7/6}$

19. $f(x) = x^{5/6}$

20. The demand function for a firm's product is given by
$$p = 6000 - 2\sqrt{q}.$$
Use the second derivative test to find the firm's maximum revenue.

A function has the following graph:

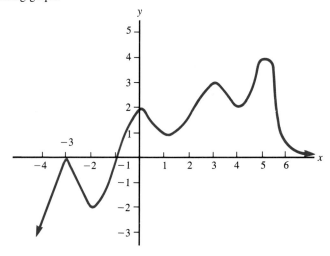

In Exercises 21 through 28, find the absolute maximum and absolute minimum of the function on the given interval.

21. $[0, 5]$ 22. $[0, 3]$ 23. $[-2, 5]$ 24. $[-2, 3]$
25. $[0, +\infty)$ 26. $[5, +\infty)$ 27. $(-\infty, -2]$ 28. $(-\infty, 0]$

29. Find the absolute maximum and absolute minimum of the function defined by
$$f(x) = x^4 - 32x^2$$
on
(a) the x-axis (b) the interval $[-1, 1]$
(c) the interval $[-5, 6]$ (d) the interval $[1, 6]$

30. A firm's profit function is given by
$$P(x) = -x^3 + 150x^2 - 4800x.$$
Find its maximum profit if it can produce and sell at most 50 units; an unlimited number of units.

31. A car dealership sells 400 new cars per year. Each new car costs $12,000, it costs the dealership $1000 to order cars, and is costs $2000 to carry a car in inventory for a year. Assuming that the average inventory of cars is one half the number ordered, find the optimal reorder quantity and the minimum annual inventory cost for the dealership.

32. Find the dimensions of a cylindrical container with a circular base and top if its volume is to be maximized and its surface area is to be 6 square feet. ($V = \pi r^2 h$ and $S = 2\pi r^2 + 2\pi rh$.)

In Exercises 33 through 36, use the curve-sketching procedure to sketch the graph of the function.

33. $f(x) = -2x^2 + 11x + 14$ 34. $f(x) = -x^3 + 6x^2 + 96x$
35. $f(x) = \dfrac{x^2 - 4}{x^2 - 9}$ 36. $f(x) = \dfrac{x^3 + 1}{2x^2}$

37. A pollster finds that t months after inauguration y percent of the voters approve of the job that a governor is doing, where
$$y = 0.01\left(-\frac{t^3}{3} + 21t^2 - 360t + 6000\right), \quad 0 \leq t \leq 46.$$
Sketch the graph of this function and analyze the governor's popularity.

38. The cost of desalinating sea water to produce fresh water depends on the proportion of salt removed from the sea water. Suppose that the cost of removing proportion p of the salt from sea water is y dollars per gallon, where
$$y = \frac{0.20p}{1 - p}, \quad 0 \leq p < 1.$$
Graph and analyze this function.

39. The neutron radiation given off by a cold-fusion experiment was measured to be y times the normal background radiation, where
$$y = \frac{261t - 30t^2}{30t + 100} + 1, \quad 0 \leq t \leq 8,$$
t in hours since the start of the experiment. Graph this function and analyze the radiation.

ADDITIONAL TOPICS

Here are some additional topics not covered in the text that you might wish to investigate on your own.

1. Rolle's Theorem says the following:

    If f is continuous on $[a, b]$ and f' exists on (a, b), and if $f(a) = f(b) = 0$, then there is some number c in (a, b) such that $f'(c) = 0$.

    Find out what Rolle's Theorem says about the graph of f and find out how it is proved.

2. The Mean Value Theorem for derivatives says the following:

    If f is continuous on $[a, b]$ and f' exists on (a, b), there is some number c in (a, b) such that $f(b) = f(a) + f'(c)(b - a)$.

    Find out what the Mean Value Theorem says about the graph of f and find out how it is proved.

3. The Extended Mean Value Theorem says the following:

    If f and f' are continuous on $[a, b]$ and f'' exists on (a, b), then there is some number c in (a, b) such that

    $$f(b) = f(a) + f'(a)(b - a) + \frac{f''(c)}{2}(b - a)^2.$$

    Find out how the Extended Mean Value Theorem is proved.

4. If f' exists in some interval containing a, then the **linear approximation of f** for x near a is

    $$f(x) \approx f(a) + f'(a)(x - a).$$

    Show how this approximation can be derived. If f'' also exists and is continuous in the interval, then the error of the linear approximation is $E(x)$, where

    $$E(x) = f(a) + f'(a)(x - a) - f(x).$$

    Show that

    $$|E(x)| \leq 0.5 \text{ Max } \{|f''(c)| \mid c \text{ between } a \text{ and } x\}(x - a)^2.$$

    Demonstrate how the linear approximation of f and the bound on its error are used.

5. Cauchy's form of the Mean Value Theorem says the following:

    If f and g are continuous on $[a, b]$ and f' and g' exist on (a, b), and if $g'(x) \neq 0$ for any x in (a, b), then there is some number c in (a, b) such that

    $$\frac{f(b) - f(a)}{g(b) - g(a)} = \frac{f'(c)}{g'(c)}.$$

    Find out the geometric meaning of Cauchy's form of the Mean Value Theorem and find out how it is proved.

6. L'Hôpital's Rule says the following:

    If f and g are continuous on $[a, b]$ and f' and g' exist on (a, b), except possibly at r, and if

    $$\lim_{x \to r} f(x) = \lim_{x \to r} g(x) = 0,$$

then

$$\lim_{x \to r} \frac{f(x)}{g(x)} = \lim_{x \to r} \frac{f'(x)}{g'(x)},$$

provided the limit on the right-hand side exists.

Find out how L'Hôpital's Rule is proved. L'Hôpital's Rule has several extensions: its conclusion holds when

$$\lim_{x \to r} f(x) = \pm \infty \quad \text{and} \quad \lim_{x \to r} g(x) = \pm \infty$$

and when the number r is replaced by $\pm \infty$. Show how L'Hôpital's Rule and its extensions are used to evaluate limits.

SUPPLEMENT: QUALITATIVE SOLUTIONS OF DIFFERENTIAL EQUATIONS

A **differential equation** is an equation involving derivatives. For instance, suppose y is a function of t; then

$$\frac{dy}{dt} = 2t,$$

or

$$y' = 2t,$$

is a differential equation for y. A **solution** of a differential equation is a function that satisfies the equation. For example, the function defined by $y = t^2 + c$, where c is any number, is a solution of the differential equation $y' = 2t$. This is so because if $y = t^2 + c$, then

$$y' = \frac{d}{dt}(t^2 + c) = 2t + 0 = 2t.$$

Differential equations are important because in many applications the rate of change (= derivative) of a function is known, but the function itself is not known. To see how this can occur, suppose a biologist studying an animal population finds that the size of the population is increasing at a constant rate of 2% per year. If $y = y(t)$ denotes the size of the population at time t, t in years, then y' is the rate of change of the population, and hence (as we noted in Section 3.3), y'/y is the *proportional* rate of change of the population. But to say that the size of the population is increasing at 2% per year is the same as saying that its proportional rate of change is 0.02, so

$$\frac{y'}{y} = 0.02,$$

or

$$y' = 0.02y.$$

Applications of Differentiation

Thus the biologist's knowledge of the growth rate of the population translates into a differential equation concerning the function that models population size.

We do not intend to consider here the question of how one finds an explicit solution of a differential equation. Instead we wish to demonstrate how it is often possible to sketch the graph of a solution to a differential equation *without* finding the solution explicitly. This procedure is referred to as finding a **qualitative solution** to the differential equation. We illustrate with an example.

EXAMPLE 1   Suppose that a biologist's knowledge of the growth rate of a population translates into the differential equation

$$y' = 0.02y,$$

as before. Suppose also that the biologist knows that the population size will be positive for all $t \geq 0$, that is, that $y > 0$ on the interval $[0, +\infty)$. Then $y > 0$ on $[0, +\infty)$ implies that

$$y' = 0.02y > 0$$

on $[0, +\infty)$. But if a function has a positive derivative on an interval, it is increasing there. Therefore we conclude that

$$y \text{ is increasing on } [0, +\infty).$$

We can also look at the second derivative of y:

$$y' = 0.02y$$

implies that

$$y'' = 0.02y'.$$

Since we already know that $y' > 0$ on $[0, +\infty)$, it follows that $y'' > 0$ on $[0, +\infty)$. But if a function has a positive second derivative on an interval, it is concave upward there. Hence

$$y \text{ is concave upward on } [0, +\infty).$$

Thus we know that $y(0) > 0$ (because $y > 0$ for all $t \geq 0$) and that y is increasing and concave upward on $[0, +\infty)$. Therefore the graph of y must be as shown in Figure 4–34. This graph is the qualitative solution of the differential equation. ☐

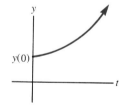

FIGURE 4–34

Qualitative solutions of differential equations are important because often we do not require an explicit solution, but only need to know how the solution behaves in a general way: is it increasing or decreasing, concave upward or downward, does it have maxima, minima, or points of inflection, does it approach a limit, and so on.

EXAMPLE 2   Again let $y = y(t)$ be the size of a population at time t, and suppose that a biologist's knowledge of the population results in the differential equation

$$y' = k(L - y),$$

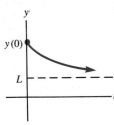

FIGURE 4–35

where k and L are positive constants and $y > L$ for $t \geq 0$. Then since $k > 0$ and $L - y < 0$ for $t \geq 0$, we have $y' < 0$ on $[0, +\infty)$. Hence,

y is decreasing on $[0, +\infty)$.

Now, $y'' = -ky'$, and since $k > 0$ and $y' < 0$ for $t \geq 0$, it follows that $y'' > 0$ on $[0, +\infty)$. Thus

y is concave upward on $[0, +\infty)$.

Therefore the graph of y must be as in Figure 4–35. Note that the size of the population is decreasing. Note also that y decreasing, concave upward, and greater than L implies that the population size declines toward L as a lower bound. □

EXAMPLE 3 Again let $y = y(t)$ be the size of a population at time t. Suppose a biologist's knowledge of the population results in the differential equation

$$y' = ky(L - y),$$

where k and L are positive numbers. Assume also that

$$0 < y < L$$

for all $t \geq 0$, and that $y(0) < L/2$. Let us find the qualitative solution of the differential equation on $[0, +\infty)$.

Since k, y, and $L - y$ are all positive on $[0, +\infty)$, so is y'. Hence

y is increasing on $[0, +\infty)$.

By the product rule,

$$y'' = ky(-y') + ky'(L - y),$$

or

$$y'' = ky'(L - 2y).$$

Since k and y' are positive on $[0, +\infty)$, we see that

$$y'' > 0 \quad \text{if } y < L/2$$

and

$$y'' < 0 \quad \text{if } y > L/2.$$

Hence,

y is concave upward when $y < L/2$;

y is concave downward when $y > L/2$.

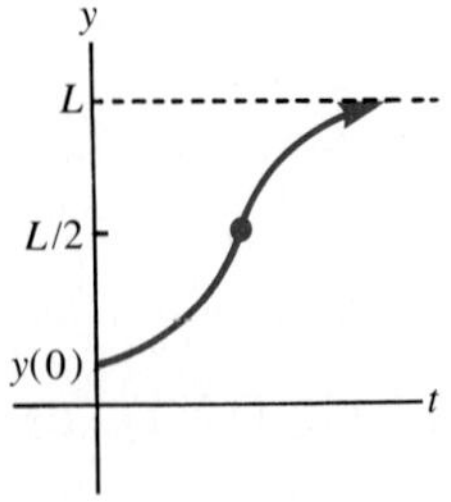

Putting all this information together, we obtain the qualitative solution shown in Figure 4–36. Notice the point of inflection: this tells us that population size increases at an increasing rate until it reaches $L/2$, and thereafter it increases at a decreasing rate. Note also that y increasing, concave downward, and less than L implies that y approaches L from below. Thus in the long run the population size grows toward the limiting value L. □

FIGURE 4–36

SUPPLEMENTARY EXERCISES

1. A biologist models the growth of an animal population as a function of time t by means of the differential equation
$$y' = ky, \quad k > 0, \quad y > 0, \quad t \geq 0.$$
Sketch the qualitative solution of the differential equation and analyze the growth of the population.

2. Repeat Exercise 1 if the model is
$$y' = -ky, \quad k > 0, \quad y > 0, \quad t \geq 0,$$
and $\lim\limits_{t \to +\infty} y = L$.

3. Repeat Exercise 1 if the model is
$$y' = -ky, \quad k > 0, \quad y > L > 0, \quad t \geq 0.$$

4. Repeat Exercise 1 if the model is
$$y' = ky^2, \quad k > 0, \quad y > 0, \quad t \geq 0.$$

5. Let $y = y(t)$ be the time between breakdowns for a supercomputer that is t months old. Suppose that
$$y' = \frac{k}{y},$$
where $k > 0$ and $y(0) > 0$. Sketch the qualitative solution of this differential equation and analyze the time between breakdowns as a function of the age of the computer.

6. The rate at which patents in the area of superconductivity are being applied for is modeled as a function of time t by the differential equation
$$y' = k(L - y), \quad k > 0, \quad L > 0, \quad 0 < y < L, \quad t \geq 0.$$
Sketch the qualitative solution of this differential equation and analyze the growth in patent applications.

7. An endemic disease responds to treatment in such a way that the rate of change of new cases per year t is given by the differential equation
$$y' = ky(L - y), \quad k > 0, \quad L > 0, \quad y > L, \quad t \geq 0.$$
Sketch the qualitative solution of this differential equation and analyze the progress of the disease.

*8. Repeat Exercise 7 if $0 < y < L$ and
   (a) $y(0) \geq L/2$ (b) $0 < y(0) < L/2$.

5

Exponential and Logarithmic Functions

In this chapter we introduce **exponential functions** and **logarithmic functions.** These functions are important because they can be used to model many situations of growth and decline. For instance, suppose a firm's annual sales were $5 million in the year it was founded and since then have been doubling every 10 years. If S denotes annual sales and t the time in years since the firm's founding, the equation

$$S = 5 \cdot 2^{0.1t}$$

expresses annual sales as a function of time. Notice that the independent variable t appears in the exponent of this equation. The function defined by the equation is thus called an **exponential function.** Exponential functions and their inverses, which are called **logarithmic functions,** are often used in the study of population growth, the growth of investments, radioactive decay, the spread of epidemics, and similar topics.

We begin this chapter with introductions to exponential and logarithmic functions and discussions of their properties. We then consider the differentiation of these functions, and finally conclude with a section devoted to some of their applications.

5.1 EXPONENTIAL FUNCTIONS

In this section we define exponential functions, discuss their characteristics, show how to graph them, and demonstrate a few of their applications. We repeatedly make use of the material on exponents presented in Section 1.1.

Exponentials

Let x be a real number. If x is an integer or a rational number, then we can calculate 2^x using the results of Section 1.1. For instance,

$$2^0 = 1, \qquad 2^1 = 2, \qquad 2^2 = 2 \cdot 2 = 4,$$

$$2^{-1} = \frac{1}{2^1} = \frac{1}{2}, \qquad 2^{-2} = \frac{1}{2^2} = \frac{1}{4}, \qquad 2^{1/2} = \sqrt{2},$$

$$2^{2/3} = (\sqrt[3]{2})^2, \qquad 2^{-1/2} = \frac{1}{\sqrt{2}}, \qquad 2^{-2/3} = \frac{1}{(\sqrt[3]{2})^2}.$$

If x is an irrational number, we can approximate x to any desired degree of accuracy by a rational number, and this allows us to approximate 2^x to any desired degree of accuracy. For example, if $x = \sqrt{3}$, successively better approximations of x are

$$\sqrt{3} \approx 1.73 \qquad \text{and} \qquad \sqrt{3} \approx 1.732,$$

so successively better approximations of $2^{\sqrt{3}}$ are

$$2^{1.73} = 2^{173/100} = 3.317278 \qquad \text{and} \qquad 2^{1.732} = 2^{1732/1000} = 3.321880.$$

In this way we can calculate $2^{\sqrt{3}}$ to any required degree of accuracy. Similar reasoning allows us to calculate 2^x when x is any irrational number.

Now that we know how to calculate 2^x for any number x, suppose we define a function f by setting

$$f(x) = 2^x$$

for all x. This function is called an **exponential function with base 2**. Its graph is shown in Figure 5–1. Note that

- The domain of the function is the set of all real numbers, and its range is the set of all positive real numbers.
- The function is one-to-one: for every real number $y > 0$, there is exactly one real number x such that $2^x = y$.
- The y-intercept of the function is $(0, f(0)) = (0, 2^0) = (0, 1)$.
- The x-axis is an asymptote for the function as $x \to -\infty$; that is,

$$\lim_{x \to -\infty} 2^x = 0.$$

Also, $\lim_{x \to +\infty} 2^x = +\infty$.

Exponential and Logarithmic Functions

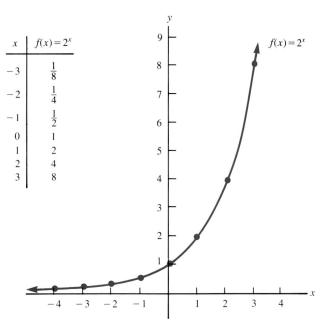

FIGURE 5-1

- The function is continuous at every value of x.

Now suppose we define a function g by setting
$$g(x) = 3 \cdot 2^{0.5x}.$$
Since we can calculate $2^{0.5x}$ for any real number x, the function g is defined for every real number x. Its graph is shown in Figure 5-2. The function g is also referred to as an exponential function with base 2.

It should be clear that what we did above using 2 as a base can be done using any positive number as a base. (We cannot use a negative number as a base because if b is negative, b^x will not be defined for all x; for instance, when b is negative, $b^{1/2} = \sqrt{b}$ is not defined.) Therefore given any number $b > 0$ and any numbers $c \neq 0$, $r \neq 0$, we can define an exponential function f by setting
$$f(x) = cb^{rx}.$$
Because $1^x = 1$ for all x, we omit the case $b = 1$.

Exponential Functions

Let b be any positive real number, $b \neq 1$, and let c and r be real numbers, with $c \neq 0$, $r \neq 0$. The function f defined by
$$f(x) = cb^{rx}$$
is an **exponential function with base b**.

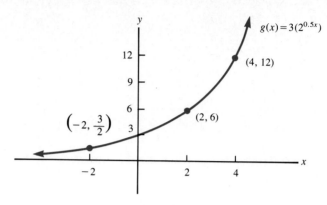

FIGURE 5-2

EXAMPLE 1 The functions f and g defined by $f(x) = 3^{2x}$ and $g(x) = 2(3^{2x})$ are both exponential with base 3. Their graphs are shown in Figure 5-3. ☐

EXAMPLE 2 The functions f and g defined by $f(x) = 2^{-x/2}$ and $g(x) = (\frac{1}{2})^{-x/2}$ are exponential with base 2 and base $\frac{1}{2}$, respectively. Their graphs are shown in Figure 5-4. ☐

Consider the exponential function defined by

$$y = (\tfrac{1}{2})^{3x}.$$

Since $\frac{1}{2} = 2^{-1}$, we may rewrite the equation as

$$y = (2^{-1})^{3x}$$

or

$$y = 2^{-3x}.$$

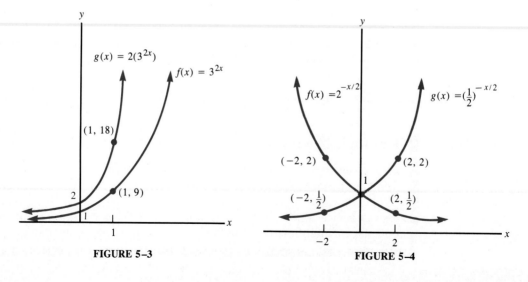

FIGURE 5-3   **FIGURE 5-4**

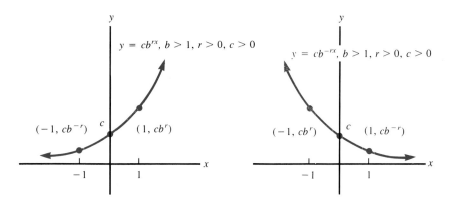

FIGURE 5–5
The Graphs of Exponential Functions

This same change of base can be carried out for any exponential function whose base is less than 1: if $y = c(1/b)^{rx}$, where $b > 1$, then $y = c(b^{-1})^{rx} = cb^{-rx}$. Hence if we wish we can always write our exponential functions using bases greater than 1. Figure 5–5 shows the graphs of $y = cb^{rx}$ and $y = cb^{-rx}$ when $b > 1$, $r > 0$, and $c > 0$. Note the following:

- The functions defined by $y = cb^{rx}$ and $y = cb^{-rx}$ are continuous and one-to-one.
- Their y-intercepts are $(0, c)$.
- $\lim\limits_{x \to +\infty} cb^{rx} = +\infty$, $\lim\limits_{x \to -\infty} cb^{rx} = 0$.
- $\lim\limits_{x \to +\infty} cb^{-rx} = 0$, $\lim\limits_{x \to -\infty} cb^{-rx} = +\infty$.

As we have previously remarked, exponential functions are often used to model situations of growth or decline. We illustrate with examples.

EXAMPLE 3 The average number of people employed by a state's Department of Motor Vehicles in year t is y, where

$$y = 1200 \cdot 2^{0.1t}.$$

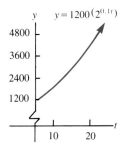

FIGURE 5–6

Here $t = 0$ represents the year 1989. The average number of employees in 1989 was therefore

$$y = 1200 \cdot 2^{0.1(0)} = 1200 \cdot 1 = 1200.$$

Figure 5–6 shows the graph of this exponential function. Notice that the average number of employees is growing at a greater and greater rate.

Suppose we wish to predict the average number of employees the Department of Motor Vehicles will have in 1992. In 1992, $t = 3$, so the average number of employees y will be

$$y = 1200 \cdot 2^{0.1(3)} = 1200 \cdot 2^{0.3}.$$

Using a calculator with an x^y key, we find that $2^{0.3} \approx 1.2311$. Therefore the average number of employees in 1992 will be approximately

$$y = 1200(1.2311) \approx 1477.$$

When will the average number of employees be 2400? If y is to be equal to 2400, we must have

$$2400 = 1200 \cdot 2^{0.1t}.$$

Dividing both sides of this equation by 1200, we obtain

$$2 = 2^{0.1t}$$

or

$$2^1 = 2^{0.1t}.$$

But because the exponential function defined by $y = 1200 \cdot 2^{0.1t}$ is one-to-one, $2^1 = 2^{0.1t}$ implies that

$$1 = 0.1t.$$

Hence $t = 10$ and thus the average number of employees will be 2400 in 1999. □

EXAMPLE 4 The number of the sea otters still alive t months after a marine oil spill has contaminated their environment is y, where

$$y = 450 \cdot 3^{-0.6t}.$$

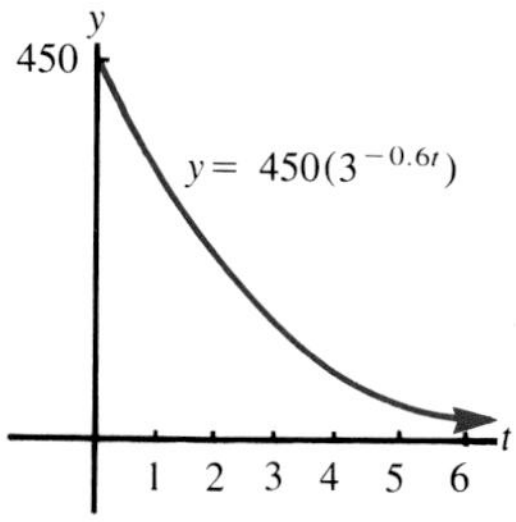

Thus the number of sea otters alive when the oil spill occurred was

$$y = 450 \cdot 3^{-0.6(0)} = 450 \cdot 1 = 450.$$

Figure 5–7 shows the graph of this exponential function. Notice that the number of otters remaining alive declines toward zero.

Five months after the oil spill, the number of otters still alive will be

$$y = 450 \cdot 3^{-0.6(5)} = 450 \cdot 3^{-3} = 450(1/27) \approx 17.$$

FIGURE 5–7

How long will it take for two-thirds of the otters to die? Since the number of otters alive at the time of the spill was 450, this is equivalent to asking how long it will take until $(1/3)450 = 150$ remain alive. But if $y = 150$, we have

$$150 = 450 \cdot 3^{-0.6t}.$$

Dividing both sides of this equation by 450, we obtain

$$\frac{1}{3} = 3^{-0.6t}$$

or

$$3^{-1} = 3^{-0.6t}.$$

Therefore, since the exponential function is one-to-one,

$$-1 = -0.6t,$$

so
$$t = \frac{1}{0.6} = \frac{5}{3}.$$

Hence it will take $1\frac{2}{3}$ months from the time of the spill until two-thirds of the otters have died.

As another application of exponential functions, let us consider the topic of **compound interest.** If an amount of money is invested, the investor receives a fee for allowing the borrower to use the money. The amount invested is the **principal,** and the fee is the **interest earned** on the principal. If the interest earned during a period depends not only on the principal but also on interest previously earned, we say that the interest is **compounded.** For instance, suppose $10,000 is invested at an interest rate of 8 percent per year compounded quarterly (every 3 months). The stated or **nominal** rate, here 8 percent, is an annual rate; if interest is to be compounded quarterly, then it will be compounded four times per year, so the interest rate for each three-month compounding period will be (8 percent)/4 = 2 percent. Therefore

- At the end of the first quarter the investment will be worth

$$\$10{,}000 + 0.02(\$10{,}000) = \$10{,}200.$$

- At the end of the second quarter the investment will be worth

$$\$10{,}200 + 0.02(\$10{,}200) = \$10{,}404.$$

- At the end of the third quarter the investment will be worth

$$\$10{,}404 + 0.02(\$10{,}404) = \$10{,}612.08.$$

- At the end of the fourth quarter the investment will be worth

$$\$10{,}612.08 + 0.02(\$10{,}612.08) = \$10{,}824.32.$$

Let us develop a formula for the worth of an investment of P dollars at the end of t years if the money earns interest at i percent per year compounded m times per year. The interest rate for each compounding period will then be (i percent)/m, or, in decimal form, $i/100m$. Let $r = i/100m$. Then

- At the end of the first compounding period the investment will be worth

$$P + rP = P(1 + r)$$

dollars.

- At the end of the second compounding period the investment will be worth

$$P(1 + r) + rP(1 + r) = P(1 + r)(1 + r) = P(1 + r)^2$$

dollars.

- At the end of the third compounding period the investment will be worth

$$P(1 + r)^2 + rP(1 + r)^2 = P(1 + r)^3$$

dollars.

Thus we can see that at the end of k compounding periods the investment will be worth

$$P(1 + r)^k$$

dollars. Therefore at the end of t years, when the money will have earned interest for mt compounding periods, the investment will be worth

$$P(1 + r)^{mt}$$

dollars. This is the compound interest formula. □

> **The Compound Interest Formula**
>
> Let principal P be invested at an interest rate of i percent per year compounded m times per year. Then $r = i/100m$ is the interest rate per compounding period in decimal form. At the end of t years the investment will be worth A dollars, where
>
> $$A = P(1 + r)^{mt}.$$

The amount A is called the **future value** of the principal. Note that A is an exponential function of t with base $1 + r$.

EXAMPLE 5 Suppose $1000 is invested at 6 percent per year compounded semiannually. How much will the investment be worth at the end of 3 years?

Here $P = \$1000$, i percent $= 6$ percent, $m = 2$, and $t = 3$. Therefore

$$r = \frac{6}{100(2)} = 0.03,$$

and

$$A = \$1000(1 + 0.03)^{2 \cdot 3} = \$1000(1.03)^6 = \$1000(1.194052) = \$1194.05,$$

rounded to the nearest cent. □

EXAMPLE 6 Suppose $1000 is invested at 6 percent per year compounded quarterly. How much will the investment be worth at the end of 3 years?

Here $P = \$1000$, i percent $= 6$ percent, $m = 4$, and $t = 3$. Therefore

$$r = \frac{6}{100(4)} = 0.015,$$

and

$$A = \$1000(1 + 0.015)^{4 \cdot 3} = \$1000(1.015)^{12} = \$1000(1.195618) = \$1195.62,$$

rounded to the nearest cent. □

Note that the only difference between the situation of Example 5 and that of Example 6 is the number of times interest is compounded each year, and that the more frequent compounding of Example 6 results in a greater worth for the investment at the end of the three-year period. This behavior holds in general: for a fixed annual interest rate of i percent, the more frequently interest is compounded, the faster it accumulates. Therefore for a fixed principal P, a fixed annual interest rate of i

percent, and a fixed number of years t, the future value A will depend on the number of times m that interest is compounded each year: the larger m is, the larger A will be.

The amount P that must be invested now in order to have a specified amount A in the future is called the **present value of A**. Since the compound interest formula relates the future value A to the present value P, we need only solve the formula for P to obtain the present value formula.

> **The Present Value Formula**
>
> Let A be an amount that is due t years in the future. Suppose money is worth $i\%$ per year compounded m times per year, and let $r = i/100m$. Then the present value P of amount A is given by
> $$P = A(1 + r)^{-mt}.$$

Thus present value P is an exponential function of t with base $1 + r$.

The process of finding the present value of a future amount is often referred to as **discounting** the future amount.

EXAMPLE 7   Let us find the present value of \$10,000 due 5 years from now if money is worth 6% per year compounded quarterly.

Here the future value $A = \$10{,}000$, $i = 6$, $m = 4$, and $t = 5$, so $r = 0.015$ and
$$P = \$10{,}000(1.015)^{-20}$$
$$= \frac{\$10{,}000}{(1.015)^{20}} = \frac{\$10{,}000}{1.346855} = \$7424.70.$$

Thus if money is worth 6% per year compounded quarterly the present value of \$10,000 due 5 years from now is \$7424.70; another way to express this is to say that \$10,000 discounted for 5 years at 6% per year compounded quarterly is \$7424.70. In either case, what is meant is that \$7424.70 invested now at 6% per year compounded quarterly will amount to \$10,000 five years from now, or, said another way, if a return on investment of 6% per year compounded quarterly is required, then \$10,000 due 5 years from now is worth \$7424.70 today. □

There is a special number denoted by e that is very important in the study of exponential functions and their applications. The number e is defined as a limit at infinity.

> **The Number e**
>
> $$e = \lim_{x \to +\infty} \left(1 + \frac{1}{x}\right)^x$$

To verify that this limit exists and to get an idea of the value of e, suppose we use a calculator with an x^y key to calculate $(x + 1/x)^x$ for some values of x:

x	$\left(1 + \dfrac{1}{x}\right)^x$
10	2.5937425
100	2.7048138
1000	2.7169239
10000	2.7181459
100000	2.7182682
1000000	2.7182805
10000000	2.7182817

Thus it appears that the limit does exist and is approximately equal to 2.71828. In fact, the number e is irrational and to 11 significant digits,

$$e = 2.7182818285.$$

Since e is positive, we may use it to construct exponential functions with base e. Any function whose defining equation is of the form

$$y = ce^{rx},$$

where c and r are nonzero real numbers, is thus an **exponential function with base** e.

EXAMPLE 8 Let us graph the exponential function defined by $y = e^x$. To do this, we must be able to evaluate e^x for any real number x. Many calculators have an exp or e^x key by means of which such evaluation may be carried out. Also, Table 1 in the back of the book gives values of e^x for selected values of x. Thus, using either a calculator or Table 1, we find that

$$e^{-2.0} \approx 0.1353, \qquad e^{-1.5} \approx 0.2231,$$
$$e^{-1.0} \approx 0.3679, \qquad e^{-0.5} \approx 0.6065,$$
$$e^{0.5} \approx 1.6487, \qquad e^{1.0} \approx 2.7183,$$
$$e^{1.5} \approx 4.4817, \qquad e^{2.0} \approx 7.3891,$$

and so on. (Check these.) Also, of course, $e^0 = 1$. We use these values of e^x to graph $y = e^x$. The graph is shown in Figure 5–8. ☐

EXAMPLE 9 A radioactive isotope decays exponentially according to the equation

$$y = 100e^{-0.25t}.$$

Here t is time in years and y is the amount of the isotope present, in grams, at time t. Note that the amount of the isotope present initially is

$$y = 100e^{-0.25(0)} = 100e^0 = 100$$

grams, and that the amount present declines toward zero. See Figure 5–9.

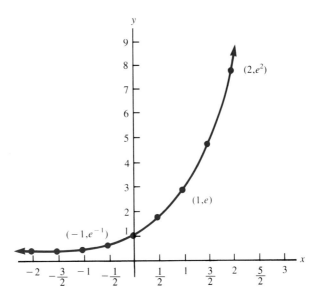

FIGURE 5–8
The Graph of $y = e^x$

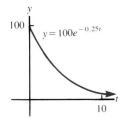

FIGURE 5–9

Let us find the amount of the isotope present at the end of 10 years. It is

$$y = 100e^{-0.25(10)} = 100e^{-2.5}$$

grams. Using either a calculator with an e^x key or Table 1, we find that $e^{-2.5} \approx 0.08208$. Hence the amount of the isotope present at the end of 10 years is approximately

$$y = 100(0.08208) = 8.208$$

grams.

Every radioactive isotope has a half-life: this is the time it takes for one-half of any amount of the isotope to decay. Let us find the half-life of the isotope of this example.

Since there are 100 grams of the isotope present initially, its half-life may be found by determining the time it will take until only 50 grams of it remain: but this can be done by substituting 50 for y in the decay equation and then solving

$$50 = 100e^{-0.25t}$$

for t. To solve this equation for t, we divide both sides of it by 100 to obtain

$$\frac{1}{2} = e^{-0.25t}$$

and then use the fact that

$$e^{-0.6931} \approx \frac{1}{2}$$

(check this) to write

$$e^{-0.6931} = e^{-0.25t}.$$

Because exponential functions are one-to-one, it then follows that

$$-0.6931 = -0.25t$$

and thus that

$$t = \frac{-0.6931}{-0.25} = 2.7724,$$

approximately. Hence the half-life of the isotope is approximately 2.77 years. ☐

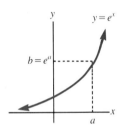

FIGURE 5–10

Why do we want to use e as a base for exponential functions? The answer is that e occurs naturally when many phenomena are modeled by exponential functions; we shall see an example of this in a moment when we discuss continuous compounding of interest. Also, as we will discover later in this chapter, using e as the base for exponential functions simplifies the process of differentiating such functions. Because of these facts, there is a tendency to use base e in preference to other bases. This is not a limitation, for if b is any base, the fact that $y = e^x$ has the positive real numbers as its range and is one-to-one implies that there is a unique number a such that $b = e^a$. (See Figure 5–10.) Hence the exponential function $y = cb^{rx}$ may be written as $y = ce^{arx}$.

EXAMPLE 10 Let us rewrite the exponential equation

$$y = 3 \cdot 2^{0.3t}$$

using e as a base. Since

$$e^{0.6931} \approx 2,$$

we have (approximately)

$$y = 3(e^{0.6931})^{0.3t}$$

or

$$y = 3e^{0.20793t}. \quad \square$$

Earlier in this section we remarked that for a fixed principal, nominal interest rate, and time, more frequent compounding results in a larger future value. For instance, suppose we invest $100,000 at 12% per year for one year, but vary the number of times that interest is compounded during the year:

Compounding	m	$A = \$100{,}000(1 + 0.12/m)^m$
Annually	1	$112,000.00
Semiannually	2	$112,360.00
Quarterly	4	$112,550.88
Monthly	12	$112,682.50
Daily	365	$112,747.46
Hourly	8760	$112,749.60

The table shows that more frequent compounding does indeed increase the interest earned, but it also seems to show that as the number of compounding periods per year increases, the additional interest earned gets smaller and smaller. What happens if we compound interest at every instant of time? This is the most frequent compounding possible, and is known as **continuous compounding.** Under continuous compounding, m, the number of compounding periods per year, becomes infinite, and we have

$$A = \lim_{m \to +\infty} P(1 + r)^{mt} = P \cdot \lim_{m \to +\infty} \left(1 + \frac{i/100}{m}\right)^{mt}.$$

For instance, if $P = \$1$, $i\% = 100\%$, and $t = 1$ year, then

$$A = \lim_{m \to +\infty} \left(1 + \frac{1}{m}\right)^m.$$

But by definition,

$$\lim_{m \to +\infty} \left(1 + \frac{1}{m}\right)^m = e.$$

More generally, it can be shown (though we shall not do so) that for any number a,

$$\lim_{m \to +\infty} \left(1 + \frac{a}{m}\right)^{mt} = e^{at}.$$

Therefore if we take $a = i/100$, we have

$$A = P \cdot \lim_{m \to +\infty} \left(1 + \frac{i/100}{m}\right)^{mt} = Pe^{(i/100)t},$$

and we have developed the formula for continuous compounding of interest.

The Formula for Continuous Compounding of Interest

If principal P is invested at $i\%$ per year compounded continuously, then $r = i/100$ is the annual interest rate expressed in decimal form, and the future value A of the investment at the end of t years is

$$A = Pe^{rt}.$$

CHAPTER 5

EXAMPLE 11 If $1000 is invested at 6% per year compounded continuously, then at the end of three years the investment will be worth

$$A = \$1000e^{(0.06)3} = \$1000e^{0.18}.$$

Using a calculator with an e^x key, we find that

$$e^{0.18} = 1.1972174.$$

Thus

$$A = \$1000(1.1972174) = \$1197.22,$$

to the nearest cent. □

It is important to be aware of the fact that for a fixed nominal interest rate, compounding continuously makes the future value grow as fast as possible. Thus, in Example 11, if the interest was compounded m times per year for any finite value of m, the future value of the $1000 at the end of 3 years would be less than $\$1000e^{0.18}$.

We can also apply the concept of present value when interest is compounded continuously:

The Present Value Formula for Continuous Compounding

If future value A is discounted at $i\%$ per year compounded continuously for t years and $r = i/100$, then the present value P of A is

$$P = Ae^{-rt}.$$

EXAMPLE 12 The present value of $25,000 discounted at 10% per year compounded continuously for five years is

$$P = \$25{,}000e^{-0.1(5)}$$
$$= \$25{,}000e^{-0.5}$$
$$= \$25{,}000(0.6065307)$$
$$= \$15{,}163.27. \quad \square$$

■ *Computer Manual: Programs LIMIT, INFLIMIT, FNCTNTBL, INTEREST, PRSNTVAL*

5.1 Exercises

Exponentials

In Exercises 1 through 16, graph the given exponential function.

1. $y = 3^x$
2. $y = 3^{-x}$
3. $y = 2(5)^x$
4. $y = 3(5)^{-x}$
5. $y = 3^{x/2}$
6. $y = 5^{-x/2}$
7. $y = 4^{2x/3}$
8. $y = 5^{-3x/2}$
9. $y = 2(3)^{-x/3}$
10. $y = \left(\dfrac{1}{2}\right)^x$
11. $y = -5\left(\dfrac{2}{5}\right)^{x/2}$
12. $y = 3\left(\dfrac{4}{3}\right)^{-2x}$
13. $y = 2^{0.2x}$
14. $y = 5^{-0.8x}$
15. $y = 2(3)^{-0.3x}$
16. $y = 4(0.6)^{0.8x}$

17. A radioactive isotope decays according to the equation
$$y = 300 \cdot 2^{-t/25}.$$
Here y is the amount of the isotope, in grams, present at time t, t in years.
   (a) Graph the exponential function defined by this equation.
   (b) What amount of the isotope is present at the end of 12.5 years?
   (c) What is the half-life of the isotope?
   (d) Find and interpret $\lim\limits_{t \to +\infty} y$.

*18. Suppose the half-life of a radioactive isotope is 12 years and you start with 100 grams of the isotope. Find a model that describes the amount of the isotope present at the end of t years. (Hint: Use a model of the form $y = c \cdot 2^{rt}$, and substitute 0 and 12 for t to find c and r.)

19. How much will you have at the end of 1 year if you invest $15,000 at 8% per year compounded annually?

20. How much will you owe at the end of 1 year if you borrow $20,000 at 12% per year compounded monthly?

21. How much will you owe at the end of two years if you borrow $8000 at 6% per year compounded semiannually?

22. How much will you have at the end of 5 years if you invest $50,000 at 8% per year compounded quarterly?

23. You borrow $7000 at 10% per year compounded quarterly. If you pay back the loan at the end of 6 years, how much will you pay?

24. You invest $22,000 at 12% per year compounded monthly. How much will you have at the end of 3 years?

25. Find the present value of $12,000 due 4 years from now if money is worth 6% per year compounded annually.

26. How much should you invest now at 8% per year compounded semiannually in order to have $10,000 five years from now?

27. Find the present value of $30,000 due 3 years from now if money is worth 8% per year compounded quarterly.

28. How much should you invest now at 6% per year compounded monthly in order to have $14,000 three years and four months from now?

29. Discount $50,000 for 3 years at 12% per year compounded monthly.

Management

30. A firm's sales are given by $S = 5 \cdot 2^{0.1t}$, where S is annual sales in millions of dollars and t is time in years since the firm was founded.
    (a) Graph this exponential function.
    (b) Find the firm's annual sales 20 years after it was founded; 25 years after it was founded.
    (c) How long did it take the firm to double its annual sales?

*31. A company's average cost per unit was $22.50 in 1985 and has been doubling every 8 years. Find a model that describes the company's average cost per unit as a function of time, and use it to predict the average cost per unit in 1993. (Hint: use a model of the form $y = c \cdot 2^{rt}$.)

*32. Repeat Exercise 31 if the company's average cost per unit has been tripling every 10 years.

The **double-declining balance** method of depreciation depreciates an asset in year t of its useful life by the amount

$$D = \frac{2C}{n}\left(1 - \frac{2}{n}\right)^{t-1}.$$

Here n is the number of years of useful life the asset has when new, C is its original cost, and D is the amount of its depreciation in year t, $1 \le t \le n$. Exercises 33 through 35 make use of this depreciation formula.

33. A car that cost $25,000 new has a useful life of five years. Write an expression for the amount of its depreciation in year t of its life using the double-declining balance method, and find the amount of its depreciation during the third and fifth years of its life.

34. Fill in the following depreciation schedule for the car of Exercise 33:

Year	Depreciation	Value of Car at End of Year
1	$10,000	$15,000
2		
3		
4		
5		

35. An apartment building cost $1.2 million to construct and has a useful life of 30 years. Find an expression for the amount of its depreciation in year t of its useful life using the double-declining balance method, and find the amount of its depreciation in the first, tenth, and twenty-fifth years of its life.

*36. Alpha Corporation is considering the purchase of a new computer. The firm has three models to choose from: X, Y, and Z. Model X will cost $50,000 now and will save Alpha $70,000 two years from now. Model Y will cost $75,000 now and save $100,000 three years from now. Model Z will cost $100,000 now and will save $150,000 four years from now. Alpha will not make any investment unless it returns at least 10% per year compounded quarterly. What should Alpha do? (Hint: find the present value of each model's savings and compare it with the model's cost.)

*37. Beta Company can invest in one or more of 3 new product lines. Product line A will cost Beta $1.5 million and will return $2.8 million four years later. Product line B will cost $1.8 million and will return $3.3 million six years later. Product line C will cost $2.4 million and will return $4.3 million eight years later. If Beta requires a return of at least 10% per year compounded annually on any investment, which, if any, of the product lines should it invest in?

Life Science

38. The number of bacteria in a culture t hours after the start of an experiment was y thousand, where $y = 10 \cdot 2^{0.2t}$.
  (a) Graph the exponential function defined by $y = 10 \cdot 2^{0.2t}$. How many bacteria were in the culture at the start of the experiment?
  (b) How many bacteria were in the culture 5 hours after the start of the experiment?

The Number e

In Exercises 39 through 46, find the given power of e.

39. $e^{0.06}$ **40.** e^4 **41.** $e^{0.75}$ **42.** $e^{-0.14}$ **43.** e^{-1}
44. $e^{-5.6}$ **45.** $e^{-3.5}$ **46.** $e^{9.9}$

In Exercises 47 through 56, graph the given exponential function.

47. $y = e^{-x}$ **48.** $y = e^{2x}$ **49.** $y = e^{-2x}$ **50.** $y = 2e^{2x}$
51. $y = e^{x/2}$ **52.** $y = -5e^{3x/2}$ **53.** $y = 1.4e^{0.5x}$ **54.** $y = 2e^{-0.8x}$
55. $y = 3e^{-0.05x}$ **56.** $y = 5.2e^{1.25x}$

In Exercises 57 through 62, rewrite the exponential function using base e. You may use the facts that

$$e^{0.6931} \approx 2 \quad \text{and} \quad e^{1.0986} \approx 3.$$

57. $y = 3 \cdot 2^{0.4t}$ **58.** $y = 5 \cdot 3^{-2x}$ **59.** $y = (1/2)^{3x}$
60. $y = -4(1/3)^{-1.2t}$ **61.** $y = 2(1/9)^{4x}$ **62.** $y = 6^{-3t}$

63. A radioactive isotope decays according to the equation

$$y = 200e^{-0.05t},$$

where t is in years and y is the amount of the isotope remaining, in grams, at time t.

  (a) Graph this exponential function.
  (b) What amount of the isotope is present at the end of 10 years?
  (c) What is the half-life of the isotope? (Hint: $e^{-0.6931} \approx 1/2$.)
  (d) Find and interpret $\lim_{t \to +\infty} y$.

64. Repeat Exercise 63 if $y = 200e^{-1.2t}$.

65. How much will you have at the end of one year if you invest $10,000 at 8% per year compounded continuously?

66. How much will you have at the end of 3.5 years if you invest $45,000 at 10.5% per year compounded continuously?

67. If you borrow $15,000 at 6% per year compounded continuously, how much will you owe at the end of 5 years?

68. If you borrow $8000 at 12.2% per year compounded continuously, how much will you owe at the end of 6.25 years?

69. Make a table showing the future value of $10,000 invested for 1 year at 6% per year compounded annually, semiannually, quarterly, monthly, and continuously.

70. Make a table showing the future value of $1 million invested for 4 years at 12% per year compounded annually, semiannually, quarterly, monthly, and continuously.

71. How much must be invested now at 10% per year compounded continuously in order to have $10,000 three years from now?

72. How much must be invested now at 6.5% per year compounded continuously in order to have $25,000 six years from now?

73. Find the present value of $100,000 discounted for 8 years at 5% per year compounded continuously.

74. Find the present value of $40,000 discounted for 6.4 years at 8.5% per year compounded continuously.

75. Make a table showing the present value of $10,000 discounted for 1 year at 9% per year compounded annually, semiannually, quarterly, monthly, and continuously.

76. Make a table showing the present value of $1 million discounted for 5 years at 6% per year compounded annually, semiannually, quarterly, monthly, and continuously.

Management

77. A firm's cost function is given by $C(x) = 300e^{0.04x}$, where cost C is in thousands of dollars and x is in thousands of units.
    (a) Graph the firm's cost function. What is its fixed cost?
    (b) What is the firm's cost if it makes 10 thousand units?

78. An oil field will produce oil at a rate of $y = 40e^{-0.3t}$ million barrels per year t years after it is opened.
    (a) Graph the exponential function defined by this equation. At what rate will the field produce oil when it is opened?
    (b) At what rate will the field produce oil 5 years after it is opened?
    (c) Find and interpret $\lim_{t \to +\infty} y$.

79. Use the fact that $e^{0.6931} \approx 2$ to express the sales of the firm of Exercise 30 as an exponential function with base e.

80. Use the fact that $e^{1.0986} \approx 3$ to express the average cost per unit function of Exercise 32 as an exponential function with base e.

*81. A company can purchase either (or neither) of two new telephone systems. System X will cost $100,000 and save $200,000 three years from now, while system Y will cost $150,000 and save $375,000 five years from now. If the company has a policy that no investment will be made unless it returns at least 12% per year compounded continuously, which system will it purchase? (Hint: compare the present value of each system's savings to its cost.)

*82. Repeat Exercise 81 if the company requires a return of at least 20% per year compounded continuously.

*83. Repeat Exercise 81 if system X will save $200,000 three and one-half years from now and the company requires a return of 20% per year compounded continuously.

Life Science

84. An article by Simons, Sherrod, Collopy, and Jenkins ("Restoring the Bald Eagle," *American Scientist*, May–June, 1988) indicates that y percent of bald eagle colonies that are founded by x individual eagles will become extinct, where $y = 107e^{-0.082x}$.
  (a) Graph the function defined by $y = 107e^{-0.082x}$.
  (b) What percentage of colonies founded by 2 eagles become extinct? What percentage of those founded by 20 eagles become extinct?

85. In the early years of the AIDS epidemic, the number of diagnosed cases was approximately y, where $y = 100e^{0.46t}$, with t being time in years since 1976.
  (a) Graph the function defined by $y = 100e^{0.46t}$.
  (b) If the epidemic had continued to be described by this exponential function, how many cases would there have been by 1991? by 1996?

5.2 LOGARITHMIC FUNCTIONS

Since every exponential function defined by an equation of the form $y = b^x$ is one-to-one, it has an inverse function, called the **logarithmic function to the base** b. Logarithmic functions are important: as we shall see later in this chapter, they can be used to model some growth situations, and they possess several properties that make them useful in computation. In this section we introduce logarithmic functions, state the properties of logarithms, and show how to use the properties of logarithms to solve exponential equations.

Logarithms

In Section 1.3 we introduced the notion of the **inverse function** of a one-to-one function. Recall that if f is a one-to-one function we define its inverse function f^{-1} by setting $f^{-1}(u) = v$ if and only if $u = f(v)$. See Figure 5–11. Thus f^{-1} is a function whose domain is the range of f and whose range is the domain of f.

Since the exponential function defined by $f(v) = b^v$ is one-to-one, with domain the set of real numbers and range the set of positive real numbers, its inverse function f^{-1} exists and has domain the set of positive real numbers and range the set of real

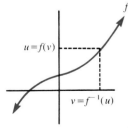

FIGURE 5–11

numbers; furthermore, f^{-1} is defined by setting $f^{-1}(u) = v$ if and only if $u = b^v$. This inverse function f^{-1} is called the **logarithmic function to the base b,** and is denoted by $\log_b$: $f^{-1}(u) = \log_b u$. Thus $\log_b u = v$ if and only if $u = b^v$. Replacing u and v by x and y, we have the following definition of logarithmic functions.

The Logarithmic Function to the Base b

Let b be a positive real number, $b \neq 1$. The inverse function of the exponential function defined by $y = b^x$ is called the **logarithmic function to the base b,** and is denoted by $\log_b$ (read as "log to the base b"). The logarithmic function to the base b is defined by

$$y = \log_b x \quad \text{if and only if} \quad x = b^y.$$

The domain of the logarithmic function to the base b is the set of positive real numbers and its range is the set of real numbers.

In Section 1.3 we noted that the graph of f^{-1} is that of f reflected about the line $y = x$. Therefore the graph of $y = \log_b x$ is that of $y = b^x$ reflected about the line $y = x$. Figure 5–12 shows the graph of the logarithmic function to the base b when $b > 1$ and when $0 < b < 1$. Notice that

- The function defined by $y = \log_b x$ is continuous and one-to-one.
- Its x-intercept is $(1, 0)$.

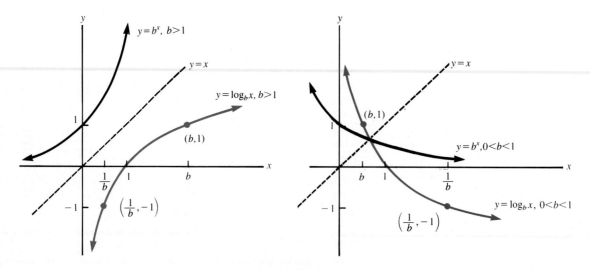

FIGURE 5–12
The Graphs of Logarithmic Functions

Exponential and Logarithmic Functions

- If $b > 1$, $\lim\limits_{x \to +\infty} \log_b x = +\infty$, $\lim\limits_{x \to 0^+} \log_b x = -\infty$.
- If $0 < b < 1$, $\lim\limits_{x \to +\infty} \log_b x = -\infty$, $\lim\limits_{x \to 0^+} \log_b x = +\infty$.

Before we proceed to some examples, we wish to emphasize that the key to understanding logarithms lies in the definition of $y = \log_b x$. Remember,

$$y = \log_b x \quad \text{means} \quad x = b^y.$$

EXAMPLE 1

$$\log_2 8 = 3 \quad \text{because } 2^3 = 8,$$
$$\log_3 81 = 4 \quad \text{because } 3^4 = 81,$$
$$\log_2 (\tfrac{1}{2}) = -1 \quad \text{because } 2^{-1} = \tfrac{1}{2},$$
$$\log_3 (\tfrac{1}{27}) = -3 \quad \text{because } 3^{-3} = \tfrac{1}{27},$$
$$\log_{1/5} 25 = -2 \quad \text{because } (\tfrac{1}{5})^{-2} = 25. \quad \square$$

EXAMPLE 2 $\log_{10} 10 = 1$, $\log_{10} 100 = 2$, $\log_{10} 1000 = 3$, $\log_{10} 10{,}000 = 4$, $\log_{10} 0.1 = -1$, $\log_{10} 0.01 = -2$, $\log_{10} 0.001 = -3$. $\square$

EXAMPLE 3 The graphs of $y = \log_2 x$ and $y = \log_{10} x$ are shown on the same set of axes in Figure 5–13. $\square$

Let b be a positive real number, $b \neq 1$. Then
- $\log_b 1 = 0$ because $1 = b^0$;
- $\log_b b = 1$ because $b = b^1$;
- $\log_b b^r = r$ because $b^r = b^r$;
- $b^{\log_b x} = x$ because if $\log_b x = y$, then $x = b^y$, so $b^{\log_b x} = b^y = x$.

The preceding identities are just some of the **properties of logarithms.**

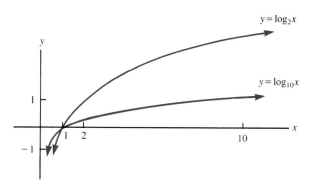

FIGURE 5–13

The Properties of Logarithms

Let b be a positive real number, $b \neq 1$. Let r be any real number and x, x_1, and x_2 be positive numbers. Then

1. $\log_b 1 = 0$.
2. $\log_b b^r = r$; in particular, $\log_b b = 1$.
3. $b^{\log_b x} = x$.
4. $\log_b(x_1 x_2) = \log_b x_1 + \log_b x_2$.
5. $\log_b(x_1/x_2) = \log_b x_1 - \log_b x_2$.
6. $\log_b x^r = r \cdot \log_b x$.
7. if $\log_b x_1 = \log_b x_2$, then $x_1 = x_2$.

Properties 1, 2, and 3 were proved above; property 7 follows from the fact that $y = \log_b x$ is a one-to-one function; and properties 4, 5, and 6 follow from the properties of exponents of Section 1.1 and the fact that $y = \log_b x$ means that $x = b^y$.

EXAMPLE 4

$$\log_2 1 = 0, \quad \log_2 2 = 1;$$

$$\log_3 9 = \log_3 3^2 = 2, \quad \log_5(1/625) = \log_5 5^{-4} = -4;$$

$$7^{\log_7 38} = 38;$$

if $\log_{10} 2x = \log_{10} 44$, then $2x = 44$, so $x = 22$. ☐

EXAMPLE 5

Given that $10^{0.3010} = 2$, $10^{0.4771} = 3$, and $10^{0.6990} = 5$, so that $\log_{10} 2 = 0.3010$, $\log_{10} 3 = 0.4771$, and $\log_{10} 5 = 0.6990$, we use the properties of logarithms to find:

$$\log_{10} 6 = \log_{10}(2 \cdot 3) = \log_{10} 2 + \log_{10} 3 = 0.3010 + 0.4771 = 0.7781;$$

$$\log_{10} 8 = \log_{10} 2^3 = 3 \log_{10} 2 = 3(0.3010) = 0.9030;$$

$$\log_{10} 45 = \log_{10}(3^2 \cdot 5) = 2 \log_{10} 3 + \log_{10} 5 = 2(0.4771) + 0.6990$$
$$= 1.6532;$$

$$\log_{10}(2/3) = \log_{10} 2 - \log_{10} 3 = 0.3010 - 0.4771 = -0.1761;$$

$$\log_{10}(75/4) = \log_{10}(3 \cdot 5^2) - \log_{10} 2^2 = \log_{10} 3 + 2 \log_{10} 5 - 2 \log_{10} 2$$
$$= 0.4771 + 2(0.6990) - 2(0.3010)$$
$$= 1.2731. \quad ☐$$

EXPONENTIAL AND LOGARITHMIC FUNCTIONS   307

Just as there is a tendency when working with exponential functions to use base e, so there is a tendency when working with logarithms to use base e. The logarithm to the base e is called the **natural logarithm** and is written ln. Thus, ln 2 means $\log_e 2$, ln 5 means $\log_e 5$, and so on. Many calculators have an ln key, which allows calculation of ln x for positive values of x. Also, Table 2 of the appendix is a table of values of ln x for selected values of x.

EXAMPLE 6   We find some natural logs (check these using a calculator or Table 2):
ln 0.5 = -0.6931,   ln 2 = 0.6931,   ln 5 = 1.6094,
ln 10 = 2.3026,   ln ($\frac{8}{25}$) = -1.1394.   □

EXAMPLE 7   Figure 5–14 shows the graph of $y = \ln x$.   □

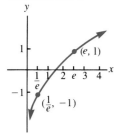

FIGURE 5–14
The Graph of $y = \ln x$

In mathematics the natural logarithm is far more common than logarithms to other bases; hence from now on we will be using natural logarithms almost exclusively. However, logarithms to bases other than e are sometimes encountered in business and the sciences. Logarithms to the base 10 are called **common logarithms;** they are well adapted to computation and are therefore frequently encountered in applications. Also, logarithms to the base 2 often appear in computer work. In a moment we will demonstrate how to change from one base to another.

In Section 5.1 we observed that it is possible to write any exponential function using base e because for any positive number b, $b \neq 1$, it is always possible to find a number r such that $e^r = b$. More generally, given any two positive numbers a and b, neither of them equal to 1, we can always find a number r such that $a^r = b$: since $a^{\log_a b} = b$, we may take $r = \log_a b$.

The Change-of-Base Formula for Exponentials

Let a, b be positive numbers not equal to 1. Then
$$a^{\log_a b} = b.$$

EXAMPLE 8   Let us rewrite the exponential function defined by $y = 2 \cdot 5^{3x}$ in terms of base e. To do so, we must find a number r such that $e^r = 5$; using the change-of-base formula with $a = e$ and $b = 5$, we see that $r = \log_e 5 = \ln 5 = 1.6094$. Hence
$$y = 2 \cdot 5^{3x} = 2(e^{1.6094})^{3x} = 2e^{4.8282x}.$$ □

Just as it is sometimes convenient to change bases in exponential functions, so too it is sometimes useful to change bases in logarithmic functions. Thus let a and b be positive numbers, neither of which is equal to 1, and suppose we wish to express $\log_b x$ in terms of $\log_a x$. If
$$\log_b x = y$$
so that
$$x = b^y,$$

then the change-of-base formula for exponentials tells us that
$$x = (a^{\log_a b})^y = a^{y(\log_a b)}.$$
But
$$x = a^{y(\log_a b)}$$
implies that
$$\log_a x = y \log_a b,$$
and hence that
$$y = \frac{\log_a x}{\log_a b}.$$

The Change-of-Base Formula for Logarithms

If a, b are positive numbers not equal to 1, then
$$\log_b x = \frac{\log_a x}{\log_a b}.$$

EXAMPLE 9

Suppose we wish to rewrite the logarithmic function $y = \log_{10} x$ using base e. Taking $a = e$ and $b = 10$ in the change-of-base formula, we have
$$y = \log_{10} x = \frac{\log_e x}{\log_e 10} = \frac{\ln x}{\ln 10} \approx \frac{\ln x}{2.3026}. \quad \square$$

Solving Exponential Equations

One of the most important uses of logarithms occurs in the solving of **exponential equations:** these are equations in which the quantity to be determined appears as an exponent. The method of solution is to take the logarithm of both sides of the equation and then use the fact that $\log_b A^x = x \cdot \log_b A$ to "get x out of the exponent." In doing this we could use any base b, but as a matter of convenience we will use the natural logarithm. We illustrate with examples.

EXAMPLE 10

Let us solve the exponential equation
$$2^{5x} = 12$$
for x. Taking the natural logarithm of both sides of the equation, we have
$$\ln 2^{5x} = \ln 12,$$
so
$$5x(\ln 2) = \ln 12,$$

or
$$x = \frac{\ln 12}{5(\ln 2)}.$$

Since $\ln 12 = 2.4849$ and $\ln 2 = 0.6931$,
$$x = \frac{2.4849}{5(0.6931)} = 0.717. \quad \square$$

EXAMPLE 11 The decay of a radioactive isotope is described by the exponential function
$$y = 200e^{-0.06t},$$
where y is the amount of the isotope, in grams, present at time t, with t in years. Let us find the half-life of the isotope.

To do so, we must determine how long it will take for one-half of the original amount present to decay. Since 200 grams are present initially (why?), we must find out how long it will take until there are only 100 grams left; in other words, we must solve the exponential equation
$$100 = 200\, e^{-0.06t}$$
for t. Simplifying the equation by dividing both sides by 200, we have
$$\tfrac{1}{2} = e^{-0.06t}.$$
Taking the natural logarithm of both sides, we obtain
$$\ln 0.5 = \ln e^{-0.06t},$$
or
$$-0.6931 = -0.06t(\ln e) = -0.06t(1) = -0.06t.$$
Hence
$$t = \frac{0.6931}{0.06} = 11.55 \text{ years.} \quad \square$$

EXAMPLE 12 If \$10,000 is invested at 8% per year compounded continuously, how long will it be until the investment is worth \$30,000?

We use the formula $A = Pe^{rt}$ with $P = \$10{,}000$, $r = 0.08$, and $A = \$30{,}000$. Thus we have
$$30{,}000 = 10{,}000 e^{0.08t},$$
or
$$3 = e^{0.08t}.$$
Taking the natural logarithm of both sides of this last equation, we find that
$$\ln 3 = 0.08t(\ln e) = 0.08t.$$
Therefore

$$t = \frac{\ln 3}{0.08} = \frac{1.0986}{0.08} = 13.73$$

years. ∎

EXAMPLE 13 What interest rate, compounded quarterly, must you obtain if an investment of $5000 is to grow to $20,000 in 12 years?

We use the compound interest formula $A = P(1 + r)^{mt}$ with $P = \$5000$, $m = 4$, $t = 12$, and $A = \$20,000$. Thus we have

$$20{,}000 = 5000(1 + r)^{4(12)},$$

or

$$4 = (1 + r)^{48}.$$

Taking the natural logarithm of both sides of this last equation, we find that

$$\ln 4 = 48 \cdot \ln(1 + r).$$

Hence

$$\ln(1 + r) = \frac{\ln 4}{48} = \frac{1.3863}{48} = 0.0289.$$

Therefore

$$e^{\ln(1 + r)} = e^{0.0289},$$

or

$$1 + r = 1.0293,$$

and

$$r = 0.0293.$$

Since r is the quarterly interest rate, the annual rate is

$$4r = 4(0.0293) = 0.1172,$$

or 11.72% per year compounded quarterly. ∎

■ *Computer Manual: Programs LIMIT, INFLIMIT, FNCTNTBL, NTRSTYRS, NTRSTRAT*

5.2 EXERCISES

Logarithms

In Exercises 1 through 4, evaluate the given expressions.

1. $\log_2 16$, $\log_2 128$, $\log_2(1/4)$, $\log_{1/2}(1/4)$, $\log_{1/2} 8$
2. $\log_3 81$, $\log_3 243$, $\log_3(1/243)$, $\log_{1/3}(1/9)$, $\log_{1/3} 27$

3. $\log_5 1$, $\log_5 5$, $\log_{1/5} 1$, $\log_{1/5}(1/5)$, $\log_{1/5} 5$

4. $\log_{10} 100{,}000$, $\log_{10} 1{,}000{,}000$, $\log_{10} 0.0000001$, $\log_{10} 0.000001$, $\log_{0.1} 100{,}000$

In Exercises 5 through 24, use the properties of logarithms to evaluate the given expression.

5. $\log_2 2^{16}$
6. $\log_5(625 \cdot 125)$
7. $2^{\log_2 6}$
8. $\ln e^{-2}$
9. $\log_2(64 \cdot 128)$
10. $\ln \sqrt{e}$
11. $\log_5 \dfrac{5^{14}}{5^{17}}$
12. $\log_3 \dfrac{81}{729}$
13. $3^{\log_3 5.1}$
14. $\log_3 81^{15}$
15. $\ln e^{4.25}$
16. $e^{\ln 5}$
17. $2^{3 \cdot \log_2 5}$
18. $5^{4 \cdot \log_5 3}$
19. $e^{\log_2 8}$
20. $\log_{10} 100^{100}$
21. $e^{-2 \cdot \ln 1.5}$
22. $\dfrac{\log_2 64}{\log_2 8}$
23. $\dfrac{\log_5 125}{\ln (1/e)}$
24. $e^{\ln e^2}$

In Exercises 25 through 28, sketch the graph of the given function.

25. $y = \log_3 x$
26. $y = 2 \cdot \log_5 x$
27. $y = \log_{1/3} x$
28. $y = \ln x^2$

In Exercises 29 through 32, change the base of the given exponential function to base e.

29. $y = 10^x$
30. $y = 2^{-x}$
31. $y = 2 \cdot 7^{4x}$
32. $y = 3(1.25)^{-2x}$

In Exercises 33 through 36, change the base of the given logarithmic function to base e.

33. $y = \log_2 x$
34. $y = 2 \cdot \log_3 x$
35. $y = 5 \cdot \log_5 x^{-2}$
36. $y = 4 \cdot \log_{11} x^3$

The intensity of earthquakes is measured by the **Richter scale r**, where

$$r = \log_{10} \frac{x}{x_0}.$$

Here x_0 represents the intensity of a benchmark quake and x is the intensity of a given quake in multiples of x_0. Thus, if a quake is 10 times as intense as the benchmark quake, then $x = 10 x_0$. Exercises 37 through 39 refer to the Richter scale.

37. Find the Richter scale measurement r for an earthquake that is
    (a) 100
    (b) 10,000
    (c) 10,000,000
    times as intense as the benchmark quake.

38. Suppose an earthquake measures 5 on the Richter scale. How much more intense is it than the benchmark quake?

39. The most destructive earthquakes measure between 8.0 and 9.0 on the Richter scale. How much more intense than the benchmark quake is such a quake?

Management

Exercises 40 through 45 were suggested by material in Coleman, *Introduction to Mathematical Sociology*, Free Press of Glencoe, 1984. They are concerned with modeling consumer purchases as a function of consumer income. For any particular consumer good, let $p(x)$ denote the proportion of consumers whose income is x who do *not* purchase the good within one year's time. One possible model for the relationship between $p(x)$ and x is

$$\ln p(x) = \begin{cases} 0, & 0 \le x < c, \\ -a(x - c), & x \ge c. \end{cases}$$

Here $a > 0$, $c \ge 0$, and a and c depend on the consumer good.

*40. Solve
$$\ln p(x) = \begin{cases} 0, & 0 \leq x < c, \\ -a(x - c), & x \geq c, \end{cases}$$
for $p(x)$ in terms of x.

*41. Use the result of Exercise 40 to graph p as a function of x. Interpret your graph in terms of the proportion of consumers who will or will not purchase the good at a particular income.

*42. Find the percentage of consumers who have incomes of \$25,000 who will not purchase a VCR within one year if $c = \$20,000$ and $a = 0.05$ for VCRs. (Measure income x in thousands of dollars.)

*43. Repeat Exercise 42 for consumers who have incomes of \$15,000, \$30,000, and \$50,000.

*44. Suppose that for a particular consumer good, the constant a remains fixed but the constant c increases. What can you say about consumer purchases of the good and their relation to consumer income?

*45. Answer the question of Exercise 44 if c remains fixed but a increases.

Solving Exponential Equations

In Exercises 46 through 55, use logarithms to solve the given exponential equation.

46. $e^{2x} = 5$ **47.** $e^{-3x} = 11$ **48.** $e^{-2.25x} = 1.9$ **49.** $e^{0.4x} = 2$
50. $e^{-2.4x} = 0.06$ **51.** $2^x = 3$ **52.** $3^{-x/2} = 12$ **53.** $5^{x/3} = 9$
54. $(1.3)^{2x} = 11.5$ **55.** $(2.2)^{0.5x} = 10.2$

In Exercises 56 through 61, change the base of the given function as indicated.

56. $y = e^x$ to base 10 **57.** $y = 10^{-2x}$ to base 2 **58.** $y = 4 \cdot 3^{1.5x}$ to base 5
59. $y = \ln x$ to base 3 **60.** $y = 3 \cdot \log_2 x$ to base 10 **61.** $y = 4 \cdot \log_5 x^2$ to base 11

62. If inflation persists at 10% per year, in t years the purchasing power of \$1, in terms of today's dollar, will be
$$y = e^{-0.1t}$$
dollars. Under these conditions, how long will it take for money to lose one-half of its value? 99% of its value?

63. The decay of a radioactive isotope is described by the equation
$$y = 1000e^{-0.02t},$$
t in days, y in grams. What is the half-life of the isotope? How long will it take for 90% of the isotope to decay?

64. The decay of a radioactive isotope is described by the equation
$$y = ce^{-0.15t},$$
t in years. What is the half-life of the isotope? How long will it take for 30% of it to decay? How long will it take for 99% of it to decay?

65. How long will it take for an investment of \$12,000 to double if it earns 9% per year compounded continuously?

66. How long will it take for an investment of $25,000 to grow to $40,000 if it earns 11.5% per year compounded continuously?

67. How long will it take for an investment of $7000 to triple if it earns 6% per year compounded quarterly?

68. How long will it take for a debt of $20,000 to grow to $35,000 if interest is added to the debt at 12% per year compounded monthly?

69. What annual interest rate must be earned if $8000 is to grow to $20,000 in five years under continuous compounding?

70. What annual interest rate must be earned if $8000 is to grow to $16,000 in five years under continuous compounding?

71. What annual interest rate i must be earned if $22,000 is to increase 10-fold in 30 years at i% per year compounded quarterly?

72. If a future value of $10,000 is discounted for four years at a rate compounded annually and its present value is $6500, what is the rate?

Management

73. If a firm can invest $500,000 in a new product and receive $750,000 in return three years later, what interest rate is it earning on its investment, assuming continuous compounding?

74. Eta Associates takes a six-month loan of $100,000 discounted at a continuously compounded rate. If Eta receives $95,000, what was the rate?

75. A firm's annual sales are given by the equation $S = 200e^{0.05t}$, where t is in years with $t = 0$ representing 1985 and annual sales S is in millions of dollars. When will annual sales attain a level of $1 billion?

76. If an airplane manufacturer has x thousand employees, it can build a plane in y worker-hours, where $y = x^{2.78}$. How many employees would be needed to build a plane in exactly 10,000 worker-hours?

Life Science

77. The number of bacteria in a culture t hours after the start of an experiment is y thousand, where $y = 120 \cdot 2^{-0.25t}$. How long will it take until the number of bacteria in the culture is 10% of the number present at the start of the experiment?

78. A patient is injected with a drug and t hours later $y = 100b^{-t/12}$ percent of the drug remains in the patient's body. If 50% of the drug is eliminated in 9 hours, what is b?

Social Science

79. The population of the earth t years after 1988 is forecast to be y billion people, where $y = 5.0562e^{0.015t}$. How long will it take for the population to reach 100 billion? 1 trillion (1000 billion)?

Internal Rate of Return

Exercises 80 through 84 were suggested by material in Higgins, *Analysis for Financial Management*, Irwin, 1984. They are concerned with the **internal rate of return** of an investment. The internal rate of return (IRR) of an investment is defined as follows:

the IRR is the annually compounded discount rate that makes the net present value of the investment zero. (The **net present value** of an investment is its present value minus its cost.)

*80. Find the IRR of an investment that will pay $100,000 four years from now and costs $75,000 now.

*81. Find a formula for calculating the IRR of an investment that costs P dollars now and will pay A dollars t years from now.

*82. If a firm invests $400,000 now in a new product, it will receive $500,000 four years from now. Use the formula of Exercise 81 to find the IRR of this investment. Suppose the firm must borrow the $400,000 at $i\%$ per year compounded annually; should they do so if $i = 5$? if $i = 7$?

*83. If a firm can invest $250,000 now in a new billing system it will save $340,000 three years from now. If the firm must borrow the $250,000 at $i\%$ per year compounded *continuously,* for what values of i will it make sense for the company to borrow the money?

*84. A firm has three investment opportunities, A, B, and C. If A costs $350,000 and will pay $500,000 in three and one-half years, B costs $400,000 and will pay $600,000 in four years, and C costs $460,000 and will pay $730,000 in five and one-quarter years, use their IRRs to rank the investments in order of desirability.

Average Daily Yield

Exercises 85 through 87 were suggested by material in Stigum, *Money Market Calculations,* Dow Jones-Irwin, 1981. They are concerned with the **average daily yield** of an investment. Short-term securities are usually purchased at a discount and redeemed for the face value of the security. For instance, an investor might purchase a 90-day note with a face value of $1000 for $950; then 90 days after the note was issued it matures, at which time the investor can redeem it for $1000. If P is the price paid by the investor for a security and F is its face value, then P is considered to be the present value of F discounted at a rate that is compounded daily: thus

$$P = F(1 + r)^{-k},$$

where k is the number of days from the purchase of the security to its redemption and $100r\%$ is the **average daily yield** of the security.

*85. An investor pays $950 for a note that will be redeemed 90 days later for $1000. Find the average daily yield of the note.

*86. Find a formula for calculating the average daily yield of a security that is purchased for P, has a face value of F, and will be redeemed in k days.

*87. Use the result of Exercise 86 to find the average daily yield of a security whose face value is $1000 and which is purchased for $982 for redemption 60 days later.

5.3 DIFFERENTIATION OF LOGARITHMIC FUNCTIONS

In Section 5.2 we introduced logarithmic functions and discussed some of their properties and uses. The graph of $y = \ln x$ suggests that there is a unique tangent line at each point on the curve (Figure 5–15), and hence that the derivative of the natural logarithm function must exist for all $x > 0$. In this section we show how to differentiate the natural logarithm and all other logarithmic functions as well.

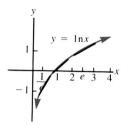

FIGURE 5–15

Differentiation of Logarithmic Functions

Let $f(x) = \ln x$. To find $f'(x)$, we must take the limit as h approaches 0 of the difference quotient

$$\frac{f(x + h) - f(x)}{h} = \frac{\ln(x + h) - \ln x}{h}.$$

We rewrite the different quotient using the properties of logarithms:

$$\frac{\ln(x + h) - \ln x}{h} = \frac{1}{h}[\ln(x + h) - \ln x]$$

$$= \frac{1}{h} \ln \frac{x + h}{x}$$

$$= \ln \left(1 + \frac{h}{x}\right)^{1/h}.$$

Now let $m = x/h$. Then

$$\frac{\ln(x + h) - \ln x}{h} = \ln \left(1 + \frac{1}{m}\right)^{m/x} = \ln \left[\left(1 + \frac{1}{m}\right)^{m}\right]^{1/x}.$$

But by definition,

$$\lim_{m \to +\infty} \left(1 + \frac{1}{m}\right)^{m} = e,$$

and it is possible to use the continuity of the natural logarithm function to show that

$$\lim_{m \to +\infty} \left[\left(1 + \frac{1}{m}\right)^{m}\right]^{1/x} = \ln \left[\lim_{m \to +\infty} \left(1 + \frac{1}{m}\right)^{m}\right]^{1/x} = \ln e^{1/x}.$$

(The details of the argument are rather technical, and we omit them.) Thus, since $m = x/h \to +\infty$ as $h \to 0$, we have

$$f'(x) = \lim_{h \to 0} \frac{\ln(x + h) - \ln x}{h}$$

$$= \lim_{m \to +\infty} \ln \left[\left(1 + \frac{1}{m}\right)^{m}\right]^{1/x}$$

$$= \ln e^{1/x}$$

$$= \frac{1}{x} \ln e$$

$$= \frac{1}{x}.$$

> **The Derivative of ln x**
>
> If $f(x) = \ln x$, then
> $$f'(x) = \frac{1}{x}.$$

EXAMPLE 1 If $f(x) = 2 \cdot \ln x$, then $f(x) = 2(1/x) = 2/x$. ∎

EXAMPLE 2 If $f(x) = x \ln x$, then by the product rule,
$$f'(x) = x(\ln x)' + (x)' \ln x = x\frac{1}{x} + 1 \cdot \ln x = 1 + \ln x. \quad \blacksquare$$

Now suppose that $u = g(x) > 0$ and consider $\frac{d}{dx}(\ln u)$. By the chain rule
$$\frac{d}{dx}(\ln u) = \frac{d}{du}(\ln u)\frac{du}{dx} = \frac{1}{u}\frac{du}{dx} = \frac{1}{g(x)}\frac{d}{dx}(g(x)).$$

Therefore we have

> **The Derivative of ln g(x)**
>
> If $f(x) = \ln g(x)$, then
> $$f'(x) = \frac{g'(x)}{g(x)}.$$

EXAMPLE 3 If $f(x) = \ln(x^2 + 3x + 1)$, then
$$f'(x) = \frac{(x^2 + 3x + 1)'}{x^2 + 3x + 1} = \frac{2x + 3}{x^2 + 3x + 1}. \quad \blacksquare$$

EXAMPLE 4 Let us use the curve-sketching procedure to draw the graph of the function defined by
$$f(x) = x \ln x$$
over the interval $[e^{-2}, +\infty)$.
From Example 2, $f'(x) = 1 + \ln x$. Setting $f'(x) = 0$ and solving for x yield
$$1 + \ln x = 0,$$
or
$$\ln x = -1,$$

and hence
$$e^{\ln x} = e^{-1}.$$
Since $e^{\ln x} = x$, we have
$$x = e^{-1}.$$
Thus $x = e^{-1}$ is the only critical value. Since e^{-2} is the interval $[e^{-2}, e^{-1})$, we may use it as a test value. Then
$$f'(e^{-2}) = 1 + \ln e^{-2} = 1 - 2 = -1$$
shows that $f'(x) < 0$ on the interval $[e^{-2}, e^{-1})$. Similarly, since e is in the interval $(e^{-1}, +\infty)$, we may use it as a test value. Then
$$f'(e) = 1 + \ln e = 1 + 1 = 2$$
implies that $f'(x) > 0$ on the interval $(e^{-1}, +\infty)$. Hence we have the diagram

and there is a relative minimum at $(e^{-1}, f(e^{-1})) = (e^{-1}, -e^{-1})$.
The second derivative is
$$f''(x) = \frac{1}{x},$$
which is always positive for $x \geq e^{-2}$. Therefore the function is concave upward on the interval $[e^{-2}, +\infty)$. Since 0 is not in the interval $[e^{-2}, +\infty)$, the function has no y-intercept. If
$$x \ln x = 0,$$
then $x \neq 0$ implies that
$$\ln x = 0,$$
and hence that
$$x = 1.$$
Therefore the only x-intercept is the point $(1, 0)$. The function has no asymptotes. Its graph is shown in Figure 5–16. □

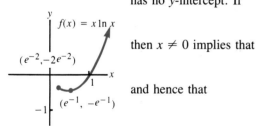

FIGURE 5–16

What about the derivative of $f(x) = \log_b g(x)$, where $b \neq e$? Using the change-of-base formula for logarithms (Section 5.2) with $a = e$, we see that
$$f(x) = \frac{\ln g(x)}{\ln b} = \frac{1}{\ln b} \ln g(x).$$
Hence
$$f'(x) = \frac{1}{\ln b} (\ln g(x))' = \frac{1}{\ln b} \frac{g'(x)}{g(x)}.$$

The Derivative of $\log_b g(x)$

If $f(x) = \log_b g(x)$, then

$$f'(x) = \frac{1}{\ln b} \frac{g'(x)}{g(x)}.$$

Compare the formula for the derivative of $\ln g(x)$ with that for the derivative of $\log_b g(x)$, $b \neq e$; the simplicity of the former is one reason why base e is preferred to other bases. Also note that if $b = e$, the formula for the derivative of $\log_b g(x)$ reduces to that for the derivative of $\ln g(x)$.

EXAMPLE 5 If $f(x) = \log_2(x^2 + 1)$, then

$$f'(x) = \frac{1}{\ln 2} \frac{(x^2 + 1)'}{x^2 + 1} = \frac{2x}{(\ln 2)(x^2 + 1)}. \quad \square$$

Logarithmic Differentiation

Sometimes it is convenient, or even necessary, to take the logarithm of both sides of an expression before differentiating it. We illustrate with an example.

EXAMPLE 6 Let us differentiate $f(x) = x^2(x^2 + 1)^2(x^4 + 1)^4$. We could do this by repeatedly using the product rule and the chain rule, but we prefer to take the natural logarithm of both sides of the equation, differentiate the result, and solve for $f'(x)$. We have

$$\ln f(x) = \ln x^2(x^2 + 1)^2(x^4 + 1)^4,$$

or

$$\ln f(x) = \ln x^2 + \ln(x^2 + 1)^2 + \ln(x^4 + 1)^4,$$

which simplifies to

$$\ln f(x) = 2 \ln x + 2 \ln(x^2 + 1) + 4 \ln(x^4 + 1).$$

Now we differentiate both sides of the last equation to obtain

$$\frac{f'(x)}{f(x)} = \frac{2}{x} + 2\left(\frac{2x}{x^2 + 1}\right) + 4\left(\frac{4x^3}{x^4 + 1}\right),$$

so that

$$f'(x) = f(x)\left(\frac{2}{x} + \frac{4x}{x^2 + 1} + \frac{16x^3}{x^4 + 1}\right),$$

or

$$f'(x) = x^2(x^2 + 1)^2(x^4 + 1)^4 \left(\frac{2}{x} + \frac{4x}{x^2 + 1} + \frac{16x^3}{x^4 + 1}\right). \quad \square$$

EXPONENTIAL AND LOGARITHMIC FUNCTIONS

The technique used in Example 6 is called **logarithmic differentiation.** It is often useful when the function to be differentiated is the product or quotient of several factors.

Sometimes logarithmic differentiation is the only way to proceed.

EXAMPLE 7 Let f be given by $f(x) = x^x$. None of our rules of differentiation apply to this function: it is neither of the form x^r, where r is a fixed real number, nor of the form b^x, where b is a fixed real number. In order to find $f'(x)$, we must use logarithmic differentiation:

$$\ln f(x) = \ln x^x,$$

so

$$\ln f(x) = x \ln x.$$

Differentiating both sides of this last equation, we obtain

$$\frac{f'(x)}{f(x)} = x \cdot \frac{1}{x} + 1 \cdot \ln x.$$

Therefore

$$f'(x) = f(x)[1 + \ln x] = x^x[1 + \ln x]. \quad \square$$

■ Computer Manual: Program DFQUOTNT

5.3 EXERCISES

Differentiation of Logarithmic Functions

In Exercises 1 through 22, find $f'(x)$.

1. $f(x) = 3 \ln x$
2. $f(x) = -4 \ln x$
3. $f(x) = \ln(7x + 2)$
4. $f(x) = \ln(4 - 2x)$
5. $f(x) = \ln(x^2 + 5x)$
6. $f(x) = \ln(x^2 + 5x - 3)$
7. $f(x) = \ln(x^4 + 2x^2 + 200)^4$
8. $f(x) = 2 \ln\sqrt{x^2 + 1}$
9. $f(x) = 2x \ln 3x$
10. $f(x) = x^2 \ln 2x$
11. $f(x) = x^3 \ln x^2$
12. $f(x) = (x^2 + 1) \ln (x^2 + 1)$
13. $f(x) = \ln \dfrac{2x + 1}{3x + 2}$
14. $f(x) = \ln \dfrac{x^2 - 1}{x^2 + 1}$
15. $f(x) = \dfrac{\ln 5x}{\ln 6x}$
16. $f(x) = \dfrac{\ln(x^2 + 4)}{x^2 + 4}$
17. $f(x) = (\ln x)^3$
18. $f(x) = [\ln(x^5 + 1)]^4$
19. $f(x) = \sqrt{\ln(x^3 + 3x^2 + 4x + 3)}$
20. $f(x) = \ln(\ln x)$
21. $f(x) = \ln(\ln(\ln x))$
22. $f(x) = \ln|x|$

In Exercises 23 through 34, use the curve-sketching procedure to draw the graph of the given function.

23. $y = \ln(2x + 3)$
24. $y = \ln(4x - 12)$
25. $y = \ln(ax + b)$, $a > 0$, $b > 0$
26. $y = x - \ln x$

27. $f(x) = x \ln 2x$ on $[e^{-3}, +\infty)$
28. $f(x) = x \ln x^2$ on $[e^{-4}, +\infty)$
29. $y = x^2 \ln x$ on $[e^{-2}, +\infty)$
30. $y = x^3 \ln x$ on $[e^{-5}, +\infty)$
31. $f(x) = (\ln x)^2$
32. $y = \dfrac{\ln x}{x}$
33. $y = \dfrac{\ln x}{x^2}$
34. $y = \dfrac{x}{\ln x}$

In Exercises 35 through 42, find $f'(x)$.

35. $f(x) = \log_2 x$
36. $f(x) = \log_3 2x$
37. $f(x) = \log_7 2x^3$
38. $f(x) = \log_{11}(3x + 5)$
39. $f(x) = 4 \log_3(x^2 + 4x)$
40. $f(x) = x \log_2 x$
41. $f(x) = x^2 \log_{10}(2x - 1)$
42. $f(x) = \log_2(2x + 3) \cdot \log_3 x^2$

Management

43. A firm's annual sales are given by the equation

$$S = \ln(10t + 8),$$

where t is in years, with $t = 0$ representing 1988, and sales S are in millions of dollars. Forecast the rate at which sales will be changing in 1992; in 2010.

44. A firm's revenue function is given by the equation

$$R = \ln(25x + 1),$$

where x is in thousands of units and R is in millions of dollars. Find and interpret the firm's marginal revenue when x is 10 thousand and when x is 100 thousand. Then sketch the graph of the revenue function and analyze revenue as x increases.

45. If steel production in year t was y million tons, where

$$y = 96 \ln (\sqrt{t} + 3)$$

and $t = 0$ represents 1955, find the rate at which production was changing in 1970, 1980, and 1990. Then sketch the graph of the function and analyze steel production as t increases.

46. A firm's cost function is

$$C(x) = \ln\sqrt{x^2 + 2}$$

and its revenue function is

$$R(x) = \ln(100x + 1),$$

both valid for $x \geq 1$. Here cost and revenue are in millions of dollars. Find the firm's maximum profit.

47. The value of all goods produced by an industry in year t is y million dollars, where $y = 1000[\ln(0.1t + 0.1)/(t + 1)] + 20$, with $t = 0$ representing the year 1950. Find the maximum value of the industry's production and the year in which it occurred.

Life Science

48. The concentration of radioactive iodine in seawater increased during the period 1945–1965 due to aboveground testing of atomic weapons. Suppose the concentration of such iodine in year t, where $t = 0$ represents 1945, was

$$y = 3 \ln(5 + t)$$

parts per billion. Find the rate at which the iodine concentration was changing in 1955 and 1960.

49. A lake that is being deacidified has its pH y described by
$$y = 2\ln(3t + 6).$$
Here t is in years, with $t = 0$ representing the start of the deacidification process. Find the rate at which the deacidification process is proceeding 5 and 10 years after its start. Then sketch the graph of the function and analyze the deacidification process as a function of time. (Deacidification ceases when pH $= 7$.)

Social Science

50. The amount spent on social services by a state is y million dollars, where y depends on the unemployment rate $x\%$ according to the equation
$$y = 90\ln(10x + 50).$$
Find the rate at which spending will be changing when the unemployment rate is 5%.

51. The number of people working in a state's bureaucracy is modeled by
$$y = 1.5\ln(60t^{2/5} + 20),$$
where y is the number of people in thousands and t is time in years, with $t = 0$ representing 1985. Find the rate at which the bureaucracy will be growing in 1995. Then sketch the graph of the function and analyze the growth of the bureaucracy as a function of time.

Nuclear Testing

52. According to Lynn Sykes and Dan Davis ("The Yields of Soviet Strategic Warheads," *Scientific American*, January 1987), if x is the yield of an underground nuclear explosion, in kilotons, and y is the magnitude of the surface seismic waves generated by the explosion, then
$$y = \log_{10}(114.4x + 1144.5).$$
Sketch the graph of this function and analyze the magnitude of the seismic waves as a function of the yield of the explosion.

Geology

53. The intensity of earthquakes is measured by the Richter scale r, where
$$r = \log_{10}\frac{x}{x_0}.$$
Here x_0 represents the intensity of a benchmark quake and x is the intensity of a given quake in multiples of x_0. Find the rate of change of the Richter scale measurement as a function of the intensity of a quake.

Logarithmic Differentiation

In Exercises 54 through 66, use logarithmic differentiation to find $f'(x)$.

54. $f(x) = x^2(x^4 + 2)^4$

55. $f(x) = x(x^3 + 1)^3(x^5 + 1)^5$

56. $f(x) = \dfrac{(x^2 + 3)^4}{(x^4 - 2)^7}$

57. $f(x) = \dfrac{(x^3 + 1)^2(x^2 + 1)^3}{(x^2 + 4x + 4)^4}$

58. $f(x) = \dfrac{(x^6 + 1)^3(x^9 - 2x + 1)^4}{(x^5 + 2)^5(x^8 - 7x + 1)^3}$

59. $f(x) = x^{2x}$
60. $f(x) = (2x)^x$
61. $f(x) = x^{-x}$
62. $f(x) = (2x)^{-2x}$
63. $f(x) = (x^2)^{x+1}$
64. $f(x) = (x + 1)^x$
65. $f(x) = (\ln x)^x$
66. $f(x) = x^{\ln x}$

5.4 DIFFERENTIATION OF EXPONENTIAL FUNCTIONS

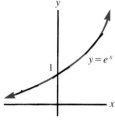

FIGURE 5–17

In the previous section we showed how to differentiate logarithmic functions. Now we are ready to consider exponential functions. The graph of $y = e^x$ suggests that there is a unique tangent line at each point on the curve (Figure 5–17), and hence that the derivative of the exponential function with base e must exist for all x. In this section we discuss the differentiation of this and other exponential functions.

We can use our knowledge about the derivative of the natural logarithm to develop a rule for the derivative of e^x: if we differentiate both sides of the identity

$$x = \ln e^x$$

with respect to x, we have

$$1 = \frac{(e^x)'}{e^x}.$$

Thus

$$(e^x)' = e^x.$$

The Derivative of e^x

If $f(x) = e^x$, then

$$f'(x) = e^x.$$

EXAMPLE 1 If $f(x) = 3e^x$, then $f'(x) = 3(e^x)' = 3e^x$. ☐

If $f(x) = e^{g(x)}$, let $u = g(x)$. Then by the chain rule,

$$\frac{df}{dx} = \frac{d}{dx}(e^u) = \frac{d}{du}(e^u)\frac{du}{dx} = e^u \frac{du}{dx} = e^{g(x)}g'(x).$$

The Derivative of $e^{g(x)}$

If $f(x) = e^{g(x)}$, then

$$f'(x) = e^{g(x)}g'(x).$$

EXPONENTIAL AND LOGARITHMIC FUNCTIONS

EXAMPLE 2 If $f(g) = e^{-3x}$, then
$$f'(x) = e^{-3x}(-3x)' = e^{-3x}(-3) = -3e^{-3x}. \quad \square$$

EXAMPLE 3 If $f(x) = 2e^{x^2+x}$, then
$$f'(x) = 2e^{x^2+x}(x^2 + x)' = 2(2x + 1)e^{x^2+x}. \quad \square$$

EXAMPLE 4 If $f(x) = xe^{-x}$, then by the product rule,
$$f'(x) = x(e^{-x})' + (x)'e^{-x} = xe^{-x}(-1) + 1 \cdot e^{-x} = e^{-x}(1 - x). \quad \square$$

EXAMPLE 5 Find the equation of the line tangent to the graph of
$$y = \frac{e^x - e^{-x}}{e^x + e^{-x}}$$
at $x = 0$. By the quotient rule,
$$\frac{dy}{dx} = \frac{(e^x + e^{-x})\frac{d}{dx}(e^x - e^{-x}) - (e^x - e^{-x})\frac{d}{dx}(e^x + e^{-x})}{(e^x + e^{-x})^2}$$
$$= \frac{(e^x + e^{-x})(e^x + e^{-x}) - (e^x - e^{-x})(e^x - e^{-x})}{(e^x + e^{-x})^2}$$
$$= \frac{(e^{2x} + 2 + e^{-2x}) - (e^{2x} - 2 + e^{-2x})}{(e^x + e^{-x})^2}$$
$$= \frac{4}{(e^x + e^{-x})^2}.$$

Therefore
$$\left.\frac{dy}{dx}\right|_{x=0} = \frac{4}{(1 + 1)^2} = 1.$$

Thus, since $y = 0$ when $x = 0$ (check this), the equation of the tangent line at $x = 0$ is $y = x$. $\square$

EXAMPLE 6 Let us sketch the graph of the function defined by
$$f(x) = xe^{-x}.$$
Since $f'(x) = e^{-x}(1 - x)$ by Example 4, setting $f'(x) = 0$ yields
$$e^{-x}(1 - x) = 0.$$
Since e to any power is positive, this implies that $x = 1$. Furthermore, when $x < 1$, $f'(x) = e^{-x}(1 - x)$ is positive, and when $x > 1$, $f'(x) = e^{-x}(1 - x)$ is negative. Hence the diagram of the first derivative test looks like this:

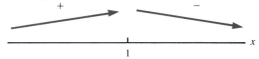

Therefore there is a relative maximum at $(1, f(1)) = (1, e^{-1})$.

The second derivative is

$$f''(x) = e^{-x}(1-x)' + (e^{-x})'(1-x)$$
$$= -e^{-x} - e^{-x}(1-x)$$
$$= e^{-x}(x-2).$$

Hence, setting the second derivative equal to zero and solving for x will yield $x = 2$. Since $f''(x) < 0$ when $x < 2$ and $f''(x) > 0$ when $x > 2$, the concavity diagram looks like this:

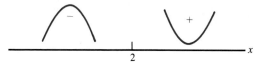

Therefore there is a point of inflection at $(2, f(2)) = (2, 2e^{-2})$.

The point $(0, 0)$ is the only intercept. It is easy to check that as $x \to +\infty$, the value of xe^{-x} approaches 0, and that as $x \to -\infty$, the value of xe^{-x} approaches $-\infty$. Thus

$$\lim_{x \to +\infty} xe^{-x} = 0 \quad \text{and} \quad \lim_{x \to -\infty} xe^{-x} = -\infty.$$

Hence the x-axis is an asymptote as $x \to +\infty$. The graph of the function is shown in Figure 5–18. ∎

EXAMPLE 7 Let us sketch the graph of $y = e^{-x^2}$. We have

$$\frac{dy}{dx} = -2xe^{-x^2}.$$

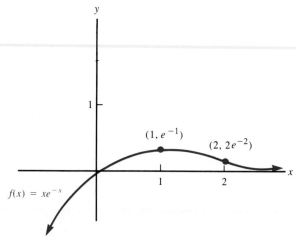

FIGURE 5–18

Setting the derivative equal to zero and solving for x yield $x = 0$. The diagram of the first derivative test looks like this:

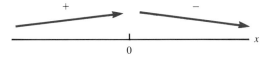

(Check this.) Hence the function has a relative maximum at $(0, 1)$.
The second derivative is

$$\frac{d^2y}{dx^2} = -2e^{-x^2} + 4x^2 e^{-x^2} = -2e^{-x^2}(1 - 2x^2),$$

so setting it equal to 0 and solving for x yield $x = \pm 1/\sqrt{2}$. The diagram of the second derivative test looks like this:

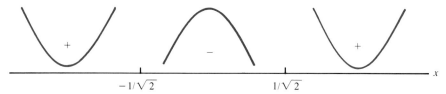

(Check this.) Hence there are points of inflection at $(\pm 1/\sqrt{2}, e^{-1/2})$. It is easy to check that $\text{Lim}_{x \to \pm \infty} e^{-x^2} = 0$. The graph is shown in Figure 5–19. □

EXAMPLE 8 Suppose sales of electronic calculators t years after their introduction were y million units, where

$$y = \frac{100}{1 + 49e^{-0.4t}}.$$

Let us find the rate at which sales were changing when $t = 10$.

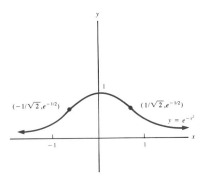

FIGURE 5–19

If we rewrite the equation in the form

$$y = 100(1 + 49e^{-0.4t})^{-1},$$

the general power rule shows that

$$\frac{dy}{dt} = -100(1 + 49e^{-0.4t})^{-2}(49(-0.4)e^{-0.4t}),$$

or

$$\frac{dy}{dt} = \frac{1960e^{-0.4t}}{(1 + 49e^{-0.4t})^2}.$$

Hence

$$\left.\frac{dy}{dt}\right|_{t=10} = \frac{1960e^{-4}}{(1 + 49e^{-4})^2} = 9.971.$$

Therefore when $t = 10$ sales were increasing at a rate of 9,971,000 units per year. □

What about the derivative of $f(x) = b^{g(x)}$, $b \neq e$? Since $e^{\ln b} = b$, we have

$$f(x) = (e^{\ln b})^{g(x)} = e^{(\ln b)g(x)}.$$

Therefore

$$f'(x) = e^{(\ln b)g(x)}((\ln b)g(x))' = e^{(\ln b)g(x)}(\ln b)g'(x),$$

which we may rewrite as

$$f'(x) = b^{g(x)}(\ln b)g'(x).$$

The Derivative of $b^{g(x)}$

If $f(x) = b^{g(x)}$, then

$$f'(x) = (\ln b)b^{g(x)}g'(x).$$

The simplicity of the formula for the derivative of $e^{g(x)}$, as compared with the formula for the derivative of $b^{g(x)}$, $b \neq e$, is another reason why base e is preferred to other bases. Note also that if $b = e$, the formula for the derivative of $b^{g(x)}$ reduces to that for the derivative of $e^{g(x)}$.

EXAMPLE 9   If $f(x) = 3^{2x}$, then

$$f'(x) = (\ln 3)3^{2x}(2x)' = 2(\ln 3)3^{2x}. \quad \square$$

■ *Computer Manual: Program DFQUOTNT*

5.4 EXERCISES

In Exercises 1 through 24, find $f'(x)$.

1. $f(x) = e^{2x}$
2. $f(x) = 3e^x$
3. $f(x) = 2e^{-x}$
4. $f(x) = -12e^{-x/2}$
5. $f(x) = \frac{1}{2}(e^x - e^{-x})$
6. $f(x) = \frac{1}{2}(e^x + e^{-x})$
7. $f(x) = 5e^{x^2+3x+4}$
8. $f(x) = 2e^{x^2-2x+7}$
9. $f(x) = 2e^{x^3}$
10. $f(x) = 4e^{\sqrt{x^2+1}}$
11. $f(x) = e^{3(\ln x)}$
12. $f(x) = e^{e^x}$
13. $f(x) = 2xe^{-4x}$
14. $f(x) = 3x^2 e^{2x}$
15. $f(x) = \frac{e^x}{x}$
16. $f(x) = \frac{2e^{-2x}}{x^2+1}$
17. $f(x) = e^x \ln x$
18. $f(x) = e^{-x} \ln x$
19. $f(x) = e^{2x} \ln 3x$
20. $f(x) = \frac{e^x}{\ln x}$
21. $f(x) = e^{ex}$
22. $f(x) = (e^x)^{e^x}$
23. $f(x) = x^e$
24. $f(x) = x^{e^x}$

In Exercises 25 through 40, use the curve-sketching procedure to draw the graph of the function.

25. $f(x) = e^{4x}$
26. $y = 2e^{-3x}$
27. $f(x) = xe^x$
28. $f(x) = 2xe^{3x}$
29. $y = x^2 e^x$
30. $y = x^2 e^{-x}$
31. $y = x^3 e^{-3x}$
32. $y = \frac{e^x}{x}$
33. $y = e^{-x}\sqrt{x}$
34. $y = \frac{1}{2}(e^x - e^{-x})$
35. $y = \frac{1}{2}(e^x + e^{-x})$
36. $y = \frac{e^x - e^{-x}}{e^x + e^{-x}}$
37. $y = 1 - e^{-x}$
38. $y = 1 + e^{-x}$
39. $y = \frac{1}{1 + e^{-x}}$
40. $y = xe^{-x^2}$

41. The function f defined by

$$f(x) = \frac{1}{\sqrt{2\pi}} e^{-x^2/2}$$

is important in statistics. Use the curve-sketching procedure to graph this function.

42. The function f defined by

$$f(x) = \frac{1}{\sigma\sqrt{2\pi}} e^{-(x-\mu)^2/2\sigma^2},$$

where μ and σ are real numbers, $\sigma > 0$, is important in statistics. Use the curve-sketching procedure to graph this function.

In Exercises 43 through 50, find $f'(x)$.

43. $f(x) = 2^x$
44. $f(x) = 3^{-5x}$
45. $f(x) = 3(5^{-2x})$
46. $f(x) = 10^{2x^2+3}$
47. $f(x) = x^2 5^x$
48. $f(x) = 2^{-3x}(7^{x^2+2x})$
49. $f(x) = 3^{2x}/x$
50. $f(x) = 4^{-x}/x^2$

51. If $10,000 is invested at 8% per year compounded continuously, find the rate at which the investment is increasing at the end of 5 years; at the end of 10 years.

52. If $25,000 is invested at 6% per year compounded quarterly, find the rate at which the investment is increasing at the end of four years; at the end of eight years.

53. If amount P is invested at $i\%$ per year compounded continuously, find the rate at which the investment is changing at the end of t years.

54. If amount P is invested at $i\%$ per year compounded m times per year, find the rate at which the investment is changing at the end of t years.

55. The amount of a radioactive isotope present t minutes after 10 grams of it are created in a reactor is $y = 100e^{-0.2t}$ grams. Find the rate at which the amount of isotope present is changing when $t = 0$ and when $t = 5$.

56. A radioactive isotope has a half-life of 30 days.
    (a) Find an equation that describes the amount of the isotope present as an exponential function of time. Use base 2.
    (b) Use the equation of part (a) to find the rate at which the amount of isotope present is changing when $t = 0$ and when $t = 60$.

Management

57. A firm's profit in year t is given by $P(t) = 100,000e^{-0.2t}$, where $t = 0$ represents 1985. Find the rate at which the firm's profit will be changing in 1990 and in 1995.

58. The supply function for a commodity is given by $q = e^{0.025p} - 1$ and the demand function by $q = 1000 - 2e^{-0.06p}$. Graph the supply and demand functions on the same set of axes and use the graph to estimate the equilibrium price.

59. A company's average cost per unit is given by
$$f(x) = 2 + 3e^{-0.1x},$$
and its average revenue per unit by
$$g(x) = 5 - e^{-0.2x}.$$
Find the rate of change of average cost per unit, average revenue per unit, and average profit per unit when $x = 10$. Then graph these functions and analyze average cost per unit and average revenue per unit.

60. The worker-hours needed to produce the xth unit of a product are y, where
$$y = 2000(2^{-0.2x}) + 1000, \quad x \geq 1.$$
Find the rate at which production time is changing when $x = 1, 5,$ and 25.

61. (This exercise and the next were suggested by material in Budnick, McLeavey, and Mojena, *Principles of Operations Research for Management*, second edition, Irwin, 1988.) Slice-n-Dice, Inc., advertises its "slicer-dicer" kitchen implement only on television. The firm comes into a city, advertises the slicer-dicer extensively on lo-

cal television for a few weeks, and then, when the market is exhausted, moves on to another city. Suppose that a city contains 300,000 potential customers for the slicer-dicer and that the proportion of them who will buy it in response to the ad campaign is $y = 1 - e^{-0.14t}$, where t is the length of the campaign in days. Suppose also that it costs Slice-n-Dice $200,000 to set up the ad campaign and $10,000 per day to run it, and that the firm makes a profit of $5 on each slicer-dicer sold. How long should the ad campaign run in order to maximize Slice-n-Dice's profits?

*62. Refer to Exercise 61: suppose that a city contains c potential customers, that it costs Slice-n-Dice $F to set up the ad campaign and $v per day to run it, that the firm's profit is $m per slicer-dicer sold, and that the proportion of customers in the city who will buy the slicer-dicer in response to the ad campaign is $y = 1 - e^{-rt}$, $r > 0$, t the length of the ad campaign in days. Find the value of t that will maximize Slice-n-Dice's profits.

Life Science

63. The concentration of medication in a patient's blood t days after going off the medication is y micrograms per milliliter, where $y = 10(3^{-0.4t})$. Find the concentration and the rate at which it is changing five days after going off the medication.

64. The pollutants in the atmosphere near a power plant will be y parts per million t years from now, where $y = 30 + 125e^{-0.06t}$. Find the rate of change of the pollutants now, and also 4 and 12 years from now. Then graph this function and analyze the plant's pollution as a function of time.

65. The number of bacteria in a culture t hours after an experiment begins will be y thousand, where $y = 100 - 10(5^{-0.1t})$. Find the rate of change of the bacteria population when the experiment begins and also 12 and 24 hours later.

66. A flu epidemic hits a town and t days later y of the inhabitants have contracted the disease, where
$$y = \frac{5000}{1 + 4999e^{-0.5t}}.$$
Find the rate at which the epidemic is proceeding when $t = 0$, 17, and 25.

67. The concentration of a drug in a patient's bloodstream t minutes after injection is $y = 10te^{-0.1t}$ milligrams per milliliter. Graph this function for $t \geq 0$, and find the maximum concentration of the drug and when this maximum occurs.

Social Science

68. The growth of a state bureaucracy is described by the equation $y = 225e^{0.005t}$. Here y is the size of the bureaucracy, in thousands, and t is time in years with $t = 0$ representing 1985. Find the rate at which the size of the bureaucracy was changing in 1985 and the rate at which it will be changing in 1995.

69. A college student starts a rumor about another student, and t hours later y of the students at the college have heard it, where
$$y = \frac{2000}{1 + 1999e^{-1.1t}}.$$
Find the rate at which the rumor spreads when $t = 0$, 5, and 10 hours.

70. A state's aid to education t years from now is forecast to be $y = 200te^{-0.05t}$ million dollars. Find the maximum aid to education and when it will occur. Then graph the function and analyze aid to education as a function of time.

Optimum Holding Period

Exercises 71 through 73 were suggested by material in Budnick, McLeavey, and Mojena, *Principles of Operations Research for Management*, second edition, Irwin, 1988. They are concerned with the optimum holding period for an investment.

*71. Suppose you are considering investing $10,000 in real estate and you estimate that the investment will be worth

$$I(t) = 10,000(1 + t)$$

dollars t years from now. You eventually intend to sell the real estate for a gain, and you would like to determine how long you should hold it before selling. It is your policy to discount future values at 10% per year compounded continuously. Should you buy the real estate? If so, how long should you hold onto it before selling? (*Hint:* Maximize the net present value (= present value − cost) of $I(t)$.)

*72. Repeat Exercise 71 if t years from now the investment will be worth $I(t) = 10,000\sqrt{t}$ and you discount future values at 12% per year compounded continuously.

*73. Suppose an investor pays C dollars for an asset that will be worth $I(t)$ dollars t years from now, where $I(t)$ is a continuous and increasing function of t. Suppose also that the investor discounts the future worth of the investment at $100r\%$ per year compounded continuously. Find the optimum holding period for the asset.

5.5 EXPONENTIAL GROWTH AND DECLINE

We have seen throughout this chapter that exponential functions can be used to model many situations involving growth and decline. In this section we discuss some useful exponential models and also briefly consider a logarithmic growth model.

Exponential Growth and Decline

A function defined by an equation of the form $y = ce^{rt}$ has the property that its rate of change is directly proportional to y:

$$\frac{dy}{dt} = rce^{rt} = ry.$$

Conversely, it is a fact from the theory of differential equations that functions of this form are the only ones having this property. It therefore follows that if the rate of change of y is directly proportional to y, then $y = ce^{rt}$, for some numbers c and r.

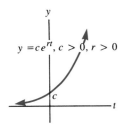

FIGURE 5–20
Exponential Growth

Models of Growth and Decline

If a quantity changes in such a way that its rate of change at time t is directly proportional to the amount of it present at time t, then it can be modeled by means of an equation of the form

$$y = ce^{rt}.$$

If the quantity increases with time, the model is said to be one of **exponential growth;** if the quantity decreases with time, the model is said to be one of **exponential decline** or **exponential decay.**

Figure 5–20 shows the graph of an exponential growth model when $c > 0$, and Figure 5–21 shows the graph of an exponential decline model when $c > 0$.

We illustrate the construction of exponential growth and decline models with examples.

EXAMPLE 1

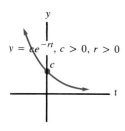

FIGURE 5–21
Exponential Decline

Suppose we have reason to believe that the rate of population growth for a caribou herd depends directly on the size of the herd. Suppose also that there were 10,000 caribou in the herd in 1987 and 11,000 in 1989, so that the size of the herd is increasing. We may then model the size of the herd by an exponential growth function whose defining equation is of the form

$$y = ce^{rt}.$$

To complete the model, we must find numbers c and r such that the resulting function describes the size of the herd.

If we measure time in years, with $t = 0$ representing 1987, we must have

$$y = 10{,}000 \quad \text{when } t = 0 \quad \text{and} \quad y = 11{,}000 \quad \text{when } t = 2.$$

Substituting 0 for t and 10,000 for y in the model yields

$$10{,}000 = ce^{r(0)} = ce^0 = c \cdot 1 = c.$$

Therefore $c = 10{,}000$, so

$$y = 10{,}000 e^{rt}.$$

To find r, we substitute 2 for t and 11,000 for y to obtain

$$11{,}000 = 10{,}000 e^{r(2)} = 10{,}000 e^{2r}.$$

Solving the exponential equation

$$11{,}000 = 10{,}000 e^{2r}$$

for r, we have

$$\frac{11{,}000}{10{,}000} = e^{2r},$$

or

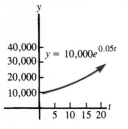

FIGURE 5-22

so

or

$$\therefore 1 = e^{2r},$$

$$\ln 1.1 = \ln e^{2r} = 2r,$$

$$0.095310 = 2r.$$

Thus $r = 0.05$, rounded to two decimal places, and our model is

$$y = 10{,}000 e^{0.05t}.$$

The graph of this function is shown in Figure 5–22.

Now we can use the model for prediction. For instance, if we wish to predict the size of the herd in 1997, we need only substitute 10 for t in the model:

$$y = 10{,}000 e^{0.05(10)} = 10{,}000 e^{0.5} = 10{,}000(1.6487) = 16{,}487.$$

Similarly, if we wish to predict how long after 1987 it will take for the size of the herd to grow to 20,000 caribou, then we must substitute 20,000 for y and solve the exponential equation

$$20{,}000 = 10{,}000 e^{0.05t}$$

for t. Dividing both sides of this last equation by 10,000 and then taking the natural logarithm of both sides, we obtain

$$\ln 2 = 0.05t(\ln e),$$

or

$$\ln 2 = 0.05t.$$

Therefore

$$t = \ln 2 / 0.05 \approx 13.9 \text{ years.} \quad \square$$

EXAMPLE 2 Radioactive decay of an isotope is known to proceed at a rate that is directly proportional to the amount of the isotope present, and the amount of the isotope present decreases with time. Therefore radioactive decay can be modeled by exponential decline functions defined by equations of the form

$$y = ce^{rt}.$$

For instance, the radioactive isotope of uranium known as U^{235} has a half-life of 12 million years. (Thus any amount of U^{235} will decrease by one half in 12 million years.) Let us measure time in millions of years. When $t = 0$,

$$y = ce^{r(0)} = ce^0 = c \cdot 1 = c,$$

so U^{235} is initially present in amount c; but since its half-life is 12 million years, when $t = 12$ we must have $y = c/2$:

$$\frac{1}{2}c = ce^{r(12)} = ce^{12r}.$$

Solving this exponential equation for r, we find that

$$\frac{1}{2} = e^{12r},$$

or

$$\ln \frac{1}{2} = \ln e^{12r},$$

or

$$-0.693147 = 12r,$$

so $r = -0.06$, rounded to two decimal places. Therefore the model is

$$y = ce^{-0.06t}.$$

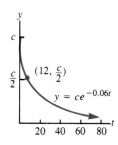

FIGURE 5–23

The graph is shown in Figure 5–23.

We see that at the end of 50 million years approximately 5% of the original amount of U^{235} will be left, because when $t = 50$ we have

$$y = ce^{-0.06(50)} = ce^{-3} = c(0.049787) \approx 0.05c.$$

How long will it be until only 1 percent of the original amount of U^{235} remains? One percent of the original amount is $0.01c$, and if we set $y = 0.01c$ in the model and solve

$$0.01c = ce^{-0.06t}$$

for t, we find that $t \approx 76.8$ million years. (Check this.) ☐

Limited Growth and Decline

Exponential growth models of the type discussed above are often unrealistic because they assume unlimited growth. For instance, the size of the caribou herd of Example 1 certainly cannot grow without limit, since, for one thing, there will be only a certain amount of food available for the caribou. Thus, if we wish to make our models of exponential growth realistic, it will sometimes be necessary to allow them to reflect limited growth. Similar remarks apply to situations of limited decline.

To see what a model for limited exponential growth might be like, consider a function defined by an equation of the form

$$y = a - ce^{-rt}, \qquad a > c > 0, \quad r > 0$$

on $[0, +\infty)$. Figure 5–24 shows the graph of such a function. (See Exercise 23.) Note that

- As t increases, y grows.
- As t increases, the rate at which y grows decreases.
- As $t \to +\infty$, growth levels off and y approaches the limiting value a from below.

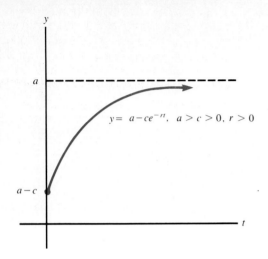

FIGURE 5–24
Limited Growth

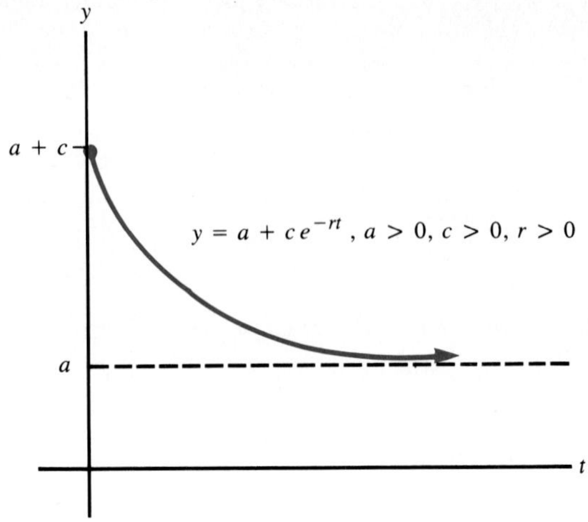

FIGURE 5–25
Limited Decline

Functions of this type are often used to model situations of limited growth when the rate of growth decreases with time and some limiting value is approached from below. Similarly, functions defined by equations of the form

$$y = a + ce^{-rt}, \quad a > 0, \quad c > 0, \quad r > 0$$

on $[0, +\infty)$ are often used to model situations of limited decline. Figure 5–25 shows the graph of such a function. (See Exercise 24.)

Limited Growth and Decline

1. Suppose that as time increases, a quantity increases, its rate of growth decreases, and its growth levels off and approaches a limiting value $a > 0$ from below. Then the amount of the quantity present at time t can be modeled by an equation of the form

$$y = a - ce^{-rt}, \quad a > c > 0, \quad r > 0.$$

2. Suppose that as time increases, a quantity decreases, its rate of decline decreases, and its decline levels off and approaches a limiting value $a > 0$ from above. Then the amount of the quantity present at time t can be modeled by an equation of the form

$$y = a + ce^{-rt}, \quad c > 0, \quad r > 0.$$

We illustrate limited growth and decline with examples.

EXAMPLE 3 Consider again the caribou herd of Example 1. As before, suppose the size of the herd is 10,000 in 1987 and 11,000 in 1989, but now we will assume that as time goes on the growth of the herd will slow down and its size will be limited by the food supply to no more than 15,000. If we model the size of the herd by a function whose defining equation is of the form

$$y = a - ce^{-rt},$$

where $t = 0$ represents 1987, then the limiting value $a = 15{,}000$, so the model is

$$y = 15{,}000 - ce^{-rt}.$$

Since $y = 10{,}000$ when $t = 0$, we have

$$10{,}000 = 15{,}000 - ce^{-r(0)} = 15{,}000 - c.$$

Therefore $c = 5000$, and the model is

$$y = 15{,}000 - 5000e^{-rt}.$$

Finally, since $y = 11{,}000$ when $t = 2$, we have

$$11{,}000 = 15{,}000 - 5000e^{-r(2)},$$

or

$$-4000 = -5000e^{-2r},$$

or

$$0.80 = e^{-2r}.$$

Taking the natural logarithm of both sides of this last equation and solving for r give $r = 0.11$, rounded to two decimal places. (Check this.) Thus the final model is

$$y = 15{,}000 - 5000e^{-0.11t}.$$

The graph is shown in Figure 5–26. With this model our prediction for the size of the herd in 1997 is

$$y = 15{,}000 - 5000e^{-0.11(10)} = 15{,}000 - 5000e^{-1.1} = 13{,}336. \quad \square$$

An important application of limited growth is the **learning curve.** Learning curves relate productivity to experience. Typically, productivity increases as employees become more experienced, since with experience they become more efficient. Productivity cannot grow without limit, however, for even perfect efficiency would result in a certain finite productivity. Thus the rate of productivity growth slows down as productivity increases toward some limit, and hence productivity can be described by a model of the form

$$y = a - ce^{-rt}.$$

EXAMPLE 4 Data collected at a plant that makes microwave carts show that a worker who has no experience can assemble carts at a rate of approximately 200 per week, whereas

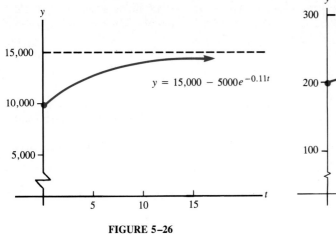

FIGURE 5-26

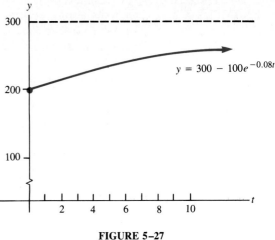

FIGURE 5-27

a worker who has five weeks of experience can assemble them at a rate of approximately 233 per week. It is also known that the most experienced and efficient workers can approach an assembly rate of 300 carts per week. Let us find the learning curve for a typical worker at the plant.

We use the model

$$y = a - ce^{-rt},$$

where t is worker experience, measured in weeks, and y is carts assembled per week. We may take the limiting value a to be 300, so our model is

$$y = 300 - ce^{-rt}.$$

We know that $y = 200$ when $t = 0$, and $y = 233$ when $t = 5$; hence

$$200 = 300 - ce^{-r(0)} = 300 - c,$$

so $c = 100$. But then

$$y = 300 - 100e^{-rt},$$

and the fact that $y = 233$ when $t = 5$ tells us that

$$233 = 300 - 100e^{-5r},$$

or

$$-67 = -100e^{-5r},$$

or

$$0.67 = e^{-5r}.$$

Taking the natural logarithm of both sides of this last equation and solving for r, we find that $r = 0.08$, to two decimal places. Hence the learning curve is

$$y = 300 - 100e^{-0.08t}.$$

Figure 5-27 shows the graph of this learning curve.

Now suppose that a worker must assemble at least 250 carts per week in order to cover the cost of his or her salary. Since $y \approx 247$ when $t = 8$ and $y \approx 251$ when $t = 9$, a worker cannot be expected to cover the cost of his or her salary until the ninth week on the job. ◻

To illustrate a limited decline model, let us consider learning curves from a different viewpoint. In Example 4 we were interested in productivity as a function of experience, but sometimes it is more important to consider production time as a function of the number of units produced. What typically occurs is that each unit can be produced in slightly less time than the preceding one because production becomes more efficient with experience. Even with perfect efficiency, however, there must be some minimum time necessary to produce a unit. Hence the time required to produce a unit will approach some limiting value from above.

EXAMPLE 5 An airplane manufacturer about to begin production of a new plane estimates that it will take 5000 worker-hours to build the first plane, 4600 worker-hours to build the second one, and that it would take 2800 worker-hours to build a plane under conditions of perfect efficiency. Let us find the learning curve that describes the number of worker-hours needed to build the xth plane, $x \geq 1$.

Our model will be of the form
$$y = a + ce^{-rx}.$$
Since the limiting value is 2800 worker-hours, we may take $a = 2800$, so that
$$y = 2800 + ce^{-rx}.$$
When $x = 1$, $y = 5000$, and when $x = 2$, $y = 4600$. Therefore
$$5000 = 2800 + ce^{-r}$$
and
$$4600 = 2800 + ce^{-2r},$$
or
$$2200 = ce^{-r}$$
and
$$1800 = ce^{-2r}.$$
We divide the first equation by the second to obtain
$$\frac{2200}{1800} = \frac{ce^{-r}}{ce^{-2r}} = e^r.$$
Solving for r yields $r = 0.20$, and then
$$2200 = ce^{-0.20}$$
implies that $c = 2687$. Therefore our learning curve is
$$y = 2800 + 2687e^{-0.20x}.$$

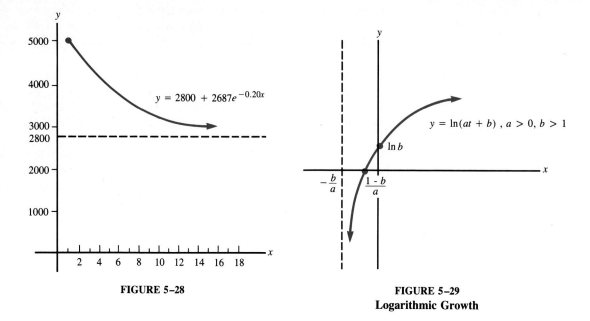

FIGURE 5–28

FIGURE 5–29
Logarithmic Growth

The graph is shown in Figure 5–28. If we wished, we could estimate the total number of worker-hours required to build the first 25 planes, say, by substituting $x = 1, 2, \ldots, 25$ into the model and summing the resulting y-values. ◻

Logarithmic Growth

Functions defined by equations of the form

$$y = \ln(at + b), \quad a > 0, \quad b > 1,$$

are sometimes used to model situations of growth in which growth is unlimited but the rate of growth is decreasing. Figure 5–29 shows the graph of such a function. (See Exercise 53.) Note that as t increases, y increases at a decreasing rate and $\lim_{t \to +\infty} y = +\infty$. A situation that can be modeled by such a function is said to be one of **logarithmic growth.**

EXAMPLE 6   The amount of steel produced in a certain country has been increasing at a decreasing rate. The amount produced was 4 million tons in 1986 and 6 million tons in 1988. Let us describe the country's steel production with a model of the form

$$y = \ln(at + b), \quad a > 0, \quad b > 1,$$

where y is in millions of tons of steel and t is in years, with $t = 0$ representing 1986.

When $t = 0$, $y = 4$, so

$$4 = \ln(a(0) + b) = \ln b.$$

Exponential and Logarithmic Functions

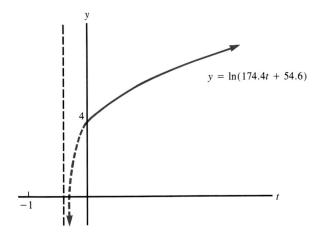

FIGURE 5–30

But

$$4 = \ln b$$

implies that

$$e^4 = e^{\ln b} = b.$$

Similarly, since $y = 6$ when $t = 2$, we have

$$6 = \ln(2a + b) = \ln(2a + e^4),$$

so

$$e^6 = e^{\ln(2a + e^4)} = 2a + e^4,$$

which shows that

$$a = \frac{1}{2}(e^6 - e^4).$$

Converting a and b to decimal form, we have

$$a = \frac{1}{2}(e^6 - e^4) \approx 174.4, \qquad b = e^4 \approx 54.6.$$

Therefore our model is

$$y = \ln(174.4t + 54.6).$$

The graph is shown in Figure 5–30. ∎

■ *Computer Manual: Programs EXPGRWTH, LIMGRWTH, LOGGRWTH*

5.5 EXERCISES

Exponential Growth and Decline

1. The rate of growth of an investment is directly proportional to the value of the investment. Suppose the original amount of the investment was $10,000 and that five years later the investment had a value of $14,918. Find a model that describes the value of the investment as a function of time.

2. The rate of growth of an investment is directly proportional to the value of the investment. Suppose that the original amount invested was $5000 and that eight years later the value of the investment had doubled. Find
   (a) a model that describes the value of the investment as a function of time;
   (b) the value of the investment 12 years after its inception;
   (c) how long it will be until the investment is worth $30,000.

*3. The rate of growth of an investment is directly proportional to the value of the investment. Suppose the original amount invested was P_o dollars and that n years later the investment was worth P_n dollars. Find a model that describes the worth of the investment as a function of time.

4. A radioactive isotope has a half-life of 500 years. Find a model that describes the amount of the isotope present as a function of time.

5. A radioactive isotope has a half-life of 16 days. Find
   (a) a model that describes the amount of the isotope present as a function of time;
   (b) the percentage of the original amount remaining at the end of 10 days;
   (c) how long it will take for 90% of the isotope to decay.

6. Ten grams of a radioactive isotope were created in a reactor and 4 minutes later 7.5 grams were left. Find a model that describes the amount of the isotope present as a function of time, and then find the half-life of the isotope.

*7. If a radioactive isotope has a half-life of T years, find a model that describes the amount of the isotope present as a function of time.

8. Inflation causes the purchasing power of money to decline at a rate proportional to the purchasing power. Suppose inflation persists at an annual rate of 10%. Find a function that describes the purchasing power of a dollar as a function of time. Then find the purchasing power of a dollar after 10 years of 10% inflation; after 15 years of such inflation.

Radiocarbon Dating

The element carbon has two isotopes: C^{14}, which is radioactive with a half-life of 5730 years, and C^{12}, which is not radioactive. In living organisms, the C^{14} incorporated into tissue decays but is replaced by C^{14} from the atmosphere in such a way that the ratio of C^{14} to C^{12} in the tissue remains constant. In dead tissue, however, the C^{14} decays and is not replaced, so the C^{14} to C^{12} ratio decreases at a rate proportional to the ratio. By comparing the C^{14} to C^{12} ratio in dead tissue to that in living tissue, it is thus possible to determine how long ago an organism died. Exercises 9 and 10 concern this technique of **radiocarbon dating.**

*9. Let y be the value of the C^{14} to C^{12} ratio in tissue t years after the tissue has died.
   - **(a)** Find an exponential model that describes y as a function of time. (*Hint:* The ratio will decrease by one-half in 5730 years.)
   - **(b)** Find the age of wheat grains found in an Egyptian tomb if the C^{14} to C^{12} ratio for the grains is 75% of today's ratio.
   - **(c)** Date an archaeological site if charcoal fragments found at the site have a C^{14} to C^{12} ratio that is 20% of today's ratio.

*10.
   - **(a)** Find an equation that describes the time T since an organism died as a function of the C^{14} to C^{12} ratio of its tissue.
   - **(b)** Use the result of part (a) to date a papyrus whose C^{14} to C^{12} ratio is 82% of today's ratio.

Management

11. The annual sales of Ampersand, Inc. have been increasing at a rate proportional to sales. They were $100,000 in 1985 and $739,000 in 1990. Find
   - **(a)** a function that describes Ampersand's annual sales;
   - **(b)** the firm's sales in 1992;
   - **(c)** when Ampersand's sales will be $15 million per year.

12. Battledore Company stopped advertising its product on January 1, and since then sales have been declining at a rate proportional to sales. January sales were 50,000 units, and February sales were 45,000 units. Find
   - **(a)** a function that describes Battledore's monthly sales;
   - **(b)** the company's sales in March;
   - **(c)** when sales will be 20,000 units per month.

13. The annual profits of Grimditch Associates were $2.5 million in 1980 and $1.4 million in 1983. If profits decrease at a rate proportional to profits, find
   - **(a)** a function that describes Grimditch's annual profits;
   - **(b)** the firm's profits in 1990;
   - **(c)** when Grimditch's annual profits will be $250,000.

14. Dynamic Corporation's revenues have been increasing at a rate proportional to revenue. Suppose they have been doubling every 10 years and were $1 million in 1985. Predict Dynamic's 1994 revenues. When will Dynamic's annual revenues attain the level of $5.5 million per year?

15. The average cost per unit for Elwin Company was $80 in 1980, and it has been decreasing at a rate proportional to itself, declining by 50% every 20 years. When will Elwin's average cost per unit be $30?

16. The quantity demanded of a commodity was 95,123 units when the price was $1 per unit, and 90,484 units when the price was $2 per unit. Assuming that the rate of change of quantity demanded is proportional to quantity demanded, find the demand function for the commodity and also find the quantity demanded when the price is $5 per unit.

17. The quantity supplied of the commodity of Exercise 16 was 20,816 units when the price was $1 per unit, and 23,470 units when the price was $4 per unit. Assuming that the rate of change of quantity supplied is proportional to quantity supplied, find the supply function for the commodity and then find its equilibrium price.

Life Science

18. Suppose that 2 hours after the beginning of an experiment there are 40,000 bacteria in a culture and that 5 hours after the beginning of the experiment there are 50,000 bacteria in the culture. Assuming that the rate of change of the bacteria population is proportional to its size, find a function that describes the number of bacteria present as a function of the time since the experiment began.

19. Biologists are attempting to revive a lake that has suffered from acid rain. It is estimated that there are currently 80,000 fish in the lake, and that over the next 5 years this number will increase to 120,000 as the lake continues to revive. If the fish population increases at a rate proportional to its size, find the number of fish the lake will support when the revivification effort is complete 12 years from now.

20. The number of cases of a new disease was 200 in month 0 and has been increasing at a rate proportional to the number of cases, tripling every 20 months. Find a model for the number of cases as a function of time.

21. The *Physician's Desk Reference* (Medical Economics Co., 1984) indicates that when administering a certain anesthetic to a patient, the vapor concentration should depend on the elapsed time since anesthesia began according to the following table:

Elapsed time (minutes)	2	4	12	24
Vapor concentration	2%	1.5%	0.475%	0.085%

Assuming that the rate of change of the vapor concentration is proportional to the vapor concentration, find a function that fits this set of data and use it to estimate the correct vapor concentration 10 minutes after anesthesia begins.

Social Science

22. A psychologist teaches subjects a complex sequence of tasks. Upon testing the subjects 10 days later, it is found that the typical subject can perform only 20% of the tasks correctly. Assuming that the rate of change of performance is proportional to performance, find a model that describes the percentage of the tasks a typical subject can perform as a function of time.

Limited Growth and Decline

23. Use the curve-sketching procedure to sketch the graph of
$$y = a - ce^{-rt}, \quad a > c > 0, \quad r > 0.$$

24. Use the curve-sketching procedure to sketch the graph of
$$y = a + ce^{-rt}, \quad a > 0, \quad c > 0, \quad r > 0.$$

Management

25. A worker with no experience can produce camshafts at a rate of 60 per week; with 1 week of experience he can produce them at a rate of 95 per week. The most experienced workers can approach a production rate of 200 camshafts per week. Find the learning curve and sketch its graph. How long will it be until a worker can produce camshafts at a rate of 140 per week?

26. Repeat Exercise 25 if a worker with one week of experience can produce camshafts at the rate of only 75 per week.

27. When it has just opened, a new plant will produce refrigerators at a rate of 20,000 per year; two months after it has opened, it will produce them at a rate of 40,000 per year. Its designed maximum rate of production is 100,000 refrigerators per year. Find the learning curve for the plant.

28. This exercise was suggested by material in McGann and Russell, *Advertising Media,* second edition, Irwin, 1988. It concerns advertising learning curves. Consumers who are exposed to an advertisement repeatedly will absorb more and more of the information content of the ad, and the amount they absorb follows an advertising learning curve of the form

$$y = 1 - ce^{-rx}, \quad r > 0.$$

Here x is the number of exposures to the advertisement and y is the proportion of the total information the ad contains that has been absorbed by the consumer. Suppose that a consumer who has seen an ad once absorbs 50% of its information content. (Of course, a consumer who has seen an ad 0 times absorbs 0% of its information content.) Find and graph the advertising learning curve for the ad.

*29. A manufacturing plant can fabricate units at the rate of 70,000 per year by the end of its first month of operation, 80,000 per year by the end of its second month of operation, and 88,000 per year by the end of its third month of operation. Find the learning curve for the plant. What is the maximum fabrication rate of the plant?

30. With perfect efficiency, an airplane manufacturer could build a plane using only 40,000 worker-hours. However, the first plane actually built required 70,000 worker-hours, and the second required 62,000 worker-hours. Find and graph the manufacturer's learning curve.

31. A computer manufacturer is introducing a new model of computer and estimates that it will take 5600 worker-hours to build the first unit and 5400 to build the second. The manufacturer also estimates that even with perfect efficiency it would take 4000 worker-hours to build a unit. Find the manufacturer's learning curve and draw its graph. How many worker-hours will it take to make the first four units?

32. An auto manufacturer knows that a newly hired worker who has no training will take an average of 4.0 minutes to install a steering wheel on a car, but that if the worker is given 1 hour of training, the average installation time falls to 3.4 minutes. The minimum possible time in which a steering wheel can be installed is 2.2 minutes. Find a limited decline model that describes average installation time as a function of training time and draw its graph. If the assembly line runs at such a speed that steering wheels must be installed in 3 minutes or less, how much training must new employees receive?

*33. Suppose a manufacturer does not know how many worker-hours would be required to build a unit with perfect efficiency, but estimates that it will take 4000 worker-hours to build the first unit, 3500 to build the second, and 3200 to build the third. Find the learning curve. What is the minimum time required to build a unit under conditions of perfect efficiency?

Life Science

34. The number of bacteria in a culture is limited to 40,000 by the amount of nutrients present. Suppose that 2000 bacteria are placed in the culture to begin with, and that 5 hours later there are 38,000 of them. Find a limited growth model that describes the number of bacteria in the culture and sketch its graph. How long does it take for the original number of bacteria to double?

35. A pond is stocked with 60 fish, and one year later contains 150 fish. The pond can support a maximum of 1200 fish. Find a limited growth model for the size of the fish population. How many fish will the pond contain 10 years after it was stocked?

36. A farmer's crop is ready to be harvested. If it is harvested today, all of it will be saved, but if it rains and harvesting must be delayed, a smaller percentage will be saved. It is known that if harvesting is delayed one day, only 70% of the crop will be saved and that the minimum percentage salvageable under any conditions is 40%. Find a limited decline model that describes the percentage of the crop that will be saved as a function of the delay in harvesting. How much will be salvaged if harvesting is delayed for four days?

37. The average concentration of ozone in the air of a large city was 200 parts per million in 1988 and 160 ppm in 1990. The minimum possible concentration is 100 ppm. Find a limited decline model for the concentration of ozone as a function of time.

*38. A patient is injected with a drug. At the moment of injection the concentration of the drug in the patient's blood is 0 ppm, 1 minute later it is 24 ppm, and 2 minutes after injection it is 40 ppm. Find a limited growth model that describes the concentration of the drug. What is the maximum concentration possible?

Exercises 39 and 40 were suggested by material on pages 81–84 of the December 1986, *Scientific American*. They concern drug toxicity.

39. A certain drug is toxic to cancer cells and also to normal cells in such a way that a dosage of x units will kill $y_c\%$ of the cancer cells and $y_n\%$ of the normal cells. Suppose that experimentation shows that a dosage of 30 units of the drug kills 77.7% of the cancer cells and 0% of the normal cells, while a dosage of 40 units kills 86.5% of the cancer cells and 39.5% of the normal cells. Assume that as the dosage x increases, both y_c and y_n will approach 100%.
   (a) Find a limited growth model for y_c.
   (b) Find a limited growth model for y_n.
   (c) Graph these functions on the same set of axes. Is it possible for the drug to kill 90% of the cancer cells without killing any normal cells?

40. Repeat Exercise 39 if experimentation shows that a dosage of 50 units will kill 91.8% of the cancer cells and 0% of the normal cells, while a dosage of 60 units will kill 95.0% of the cancer cells and 39.5% of the normal cells.

Social Science

41. The percentage of city voters who consider themselves independents was 22% in 1970 and 29% in 1980. Find a limited growth model that describes the percentage of independents as a function of time, if the maximum possible percentage who could consider themselves independents is 100%. If this model is valid, when will a majority of the voters consider themselves independents?

Computer Science

*42. An article by James Meindl ("Chips for Advanced Computing," *Scientific American*, October 1987) offers two predictions for the maximum number of components that will be placed on computer chips in the future. Both predictions assume that the maximum number of components per chip can be modeled by a limited growth function. The first prediction assumes that the maximum number of components per chip will ultimately approach 100 billion, the second that it will ultimately approach 1 billion. Both predictions imply that in 1992 the maximum number of components per chip will be approximately 500 million. The maximum number of components per chip in 1987 was approximately 1 million.

(a) Find the model for Meindl's first prediction.
(b) Find the model for Meindl's second prediction.
(c) Graph the functions of parts (a) and (b) on the same set of axes. If the first prediction holds, when will the maximum number of components per chip reach 1 billion?

Nuclear Strategy

Exercises 43 and 44 were suggested by an article of von Hippel, Levi, Postol, and Daugherty ("Civilian Casualties from Counterforce Attacks," *Scientific American,* September 1988). They concern casualties under nuclear attack. The article indicates that the approximate percentage of the U.S. population that would be killed by a nuclear attack of x hundred megatons would be between y_L and y_U, and the approximate percentage of the Soviet population that would be killed by such an attack would be between z_L and z_U, where y_L, y_U, z_L, and z_U can be described by limited growth models. The models imply that a 100-megaton attack would kill between 8.2% and 12.7% of the U.S. population and between 11.1% and 16.7% of the Soviet population, while a 500-megaton attack would kill between 34.9% and 49.3% of the U.S. population and between 44.6% and 59.9% of the Soviet population. Of course, as x increases, y_L, y_U, z_L, and z_U all approach 100.

*43. (a) Find the limited growth model for y_L.
(b) Find the limited growth model for y_U.
(c) Graph the functions of parts (a) and (b) on the same set of axes. Under the doctrine of "mutual assured destruction" (MAD), the capability of mounting an attack of 200 megatons will deter aggression. According to these models, what percentage of the U.S. population would such an attack kill?

*44. (a) Find the limited growth model for z_L.
(b) Find the limited growth model for z_U.
(c) Graph y_L, y_U, z_L, and z_U on the same set of axes. What percentage of the Soviet population would a 200-megaton attack kill?

Learning Curves Revisited

Exercise 45 through 52 were suggested by material in Chase and Aquilano, *Production and Operations Management,* fourth edition, Irwin, 1985. They are concerned with learning curves. In industry, learning curves are sometimes defined differently than we have done in this section. Production managers often refer to a "90% learning curve" or an "80% learning curve," for example. A 90% learning curve is one with the property that every time the unit number doubles, the worker-hours required to produce the unit decrease by 10%. For instance, if production follows a 90% learning curve and it takes 1000 worker-hours to produce the first unit, then it will take

$0.9(1000) = 900$ worker-hours to produce the second unit,

$0.90(900) = 810$ worker-hours to produce the fourth unit,

$0.90(810) = 729$ worker-hours to produce the eighth unit,

and so on. Similarly, for an 80% learning curve, 1000 worker-hours for the first unit would lead to $0.80(1000) = 800$ for the second, $0.80(800) = 640$ for the fourth, and so on. When a learning curve is defined in this way, the worker-hours y required to produce the xth unit, $x \geq 1$, are given by an equation of the form

$$y = cx^r$$

for some real numbers c and r. The use of such an equation implies that as the number of units x increases, the worker-hours required to produce the xth unit approach 0.

*45. Find the worker-hours required to produce the second, fourth, eighth, and sixteenth unit if the first unit requires 10,000 worker-hours and production follows an 85% learning curve.

*46. Graph an 85% learning curve if the first unit requires 10,000 worker-hours. Note that as x increases y approaches 0.

*47. Find the equation $y = cx^r$ that describes a 90% learning curve if the first unit produced requires 10,000 worker-hours.

*48. Suppose production follows a $100p$% learning curve and the first unit produced requires c_0 worker-hours. Find and graph a function that describes the number of worker-hours required to build the xth unit, $x \geq 1$.

*49. Use the result of Exercise 48 to find an equation that describes an 88% learning curve if it takes 5000 worker-hours to produce the first unit. How many worker-hours will it take to produce the tenth unit?

How can a production manager tell if production follows a $100p$% learning curve for some p? Note that if

$$y = cx^r,$$

then

$$\ln y = r \cdot \ln x + \ln c.$$

If we set $Y = \ln y$ and $X = \ln x$, the equation

$$\ln y = r \cdot \ln x + \ln c$$

becomes

$$Y = rX + \ln c,$$

which is a linear equation in X and Y. Therefore the production manager can proceed as follows:

1. Collect data on the number of hours y_i it takes to produce unit x_i.
2. Let $Y_i = \ln y_i$ and $X_i = \ln x_i$.
3. Determine whether the points (X_i, Y_i) lie on a line $Y = mX + b$; if they do, production follows a $100p$% learning curve $y = cx^r$, where $c = e^b$, $r = m$, and $p = 2^r$.

*50. Justify statement 3 above.

*51. Suppose a production manager collects the following data concerning the worker-hours required to produce selected units:

Unit Number	Worker-Hours Required
2	4000.0
5	2977.8
7	2672.1
12	2245.8

Show that production follows a learning curve and find its equation $y = cx^r$. What percentage learning curve is it?

*52. Repeat Exercise 51 using the following data:

Unit Number	Worker-Hours Required
6	3200
10	2581
11	2479
15	2176

Logarithmic Growth

53. Use the curve-sketching procedure to sketch the graph of
$$y = \ln(at + b), \quad a > 0, \quad b > 1.$$

54. Annual aluminum production in a country was 7 million tons in 1950 and 10 million tons in 1980. Find a model that describes aluminum production in the country and predict
   (a) production in 1990;
   (b) the year in which production will reach 11 million tons.

Management

55. A firm's sales were $1.5 million when it spent $200,000 on advertising and $1.6 million when it spent $250,000 on advertising. Find a logarithmic growth model for sales as a function of the amount spent on advertising.

56. A firm's annual profits have been growing logarithmically; they were $2.1 million in 1988 and $2.9 million in 1989. Find and graph the function that models the firm's annual profits as a function of time. What will the firm's profit be in 1995? When will profits attain an annual level of $5 million?

57. When a company sold 1000 units of its product, its revenue was $3,135,000; when it sold 2000 units, its revenue was $3,761,000. If the company's revenues are a logarithmic function of the number of units sold, find the revenue function. What would the company's revenue be if it sold 10,000 units? How many units would it have to sell in order to have revenue of $10,000,000?

Life Science

58. If the concentration of radioactive iodine in sea water was 4.8 parts per million in 1945 and 8.1 ppm in 1955, find a logarithmic growth model that describes the concentration as a function of time. The increase in radioactive iodine was largely due to aboveground testing of nuclear weapons; if damage to some forms of sea life occurs at a concentration of 9 ppm, when would such damage have begun if aboveground atomic testing had not been halted in the early 1960s?

59. The concentration of nitrous oxide in the air of a large city is growing logarithmically. If the concentration was 2.0 ppb in 1988 and 2.2 ppb in 1989, find a function that describes the concentration as a function of time.

60. A lake that is being deacidified has its pH y increasing logarithmically. If the pH at the start of the deacidification process was 2.1 and three years later it was 2.7, find and graph a model for y as a function of time. How long will it take until the water in the lake is neutral (pH = 7)?

Social Science

61. A state's aid to higher education is growing logarithmically, and was \$1.2 billion when there were 450,000 students in the system and \$1.5 billion when there were 500,000 students in the system. Find a function that describes the amount of such aid as a function of the number of students and predict the amount of such aid when there are 600,000 students in the system.

62. The number of people employed in a state's bureaucracy is growing logarithmically; in 1985 it was 449,000 and in 1988 it was 472,000. Find a function that describes the number of people employed in the bureaucracy as a function of time. When will the bureaucracy number 500,000?

Nuclear Testing

*63. According to an article by Lynn Sykes and Dan Davis ("The Yields of Soviet Strategic Warheads," *Scientific American*, January 1987), if x is the yield of an underground nuclear explosion, in kilotons, and y is the magnitude of the surface seismic waves generated by the explosion, then y is a logarithmically increasing function of x; furthermore, when $x = 10$, $y = 3.36$, and when $x = 25$, $y = 3.60$.
  (a) Find a model that describes y as a function of x.
  (b) What is the seismic magnitude of a 50-kiloton explosion?
  (c) Suppose the underground test of an atomic weapon can be detected by the seismic waves it generates if the magnitude of the waves is at least 3.2. What is the size of the smallest detectable underground explosion?

SUMMARY

This summary consists of terms and symbols whose meaning you should know, facts you should know, and techniques and methods you should be able to employ.

Section 5.1

Exponential function with base b: $y = cb^{rx}$, $b > 0$, $b \neq 1$. Graphing exponential functions with base b. Domain, range, continuity, one-to-oneness, intercepts, limits at infinity of exponential functions. Rewriting an exponential function with base less than 1 as an exponential function with base greater than 1. Using exponential models. Compound interest. The compound interest formula:

$A = P(1 + r)^{mt}$. Present value. The present value formula: $P = A(1 + r)^{-mt}$. The number e. Finding powers of e. Graph of $y = e^x$. Why use e as a base? Every exponential function with base not equal to e can be rewritten as an exponential function with base e. Continuous compounding of interest. The formula for continuous compounding of interest: $A = Pe^{rt}$. The present value formula for continuously compounded interest: $P = Ae^{-rt}$.

Section 5.2

Inverse function of a one-to-one function. Logarithmic function to the base b: $y = \log_b x$, $b > 0$, $b \neq 1$ as the inverse function of $y = b^x$. $y = \log_b x$ means $x = b^y$. Graphing $y = \log_b x$. Domain, range, continuity, one-to-oneness, intercepts, limits at infinity of logarithmic functions. The properties of logarithms. Using the properties of logarithms. The natural logarithmic function: $y = \ln x$. Finding values of $\ln x$. The graph of $y = \ln x$. Common logarithms. The change-of-base formula for exponentials. Using the change-of-base formula for exponentials to change the bases of exponential functions. The change-of-base formula for logarithms. Using the change-of-base formula for logarithms to change the bases of logarithmic functions. Solving exponential equations by taking logarithms of both sides of the equation.

Section 5.3

The existence of the derivative of $y = \ln x$ at every positive number x. The derivative of $\ln x$. The derivative of $\ln g(x)$. Finding derivatives of functions involving the natural logarithm. Sketching the graphs of functions involving the natural logarithm. The derivative of $\log_b g(x)$. Finding derivatives of functions involving logarithms to the base b, $b \neq e$. Logarithmic differentiation.

Section 5.4

The existence of the derivative of $y = e^x$ for all x. The derivative of e^x. The derivative of $e^{g(x)}$. Finding derivatives of functions that involve powers of e. Sketching the graphs of functions that involve powers of e. The derivative of $b^{g(x)}$. Finding derivatives of functions that involve b to a power, $b \neq e$.

Section 5.5

The rate of change of $y = ce^{rt}$ is directly proportional to y. A quantity whose rate of change with respect to time t is directly proportional to the amount present at time t can be modeled by an equation of the form $y = ce^{rt}$. Models of exponential growth: $y = ce^{rt}$, $r > 0$. Models of exponential decline: $y = ce^{-rt}$, $r > 0$. Graphs of exponential growth and exponential decline models. Creating and using models of exponential growth and exponential decline. Limited and unlimited growth. Limited growth model $y = a - ce^{-rt}$, $a > c > 0$, $r > 0$. Limited and limited decline models. Creating and using models of limited growth and limited decline: learning curves. Logarithmic growth model $y = \ln(at + b)$, $a > 0, b > 1$. Graph of logarithmic growth model. Creating and using logarithmic growth models.

REVIEW EXERCISES

In Exercises 1 through 4, graph the given function.

1. $y = 7^{2x}$
2. $y = 2\left(\dfrac{1}{3}\right)^{-2x}$
3. $y = 3e^{-x/2}$
4. $y = 2 + e^{2x}$

5. Show that $e = \lim\limits_{x \to 0^+} (1 + x)^{1/x}$.

6. A firm's average cost per unit was \$12 in the year the firm was founded and has been tripling every 20 years. Find an exponential equation that describes the average cost per unit as a function of time. Use base 3.

7. How much will you have in seven years if you invest \$4000 at 6.5% per year compounded semiannually? compounded continuously?

8. What is the present value of \$54,000 discounted for 12 years at 11% per year compounded quarterly? compounded continuously?

In Exercises 9 through 20, evaluate the given expression.

9. $\log_7 7$
10. $\log_7 1$
11. $\log_7 49^3$
12. $\log_7 7^{200}$
13. $\log_7 \dfrac{1}{343^2}$
14. $\log_7 7^{-35}$
15. $\log_7 7^{\log_7 22}$
16. $\ln e^{5.2}$
17. $e^{\ln 4.2}$
18. $e^{-2\ln 5}$
19. $\log_7 \dfrac{7^4}{7^9}$
20. $\log_7 (7^{10} \cdot 7^3)$

In Exercises 21 and 22, use only the properties of logarithms and the facts that
$$\log_{10} 2 = 0.3010, \quad \log_{10} 7 = 0.8451, \quad \text{and} \quad \log_{10} 11 = 1.0414$$
to evaluate the given expression.

21. $\log_{10} 56$
22. $\log_{10} \dfrac{392}{1331}$

In Exercises 23 through 26, graph the given function.

23. $y = \log_5 x$
24. $y = \log_5 x^3$
25. $y = \ln 3x$
26. $y = 2 \ln 1.5x$

In Exercises 27 through 30, rewrite the given function using the given base.

27. $f(x) = 7^{2x}$, using base e
28. $y = 10^{-3x}$, using base 7
29. $y = 2 \log_{13} 3x$, using base e
30. $g(x) = \log_8 x$, using base 5

31. Loudness is measured on a logarithmic scale known as the **decibel scale.** This is done by comparing the loudness x of a given sound with the loudness x_0 of a threshold sound (one that is barely audible); the decibel measure of x is then defined to be

$$db(x) = \log_{10}\frac{x}{x_0}.$$

Find the loudness in decibels of a sound that is
(a) 10 (b) 1000 (c) 1,000,000 (d) 2,500,000
times as loud as the threshold sound.
(e) A large passenger jet at full throttle produces a noise of approximately 14 db. How much louder is this than the threshold sound?

In Exercises 32 through 35, solve the given equation for x.

32. $e^{-3x} = 5$ **33.** $e^{2.3x} = 4.2$ **34.** $5^{-3x} = 40$ **35.** $(1.1)^{0.02x} = 0.65$

36. Suppose the number of fish still alive in a pond t days after it is contaminated by an oil spill is given by
$$y = 2000e^{-0.1t}.$$
(a) Graph this function.
(b) How many fish are still alive five days after the oil spill?
(c) How long will it be until half the fish have died?

37. A radioactive isotope has a half-life of eight months. Find an exponential function with base 2 that describes the amount of the isotope present as a function of time.
(a) What percentage of the isotope decays in two years?
(b) How long does it take for 75% of the isotope to decay?

38. What interest rate, compounded continuously, must be earned if $10,000 is to grow to $50,000 in 30 years?

39. What interest rate, compounded monthly, must be earned if $10,000 is to grow to $50,000 in 30 years?

40. How long will it take for $35,000 to triple at 10% per year compounded continuously?

In Exercises 41 through 54, find $f'(x)$.

41. $f(x) = \ln(x^3 + 3x - 2)$ **42.** $f(x) = \ln(x^2 + 1)^2$ **43.** $f(x) = \ln\frac{x^2}{2x - 1}$

44. $f(x) = \dfrac{\ln x^2}{\ln(2x - 1)}$ **45.** $f(x) = 2e^{5x}$ **46.** $f(x) = 4x^2 e^{-x}$

47. $f(x) = e^{x^2}\ln x^2$ **48.** $f(x) = \dfrac{e^x - 1}{e^x + 1}$ **49.** $f(x) = e^{e^{-2x}}$

50. $f(x) = x^{-x^2}$ **51.** $f(x) = \log_2(4x - 3)$ **52.** $f(x) = e^x \log_{10} e^{-2x}$

53. $f(x) = 4 \cdot 2^{-5x}$ **54.** $f(x) = x^2 10^{-3x}$

In Exercises 55 through 60, use the curve-sketching procedure to graph the given function.

55. $y = x + \ln x$ **56.** $y = x^4 \ln x$ **57.** $y = x^{-3} \ln x$

58. $y = x + e^{-x}$ **59.** $y = \dfrac{e^x + e^{-x}}{e^x - e^{-x}}$ **60.** $y = x^2 e^{-x^2}$

61. The money spent on social services by a state government in year t is
$$y = [\ln(3t + 2)]^{1/2}$$

million dollars. Here $t = 0$ represents 1987. Find the rate at which spending on social services was changing in 1989.

62. It is estimated that the concentration of pollutants in a lake t years from now will be
$$y = 200e^{-0.2t}$$
parts per million. Find the rate at which the concentration of pollutants in the lake will be changing five years from now.

63. Find the rate at which the size of a bacteria population is changing 2 hours after the start of an experiment if the number of bacteria present t hours after the start of the experiment is $y = 20,000 \cdot 3^{-0.45t} + 100,000$.

64. Under a runaway greenhouse effect, the earth's average temperature would increase at a rate directly proportional to the average temperature. If the average temperature of the earth is now 59.5°F and if under a runaway greenhouse effect it would be 77.4°F 100 years from now, find a model for the average temperature under a runaway greenhouse effect. Then find
    (a) the average temperature 25 years from now;
    (b) how long it would take for the average temperature to reach 120°F.

65. Insecticides have caused an exponential decline in the number of waterfowl in a wetlands region. Four years ago there were 2000 waterfowl in the wetlands, while today there are only 1340 of them. Find an exponential decline model that describes the number of waterfowl in the wetlands as a function of time.

66. According to an article by Peter Moretti and Louis Divone ("Modern Windmills," *Scientific American*, June 1986), the ideal efficiency for rotor-type windmills is approximately given by the equation
$$y = 0.593 - 0.4e^{-0.73x}.$$
Here x is the ratio of the speed of the rotor tips to the speed of the wind and y is the so-called power coefficient of the windmill: the power coefficient measures the amount of power the windmill extracts from the wind.
    (a) Graph this function.
    (b) Suppose the speed of the rotor tips increases while the wind speed remains constant. What can you say about the power that the windmill will extract from the wind?
    (c) Same question as in part (b) if the speed of the wind increases while the speed of the rotor tips remains constant.

67. A newly hired employee can produce at the rate of 1600 units per month, whereas one with one month's experience can produce at the rate of 1700 units per month. The maximum production with perfect efficiency is 2400 units per month. Find and graph the learning curve. When will an employee be producing at the rate of 2000 units per month?

68. It takes 400 worker-hours to produce the first unit of a new computer chip and 380 worker-hours to produce the second unit. The minimum time needed to produce a chip is estimated to be 250 worker-hours. Find and graph the learning curve. How long will it take to produce the first four chips? When will production time per chip fall below 300 worker-hours?

69. The pollen count during August followed a limited growth model, with the maximum daily count y being expressed as a function of the number of days since August 1. If the count on August 3 was 118 and on August 6 was 125, and if the

maximum possible count is 200, find the model and find out when the count reached 150.

70. It takes a rat 350 seconds to negotiate a maze on its first attempt, 335 seconds on its second attempt, and 322 seconds on its third attempt. If the time the rat takes to get through the maze follows a limited decline model, find the model.

71. A firm's profits are growing logarithmically, and were $100,000 in 1980 and $150,000 in 1990. Predict the firm's profits in 1995 and the year in which they will reach $250,000.

ADDITIONAL TOPICS

Here are some suggestions for topics not covered in the text that you might want to investigate on your own.

1. In Section 5.2 we mentioned that 4, 5, and 6 of the properties of logarithms follow from the properties of exponents introduced in Section 1.1. Use the definition of logarithm and the properties of exponents to prove 4, 5, and 6 of the properties of logarithms.

2. Let f and g be functions. If

$$\lim_{x \to +\infty} \frac{f(x)}{g(x)} = +\infty,$$

then f **grows faster than** g. If

$$\lim_{x \to +\infty} \frac{f(x)}{g(x)} = 0,$$

then f **grows more slowly than** g. If

$$\lim_{x \to +\infty} \frac{f(x)}{g(x)} = L,$$

where $L \neq 0$ is a finite number, then f and g **grow at the same rate.**

  (a) If f and g are polynomial functions, what conditions on them will cause them to grow at the same rate? What conditions will cause one of them to grow faster than the other?

  (b) If f is an exponential function defined by $f(x) = e^{rx}$, $r > 0$, if g is a polynomial function, and if h is a logarithmic function defined by $h(x) = (\ln x)^s$, $s > 0$, show that f grows faster than g and that g grows faster than h. (*Hint:* You may wish to use L'Hôpital's Rule; see Additional Topics, chapter 4, number 6.)

3. The number defined as the limit of $(1 + 1/x)^x$ as $x \to +\infty$ is denoted by e in honor of the Swiss mathematician Leonhard Euler (1707–1783). Euler (pronounced ''oiler'') was one of the most prolific mathematicians of all time: his output of mathematical discoveries and results was immense. Find out about Euler's life and some of his contributions to mathematics.

4. Logarithms were discovered by the Scottish mathematician and inventor John Napier (1550–1617), who also labored for many years constructing the first tables of logarithms. The availability of such tables simplified many calculations and had a profound influence on the development of the sciences, astronomy in particular. Find out

about the life of Napier and the early applications of logarithms by the great Danish astronomer Tycho Brahe and others. (A good place to start is H. W. Turnbull's article "The Great Mathematicians," which appears in volume 1 of *The World of Mathematics,* James R. Newman, ed., Simon & Schuster, 1956.)

SUPPLEMENT: S-CURVES

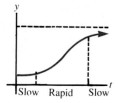

FIGURE 5–31
S-Curve

Adequate modeling of some situations involving limited growth or decline may require the use of more complicated exponential functions than the ones we introduced in Section 5.5. For instance, the growth of an individual company or industry often follows an **S-curve,** such as that in Figure 5–31.

Notice how the S-curve has an initial period of slow growth, then a period of rapid growth, and finally another period of slow growth. Such behavior is fairly common: a new company will typically grow quite slowly during the first few years of its existence, then "take off" and experience a period of very rapid growth. Eventually, however, the company matures and the growth slows down again. Thus the company's **life cycle** can be described by an S-curve. For this reason, S-curves are sometimes called **life-cycle curves.**

One function that has an S-curve as its graph is the **logistic function**

$$y = \frac{a}{1 + be^{-rt}}, \quad a > 0, \quad b > 1, \quad r > 0.$$

The graph of the logistic function on $[0, +\infty)$ is shown in Figure 5–32. (See Exercise 1.) The larger the value of r, the steeper the curve will be. (See Exercise 16.) Notice the point of inflection at $((\ln b)/r, a/2)$; to the left of this point of inflection, the rate of growth is increasing, while to the right of the point of inflection, the rate

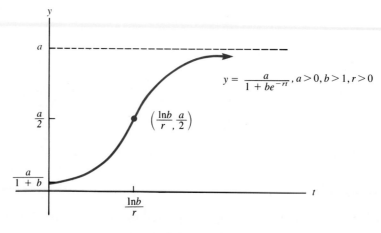

FIGURE 5–32
The Logistic Function

of growth is decreasing. (Why?) Thus the point of inflection marks the place where the *rate* of growth stops increasing and begins to decrease.

EXAMPLE 1   A new computer company was started and t years later its annual sales were S million dollars, where

$$S = \frac{20}{1 + 19e^{-0.6t}}.$$

The graph of this function is shown in Figure 5–33. Initially the firm's annual sales were running at a level of $1 million per year. Sales increased slowly for the first several years, then increased rapidly for several more, and finally settled down, approaching a level of $20 million per year. The rate at which sales grew increased during the first 4.9 years of the company's existence and decreased thereafter.

What were annual sales four years after the company was founded? The answer is

$$S = \frac{20}{1 + 19e^{-0.6(4)}} \approx 7.3$$

million dollars. Now let us find how many years it took until sales attained an annual level of $16 million. To do so, we must solve

$$16 = \frac{20}{1 + 19e^{-0.6t}}$$

for t. We have

$$1 + 19e^{-0.6t} = \frac{20}{16} = 1.25,$$

or

$$e^{-0.6t} = \frac{1.25 - 1}{19} = \frac{1}{76}.$$

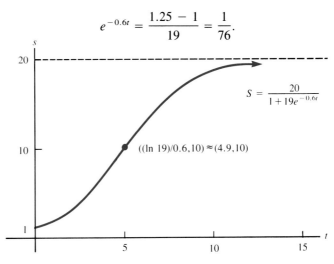

FIGURE 5–33

Thus
$$\ln e^{-0.6t} = \ln(1/76) = -\ln 76,$$
and hence
$$t = \frac{\ln 76}{0.6} \approx 7.2$$
years. □

Another function that has an S-curve as its graph is the **Gompertz function**
$$y = ae^{-be^{-rt}}, \quad a > 0, \quad b > 1, \quad r > 0.$$
The graph of the Gompertz function on $[0, +\infty)$ is shown in Figure 5–34. (See Exercise 7.) The larger the value of r, the steeper the curve will be. (See Exercise 17.) Note the point of inflection at $((\ln b)/r, a/e)$: to the left of this point, the rate of growth is increasing, while to the right of it the rate of growth is decreasing.

EXAMPLE 2 A new video game is placed in arcades and t weeks later the revenue from the game is running at a level of y thousand dollars per week, where
$$y = 25e^{-5e^{-0.4t}}.$$
The graph of this function is shown in Figure 5–35. Initially revenue was running at a level of approximately $170 per week, and it remained low for the first few weeks. Weekly revenues from the game then increased rapidly, however, before finally slowing down and approaching $25,000 per week. The rate at which revenues grew increased during the first four weeks and decreased thereafter.

Suppose we wish to find the revenue five weeks after the installation of the game. We have
$$y = 25e^{-5e^{-0.4(5)}} = 25e^{-5e^{-2}} = 25e^{-5(0.1353)} \approx 12.7$$
thousand dollars.

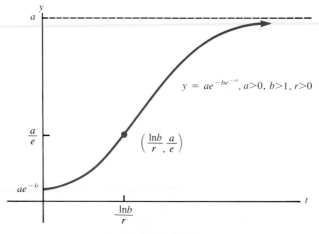

FIGURE 5–34
The Gompertz Function

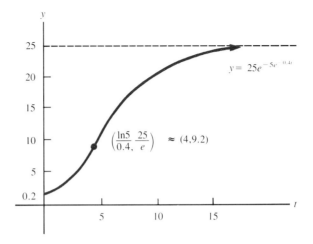

FIGURE 5-35

Now suppose we wish to find out how long it would take for revenues to attain the level of $24,000 per week. Then we must solve

$$24 = 25e^{-5e^{-0.4t}}$$

for t. Thus

$$\ln(24/25) = -5e^{-0.4t},$$

so

$$\frac{-0.0408}{-5} = e^{-0.4t},$$

and hence

$$\ln(0.0408/5) = -0.4t.$$

Therefore

$$t \approx 12.0 \text{ weeks.} \quad \square$$

Both the logistic function and the Gompertz function have S-curves as their graphs. The difference between them is that, for the same values of a, b, and r, the Gompertz curve increases more slowly than the logistic curve. See Figure 5-36.

Finally, we remark that sometimes decline follows a reverse S-curve, as in Figure 5-37. One function that has such a graph is the **sigmoid function**

$$y = a - b(1 - e^{-rt})^n, \quad a > b > 0, \quad r > 0, \quad n \in \mathbb{N}, \quad n > 1.$$

The graph of the sigmoid function is shown in Figure 5-38. (See Exercise 11.) The smaller n is, the steeper the curve will be. (See Exercise 18.) Notice the point of

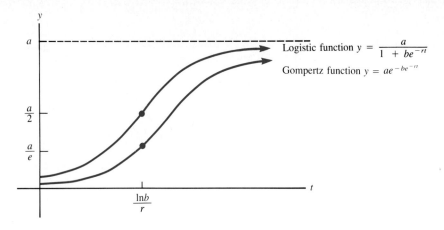

FIGURE 5–36

inflection at $((\ln n)/r, a - b(1 - 1/n)^n)$: the rate of decline increases to the left of the point of inflection and decreases to its right.

EXAMPLE 3 A patient is given a temperature-reducing drug and t hours later has a temperature of

$$y = 103 - 4.4(1 - e^{-0.4t})^3$$

degrees Fahrenheit. The graph of this function is shown in Figure 5–39. Notice that the drug had little effect on the patient's temperature during the first hour, but

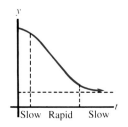

FIGURE 5–37
Reverse S-Curve

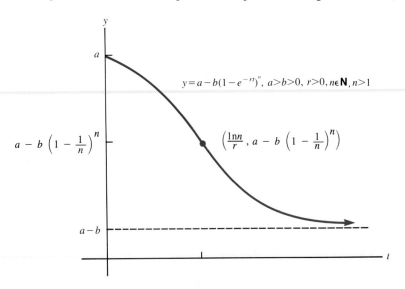

FIGURE 5–38
The Sigmoid Function

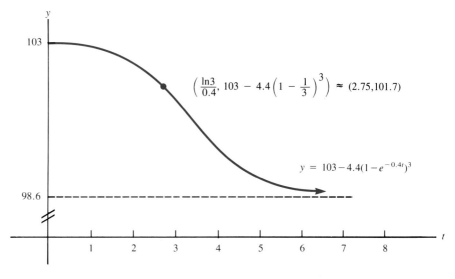

FIGURE 5–39

that it then brought the temperature down quite rapidly to a near-normal level before its effect began to wear off. ☐

■ *Computer Manual: Program ESSCURVE*

SUPPLEMENTARY EXERCISES

1. Use the curve-sketching procedure to sketch the graph of
$$y = \frac{a}{1 + be^{-rt}}, \quad a > 0, \quad b > 1, \quad r > 0$$
on $[0, +\infty)$.

2. Sales of electronic calculators are described by the logistic function
$$y = \frac{100}{1 + 49e^{-0.4t}},$$
where y is annual sales of calculators in millions and t is time in years, with $t = 0$ representing 1970.
   (a) Sketch the graph of this function.
   (b) What were calculator sales in 1970? in 1975? in 1985?
   (c) What will sales be in 1995?
   (d) When did the rate of sales growth begin to slow down? What were sales at this time?
   (e) When were sales running at a level of 80 million per year?

3. The manufacturing capacity of a nation is given by
$$y = \frac{450}{1 + 300e^{-0.06t}}.$$

Here y is the volume of manufactured goods produced in the country in year t, y in billions of 1990 dollars and $t = 0$ representing A.D. 1800.
(a) Sketch the graph of this function.
(b) What was the manufacturing capacity of the nation in A.D. 1800? What will it be in the year 2000?
(c) When did the rate of growth of the manufacturing capacity begin to slow down?

4. An article on energy-efficient buildings by Arthur Rosenfeld and David Hafemeister ("Energy Efficient Buildings," *Scientific American*, April 1988) suggests that the total savings over the period 1975–2050 of making appliances, homes, and commercial buildings energy efficient would be approximately

$$y = \frac{100}{1 + 99e^{-0.21t}}$$

billion 1985 dollars. Here t is in years with $t = 0$ representing 1975.
(a) Sketch the graph of this function.
(b) When would the rate at which savings increase begin to slow down? What would total savings be at this time?
(c) When would total savings amount to $75 billion?

5. In a town of 12,000 people the spread of an influenza epidemic is described by the function

$$y = \frac{5000}{1 + 4999e^{-0.5t}},$$

where t is time in days since the first case of flu appeared and y is the number of people who contracted the flu on or before day t.
(a) Sketch the graph of this function.
(b) When did the epidemic begin to slow down?
(c) What proportion of the town's inhabitants contracted the flu?

6. Two hundred students live in a college dormitory. One of them starts a rumor and t hours later y of them have heard it, where

$$y = \frac{200}{1 + 199e^{-1.1t}}.$$

(a) Sketch the graph of this function.
(b) How long will it take until half of the dorm residents have heard the rumor?
(c) Assuming that the resident who is the subject of the rumor will be the last one to hear it, how long will it take until everyone else has heard it?

7. Use the curve-sketching procedure to sketch the graph of

$$y = ae^{-be^{-rt}}, \quad a > 0, \quad b > 1, \quad r > 0.$$

8. The growth of a cell colony inside a container is described by the Gompertz curve

$$y = 100e^{-2e^{-0.02t}}.$$

Here y is the percentage of the volume of the container that the colony fills after t days of growth.
(a) Sketch the graph of this function.
(b) What percentage of the container did the colony fill at the start of its growth? after 5 days of growth? after 10 days of growth?
(c) When did the colony's rate of growth begin to slow? What percentage of the container did it fill at this time?
(d) How long did it take for the colony to fill 90% of the container?

9. The manufacturing capacity of a nation is given by
$$y = 450e^{-300e^{-0.06t}},$$
where y is in billions of 1990 dollars and t is in years with $t = 0$ representing A.D. 1800. Repeat Exercise 3 above using this function and compare your results with those of Exercise 3.

10. Repeat Exercise 6 above if
$$y = 200e^{-199e^{-1.1t}}$$
and compare your results with those of Exercise 6.

11. Use the curve-sketching procedure to sketch the graph of
$$y = a - b(1 - e^{-rt})^n, \quad a > b > 0, \quad r > 0, \quad n \in \mathbf{N}, \quad n > 1$$
on $[0, +\infty)$.

12. Annual sales y of the buggy-whip industry are given by
$$y = 400 - 350(1 - e^{-0.1t})^3,$$
where y is in millions of 1989 dollars and t is in years since A.D. 1900.
   (a) Sketch the graph of this function.
   (b) What were annual sales of the industry in 1900? in 1910? in 1925?
   (c) When did annual sales fall to a level of $100 million?

13. The effects of a painkiller are modeled by the equation
$$y = 100[1 - (1 - e^{-0.4t})^2],$$
where t is time in minutes since the pain-killer was taken and y is a discomfort index, with $y = 0$ indicating no discomfort.
   (a) Sketch the graph of this function.
   (b) What was the discomfort index at the time the painkiller was taken? What was it 5 minutes later? ten minutes later?
   (c) How long did it take for the discomfort index to fall to 10?

14. Repeat Exercise 13 if
$$y = 100[1 - (1 - e^{-0.4t})^4],$$
and compare your results with those of Exercise 13.

15. The sales of fad items can often be described by the combination of a Gompertz function and a sigmoid function. Thus, suppose that sales of a fad item are given by
$$y = \begin{cases} 200e^{-10e^{-0.5t}}, & 0 \leq t \leq 6, \\ 121.6[1 - (1 - e^{-0.5(t-6)})^2], & 6 < t \leq 12. \end{cases}$$
Graph this function and analyze the sales of the item.

16. Let a, b, r, and s be positive numbers, with $b > 1$ and $r > s$. Graph the logistic functions
$$y = \frac{a}{1 + be^{-rx}} \quad \text{and} \quad y = \frac{a}{1 + be^{-sx}}$$
on the same set of axes. This shows that the larger r is, the steeper the curve.

17. Let a, b, r, and s be positive numbers, with $b > 1$ and $r > s$. Graph the Gompertz functions
$$y = ae^{-be^{-rx}} \quad \text{and} \quad y = ae^{-be^{-sx}}$$
on the same set of axes. This shows that the larger r is, the steeper the curve.

18. Let a, b, and r be positive numbers, with $a > b$, and let $n > m > 1$. Graph the sigmoid curves
$$y = a - b(1 - e^{-rt})^n \quad \text{and} \quad y = a - b(1 - e^{-rt})^m$$
on the same set of axes. This shows that the smaller n is, the steeper the curve.

6

Integration

In Chapters 2 through 5 we were concerned with the problem of finding the rate of change of a function. We saw that the rate of change of a function is the derivative of the function, and we have exploited this fact in many interesting and important ways. Often, however, it is not the function that is known and its rate of change that must be found, but rather the rate of change that is known and the function that must be found. For instance,

- A biologist might know the rate of population growth for a species and wish to find a function that describes the size of its population.
- A physicist might know the velocity of a particle and want to find a function that describes its position.
- A manufacturer might know the marginal cost function for a product and wish to find its cost function.

In cases like these, instead of starting with a function f and differentiating it to find its rate of change f', we start with a function f that describes a rate of change and **antidifferentiate** it to find a function F such that $F' = f$. Such a function F is called an **antiderivative** of f.

We begin this chapter with two sections devoted to antiderivatives. In these sections we consider the set of all antiderivatives of a function: this set is known as the **indefinite integral** of the function. We state the rules of indefinite integrals and demonstrate some techniques for finding indefinite integrals. In the third section we introduce the **definite integral** of a function: the definite integral is a number that gives the total change of an antiderivative of the function over an interval. We then show how the definite integral may be used to find areas, and conclude by defining the definite integral as the limit of a certain type of sum.

6.1 THE INDEFINITE INTEGRAL

We have discussed the motivation for antidifferentiation. In this section we consider the mechanics of the antidifferentiation process, introduce the concept of the indefinite integral of a function, and discuss some properties of the indefinite integral.

> **The Antiderivative of a Function**
>
> If f and F are functions such that
> $$F'(x) = f(x)$$
> for all x in the domain of f, then $F(x)$ is said to be an **antiderivative** of $f(x)$.

EXAMPLE 1 Let f be defined by $f(x) = 2x + 3$ and F by $F(x) = x^2 + 3x$. Then $F(x)$ is an antiderivative of $f(x)$, because
$$F'(x) = 2x + 3 = f(x)$$
for all x. More generally, if $G(x) = x^2 + 3x + c$, where c is any number, then
$$G'(x) = 2x + 3 + 0 = f(x)$$
for all x, and therefore $G(x)$ is an antiderivative for $f(x)$. Thus
$$f(x) = 2x + 3$$
has many antiderivatives: each choice of the number c in
$$G(x) = x^2 + 3x + c$$
results in a different antiderivative of $f(x)$. ☐

As Example 1 suggests, if $f(x)$ has one antiderivative $F(x)$, then it has a whole *set* of antiderivatives, because if $F'(x) = f(x)$ for all x in the domain of f and c is any number, then
$$(F(x) + c)' = F'(x) + 0 = f(x)$$
for all x in the domain of f; hence $F(x) + c$ is an antiderivative for $f(x)$ also. On the other hand, if $F(x)$ and $G(x)$ are any two antiderivatives of $f(x)$, then there is some number c such that
$$G(x) = F(x) + c$$
for all x in the domain of f. To see this, set $H(x) = G(x) - F(x)$ for all x in the domain of f. Then
$$H'(x) = G'(x) - F'(x) = f(x) - f(x) = 0.$$

Hence the derivative of H is always zero, and therefore H has a horizontal tangent line at every point of its graph. But the only way this can happen is if the graph of H is itself a horizontal line. If this is so, then there is some number c such that $H(x) = c$ for every x in the domain of f. Thus $G(x) - F(x) = c$, or $G(x) = F(x) + c$ for all x in the domain of f.

> **Properties of the Antiderivative**
>
> If $F(x)$ is an antiderivative of $f(x)$, then
>
> 1. $F(x) + c$ is also an antiderivative of $f(x)$, for any number c.
> 2. Every antiderivative of $f(x)$ is of the form $F(x) + c$ for some number c.

Now we are ready to define the **indefinite integral** of a function f. This is just the set of all antiderivatives of $f(x)$.

> **The Indefinite Integral**
>
> The **indefinite integral** of a function f with respect to x, denoted by
> $$\int f(x)\,dx,$$
> is defined by
> $$\int f(x)\,dx = F(x) + c,$$
> where $F(x)$ is any antiderivative of $f(x)$ and c is a constant.

The symbol $\int$ is called an **integral sign,** and the expression $f(x)$ that follows the integral sign is called the **integrand.** The dx after the integrand is the differential of x discussed in Section 3.3; its presence tells us that the integral is taken with respect to the variable x. The constant c is referred to as the **constant of integration.** Note that to find the indefinite integral of f, we need only find *one* antiderivative $F(x)$ of $f(x)$.

EXAMPLE 2 Let us find
$$\int (2x + 3)\,dx.$$
Since $F(x) = x^2 + 3x$ is an antiderivative of $f(x) = 2x + 3$, we have
$$\int (2x + 3)\,dx = x^2 + 3x + c. \quad \square$$

EXAMPLE 3 Let us find
$$\int (3x^2 + 2x + 1)\,dx.$$

Since $F(x) = x^3 + x^2 + x$ is an antiderivative of $f(x) = 3x^2 + 2x + 1$ (why?), it follows that

$$\int (3x^2 + 2x + 1)\, dx = x^3 + x^2 + x + c. \quad \square$$

EXAMPLE 4 We claim that

$$\int (6t^5 - 6t^2 + 2t - 2)\, dt = t^6 - 2t^3 + t^2 - 2t + c.$$

The claim is true if the derivative of the right side with respect to t is the integrand. (Why?) Check that this is so. $\quad \square$

Since the process of integration is the reverse of the process of differentiation, every rule of differentiation gives rise to a corresponding rule of integration. The rules are traditionally formulated in terms of a variable u. We can prove each rule by showing that the derivative with respect to u of the right side is the integrand.

The Rules of Integration

1. For any number k,

$$\int k\, du = ku + c.$$

2. For any number $r \neq -1$,

$$\int u^r\, du = \frac{u^{r+1}}{r+1} + c.$$

3. $\int \dfrac{1}{u}\, du = \ln |u| + c.$ In other words,

$$\text{if } u > 0, \quad \int \frac{1}{u}\, du = \ln u + c;$$

$$\text{if } u < 0, \quad \int \frac{1}{u}\, du = \ln(-u) + c.$$

4. $\int e^u\, du = e^u + c.$

5. For any number k and function $f(u)$,

$$\int k f(u)\, du = k \int f(u)\, du.$$

6. For any functions $f_1(u)$ and $f_2(u)$,

$$\int [f_1(u) \pm f_2(u)]\, du = \int f_1(u)\, du \pm \int f_2(u)\, du.$$

Rule 6 remains true if the integrand is the sum or difference of more than two functions.

EXAMPLE 5

$$\int 2 \, du = 2u + c.$$

$$\int -3 \, dx = -3x + c.$$

$$\int dt = \int 1 \, dt = t + c.$$

$$\int 0 \, du = c. \quad \square$$

EXAMPLE 6

$$\int u \, du = \int u^1 \, du = \frac{u^{1+1}}{1+1} + c = \frac{u^2}{2} + c.$$

$$\int x^2 \, dx = \frac{x^3}{3} + c.$$

$$\int \sqrt{x} \, dx = \int x^{1/2} \, dx = \frac{x^{3/2}}{3/2} + c = \frac{2}{3} x^{3/2} + c.$$

$$\int t^{-3} \, dt = \frac{t^{-3+1}}{-3+1} + c = -\frac{t^{-2}}{2} + c. \quad \square$$

EXAMPLE 7

$$\int 4x^2 \, dx = 4 \int x^2 \, dx = 4 \frac{x^3}{3} + c = \frac{4}{3} x^3 + c.$$

$$\int \frac{4}{x} \, dx = 4 \int \frac{1}{x} \, dx = 4 \ln |x| + c.$$

$$\int -2e^t \, dt = -2 \int e^t \, dt = -2e^t + c. \quad \square$$

EXAMPLE 8

$$\int (u^2 + 3u) \, du = \int u^2 \, du + \int 3u \, du$$

$$= \frac{u^3}{3} + 3 \int u \, du$$

$$= \frac{u^3}{3} + \frac{3u^2}{2} + c. \quad \square$$

EXAMPLE 9

$$\int (4x^4 - 3x^{-2} + 6) \, dx = 4 \int x^4 \, dx - 3 \int x^{-2} \, dx + 6 \int dx$$

$$= 4 \frac{x^5}{5} - 3 \frac{x^{-1}}{-1} + 6x + c$$

$$= \frac{4}{5} x^5 + \frac{3}{x} + 6x + c. \quad \square$$

If $F'(x) = f(x)$ for all x in the domain of f, then $F(x)$ is an antiderivative of $f(x)$, so

$$\int f(x) \, dx = \int F'(x) \, dx = F(x) + c.$$

Thus if f is the derivative of a function F, we can find F by integrating $f(x)$ and then determining the value of the constant of integration. We illustrate with examples.

EXAMPLE 10 Suppose a firm's marginal cost function is given by

$$f(x) = 2x + 5.$$

Marginal cost is the derivative of cost, so we can find the firm's cost function by integrating $f(x)$ and then determining the value of the constant of integration. Thus

$$C(x) = \int f(x)\,dx = \int (2x+5)\,dx = x^2 + 5x + c.$$

To find c, we must know the value of $C(x)$ for some value of x. For instance, suppose the firm's fixed cost is \$10,000. Then

$$10{,}000 = C(0) = 0^2 + 5(0) + c = c,$$

so $c = 10{,}000$ and the cost function is given by

$$C(x) = x^2 + 5x + 10{,}000. \quad \square$$

EXAMPLE 11 Suppose that t hours after the start of an experiment a colony of bacteria is growing at the rate of $y = 300\sqrt{t}$ bacteria per hour and that there were 2000 bacteria in the colony 4 hours after the start of the experiment. Since y is the rate at which the size of the bacteria population changes with respect to time, it is the derivative of the function F that describes population size as a function of time. Therefore we may find F by integrating y. But it is easy to check that $200t^{3/2}$ is an antiderivative for y (do so), and hence

$$\int y\,dt = \int 300\sqrt{t}\,dt = 200t^{3/2} + c.$$

Thus $F(t) = 200t^{3/2} + c$ for some c. But since there were 2000 bacteria present when $t = 4$, it follows that

$$2000 = F(4) = 200(4^{3/2}) + c = 200(8) + c = 1600 + c,$$

and thus that $c = 400$. Therefore F is given by

$$F(t) = 200t^{3/2} + 400. \quad \square$$

■ **Computer Manual: Program INTPOLY**

6.1 EXERCISES

In Exercises 1 through 40, find the indefinite integral.

1. $\int 3\,dx$
2. $\int \sqrt{2}\,dx$
3. $\int 0\,dx$
4. $\int u^4\,du$
5. $\int x^{10}\,dx$
6. $\int t^{12}\,dt$
7. $\int t^{-4}\,dt$
8. $\int q^{-4/5}\,dq$
9. $\int \sqrt{x^3}\,dx$
10. $\int 2x\,dx$
11. $\int 2x^3\,dx$
12. $\int \dfrac{5}{x^4}\,dx$
13. $\int 3x^{-6}\,dx$
14. $\int -3\sqrt{y}\,dy$
15. $\int 2t^{-1/2}\,dt$

16. $\int 2u^{1/3}\, du$ 17. $\int \frac{2}{p}\, dp$ 18. $\int -\frac{4}{x}\, dx$ 19. $\int \frac{3}{t}\, dt,\ t<0$ 20. $\int -3e^y\, dy$

21. $\int 2e^x\, dx$ 22. $\int 12e^x\, dx$ 23. $\int (x+2)\, dx$ 24. $\int (3-2x)\, dx$

25. $\int (2v^2 - 3v + 1)\, dv$ 26. $\int (x^2 - 3x + 5)\, dx$ 27. $\int (2x^3 - 6x^2 + 18x - 3)\, dx$

28. $\int (x^3 + 3x^2 - 2x)\, dx$ 29. $\int (-4x^3 + 3x^2 - 2x + 7)\, dx$ 30. $\int (-3x^{-5} + 4x^2 + 12)\, dx$

31. $\int (6e^x + x)\, dx$ 32. $\int \left(\frac{2}{t^2} + \frac{1}{t^3} + 3\right) dt$ 33. $\int \left(7x^2 - 14x + \frac{2}{x} - \frac{5}{x^3}\right) dx$

34. $\int \left(\frac{2}{\sqrt{q}} - \frac{3}{q} + 1\right) dq,\ q>0$ 35. $\int (100x^{99} + 99x^{98} + 98x^{97} + \cdots + 2x + 1)\, dx$

36. $\int -\left(\frac{1}{t^2} + \frac{2}{t^3} + \frac{3}{t^4} + \cdots + \frac{99}{t^{100}}\right) dt$ 37. $\int (x+5)(2x-3)\, dx$

38. $\int (2x+5)^2\, dx$ 39. $\int (4-x)^3\, dx$ 40. $\int (3x+1)(x+5)^2\, dx$

41. An airplane's velocity t hours after takeoff is $v = 300t$ kilometers per hour. If the plane flies in a straight line, find the distance it travels in the 2 hours after takeoff. (*Hint:* The distance the plane has traveled at takeoff is 0 kilometers.)

42. If a rocket's acceleration t seconds after lift-off is $a(t) = 0.4t + 200$ mps², find the rocket's velocity function. (*Hint:* Its initial velocity is 0 feet per second.)

43. A rocket is launched straight up with a constant acceleration of 100 feet per second per second.
    (a) Find the rocket's velocity 10 seconds after launch.
    (b) Find its altitude at the end of 10 seconds. (*Hint:* Its initial altitude is 0 ft.)

*44. Stefan's Law says that the rate at which an object radiates energy is proportional to the fourth power of its absolute temperature: if y is the rate at which energy is radiated per unit area, then $y = kT^4$, where T is temperature in degrees Kelvin (K) and k is a constant. Find an expression for the amount of energy radiated per unit area. (*Hint:* An object will radiate no energy at $T = 0°$ K.)

Management

45. A firm's marginal cost function is given by $c(x) = 2x + 1$, its fixed cost is $5000, and its marginal revenue function is given by $r(x) = 15$. Find the firm's cost, revenue, and profit functions. (*Hint:* The revenue obtained from selling 0 units is $0.)

46. A firm's marginal cost function is given by $c(x) = 5x + 20$, the cost of making 10 units is $10,000, and its marginal revenue function is given by $r(x) = 40$. Find the firm's cost, revenue, and profit functions.

47. A firm's marginal profit function is given by $p(x) = x^2 - 100x + 1600$, and its profit from making and selling 3 units is $4341. Find the firm's profit function.

48. Suppose a firm's marginal cost function is given by
$$c(x) = 0.3x^2 + 0.5x + 2.$$
Find the firm's cost function if its fixed cost is $20,000.

49. The rate at which the quantity of a commodity demanded changes with respect to price is y units per dollar, where

$$y = -50p.$$

Find the demand function for the product if 25,000 units are demanded at a price of $100.

50. The rate at which the quantity of a product supplied to the market changes with respect to price is y units per dollar p, where

$$y = \frac{200}{p}.$$

Find the supply function for the product if 1000 units were supplied at a price of $10.

Life Science

51. A nesting pair of blue herons is introduced into a bird sanctuary where blue herons did not previously exist. It is estimated that the population of blue herons in the sanctuary will increase at the rate of y individuals per year, where $y = \sqrt{t} + 4$ and t is time in years since the species was introduced into the sanctuary. Find a function that describes the size of the blue heron population in the sanctuary in year t.

52. A population of bacteria grows at a rate of $y = 0.02e^t$ bacteria per minute. If the initial population size was one bacterium, find an equation that describes the size of the population as a function of time.

53. Ivory poachers are killing elephants in a game preserve at rate of $y = -100e^{-t}$ individuals per year t, where $t = 0$ represents the present. If there are 400 elephants in the preserve now, find a function that describes the number in the preserve as a function of time. (*Hint:* e^{-t} is an antiderivative for $-e^{-t}$.)

Social Science

54. In 1976, 68% of the voters in a state were affiliated with a political party. Since then the percentage affiliated with a party has changed at the rate of $-0.6\sqrt{t}$ percent per year, where t is in years, with $t = 0$ representing 1976. Find a function that describes the percentage of voters affiliated with a political party in year t.

55. The property tax rate in a town is expressed in terms of dollars per thousand dollars of the assessed value of the property. The tax rate has been changing at a rate of

$$y = \frac{2}{t+1}$$

dollars per thousand dollars of assessed value, and was $80 per thousand in 1989. Find a function that describes the property tax rate as a function of time. (*Hint:* $\ln|t+1|$ is an antiderivative for $(t+1)^{-1}$.)

Fick's Law

*56. Fick's Law states that if $s(t)$ is the concentration of a solute inside a cell at time t, its rate of diffusion is

$$s'(t) = \frac{kA}{V}(S - s(t)),$$

where A is the area of the cell membrane, V is the volume of the cell, S is the concentration of the solute outside the cell, and k is a constant. Find $s(t)$. (*Hint:* Write Fick's Law in the form

$$\frac{s'(t)}{S - s(t)} = \frac{kA}{V},$$

find an antiderivative for each side of this equation, and solve for $s(t)$. You may assume that $S - s(t)$ is positive for all t.)

6.2 FINDING INDEFINITE INTEGRALS

The rules of integration of Section 6.1 do not tell us how to integrate all functions of interest. For instance, none of the rules apply directly to the integrals

$$\int \sqrt{3x + 1} \, dx \quad \text{and} \quad \int x^2(x^3 + 1)^{10} \, dx.$$

In this section we discuss and illustrate two techniques for finding indefinite integrals: integration by guessing and integration by substitution.

Sometimes we can guess the general form that an integral must take, and then by differentiating this general form, find the exact expression for the integral. For instance, consider the integral

$$\int \sqrt{3x + 1} \, dx.$$

An antiderivative of $\sqrt{3x + 1}$ must have the general form $(3x + 1)^{3/2}$. However, $(3x + 1)^{3/2}$ is not quite the antiderivative we seek, because its derivative is not quite equal to $\sqrt{3x + 1}$:

$$\frac{d}{dx}(3x + 1)^{3/2} = \frac{9}{2}(3x + 1)^{1/2} = \frac{9}{2}\sqrt{3x + 1}.$$

Thus we have an unwanted constant of $9/2$. However, if we multiply $(3x + 1)^{3/2}$ by $2/9$, the unwanted constant will cancel out:

$$\frac{d}{dx}\left[\frac{2}{9}(3x + 1)^{3/2}\right] = \frac{2}{9}\frac{d}{dx}(3x + 1)^{3/2} = \frac{2}{9}\frac{9}{2}\sqrt{3x + 1} = \sqrt{3x + 1}.$$

Thus

$$\frac{2}{9}(3x + 1)^{3/2}$$

is an antiderivative of $\sqrt{3x + 1}$, so

$$\int \sqrt{3x + 1} \, dx = \frac{2}{9}(3x + 1)^{3/2} + c.$$

EXAMPLE 1 Consider the integral

$$\int \frac{1}{3x + 2} \, dx, \quad 3x + 2 > 0.$$

Since $3x + 2 > 0$, an antiderivative of $1/(3x + 2)$ must have the general form $\ln(3x + 2)$. But

$$\frac{d}{dx}(\ln(3x + 2)) = \frac{1}{3x + 2}\frac{d}{dx}(3x + 2) = \frac{3}{3x + 2}.$$

Therefore

$$\frac{d}{dx}\left(\frac{1}{3}\ln(3x + 2)\right) = \frac{1}{3}\frac{d}{dx}(\ln(3x + 2)) = \frac{1}{3}\frac{3}{3x + 2} = \frac{1}{3x + 2}.$$

Thus

$$\int \frac{1}{3x + 2}\,dx = \frac{1}{3}\ln(3x + 2) + c. \quad \square$$

EXAMPLE 2 Let us find the indefinite integral

$$\int e^{-2x}\,dx.$$

Since

$$\frac{d}{dx}(e^{-2x}) = -2e^{-2x},$$

we have

$$\int e^{-2x}\,dx = -\frac{1}{2}e^{-2x} + c. \quad \square$$

Guessing the general form of an integral does not always work, because often we cannot tell what the general form should look like. In any case, we would like to have a more systematic technique for performing integration. No technique of integration works for all integrals, but one widely used method is **integration by substitution,** which we now discuss.

Recall that in Section 3.3 we introduced the differentials dx and dy and showed that if $y = f(x)$ then $dy = f'(x)\,dx$. We shall use u rather than y as the dependent variable; thus if $u = f(x)$, then $du = f'(x)\,dx$. We can often use this relationship between du and dx to change the variable in an integral from x to u in such a way that the resulting integral in u is easy to find. The technique is called **integration by substitution,** because we substitute u for some portion of the integrand. We illustrate with an example.

EXAMPLE 3 Let us consider again the integral

$$\int \sqrt{3x + 1}\,dx.$$

Suppose we set $u = 3x + 1$. Then

$$du = (3x + 1)'\,dx = 3\,dx,$$

or

$$\frac{1}{3} du = dx.$$

If we substitute u for $3x + 1$ in the integrand and $\frac{1}{3} du$ for dx, we obtain

$$\int \sqrt{3x + 1}\, dx = \int \sqrt{u}\, \frac{1}{3} du = \frac{1}{3} \int u^{1/2}\, du$$

$$= \frac{1}{3} \frac{u^{3/2}}{3/2} + c = \frac{2}{9} u^{3/2} + c.$$

Back-substituting $3x + 1$ for u now yields

$$\int \sqrt{3x + 1}\, dx = \frac{2}{9}(3x + 1)^{3/2} + c.$$

This is the result we obtained previously by guessing. ☐

How did we know to choose $u = 3x + 1$ in Example 3? The answer is that no other choice for u simplifies the integral. For instance, if we had set $u = 3x$, then $du = 3\, dx$ and substitution would give the integral

$$\int \sqrt{u + 1}\, \frac{1}{3} du,$$

which is really no simpler than the one we started with. On the other hand, if we had let $u = \sqrt{3x + 1}$, then we would have had

$$du = \frac{3}{2\sqrt{3x + 1}}\, dx;$$

but with this expression for du, substitution does not even give an integral in the variable u. (Try it.)

EXAMPLE 4   Let us find

$$\int e^{-2x}\, dx,$$

using the technique of integration by substitution.

We set $u = -2x$. Then $du = -2\, dx$, so $-\frac{1}{2} du = dx$. Thus

$$\int e^{-2x}\, dx = \int e^u \left(-\frac{1}{2}\right) du = -\frac{1}{2} \int e^u\, du$$

$$= -\frac{1}{2} e^u + c = -\frac{1}{2} e^{-2x} + c,$$

which is the result we obtained in Example 2 by guessing. ☐

Integration by substitution is often useful in dealing with more complicated integrals, such as

$$\int x^2(x^3 + 1)^{10}\, dx.$$

To illustrate, let $u = x^3 + 1$. Then $du = 3x^2\, dx$, or $\frac{1}{3}\, du = x^2\, dx$, so

$$\int x^2(x^3 + 1)^{10}\, dx = \int (x^3 + 1)^{10}\, x^2\, dx = \int u^{10} \frac{1}{3}\, du = \frac{1}{3} \frac{u^{11}}{11} + c$$

$$= \frac{1}{33}(x^3 + 1)^{11} + c.$$

How did we know to let $u = x^3 + 1$? The answer is that we recognized part of the integrand, namely x^2, as the derivative (except for a constant factor) of another part, namely $x^3 + 1$.

Key to Integration by Substitution

Whenever part B of an integrand is the derivative (except possibly for a constant factor) of part A of the integrand, the substitution $u = $ part A will simplify the integral.

This is so because, under these conditions, part B of the integrand will be replaced by du when the substitution is performed.

EXAMPLE 5 Let us find $\int \frac{\ln x}{x}\, dx$.

We recognize $1/x$ as the derivative of $\ln x$. Therefore we make the substitution $u = \ln x$. Since $du = (1/x)\, dx$, we have

$$\int \frac{\ln x}{x}\, dx = \int (\ln x) \frac{1}{x}\, dx = \int u\, du = \frac{1}{2}u^2 + c = \frac{1}{2}(\ln x)^2 + c. \quad \square$$

EXAMPLE 6 Find $\int xe^{-x^2}\, dx$.

Except for a constant factor of -2, x is the derivative of $-x^2$. Therefore we set $u = -x^2$, so $du = -2x\, dx$, or $-\frac{1}{2}\, du = x\, dx$, and

$$\int xe^{-x^2}\, dx = \int e^{-x^2} x\, dx = \int e^u \left(-\frac{1}{2}\right) du = -\frac{1}{2}\int e^u\, du$$

$$= -\frac{1}{2} e^u + c = -\frac{1}{2} e^{-x^2} + c. \quad \square$$

INTEGRATION

We conclude this section by remarking that integration by substitution is in fact just the chain rule "in reverse." For instance, if $F(x) = (x^2 + 1)^4 + c$, then the chain rule tells us that

$$F'(x) = 4(x^2 + 1)^3 (x^2 + 1)' = 8x(x^2 + 1)^3.$$

If we now integrate $8x(x^2 + 1)^3$ by making the substitution $u = x^2 + 1$, so that $du = 2x\,dx$, we obtain

$$\int 8x(x^2 + 1)^3\,dx = \int 4(x^2 + 1)^3\,2x\,dx$$
$$= \int 4u^3\,du$$
$$= u^4 + c$$
$$= (x^2 + 1)^4 + c.$$

Thus our integration by substitution "reversed" or "undid" the chain rule.

6.2 EXERCISES

In Exercises 1 through 42, find the integral.

1. $\int \sqrt{x + 1}\,dx$
2. $\int \sqrt{2x + 5}\,dx$
3. $\int \sqrt{4x - 3}\,dx$
4. $\int (5t + 2)^{3/2}\,dt$
5. $\int (6x - 2)^{4/3}\,dx$
6. $\int \dfrac{1}{(2x - 1)^2}\,dx$
7. $\int (7x - 3)^{12}\,dx$
8. $\int (8x + 1)^9\,dx$
9. $\int \dfrac{5}{(2x + 3)^2}\,dx$
10. $\int \dfrac{3}{\sqrt{2 - 3x}}\,dx$
11. $\int 6(4 - 7x)^3\,dx$
12. $\int \dfrac{-6}{5x + 2}\,dx$
13. $\int \dfrac{2}{1 - 2x}\,dx$
14. $\int \dfrac{3}{2x + 1}\,dx$
15. $\int \dfrac{4}{8 - 3x}\,dx$
16. $\int e^{4x}\,dx$
17. $\int 2e^{3x}\,dx$
18. $\int 8e^{-2x}\,dx$
19. $\int 3e^{-7x}\,dx$
20. $\int 2x(x^2 - 2)^3\,dx$
21. $\int x(x^2 + 1)^4\,dx$
22. $\int -3x^2(x^3 + 1)^2\,dx$
23. $\int x^4(2x^5 + 10)\,dx$
24. $\int 2x\sqrt{x^2 + 1}\,dx$
25. $\int 5x\sqrt{2 - x^2}\,dx$
26. $\int 2x^2\sqrt{x^3 + 1}\,dx$
27. $\int (2x + 3)(x^2 + 3x + 1)^4\,dx$
28. $\int (x - 2)(2x^2 - 8x^5)\,dx$
29. $\int \dfrac{2x}{x^2 + 1}\,dx$
30. $\int \dfrac{2x}{3x^2 + 6}\,dx$
31. $\int \dfrac{-7t^2}{t^3 + 2}\,dt$
32. $\int \dfrac{2 - 3x^2}{x^3 - 2x}\,dx$
33. $\int 3xe^{x^2}\,dx$
34. $\int 4xe^{-x^2}\,dx$
35. $\int (2x + 1)e^{2x^2 + 2x}\,dx$
36. $\int (1 - 2x)e^{x^2 - x + 1}\,dx$
37. $\int e^x e^{e^x}\,dx$
38. $\int \dfrac{2e^{\sqrt{x}}}{\sqrt{x}}\,dx$
39. $\int \dfrac{(\ln x)^2}{x}\,dx$
40. $\int \dfrac{1}{x \ln x}\,dx$
41. $\int \dfrac{1}{x(\ln x)^2}\,dx$
42. $\int \dfrac{\sqrt{\ln x}}{x}\,dx$

Management

43. A firm's marginal cost and marginal revenue functions are given by

$$c(x) = 3x\sqrt{x^2 + 1} \quad \text{and} \quad r(x) = \frac{7}{2}x(x^2 + 1)^{3/4},$$

respectively. Find the firm's cost function and revenue function if the fixed cost is $1000.

44. A firm's marginal revenue function is given by $r(x) = 0.2x(1.2x^2 + 4)^{1/2}$. Find the firm's revenue function.

45. A firm's marginal profit function is given by

$$p(x) = \frac{x}{\sqrt{2x^2 + 9}}.$$

Find the firm's profit function if the profit from making and selling 6 units is $52.50.

46. Suppose that the amount of information that must be transmitted by the national telephone network is increasing at a rate of

$$y = (2 \cdot 10^7)e^{0.02t}$$

bits per year t, where $t = 0$ represents the current year. If the network must transmit 10^{10} bits of information this year, find a function that describes the amount of information transmitted as a function of time.

47. With no experience, a worker can assemble 100 units per week. With t weeks of experience, the worker can assemble them at the rate of

$$f(t) = 200e^{-2t}$$

units per week. Find a function that describes the worker's weekly production.

48. A firm's marginal profit function is given by $p(x) = (0.5x + 0.75)/(x^2 + 3x + 9)$, profit in thousands of dollars. Find the firm's profit function if its fixed cost is $10,000.

Life Science

49. It is estimated that the total world population of blue whales in 1978 was 250 and that the population since then has been changing at a rate of y individuals per year, where

$$y = \frac{20t}{(2t^2 + 1)^{3/2}}.$$

Here t is in years, with $t = 0$ representing 1978. Find a function that describes the size of the blue whale population in year t.

50. An animal population is growing at the rate of y thousand individuals per year t, where $y = 8t/(6t^2 + 11)$ and $t = 0$ represents 1988. Find an expression for population size as a function of time if the population in 1988 was 60,000.

51. The concentration of a drug in a patient's bloodstream was 2.5 micrograms per milliliter (µg/mL) 3 minutes after the drug was injected, and it was decreasing at a rate of $-0.5e^{-0.25t}$ (µg/mL)/minute. Find a function that describes the concentration of the drug in the patient's bloodstream as a function of time.

52. The body produces antibodies as a reaction to vaccination. Suppose that t days after a vaccination, antibodies are being produced at a rate of

$$y = \frac{200}{5t + 4}$$

per day, where y is in thousands of antibodies. Assuming there were no antibodies present before vaccination, find a function that describes antibody production t days after vaccination.

53. Air pollution in a region is decreasing at a rate of

$$y = -\frac{20}{2t + 5}$$

parts per million per year t, where $t = 0$ represents the present. If air pollution in the region currently stands at 40 ppm, find a function that describes air pollution as a function of time.

54. The concentration of smog produced by automobiles in a large city changes with the temperature at a rate given by

$$f(x) = \frac{0.006}{(0.02x + 100)^{1/2}},$$

where x is temperature in degrees Celsius and the smog concentration is in parts per million (ppm). Find a function that describes the smog concentration if 6 ppm are produced at 0°C.

6.3 THE DEFINITE INTEGRAL

The concept of the **definite integral** of a function over an interval is one of the most important of calculus. As we shall see in the remainder of this chapter and in the next, definite integrals have many uses. In this section we introduce the definite integral of a function in its aspect as the total change in any antiderivative of the function. We also demonstrate how to apply the method of integration by substitution to definite integrals.

The Definite Integral as Total Change

The **definite integral** of a function over an interval is a number that expresses the total change in any antiderivative of the function over the interval. For instance, if

$$f(x) = 2x + 1,$$

any antiderivative of $f(x)$ is of the form

$$F(x) = x^2 + x + c.$$

The total change in F over the interval $[1, 3]$ is thus

$$F(3) - F(1) = (3^2 + 3 + c) - (1^2 + 1 + c) = 12 - 2 = 10.$$

Hence we say that the definite integral of f over the interval $[1, 3]$ is equal to 10. Note that when calculating a definite integral, we can use any antiderivative, since the constant of integration subtracts out. Therefore we omit the constant of integration when working with definite integrals.

> **The Definite Integral**
>
> If f is a function defined on the interval $[a, b]$, then the **definite integral of f over $[a, b]$**, symbolized by
>
> $$\int_a^b f(x)\, dx,$$
>
> is the total change in any antiderivative of $f(x)$ over the interval. Thus if $F(x)$ is any antiderivative of $f(x)$, then
>
> $$\int_a^b f(x)\, dx = F(b) - F(a).$$
>
> The numbers a and b are referred to as the **limits of integration.**

(The foregoing is not really the *definition* of the definite integral, but rather a *property* of the definite integral. As we shall see in Section 6.5, the definition of the definite integral identifies it as the limit of a certain type of sum, and the fact that this limit may be expressed as the total change in any antiderivative of the integrand is a theorem about the definite integral. We have chosen to introduce the definite integral through this property because we wish to develop some facility in handling definite integrals before we take up their formal definition.)

EXAMPLE 1 If f is defined by $f(x) = 4x - 5$, let us find the definite integral of f over the interval $[1, 2]$. Since $F(x) = 2x^2 - 5x$ is an antiderivative for $f(x)$, we have

$$\int_1^2 (4x - 5)\, dx = F(2) - F(1)$$

$$= (2(2)^2 - 5(2)) - (2(1)^2 - 5(1))$$

$$= 1. \quad \square$$

If $F(x)$ is an antiderivative of $f(x)$, we usually write

$$\int_a^b f(x)\, dx = F(x) \Big|_a^b = F(b) - F(a)$$

when evaluating the definite integral. For instance, we could have carried out the integration of the preceding example as follows:

$$\int_1^2 (4x - 5)\, dx = (2x^2 - 5x) \Big|_1^2 = (2(2)^2 - 5(2)) - (2(1)^2 - 5(1)) = 1.$$

INTEGRATION

Note the difference between the definite integral $\int_b^a f(x)\,dx$ and the indefinite integral $\int f(x)\,dx$: the definite integral is a *number*, while the indefinite integral is a *family of functions*.

EXAMPLE 2 Let us find $\int_{-1}^{3} (x^2 + 2x - 1)\,dx$. We have

$$\int_{-1}^{3} (x^2 + 2x - 1)\,dx = \left(\frac{x^3}{3} + x^2 - x\right)\Bigg|_{-1}^{3}$$

$$= \left(\frac{3^3}{3} + 3^2 - 3\right) - \left(\frac{(-1)^3}{3} + (-1)^2 - (-1)\right)$$

$$= 15 - \frac{5}{3}$$

$$= \frac{40}{3}. \quad \square$$

EXAMPLE 3 Let us evaluate $\int_{-4}^{4} 2e^x\,dx$. We have

$$\int_{-4}^{4} 2e^x\,dx = 2e^x \Bigg|_{-4}^{4} = 2(e^4 - e^{-4}). \quad \square$$

Note that the rules for indefinite integrals

$$\int kf(x)\,dx = k\int f(x)\,dx \qquad \text{for any number } k$$

and

$$\int [f_1(x) \pm f_2(x)]\,dx = \int f_1(x)\,dx \pm \int f_2(x)\,dx$$

imply the corresponding rules for definite integrals:

$$\int_a^b kf(x)\,dx = k\int_a^b f(x)\,dx \qquad \text{for any number } k$$

and

$$\int_a^b [f_1(x) \pm f_2(x)]\,dx = \int_a^b f_1(x)\,dx \pm \int_a^b f_2(x)\,dx.$$

Now we demonstrate how the interpretation of the definite integral as total change can be exploited.

EXAMPLE 4 Suppose a city's population is growing at the rate of

$$f(t) = 12{,}000\sqrt{t}$$

persons per year, where t is in years with $t = 0$ representing 1988. Let us find the total change in population from 1988 to 1992.

Since $f(t)$ describes the rate of change of the population, its definite integral $\int_a^b f(t)\,dt$ gives the total change in the population from time $t = a$ to time $t = b$. Therefore we must evaluate

$$\int_0^4 12{,}000\sqrt{t}\,dt.$$

But

$$\int_0^4 12{,}000\sqrt{t}\,dt = 8000 t^{3/2}\Big|_0^4 = 8000(4^{3/2} - 0^{3/2}) = 64{,}000.$$

Hence, over the period 1988–1992 the city will experience a net gain in population of 64,000 people. ☐

EXAMPLE 5   A firm's marginal cost function is

$$c(x) = 2x^2 - 8x + 30.$$

The firm is currently producing six units per day. What will be the additional cost if the firm decides to produce 10 units per day?

Since marginal cost is the rate of change of cost, the total change in cost due to increasing production from $x = 6$ to $x = 10$ units is

$$\int_6^{10} (2x^2 - 8x + 30)\,dx = \frac{2}{3}x^3 - 4x^2 + 30x\Big|_6^{10}$$

$$= \left(\frac{2}{3}(10)^3 - 4(10)^2 + 30(10)\right) - \left(\frac{2}{3}(6)^3 - 4(6)^2 + 30(6)\right)$$

$$= \frac{1700}{3} - \frac{540}{3}$$

$$\approx \$386.67$$

per day. ☐

EXAMPLE 6   An investment in a new computer will save a company money at the rate of

$$s(t) = 800 - 0.5t^2$$

dollars per month, where t is time in months since the installation of the computer. Figure 6–1 shows the graph of $s(t)$. Note that the rate of savings declines with time, reaching zero when $t = 40$.

The definite integral $\int_a^b s(t)\,dt$ gives the savings due to the computer from time $t = a$ to time $t = b$. Therefore to find the total savings due to the computer, we must evaluate the integral from the time the savings begin until the time they end. But the savings begin when $t = 0$ and end when $t = 40$. Thus the total savings due to the computer are

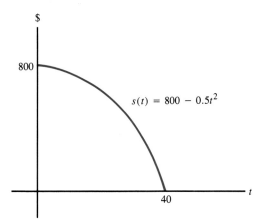

FIGURE 6–1

$$\int_0^{40} (800 - 0.5t^2)\, dt = 800t - \frac{t^3}{6} \Big|_0^{40}$$

$$= \left(800(40) - \frac{(40)^3}{6}\right) - \left(800(0) - \frac{0^3}{6}\right)$$

$$= \$21{,}333.33. \quad \square$$

Integration by Substitution for Definite Integrals

The definite integral $\int_a^b f(x)\, dx$ can be evaluated in two ways. In the first method we find an antiderivative for the integrand $f(x)$, either by guessing or by integration by substitution; then we use the antiderivative to evaluate the definite integral. For instance, to evaluate

$$\int_0^1 x(x^2 + 1)^4\, dx$$

by substitution, we let $u = x^2 + 1$; then $du = 2x\, dx$, and

$$\int x(x^2 + 1)^4\, dx = \frac{1}{2}\int u^4\, du = \frac{1}{10} u^5 + c = \frac{1}{10}(x^2 + 1)^5 + c.$$

Thus

$$\int_0^1 x(x^2 + 1)^4\, dx = \frac{1}{10}(x^2 + 1)^5 \Big|_0^1 = \frac{1}{10}(2^5 - 1^5) = \frac{31}{10}.$$

Alternatively, we can apply the method of integration by substitution directly to the definite integral. If we do so, however, we must be careful to change the limits of integration when we change the variable from x to u in the integral. Thus, in our example, since $u = x^2 + 1$, it follows that

$$\text{when } x = 0, \quad u = 0^2 + 1 = 1$$

and

$$\text{when } x = 1, \quad u = 1^2 + 1 = 2.$$

Applying the method of integration by substitution directly to the definite integral, we now have

$$\int_0^1 x(x^2 + 1)^4 \, dx = \frac{1}{2} \int_1^2 u^4 \, du = \frac{1}{10} u^5 \bigg|_1^2 = \frac{1}{10}(2^5 - 1^5) = \frac{31}{10}.$$

EXAMPLE 7 Let us find $\int_1^2 \dfrac{x + 1}{x^2 + 2x} \, dx$ by the method of substitution for definite integrals. If we let

$$u = x^2 + 2x,$$

then

$$du = 2(x + 1) \, dx.$$

Thus,

$$\text{when } x = 1, \quad u = (1)^2 + 2(1) = 3,$$
$$\text{when } x = 2, \quad u = (2)^2 + 2(2) = 8,$$

and

$$\int_1^2 \frac{x + 1}{x^2 + 2x} \, dx = \frac{1}{2} \int_3^8 \frac{du}{u} = \frac{1}{2} \ln u \bigg|_3^8$$
$$= \frac{1}{2}(\ln 8 - \ln 3) = \frac{1}{2} \ln \frac{8}{3}.$$

(Note that we did not use absolute values and write $\ln |u|$ when we integrated $1/u$ because we already knew that u was positive, since it was between 3 and 8.) ☐

■ *Computer Manual: Program DEFNTGRL*

6.3 EXERCISES

The Definite Integral as Total Change

In Exercises 1 through 22, evaluate the definite integral.

1. $\int_0^4 x \, dx$
2. $\int_1^5 x^2 \, dx$
3. $\int_{-1}^1 3x^3 \, dx$
4. $\int_{-1}^1 x^4 \, dx$
5. $\int_0^3 (x^2 - 2x + 2) \, dx$
6. $\int_2^4 (3x^2 + 1) \, dx$

7. $\int_{-1}^{1} (2x^3 + 3x^2 - 2)\, dx$ 8. $\int_{-2}^{0} (x^4 - 5x^2 + 3x - 1)\, dx$ 9. $\int_{1}^{4} \sqrt{x}\, dx$

10. $\int_{0}^{4} (3x - \sqrt{x})\, dx$ 11. $\int_{0}^{1} (x + \sqrt[3]{x})\, dx$ 12. $\int_{1}^{64} (x^{5/3} - \sqrt{x})\, dx$

13. $\int_{1}^{2} x^{-2}\, dx$ 14. $\int_{1}^{2} -\left(\frac{1}{t^2} + \frac{1}{t^3} + \frac{1}{t^4}\right) dt$ 15. $\int_{0}^{3} 3e^x\, dx$

16. $\int_{0}^{3} e^{-x}\, dx$ 17. $\int_{\ln 2}^{\ln 5} 5e^t\, dt$ 18. $\int_{1}^{e} \frac{1}{x}\, dx$

19. $\int_{-2}^{-1} \frac{4}{t}\, dt$ 20. $\int_{e^2}^{e^3} \frac{3}{x}\, dx$

*21. $\int_{-a}^{a} x^n\, dx$, $a > 0$, n an even positive integer.

*22. $\int_{-a}^{a} x^n\, dx$, $a > 0$, n an odd positive integer.

Management

23. Addlepate, Inc., has marginal cost function given by
$$c(x) = x^2 + 5x$$
and marginal revenue function given by
$$r(x) = 3x^2 + 8x,$$
where cost and revenue are in thousands of dollars and x is in thousands of units. The corporation is currently producing and selling 5000 units. What will be the additional cost and revenue if the firm increases production and sales to 8000 units?

24. Referring to Exercise 23: find the total cost and revenue generated by the first 5000 units Addlepate produces.

25. A firm's marginal profit function is given by $p(x) = -2x + 200$. If the firm is currently producing and selling 70 units, what will its additional profit be if it produces and sells another 10 units?

26. A firm's product generates revenue at the rate of
$$r = 9600t - 100t^2$$
dollars per month, where t is the number of months it has been on the market. Sketch the graph of this function and find the total revenue generated by the product over its lifetime.

27. Apex Company has marginal revenue function given by
$$r(x) = 10x^{1/2},$$
where $x \geq 1$. If Apex currently sells 25 units per day, find the additional revenue if the firm could sell 100 units per day.

28. Maintenance costs for a building t years old increase at a rate of $y = 0.5t^{3/2}$ thousand dollars per year. Find the total maintenance cost for the building over its first four years of life.

29. An investment in new equipment will save a firm $s(t)$ dollars per year, where
$$s(t) = 20{,}000 - 2000t$$

and t is measured in years since the installation of the equipment. Find the firm's total savings due to the new equipment.

Life Science

30. The pollutants in a pond are increasing at the rate of y grams per day, where
$$y = 10t + 3$$
and t is in days with $t = 0$ representing the present. Find the total amount of pollutants that will accumulate in the pond over the next 30 days.

31. A certain strain of influenza spreads in such a way that $f(t)$ persons per day contract the disease, where
$$f(t) = -5t^2 + 300t,$$
t in days. Find the number of persons who will contract the disease in the first two weeks of an epidemic that begins on day 0.

32. A fast-growing strain of bamboo grows $f(t)$ inches per day, where
$$f(t) = 6 + \frac{2}{\sqrt{t}},$$
t being the number of days since germination, and the formula is valid for $t \geq 10$. Find the amount the bamboo will grow in the two weeks from $t = 10$ to $t = 24$.

Social Science

33. In a certain metropolitan region the number of unemployed persons is increasing at a rate of y persons per year t, where
$$y = 60t^2 + 2000$$
and $t = 0$ is the present. Forecast the increase in unemployment in the region over the next two years.

Hurricanes

*34. According to an article by Kerry Emanuel ("Toward a General Theory of Hurricanes," *American Scientist*, July–August 1988), the total pressure drop from the outer wall of a hurricane to its eye is given by
$$-\int_a^c \frac{RT}{p}\, dp,$$
where p is atmospheric pressure, in millibars, a is the pressure at the outer wall and c the pressure at the eye, T is the temperature in degrees Celsius, and R is a constant. Evaluate the integral to find an expression for the total pressure drop as a function of T.

Integration by Substitution for Definite Integrals

In Exercises 35 through 50, evaluate the definite integral by the method of integration by substitution for definite integrals.

35. $\int_{\ln 2}^{\ln 3} e^{3x}\, dx$ 36. $\int_{-2}^{3} e^{-2x}\, dx$ 37. $\int_0^2 2xe^{-x^2}\, dx$ 38. $\int_1^3 xe^{x^2}\, dx$

39. $\int_0^9 \frac{e^{\sqrt{x}}}{\sqrt{x}} dx$ 40. $\int_{-1}^{e-2} \frac{6}{x+2} dx$ 41. $\int_0^4 \frac{1}{2x+1} dx$ 42. $\int_1^2 \frac{4}{3x-2} dx$

43. $\int_0^{\sqrt{2}} x(x^2-2) dx$ 44. $\int_0^4 x^2(x^3-1)^2 dx$ 45. $\int_{-1}^2 2x^2\sqrt{x^3+2} dx$ 46. $\int_0^1 \frac{2x}{x^2+5} dx$

47. $\int_1^2 \frac{3x^2+x}{2x^3+x^2} dx$ 48. $\int_0^1 \frac{3x}{x^2-4} dx$ 49. $\int_0^9 (\sqrt{x}+1)\sqrt{\frac{2}{3}x^{3/2}+x} dx$ 50. $\int_2^4 \frac{x^2+2x+1}{x^3} dx$

Management

51. A firm intends to build a new plant. It is estimated that t years after the plant is built it will be producing at the rate of

$$f(t) = 100{,}000 e^{0.01t}$$

units per year. Find the number of units the plant will produce in its first five years. How long must the plant operate if it is to produce 2 million units?

52. Able Associates has marginal profit function given by

$$p(x) = \frac{2000}{0.5x+10}.$$

Find the firm's profit from making and selling its first 50 units.

53. With t weeks of experience a worker can assemble parts at a rate of

$$y = 200 - 120 e^{-0.05t}$$

per week. Find the number of parts the worker can assemble over weeks 0 through 5; over weeks 5 through 10.

54. Suppose it takes y worker-hours to build the xth unit of a product, where

$$y = 4000 + 2000 e^{-0.1x}, \quad x \geq 1.$$

Find the total worker-hours needed to build units 1 through 5; to build units 1 through 100.

55. Investment in new industrial robots saves a firm money at the rate of

$$S = 100(1 - e^{-0.02t})$$

thousand dollars per year t years after purchase. Find the firm's total savings due to the robots over the period from two to five years after they were purchased.

56. Suppose that platinum is being mined at the rate of

$$y = \frac{2000}{0.5t+1}$$

metric tons per year t, where $t = 0$ represents the present. Find the total amount of platinum that will be mined over the next two years; over the following three years.

57. Suppose that a natural gas field is increasing production at the rate of

$$y = \frac{2t}{0.2t^2+1}$$

million cubic feet per year t, where $t = 0$ represents the present. Find the total amount of gas that will be produced over the next five years; over the period $t = 10$ to $t = 15$.

Life Science

58. A patient receives a solution intravenously at a rate of
$$f(t) = 10.8e^{0.06t}$$
cubic centimeters per hour, t in hours. Find the amount of solution the patient receives during the first 24 hours of treatment.

59. An animal population is increasing at a rate of
$$y = \frac{200}{t+1} + 20$$
individuals per year t. Find the total population gain from time $t = 0$ to time $t = 10$.

60. The size of a bacteria population is changing at the rate of $y = 50te^{-t^2/32}$ thousand individuals per hour t, where $t = 0$ represents the present. Find the total change in the population over the next 2 hours.

61. Cases of a new disease are increasing at a rate of
$$y = 1200t/(0.5t^2 + 800)$$
cases per week, where t is in weeks since the beginning of the year. Find the total number of new cases during the tenth week of the year ($t = 9$ to $t = 10$); during the fiftieth week of the year.

Social Science

62. A politician's campaign manager estimates that voters who support the politician can be registered at a rate of y individuals per day t, where
$$y = 2e^{0.25t},$$
with $t = 0$ being the present. How many voters can be registered over the next 30 days?

63. Research shows that people who learn a foreign language but never use it tend to forget their vocabulary at the rate of $100y\%$ per year, where
$$y = \frac{0.20}{t+1}.$$
Here t is time in years since the language was last used. How much of a language will be forgotten if it is not used for 10 years?

6.4 THE DEFINITE INTEGRAL AND AREA

Let f be a function continuous and nonnegative over the interval $[a, b]$, as in Figure 6–2, and consider the shaded region bounded by the graph of f, the x-axis, and the lines $x = a$ and $x = b$. This region has an area, which we call the **area under f from a to b**. This area and the definite integral are related as follows: the area under f from a to b is the definite integral of f from a to b:

INTEGRATION

The Definite Integral as Area

If f is continuous and nonnegative over the interval $[a, b]$, then the area under f from a to b is given by the definite integral

$$\int_a^b f(x)\, dx.$$

This remarkable fact is the subject of this section. We demonstrate why it is true and show how to use it to find the area of planar regions.

The Area under a Curve

Let us see why the area under f from a to b is the definite integral of f from a to b. If f is continuous and nonnegative over the interval $[a, b]$, as in Figure 6–2, then associated with f is an **area function** A, which is defined as follows: for each number x in the interval $[a, b]$,

$$A(x) = \text{the area under } f \text{ from } a \text{ to } x.$$

See Figure 6–3. Note that

$$A(b) = \text{the area under } f \text{ from } a \text{ to } b$$

and that

$$A(a) = \text{the area under } f \text{ from } a \text{ to } a = 0.$$

Now we will show that $A(x)$ is an antiderivative for $f(x)$. To do this, we must show that $A'(x) = f(x)$ for every number x in $[a, b]$. But

$$A'(x) = \lim_{h \to 0} \frac{A(x + h) - A(x)}{h},$$

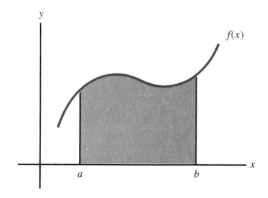

FIGURE 6–2
The Area under f from $x = a$ to $x = b$

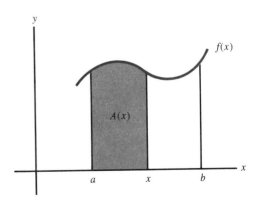

FIGURE 6–3

if the limit exists. However, if h is small enough so that $x + h$ is in the interval, then $A(x + h) - A(x)$ is just the area under f from x to $x + h$. See Figure 6–4. But as Figure 6–5 demonstrates, the area under f from x to $x + h$ can be approximated by the area of a rectangle whose height is $f(x + h)$ and whose base has length $x + h - x = h$; thus

$$A(x + h) - A(x) \approx hf(x + h).$$

Furthermore, it is clear that as $h \to 0$, the approximation becomes better and better. Therefore if we rewrite the approximation as

$$\frac{A(x + h) - A(x)}{h} \approx f(x + h)$$

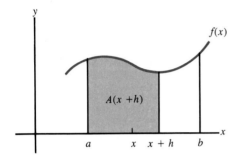

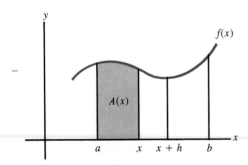

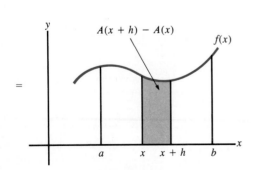

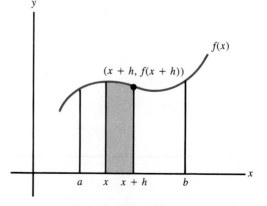

FIGURE 6–4

FIGURE 6–5

INTEGRATION

and take the limit as $h \to 0$, we obtain

$$\lim_{h \to 0} \frac{A(x + h) - A(x)}{h} = \lim_{h \to 0} f(x + h).$$

But since f is continuous,

$$\lim_{h \to 0} f(x + h) = f(x)$$

by the second limit definition of continuity. Therefore

$$A'(x) = \lim_{h \to 0} \frac{A(x + h) - A(x)}{h} = \lim_{h \to 0} f(x + h) = f(x),$$

which shows that $A(x)$ is an antiderivative for $f(x)$.

Now suppose that $F(x)$ is any antiderivative for $f(x)$ on $[a, b]$. Then, as we saw in Section 6–1, there is some number c such that

$$F(x) = A(x) + c$$

for all x in $[a, b]$. But then

$$F(a) = A(a) + c = 0 + c = c,$$

so

$$F(x) = A(x) + F(a)$$

for all x in $[a, b]$. This in turn implies that when $x = b$,

$$A(b) = F(b) - F(a).$$

But $A(b)$ is the area under f from a to b, and

$$\int_a^b f(x)\, dx = F(b) - F(a).$$

Therefore

$$\int_a^b f(x)\, dx = A(b),$$

so the area under f from a to b is the definite integral of f from a to b.

Finding Areas

The relationship between the definite integral and area makes it easy to find the area of many regions in the xy-plane. We illustrate with examples.

EXAMPLE 1   Let $f(x) = x$. We will find the area under f from 0 to 1. As Figure 6–6 shows, this is the area of a triangular region whose base has length 1 and whose height is also 1. Since the area of a triangle is one-half the length of its base times its height, the area in question must be $\frac{1}{2}$. We can obtain the same result by using the definite integral, since the area in question is equal to

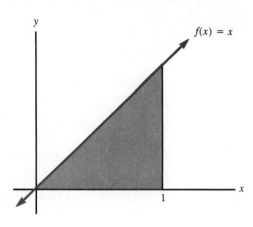

FIGURE 6–6   **FIGURE 6–7**

$$\int_0^1 f(x)\,dx = \int_0^1 x\,dx = \frac{1}{2}x^2 \Big|_0^1 = \frac{1}{2}(1^2 - 0^2) = \frac{1}{2}. \quad \square$$

EXAMPLE 2 Let $f(x) = x^2$. We find the area of the region shaded in Figure 6–7, which is the area under f from -1 to 2, and thus is equal to

$$\int_{-1}^2 x^2\,dx = \frac{1}{3}x^3 \Big|_{-1}^2 = \frac{1}{3}(2^3 - (-1)^3) = 3. \quad \square$$

If the graph of a function f lies below the x-axis from $x = a$ to $x = b$, as in Figure 6–8, then the definite integral $\int_a^b f(x)\,dx$ will be a negative number; in such a case the area of the region that lies between the graph and the x-axis from a to b (the shaded region in Figure 6–8) is equal to the absolute value of the definite integral.

EXAMPLE 3 Suppose $f(x) = x^2 - 3x$ and we want to find the area of the region bounded by the graph of f and the x-axis from $x = 0$ to $x = 3$. This is the shaded region in Figure 6–9. Since the region lies below the x-axis, the definite integral is negative:

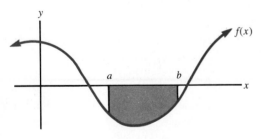

FIGURE 6–8

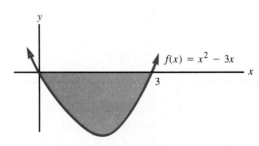

FIGURE 6–9

INTEGRATION

$$\int_0^3 f(x)\,dx = \int_0^3 (x^2 - 3x)\,dx$$

$$= \frac{1}{3}x^3 - \frac{3}{2}x^2 \Big|_0^3$$

$$= \frac{1}{3} 3^3 - \frac{3}{2} 3^2$$

$$= -\frac{9}{2}.$$

Therefore the region has area equal to $|-\frac{9}{2}| = \frac{9}{2}$. □

If we need to find the area of a region part of which lies above the *x*-axis and part below, we can find the areas of the separate parts of the region and add them.

EXAMPLE 4   Let $f(x) = x^3 - 3x^2 + 2x$. Suppose we need to find the area of the shaded region in Figure 6–10. The area of the part of the region that is above the *x*-axis is

$$\int_0^1 (x^3 - 3x^2 + 2x)\,dx = \left(\frac{1}{4}x^4 - x^3 + x^2\right)\Big|_0^1 = \frac{1}{4}.$$

The area of the part of the region that is below the *x*-axis is the absolute value of

$$\int_1^2 (x^3 - 3x^2 + 2x)\,dx = \left(\frac{1}{4}x^4 - x^3 + x^2\right)\Big|_1^2 = -\frac{1}{4}.$$

Thus the area of this region is $|-\frac{1}{4}| = \frac{1}{4}$. The area of the entire region is therefore $\frac{1}{4} + \frac{1}{4} = \frac{1}{2}$. □

The preceding examples show that when using the definite integral to find the area of a planar region, it is essential before doing the integration to determine

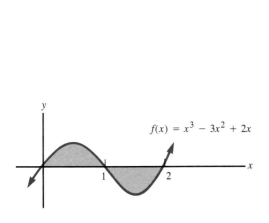

FIGURE 6–10

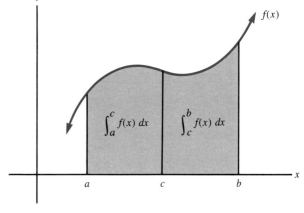

FIGURE 6–11

whether the region lies above the x-axis, below it, or partly above it and partly below it. Thus careful curve sketching is required *before* any integration is performed.

Considering the definite integral as an area leads to many interesting consequences. For instance, it is true that for any number a,

$$\int_a^a f(x)\,dx = 0.$$

This is so because, as noted previously, the area under f from a to a must be zero. Similarly, if c is any number such that $a < c < b$, then

$$\int_a^b f(x)\,dx = \int_a^c f(x)\,dx + \int_c^b f(x)\,dx,$$

because, as Figure 6–11 shows, the area under f from a to b is equal to the area under f from a to c plus the area under f from c to b.

Another consequence of regarding the definite integral as an area is that total change can be represented by area. For instance, we have seen (Example 5, Section 6.3) that since marginal cost is the rate of change of cost, the definite integral of marginal cost from a to b is the total change in cost over the interval $[a, b]$; but this integral is also the area under the marginal cost curve from a to b. Hence, total change in cost is represented by the area of a region under the marginal cost curve. The same reasoning applied to any rate-of-change function leads to the following conclusion:

Total Change as Area

If f describes the rate of change of some quantity Q over the interval $[a, b]$, then the total change in Q over $[a, b]$ is represented by the area under f from a to b.

See Figure 6–12. This is an important fact to remember, because in business, economics, and the sciences, arguments concerning total change are often presented in terms of areas under some rate-of-change function.

The Area Bounded by Two Curves

Figure 6–13 shows a region bounded by the graphs of the functions f and g and the vertical lines $x = a$ and $x = b$. The area of such a region is referred to as the **area bounded by the curves f and g from $x = a$ to $x = b$**. As Figure 6–14 indicates, we can find the area of this region by subtracting the area under the lower curve from the area under the upper one. Since the area under the lower curve in Figure 6–14 is $\int_a^b g(x)\,dx$ and that under the upper curve is $\int_a^b f(x)\,dx$, the area bounded by the curves is

$$\int_a^b f(x)\,dx - \int_a^b g(x)\,dx = \int_a^b [f(x) - g(x)]\,dx.$$

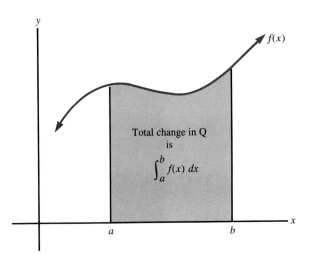

FIGURE 6–12
f Describes Rate of Change of *Q*

FIGURE 6–13
The Area Bounded by *f* and *g* from $x = a$ to $x = b$.

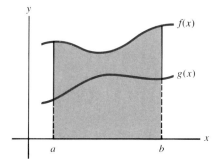 −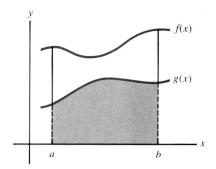

=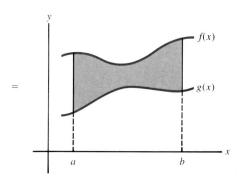

FIGURE 6–14
$$\int_a^b f(x)\, dx - \int_a^b g(x)\, dx = \int_a^b [f(x) - g(x)]\, dx$$

Finding the Area of a Region Bounded by Two Curves

To find the area of a region bounded by two curves from $x = a$ to $x = b$: Determine which of the curves is the upper and which the lower over the interval $[a, b]$. If f is the upper and g the lower, the area bounded by the curves is

$$\int_a^b [f(x) - g(x)]\, dx.$$

Figure 6-14 pictured the curves as nonnegative, but this is not necessary: the procedure for finding the area bounded by two curves works even if part or all of the region between them lies below the x-axis. However, it *is* necessary that in the integral $\int_a^b [f(x) - g(x)]\, dx$, f must be the upper curve and g the lower one.

EXAMPLE 5 Let us find the area of the shaded region in Figure 6-15. From the figure, we see that the upper curve over the interval $[-1, 1]$ is $y = x^2 + 1$, while the lower one is $y = -x^2$. Hence the area of the region bounded by the curves from $x = -1$ to $x = 1$ is

$$\int_{-1}^{1} [(x^2 + 1) - (-x^2)]\, dx = \int_{-1}^{1} [2x^2 + 1]\, dx$$

$$= \frac{2}{3}x^3 + x \Big|_{-1}^{1}$$

$$= \frac{10}{3}. \quad \square$$

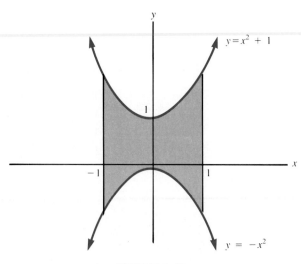

FIGURE 6-15

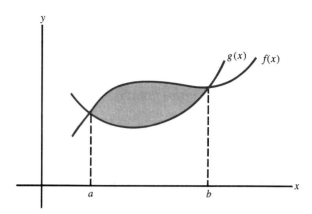

FIGURE 6–16
The Area Bounded by *f* and *g*

Suppose we wish to find the area of a region that is completely enclosed by the graphs of two functions *f* and *g*, such as the area in Figure 6–16, for instance. The area of such a region is referred to as the **area bounded by *f* and *g*,** and it is equal to

$$\int_a^b [f(x) - g(x)] \, dx.$$

The limits of integration *a* and *b* may be found by setting $f(x) = g(x)$ and solving for *x*.

EXAMPLE 6   Let us find the area of the region bounded by *p* and *q*, where

$$p(x) = 8x - x^2 \quad \text{and} \quad q(x) = x^2 - 4x + 10.$$

Since we must know which curve is the upper one and which the lower one, we sketch the graphs of these functions. See Figure 6–17. To find the points where the curves intersect, we set $p(x) = q(x)$ and solve for *x*:

$$8x - x^2 = x^2 - 4x + 10,$$

so

$$2x^2 - 12x + 10 = 0,$$

or

$$2(x - 1)(x - 5) = 0.$$

Therefore the curves intersect at $(1, p(1)) = (1, 7)$ and $(5, p(5)) = (5, 15)$. Since *p* is the upper curve and *q* the lower one over the interval [1, 5], it follows that the area of the region in question is equal to

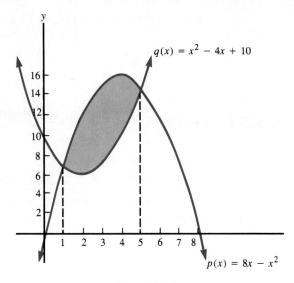

FIGURE 6-17   **FIGURE 6-18**

$$\int_1^5 [p(x) - q(x)]\, dx = \int_1^5 [(8x - x^2) - (x^2 - 4x + 10)]\, dx$$

$$= \int_1^5 (-2x^2 + 12x - 10)\, dx$$

$$= -\frac{2}{3}x^3 + 6x^2 - 10x \bigg|_1^5$$

$$= \frac{64}{3}. \quad \square$$

Our procedure for finding the area bounded by two curves requires that we subtract the lower curve from the upper one. But which curve is the lower and which is the upper may change from interval to interval. Again, careful curve sketching is essential.

EXAMPLE 7 Find the area between the curves

$$y = x^2 \quad \text{and} \quad y = x^3 - 2x.$$

Setting the equations equal to each other and solving for x yields

$$x^2 = x^3 - 2x,$$

or

$$x^3 - x^2 - 2x = 0,$$

or

$$x(x + 1)(x - 2) = 0,$$

INTEGRATION

so the points of intersection are $(0, 0)$, $(-1, 1)$, and $(2, 4)$. See Figure 6–18. Note that over the interval $[-1, 0]$ the upper curve is $y = x^3 - 2x$, but over the interval $[0, 2]$ the upper curve is $y = x^2$. Therefore the area of the region bounded by the two curves is

$$\int_{-1}^{0} [(x^3 - 2x) - x^2] \, dx + \int_{0}^{2} [x^2 - (x^3 - 2x)] \, dx$$

$$= \left(\frac{1}{4}x^4 - \frac{1}{3}x^3 - x^2 \right) \Big|_{-1}^{0} + \left(-\frac{1}{4}x^4 + \frac{1}{3}x^3 + x^2 \right) \Big|_{0}^{2}$$

$$= 0 - \left(\frac{1}{4} + \frac{1}{3} - 1 \right) + \left(-\frac{16}{4} + \frac{8}{3} + 4 \right) - 0$$

$$= \frac{37}{12}. \quad \square$$

■ *Computer Manual: Program DEFNTGRL*

6.4 EXERCISES

The Area under a Curve

In Exercises 1 through 16, find the area under f from a to b.

1. $f(x) = x^2 + 3x + 5$, $a = 1$, $b = 3$
2. $f(x) = x^2 + 3x + 5$, $a = -1$, $b = 3$
3. $f(x) = x^2 - 2x + 3$, $a = 0$, $b = 2$
4. $f(x) = x^2 - 2x + 3$, $a = 2$, $b = 4$
5. $f(x) = x^2 - 2x + 3$, $a = 0$, $b = 4$
6. $f(x) = -x^3 + 2x^2 + 15x$, $a = -3$, $b = 0$
7. $f(x) = -x^3 + 2x^2 + 15x$, $a = -3$, $b = 5$
8. $f(x) = x^3 - 3x^2 - 6x + 8$, $a = -2$, $b = 1$
9. $f(x) = x^3 - 3x^2 - 6x + 8$, $a = 1$, $b = 4$
10. $f(x) = x^3 - 3x^2 - 6x + 8$, $a = -2$, $b = 4$
11. $f(x) = \frac{1}{x}$, $a = \frac{1}{2}$, $b = e$
12. $f(x) = e^{2x}$, $a = 0$, $b = 4$
13. $f(x) = e^{-3x}$, $a = 0$, $b = 1$
14. $f(x) = \frac{2}{x + 1}$, $a = 0$, $b = 3$
15. $f(x) = (\ln x)/x$, $a = e^2$, $b = e^3$
16. $f(x) = xe^{-x^2}$, $a = 0$, $b = 3$

Management

17. The typical worker in a jewelry plant can produce $f(t)$ necklaces per hour, where

$$f(t) = -3t^2 + 36t,$$

t in hours.
(a) Graph this function.
(b) Interpret the number of necklaces the typical worker can produce in an 8-hour day as an area, and find this number.

18. A firm's marginal cost and marginal revenue functions are defined by
$$c(x) = x^2 - 80x + 3600 \quad \text{and} \quad r(x) = -x^2 + 100x,$$
respectively.
   (a) Graph these functions.
   (b) Suppose the firm is currently making and selling 20 units per day. Interpret as areas the additional cost and revenue that would result from doubling production and sales, and then find the additional cost and revenue.

19. The savings produced by new equipment accrue at a rate of
$$y = 10{,}000 - 400t$$
dollars per year, where t is the number of years since the equipment was purchased. Find the area of the region bounded by the graph of this function and the coordinate axes, and interpret your result.

20. The savings produced by a new computer accrue at a rate of
$$y = 20e^{-0.05t}$$
thousand dollars per month, where t is in months since the purchase of the computer. Find the area of the region bounded by the graph of this function, the y-axis, and the line $t = 10$, and interpret your result.

21. With t months of experience on the job, a worker can produce at the rate of y units per month, where
$$y = 1000 - 600e^{-0.25t}.$$
Find the area under the graph of this function from $t = 0$ to $t = 12$ and interpret your result.

22. Suppose gold is being produced at a rate of
$$y = \frac{25}{t + 0.5} + 3$$
metric tons per year t, where $t = 0$ represents the present. Find the area under this function from $t = 1$ to $t = 4$ and interpret your result.

23. The savings produced by new equipment accrue at a rate of y thousand dollars per year t, where t is years since the equipment was purchased and
$$y = \begin{cases} 50(t + 4)^{-1}, & 0 \le t \le 6, \\ 12.5 - 1.25t, & t > 6. \end{cases}$$
Find the total savings due to the new equipment.

Life Science

24. A virus grows within the human body at a rate of y organisms per hour, where
$$y = 2000e^{0.02t}$$
and t is measured in hours since infection occurred.
   (a) Graph this function.
   (b) Interpret the number of viral organisms that are produced during the first 3 hours after infection as an area and find this number.
   (c) If death results when the total number of viral organisms produced since infection reaches 1 million, how long will it be from infection until death if no treatment is given?

25. The wolf population in a certain region of Alaska is increasing at a rate of
$$f(t) = 12\sqrt{0.5t + 4}$$
individuals per year t, where $t = 0$ represents the present. Find the area under f from $t = 0$ to $t = 10$ and interpret your result.

The Area Bounded by Two Curves

In Exercises 26 through 29, find the area of the region bounded by f and g from $x = a$ to $x = b$.

26. $f(x) = x^2 + 2$, $g(x) = x^2$, $a = -2$, $b = 2$ **27.** $f(x) = 2$, $g(x) = -x^2$, $a = 0$, $b = 1$
28. $f(x) = x^3$, $g(x) = 4 - x^2$, $a = -1$, $b = 1$ **29.** $f(x) = e^x$, $g(x) = x$, $a = 1$, $b = 2$

In Exercises 30 through 37, find the area bounded by the given functions.

30. $f(x) = x$, $g(x) = x^2$ **31.** $f(x) = x$, $g(x) = x^3$
32. $f(x) = \sqrt{x}$, $g(x) = x^2$ **33.** $f(x) = -x + 4$, $g(x) = -x^2 + 4x$
34. $y = x^2 - 2x$, $y = x$ **35.** $f(x) = x^2 - 9x + 26$, $g(x) = -x^2 + 9x - 14$
36. $y = -x^2 + 4x + 96$, $y = x^2 + 6x + 12$ **37.** $y = x^4 - 2x^2$, $y = 2x^2$

In Exercises 38 through 47, find the area of the region bounded by the graphs of the given equations.

38. $y = e^{-x/2}$, $y = 0$, $x = 0$, $x = 4$ **39.** $y = 2/(x + 3)$, $y = 0$, $x = 0$, $x = 2$
40. $y = 4 + 2x$, $y = 7 - x$, $y = -3 + 4x$ **41.** $y = 2x + 5$, $y = 8 + x$, $y = 29 - 2x$
42. $y = x + 1$, $y = 1 - x$, $y = x - 1$, $y = 3 - x$
43. $y = x + 1$, $y = x - 5$, $y = 4 - 2x$, $y = 9 - x$
44. $y = x^2 - 4x + 14$, $y = -x^2 + 6x - 9$, $x = 1$, $x = 4$
45. $y = e^x$, $y = e^{-x}$, $x = -1$, $x = 1$, $y = 0$
46. $y = 12 - x^2$, $y = x$ for $x \geq 0$, $y = -4x$ for $x \leq 0$
47. $y = 12 - x^2$, $y = x$ for $x \leq 0$, $y = -4x$ for $x \geq 0$

Management

48. A new car brings in revenue for a car rental agency at a rate of
$$f(t) = 12{,}500$$
dollars per year, and its cost to the rental company increases at a rate of
$$g(t) = 5000 + 2000t$$
dollars per year. Find the area of the region bounded by the graphs of f and g and the y-axis, and interpret your result.

49. A firm's savings due to new equipment are given by
$$y = 6 + 2t$$
and the cost of maintaining the new equipment by
$$y = 2.4t,$$

where y is in thousands of dollars and t is in years since the equipment was purchased. Find the area of the region bounded by these curves and the y-axis, and interpret your result.

50. The savings due to a new telephone system accrue at a rate of
$$y = 10 + 1.2t$$
thousand dollars per month, where t is the number of months since the system was installed. Maintenance costs for the system t months after its installation are m thousand dollars, where
$$m = 0.8 + 0.4t.$$
Find the area of the region bounded by these curves, the y-axis, and the line $t = 5$, and interpret your result.

51. A firm's new drill press is saving s thousand dollars per month in raw materials cost, where $s = 150 - t^2$. Also, the cost of the labor needed to run the new press is increasing at the rate of $c = 0.5t^2$ thousand dollars per month. Here t is months since the press was installed. Graph these functions, interpret the firm's net savings as an area, and find the net savings.

52. Since time $t = 0$, a firm's rate of cash inflow has been $C = -t^2 + 12t + 160$ thousand dollars per month t.
    (a) Find the firm's total cash inflow.
    (b) If the firm's cash outflow since time $t = 0$ has been $E = t^2 + 70$ thousand dollars per month, find the firm's net cash flow.

Social Science

53. Emigrants are leaving a nation at the rate of $f(t)$ thousand persons per year, where $f(t) = 20e^{0.5t}$, and immigrants are entering the nation at the rate of $g(t)$ thousand persons per year, where $g(t) = 40e^{0.4t}$. Here $t = 0$ is 1985. Graph these functions, and then find the net population change due to the combination of emigration and immigration over the period
    (a) 1985–1990   (b) 1985–1995.

Lorenz Curves

Suppose a function L describes the distribution of income in a society in the following manner:

> for $0 \le x \le 100$, $100L(x)\%$ is the percentage of the total income earned by the $x\%$ of the population that earns the least.

For instance, if $L(0.2) = 0.05$, then 5% of the total income is earned by the 20% of the population that earns the least; if $L(0.5) = 0.3$, then 30% of the total income is earned by the 50% of the population that earns the least. Such a function L is called a **Lorenz curve**. Exercises 54 through 58 are concerned with Lorenz curves.

*54. Suppose a Lorenz curve is given by $L(x) = 0.75x^2 + 0.25x$. Find $L(0.1)$, $L(0.2)$, and $L(0.9)$, and interpret your results.

*55. Show that if L is a Lorenz curve, then $L(0) = 0$ and $L(1) = 1$.

*56. The line segment $y = x$, $0 \leq x \leq 1$, is the Lorenz curve for a society in which income is equally distributed. Why?

The graph of a typical Lorenz curve is as follows:

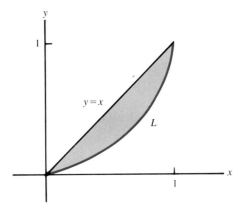

The greater the inequality of income distribution, the farther below the line $y = x$ the Lorenz curve will be, and hence the larger the shaded region will be. Economists measure inequality of income distribution by the **coefficient of inequality**

$$C = \frac{\text{Area of region bounded by } y = x \text{ and Lorenz curve } L}{\text{Area under } y = x \text{ from } x = 0 \text{ to } x = 1}.$$

Thus $0 \leq C \leq 1$, and the larger C is, the greater the inequality of income distribution.

*57. Show that for a given Lorenz curve L,

$$C = 2\int_0^1 (x - L(x))\, dx.$$

*58. Graph the Lorenz curves

$$L_1(x) = \frac{1}{2}x^2 + \frac{1}{2}x$$

and

$$L_2(x) = \frac{1}{5}x^2 + \frac{4}{5}x$$

on the same set of axes and find their coefficients of inequality. Which of them describes the more unequal distribution of income?

6.5 RIEMANN SUMS

When we discussed the definite integral in Section 6.3 we remarked that we had not actually defined it but rather had introduced it by means of one of its properties. Now it is time to define the definite integral as the limit of a certain type of sum known as a **Riemann sum.** Before doing so, however, we wish to introduce the summation notation Σ.

Summation Notation

The summation symbol Σ is the Greek capital letter sigma. It is used in mathematics as shorthand for "sum up what follows." Thus, if $x_1, x_2, \ldots, x_n$ are numbers, then by definition,

$$\sum_{i=1}^{n} x_i = x_1 + x_2 + \cdots + x_n$$

The left-hand side of the preceding identity is read as "the sum of x sub i from i equal 1 to n." Note that the sum is formed by setting i equal to 1, then 2, then 3, and so on, up to n, and adding the resulting terms. Where there is no possibility of confusion, we shall write $\Sigma\, x$ for $\Sigma_{i=1}^{n} x_i$.

EXAMPLE 1 Let $x_1 = 2$, $x_2 = 5$, $x_3 = 0$, $x_4 = 6$, and $x_5 = -1$. Then

$$\sum_{i=1}^{5} x_i = x_1 + x_2 + x_3 + x_4 + x_5 = 2 + 5 + 0 + 6 + (-1) = 12,$$

$$\sum_{i=1}^{4} x_i = x_1 + x_2 + x_3 + x_4 = 2 + 5 + 0 + 6 = 13,$$

$$\sum_{i=1}^{3} x_i = x_1 + x_2 + x_3 = 2 + 5 + 0 = 7,$$

$$\sum_{i=1}^{2} x_i = x_1 + x_2 = 2 + 5 = 7,$$

and

$$\sum_{i=1}^{1} x_i = x_1 = 2. \quad \square$$

EXAMPLE 2 Consider the summation

$$\sum_{i=1}^{4} i^2.$$

We have

$$\sum_{i=1}^{4} i^2 = 1^2 + 2^2 + 3^2 + 4^2 = 1 + 4 + 9 + 16 = 30.$$

Similarly,

$$\sum_{j=0}^{3} 2(j + 1) = 2(0 + 1) + 2(1 + 1) + 2(2 + 1) + 2(3 + 1) = 20.$$

(Note that we could have factored the 2 out of this summation before evaluating it:

$$\sum_{j=0}^{3} 2(j + 1) = 2 \sum_{j=0}^{3} (j + 1) = 2[(0 + 1) + (1 + 1) + (2 + 1) + (3 + 1)]$$
$$= 2[10] = 20).$$

INTEGRATION

Also,
$$\sum_{k=3}^{5} \frac{k+1}{k} = \frac{3+1}{3} + \frac{4+1}{4} + \frac{5+1}{5} = \frac{4}{3} + \frac{5}{4} + \frac{6}{5} = \frac{227}{60}. \quad \square$$

EXAMPLE 3 Note that
$$\underbrace{\sum_{i=0}^{6} 3 = 3 + 3 + 3 + 3 + 3 + 3}_{6 \text{ times}} = 6(3) = 18.$$

In general, if a is any number and n is a positive integer, then
$$\sum_{i=0}^{n} a = na. \quad \square$$

The Definition of the Definite Integral

Now we are ready to define the definite integral as a limit of sums. Let n be any positive integer. Suppose we divide the interval $[a, b]$ into n subintervals of equal length. Then each of the subintervals will have length Δx, where

$$\Delta x = \frac{1}{n}(\text{length of } [a, b]) = \frac{1}{n}(b - a) = \frac{b-a}{n}.$$

See Figure 6–19. Note that the endpoints of the subintervals are
$$a = a_0, a_1, a_2, \ldots, a_{n-1}, a_n = b.$$
Thus the ith subinterval is $[a_{i-1}, a_i]$, and its length is Δx.

Now let f be a continuous function defined on $[a, b]$. From each subinterval $[a_{i-1}, a_i]$ we choose a single number x_i, find the functional value $f(x_i)$ (see Figure 6–20), and form the sum

$$\sum_{i=1}^{n} f(x_i) \Delta x = f(x_1) \Delta x + f(x_2) \Delta x + \cdots + f(x_n) \Delta x.$$

This sum is called a **Riemann sum for f over $[a, b]$**. If f is nonnegative on $[a, b]$ and we construct a rectangle of height $f(x_i)$ on each subinterval $[a_{i-1}, a_i]$, the area

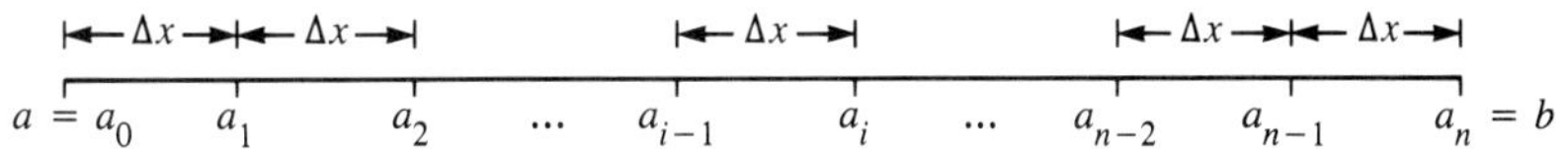

|←Δx→|←Δx→| |←Δx→| |←Δx→|←Δx→|
$a = a_0 \quad a_1 \quad a_2 \quad \ldots \quad a_{i-1} \quad a_i \quad \ldots \quad a_{n-2} \quad a_{n-1} \quad a_n = b$

FIGURE 6–19

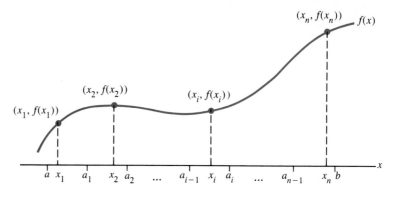

FIGURE 6–20

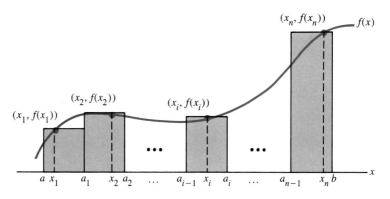

FIGURE 6–21

of the rectangle will be $f(x_i)\,\Delta x$. See Figure 6–21. Thus the Riemann sum will be the sum of the areas of all the rectangles, and hence will be an approximation to the area under f from a to b. See Figure 6–22, where the Riemann sum is the sum of the areas of the rectangles.

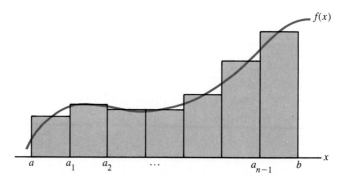

FIGURE 6–22

EXAMPLE 4 Suppose we let f be defined by $f(x) = x$, let $[a, b] = [0, 1]$ and take $n = 3$. Then

$$\Delta x = \frac{1 - 0}{3} = \frac{1}{3},$$

and

$$a_0 = 0, \quad a_1 = \frac{1}{3}, \quad a_2 = \frac{2}{3}, \quad a_3 = 1.$$

Thus a Riemann sum for f over $[0, 1]$ with $n = 3$ is a sum of the form

$$f(x_1)\frac{1}{3} + f(x_2)\frac{1}{3} + f(x_3)\frac{1}{3},$$

where x_1 may be any number in $[0, \frac{1}{3}]$, x_2 may be any number in $[\frac{1}{3}, \frac{2}{3}]$, and x_3 may be any number in $[\frac{2}{3}, 1]$. For instance, if we choose $x_1 = \frac{1}{6}$, $x_2 = \frac{2}{3}$, and $x_3 = \frac{5}{6}$, then the sum becomes

$$f\left(\frac{1}{6}\right)\frac{1}{3} + f\left(\frac{2}{3}\right)\frac{1}{3} + f\left(\frac{5}{6}\right)\frac{1}{3} = \frac{1}{6} \cdot \frac{1}{3} + \frac{2}{3} \cdot \frac{1}{3} + \frac{5}{6} \cdot \frac{1}{3} = \frac{5}{9}.$$

For this Riemann sum, we obtain the rectangles of Figure 6–23. □

Suppose we form a Riemann sum $\sum_{i=1}^{n} f(x_i) \Delta x$ for a continuous function f over an interval $[a, b]$. What will happen to this sum if we divide the interval $[a, b]$ into more and more subintervals of shorter and shorter length? That is, what will happen to the sum if we let Δx approach 0? (Note that if Δx is to approach 0, then n must approach $+\infty$.) If f is nonnegative on $[a, b]$, as $\Delta x \to 0$ the total area of the rectangles approaches the area under f from a to b (see Figure 6–24) and hence the Riemann

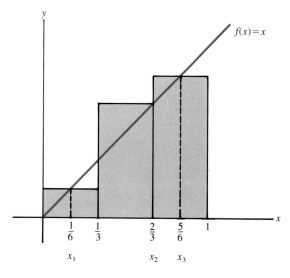

FIGURE 6–23

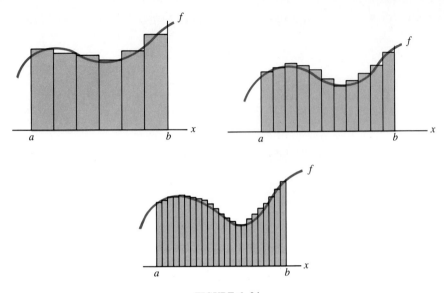

FIGURE 6–24

sum, which is the total area of the rectangles, approaches a real number. It can be shown that this conclusion holds even when f is not nonnegative on $[a, b]$:

$$\lim_{\Delta x \to 0} \sum_{i=1}^{n} f(x_i)\, \Delta x$$

is a real number, regardless of the choice of $x_1, \ldots, x_n$. The definite integral of f from a to b is defined to be this limit of Riemann sums.

Definition of the Definite Integral

Let f be a function continuous on the interval $[a, b]$. Let n be a positive integer, $\Delta x = \dfrac{b - a}{n}$, and set

$$a_0 = a, \quad a_i = a + i\Delta x \quad \text{for } i = 1, \ldots, n.$$

For $i = 1, \ldots, n$, let x_i be any number in the interval $[a_{i-1}, a_i]$. Then $\int_a^b f(x)\, dx$, the **definite integral of f from a to b,** is defined by

$$\int_a^b f(x)\, dx = \lim_{\Delta x \to 0} \sum_{i=1}^{n} f(x_i)\, \Delta x.$$

EXAMPLE 5 We illustrate the definition by using it to find

$$\int_0^1 f(x)\, dx$$

when $f(x) = x$. We have

$$\Delta x = \frac{1-0}{n} = \frac{1}{n}$$

for any integer n. Hence

$$a_0 = 0, \quad a_1 = \frac{1}{n}, \quad a_2 = \frac{2}{n}, \quad \ldots, \quad a_{n-1} = \frac{n-1}{n}, \quad a_n = 1;$$

that is, $a_i = i/n$ for $i = 1, \ldots, n$. For each i, we may choose x_i to be any number in $[a_{i-1}, a_i]$; let us choose $x_i = a_i = i/n$. Then the Riemann sum is

$$\sum_{i=1}^{n} f(x_i)\, \Delta x = \sum_{i=1}^{n} f\left(\frac{i}{n}\right) \frac{1}{n} = \sum_{i=1}^{n} \frac{i}{n} \cdot \frac{1}{n} = \frac{1}{n^2} \sum_{i=1}^{n} i$$

$$= \frac{1}{n^2}(1 + 2 + \cdots + n).$$

Now, there is a formula for the sum of the first n positive integers:

$$1 + 2 + \cdots + n = \frac{n(n+1)}{2}.$$

Therefore

$$\sum_{i=1}^{n} f(x_i)\, \Delta x = \frac{1}{n^2} \frac{n(n+1)}{2} = \frac{1}{2} \frac{n+1}{n},$$

and thus

$$\int_0^1 f(x)\, dx = \int_0^1 x\, dx$$

$$= \underset{\Delta x \to 0}{\text{Lim}} \sum_{i=1}^{n} f(x_i)\, \Delta x$$

$$= \underset{n \to +\infty}{\text{Lim}} \frac{1}{2} \frac{n+1}{n}$$

$$= \frac{1}{2} \underset{n \to +\infty}{\text{Lim}} \frac{1 + 1/n}{1}$$

$$= \frac{1}{2} \frac{1+0}{1}$$

$$= \frac{1}{2}. \quad \square$$

Note that the definition of the definite integral as a limit of Riemann sums makes the connection between the integral and area obvious. For, if f is nonnegative on $[a, b]$, the Riemann sums for f over $[a, b]$ approximate the area under f from a to b; since this approximation becomes better and better as Δx approaches 0 (see Figure 6–24 again), it follows that the area under f from a to b is the limit of the Riemann sums as $\Delta x \to 0$, and hence by definition is equal to the definite integral of f from a to b.

From the definition of the definite integral as the limit of Riemann sums it is possible to prove that if f is continuous on $[a, b]$ and $F(x)$ is an antiderivative of $f(x)$ on $[a, b]$, then

$$\int_a^b f(x)\,dx = F(b) - F(a).$$

This fact is so important that it is known as the **Fundamental Theorem of Calculus:**

> **The Fundamental Theorem of Calculus**
>
> If f is continuous on $[a, b]$ and $F(x)$ is an antiderivative of $f(x)$ on $[a, b]$, then
>
> $$\int_a^b f(x)\,dx = F(b) - F(a).$$

We introduced the definite integral in Section 6.3 by using the Fundamental Theorem of Calculus, and have until now gotten along perfectly well without knowing that the integral is in fact a limit of Riemann sums; this being the case, why do we bother to define the integral this way at all? Why not just use the Fundamental Theorem as the definition of the definite integral? There are two reasons for not doing so. The first of these is that we cannot use the Fundamental Theorem to evaluate a definite integral if we cannot find an antiderivative for its integrand. This sometimes occurs, and when it does we must fall back on the Riemann sum definition and try to evaluate the integral in the manner of Example 5. We will have more to say about this in Section 7.4. The second, and perhaps more important, reason for introducing the Riemann sum definition of the definite integral is that it allows us to use the integral as an accumulator. We illustrate the use of the integral as an accumulator by finding the average value of a function over an interval.

The Average Value of a Function

It is often useful to be able to find the average value of a function over some interval. For instance, if f is a function that describes a retailer's inventory as a function of time, we might wish to find the retailer's average inventory for a month; if g is a function that describes the temperature during a day, we might want to find the average temperature for the day.

We know that we average a collection of numbers by summing them and dividing the sum by the number of them. Now suppose f is a continuous function defined on $[a, b]$. We will show how to find the average value of f on $[a, b]$. Let M denote this average value; then M ought to be the sum of all functional values $f(x)$, for all x in $[a, b]$, divided by the number of such x's. We cannot find M directly because there are infinitely many such x's, but we can approximate M as follows: let

$$\Delta x = \frac{b-a}{n}$$

INTEGRATION

and divide $[a, b]$ into n subintervals $[a_{i-1}, a_i]$, where $a_0 = a$ and

$$a_i = a + i\Delta x$$

for $i = 1, \ldots, n$. If we choose one number x_i from each subinterval $[a_{i-1}, a_i]$ and then average the functional values of the x_i's, we will obtain an approximation for M:

$$M \approx \frac{f(x_1) + f(x_2) + \cdots + f(x_n)}{n}.$$

As we let $n \to +\infty$, the preceding approximation becomes better and better. Let us rewrite the approximation as

$$M \approx \frac{1}{b-a}\left[f(x_1)\frac{b-a}{n} + f(x_2)\frac{b-a}{n} + \cdots + f(x_n)\frac{b-a}{n}\right],$$

or

$$(b-a)M \approx \sum_{i=1}^{n} f(x_i)\,\Delta x.$$

The sum on the right is a Riemann sum for f over $[a, b]$, and since the approximation becomes exact as $\Delta x \to 0$ (that is, as $n \to +\infty$), we have

$$(b-a)M = \lim_{\Delta x \to 0} \sum_{i=1}^{n} f(x_i)\,\Delta x = \int_a^b f(x)\,dx.$$

Therefore

$$M = \frac{1}{b-a}\int_a^b f(x)\,dx.$$

The Average Value of a Function

The average value of a continuous function f over the interval $[a, b]$ is

$$M = \frac{1}{b-a}\int_a^b f(x)\,dx.$$

Note that the average value M of f over $[a, b]$ has the property that

$$(b-a)M = \int_a^b f(x)\,dx;$$

in other words, M is the value of $f(x)$ over $[a, b]$ such that a rectangle with base $[a, b]$ and height M has the same area as the region under the graph of f from $x = a$ to $x = b$. See Figure 6–25.

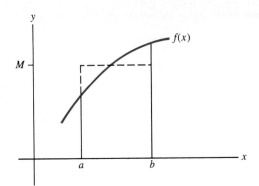

FIGURE 6–25

$$(b - a)M = \int_a^b f(x)\, dx$$

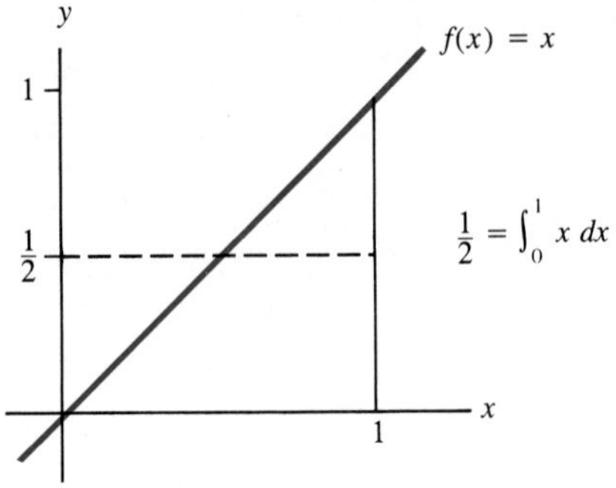

FIGURE 6–26

EXAMPLE 6 Let f be the function with defining equation $f(x) = x$. The average value of f over the interval $[0, 1]$ is

$$M = \frac{1}{1-0} \int_0^1 x\, dx = \frac{x^2}{2} \bigg|_0^1 = \frac{1}{2}.$$

Thus the rectangle of height $\frac{1}{2}$ over the interval $[0, 1]$ has the same area as the region under f from $x = 0$ to $x = 1$. See Figure 6–26. ☐

EXAMPLE 7 Over the course of each month (30 days) a retailer's inventory declines linearly from 300 units to 60 units. What is the retailer's average monthly inventory?

If we let y be the number of units in inventory on day t, $0 \le t \le 30$, then y is a linear function of t such that $y = 300$ when $t = 0$ and $y = 60$ when $t = 30$. Hence

$$y = \frac{60 - 300}{30 - 0} t + 300$$

or

$$y = -8t + 300,$$

$0 \le t \le 30$. Therefore the average number of units in inventory each month is

$$\frac{1}{30 - 0} \int_0^{30} (-8t + 300)\, dt = \frac{1}{30}(-4t^2 + 300t) \bigg|_0^{30} = \frac{1}{30}(5400) = 180. \quad \square$$

Notice the technique we used to develop the formula for the average value of a function: we approximated the quantity $(b - a)M$ by a Riemann sum and then showed that the quantity was equal to the limit of the Riemann sum as Δx approached 0, and hence that it was equal to a definite integral. This technique is what is meant by using the definite integral as an accumulator. It is a very general procedure that

INTEGRATION

can be applied to many different situations; we shall consider some of them in Section 7.1

■ *Computer Manual: Program DEFNTGRL*

6.5 EXERCISES

Summation Notation

In Exercises 1 through 4, evaluate each sum given that $x_1 = 8$, $x_2 = 4$, $x_3 = -3$, $x_4 = 10$, $x_5 = 0$, and $x_6 = 9$.

1. $\sum_{i=1}^{6} x_i$
2. $\sum_{i=1}^{3} x_i$
3. $\sum_{i=1}^{5} 2x_i$
4. $\sum_{i=1}^{4} 3x_i$

In Exercises 5 through 12, evaluate the given sum.

5. $\sum_{i=1}^{5} i$
6. $\sum_{i=2}^{5} 3i$
7. $\sum_{j=0}^{6} 4(j + 3)$
8. $\sum_{k=2}^{6} k(k + 2)$
9. $\sum_{i=1}^{3} i(2i + 1)$
10. $\sum_{i=1}^{5} \frac{i}{i + 1}$
11. $\sum_{i=1}^{3} 4$
12. $\sum_{i=1}^{100} \frac{1}{2}$

The Definition of the Definite Integral

In Exercises 13 through 24, use the method of Example 5 of this section to find the definite integral of the given function over the given interval.

13. $f(x) = x$ over $[1, 2]$
14. $f(x) = 2x$ over $[0, 2]$
15. $f(x) = x + 1$ over $[1, 3]$
16. $f(x) = 3 - x$ over $[0, 3]$
17. $f(x) = 2x - 7$ over $[4, 8]$
18. $f(x) = |x|$ over $[-1, 1]$
19. $f(x) = x^2$ over $[0, 1]$ (*Hint:* Use the formula

$$1^2 + 2^2 + \cdots + n^2 = \frac{n(n + 1)(2n + 1)}{6}.)$$

20. $f(x) = x^2$ over $[1, 3]$
21. $f(x) = x^2 + 1$ over $[1, 2]$
22. $f(x) = 2 - x^2$ over $[0, 2]$
23. $f(x) = (x - 1)^2$ over $[0, 2]$
24. $f(x) = x^2 - x$ over $[0, 1]$

The Average Value of a Function

In Exercises 25 through 36, find the average value of the given function over the given interval.

25. $f(x) = 2x + 3$, $[1, 5]$
26. $f(x) = x^2$, $[-2, 2]$
27. $f(x) = \sqrt{x}$, $[1, 4]$
28. $f(x) = x^3$, $[0, 1]$
29. $f(x) = x^3 - 3x + 2$, $[1, 2]$
30. $f(x) = e^{-x}$, $[0, 2]$

31. $f(x) = 2e^x$, $[0, 2]$
32. $f(x) = x^{-2}$, $[3, 10]$
33. $f(x) = \dfrac{1}{x}$, $[e, e^2]$

*34. $y = mx + b$, $[0, c]$

*35. $y = x^n$, n an odd positive integer, $[-a, a]$, $a > 0$.

*36. $y = x^n$, n an even positive integer, $[-a, a]$, $a > 0$.

37. The velocity of a freely falling body t seconds after the start of its fall is $v = 32t$ feet per second. Find the average velocity of a freely falling body over the first 4 seconds of its fall.

38. If a rocket is fired straight up, its altitude t minutes after launch is
$$s = -t^2 + 40t$$
kilometers. Find the rocket's average altitude during the second 5 minutes of its flight.

Management

39. A firm's monthly sales, in dollars, in month t are given by
$$S = 120,000t + 52,000,$$
where $t = 0$ represents the beginning of the year. Find the firm's average monthly sales over the first six months of the year.

40. Dour Company's annual profits are given by the equation
$$P = 10e^{0.06t},$$
where P is in millions of dollars and t is in years, with $t = 0$ representing the present. Find Dour's average annual profit over the past five years and forecast their average annual profit over the next five years.

41. A new plant produces
$$y = 0.1t^2 + 0.5t + 400$$
units per month, where t is in months since the plant began operating. Find the plant's average monthly production during its first year of operation.

42. A retailer's inventory declines linearly from 250 units at the beginning of the month to 20 units at the end of the month. Find the retailer's average inventory for the month (30 days).

43. A retailer's inventory declines linearly from 600 units to 100 units over a 90-day period. Find the retailer's average inventory over the period.

44. A retailer's inventory declines linearly from 1000 units to 250 units over a 30-day period, then declines linearly from 250 units to 25 units over the next 60 days. Find the retailer's average inventory over the 90-day period.

45. A retailer's inventory t days after the arrival of a shipment of goods is $y = 800 - 0.5t^2$ units. Find the retailer's average inventory, assuming that a new shipment of goods arrives just as inventory reaches zero.

46. A worker can assemble compact disk players at the rate of $8\sqrt{t}$ per hour, where $0 \le t \le 8$. Find the worker's average hourly assembly rate over the 8-hour workday.

47. A manufacturer's inventory of a finished product over a 90-day period is y units, where

$$y = 100 - \frac{1}{90}t^2, \quad 0 \le t \le 90.$$

Find the manufacturer's average inventory over the period.

48. The supply function for a commodity is given by
$$q = 900\sqrt{p - 1}, \quad p \ge 1.$$
Find the average quantity supplied over the price range $1 to $10.

49. The supply function for a commodity is given by
$$p = \frac{q^2}{900^2} + 1.$$
Find the average price of the commodity if the quantity supplied ranges from 0 to 300 units.

50. The demand function for a product is given by
$$q = 10{,}000e^{-0.04p}.$$
Find the average quantity demanded as the price ranges from $5 to $10 per unit.

51. Suppose that it is a publisher's policy to keep a textbook in inventory until only 100 copies of it are left and then declare it out of print. If t years after a certain book is published there are $y = 10{,}000e^{-0.8t}$ copies in the publisher's inventory and if the retail price of each copy is $39.95, find the average retail value of the publisher's inventory of the book.

52. With t weeks of experience, a worker can produce
$$y = 200 - 100e^{-0.3t}$$
units of product per week. Find the worker's average weekly productivity over his or her
   (a) first month on the job;
   (b) second month on the job;
   (c) fifth month on the job.

Life Science

53. The amount of a drug present in a patient's liver t minutes after it has been injected into the patient's bloodstream is $y = 0.35t$ grams. Find the average amount of the drug present in the patient's liver during the first 10 minutes after injection.

54. During an experiment a colony of bacteria is treated with a bactericide and t minutes later there are $y = 100{,}000 - 22{,}000t$ bacteria present. Find the average number of bacteria present during the first 3 minutes after the application of the bactericide.

55. It is estimated that as of year t the harp seal population will be $y = 4500/(2t + 1)$ individuals. Here $t = 0$ represents 1988. Estimate the average harp seal population over the period 1988–2000.

SUMMARY

This summary consists of terms and symbols whose meaning you should know, facts you should know, and techniques and methods you should be able to employ.

Section 6.1

Antiderivative of a function f: a function F such that $F'(x) = f(x)$ for all x in the domain of f. If $F(x)$ is an antiderivative of $f(x)$, so is $F(x) + c$ for every constant c. If $F(x)$ is an antiderivative of $f(x)$, every antiderivative of $f(x)$ is of the form $F(x) + c$ for some constant c. Finding antiderivatives. The indefinite integral of $f(x)$:

$$\int f(x)\,dx = F(x) + c,$$

where $F(x)$ is any antiderivative of $f(x)$. Integral sign, integrand, constant of integration. The rules of integration. Using the rules of integration to find indefinite integrals. Determining the value of the constant of integration.

Section 6.2

Integration by guessing. Integration by substitution. Using substitution to find indefinite integrals. When a substitution will simplify an integral. Integration by substitution and the chain rule.

Section 6.3

The definite integral as the total change in any antiderivative:

$$\int_a^b f(x)\,dx = F(b) - F(a),$$

$F(x)$ any antiderivative of $f(x)$ on $[a, b]$. Limits of integration. Notation:

$$\int_a^b f(x)\,dx = F(x)\Big|_a^b = F(b) - F(a)$$

Evaluating definite integrals. Using the definite integral to find total change. Integration by substitution for definite integrals. Evaluating definite integrals by substitution.

Section 6.4

The area under a continuous nonnegative function f from a to b. The area under a continuous nonnegative function f from a to b is the definite integral of f from $x = a$ to $x = b$. The area function A for a continuous nonnegative function f. The area function A is an antiderivative for f. Finding the area under a curve from a to b using the definite integral. Total change as area: if f describes the rate of change of Q over $[a, b]$, the total change in Q over $[a, b]$ is represented by the area under f from a to b. The area bounded by f and g from $x = a$ to $x = b$. If f is the upper curve and g is the lower on $[a, b]$ then the area bounded by f and g from $x = a$ to $x = b$ is the definite integral of $f(x) - g(x)$ from $x = a$ to $x = b$. Using the definite integral to find the area bounded by two curves from $x = a$ to

INTEGRATION

$x = b$. The area bounded by f and g. Finding the area bounded by two curves by finding their points of intersection and then integrating.

Section 6.5

Summation notation Σ. Riemann sum for f over $[a, b]$. If f is continuous and nonnegative on $[a, b]$, as $\Delta x \to 0$ the Riemann sums approach the area under f from a to b. The definition of the definite integral:

$$\int_a^b f(x)\, dx = \lim_{\Delta x \to 0} \sum_{i=1}^{n} f(x_i)\, \Delta x.$$

Finding definite integrals using Riemann sums. The Fundamental Theorem of Calculus: if f is continuous on $[a, b]$ and $F(x)$ is an antiderivative of $f(x)$ on $[a, b]$, then

$$\int_a^b f(x)\, dx = F(b) - F(a).$$

Why is it necessary to define the definite integral as a limit of Riemann sums? The average value of a function over an interval. Finding the average value of a continuous function over an interval.

REVIEW EXERCISES

In Exercises 1 through 20, find the indefinite integral.

1. $\int (-x^5 + x^3 + 9x^2 - 1)\, dx$

2. $\int (-4x^5 + 3x^2 - 8x + 14)\, dx$

3. $\int (1 + x + x^2/2 + x^3/6 + x^4/24 + x^5/120)\, dx$

4. $\int (u^{-1/2} + u^{-1/3} + u^{-1/4} + u^{-1/5})\, du$

5. $\int (-2x^{-2} + 3x^{4/5} + 4)\, dx$

6. $\int \left(-2t^{-3} + \dfrac{2}{t^2} + 4\sqrt{t} + 1\right) dt$

7. $\int 2e^x\, dx$

8. $\int -\dfrac{3}{x}\, dx$

9. $\int \left(\dfrac{2}{x} - 3e^x\right) dx$

10. $\int (5x^{3/5} + 6x^5 - 2x^{-1} + 3x^{-4/3} + x^{1/2} - e^{2x})\, dx$

11. $\int \sqrt{4x + 1}\, dx$

12. $\int 3x(x^2 + 4)^3\, dx$

13. $\int (x + 1)(x^2 + 2x - 8)^5\, dx$

14. $\int 5(x^2 + 2)(x^3 + 6x - 2)^{4/5}\, dx$

15. $\int \dfrac{2x}{4x^2 + 1}\, dx$

16. $\int \dfrac{6x + 2}{(3x^2 + 2x + 1)^{4/3}}\, dx$

17. $\int \dfrac{x - 2}{(x^2 - 4x + 1)^{2/3}}\, dx$

18. $\int 3x^2 e^{x^3}\, dx$

19. $\int 5xe^{-x^2}\, dx$

20. $\int \dfrac{dx}{x\sqrt{\ln x}}$

21. A firm's marginal cost and marginal revenue functions are defined by
$$c(x) = 6x + 2 \quad \text{and} \quad r(x) = 20x,$$
respectively. Find the firm's cost and revenue functions if its fixed cost is $5000.

22. During an epidemic the number of cases of influenza increases at a rate of $y = 300\sqrt{t}$ cases per day t. Find a function that describes the number of cases as a function of time if there were 2000 cases on day 4 of the epidemic.

23. An investment is made and t years later it is growing at a rate of $y = 120e^{0.12t}$ dollars per year. Find an expression that describes the worth of the investment as a function of time if the initial amount of the investment was $1000.

24. The gross national product of a country is changing at a rate of
$$y = \frac{25}{t + 1}$$
billion dollars per year t, where $t = 0$ represents 1985. Find an expression for the gross national product of the country as a function of time if the GNP in 1985 was $1000 billion.

In Exercises 25 through 36, evaluate the given definite integral.

25. $\displaystyle\int_1^3 (2 - 8x)\, dx$

26. $\displaystyle\int_{-4}^{-2} (3x^2 + 5x)\, dx$

27. $\displaystyle\int_0^2 (x^2 - 3x + 1)\, dx$

28. $\displaystyle\int_{-2}^{-1} (4x^3 - x^2 + 1)\, dx$

29. $\displaystyle\int_0^{\ln 2} e^{3x}\, dx$

30. $\displaystyle\int_0^4 \sqrt{2x + 3}\, dx$

31. $\displaystyle\int_4^9 (\sqrt{x} + x^{-2})\, dx$

32. $\displaystyle\int_{4/5}^{(e+3)/5} \frac{10}{5x - 3}\, dx$

33. $\displaystyle\int_0^1 xe^{-x^2/2}\, dx$

34. $\displaystyle\int_0^2 \frac{x^2}{x^3 + 1}\, dx$

35. $\displaystyle\int_e^{e^2} \frac{1}{x \ln x}\, dx$

36. $\displaystyle\int_e^{e^3} \frac{1}{x(\ln x)^2}\, dx$

37. The population of a town is growing at the rate of
$$f(t) = 15\sqrt{t} + 8$$
persons per year, where t is in years with $t = 0$ representing the present. What will be the change in the town's population size over the next four years?

38. A firm's marginal cost function is given by
$$c(x) = 3x + 10 + \frac{500}{x},$$
valid for $x \geq 1$. If the firm is currently producing 10 units, find the additional cost of producing 2 more units.

39. Let $f(x) = 4x^{-3}$. Find the area under f from 1 to 4.

40. Let $f(x) = \dfrac{x}{x^2 + 1}$. Find the area under f from 0 to 2.

41. Let $f(x) = -x^2 + 13x - 36$. Find the area of the region bounded by the x-axis and the graph of f.

42. Find the area of the region bounded by the graph of
$$y = (x + 2)(x - 1)(x - 3)$$
and the x-axis.

INTEGRATION

43. Find the area bounded by the x-axis and the graph of $y = x^3 - 2x^2 - 15x$.

44. Find the area of the region bounded by p and q, where
$$p(x) = -x^2 + 6x + 2 \quad \text{and} \quad q(x) = x + 2.$$

45. Find the area of the region bounded by the graphs of $y = e^x$, $y = e^{-x}$, $x = -1$, and $x = 1$.

46. It is estimated that the number of junk bonds on the market is increasing at the rate of
$$f(t) = \frac{50}{t + 2}$$
million bonds per year. Here $t = 0$ represents 1984.
(a) Find the area under f from 0 to b. What does this area represent?
(b) When will there be a total of 100 million junk bonds on the market?

47. Use Riemann sums to show that $\int_0^1 (4 - 3x)\, dx = 2.5$.

48. Use Riemann sums to show that $\int_1^2 x^3\, dx = 3.75$.
$$\left(\text{Hint: } 1^3 + 2^3 + \cdots + n^3 = \frac{n^2(n + 1)^2}{4}.\right)$$

49. Find the average value of the function f defined by $f(x) = xe^{-x^2}$ over the interval $[0, 2]$.

50. A company's annual return on investment is given by
$$f(t) = 100e^{-0.05t},$$
where $f(t)$ is in thousands of dollars and $t = 0$ represents 1985. Forecast the firm's average annual return on investment over the period
(a) 1990–1995 (b) 1999–2005

51. A merchant's inventory over a 60-day period is given by
$$y = 1000e^{-0.02t}, \qquad 0 \le t \le 60.$$
Find the average inventory over the period.

52. There were y thousand riders each day on a city's transportation system, where
$$y = -0.1t^2 + 2.4t + 45.6$$
and $0 \le t \le 30$. Find the average number of riders over the 30-day period.

ADDITIONAL TOPICS

Here are some suggestions for topics not covered in the text that you might want to investigate on your own.

1. Let f, g be continuous on $[a, b]$. Show why the following properties of the definite integral are true:

$$\int_a^b f(x)\, dx \geq 0 \qquad \text{if } f(x) \geq 0 \text{ on } [a, b]$$

$$\int_a^b f(x)\, dx \geq \int_a^b g(x)\, dx \qquad \text{if } f(x) \geq g(x) \text{ on } [a, b]$$

$$\int_a^a f(x)\, dx = 0$$

$$\int_b^a f(x)\, dx = -\int_a^b f(x)\, dx$$

$$\int_a^b f(x)\, dx = \int_a^c f(x)\, dx + \int_c^b f(x)\, dx \qquad \text{if } a \leq c \leq b$$

2. If f is continuous on $[a, b]$, let m denote the minimum value of $f(x)$ on $[a, b]$ and M the maximum value of $f(x)$ on $[a, b]$. Show that

$$m(b - a) \leq \int_a^b f(x)\, dx \leq M(b - a).$$

3. The Mean Value Theorem for integrals says that if f is continuous on $[a, b]$, there is some number c in $[a, b]$ such that

$$f(c)(b - a) = \int_a^b f(x)\, dx.$$

Find out how to prove this theorem. What is its geometric significance? What connection does it have with the average value of a function over an interval?

4. In higher mathematics functions are often defined by means of definite integrals. For instance, it is possible to define the natural logarithm function by setting

$$\ln x = \int_1^x \frac{1}{t}\, dt, \qquad x > 0.$$

Using only this definition and your knowledge of the definite integral, show that

$$\ln 1 = 0, \qquad \ln x_1 > \ln x_2 \text{ if } x_1 > x_2, \qquad \frac{d}{dx}(\ln x) = \frac{1}{x},$$

$$\ln x \text{ is continuous on } (0, +\infty), \qquad \lim_{x \to +\infty} \ln x = +\infty,$$

and

$$\lim_{x \to 0^+} \ln x = -\infty.$$

5. Riemann sums were named for the German mathematician Bernhard Riemann (1826–1866). Indeed, the definite integral we have studied in this chapter is formally known as the Riemann integral. (There are other, more general, integrals.) Besides his work with integrals, Riemann made many other important contributions to mathematics; to cite just one of these, he discovered a non-Euclidean geometry that now bears his name. Find out about Riemann's life and his contributions to mathematics.

6. The development of the definite integral was a very long process. It may properly be said to have begun in ancient Greece when the mathematicians of the time began to consider the problem of finding areas of geometric figures. Find out about the development of the definite integral. (A good place to start would be with the supplement to this chapter.)

SUPPLEMENT: THE DEVELOPMENT OF THE DEFINITE INTEGRAL

We began our study of the definite integral by introducing it as the total change in an antiderivative over an interval. We next showed how the definite integral could be used to find areas, and concluded by defining it as a limit of Riemann sums. The historical development of the definite integral followed these same steps, but in a different order: first came the problem of finding areas, then the discovery that areas could be computed as limits of what we would today call Riemann sums, and finally the realization that such limits could be evaluated by making use of the antiderivative. In this supplement we give a brief account of the historical development of the definite integral.

If integration is considered to be a type of process in which areas are obtained as the limit of sums of simpler areas (the general approach taken in Section 6.5), its beginnings go back to antiquity. The Greek mathematician Archimedes (287–212 B.C.) found the areas of many geometric figures by using a technique in which the figure under question was "filled up" with simpler figures whose areas could be easily determined. This method of integration was known as the **method of exhaustion.** We will demonstrate how the method of exhaustion may be employed to find the area P of a parabolic region such as that pictured in Figure 6–27.

We let D be the midpoint of the segment AB, and let AB serve as the base of a triangle ACB ($\triangle ACB$), where C is a point on the parabola such that CD is parallel to the axis of symmetry of the parabola. See Figure 6–28, where we have chosen C so that CD lies on the axis of symmetry. As a first approximation, we have

$$P \approx \text{area of } \triangle ACB.$$

Now we repeat the construction with E and F the midpoints of AC and CB, respectively, and G and H the points on the parabola such that GE and HF are parallel to the axis of symmetry of the parabola. See Figure 6–29. We thus obtain the triangles AGC and CHB, and

$$P \approx \text{area of } \triangle ACB + \text{area of } \triangle AGC + \text{area of } \triangle CHB.$$

Clearly, this approximation is better than the previous one. It is possible to show geometrically that

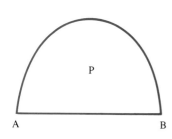 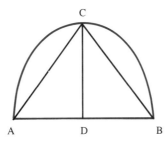 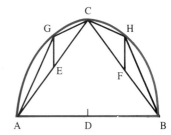

$$\text{area of } \triangle AGC + \text{area of } \triangle CHB = \frac{1}{4}(\text{area of } \triangle ACB).$$

Therefore $P = \frac{5}{4}(\text{area of } \triangle ACB)$.

Now we repeat the construction with segments AG, GC, CH, and HB as the bases of four new triangles, to obtain a better approximation:

$$P \approx \frac{5}{4}(\text{area of } \triangle ACB) + \frac{1}{4^2}(\text{area of } \triangle ACB)$$

$$= \frac{21}{16}(\text{area of } \triangle ACB).$$

As the process continues, the triangles "fill up" or "exhaust" the parabolic region, and hence the sum of their areas approaches the value of P. It can be shown that after n repetitions of the construction, the sum of the areas of the triangles is

$$\frac{4}{3}\left(1 - \frac{1}{4^n}\right)(\text{area of } \triangle ACB).$$

But since $\frac{1}{4^n}$ can be made arbitrarily small by taking n large enough, it follows that

$$P = \frac{4}{3}(\text{area of } \triangle ACB).$$

In modern terminology we would say that

$$P = \lim_{n \to +\infty} (\text{sum of the areas of the triangles})$$

$$= \lim_{n \to +\infty} \frac{4}{3}\left(1 - \frac{1}{4^n}\right)(\text{area of } \triangle ACB)$$

$$= \frac{4}{3}(1 - 0)(\text{area of } \triangle ACB)$$

$$= \frac{4}{3}(\text{area of } \triangle ACB).$$

We can use this result to determine the area of a parabolic segment by simply finding the length of the segments AB and CD of Figure 6–28. For instance, if AB has length 1 and CD has length $\frac{1}{4}$, then

$$\text{area of } \triangle ACB = \frac{1}{2}(1)\left(\frac{1}{4}\right) = \frac{1}{8},$$

and

$$P = \frac{4}{3} \cdot \frac{1}{8} = \frac{1}{6}.$$

The method of exhaustion depends heavily on the geometric properties of the region under consideration and is difficult to apply in general situations. A better

INTEGRATION

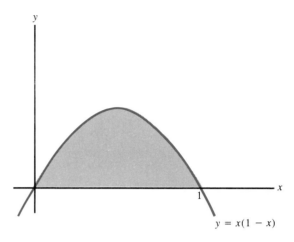

FIGURE 6–30

method for finding areas as the limits of sums of simpler areas became possible after the French mathematician René Descartes (1596–1650) introduced the cartesian coordinate system: This allowed mathematicians to regard curves as the graphs of equations and hence to consider regions bounded by curves as being defined by equations. This in turn made it possible to express the areas of such regions as limits of sums of areas of rectangles, much as we did in Section 6.5. Thus, for instance, suppose it is desired to find the area of the parabolic region bounded by the x-axis and the graph of $f(x) = x(1 - x)$. See Figure 6–30. We could proceed by subdividing the interval $[0, 1]$ into subintervals $[i/n, (i + 1)/n]$, $i = 0, 1, \ldots, n - 1$, and constructing on each subinterval a rectangle of height

$$f\left(\frac{i}{n}\right) = \frac{i}{n}\left(1 - \frac{i}{n}\right).$$

See Figure 6–31. Then the area P of the parabolic segment is approximated by the sum of the area of the rectangles:

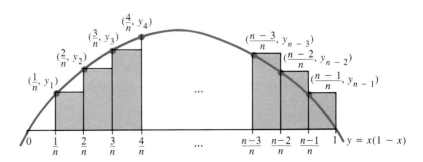

FIGURE 6–31

$$P \approx \sum_{i=0}^{n-1} \frac{i}{n}\left(1 - \frac{i}{n}\right)\frac{1}{n}.$$

But the sum on the right-hand side simplifies to

$$\frac{1}{6}\left(1 - \frac{1}{n^2}\right).$$

Since $1/n^2$ can be made arbitrarily small by taking n large enough, it follows that $P = \frac{1}{6}$. In other words, since

$$P = \lim_{n \to +\infty} (\text{sum of the areas of the rectangles}),$$

we have

$$P = \lim_{n \to +\infty} \frac{1}{6}\left(1 - \frac{1}{n^2}\right) = \frac{1}{6}(1 - 0) = \frac{1}{6}.$$

This technique of expressing the area of a region as a limit of sums of areas of rectangles is a very general method for finding areas, and it was known to mathematicians before the time of Newton and Leibniz. Thus if f is a function that is nonnegative over the interval $[a, b]$ and we *define* the definite integral

$$\int_a^b f(x)\, dx$$

to be the area under f from $x = a$ to $x = b$, then we can say that mathematicians possessed a technique for evaluating definite integrals (i.e., finding areas) before Newton and Leibniz. What was not appreciated was the fact that definite integrals could be evaluated by antidifferentiation. The realization that this is so is usually credited to the English mathematician Isaac Barrow (1630–1677), who was Newton's teacher; however, it was not until Newton and Leibniz put differentiation on a firm basis that this fact could be easily exploited.

Once the notion of the derivative of a function was formalized, mathematicians could easily show that definite integrals could be evaluated by finding antiderivatives for their integrands. The proof was as follows. By definition,

$$\int_a^x f(x)\, dx = \text{area under } f \text{ from } a \text{ to } x.$$

Set

$$A(x) = \int_a^x f(x)\, dx.$$

Then the Fundamental Theorem of Calculus can be proved by showing (just as we did in Section 6.4) that $A'(x) = f(x)$ and hence that if $F(x)$ is an antiderivative for $f(x)$,

$$\int_a^b f(x)\, dx = F(b) - F(a).$$

Since $A'(x) = f(x)$, the Fundamental Theorem of Calculus is often written as

$$\frac{d}{dx} \int_a^x f(x) \, dx = f(x).$$

In this form the Fundamental Theorem of Calculus deserves its name, for it makes explicit the relationship between the two great concepts of calculus—the derivative and the integral—by showing that they are inverse operations in the same sense that addition and subtraction are inverse operations: each can be used to undo the other. Thus the Fundamental Theorem is the cord that ties the two subjects of differential calculus and integral calculus together into one coherent theory.

Suggestion

If you would like to read more about the development of the integral, here are some books to consult:

Bell, E. T., *Men of Mathematics,* Simon and Schuster, New York, 1937.

Boyer, Carl B., *History of the Calculus and Its Conceptual Development,* Dover, New York, 1959.

Boyer, Carl B., *A History of Mathematics,* Wiley, New York, 1968.

Eves, Howard, *An Introduction to the History of Mathematics,* Holt, Rinehart, and Winston, New York, 1961.

Eves, Howard, *Great Moments in Mathematics before 1650,* Dolciani Mathematical Expositions, No. 5, The Mathematical Association of America, 1980.

Eves, Howard, *Great Moments in Mathematics after 1650,* Dolciani Mathematical Expositions, No. 7, The Mathematical Association of America, 1983.

7

More about Integration

In this chapter we consider some important applications and techniques of integration. We answer such questions as

- How much do consumers and producers save when the price of a product is at equilibrium?
- If income flows into a firm at a certain rate and is reinvested at another rate, how much will it be worth over a given period of time?
- How can we integrate functions defined by equations such as

$$f(x) = x^2 e^{-x} \quad \text{and} \quad g(x) = \sqrt{x^2 + 9}?$$

- What is the meaning of an integral such as $\int_{1}^{+\infty} \frac{1}{x^2}\,dx$?
- How do we evaluate

$$\int_{a}^{b} f(x)\,dx$$

when we cannot find an antiderivative for $f(x)$?

The chapter begins with a demonstration of how we can use the integral as accumulator to find producers' surplus, consumers' surplus, and the future value of an income stream. We next consider the integration technique known as integration by parts and then discuss how to use a table of integrals. This is followed by a section in which we expand the notion of the definite integral to encompass so-called improper integrals: these are integrals for which the interval of integration is infinite. The final section is devoted to the topic of numerical integration, which considers methods for approximating the value of definite integrals.

7.1 CONSUMERS' SURPLUS, PRODUCERS' SURPLUS, AND STREAMS OF INCOME

Recall that in Section 6.5 we defined the definite integral of a continuous function f over an interval $[a, b]$ to be the limit of its Riemann sums over the interval: if

$$\Delta x = \frac{b-a}{n}, \quad a_0 = a, \quad \text{and} \quad a_i = a + i\Delta x \quad \text{for} \quad i = 1, \ldots, n,$$

then any sum of the form $\sum_{i=1}^{n} f(x_i) \Delta x$, where x_i is in $[a_{i-1}, a_i]$, is a **Riemann sum** for f over $[a, b]$, and

$$\int_a^b f(x)\, dx = \lim_{\Delta x \to 0} \sum_{i=1}^{n} f(x_i)\, \Delta x.$$

We also remarked that as a consequence of this, any quantity that is the limit of Riemann sums as Δx approaches 0 is a definite integral:

The Integral as an Accumulator

If a quantity Q can be approximated by a Riemann sum for some function f over some interval $[a, b]$, and if the approximation becomes exact as $\Delta x \to 0$, then

$$Q = \int_a^b f(x)\, dx.$$

In this section we use the integral as an accumulator to calculate producers' surplus, consumers' surplus, and the future value of a stream of income. We also briefly discuss a general procedure for setting up the integral as an accumulator.

Producers' Surplus and Consumers' Surplus

Let the equation $p = s(q)$ define a supply function for some commodity: here q is the number of units of the commodity supplied to the market, and p is the resulting unit price. Similarly, let $p = d(q)$ define the demand function for the commodity. See Figure 7–1. Note that the market equilibrium quantity for the commodity is q_E, and its market equilibrium price is p_E.

For the moment, let us consider only the supply curve. Some producers would be willing to supply the commodity at prices lower than the equilibrium price p_E;

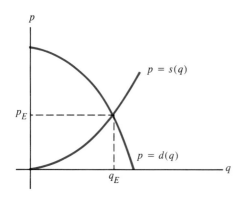

FIGURE 7–1

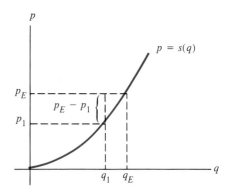

FIGURE 7–2

these producers gain when the price is at equilibrium. For instance, Figure 7–2 indicates that some producers are willing to supply quantity q_1 of the commodity at price $p_1 < p_E$; when the price is at equilibrium, these producers gain an amount equal to $(p_E - p_1)q_1$. The total amount that producers as a group gain in this manner when the price is at equilibrium is called **producers' surplus.** We will use the integral as an accumulator to find producers' surplus.

Let us divide the interval $[0, q_E]$ on the q-axis into n subintervals, each of length Δq. For each i we select an arbitrary number q_i in the ith subinterval (see Figure 7–3); on this subinterval Δq units will be supplied at the approximate price of $s(q_i)$. Thus the total gained by producers over the ith subinterval (the area of the shaded region in Figure 7–3) is approximately $[p_E - s(q_i)]\Delta q$ (the area of the rectangle bounded by the solid lines in Figure 7–3). Hence

$$\text{Producers' surplus} \approx [p_E - s(q_1)]\Delta q + \cdots + [p_E - s(q_n)]\Delta q$$

on $[0, q_E]$. But the sum on the right-hand side above is a Riemann sum for $p_E - s(q)$ over $[0, q_E]$, and since the approximation becomes exact as $\Delta x \to 0$

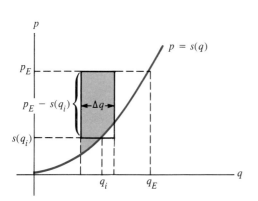

FIGURE 7–3

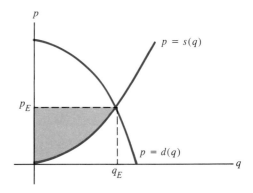

FIGURE 7–4
Producers' Surplus

(that is, since producers' surplus is the limit of such Riemann sums as $\Delta x \to 0$), it follows that producers' surplus is the definite integral of $p_E - s(q)$ from 0 to q_E.

> **Producers' Surplus**
>
> If $p = s(q)$ defines a supply function for a commodity and (q_E, p_E) is the commodity's market equilibrium point, then producers' surplus for the commodity is
>
> $$\int_0^{q_E} [p_E - s(q)] \, dq.$$

Producers' surplus is thus the area of the region shaded in Figure 7–4.

EXAMPLE 1 Suppose the supply function for a commodity is defined by

$$s(q) = 0.009q^2$$

and its demand function by

$$d(q) = 30 - 0.003q^2.$$

Solving the equation $s(q) = d(q)$ for q yields

$$0.009q^2 = 30 - 0.003q^2,$$

or

$$q^2 = \frac{30}{0.012} = 2500.$$

Therefore $q_E = 50$ and $p_E = s(50) = 22.50$. See Figure 7–5. Thus we have

$$\text{Producers' surplus} = \int_0^{q_E} [p_E - s(q)] \, dq$$

$$= \int_0^{50} [22.50 - 0.009q^2] \, dq$$

$$= \$750. \quad \square$$

Analogous to producers' surplus is **consumers' surplus.** Figure 7–6 depicts the situation. Here the equation $p = d(q)$ defines the demand function. Consumers who would be willing to purchase the commodity at a higher price than p_E save when the price is at p_E. For example, if some consumers would have purchased q_1 units at price $p_1 > p_E$, as in the figure, then they save an amount equal to $(p_1 - p_E)q_1$ when the price is at equilibrium. **Consumers' surplus** is the total amount saved by consumers as a group when the price is at equilibrium. Just as we did for producers' surplus, let us divide the interval $[0, q_E]$ into n subintervals, each of length Δq. On the ith subinterval we choose a number q_i and note that the

More about Integration   431

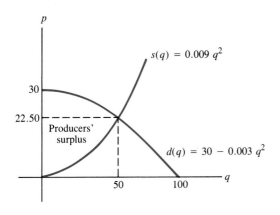

FIGURE 7–5   **FIGURE 7–6**

total gained by consumers over the ith subinterval (the shaded region in Figure 7–7) is approximated by $[d(q_i) - p_E] \Delta q$ (the area of the rectangle bounded by the solid lines in Figure 7–7). Thus

$$\text{Consumers' surplus} \approx \sum_{i=1}^{n} [d(q_i) - p_E] \Delta q,$$

and since this approximation by Riemann sums becomes exact as $\Delta x \to 0$, it follows that consumers' surplus is the definite integral of $d(q) - p_E$ over $[0, q_E]$.

Consumers' Surplus

If $p = d(q)$ defines a demand function for a commodity and (q_E, p_E) is the commodity's market equilibrium point, then consumers' surplus for the commodity is

$$\int_0^{q_E} [d(q) - p_E] \, dq.$$

Thus consumers' surplus is the area of the region shaded in Figure 7–8.

EXAMPLE 2   Let $s(q) = 0.009q^2$ and $d(q) = 30 - 0.003q^2$, as in Example 1. Then

$$\text{Consumers' surplus} = \int_0^{q_E} [d(q) - p_E] \, dq$$

$$= \int_0^{50} [30 - 0.003q^2 - 22.50] \, dq$$

$$= \$250.$$

See Figure 7–9 on page 433. ☐

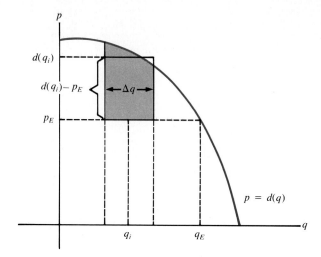

FIGURE 7–7

Income Streams

Any firm continually receives income from its business activities. This is referred to as the firm's **stream of income.** Most of the income that a firm receives is reinvested, either to keep the business going or in other ways. For purposes of analysis it is often convenient to assume that income flows into the firm continuously over the course of a year and that it is immediately reinvested at a certain interest rate that is compounded continuously. Thus we assume that at time t income is flowing into the firm at the rate of $P(t)$ dollars per year, where P is some continuous function of t, and that this income is immediately reinvested at $100r\%$ per year

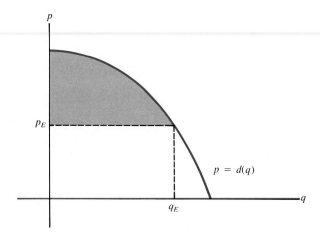

FIGURE 7–8
Consumers' Surplus

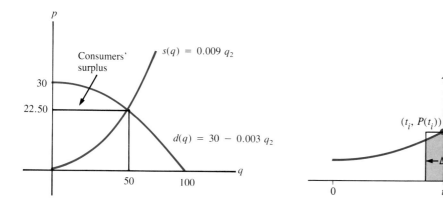

FIGURE 7–9 **FIGURE 7–10**

compounded continuously. We now ask the following question: if the firm continues to receive and reinvest income in this manner, what will be the value of its income stream over the next b years? To answer this question, we will again use the integral as an accumulator.

Suppose we divide the time interval $0 \leq t \leq b$ into n subintervals, each of length Δt, and select a number t_i in each subinterval. See Figure 7–10. Note that the rate at which income is flowing into the firm at time $t = t_i$ is $P(t_i)$, and that this money is reinvested at $100r$ percent per year compounded continuously for $b - t_i$ years. Hence by the formula for continuously compounded interest of Section 5.1, the money received by the firm at time $t = t_i$ will grow to

$$P(t_i)e^{r(b-t_i)}$$

dollars at the end of the b-year period. Since the total amount of income flowing into the firm over the ith subinterval (the shaded region in Figure 7–10) is approximately

$$P(t_i)\,\Delta t$$

(the area of the rectangle over the ith subinterval), and since this amount will accumulate at interest for approximately $b - t_i$ years, the future value at the end of the b-year period of the income received over the ith subinterval is approximately

$$[P(t_i)\,\Delta t]e^{r(b-t_i)}.$$

This approximation will become exact as $\Delta t \to 0$. Thus the total future value of the income stream at the end of b years can be approximated by a Riemann sum

$$P(t_1)e^{r(b-t_1)}\,\Delta t + \cdots + P(t_n)e^{r(b-t_n)}\,\Delta t,$$

and the approximation will become exact as $\Delta t \to 0$. Therefore we have the following result for the future value of a stream of income.

> **The Future Value of a Stream of Income**
>
> If income flows in continuously at a rate of $P(t)$ dollars per year and is reinvested at $100r\%$ per year compounded continuously, then the future value of the income stream at the end of b years is
>
> $$\int_0^b P(t)e^{r(b-t)}\, dt.$$

EXAMPLE 3 The earnings of Zymax, Inc. are currently flowing into the firm at a constant rate of $100,000 per year and are increasing at the rate of 6% per year. If we assume that earnings flow into the company continuously over the course of a year and are immediately reinvested at 6% per year compounded continuously, then we have $P(t) = \$100,000$ and $r = 0.06$. Thus the value of Zymax's earnings stream over the next four years will be

$$\int_0^b P(t)e^{r(b-t)}\, dx = \int_0^4 100{,}000 e^{0.06(4-t)}\, dt$$

$$= \int_0^4 100{,}000 e^{0.24} e^{-0.06t}\, dt$$

$$= 100{,}000 e^{0.24} \int_0^4 e^{-0.06t}\, dt$$

$$= 100{,}000 e^{0.24} \left[\frac{e^{-0.06t}}{-0.06} \bigg|_0^4 \right]$$

$$= 100{,}000 e^{0.24} \left[\frac{e^{-0.24} - 1}{-0.06} \right]$$

$$= -\frac{100{,}000}{0.06}(1 - e^{0.24})$$

$$= \$452{,}082. \quad \square$$

EXAMPLE 4 A new computer will save a company money at the rate of $P(t) = 2t$ thousand dollars per year, where t is in years, with $t = 0$ representing the present. Find the value of this stream of savings over the next 10 years if the company reinvests the savings at 10% per year compounded continuously. The answer is given by the integral

$$\int_0^{10} 2te^{0.10(10-t)}\, dt = \int_0^{10} 2te^{1-0.10t}\, dt$$

$$= 2e \int_0^{10} te^{-0.10t}\, dt.$$

But, as we will show in the next section,

$$\int ue^{au}\,du = \frac{ue^{au}}{a} - \frac{e^{au}}{a^2} + c.$$

(Check this by differentiating the right side with respect to u.) Thus

$$\int_0^{10} 2te^{0.10(10-t)}\,dt = 2e\left[\frac{te^{-0.10t}}{-0.10} - \frac{e^{-0.10t}}{(-0.10)^2}\right]\Bigg|_0^{10}$$

$$= 2e\left[\left(-\frac{10e^{-1}}{0.10} - \frac{e^{-1}}{0.01}\right) - \left(0 - \frac{1}{0.01}\right)\right]$$

$$= 2e[-100e^{-1} - 100e^{-1} + 100]$$

$$= 200e - 400$$

$$= 143{,}656,$$

or \$143,656. □

Using the Integral as an Accumulator

In this section we have shown how to use the integral as an accumulator to find three different quantities: producers' surplus, consumers' surplus, and the value of a stream of income. If we look back at these examples, we can see that it was not actually necessary to write out the Riemann sums in each case. The essential steps in setting up a definite integral to find the value of some quantity Q over an interval [a, b] are as follows.

Procedure for Setting Up the Integral as an Accumulator

1. Divide the interval [a, b] into subintervals, each of which has length Δx.

2. Approximate the value of Q over a typical subinterval by an expression of the form $f(x)\,\Delta x$ in such a way that the approximation will become exact as $\Delta x \to 0$.

3. Then the value of Q over [a, b] is $\int_a^b f(x)\,dx$.

There are many situations other than those considered here in which this procedure can be profitably employed. Some of these are introduced in the exercises.

■ *Computer Manual: Program DEFNTGRL*

7.1 EXERCISES

Producers' Surplus and Consumers' Surplus

In Exercises 1 through 10, find producers' surplus and consumers' surplus using the given supply and demand functions.

1. supply: $p = 0.06q$
   demand: $p = -0.94q + 220$
2. supply: $p = 0.5q$
   demand: $p = 10 - 0.05q^2$
3. supply: $p = \sqrt{q}$
   demand: $p = -2q + 55$
4. supply: $p = 1.6(q + 1)^2$
   demand: $p = 200/(q + 1)$
5. supply: $p = q^2/64$
   demand: $p = 10 - 3q^2/128$
6. supply: $p = 5e^{0.8q}$
   demand: $p = 20e^{-0.2q}$
7. supply: $p = \sqrt{q + 5}$
   demand: $p = 27/(q + 5)$
8. supply: $p = 4(q - 4)/3$
   demand: $p = 400/(q + 1)$
9. supply: $q = 0.05p + 0.25$
   demand: $q = 52.5 - 0.5p$
10. supply: $q = \ln p$
    demand: $q = \ln(100/p)$

Income Streams

11. Find the value over its first five years of a continuous stream of income of $10,000 per year reinvested at 6% per year compounded continuously.

12. Repeat Exercise 11 if the income is reinvested at 10% per year compounded continuously.

13. A firm's earnings are flowing in continuously at a constant rate of $2.5 million per year and are reinvested at 10% per year compounded continuously. Find
    (a) the value of the firm's earnings over the next two years;
    (b) how long it will be until the value of the firm's earnings reaches $20 million.

*14. Show that if a stream of income is constant at $P per year and is reinvested at 100r% per year compounded continuously, then its value over the first b years is
$$\frac{P}{r}(e^{rb} - 1).$$
Use this formula to find the value over its first 20 years of a continuous stream of income of $100,000 per year reinvested at 16% per year compounded continuously.

15. A firm's earnings are flowing in continuously at $0.5t$ million dollars per year and are reinvested at 6% per year compounded continuously. Find the value of the firm's stream of earnings over its first five years. You may use the integration formula for ue^{au} that was stated in Example 4 of this section.

16. Find the value at the end of its first three years of a continuous stream of income that flows in at a rate of $P(t) = 1 + 0.2t$ million dollars per year and is reinvested at 9% per year compounded continuously.

17. Find the value at the end of its first six years of a continuous stream of income that flows in at a rate of $100 + 5t$ thousand dollars per year and is reinvested at 7% per year compounded continuously.

18. If a firm is generating profits at the rate of
$$P(t) = 20e^{0.01t}$$

million dollars per year, where t is in years and $t = 0$ is the present, find the value of the firm's stream of profits over the next three years if all profits are reinvested at 12% per year compounded continuously.

19. A firm's profits are being earned continuously at a rate of $50e^{0.01t}$ million dollars per year, where $t = 0$ represents the present. Find the value of the firm's stream of profits over the next five years if all profits are reinvested at
   (a) 8% per year compounded continuously;
   (b) 12% per year compounded continuously.

20. Suppose the earnings of a firm are flowing in continuously at a rate of $2.5 + 0.1t$ million dollars per year, where $t = 0$ represents the present.
   (a) Find a general formula for the value of the firm's earnings over the next b years.
   (b) Use the formula of part (a) to find the value of the firm's earnings over the next two years.
   (c) Use the formula of part (a) to find approximately how long it will take for earnings to reach $20 million.

21. A stream of savings is reinvested at 8% per year compounded continuously. If the savings accrue continuously at the rate of $20,000 per year, how long will it take for the value of the stream of savings to reach $250,000?

22. It is estimated that a company's investment in industrial robots will save it

$$P(t) = 200 + 12t$$

thousand dollars per year over the next 10 years, where t is time in years with $t = 0$ representing the present. Find the value of the stream of savings due to the robots if the company reinvests the savings at 10% per year compounded continuously.

23. A firm's investment in a new inventory control system will save it $P(t) = 40e^{-0.05t}$ thousand dollars per year for the next five years. If the savings are reinvested at 12% per year compounded continuously, find the value of the stream of savings.

*24. Show that if a continuous stream of income flows in at a rate of $P(t)$ dollars per time period and is *not* reinvested, then the total income produced by the stream from time $t = a$ to time $t = b$ is equal to

$$\int_a^b P(t)\, dt.$$

Use this result to find the total income over its first 10 years of a continuous stream of income that flows in at a rate of $10,000 per year.

*25. Use the result of Exercise 24 to find the total income over its first five years of a continuous stream of income that flows in at a rate of $4t$ thousand dollars per year t.

*26. Suppose a continuous stream of income flows in at a rate of $10e^{0.2t}$ thousand dollars per year t. Use the result of Exercise 24 to find the total income produced by the stream from time $t = 2$ to time $t = 6$.

Using the Integral as an Accumulator

Management

*27. **Carrying Cost.** Suppose that it costs $\$p$ to carry one unit in inventory for one time period and that the number of units in inventory at time t is given by an inventory function $I(t)$. Then the total cost of carrying the inventory from time $t = 0$ to time $t = b$ is

$$\int_0^b pI(t)\, dt.$$

Use the procedure for setting up the integral as an accumulator to show that this is so.

*28. Refer to Exercise 27. If it costs a store \$4 to carry one unit in inventory for one month and if the number of units in inventory at time t is given by

$$I(t) = 40{,}000 - 2000t,$$

where t is in months with $t = 0$ representing the present, find the store's total carrying cost for the next year.

*29. Suppose it costs a manufacturer \$0.20 to carry a unit in inventory for one week and that the number of units in inventory as of week t is given by

$$I(t) = 12{,}000 - 150t^2,$$

where $t = 0$ represents the present. Use the result of Exercise 27 to find the manufacturer's total carrying cost over the next four weeks.

Life Science

*30. Survival-Renewal Functions. Suppose that $p(t)$ is the size of a population at time t, $t \geq 0$. Let $f(t)$ be the proportion of the original population that survives at time t; the function f is called a **survival function.** Let $g(t)$ denote the number of individuals added to the population at time t; the function g is called a **renewal function.** Use the integral as an accumulator to show that the size of the population at time $t = b$ is

$$p(b) = p(0)f(b) + \int_0^b g(t)f(b - t)\, dt.$$

*31. Use the result of Exercise 30 to find the size of a caribou herd six months from now if there are 2000 caribou in the herd now, the proportion of the 2000 who will survive t months from now is given by

$$f(t) = \frac{1}{1 + 0.02t},$$

and the number of caribou added to the herd is constant at 100 per month.

*32. Use the result of Exercise 30 to find the number of rabbits you will have one year from now if you start with 100 rabbits, the proportion of the original population that survives as of month t is given by

$$f(t) = e^{-0.1t},$$

and the number of rabbits being born at time t is given by $g(t) = 20t$.

*33. Epidemiology. According to an article by Peter J. Denning ("The Science of Computing: Computer Models of AIDS Epidemiology," *American Scientist*, July–August 1987) the relationship between infection by the human immunodeficiency virus (HIV) and the development of the disease AIDS is as follows: Let $H(t)$ denote the number of people who have been infected by the virus as of year t, let $A(t)$ denote the number who have AIDS as of year t, and let $I(x)$ denote the proportion of HIV-infected individuals whose incubation period for AIDS is x years.

Then
$$A(b) = \int_0^b I(t)H(b-t)\,dt.$$

Use the integral as an accumulator to derive the formula for $A(b)$.

*34. Refer to Exercise 33: If $I(t) = 0.3e^{-0.3t}$ and $H(t) = e^{0.01t}$, $H(t)$ in hundreds of thousands of individuals, find $A(5)$, $A(10)$, and $A(20)$, and interpret your results.

*35. Refer to Exercise 33: If $I(t) = 0.2e^{-0.2t}$ and $H(t) = e^{0.01t}$, find $A(5)$, $A(10)$, and $A(20)$, and compare your results with those of Exercise 34.

*36. Refer to Exercise 33: Some models of the AIDS epidemic assume that $I(t) = 0.125e^{-0.125t}$ and $H(t) = e^{0.01t}$. Find and graph $A(x)$, $x \geq 0$, and interpret your result.

*37. Blood Flow. Poiseuille's Law states that the velocity of blood in an artery depends on the distance of the blood from the center of the artery according to the equation
$$v(x) = c(b^2 - x^2).$$

Here x is the distance from the central axis of the artery, b is the radius of the artery, c is a constant, and $v(x)$ is the velocity of the blood in cubic centimeters per second. See the figure.

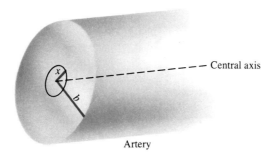

Artery

Use the integral as accumulator to show that the rate of blood flow through the artery is
$$2\pi c \int_0^b (b^2 x - x^3)\,dx = \frac{\pi c b^4}{2}$$

cubic centimeters per second. (*Hint:* Divide the artery into rings of width Δx and choose a number x_i in each ring; the approximate area of the ith ring is $2\pi x_i\,\Delta x$, and the approximate speed of blood in the ith ring is the area of the ring times $v(x_i)$.)

*38. Use the result of Exercise 37 to find the rate at which blood flows through an artery of radius 1 centimeter; of radius 2 centimeters.

Social Science

*39. A religious cult currently has 10,000 adherents. The proportion of them who will remain members of the cult t weeks from now is given by
$$f(t) = e^{-0.05t},$$

and the number of new members recruited during week t is given by
$$g(t) = 1000e^{-0.05t}.$$
Find the size of the cult one year (52 weeks) from now. (*Hint:* See Exercise 30.)

***40.** A prison has space for 3500 inmates and currently has a population of 3000 inmates. The proportion of inmates who will remain in the prison t months from now is given by
$$f(t) = \frac{1}{1 + 0.005t},$$
and new prisoners are arriving at the rate of 40 per month. Will the prison be overcrowded 20 months from now? 21 months from now? (*Hint:* See Exercise 30.)

***41. Population Density.** Suppose that the population x kilometers from the center of a region is $f(x)$ persons per square kilometer. Show that the total population within b miles of the center of the region is
$$2\pi \int_0^b xf(x)\, dx.$$
(*Hint:* Divide the region into circular rings as in Exercise 37.)

***42.** Refer to Exercise 41. Find the total population within 10 kilometers of the center of a city if there are $f(x) = 1000e^{-0.1x}$ persons per square kilometer x kilometers from the center.

7.2 INTEGRATION BY PARTS AND TABLES OF INTEGRALS

To this point we have discussed only two methods for finding the indefinite integral of a function: integration by guessing an antiderivative and integration by substitution. No technique of integration is applicable to all integrals, and over the years many different techniques have been developed to handle different types of indefinite integrals. Most of these techniques of integration are beyond the scope of this book, but two of them, **integration by parts** and integration using tables of integrals, are so useful that we must present them. In this section we discuss integration by parts and demonstrate how to perform integration with the aid of a table of integrals.

Integration by Parts

If u and v are functions of x, the product rule for differentiation states that
$$\frac{d}{dx}(uv) = u\frac{dv}{dx} + v\frac{du}{dx}.$$
If we multiply both sides of this equation by the differential dx, we may write it in differential form as
$$d(uv) = u\, dv + v\, du,$$
or
$$u\, dv = d(uv) - v\, du.$$

MORE ABOUT INTEGRATION

Integrating both sides of this last equation, we obtain

$$\int u \, dv = \int d(uv) - \int v \, du.$$

But uv is an antiderivative for $d(uv)$, so we have

$$\int u \, dv = uv - \int v \, du.$$

This is known as the **integration by parts** formula.

Integration by Parts Formula

$$\int u \, dv = uv - \int v \, du.$$

EXAMPLE 1 Let us use the integration by parts formula to find

$$\int xe^{-x} \, dx.$$

We set

$$u = x \quad \text{and} \quad dv = e^{-x} \, dx.$$

Then

$$du = dx \quad \text{and} \quad v = \int e^{-x} \, dx = -e^{-x},$$

so

$$\int xe^{-x} \, dx = \int u \, dv$$

$$= uv - \int v \, du$$

$$= x(-e^{-x}) - \int (-e^{-x}) \, dx$$

$$= -xe^{-x} + \int e^{-x} \, dx$$

$$= -xe^{-x} - e^{-x} + c.$$

(Note that when we found v we did not include the constant of integration and write

$$v = \int e^{-x} \, dx = -e^{-x} + c;$$

instead, we simply postponed adding the constant of integration until the final step.) ∎

EXAMPLE 2 How did we know in Example 1 to let $u = x$ and $dv = e^{-x} \, dx$? If we had tried the other possibility,

$$u = e^{-x} \quad \text{and} \quad dv = x \, dx,$$

then we would have obtained

$$du = -e^{-x}\,dx \quad \text{and} \quad v = \int x\,dx = \frac{x^2}{2},$$

and thus

$$\int xe^{-x}\,dx = \int u\,dv$$
$$= uv - \int v\,du$$
$$= \frac{x^2}{2}e^{-x} - \int \left(-\frac{x^2}{2}e^{-x}\right)dx.$$

But the integral on the right is more complicated than the one we started with. Therefore this choice of u and dv is not a useful one. ☐

Integration by parts can often be used to convert an integral that cannot be handled into one that can, as in Example 1. Examples 1 and 2 together suggest the following:

Keys to Integration by Parts

When performing integration by parts:

1. dv must include the differential dx of the original integral.

2. dv must be chosen so that it can be integrated to find v.

3. The integral $\int v\,du$ should be easy to find, or at least simpler than the integral $\int u\,dv$.

EXAMPLE 3 Let us use integration by parts to find

$$\int \ln x\,dx.$$

If we choose

$$u = \ln x \quad \text{and} \quad dv = dx,$$

then

$$du = \frac{1}{x}\,dx \quad \text{and} \quad v = \int dx = x,$$

so

$$\int v\,du = \int x\frac{1}{x}\,dx = \int 1\,dx,$$

which is certainly easy to find. Hence this is a promising choice for u and dv. We have

$$\int \ln x \, dx = \int u \, dv$$
$$= uv - \int v \, du$$
$$= x \ln x - \int 1 \, dx$$
$$= x \ln x - x + c. \quad \square$$

Sometimes it is necessary to use integration by parts more than once.

EXAMPLE 4 Let us find

$$\int x^2 e^{-x} \, dx.$$

We let

$$u = x^2 \quad \text{and} \quad dv = e^{-x} \, dx.$$

Then

$$du = 2x \, dx \quad \text{and} \quad v = \int e^{-x} \, dx = -e^{-x},$$

so

$$\int v \, du = \int (-2xe^{-x}) \, dx = -2 \int xe^{-x} \, dx.$$

But $\int xe^{-x} \, dx$ is simpler than $\int x^2 e^{-x} \, dx$; in fact, we know how to integrate xe^{-x} from Example 1. Therefore

$$\int x^2 e^{-x} \, dx = \int u \, dv$$
$$= uv - \int v \, du$$
$$= -x^2 e^{-x} - \int (-2xe^{-x}) \, dx$$
$$= -x^2 e^{-x} + 2 \int xe^{-x} \, dx.$$

But from Example 1,

$$\int xe^{-x} \, dx = -xe^{-x} - e^{-x} + c.$$

Thus

$$\int x^2 e^{-x} \, dx = -x^2 e^{-x} + 2(-xe^{-x} - e^{-x}) + c$$
$$= -e^{-x}(x^2 + 2x + 2) + c. \quad \square$$

When using integration by parts with definite integrals, the term uv must be evaluated at the limits of integration.

EXAMPLE 5 Let us evaluate

$$\int_0^2 xe^{-x}\,dx$$

using integration by parts. As in Example 1, we take

$$u = x \quad \text{and} \quad dv = e^{-x}\,dx,$$

so

$$du = dx \quad \text{and} \quad v = \int e^{-x}\,dx = -e^{-x}.$$

Thus

$$\int_0^2 xe^{-x}\,dx = -xe^{-x}\bigg|_0^2 + \int_0^2 e^{-x}\,dx$$

$$= (-2e^{-2}) - (-0e^{-0}) - \left[e^{-x}\bigg|_0^2\right]$$

$$= -2e^{-2} - [e^{-2} - e^{-0}]$$

$$= 1 - 3e^{-2}. \quad \square$$

Integral Tables

Many integrals cannot be found using only the techniques we have presented. For example, the integrals

$$\int \sqrt{x^2 - 4}\,dx \quad \text{and} \quad \int \frac{1}{4x^2 - 9}\,dx$$

cannot be determined by any method we have discussed so far. Fortunately there are tables of integrals that show how to integrate these and many other functions. Mathematical handbooks contain extensive tables of integrals; the integration formulas in the back endpapers of this book form an abbreviated version of such an integral table.

EXAMPLE 6 Let us use the integration formulas in the back endpapers to find

$$\int \sqrt{x^2 - 4}\,dx.$$

We search through the list of formulas until we find one in which the integrand has the form of our integrand $\sqrt{x^2 - 4}$. Formula 8, which tells us that

$$\int \sqrt{u^2 - a^2}\,du = \frac{1}{2}[u\sqrt{u^2 - a^2} - a^2 \ln|u + \sqrt{u^2 - a^2}|] + c,$$

has an integrand of the proper form if we take

$$u = x \quad \text{and} \quad a = 2.$$

More about Integration

Note that with the substitutions $u = x$ (so that $du = dx$) and $a = 2$ the integral of formula 8 exactly matches our integral. Thus,

$$\int \sqrt{x^2 - 4}\, dx = \int \sqrt{u^2 - a^2}\, du$$
$$= \frac{1}{2}[u\sqrt{u^2 - a^2} - a^2 \ln|u + \sqrt{u^2 - a^2}|] + c,$$
$$= \frac{1}{2}[x\sqrt{x^2 - 4} - 4 \ln|x + \sqrt{x^2 - 4}|] + c. \quad \square$$

When we match an integral with a formula from a table of integrals, the entire integral, including the differential, must match the formula in the table exactly. This sometimes requires making a substitution or rewriting the integral.

EXAMPLE 7 Let us find

$$\int \frac{1}{4x^2 - 9}\, dx.$$

The integrand matches that of formula 11 in the endpapers if we take

$$u = 2x \quad \text{and} \quad a = 3.$$

But if $u = 2x$, then

$$du = 2\, dx \quad \text{or} \quad \frac{1}{2}du = dx.$$

Thus,

$$\int \frac{1}{4x^2 - 9}\, dx = \int \frac{1}{u^2 - 3^2} \frac{1}{2}\, du$$
$$= \frac{1}{2}\int \frac{1}{u^2 - 3^2}\, du$$
$$= \frac{1}{2}\left[\frac{1}{2 \cdot 3} \ln\left|\frac{u - 3}{u + 3}\right|\right] + c$$
$$= \frac{1}{12} \ln\left|\frac{2x - 3}{2x + 3}\right| + c. \quad \square$$

EXAMPLE 8 Let us find

$$\int \frac{x}{4x^2 + 12x + 9}\, dx.$$

This integral does not appear to match any of those in the endpapers. However,

$$4x^2 + 12x + 9 = (2x + 3)^2,$$

and

$$\int \frac{x}{(2x + 3)^2}\, dx$$

matches formula 16 exactly if $u = x$ (so that $du = dx$), $a = 2$, and $b = 3$. Therefore

$$\int \frac{x}{4x^2 + 12x + 9} dx = \int \frac{x}{(2x + 3)^2} dx$$

$$= \int \frac{u}{(2u + 3)^2} du$$

$$= \frac{1}{4}\left[\ln|2u + 3| + \frac{3}{2u + 3}\right] + c$$

$$= \frac{1}{4}\left[\ln|2x + 3| + \frac{3}{2x + 3}\right] + c. \quad \square$$

■ *Computer Manual: Program DEFNTGRL*

15.2 EXERCISES

Integration by Parts

In Exercises 1 through 22, integrate by parts.

1. $\int xe^x \, dx$
2. $\int xe^{2x} \, dx$
3. $\int x^2 e^x \, dx$
4. $\int x^3 e^x \, dx$
5. $\int_0^1 x^3 e^{-x} \, dx$
6. $\int x^3 e^{-x^2} \, dx$
7. $\int x \ln x \, dx$
8. $\int_1^e x^2 \ln x \, dx$
9. $\int x^3 \ln x \, dx$
10. $\int \sqrt{x} \ln x \, dx$
11. $\int (\ln x)^2 \, dx$
12. $\int x(\ln x)^2 \, dx$
13. $\int (\ln x)^3 \, dx$
14. $\int_e^{e^2} \frac{\ln x}{x^2} \, dx$
15. $\int_1^2 x(x + 2)^5 \, dx$
16. $\int x^2(x + 1)^{3/2} \, dx$
17. $\int_1^2 x(2x + 1)^3 \, dx$
18. $\int_0^1 x^2(x + 3)^4 \, dx$
19. $\int x^2 \sqrt[3]{x + 2} \, dx$
20. $\int \frac{x}{\sqrt{x + 1}} \, dx$
21. $\int_{-1}^4 \frac{x^2}{\sqrt{x + 5}} \, dx$
22. $\int e^{\sqrt{x}} \, dx$ (*Hint:* Let $u^2 = x$.)

Management

23. A firm's marginal cost function is given by
$$c(x) = x\sqrt{x + 1}.$$
Find the firm's cost function if its fixed cost is $0.

24. A well produces oil at a rate of $20{,}000te^{-0.2t}$ barrels per year, where t is in years with $t = 0$ representing the present. Find the total amount of oil the well will produce over the next two years.

25. The supply function for a product is given by $p = 2 \ln(q + 1)$ and the demand function by $p = 50 - 3 \ln(q + 1)$. Find producers' surplus and consumers' surplus for the product.

Life Science

26. Suppose that the amount of CO_2 in the atmosphere is increasing at a rate of $t^2\sqrt{t+1}$ parts per million per decade. Here t is in decades, with $t = 0$ representing the present. Find out by how much the total concentration of CO_2 will increase over the next decade.

27. The concentration of a drug in a patient's bloodstream is

$$y = \frac{2\ln(t+3)}{(t+3)^2}$$

milligrams per milliliter t minutes after injection. Find the average concentration of the drug over the first 10 minutes after injection.

Integral Tables

In Exercises 28 through 49, use the integration formulas in the back endpapers to find the integral.

28. $\displaystyle\int \frac{1}{x^2 - 16}\,dx$

29. $\displaystyle\int \frac{1}{x\sqrt{81-x^2}}\,dx$

30. $\displaystyle\int \frac{2}{9-x^2}\,dx$

31. $\displaystyle\int_1^2 \frac{3}{x\sqrt{x^2+16}}\,dx$

32. $\displaystyle\int 2^{3x}\,dx$

33. $\displaystyle\int 5^{-3x}\,dx$

34. $\displaystyle\int x\sqrt{1-x}\,dx$

35. $\displaystyle\int \frac{x}{3x+2}\,dx$

36. $\displaystyle\int \frac{1}{x\sqrt{25-4x^2}}\,dx$

37. $\displaystyle\int \frac{1}{3x\sqrt{4+9x^2}}\,dx$

38. $\displaystyle\int \frac{1}{8-2x^2}\,dx$

39. $\displaystyle\int \frac{1}{x^2+3x}\,dx$

40. $\displaystyle\int \frac{2}{\sqrt{9x^2-0.25}}\,dx$

41. $\displaystyle\int \frac{3x}{36x^2+60x+25}\,dx$

42. $\displaystyle\int_0^1 \frac{2x}{x^2+2x+1}\,dx$

43. $\displaystyle\int \frac{3}{2x^2+6x}\,dx$

44. $\displaystyle\int 4\sqrt{x^2+5}\,dx$

45. $\displaystyle\int \sqrt{2x^2-1}\,dx$

46. $\displaystyle\int \sqrt{4x^2+9}\,dx$

47. $\displaystyle\int \frac{1}{3x\sqrt{2-4x^2}}\,dx$

48. $\displaystyle\int \frac{4}{\sqrt{2x^2-1}}\,dx$

49. $\displaystyle\int \log_2 2x\,dx$

Management

50. The supply function for a product is given by

$$p = \frac{5q}{50-q}$$

and the demand function by

$$p = \frac{2000-45q}{50-q}$$

Find producers' and consumers' surplus for the product.

51. A firm's marginal revenue function is given by

$$r(x) = \frac{20}{\sqrt{4x^2-1}}, \quad x \geq 1.$$

Find the firm's revenue function if the revenue from selling one unit is $76.93.

Life Science

52. A desert is encroaching on arable land at a rate of

$$f(t) = \frac{100}{\sqrt{t^2 + 4}}$$

square miles per year t, where $t = 0$ represents the present. Find the number of square miles of arable land the desert will claim over the next 10 years.

Social Science

53. A student who studies t hours absorbs y new concepts per hour, where

$$y = \frac{8t}{0.25t^2 + 2t + 4}.$$

Find the total number of new concepts absorbed in 4 hours of study.

7.3 IMPROPER INTEGRALS

Sometimes it is necessary to consider areas of regions that extend infinitely far to the left or right along the x-axis. For instance, the shaded region in Figure 7–11 is meant to extend infinitely far to the right along the x-axis. As we shall see, the area of such a region may or may not be finite. In this section we show how to determine if the area of such a region is finite, and how to find it if it is. In so doing, we use **improper integrals:** These are definite integrals that have at least one infinite limit of integration.

(There is another type of improper integral in which the limits of integration are finite but the integrand becomes infinite at some point of the interval of integration. We will not consider this type of improper integral.)

Consider the function f defined by $f(x) = e^{-x}$. The region under the graph of f above the x-axis and to the right of $x = 0$ (Figure 7–12) is important in probability and statistics. Let us see if this region has a finite area.

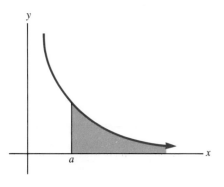

FIGURE 7–11

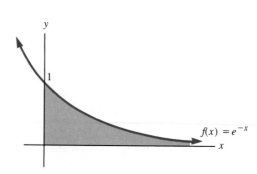

FIGURE 7–12

MORE ABOUT INTEGRATION

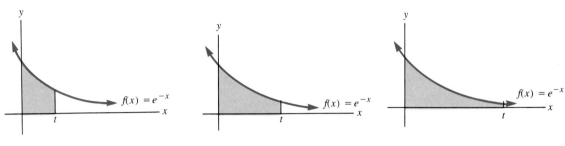

FIGURE 7–13

If t is any positive number, then the area under $f(x) = e^{-x}$ from 0 to t is certainly finite, and we know that it is equal to

$$\int_0^t e^{-x}\, dx.$$

As $t \to +\infty$, the area under $f(x)$ from 0 to t will approach the area of the region in question. See Figure 7–13. Thus, if

$$\operatorname*{Lim}_{t \to +\infty} \int_0^t e^{-x}\, dx$$

exists, then the region in question has area equal to the value of the limit; if the limit does not exist, then the area of the region is not finite. Now,

$$\operatorname*{Lim}_{t \to +\infty} \int_0^t e^{-x}\, dx = \operatorname*{Lim}_{t \to +\infty} (-e^{-x})\Big|_0^t = \operatorname*{Lim}_{t \to +\infty} (-e^{-t} + 1);$$

but, as Figure 7–12 shows,

$$\operatorname*{Lim}_{t \to +\infty} e^{-t} = 0,$$

so

$$\operatorname*{Lim}_{t \to +\infty} \int_0^t e^{-x}\, dx = 0 + 1 = 1.$$

Therefore the shaded region of Figure 7–12 has finite area equal to 1.

It is customary to write a limit such as

$$\operatorname*{Lim}_{t \to +\infty} \int_0^t e^{-x}\, dx$$

as an integral:

$$\operatorname*{Lim}_{t \to +\infty} \int_0^t e^{-x}\, dx = \int_0^{+\infty} e^{-x}\, dx.$$

The integral on the right is called an **improper integral** because one of its limits of integration is infinite.

Improper Integrals

The **improper integrals**

$$\int_a^{+\infty} f(x)\,dx \quad \text{and} \quad \int_{-\infty}^b f(x)\,dx$$

are defined by

$$\int_a^{+\infty} f(x)\,dx = \lim_{t \to +\infty} \int_a^t f(x)\,dx$$

and

$$\int_{-\infty}^b f(x)\,dx = \lim_{t \to -\infty} \int_t^b f(x)\,dx.$$

The improper integrals are said to **converge** if the limits on the right side exist; otherwise they are said to **diverge**.

EXAMPLE 1 For the improper integral

$$\int_2^{+\infty} 2x^{-3}\,dx$$

we have

$$\int_2^{+\infty} 2x^{-3}\,dx = \lim_{t \to +\infty} \int_2^t 2x^{-3}\,dx$$

$$= \lim_{t \to +\infty} (-x^{-2})\Big|_2^t$$

$$= \lim_{t \to +\infty} \left(-\frac{1}{t^2} + \frac{1}{4}\right)$$

$$= 0 + \frac{1}{4}$$

$$= \frac{1}{4}.$$

Therefore the improper integral

$$\int_2^{+\infty} 2x^{-3}\,dx$$

converges, and the shaded region of Figure 7–14 has an area of $\frac{1}{4}$. ☐

EXAMPLE 2 Let us evaluate the improper integral

$$\int_{-\infty}^0 e^x\,dx.$$

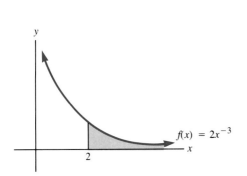

FIGURE 7–14 **FIGURE 7–15**

See Figure 7–15. We have

$$\int_{-\infty}^{0} e^x \, dx = \lim_{t \to -\infty} \int_{t}^{0} e^x \, dx = \lim_{t \to -\infty} e^x \Big|_{t}^{0} = \lim_{t \to -\infty} (1 - e^t).$$

But as Figure 7–15 shows, $\lim_{t \to -\infty} e^t = 0$, so

$$\int_{-\infty}^{0} e^x \, dx = 1 - 0 = 1.$$

Hence the integral converges, and the shaded region in Figure 7–15 has an area of 1. □

EXAMPLE 3 The improper integral

$$\int_{1}^{+\infty} \frac{1}{x} \, dx$$

diverges, because

$$\int_{1}^{+\infty} \frac{1}{x} \, dx = \lim_{t \to +\infty} \int_{1}^{t} \frac{1}{x} \, dx = \lim_{t \to +\infty} \ln x \Big|_{1}^{t} = \lim_{t \to +\infty} (\ln t),$$

and $\lim_{t \to +\infty} (\ln t) = +\infty$. Therefore the shaded area in Figure 7–16 is not finite. □

We may also define an improper integral having two infinite limits of integration by setting

$$\int_{-\infty}^{+\infty} f(x) \, dx = \int_{-\infty}^{a} f(x) \, dx + \int_{a}^{+\infty} f(x) \, dx$$

for any number a. The integral on the left converges if and only if both of the integrals on the right converge.

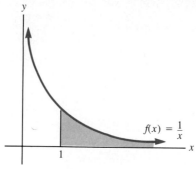

FIGURE 7–16

EXAMPLE 4   The improper integral

$$\int_{-\infty}^{+\infty} e^{-x}\, dx$$

diverges, because

$$\int_{-\infty}^{+\infty} e^{-x}\, dx = \int_{-\infty}^{0} e^{-x}\, dx + \int_{0}^{+\infty} e^{-x}\, dx,$$

and

$$\int_{-\infty}^{0} e^{-x}\, dx = \lim_{t \to -\infty} \int_{t}^{0} e^{-x}\, dx = \lim_{t \to -\infty} (-1 + e^{-t})$$

diverges. (Why?) ☐

7.3 EXERCISES

In Exercises 1 through 30, determine whether the given integral converges or diverges; if it converges, find its value and interpret your result as an area.

1. $\int_{1}^{+\infty} \dfrac{3}{x}\, dx$
2. $\int_{4}^{+\infty} \dfrac{1}{x}\, dx$
3. $\int_{1}^{+\infty} \dfrac{1}{x^2}\, dx$
4. $\int_{-\infty}^{1} \dfrac{1}{x^2}\, dx$
5. $\int_{2}^{+\infty} \dfrac{1}{x^3}\, dx$
6. $\int_{-\infty}^{-2} \dfrac{1}{x^3}\, dx$
7. $\int_{1}^{+\infty} \dfrac{1}{x^{1/3}}\, dx$
8. $\int_{1}^{+\infty} \dfrac{1}{x^{1.0001}}\, dx$
9. $\int_{1}^{+\infty} \dfrac{1}{x^{0.9999}}\, dx$
10. $\int_{-\infty}^{-1} \dfrac{1}{2-x}\, dx$
11. $\int_{1}^{+\infty} \dfrac{x}{x^2+1}\, dx$
12. $\int_{2}^{+\infty} \dfrac{x^2}{x^3+1}\, dx$
13. $\int_{0}^{+\infty} e^{x}\, dx$
14. $\int_{-\infty}^{0} e^{x}\, dx$
15. $\int_{-\infty}^{+\infty} e^{x}\, dx$
16. $\int_{-\infty}^{-1} e^{2x}\, dx$
17. $\int_{-1}^{+\infty} e^{-2x}\, dx$
18. $\int_{-\infty}^{+\infty} f(x)\, dx,$ where $f(x) = \begin{cases} e^{x}, & x \le 0, \\ e^{-x}, & x \ge 0 \end{cases}$
19. $\int_{0}^{+\infty} xe^{-x^2}\, dx$
20. $\int_{-\infty}^{+\infty} xe^{-x^2}\, dx$
21. $\int_{0}^{+\infty} x^2 e^{-x^3}\, dx$
22. $\int_{1}^{+\infty} \dfrac{\ln x}{x}\, dx$
23. $\int_{2}^{+\infty} \dfrac{1}{x \ln x}\, dx$
24. $\int_{2}^{+\infty} \dfrac{1}{x(\ln x)^2}\, dx$
*25. $\int_{1}^{+\infty} \dfrac{1}{x^r}\, dx, \quad 0 \le r \le 1$
*26. $\int_{1}^{+\infty} \dfrac{1}{x^r}\, dx, \quad r > 1$

*27. $\displaystyle\int_a^{+\infty} \frac{1}{x^r}\,dx, \quad 0 \le r \le 1, \quad a > 0$

*28. $\displaystyle\int_a^{+\infty} \frac{1}{x^r}\,dx, \quad r > 1, \quad a > 0$

*29. $\displaystyle\int_2^{+\infty} \frac{1}{x(\ln x)^r}\,dx, \quad r > 1$

*30. $\displaystyle\int_2^{+\infty} \frac{1}{x(\ln x)^r}\,dx, \quad 0 \le r \le 1$

Management

31. A firm's marginal cost function is given by
$$c(x) = \frac{20{,}000}{x^{3/2}}, \quad x \ge 1.$$
Find the firm's total cost if it makes an unlimited number of units.

32. A firm's marginal cost function is given by
$$c(x) = \frac{100{,}000}{x^{5/4}}, \quad x \ge 1.$$
Find the firm's total cost if it makes an unlimited number of units.

33. A firm's marginal profit function is given by $p(x) = 200x^{-1.1}$, profit in thousands of dollars, $x \ge 1$. Find the firm's profit if it can make and sell an unlimited number of units.

34. If a well produces oil at a rate of $10{,}000e^{-0.2t}$ barrels per year t, where $t = 0$ represents the present, find the total amount of oil the well could produce in the future.

35. Repeat Exercise 34 if the rate is $10{,}000te^{-0.2t}$ barrels per year. (*Hint:* If $r > 0$, the graph of $y = te^{-rt}$ shows that te^{-rt} approaches 0 as t approaches $+\infty$.)

36. Investment income is flowing in continuously to a blind trust at a rate of $25{,}000e^{-0.02t}$ dollars per year t. Find the value of this stream of income if it lasts forever.

37. If a stream of income flows into a firm continuously at a rate of $50e^{-0.25t}$ thousand dollars per year, find its total future value.

38. Repeat Exercise 37 if the rate is $50t^2e^{-0.25t}$ thousand dollars per year.

Life Science

39. It is estimated that the total amount of pollutants released into the atmosphere worldwide by automobiles is y million tons per decade t, where
$$y = 200(t + 2)^{-2}.$$
Here $t = 0$ represents the present. Find the total amount of pollutants that will be released by automobiles in the future.

40. Repeat Exercise 39 if $y = 200t(t + 2)^{-2}$.

7.4 NUMERICAL INTEGRATION

There are situations when it is necessary to evaluate $\int_a^b f(x)\,dx$ and it is not possible to find a simple expression for the antiderivative $F(x)$ of $f(x)$. For instance, there is no simple formula or defining equation for the antiderivative of

$$f(x) = e^{-x^2};$$

nevertheless, in statistics we must be able to evaluate integrals of the form

$$\int_a^b e^{-x^2} \, dx.$$

In a case like this we can use **numerical integration** to approximate the value of the integral. Numerical integration can be used to approximate a definite integral to any degree of accuracy. It is also useful when the integrand is defined by a collection of data points rather than by a functional expression.

Methods of numerical integration have become increasingly important with the advent of the computer. In this section we discuss and illustrate three of the most common methods: the midpoint rule, the trapezoidal rule, and Simpson's rule.

The Midpoint Rule

We saw in Section 6.5 that the definite integral $\int_a^b f(x)\,dx$ is the limit of Riemann sums:

$$\int_a^b f(x)\,dx = \lim_{\Delta x \to 0} \sum_{i=1}^{n} f(x_i)\,\Delta x,$$

where $\Delta x = \dfrac{b-a}{n}$, $a_0 = a$, $a_i = a + i\,\Delta x$ for $i = 1, \ldots, n$, and x_i may be chosen to be any number in the interval $[a_{i-1}, a_i]$. Thus

$$\int_a^b f(x)\,dx \approx \sum_{i=1}^{n} f(x_i)\,\Delta x,$$

and the approximation becomes exact as $\Delta x \to 0$, that is, as $n \to +\infty$. Since we are free to choose x_i to be any number in $[a_{i-1}, a_i]$, we may choose it to be the midpoint of this interval, which is

$$a_{i-1} + \frac{1}{2}\Delta x = a + (i-1)\,\Delta x + \frac{1}{2}\Delta x = a + \frac{2i-1}{2}\,\Delta x.$$

Hence

$$\int_a^b f(x)\,dx \approx \left[\sum_{i=1}^{n} f\!\left(a + \frac{2i-1}{2}\Delta x\right)\right]\Delta x;$$

that is,

$$\int_a^b f(x)\,dx \approx \left[f\!\left(a + \frac{1}{2}\Delta x\right) + f\!\left(a + \frac{3}{2}\Delta x\right) + \cdots + f\!\left(a + \frac{2n-1}{2}\Delta x\right)\right]\Delta x.$$

This is known as the **midpoint rule** for approximating definite integrals. The approximation is the area of the shaded region in Figure 7–17.

More about Integration

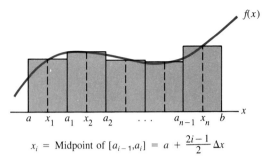

x_i = Midpoint of $[a_{i-1}, a_i] = a + \dfrac{2i-1}{2}\Delta x$

FIGURE 7–17

The Midpoint Rule

If n is a positive integer and $\Delta x = \dfrac{b-a}{n}$, then

$$\int_a^b f(x)\,dx \approx \left[\sum_{i=1}^n f\!\left(a + \dfrac{2i-1}{2}\Delta x\right)\right]\Delta x.$$

The approximation becomes exact as $n \to +\infty$.

EXAMPLE 1 Let us use the midpoint rule to approximate the definite integral

$$\int_1^2 x^2\,dx$$

using $n = 5$ rectangles. Then $a = 1$, $b = 2$, and $\Delta x = (2-1)/5 = \tfrac{1}{5}$. Since $f(x) = x^2$ and

$$a + \dfrac{2i-1}{2}\Delta x = 1 + \dfrac{2i-1}{2}\cdot\dfrac{1}{5} = 1 + \dfrac{2i-1}{10}$$

for $i = 1, 2, 3, 4$, and 5, the midpoint rule says that

$$\int_1^2 x^2\,dx \approx \left[\sum_{i=1}^5 \left(1 + \dfrac{2i-1}{10}\right)^2\right]\dfrac{1}{5}.$$

We make a table:

i	$1 + \dfrac{2i-1}{10}$	$\left(1 + \dfrac{2i-1}{10}\right)^2$
1	$1 + 1/10 = 11/10$	$(11/10)^2 = 121/100$
2	$1 + 3/10 = 13/10$	$(13/10)^2 = 169/100$
3	$1 + 5/10 = 15/10$	$(15/10)^2 = 225/100$
4	$1 + 7/10 = 17/10$	$(17/10)^2 = 289/100$
5	$1 + 9/10 = 19/10$	$(19/10)^2 = 361/100$
		$1165/100$

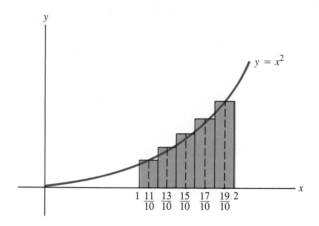

FIGURE 7–18

Therefore

$$\int_1^2 x^2 \, dx \approx \left[\frac{1165}{100} \right] \frac{1}{5} = 2.330.$$

See Figure 7–18, where the shaded area is 2.330.
Since

$$\int_1^2 x^2 \, dx = \frac{1}{3} x^3 \Big|_1^2 = \frac{7}{3} = 2.333333$$

to six decimal places, our approximation of 2.330 is accurate to two decimal places. We could obtain a better approximation to the integral by using more rectangles, that is, by using a larger value of n. ☐

The virtue of the midpoint rule is its simplicity; however, it often does not produce a good approximation for a definite integral unless the value of n is very large. Therefore other methods of numerical integration are usually preferable.

The Trapezoidal Rule

Since the definite integral of f from a to b may be interpreted as the area under f from a to b, any approximation to this area, whether it uses rectangles or some other means, is also an approximation to the integral. The **trapezoidal rule** approximates areas, and hence definite integrals, by means of trapezoids rather than rectangles. Figure 7–19 shows how this is done. The area under the graph of f from a to b, which is the definite integral of f from $x = a$ to $x = b$, is approximated by the sum of the areas of n trapezoids. As before, we set $\Delta x = (b - a)/n$ and partition the interval $[a, b]$ into n subintervals $[a, a_1], [a_1, a_2], \ldots, [a_{n-1}, b]$, where $a_i = a + i \Delta x$ for each i.

More about Integration

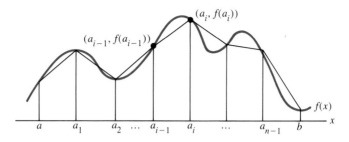

FIGURE 7–19

If a trapezoid has bases of length b_1 and b_2 and an altitude of length h, then its area is $(b_1 + b_2)h/2$. The ith trapezoid in Figure 7–19 has as its bases the line segments from $(a_{i-1}, 0)$ to $(a_{i-1}, f(a_{i-1}))$ and from $(a_i, 0)$ to $(a_i, f(a_i))$; its altitude is the interval $[a_{i-1}, a_i]$ on the x-axis. Therefore the area of the ith trapezoid is

$$\frac{1}{2}(f(a_{i-1}) + f(a_i))\, \Delta x.$$

The sum of the areas of all n trapezoids is thus

$$\frac{1}{2}[f(a) + f(a_1)]\, \Delta x + \frac{1}{2}[f(a_1) + f(a_2)]\, \Delta x + \cdots$$

$$\cdots + \frac{1}{2}[f(a_{n-2}) + f(a_{n-1})]\, \Delta x + \frac{1}{2}[f(a_{n-1}) + f(b)]\, \Delta x,$$

which simplifies to

$$\left[\frac{1}{2}f(a) + f(a_1) + f(a_2) + \cdots + f(a_{n-1}) + \frac{1}{2}f(b)\right]\Delta x,$$

or

$$\left[\frac{1}{2}f(a) + \frac{1}{2}f(b) + \sum_{i=1}^{n-1} f(a + i\, \Delta x)\right]\Delta x.$$

(Check this.) Therefore we have

The Trapezoidal Rule

If n is a positive integer and $\Delta x = \dfrac{b - a}{n}$, then

$$\int_a^b f(x)\, dx \approx \left[\frac{1}{2}f(a) + \frac{1}{2}f(b) + \sum_{i=1}^{n-1} f(a + i\, \Delta x)\right]\Delta x.$$

This approximation becomes exact as $n \to +\infty$.

EXAMPLE 2 Use the trapezoidal rule to approximate

$$\int_1^2 \frac{1}{x} dx$$

with $n = 5$ trapezoids. Then $a = 1$, $b = 2$, and $\Delta x = (2 - 1)/5 = 1/5$, so the trapezoidal rule says that

$$\int_1^2 f(x)\, dx \approx \left[\frac{1}{2}f(1) + \frac{1}{2}f(2) + \sum_{i=1}^{4} f\left(1 + i\frac{1}{5}\right)\right]\frac{1}{5}.$$

But since $f(x) = 1/x$, we have

$$\int_1^2 \frac{1}{x} dx \approx \left[\frac{1}{2}\frac{1}{1} + \frac{1}{2}\frac{1}{2} + \frac{1}{1 + 1/5} + \frac{1}{1 + 2/5} + \frac{1}{1 + 3/5} + \frac{1}{1 + 4/5}\right]\frac{1}{5}$$

$$= \left[\frac{1}{2}\frac{1}{1} + \frac{1}{2}\frac{1}{2} + \frac{1}{6/5} + \frac{1}{7/5} + \frac{1}{8/5} + \frac{1}{9/5}\right]\frac{1}{5}$$

$$= \left[\frac{1}{2} + \frac{1}{4} + \frac{5}{6} + \frac{5}{7} + \frac{5}{8} + \frac{5}{9}\right]\frac{1}{5}$$

$$= 0.6956349.$$

Since

$$\int_1^2 \frac{1}{x} dx = \ln x \Big|_1^2 = \ln 2 - \ln 1 = \ln 2 = 0.693147,$$

our approximation using the trapezoidal rule is accurate to two decimal places. We could obtain a more accurate approximation by using a larger value of n. □

In Section 6.4 we remarked that the total change in a quantity Q can be thought of as an area under the graph of the function that describes the rate of change of Q. With the aid of numerical integration we can exploit this fact to estimate total change when the function f is unknown, provided some of its functional values have been determined. We illustrate with an example.

EXAMPLE 3 Suppose a firm wishes to find the total cost of increasing its daily production from 100 units per day to 125 units per day. This cost is equal to

$$\int_{100}^{125} c(x)\, dx,$$

where $c(x)$ denotes the marginal cost of making x units. Suppose also that the firm does not know its marginal cost function, but only has information on its value at selected points, as follows:

Units Produced	Marginal Cost
100	36
105	28
110	26
115	30
120	36
125	44

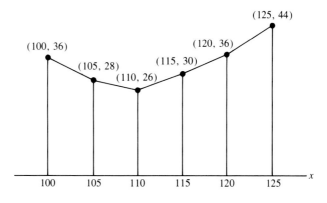

FIGURE 7–20

See Figure 7–20. Then by the trapezoidal rule with $a = 100$, $b = 125$, and $\Delta x = 5$, we have

$$\int_{100}^{125} c(x)\,dx \approx \left[\frac{1}{2} c(100) + \frac{1}{2} c(125) + c(105) + c(110) + c(115) + c(120) \right] 5$$

$$= \left[\frac{1}{2}(36) + \frac{1}{2}(44) + 28 + 26 + 30 + 36 \right] 5$$

$$= 800$$

dollars. ☐

Simpson's Rule

Another widely used method of numerical integration is **Simpson's rule,** which approximates the definite integral by areas of parabolic regions. Figure 7–21 shows how this is done. As usual,

$$\Delta x = (b - a)/n \quad \text{and} \quad a_i = a + i\,\Delta x$$

for each i. Here n must be an even positive integer. It can be shown that the area of the region bounded by $x = a_{i-1}$, $x = a_{i+1}$, the x-axis, and the parabolic arc

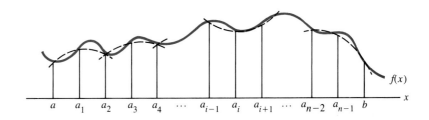

FIGURE 7–21

that passes through the points $(a_{i-1}, f(a_{i-1}))$, $(a_i, f(a_i))$, and $(a_{i+1}, f(a_{i+1}))$ on the graph of f is

$$[f(a_{i-1}) + 4f(a_i) + f(a_{i+1})]\frac{\Delta x}{3}.$$

(See Exercise 46 of Exercises 7.4.) The sum of all such terms simplifies to

$$[f(a) + 4f(a_1) + 2f(a_2) + 4f(a_3) + \cdots + 2f(a_{n-2}) + 4f(a_{n-1}) + f(b)]\frac{\Delta x}{3}.$$

(Check this.) Therefore Simpson's rule may be stated as follows:

Simpson's Rule

If n is an even positive integer, $\Delta x = \dfrac{b-a}{n}$, and $a_i = a + i\,\Delta x$ for each i, then

$$\int_a^b f(x)\,dx \approx [f(a) + 4f(a_1) + 2f(a_2) + \cdots$$

$$\cdots + 2f(a_{n-2}) + 4f(a_{n-1}) + f(b)]\frac{\Delta x}{3}.$$

This approximation becomes exact as $n \to +\infty$.

EXAMPLE 4 Use Simpson's rule with $n = 4$ subintervals to approximate

$$\int_1^2 \frac{1}{x}\,dx.$$

Then $a = 1$, $b = 2$, and $\Delta x = \frac{1}{4}$, so

$$a = 1,\quad a_1 = \frac{5}{4},\quad a_2 = \frac{3}{2},\quad a_3 = \frac{7}{4},\quad \text{and}\quad b = 2.$$

Thus

$$\int_1^2 \frac{1}{x}\,dx \approx \left[\frac{1}{1} + 4\frac{1}{5/4} + 2\frac{1}{3/2} + 4\frac{1}{7/4} + \frac{1}{2}\right]\frac{1/4}{3} = 0.693254,$$

to six decimal places. Since

$$\int_1^2 \frac{1}{x}\,dx = \ln x \Big|_1^2 = \ln 2 - \ln 1 = \ln 2 = 0.693147,$$

our approximation using Simpson's rule is accurate to three decimal places. As usual, a larger value of n will produce a better approximation. □

■ *Computer Manual: Program NUMERINT*

7.4 EXERCISES

The Midpoint Rule

In Exercises 1 through 10, use the midpoint rule with the given value of n to approximate the given definite integral. In Exercises 1 through 6, compare your answer with the actual value of the integral.

1. $\int_1^5 2x\,dx$, $n = 4$
2. $\int_1^5 2x\,dx$, $n = 6$
3. $\int_1^2 x^2\,dx$, $n = 6$
4. $\int_1^2 x^2\,dx$, $n = 8$
5. $\int_2^4 \frac{1}{x}\,dx$, $n = 5$
6. $\int_2^4 \frac{1}{x}\,dx$, $n = 6$
7. $\int_1^2 \frac{1}{1+\sqrt{x}}\,dx$, $n = 8$
8. $\int_0^1 \sqrt{1+x^2}\,dx$, $n = 8$
9. $\int_0^{1/2} \frac{1}{\sqrt{1-x^2}}\,dx$, $n = 10$
10. $\int_2^4 e^{\sqrt{x}}\,dx$, $n = 10$

11. Use the midpoint rule with $n = 4$ to approximate
$$\int_0^1 \frac{1}{x^2+1}\,dx.$$
The actual value of this integral is $\pi/4$. How does your approximation compare with the actual value?

12. Use the midpoint rule with $n = 4$ to approximate
$$\int_0^1 e^{-x^2}\,dx.$$

13. A building lot is bordered by a stream:

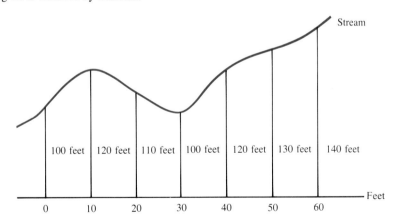

Estimate the area of the lot using the midpoint rule.

The Trapezoidal Rule

In Exercises 14 through 19, use the trapezoidal rule with the given value of n to approximate the given definite integral. In each case compare your answer with the actual value of the integral.

14. $\int_0^2 x\, dx$, $n = 5$
15. $\int_1^2 x^2\, dx$, $n = 6$
16. $\int_1^2 x^2\, dx$, $n = 8$
17. $\int_1^2 x^2\, dx$, $n = 12$
18. $\int_1^2 \frac{1}{x}\, dx$, $n = 5$
19. $\int_1^2 \frac{1}{x}\, dx$, $n = 10$

20. Use the trapezoidal rule with $n = 8$ to approximate the value of
$$\int_1^2 \frac{1}{1 + \sqrt{x}}\, dx.$$
Compare your answer with that of Exercise 7.

21. Use the trapezoidal rule with $n = 8$ to approximate the value of
$$\int_0^1 \sqrt{1 + x^2}\, dx.$$
Compare your answer with that of Exercise 8.

22. Use the trapezoidal rule with $n = 10$ to approximate the value of
$$\int_0^{1/2} \frac{1}{\sqrt{1 - x^2}}\, dx.$$
Compare your answer with that of Exercise 9.

23. Use the trapezoidal rule with $n = 10$ to approximate the value of
$$\int_2^4 e^{\sqrt{x}}\, dx.$$
Compare your answer with that of Exercise 10.

24. Use the trapezoidal rule with $n = 4$ to approximate the value of
$$\int_0^1 \frac{1}{x^2 + 1}\, dx.$$
Compare your answer with that of Exercise 11.

25. Use the trapezoidal rule with $n = 4$ to approximate the value of
$$\int_0^1 e^{-x^2}\, dx.$$
Compare your answer with that of Exercise 12.

*26. Show that using the trapezoidal rule to approximate
$$\int_a^b (mx + c)\, dx$$
with any n yields the exact value of the integral.

Management

27. A firm estimates its marginal cost at various production levels to be as follows:

Production (units)	10	12	14	16	18	20
Marginal Cost ($)	200	202	204	207	212	220

Use these data and the trapezoidal rule to approximate the firm's cost if it increases production from 10 to 20 units.

28. A firm's marginal profit at various production levels is as follows:

Production (units)	100	110	120	130	140	150	160
Marginal Profit ($)	48	32	12	−2	−22	−35	−45

Use the trapezoidal rule to approximate the change in profit if production is increased from 100 to 160 units.

29. An efficiency expert sampled a new worker's daily production at various times during the worker's first three weeks on the job. The result was the following graph:

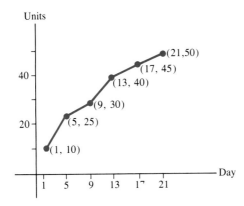

Use the trapezoidal rule to approximate the area under the graph from $x = 1$ to $x = 21$ and interpret your result.

Life Science

30. All ponds that have a surface area of more than 1 acre (43,560 square feet) qualify for stocking by the state fish and game department. Use the trapezoidal rule to determine whether or not the following pond qualifies for stocking (all figures are in feet):

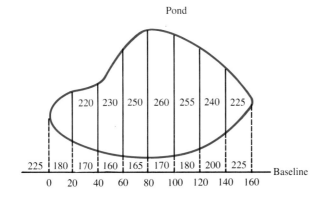

31. The concentration of a drug in a patient's bloodstream was sampled once each day, and the following graph was prepared:

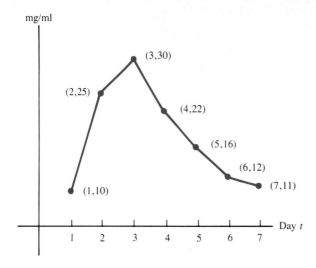

Use the trapezoidal rule to approximate the average concentration of the drug in the patient's bloodstream.

32. The rate at which an enzyme is produced by a biochemical reaction is as follows:

Time (secs)	0.0	0.1	0.2	0.3	0.4	0.5	0.6	0.7	0.8	0.9	1.0
Production (grams/sec)	0.0	0.2	0.8	1.4	2.9	4.5	3.2	1.7	0.9	0.3	0.1

Use the trapezoidal rule to estimate the total amount of the enzyme produced in 1 second.

Social Science

33. A psychologist measures the rate at which college freshmen can master new concepts in mathematics as a function of the number of weeks they have spent in college mathematics courses and finds the following:

Number of Weeks of Math Completed	5	10	15	20	25
Number of New Concepts Mastered per Week	2	6	10	13	14

Use the trapezoidal rule to approximate the total number of new concepts mastered from week 5 through week 25.

Simpson's Rule

In Exercises 34 through 39, use Simpson's rule with the given value of n to approximate the given definite integral. In each case compare your answer with the actual value of the integral.

34. $\int_{1}^{5} 2x\, dx$, $n = 4$

35. $\int_{1}^{2} x^2\, dx$, $n = 8$

36. $\int_{0}^{1} x^3\, dx$, $n = 4$

37. $\int_{0}^{1} x^3\, dx$, $n = 8$

38. $\int_{1}^{2} \frac{1}{x}\, dx$, $n = 6$

39. $\int_{1}^{2} \frac{1}{x}\, dx$, $n = 10$

40. Use Simpson's rule with $n = 8$ to approximate the value of
$$\int_1^2 \frac{1}{1+\sqrt{x}}\,dx.$$
Compare your answer with those of Exercises 7 and 20.

41. Use Simpson's rule with $n = 8$ to approximate the value of
$$\int_0^1 \sqrt{1+x^2}\,dx.$$
Compare your answer with those of Exercises 8 and 21.

42. Use Simpson's rule with $n = 10$ to approximate the value of
$$\int_0^{1/2} \frac{1}{\sqrt{1-x^2}}\,dx.$$
Compare your answer with those of Exercises 9 and 22.

43. Use Simpson's rule with $n = 10$ to approximate the value of
$$\int_2^4 e^{\sqrt{x}}\,dx.$$
Compare your answer with those of Exercises 10 and 23.

44. Use Simpson's rule with $n = 4$ to approximate the value of
$$\int_0^1 \frac{1}{x^2+1}\,dx.$$
Compare your answer with those of Exercises 11 and 24.

45. Use Simpson's rule with $n = 4$ to approximate the value of
$$\int_0^1 e^{-x^2}\,dx.$$
Compare your answer with those of Exercises 12 and 25.

*46. Let $f(x) = ax^2 + bx + c$ and $a_2 = a_1 + \Delta x$, $a_3 = a_1 + 2\Delta x$, $\Delta x > 0$. Show that
$$\int_{a_1}^{a_3} f(x)\,dx = [f(a_1) + 4f(a_2) + f(a_3)]\frac{\Delta x}{3}.$$

Management

47. Acme Computer Company estimates that the number of worker-hours that will be required to build the xth unit of its new computer is as follows:

Unit Number	1	4	7	10	13
Worker-hours	4000	3200	3000	2850	2750

Use Simpson's rule to approximate the total number of worker-hours that will be required to build units 1 through 13.

48. Repeat Exercise 13 using Simpson's rule and compare your result with that of Exercise 13.

49. Repeat Exercise 28 using Simpson's rule and compare your result with that of Exercise 28.

Life Science

50. The following graph shows the rate at which pollutants entered a lake at various times during one day of sampling:

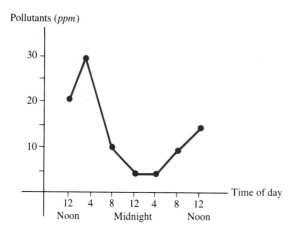

Use Simpson's rule to approximate the area under the curve and interpret your result.

51. Repeat Exercise 30 using Simpson's rule and compare your result with that of Exercise 30.

52. Repeat Exercise 32 using Simpson's rule and compare your result with that of Exercise 32.

SUMMARY

This summary consists of terms and symbols whose meaning you should know, facts you should know, and techniques and methods you should be able to employ.

Section 7.1

Riemann sum. The definite integral as a limit of Riemann sums. The definite integral as an accumulator. Producers' surplus. The formula for producers' surplus. Finding producers' surplus. Consumers' surplus. The formula for consumers' surplus. Finding consumers' surplus. Stream of income. Formula for finding the future value of a continuous stream of income reinvested at a continuously compounded interest rate. Finding the future value of a stream of income. Method for setting up the integral as an accumulator. Setting up the integral as an accumulator,

Section 7.2

The integration by parts formula: $\int u\, dv = uv - \int v\, du$. The keys to integration by parts. Using integration by parts to find integrals. Integral tables. Using integral tables to find integrals.

Section 7.3

Improper integrals:

$$\int_a^{+\infty} f(x)\, dx = \lim_{t \to +\infty} \int_a^t f(x)\, dx$$

$$\int_{-\infty}^b f(x)\, dx = \lim_{t \to -\infty} \int_t^b f(x)\, dx.$$

Convergence, divergence of improper integrals of the above forms. Determining whether improper integrals converge or diverge, and evaluating them when they converge. Interpreting the value of a convergent improper integral as an area. Improper integrals of the form $\int_{-\infty}^{+\infty} f(x)\, dx$. Convergence, divergence of such integrals.

Section 7.4

Numerical integration. The need for numerical integration. The midpoint rule as approximation by rectangles. The formula for the midpoint rule. Using the midpoint rule to approximate definite integrals. The trapezoidal rule as approximation by trapezoids. The formula for the trapezoidal rule. Using the trapezoidal rule to approximate definite integrals. Simpson's rule as approximation by parabolic segments. The formula for Simpson's rule. Using Simpson's rule to approximate definite integrals.

REVIEW EXERCISES

In Exercises 1 through 3, find producers' surplus and consumers' surplus for the given supply and demand functions.

1. supply: $p = 5q + 50$
   demand: $p = -2q + 120$

2. supply: $p = 200e^{0.01q}$
   demand: $p = 1000e^{-0.02q}$

3. supply: $q = 15p$
   demand: $q = 100 - p^2$

4. A firm's earnings are flowing in at the rate of $50 million per year and are compounding continuously at 15% per year. Find the value of the firm's earnings over the next 10 years.

5. Repeat Exercise 4 if the firm's earnings are flowing in at a rate of $50e^{0.01t}$ million dollars per year.

6. Repeat Exercise 4 if the firm's earnings are flowing in at a rate of $0.5t + 50$ million dollars per year.

7. Let f be a function that is positive over the interval $[a, b]$. If we revolve the graph of f about the x-axis, we obtain a solid known as a **solid of revolution:**

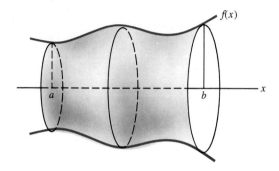

A small subinterval of $[a, b]$ of length Δx thus generates a portion of the solid of revolution whose volume can be approximated by that of a circular disk of radius $f(x)$ and width Δx:

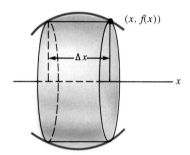

The volume of such a circular disk is $\pi[f(x)]^2 \Delta x$. Use the integral as an accumulator to show that the volume of the solid of revolution is given by

$$\pi \int_a^b [f(x)]^2 \, dx.$$

In Exercises 8 and 9, use the result of Exercise 7 to find the volume of the solid of revolution obtained by revolving the given function about the x-axis over the given interval:

8. $f(x) = x^2$, $[1, 2]$ 9. $f(x) = e^{-x}$, $[0, 2]$

In Exercises 10 through 17, perform the indicated integration.

10. $\int 4xe^{-4x}\, dx$

11. $\int 3x(x + 5)^6\, dx$

12. $\int_1^2 x^{3/2} \ln x\, dx$

13. $\int x^3(x + 1)^{4/3}\, dx$

14. $\int \dfrac{1}{2x^2 + 4x}\, dx$

15. $\int \dfrac{1}{16 - 4x^2}\, dx$

16. $\displaystyle\int_{1/\sqrt{2}}^{\sqrt{2}} \frac{1}{2x\sqrt{8-2x^2}}\,dx$ **17.** $\displaystyle\int \frac{2x}{x^2+4x+4}\,dx$

In Exercises 18 through 23, determine whether the given improper integral converges or diverges, and find its value if it converges.

18. $\displaystyle\int_0^{+\infty} e^{-kx}\,dx, \quad k > 0$ **19.** $\displaystyle\int_3^{+\infty} \frac{2x^2}{x^3+1}\,dx$

20. $\displaystyle\int_{-\infty}^{-2} x^{-4}\,dx$ **21.** $\displaystyle\int_1^{+\infty} \frac{(\ln x)^2}{x}\,dx$

22. $\displaystyle\int_0^{+\infty} x^r e^{-x^{r+1}}\,dx, \quad r > -1$ **23.** $\displaystyle\int_{-\infty}^{+\infty} f(x)\,dx, \quad \text{where } f(x) = \begin{cases} e^{2x} & \text{for } x \le 0, \\ e^{-3x} & \text{for } x > 0. \end{cases}$

24. Use the midpoint rule with $n = 6$ to approximate

$$\int_0^1 (4 - x^4)\,dx$$

and compare your result with the actual value of the integral.

25. Repeat Exercise 24 using the trapezoidal rule.

26. Repeat Exercise 24 using Simpson's rule.

27. A criminologist sampled the police records of a city and found the following:

Day	1	5	9	13	17	21	25
Crimes per day	125	128	136	132	130	124	118

Use the trapezoidal rule to estimate the total number of crimes reported over the 25-day period.

28. Repeat Exercise 27 using Simpson's rule and compare your answer with that of Exercise 27.

29. A beaver dam has flooded the region shown below (figures in meters):

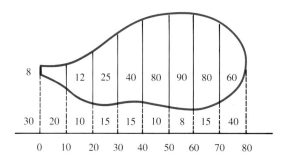

Use the trapezoidal rule to estimate the area flooded.

30. Repeat Exercise 29 using Simpson's rule and compare your answer with that of Exercise 29.

ADDITIONAL TOPICS

Here are some suggestions for topics not covered in the text that you might want to investigate on your own.

1. In Exercise 7 of the Review Exercises for this chapter we outlined a method of using the integral as an accumulator to find the volume of a solid of revolution. This method is known as the **disk method,** because it depends on approximating the volume of the solid by the volumes of disks. Another method of using the integral as an accumulator to find the volume of a solid of revolution is known as the **shell method.** Find out how and why the shell method works.

2. The length of a curve between any two points on it is called the **arc length** of the curve between the points. Under certain conditions arc length can be found by using the integral as an accumulator. Find out how this is done and why it works.

3. A **surface of revolution** is a surface that is obtained by revolving a curve about an axis. Under certain conditions the surface area of a surface of revolution can be found by using the integral as an accumulator. Find out how this is done and why it works.

4. In physics, work is defined as force applied times the distance through which it is applied. Find out how the integral as an accumulator can be used to calculate work.

5. A technique of integration that is very useful in integrating rational functions is the method of **partial fractions.** Find out how the method of partial fractions works.

6. We mentioned at the beginning of Section 7.3 that integrals for which the integrand becomes infinite at some point of the interval of integration are also called improper integrals. Thus integrals such as

$$\int_{-1}^{1} \frac{1}{x}\,dx \quad \text{and} \quad \int_{1}^{2} \frac{1}{x-1}\,dx$$

are improper integrals of this type: the first because $1/x$ becomes infinite at 0 and the second because $1/(x-1)$ becomes infinite at 1. Find out how this type of improper integral is integrated.

7. When using either the trapezoidal rule or Simpson's rule to approximate the value of a definite integral, it is possible to make estimates for the error of the approximation. Find out about the trapezoidal error estimate and Simpson's error estimate and how they are used.

SUPPLEMENT: THE PRESENT VALUE OF A STREAM OF INCOME

In Section 5.1 we introduced the notion of the present value of an amount due in the future. We can apply the present value concept to streams of income such as those discussed in Section 7.1. In this supplement we show how to find the present value of a stream of income, and illustrate some of the uses of this concept.

Suppose a stream of income flows into a firm continuously, arriving at the rate of $I(t)$ dollars per year at time t. Suppose also that money is worth $100s\%$ per year compounded continuously; that is, that $100s$ is the continuous discount rate. We will find an expression for the present value of this income stream over the interval $[0, b]$.

If we divide the time interval $[0, b]$ into small subintervals, each of width Δt, then the amount of income that flows in during any particular subinterval is approximately $I(t)\,\Delta t$ dollars. In Section 5.1 we saw that the present value P of an amount A due t years in the future when money is worth $100r\%$ per year compounded continuously is given by

$$P = Ae^{-rt}.$$

Therefore the present value of the amount $I(t)\,\Delta t$ that will arrive during the subinterval and is to be discounted at $100s\%$ per year is equal to

$$[I(t)\,\Delta t]e^{-st}$$

dollars. Using the integral as an accumulator, we thus find that the total present value of the stream of income over the time period $[0, b]$ is

$$\int_0^b I(t)e^{-st}\,dt.$$

The Present Value of a Stream of Income

If a stream of income flows in continuously, arriving at the rate of $I(t)$ dollars per year at time t, and if its future value is discounted at $100s\%$ per year compounded continuously, then the present value of the stream of income over the next b years is

$$\int_0^b I(t)e^{-st}\,dt.$$

EXAMPLE 1 Suppose that each year Blink, Inc., earns $100,000 in revenue. What is the present value of Blink's revenue stream over the next four years if Blink discounts future revenues at 8% per year compounded continuously? The answer is

$$\int_0^4 100{,}000\,e^{-0.08t}\,dt = -\frac{100{,}000}{0.08}\left(e^{-0.08t}\Big|_0^4\right)$$

$$= -\frac{100{,}000}{0.08}(e^{-0.32} - 1)$$

$$= \$342{,}313.70. \quad \square$$

The present value of a stream of income discounted over all future time is called its **capital value.**

EXAMPLE 2 Suppose Blink, Inc. of the previous example expects its revenue to continue to be $100,000 per year forever. Then the capital value of Blink's revenue stream is

$$\int_0^{+\infty} 100{,}000 e^{-0.08t}\, dx = \lim_{b \to +\infty} \int_0^b 100{,}000 e^{-0.08t}\, dt$$

$$= -\frac{100{,}000}{0.08} \lim_{b \to +\infty} (e^{-0.08b} - 1)$$

$$= -1{,}250{,}000(0 - 1)$$

$$= \$1{,}250{,}000. \quad \square$$

Another viewpoint of the present value of an income stream takes into account the fact that the money received can be invested. Thus, suppose that income is currently arriving continuously at the rate of P dollars per year, that it compounds continuously at $100r\%$ per year, and that it is discounted at $100s\%$ per year. Then at time t in the future the amount P will have grown to

$$A(t) = Pe^{rt}$$

dollars. But the present value of the amount $A(t)$ is equal to

$$A(t)e^{-st} = Pe^{rt}e^{-st} = Pe^{(r-s)t}.$$

Therefore by the usual argument using the integral as accumulator, the present value of the income stream over the next b years is

$$\int_0^b Pe^{(r-s)t}\, dt.$$

The Present Value of an Invested Stream of Income

If a stream of income is currently flowing in at a rate of P dollars per year, and if this income is invested at $100r\%$ per year compounded continuously and its future value is discounted at $100s\%$ per year compounded continuously, then the present value of the income received over the next b years is

$$\int_0^b Pe^{(r-s)t}\, dt.$$

EXAMPLE 3 Suppose that a firm's profits are currently flowing in at a rate of $1 million per year and that it reinvests all profits. If the reinvested profits earn 8% per year compounded continuously and if the firm discounts future profits at 12% per year, then the present value of its profits over the next five years, in millions of dollars, will be

$$\int_0^5 1e^{(0.10-0.12)t}\,dt = \int_0^5 e^{-0.02t}\,dt$$

$$= -\frac{e^{-0.02t}}{0.02}\bigg|_0^5$$

$$= -\frac{e^{-0.1}-1}{0.02}$$

$$\approx 4.758,$$

or approximately $4,758,000. □

SUPPLEMENTARY EXERCISES

1. If profits are flowing into Aurora Corporation at a rate of 2 million dollars per year, and if Aurora discounts future profits at 10% per year compounded continuously, find the present value of Aurora's future profits over the next 6 years.

2. Refer to Exercise 1: Find the capital value of Aurora's profits.

3. A firm's revenues are flowing in at a rate of $I(t) = 2 + 5t$ million dollars per year. If the firm discounts future revenues at 7% per year compounded continuously, find the present value of the firm's stream of revenues over
   (a) the next 4 years   (b) the next 10 years

4. Refer to Exercise 3: Find the capital value of the firm's revenues.

5. Suppose Borealis Company expects to be earning profits at the rate of
   $$I(t) = 1 + e^{0.05t}$$
   million dollars per year t years from now. If the company discounts future earnings at 8% per year compounded continuously, find the present value of its expected earnings
   (a) over the next 10 years   (b) over the next 50 years   (c) forever

6. An oil well brings in revenues of $10te^{-0.02t}$ thousand dollars per year t. Using a discount rate of 6% per year compounded continuously, find
   (a) the present value of the revenue stream from the well over the next five years;
   (b) the capital value of the well.

7. Camran, Ltd. is currently earning profits at the rate of $5 million per year. Camran reinvests its profits at 9% per year compounded continuously and discounts future profits at 10% per year compounded continuously. Find the present value of Camran's profits
   (a) over the next three years   (b) forever

8. Do Exercise 7 if Camran's discount rate per year is
   (a) 20%   (b) 9%

9. You are considering buying some rental property as an investment. You know that you can rent the property now for $10,000 per year and that the rent will increase at 6% per year compounded continuously. If you discount future income at 10%

per year compounded continuously, find the present value of the rent you will receive if you keep the property

    (a) 10 years   (b) 40 years   (c) forever

10. Refer to Exercise 9. How long would you have to hold the property to receive rent having a present value of $100,000?

*11. Refer to Exercise 9. Suppose you intend to hold the property for 10 years and then sell it.

    (a) If you could sell it for $100,000, what is the maximum amount that you would be willing to pay for it now?

    (b) If the property will cost you $150,000, how much will you have to sell it for in order to earn more, on a present value basis, than it costs?

12. A certain stock pays a dividend of $14 per share, and this is increasing at a rate of 8% per year compounded continuously. If future income is discounted at 16% per year compounded continuously, find the present value of all future dividends you will receive if you buy the stock and hold it for

    (a) 5 years   (b) 50 years   (c) forever

*13. Refer to Exercise 12.

    (a) Would you be willing to pay $100 per share for the stock now if you knew you could sell it five years from now for $100? If you could sell it 5 years from now for $50?

    (b) Suppose the stock costs $75 a share now. How much would you have to sell it for five years from now in order to earn more, on a present value basis, than it costs?

8

Functions of Several Variables

The functions we have studied up until now have been **functions of one variable**, that is, functions with just one independent variable. However, in many applications we must consider functions with more than one independent variable. For instance, suppose a firm produces two products. Its total cost C will then depend on the number of units x of the first product *and* the number of units y of the second product that it makes, and might be given by an equation such as

$$C = 20x + 30y + 50{,}000.$$

This equation defines C as a function of the two independent variables x and y. Similarly, a farmer's crop yield Y might depend on the amount q of fertilizer and the amount w of weedkiller used on the crop, the amount r of rainfall the crop receives, and the time t that the farmer devotes to cultivating the crop, according to the equation.

$$Y = 0.5q + 2.2w + 2\sqrt{r} + 1.2t^2.$$

This equation defines crop yield Y as a function of the four independent variables q, w, r, and t. Functions such as these are called **functions of several variables.**

In this chapter we study functions of several variables. We begin by illustrating how functions of several variables are defined and interpreted, introduce the three-dimensional cartesian coordinate system, and show how functions of two variables may be graphed as surfaces in three-dimensional space. Next we discuss derivatives of functions of several variables, show how these derivatives may be interpreted, and demonstrate how to find relative maxima and minima for functions of two variables. We also present a method by means of which functions of several variables can be maximized or minimized subject to constraints. The final section of the chapter is devoted to integrating functions of two variables.

8.1 FUNCTIONS OF MORE THAN ONE VARIABLE

In this section we introduce functions of several variables and briefly discuss how functions of two variables may be graphed using the three-dimensional cartesian coordinate system.

An equation such as

$$f(x, y) = x^2 + 2xy + 5y$$

defines f as a function of two independent variables x and y. The domain of such a function f is a set of ordered pairs (a, b), where a and b are numbers. We evaluate the function at the ordered pair (a, b) by substituting a for x and b for y in the defining equation.

EXAMPLE 1 Let

$$f(x, y) = x^2 + 2xy + 5y.$$

Then

$$f(0, 0) = 0^2 + 2(0)(0) + 5(0) = 0,$$
$$f(2, 3) = 2^2 + 2(2)(3) + 5(3) = 31,$$
$$f(-1, 2) = (-1)^2 + 2(-1)(2) + 5(2) = 7,$$
$$f(4, y) = 4^2 + 2(4)y + 5y = 16 + 13y,$$
$$f(x, 1) = x^2 + 2x(1) + 5(1) = x^2 + 2x + 5,$$
$$f(x + 1, b) = (x + 1)^2 + 2(x + 1)b + 5b$$
$$= x^2 + 2x + 1 + 2xb + 2b + 5b$$
$$= x^2 + 2(1 + b)x + 7b + 1,$$

and so on. Since we may substitute any number for x and any number for y in the defining equation for f, the domain of f is the set of all ordered pairs (x, y). □

Similarly, a function defined by an equation such as

$$f(x, y, z) = \frac{x - y}{y - z}$$

is a function of three variables. The domain of such a function is a set of ordered triples (a, b, c), where a, b, and c are numbers. We evaluate the function at the ordered triple (a, b, c) by substituting a for x, b for y, and c for z in its defining equation.

EXAMPLE 2

Let
$$f(x, y, z) = \frac{x - y}{y - z}.$$

Then
$$f(1, 1, 0) = \frac{1 - 1}{1 - 0} = 0,$$

and
$$f(2, -1, 3) = \frac{2 - (-1)}{-1 - 3} = -\frac{3}{4}.$$

However, $f(1, 2, 2)$ is not defined. (Why not?) More generally, notice that
$$f(x, y, z) = \frac{x - y}{y - z}$$

is not defined if and only if $y = z$. Thus the domain of this function is
$$\{(x, y, z) | y \neq z\}. \quad \square$$

Functions of several variables can also be defined by equations that relate a dependent variable to two or more independent variables.

EXAMPLE 3

Let
$$z = xe^x + \ln y.$$

Then

when $x = 0$ and $y = 1$, $\quad z = 0e^0 + \ln 1 = 0 + 0 = 0;$

when $x = 1$ and $y = 1$, $\quad z = 1e^1 + \ln 1 = e + 0 = e;$

when $x = -2$ and $y = 2$, $\quad z = -2e^{-2} + \ln 2;$

and so on. Note that since $\ln y$ is not defined when $y \leq 0$, the function defined by this equation has domain $\{(x, y) | y > 0\}. \quad \square$

Functions of four or more variables are defined and evaluated in a manner similar to that illustrated in the preceding examples.

As we mentioned at the beginning of this chapter, functions of several variables arise naturally in many situations. We illustrate with examples.

EXAMPLE 4

Suppose that a store sells two products, X and Y, and that the sales price of one product is not affected by the sales price of the other. If each unit of product X sells for \$15 and each unit of product Y for \$20, then the store's revenue from selling x units of X and y units of Y is
$$R(x, y) = 15x + 20y$$

dollars. $\quad \square$

EXAMPLE 5   Tropicoid Corporation makes two models of its instant camera, the regular and the deluxe. The company's cost if it makes x units of the regular model and y units of the deluxe model is

$$C(x, y) = 40x + 60y + 0.01xy + 20{,}000$$

dollars. Note that the company's fixed cost is therefore

$$C(0, 0) = 20{,}000$$

dollars.

Now suppose Tropicoid makes a fixed number, say 1000 units, of the regular model; then its cost will be

$$C(1000, y) = 40(1000) + 60y + 0.01(1000)y + 20{,}000 = 60{,}000 + 70y$$

dollars. Therefore if production of the regular model is held fixed at 1000 units, each additional unit of the deluxe model made will cost $70. Similarly, if production of the regular model is held fixed at 2000 units, then

$$C(2000, y) = 100{,}000 + 80y,$$

which tells us that each additional unit of the deluxe model produced will cost $80. On the other hand, if production of the deluxe model is held fixed at 2000 units, then

$$C(x, 2000) = 140{,}000 + 60x$$

shows that each additional unit of the regular model produced will cost $60. ☐

EXAMPLE 6   Suppose a company sells both electric heaters and kerosene heaters. Let p_E be the unit price of the electric heaters and p_K the unit price of the kerosene heaters. The quantity demanded of either product may depend on both the price of that product and the price of the other; thus, if E is the quantity of electric heaters demanded, then E may depend on both p_E and p_K. For example we might have

$$E = -100p_E + 500p_K + 10{,}000.$$

Then if electric heaters sell for $40 each and kerosene heaters for $60 each, the quantity of electric heaters demanded will be

$$E = -100(40) + 500(60) + 10{,}000 = 36{,}000$$

units.

Let us analyze the demand function

$$E = -100p_E + 500p_K + 10{,}000.$$

If the price p_K of kerosene heaters remains fixed at, say, $p_K = b$, then the quantity of electric heaters demanded is

$$E = -100p_E + 500b + 10{,}000.$$

Therefore each $1 rise in the price p_E of electric heaters will result in a decrease of 100 in the quantity of electric heaters demanded by consumers. In other words, if the price of kerosene heaters remains fixed, a rise in the price of electric heaters leads to a decrease in the quantity of electric heaters demanded. On the other hand, if the price of electric heaters is fixed at, say, $p_E = a$, then

$$E = -100a + 500p_K + 10{,}000,$$

which shows that each increase of $1 in the price p_K of kerosene heaters leads to an increase of 500 in the quantity of electric heaters demanded. Thus, if the price of electric heaters remains fixed while the price of kerosene heaters goes up, the quantity of electric heaters demanded will increase. ☐

Functions of two variables can be graphed using the **three-dimensional cartesian coordinate system.** (Functions of three or more variables cannot be graphed.) The three-dimensional cartesian coordinate system is depicted in Figure 8–1: the x-, y-, and z-axes are perpendicular and intersect at the **origin;** their positive portions are shown as solid lines, their negative portions as dashed lines. Because of the difficulties with perspective that arise when the negative portions of the axes are shown, diagrams are sometimes restricted to the region where all three axes are nonnegative; this region, called the **first octant,** is shown in Figure 8–2.

Just as any ordered pair *(a, b)* of numbers can be plotted as a point on the two-dimensional cartesian coordinate system, so any ordered triple *(a, b, c)* of numbers can be plotted as a point in three-dimensional space using the three-dimensional cartesian coordinate system (Figure 8–3).

EXAMPLE 7 In Figure 8–4 we have plotted the origin (0, 0, 0) and the points

(3, 0, 0),   (0, 2, 0),   (0, 0, 4),   (3, 2, 0),   (0, 2, 3),
(1, 0, 1),   (3, 0, 4),   (3, 2, 4),   (0, 2, 4),   (1, 2, 2). ☐

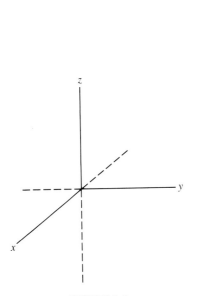

FIGURE 8–1
The Cartesian Coordinate System

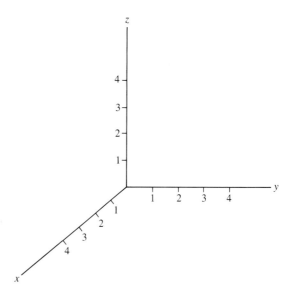

FIGURE 8–2
The First Octant

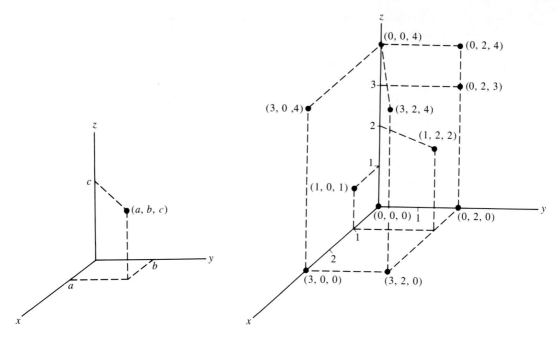

FIGURE 8–3 **FIGURE 8–4**

The set of all points of the form $(x, y, 0)$ is called the **xy-plane,** the set of all points of the form $(x, 0, z)$ the **xz-plane,** and the set of all points of the form $(0, y, z)$ the **yz-plane.** In Example 7, $(3, 2, 0)$ and $(0, 2, 0)$ are points in the *xy*-plane, $(3, 0, 4)$ and $(3, 0, 0)$ are points in the *xz*-plane, and $(0, 2, 4)$ and $(0, 2, 0)$ are points in the *yz*-plane. The *xy*-, *xz*-, and *yz*-planes are known collectively as the **coordinate planes.**

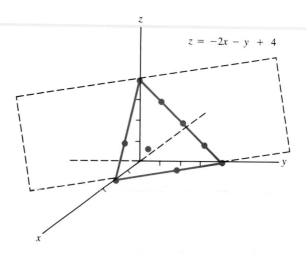

FIGURE 8–5

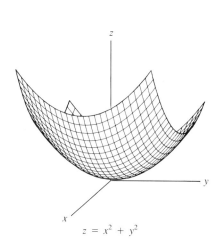

$z = x^2 + y^2$

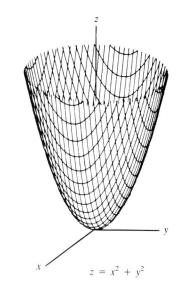

$z = x^2 + y^2$

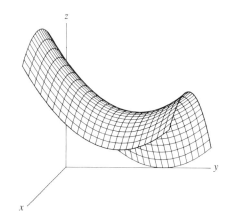

$z = 4 + (y-2)^2 - (x-2)^2$

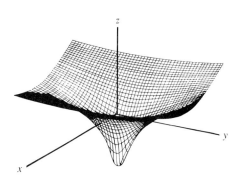

$z = \ln(x^2 + 2y^2)$

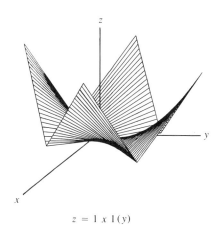

$z = 1 \, x \, 1 \, (y)$

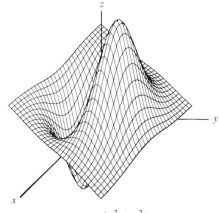

$z = ye - (x^2 + y^2)$

FIGURE 8–6

484   CHAPTER 8

We can use the three-dimensional cartesian coordinate system to graph any function f of two variables. The technique is to set $z = f(x, y)$, give values to x and y, calculate $z = f(x, y)$, and plot the resulting point (x, y, z). The graph will be a surface.

EXAMPLE 8   Let us graph the function whose defining equation is

$$f(x, y) = -2x - y + 4.$$

We set

$$z = -2x - y + 4.$$

Then we have

x	y	$z = -2x - y + 4$	(x, y, z)
0	0	4	(0, 0, 4)
1	0	2	(1, 0, 2)
2	0	0	(2, 0, 0)
0	1	3	(0, 1, 3)
1	1	1	(1, 1, 1)
0	2	2	(0, 2, 2)
1	2	0	(1, 2, 0)
0	3	1	(0, 3, 1)
0	4	0	(0, 4, 0)

The graph of f is the plane depicted in Figure 8–5. ◻

We do not intend to place much emphasis on graphing functions of two variables: their graphs are often quite difficult to draw, and today computers can do the graphing more quickly and with better artistic results than most people can. By way of example, Figure 8–6 presents several computer-generated graphs of functions of two variables.

■ *Computer Manual: Program TWOVARS*

8.1   EXERCISES

In Exercises 1 through 10, find the domain of the function f and the indicated functional values, if possible.

1. $f(x, y) = x^2 + 2xy + y^2$, $f(0, 0), f(1, 3)$

2. $f(x, y) = \dfrac{2x}{y + 1}$, $f(-2, -3), f(4, -1)$

3. $f(p, q) = \sqrt{p} + 2\sqrt{q}$, $f(0, 4)$, $f(9, 16)$

4. $f(x, y) = \dfrac{2x^2 - 3y^2}{x + y}$, $f(1, 1), f(4, 0)$

5. $f(u, v) = \sqrt{2uv} + 1$, $f(0, 0)$, $f(3, -2)$

6. $f(x, y) = e^{-xy} + e^{x/y}$, $f(1, 0), f(2, 2)$

7. $f(x, y, z) = x^2 + y^2 + z^2$, $f(1, 2, -1), f(0, 2, 5)$

8. $f(r, s, t) = \dfrac{1}{r} + \dfrac{2}{s^2 - t^2}$, $f(1, 2, 3), f(-2, 2, 2)$

9. $f(x, y, z) = \ln x + 2 \ln y + 3 \ln z$, $f(1, 1, 1), f(e, e^2, e^3)$
10. $f(x, y, z, w) = \ln |x| + \ln |y| + \ln |z| + \ln |w|$, $f(-1, 1, 1, 1)$, $f(e, -e, -e, e)$
11. Let $z = x^2 + 2x^3y - 3y^2$. Find z when $x = 0$ and $y = 2$; when $x = -2$ and $y = -1$. What is the domain of this function?
12. Let $z = 2x^2 + \dfrac{x}{y} + 2$. Find z when $x = 0$ and $y = 1$; when $x = 2$ and $y = 2$. What is the domain of this function?
13. Let $w = u^2 - uv + (uv)^{-1}$. Find w when $u = 1$ and $v = 2$; when $u = -1$ and $v = -2$; when $u = v = a$. What is the domain of this function?
14. Let $w = e^{z-2x+3y}$. Find w when $x = -1$, $y = z = 2$; when $x = 0$, $y = a$, and $z = -a$. What is the domain of this function?
15. Let $u = z(x^2 - y^2)^{1/2}$. Find u when $x = 1$, $y = 2$, and $z = 3$; when $x = 3$, $y = 2$, and $z = 1$; when $x = a$, $y = 0$, and $z = b$. When is the domain of this function?
16. Let $q = (2x - 3y)^{1/2} + z^2(w - x)^{-1}$. Find q when $x = y = z = w = 1$; when $x = 1$, $y = -1$, $z = 2$, and $w = 3$; when $x = t/2$, $y = t/3$, $z = t$, and $w = -t/2$. What is the domain of this function?
17. A stockbroker charges a commission of

$$c(x, y) = 25 + 0.15x + 0.005xy$$

dollars on any trade involving x shares of stock selling at y dollars per share. Find the stockbroker's commission on a trade involving 100 shares selling at $20 per share; on 400 shares selling at $60 per share.

Management

18. A refinery produces two grades of gasoline, regular and premium. If the cost of producing one grade does not affect the cost of producing the other, if it costs $0.40 to produce each gallon of regular and $0.50 to produce each gallon of premium, and if the refinery's fixed cost is $100,000 per day, find
    (a) its cost function;
    (b) the cost of making 20,000 gallons of each grade of gasoline.
19. A firm makes two products, A and B. The profit contribution of one product is not affected by that of the other. If each unit of product A contributes $60 toward profit, each unit of product B contributes $75 toward profit, and the fixed cost is $30,000, find the profit function and the firm's profit when it makes and sells 300 units of A and 500 of B.
20. A store sells white paint and blue paint. If the prices at which the two colors are sold do not affect each other, if x is the price per gallon of the white paint and y the price per gallon of the blue paint, and if the quantity of white paint demanded is

$$W = 400 - 6x + 2y$$

gallons and that of blue paint is

$$B = 100 + 4x - 10y$$

gallons, find

(a) the function that describes the store's revenue from paint sales;
(b) the store's revenue if it sells the white paint for $5 per gallon and the blue for $7 per gallon.

Suppose the demand function for product A is given by
$$A(p_A, p_B) = 200 - 8p_A + 3p_B,$$
where p_A is the unit price of A and p_B the unit price of another product B.
Exercises 21 through 24 refer to this product

21. Find the quantity of A demanded if its price is $10 per unit and the price of B is $12 per unit.
22. If the unit price of A is fixed at a dollars, what can you say about the quantity of A demanded?
23. If the unit price of B is fixed at b dollars, what can you say about the quantity of A demanded?
24. How can the quantity of A demanded increase? How can it decrease?

A firm makes two products, X and Y. Its cost function is given by
$$C(x, y) = 2000 + 40x + 20y$$
and its revenue function by
$$R(x, y) = 12x\sqrt{y},$$
where x is the number of units of X produced and sold and y the number of units of Y produced and sold. Exercises 25 through 33 refer to this situation.

25. What is the firm's fixed cost?
26. What is the cost if 150 units of X and 100 units of Y are produced?
27. What can you say about the cost if the number of units of X produced is held fixed? If the number of units of Y produced is held fixed?
28. Find a function that describes the firm's average cost per unit, and find the average cost per unit when 150 units of X and 100 units of Y are produced.
29. What can you say about the average cost per unit when production of X is held fixed at 150 units? When production of Y is held fixed at 100 units?
30. Find the firm's profit function and its profit if it makes and sells 150 units of X and 100 units of Y.
31. Find a function that describes the firm's average profit per unit, and find its average profit per unit if it makes and sells 150 units of X and 100 units of Y.
32. What can you say about the firm's profit when production and sales of X are held fixed at 150 units? When production and sales of Y are held fixed at 100 units?
33. What can you say about the firm's average profit per unit when production and sales of X are held fixed at 150 units? When production and sales of Y are held fixed at 100 units?

A firm makes three products, X, Y, and Z. Its cost function is given by
$$C(x, y, z) = 25x + 30y + 32z + 40{,}000,$$
where x is the number of units of X, y the number of units of Y, and z the number of units of Z made. Exercises 34 through 36 refer to this situation.

34. Find the cost of producing 50 units of X, 60 units of Y, and 40 units of Z.
35. What can you say about the firm's cost if production of X and Y are held fixed at 50 and 60 units, respectively?
36. Find a function that describes the firm's average cost per unit and find the firm's average cost per unit when $x = 50$, $y = 60$, and $z = 40$.

The supply function for commodity A is

$$S = (8p_1 + 2p_2 + 1)/(3p_3 + 1),$$

where S is in millions of units, p_1 is the unit price of A, and p_2, p_3 are unit prices of other commodities. Exercises 37 through 40 refer to this situation.

37. Find the quantity of commodity A supplied if $p_1 = \$10$, $p_2 = \$5$, and $p_3 = \$4$.
38. What can you say about the quantity of A supplied if p_1 and p_2 remain fixed at $10 and $4, respectively?
39. How can the quantity of A supplied increase?
40. How can the quantity of A supplied decrease?

Exercises 41 through 45 were suggested by material in Niebel, *Motion and Time Study*, eighth edition, Irwin, 1988. They concern the number of machines an operator can tend. Suppose an operator tends several identical machines, moving from one to another loading and unloading them. Let

s = service time (loading time plus unloading time) per machine;
m = machine time (running time) per machine;
t = average travel time (time spent traveling from one machine to another).

The number of machines the operator can be assigned to is then given by the **synchronous service function** f defined by

$$f(s, m, t) = \left[\frac{s + m}{s + t}\right].$$

Here $\left[\dfrac{s + m}{s + t}\right]$ denotes the largest integer less than or equal to $\dfrac{s + m}{s + t}$.

*41. What is the domain of the synchronous service function?
*42. Find the number of machines an operator can tend if loading time is 1 minute, running time is 5 minutes, unloading time is 30 seconds, and average travel time is 15 seconds.
*43. Repeat Exercise 42 if loading time is 2 minutes, running time is 10 minutes, no unloading is necessary, and average travel time is 20 seconds.

If w_1 is the operator's wage rate and w_2 is the cost of running the machine, then the total cost of one cycle (loading, running, unloading, travel to next machine) for one machine will be

$$C = \frac{(s + m)(w_1 + w_2 f(s, m, t))}{f(s, m, t)}.$$

*44. Suppose that loading time is 2 minutes, running time is 12 minutes, unloading time is 1 minute, travel time averages 30 seconds, the operator earns $12 per hour, and it costs $72 per hour to run a machine. Find the cost of one cycle for one machine.
*45. Repeat Exercise 44 if the loading time is 2 minutes 15 seconds, the running time is 3 minutes 30 seconds, the unloading time is 45 seconds, there is no travel re-

quired, the operator earns $9 per hour, and it costs $240 per hour to run the machine.

Life Science

46. Let x denote the number of dogs, y the number of mice, and z the number of cats in a neighborhood. If
$$z = 10 - 2x + 0.2y,$$
   (a) Find the number of cats in the neighborhood if there are 6 dogs and 40 mice there.
   (b) What can you say about the cat population if the dog population of the neighborhood remains constant? If the mouse population remains constant?

The approximate time it will take for a patient to recover from pneumonia is
$$f(t, d) = \frac{1.2t}{d}$$
days, where t is the patient's age in years, $t \geq 21$, and d is the patient's daily dosage of penicillin in millions of units. Exercises 47 through 50 refer to this situation.

47. Find the approximate recovery time for a patient who is 40 years old and takes 4 million units of penicillin per day.

48. What can you say about the recovery time for patients who are all the same age?

49. What can you say about the recovery time for patients all of whom take the same daily dosage of penicillin?

50. How can recovery time increase; how can it decrease?

Exercises 51 through 55 were suggested by material in Niebel, *Motion and Time Study*, eighth edition, Irwin, 1988. They concern the contrast between an object and its background. This contrast is given by the equation
$$C = \frac{L - D}{L},$$
where D is the luminance of the darker of the object and its background, L is the luminance of the lighter of the object and its background, both measured in lamberts, and C is the contrast.

51. Find the contrast if $D = 10$ and $L = 20$ lamberts.

52. Find C if $D = 20$ and $L = 10$ lamberts.

53. Find C when D is fixed at D_0. What happens to C as L increases?

54. Find C when L is fixed at L_0. What happens to C as D increases?

55. The larger the value of C, the better the contrast. How can the contrast be increased?

Social Science

Annual housing starts H in a community depend on average weekly income Y and the average annual percentage rate I for mortgage loans according to the equation
$$H = 200 + 0.2Y - 0.65I^2.$$
Exercises 56 through 58 refer to this situation.

56. What will housing starts be if average income is $250 per week and the average annual percentage for mortgage loans is 12%?

57. What can you say about housing starts if I remains stable at 12% per year?

58. What can you say about housing starts if Y remains fixed at $250 per week?

8.2 PARTIAL DERIVATIVES

Just as a function of one variable may have a first derivative with respect to its independent variable, so a function of more than one variable may have several first derivatives, one with respect to each of its independent variables. These derivatives, in turn, may themselves have derivatives with respect to each independent variable, and so on. Such derivatives of functions of several variables are known as **partial derivatives,** and the process by which they are found is **partial differentiation.**

Partial derivatives are important for many reasons, not the least of which is that they can be used to find the rate of change of a function of several variables with respect to any one of its variables while the others remain fixed. Thus, for instance, if $C = C(x, y)$ is a cost function, the partial derivative of C with respect to x will tell us (approximately) how much total cost C will change as x increases by one unit while y remains the same. This section is devoted to showing how to find partial derivatives; we postpone further discussion of their applications until the following sections.

First Partial Derivatives

A function of two variables can have two first partial derivatives, one with respect to each of its independent variables.

First Partial Derivatives

Let f be a function of the two variables x and y.

1. We obtain the **first partial derivative of f with respect to x,** denoted by

$$f_x \quad \text{or} \quad \frac{\partial f}{\partial x},$$

by differentiating f with respect to x while treating y as a constant.

2. We obtain the **first partial derivative of f with respect to y,** denoted by

$$f_y \quad \text{or} \quad \frac{\partial f}{\partial y},$$

by differentiating f with respect to y while treating x as a constant.

Partial differentiation for functions of more than two variables is carried out in a similar manner.

EXAMPLE 1 Let us find the first partial derivatives f_x and f_y for the function f defined by

$$f(x, y) = x^2 + y^2 + 2x + 3y.$$

To find f_x, we must differentiate f with respect to x while treating y as a constant. To remind ourselves that y is to be treated as a constant, we replace y by c in $f(x, y)$:

$$f(x, c) = x^2 + c^2 + 2x + 3c,$$

or

$$f(x, c) = x^2 + 2x + (c^2 + 3c);$$

differentiating with respect to x, we obtain

$$f_x = 2x + 2 + 0 = 2x + 2.$$

Similarly, to find f_y we replace x by c, obtaining

$$f(c, y) = y^2 + 3y + (c^2 + 2c),$$

and differentiate with respect to y to find that

$$f_y = 2y + 3. \quad \square$$

It is not really necessary for us to replace x and y with c when finding partial derivatives. Instead, we need merely keep in mind that when we are differentiating f to find f_x, y is to be treated as a constant, and when we are differentiating f to find f_y, x is to be treated as a constant.

EXAMPLE 2 Let us find f_x and f_y if f is defined by

$$f(x, y) = 2x - 5y + x^2y^3.$$

Differentiating with respect to x while treating y as a constant, we have

$$f_x = 2(1) - 5(0) + 2xy^3 = 2 + 2xy^3.$$

Similarly, differentiating with respect to y while treating x as a constant, we obtain

$$f_y = 2(0) - 5(1) + x^2(3y^2) = -5 + 3x^2y^2. \quad \square$$

EXAMPLE 3 Let us find the partial derivatives $\partial g/\partial x$, $\partial g/\partial y$, and $\partial g/\partial z$ for the function g of three variables defined by

$$g(x, y, z) = 5x + 3xy - 2x^2z + y^2.$$

To find $\partial g/\partial x$, we differentiate with respect to x while treating y and z as constants:

$$\frac{\partial g}{\partial x} = 5(1) + 3(1)y - 2(2x)z + 0 = 5 + 3y - 4xz.$$

Similarly, to find $\partial g/\partial y$, we differentiate with respect to y while treating x and z as constants:

$$\frac{\partial g}{\partial y} = 5(0) + 3x(1) - 2(0) + 2y = 3x + 2y.$$

To find $\partial g/\partial z$, we differentiate with respect to z while treating x and y as constants:

$$\frac{\partial g}{\partial z} = 5(0) + 3(0) - 2x^2(1) + 0 = -2x^2. \quad \square$$

EXAMPLE 4   Let us find $f_x(1, 1)$ and $f_y(2, 1)$ if

$$f(x, y) = x^2 - 7x^2y^3 + y + 1.$$

Differentiating with respect to x and treating y as a constant, we have

$$f_x = 2x - 14xy^3.$$

Therefore

$$f_x(1, 1) = 2(1) - 14(1)(1)^3 = -12.$$

Similarly,

$$f_y = -21x^2y^2 + 1,$$

so

$$f_y(2, 1) = -21(2)^2(1)^2 + 1 = -83. \quad \square$$

EXAMPLE 5   Let

$$h(x, y, z, w) = (2x + 3y^2 + 4z^3 + 5w^4)^{10}.$$

Then

$$h_x = 10(2x + 3y^2 + 4z^3 + 5w^4)^9(2)$$
$$= 20(2x + 3y^2 + 4z^3 + 5w^4)^9,$$
$$h_y = 10(2x + 3y^2 + 4z^3 + 5w^4)^9(6y)$$
$$= 60y(2x + 3y^2 + 4z^3 + 5w^4)^9,$$
$$h_z = 10(2x + 3y^2 + 4z^3 + 5w^4)^9(12z^2)$$
$$= 120z^2(2x + 3y^2 + 4z^3 + 5w^4)^9,$$

and

$$h_w = 10(2x + 3y^2 + 4z^3 + 5w^4)^9(20w^3)$$
$$= 200w^3(2x + 3y^2 + 4z^3 + 5w^4)^9.$$

(Check this.) $\square$

Notice that we have shown how to *find* first partial derivatives, but we have not *defined* them. First partial derivatives are defined by difference quotients in a manner analogous to the way we used the difference quotient to define the derivative of a function of one variable in Section 2.3. For the sake of completeness, we give these definitions here for a function of two variables.

Definition of Partial Derivatives

Let f be a function of two variables x and y. Then

$$f_x(x, y) = \lim_{h \to 0} \frac{f(x + h, y) - f(x, y)}{h}$$

and

$$f_y(x, y) = \lim_{h \to 0} \frac{f(x, y + h) - f(x, y)}{h},$$

provided these limits exist.

For our purposes, knowing how to find partial derivatives will be sufficient; we will not need to use their definitions.

Higher Partial Derivatives

Just as a function of one variable may have a first derivative, a second derivative, and so on, a function of several variables may have first partial derivatives, second partial derivatives, and so on. The notation for second partial derivatives of a function of two variables is as follows:

Second Partial Derivatives

Let f be a function of two variables x and y with first partial derivatives f_x and f_y. Then the second partial derivatives of f are f_{xx}, f_{xy}, f_{yx}, and f_{yy}, where

1. f_{xx} is the partial derivative of f_x with respect to x;
2. f_{xy} is the partial derivative of f_x with respect to y;
3. f_{yx} is the partial derivative of f_y with respect to x;
4. f_{yy} is the partial derivative of f_y with respect to y.

Note that if we read the subscripts that label a second partial derivative from left to right, they tell us how to find the second partial; for instance, the notation

FUNCTIONS OF SEVERAL VARIABLES

f_{xy} tells us that we are to find the first partial of f with respect to x and then differentiate it with respect to y.

EXAMPLE 6 Let f be defined by

$$f(x, y) = 5x + y^2 + 2x^3y^4.$$

Then

$$f_x = 5 + 6x^2y^4,$$

so

$$f_{xx} = 12xy^4 \quad \text{and} \quad f_{xy} = 24x^2y^3.$$

Similarly,

$$f_y = 2y + 8x^3y^3$$

and thus

$$f_{yx} = 24x^2y^3 \quad \text{and} \quad f_{yy} = 2 + 24x^3y^2. \quad \square$$

Notice that the **mixed partials** f_{xy} and f_{yx} in Example 6 were the same. This is not the case for all functions, but it will be true for all the functions we use in this book.

The extension of the concept of second partial derivatives to functions of more than two variables is straightforward.

EXAMPLE 7 Let

$$g(x, y, z) = x^2y^3 + xyz + z^2.$$

Then

$$g_x = 2xy^3 + yz, \quad g_y = 3x^2y^2 + xz,$$

and

$$g_z = xy + 2z.$$

Therefore there are nine second partial derivatives:

$$\begin{array}{lll} g_{xx} = 2y^3, & g_{xy} = 6xy^2 + z, & g_{xz} = y, \\ g_{yx} = 6xy^2 + z, & g_{yy} = 6x^2y, & g_{yz} = x, \\ g_{zx} = y, & g_{zy} = x, & g_{zz} = 2. \end{array}$$

Notice that mixed partials that have the same subscripts are equal:

$$g_{xy} = g_{yx}, \quad g_{xz} = g_{zx}, \quad \text{and} \quad g_{yz} = g_{zy}. \quad \square$$

An alternative notation for second partial derivatives is as follows:

$$f_{xx} = \frac{\partial}{\partial x}\left(\frac{\partial f}{\partial x}\right) = \frac{\partial^2 f}{\partial x^2}, \quad f_{xy} = \frac{\partial}{\partial y}\left(\frac{\partial f}{\partial x}\right) = \frac{\partial^2 f}{\partial y \, \partial x},$$

$$f_{yx} = \frac{\partial}{\partial x}\left(\frac{\partial f}{\partial y}\right) = \frac{\partial^2 f}{\partial x \, \partial y}, \quad f_{yy} = \frac{\partial}{\partial y}\left(\frac{\partial f}{\partial y}\right) = \frac{\partial^2 f}{\partial y^2}.$$

Notice that when we use this notation, it is the order of the variables in the denominator from right to left that tells us how to find the second partial:

$$\frac{\partial^2 f}{\partial x\, \partial y} \text{ is the partial derivative of } \frac{\partial f}{\partial y} \text{ with respect to } x.$$

EXAMPLE 8 Let

$$f(x, y) = e^y + xy^2.$$

Then

$$f_y = e^y + 2xy$$

and

$$\frac{\partial^2 f}{\partial x\, \partial y} = \frac{\partial}{\partial x}(f_y) = \frac{\partial}{\partial x}(e^y + 2xy) = 2y. \quad \square$$

EXAMPLE 9 Let

$$z = x^2 + \ln xy.$$

Then

$$\frac{\partial z}{\partial x} = 2x + \frac{1}{xy}y = 2x + \frac{1}{x},$$

so

$$\frac{\partial^2 z}{\partial x^2} = 2 - \frac{1}{x^2} \quad \text{and} \quad \frac{\partial^2 z}{\partial y\, \partial x} = 0. \quad \square$$

Third and higher-order partial derivatives are found in the obvious manner.

EXAMPLE 10 Find the third partial derivative f_{xyx} for

$$f(x, y) = 3y - 2xy^2 + y \ln x.$$

The notation f_{xyx} tells us that we must differentiate first with respect to x, then with respect to y, and finally with respect to x again. Thus,

$$f_x = -2y^2 + \frac{y}{x},$$

$$f_{xy} = -4y + \frac{1}{x},$$

and

$$f_{xyx} = -\frac{1}{x^2}. \quad \square$$

8.2 EXERCISES

First Partial Derivatives

1. Find f_x and f_y if $f(x, y) = 5x - 7y + 12$.
2. Find f_x and f_y if $f(x, y) = 6x + 12y^2 - 3$.
3. Find f_x and f_y if $f(x, y) = 8xy - 2y^2$.
4. Find f_x and f_y if $f(x, y) = x^2y^3 - 2xy^2 - 1$.
5. Find $\dfrac{\partial f}{\partial x}$ and $\dfrac{\partial f}{\partial y}$ if $f(x, y) = 3x + \dfrac{2x^2}{y^2} + 6$.
6. Find $\dfrac{\partial f}{\partial x}$ and $\dfrac{\partial f}{\partial y}$ if $f(x, y) = 5x^3 + 7y^4 - 3x^2y^2 + 5xy - 2x + y$.
7. Find g_x and g_y if $g(x, y) = \sqrt{2xy}$.
8. Find $\dfrac{\partial z}{\partial x}$ and $\dfrac{\partial z}{\partial y}$ if $z = \sqrt{x^2 - y^2}$.
9. Find $\dfrac{\partial h}{\partial u}$ and $\dfrac{\partial h}{\partial v}$ if $h(u, v) = (u^2 + uv + v^3)^4$.
10. Find f_r and f_s if $f(r, s) = e^{2r} - \ln s$.
11. Find $\dfrac{\partial z}{\partial x}$ and $\dfrac{\partial z}{\partial y}$ if $z = x \ln y^2$.
12. Find $\dfrac{\partial s}{\partial p}$ and $\dfrac{\partial s}{\partial q}$ if $s = 2p^2q^4 - 5pq + 2$.
13. Find $\dfrac{\partial w}{\partial x}$ and $\dfrac{\partial w}{\partial y}$ if $w = \dfrac{2x - y}{x + y}$.
14. Find h_x and h_y if $h(x, y) = \ln \dfrac{x + y}{x - y}$.
15. Find f_x, f_y, and f_z if $f(x, y, z) = xz + x^2z^2 + yz^3 - 5y$.
16. Find h_x, h_y, and h_z if $h(x, y, z) = e^{2xz} - \ln yz + 1$.
17. Find $f_x(1, 3)$ and $f_y(-2, 1)$ if $f(x, y) = 2xy$.
18. Find $f_x(0, 0)$ and $f_y(1,1)$ if $f(x, y) = 2x + y + 4xy - x^2 + y^2$.
19. Find $f_x(1, 0)$ and $f_y(0, 1)$ if $f(x, y) = 2e^x - 3e^{-y}$.
20. Evaluate $\dfrac{\partial f}{\partial x}$ and $\dfrac{\partial f}{\partial y}$ at $(8, 3)$ if $f(x, y) = \sqrt{2x + 3y}$.
21. Evaluate $\dfrac{\partial z}{\partial x}$ and $\dfrac{\partial z}{\partial y}$ at $(0, e)$ if $z = e^{-x} \ln y$.
22. Find $\dfrac{\partial w}{\partial x}\bigg|_{(1,-1,1)}$ and $\dfrac{\partial w}{\partial y}\bigg|_{(1,-1,1)}$ if $w = \sqrt{x^2 + y^2} + 4xz$.
23. Find $f_x(1, -1, 2)$, $f_y(1, -1, 2)$, and $f_z(1, -1, 2)$ if $f(x, y, z) = x^2 + \dfrac{5y}{xz} + 10xz$.
24. Find $\dfrac{\partial p}{\partial u}\bigg|_{(0,0,2)}$, $\dfrac{\partial p}{\partial v}\bigg|_{(0,0,2)}$, and $\dfrac{\partial p}{\partial w}\bigg|_{(0,0,2)}$ if $p = (3u^2w^2 + 4v^2w - 2u + 3v + w + 1)^4$.

Higher Partial Derivatives

25. If $f(x, y) = 2x + 3y^2 + 1$, find f_{xx}, f_{xy}, f_{yx}, and f_{yy}.
26. If $f(x, y) = 4x^2 + 3xy - y^2$, find f_{xx}, f_{xy}, f_{yx}, and f_{yy}.
27. If $g(x, y) = x^2 + y^2 - 2xy^3$, find g_{xx}, g_{xy}, g_{yx}, and g_{yy}.
28. If $f(x, y) = 2y^3 - 3x^3 + 7x + 12y^2 + 1$, find f_{xx}, f_{xy}, f_{yx}, and f_{yy}.
29. If $f(x, y) = \dfrac{2x}{3y^2} + 1$, find f_{xx}, f_{xy}, f_{yx}, and f_{yy}.

30. If $z = y^2\sqrt{x} - 2x^3 + 3$, find $\dfrac{\partial^2 z}{\partial x^2}, \dfrac{\partial^2 z}{\partial y\, \partial x}, \dfrac{\partial^2 z}{\partial x\, \partial y}$, and $\dfrac{\partial^2 z}{\partial y^2}$.

31. If $z = \dfrac{4x^4}{y} + \dfrac{2y^2}{x^2}$, find $\dfrac{\partial^2 z}{\partial x^2}, \dfrac{\partial^2 z}{\partial y\, \partial x}, \dfrac{\partial^2 z}{\partial x\, \partial y}$, and $\dfrac{\partial^2 z}{\partial y^2}$.

32. If $f(x, y, z) = x^2 + 3xyz + 2yz + z^2$, find all second partial derivatives of f.

33. If $f(x, y, z) = ye^x + 2\ln x + x\ln z$, find all second partial derivatives of f.

34. If $f(x, y) = 10x^2 y^3 + x + 2y^3$, find f_{xx}, f_{xyx}, and f_{yyx}.

35. If $f(x, y) = e^{xy}$, find f_{xyxy} and f_{xyyy}.

36. Find f_{uvu} if $f(u, v) = u^4 + 2e^{u/v}$.

37. Find g_{pqq} if $g(p, q, r) = 1 + 3pqr^{-1} + q^3 p^{-1}$.

38. Find f_{xxy} if $f(x, y) = e^{x-2y}$.

39. Find f_{xxyz} if $f(x, y, z) = \ln xyz + 2x^3 y^2 z^3$.

40. If $z = \ln 2x + 3\ln y$, find $\dfrac{\partial^2 z}{\partial x^2}, \dfrac{\partial^3 z}{\partial x\, \partial y\, \partial x}$, and $\dfrac{\partial^3 z}{\partial x\, \partial y^2}$.

41. If $f(x, y) = 5x^2 - 3xy + y$, find $f_{xx}(1, 1), f_{xy}(2, 2), f_{yx}(3, -5)$, and $f_{yy}(-2, 7)$.

42. If $f(x, y) = 6x + 2ye^x$, find $f_{xx}(2, 0), f_{xy}(1, -1), f_{yx}(-2, 1)$, and $f_{yy}(0, 2)$.

43. If $f(x, y) = x^2 y^3 + x^4 + y^4$, find $f_{xyx}(1, 1)$ and $f_{yyx}(-1, -1)$.

44. If $f(x, y) = xy + \ln xy$, find $f_{xxyy}(2, 2)$ and $f_{xyxx}(2, 2)$.

8.3 PARTIAL DERIVATIVES AS RATES OF CHANGE

At the beginning of Section 8.2 we remarked that a first partial derivative of a function of several variables gives its rate of change with respect to one of the variables as the others remain fixed. In this section we illustrate some uses of this fact. Before considering partial derivatives as rates of change, however, we briefly

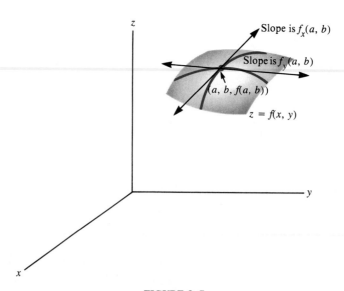

FIGURE 8–7

explain the geometric significance of the partial derivatives of a function of two variables.

The Geometric Significance of Partial Derivatives

Just as the derivative $f'(a)$ of a function of one variable is the slope of the line tangent to the graph of f at the point $(a, f(a))$, so the partial derivatives $f_x(a, b)$ and $f_y(a, b)$ of a function of two variables are slopes of tangent lines to the graph of f at the point $(a, b, f(a, b))$. To be specific:

Slopes of Tangent Lines to Surfaces

Let f be a function of the two variables x and y, and let (a, b) be a point in the domain of f.

1. If $f_x(a, b)$ exists, it is the slope of the line parallel to the xz-plane and tangent to the surface $z = f(x, y)$ at the point $(a, b, f(a, b))$.

2. If $f_y(a, b)$ exists, it is the slope of the line parallel to the yz-plane and tangent to the surface $z = f(x, y)$ at the point $(a, b, f(a, b))$.

See Figure 8–7.

EXAMPLE 1 Let f be defined by $f(x, y) = y^2 - x^2$. When $x = 1$ and $y = 2$, $f(1, 2) = 3$, so the point $(1, 2, 3)$ is on the surface $z = y^2 - x^2$. Since

$$f_x(x, y) = -2x,$$

the slope of the line tangent to the surface at $(1, 2, 3)$ and parallel to the xz-plane is

$$f_x(1, 2) = -2(1) = -2.$$

Thus, since $f_x(1, 2)$ is negative, at the point $(1, 2, 3)$ the surface bends downward in the positive x-direction. See Figure 8–8. Similarly, since

$$f_y(x, y) = 2y,$$

the slope of the line tangent to the surface at $(1, 2, 3)$ and parallel to the yz-plane is

$$f_y(1, 2) = 2(2) = 4,$$

which is positive; hence at $(1, 2, 3)$ the surface bends upward in the positive y-direction. Again, see Figure 8–8. □

As Example 1 indicates, the signs of the first partial derivatives f_x and f_y evaluated at a point tell us whether at that point the surface bends upward or downward in the positive x- and y-directions, respectively. Another way to think of this is to

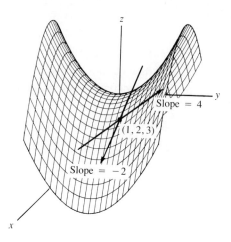

FIGURE 8–8

consider what will happen if we take a "slice" through the surface parallel to the *xz*- or *yz*-plane: in either case the result will be a curve on the surface, and the signs of the first partials will tell us whether the curve is increasing or decreasing. Thus, suppose we take a slice through the surface parallel to the *xz*-plane: if f_x is positive

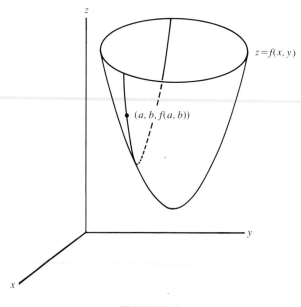

FIGURE 8–9
$f_x(a, b) > 0, f_{xx}(a, b) > 0$

at a point on the curve, then at that point the curve will be increasing in the positive x-direction; if f_x is negative, the curve will be decreasing. See Figure 8–9, where $f_x(a, b) > 0$ and the curve is increasing in the positive x-direction. Similar remarks apply to f_y and slices parallel to the yz-plane.

The signs of the second partials f_{xx} and f_{yy} evaluated at a point tell us whether at that point the surface is concave upward or downward in the x- and y-directions, respectively. For example, if we take a slice through the surface parallel to the xz-plane and f_{xx} is positive at a point on the resulting curve, then the curve is concave upward there; if f_{xx} is negative, the curve is concave downward. Again see Figure 8–9, where $f_{xx}(a, b) > 0$ and the curve is concave upward. Similar remarks apply to f_{yy} and slices parallel to the yz-plane.

EXAMPLE 2   Let $f(x, y) = y^2 - x^2$. At any point on the surface, $f_{xx} = -2$. Hence if we take a slice through any point on the surface and parallel to the xz-plane, the resulting curve will be concave downward. Similarly, $f_{yy} = 2$ at any point on the surface, so any slice parallel to the yz-plane will result in a curve that is concave upward. See Figure 8–10. ☐

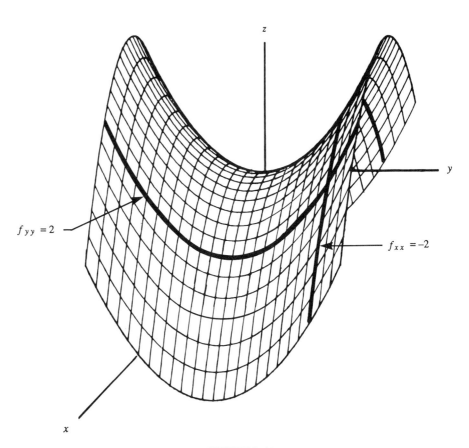

FIGURE 8–10

Partial Derivatives as Rates of Change

Consider Figure 8–7 again: as the figure indicates, the slope $f_x(a, b)$ may be interpreted as the instantaneous rate of change of $z = f(x, y)$ at the point $(a, b, f(a, b))$ in the positive x-direction; that is, $f_x(a, b)$ is the instantaneous rate of change of f with respect to x at $(a, b, f(a, b))$ when y is held fixed at the value $y = b$. Similarly, $f_y(a, b)$ is the instantaneous rate of change of f with respect to y at $(a, b, f(a, b))$ when x is held fixed at the value $x = a$. Thus for functions of two variables, partial derivatives can be interpreted as rates of change, and this remains true for functions of more than two variables.

Partial Derivatives as Rates of Change

Let f be a function of several variables, and let x be any one of the independent variables of f. Then f_x gives the **instantaneous rate of change of f with respect to x** when all the other independent variables of f are fixed. Thus, when all the other independent variables of f are fixed, a small change in x will lead to a change of approximately f_x in f.

We illustrate some applications of this point of view with examples.

EXAMPLE 3 Two bactericides are tested in combination, and it is found that when x units of bactericide X and y units of bactericide Y are used together, they will destroy

$$f(x, y) = 5x + 4y$$

million bacteria. Thus if y is held fixed, the rate of change of $f(x, y)$ with respect to x is

$$f_x(x, y) = 5.$$

The interpretation is that if the number of units y of bactericide Y is held fixed, each additional unit of bactericide X used will kill approximately 5 million bacteria.
    Similarly, since

$$f_y(x, y) = 4,$$

if the number of units of bactericide X is held fixed, then each additional unit of bactericide Y used will destroy approximately 4 million bacteria. □

EXAMPLE 4 Suppose that if a firm produces and sells x units of product A, y units of product B, and z units of product C, its profit is

$$P(x, y, z) = 2\sqrt{x} + 5y + 4\sqrt{z} + xyz$$

dollars. Then

$$P_x(x, y, z) = \frac{1}{\sqrt{x}} + yz$$

FUNCTIONS OF SEVERAL VARIABLES   501

is the firm's **marginal profit with respect to** x. If y and z are held fixed, an additional unit of product A made and sold will increase the firm's profit by approximately

$$\frac{1}{\sqrt{x}} + yz$$

dollars. For instance, if $x = 16$, $y = 10$, and $z = 25$, then the firm's profit is

$$P(16, 10, 25) = 4068$$

dollars. If y and z are held fixed at 10 and 25, respectively, then making and selling an additional unit of product A (the seventeenth one) will increase profit by approximately

$$P_x(16, 10, 25) = \frac{1}{\sqrt{16}} + (10)(25) = \$250.25.$$

Similarly, since

$$P_y(x, y, z) = 5 + xz \quad \text{and} \quad P_z(x, y, z) = \frac{2}{\sqrt{z}} + xy,$$

holding x and z fixed at 16 and 25, respectively, and producing an additional unit of B (the eleventh one) will contribute approximately

$$P_y(16, 10, 25) = 5 + (16)(25) = \$405$$

to profit; holding x and y fixed at 16 and 10, respectively, and producing an additional unit of C (the twenty-sixth one) will contribute approximately

$$P_z(16, 10, 25) = \frac{2}{\sqrt{25}} + (16)(10) = \$160.40$$

to profit. Therefore if the firm is currently making 16 units of A, 10 of B, and 25 of C and has limited resources for expanding production, it would be wise to use its resources to increase the production of B. ☐

EXAMPLE 5   Two products are said to be **competitive** if a *decrease* in the quantity demanded of one can lead to an *increase* in the quantity demanded of the other. (This occurs if the products can be substituted for each other. For instance, beef products and pork products are competitive; if the price of beef goes up while the price of pork remains fixed, the quantity of beef demanded will decrease and the quantity of pork demanded will increase, as consumers substitute pork for beef. Similarly, if the price of pork goes up while the price of beef remains unchanged, the quantity of pork demanded will decrease while the quantity of beef demanded will increase.) On the other hand, if a *decrease* in the quantity demanded of one product can lead to a *decrease* in the quantity demanded of the other, then the products are said to be **complementary**. (Automobiles and automobile tires are examples of complementary products.)

We can tell whether products are competitive or complementary by analyzing the partial derivatives of their demand functions. To see how this is done, suppose that the price of butter is b dollars per pound and that the price of margarine is m dollars per pound. Let

$$B(b, m) = 10{,}000 - 200b + 400m \quad \text{and} \quad M(b, m) = 8000 + 300b - 500m$$

be the demand functions for butter and margarine, respectively. Then

$$B_b = -200, \quad B_m = 400, \quad M_b = 300, \quad \text{and} \quad M_m = -500.$$

Now, B_b is negative, so an increase in the price b of butter while the price m of margarine is held fixed will lead to a decrease in the quantity of butter demanded. But since M_b is positive, an increase in b when m is held fixed will also lead to an increase in the quantity of margarine demanded. Similarly, since M_m is negative and B_m is positive, an increase in m while b is held fixed leads to a decrease in the quantity of margarine demanded and an increase in the quantity of butter demanded. Thus butter and margarine are competitive products. ∎

EXAMPLE 6 If a firm uses x worker-hours per week and y dollars of capital per week, it can produce

$$f(x, y) = 4x^{3/4}y^{1/4}$$

units per week. A function such as this which relates output to input by means of an expression of the form $kx^p y^{1-p}$, $0 \le p \le 1$, is called a **Cobb-Douglas production function**. In this case, if the firm has 1296 worker-hours and \$10,000 in capital available this week, it can produce

$$f(1296, 10{,}000) = 4(1296)^{3/4}(10{,}000)^{1/4} = 8640$$

units. The partial derivative f_x is called the **marginal productivity of labor**; the partial derivative f_y is the **marginal productivity of capital**. Here

$$f_x(x, y) = 3x^{-1/4}y^{1/4} \quad \text{and} \quad f_y(x, y) = x^{3/4}y^{-3/4}.$$

Thus if $x = 1296$ and $y = 10{,}000$,

$$f_x(1296, 10{,}000) = 3(1296)^{-1/4}(10{,}000)^{1/4} = 5$$

and

$$f_y(1296, 10{,}000) = (1296)^{3/4}(10{,}000)^{-3/4} = 0.216.$$

Therefore if capital remains fixed at \$10,000 per week, the availability of an additional worker-hour each week would result in an increase in production of approximately five units per week. If labor remains fixed at 1296 worker-hours per week, the availability of an additional dollar of capital each week would result in an increase in production of approximately 0.216 units per week. ∎

EXAMPLE 7 Poiseuille's Law states that the volume rate of blood flow F in an artery is given by the equation

$$F = \frac{kr^4}{L},$$

where r is the radius of the artery, L is its length, and k is a positive constant that depends on the viscosity of the blood. Since

$$\frac{\partial F}{\partial r} = \frac{4kr^3}{L},$$

when length L is fixed increasing the radius slightly will cause the volume rate of blood flow F to increase (because $4k/L$ is positive) by an amount approximately equal to $(4k/L)r^3$, that is, by an amount approximately proportional to the cube of the radius. On the other hand, since

$$\frac{\partial F}{\partial L} = -\frac{kr^4}{L^2},$$

when the radius r is held fixed increasing the length slightly will cause F to decrease (because $-kr^4$ is negative) by an amount approximately equal to $(kr^4)(1/L^2)$, that is, by an amount approximately inversely proportional to the square of the length. ☐

8.3 EXERCISES

The Geometric Significance of Partial Derivatives

In Exercises 1 through 8, find the slopes of the lines tangent to the given surface at the given point and parallel to the xz- and yz-planes. If slices are taken through the surface at the point and parallel to the xz- and yz-planes, state whether the resulting curves are increasing, decreasing, concave upward, or concave downward at the point.

1. $z = 20 - 2x - 5y$, (2, 1, 11)
2. $z = 2x^2 + 4y^2 + 10$, (2, 3, 54)
3. $z = 2xy + x^2y$, (1, −1, −3)
4. $z = xy^2 + 2x + 3y + 1$, (0, 2, 7)
5. $z = x^2y^3 + 2x - 3y$, (1, −1, 4)
6. $z = x^2 + y^2$, $(a, 1, a^2 + 1)$
7. $z = x^2$, (1, b, 1)
8. $z = e^y$, (a, b, e^b)

Partial Derivatives as Rates of Change

Management

If a firm makes and sells x units of product A and y units of product B, its cost function is given by

$$C(x, y) = 20x + 14y + 200$$

and its revenue function by

$$R(x, y) = 25x + 20y.$$

Exercises 9 through 11 refer to this firm.

9. Find the firm's marginal cost with respect to x and with respect to y. Interpret your results.
10. Find the firm's marginal revenue with respect to x and with respect to y. Interpret your results.

11. Find the firm's marginal profit with respect to x and with respect to y. Interpret your results.

If a firm makes and sells x units of product X and y units of product Y, its cost and revenue functions are given by $C(x, y) = 30x + xy + 42y + 80,000$ and $R(x, y) = 35x + 2xy + 50y$, respectively. Exercises 12 through 21 refer to this firm.

12. Find the firm's marginal cost functions.
13. Find the firm's marginal cost with respect to x and y when $x = 100$ and $y = 200$.
14. If y is held fixed at 200, what is the effect of making one more unit of X?
15. If x is held fixed at 100, what is the effect of making one more unit of Y?
16. In general, if y is held fixed and one more unit of X is produced, what happens to the cost? What if x is held fixed and one more unit of Y is produced?
17. Find the firm's marginal revenue functions.
18. Find the firm's marginal revenue with respect to x and y when $x = 100$ and $y = 200$.
19. If y is held fixed at 200, what is the effect of selling one more unit of X?
20. If x is held fixed at 100, what is the effect of selling one more unit of Y?
21. In general, if y is fixed and one more unit of X is sold, what happens to the revenue? What if x is fixed and one more unit of Y is sold?

A company that makes and sells x units of product X and y units of product Y has cost function given by
$$C(x, y) = \sqrt{xy + 25}$$
and revenue function given by
$$R(x, y) = 2x + y.$$
Exercises 22 through 25 refer to this company.

22. Find the company's marginal cost, marginal revenue, and marginal profit with respect to x and y when $x = 20$ and $y = 10$.
23. If y is held fixed at 10, what is the effect of making and selling one more unit of X? If x is held fixed at 20, what is the effect of making and selling one more unit of Y?
24. In general, if x is held fixed and one additional unit of Y is made and sold, what happens to cost, revenue, and profit?
25. Answer the question of Exercise 24 if y is held fixed and one more unit of X is made and sold.
26. If a corporation's cost and revenue functions are given by
$$C(x, y) = 5x^2 + 3y^2 + xy + 100$$
and
$$R(x, y) = 6x^2 + 5y^2,$$
respectively, find and interpret $C_x(100, 200)$, $C_y(100, 200)$; $R_x(100, 200)$, $R_y(100, 200)$; $R_x(100, 200) - C_x(100, 200)$; and $R_y(100, 200) - C_y(100, 200)$.
27. Suppose the demand function for commodity V is given by
$$V(v, w) = 20,000 - 500v + 300w$$

and that for commodity W by
$$W(v, w) = 30{,}000 + 800v - 400w,$$
where v is the unit price of V and w is the unit price of W. Find V_v, V_w, W_v, and W_w, and interpret your results. Are the commodities competitive or complementary?

28. Repeat Exercise 27 if
$$V(v, w) = 20{,}000 - 200v - 100w,$$
and
$$W(v, w) = 30{,}000 - 400v - 300w.$$

29. Suppose the demand functions for products X and Y are given by
$$f(x, y) = 200{,}000 - 200x^2 + 50xy - 10y^2$$
and
$$g(x, y) = 300{,}000 - 400y^2 + 100xy - 30x^2,$$
respectively, with x the unit price of X and y the unit price of Y. Find $f_x(10, 10)$, $f_y(10, 10)$, $g_x(10, 10)$, and $g_y(10, 10)$ and interpret your results. Are the commodities competitive at these prices?

30. Repeat Exercise 29 for $f_x(10, 20)$, $f_y(10, 20)$, $g_x(10, 20)$, and $g_y(10, 20)$.

31. Suppose the supply function for commodity X is given by
$$f(x, y) = 200x - 300y - 10{,}000$$
while that for commodity Y is given by $g(x, y) = -400x + 200y - 12{,}000$. Here x is the unit price of X and y the unit price of Y. Find f_x, f_y, g_x, and g_y and interpret your results.

32. Suppose the supply functions for products X and Y are given by
$$A(x, y) = 2x^2 + 10xy - 3y^2$$
and $B(x, y) = -5x^2 + 20xy + 8y^2$, respectively, where x is the unit price of X and y that of Y. Find A_x, A_y, B_x, B_y when $x = 5$ and $y = 10$ and interpret your results.

33. Repeat Exercise 32 using $x = 7$ and $y = 3$.

34. The value of a company's daily output is z dollars, where $z = 10x^{2/5}y^{3/5}$. Here x is labor input in worker-hours per day and y is capital input in dollars per day. Find z, z_x, and z_y when $x = 32{,}768$ and $y = 100{,}000$, and interpret your results.

35. Refer to Exercise 34: In general, what happens to production if labor input is fixed and capital input increases by \$1? if capital input is fixed and labor input increases by 1 worker-hour?

36. A firm's production function is given by
$$f(x, y) = 200\sqrt{xy},$$
where x is labor input in worker-hours per week, y is capital input in dollars per week, and output is in units per week.
   (a) Find the firm's weekly output if it uses 1600 worker-hours and \$10,000 of capital per week.
   (b) Find $f_x(1600, 10{,}000)$ and $f_y(1600, 10{,}000)$, and interpret your results.

37. Using the notation of Exercise 36, suppose the production function is defined by
$$f(x, y) = 12x^{1/4}y^{3/4}.$$
Find $f(81, 625)$, $f_x(81, 625)$, and $f_y(81, 625)$, and interpret your results.

38. Suppose a production function is given by
$$f(x, y) = xe^{0.01y},$$
where x is labor input in thousands of worker-hours per week, y is capital input in thousands of dollars per week, and output is in thousands of units per week. Evaluate the first partial derivatives of f with respect to x and y at
(a) $x = y = 50$ (b) $x = y = 100$ (c) $x = y = 200$
and interpret your results.

39. Refer to Exercise 38: What happens to production if labor input is fixed and capital input increases by $1000 per week? if capital input is fixed and labor input increases by 1000 worker-hours per week?

Exercises 40 and 41 were suggested by material in McGann and Russell, *Advertising Media*, second edition, Irwin, 1988. They are concerned with the problem of measuring the net audience reached by an advertisement when there is audience duplication because the advertisement appears in several different outlets. A model for net audience reached due to J. M. Agostini is

$$C = \frac{A^2}{A + 1.25D},$$

where

A = total audience of all outlets used,
D = sum of the duplicated audiences for all pairs of outlets used,
C = net (unduplicated) audience.

*40. Find $\partial C/\partial A$ and interpret your result.

*41. Find $\partial C/\partial D$ and interpret your result.

Life Science

42. If a farmer irrigates a field x times, weeds it y times, and fertilizes it z times during the growing season, the crop yield will be
$$w = 4x + 5y + 2z + 400$$
bushels. Find the first partial derivatives of w with respect to x, y, and z, and interpret your results.

43. The number of bacteria present in a culture t hours after an experiment has begun depends on t, on the amount of nutrient provided, and on the temperature at which the experiment is run. Suppose that
$$p = 100e^{0.2x} + 2xT + 10T^2 + 10,000t,$$
where p is the number of bacteria in the culture, x is nutrient provided in grams, and T is temperature in degrees Celsius. Evaluate the first partial derivatives of p with respect to x, T, and t when $x = 2$, $T = 20$, and $t = 3$, and interpret your results.

44. Suppose that if there are x rabbits for coyotes to prey on and y hunters to kill the coyotes, then the size of the coyote population is

$$f(x, y) = 4\frac{\sqrt{x}}{y}$$

individuals. Find $f_x(400, 2)$ and $f_y(400, 2)$ and interpret your results.

45. Refer to Exercise 44: In general, what happens to the coyote population if the number of rabbits is fixed and the number of hunters increases by one? If the number of hunters is fixed and the number of rabbits increases by one?

46. If microorganisms are supplied with p grams of nutrient and sprayed with q milliliters of a weak bactericide, there will be $w = 1000e^{p/2q}$ of them present. Find $w_p(5, 8)$ and $w_q(5, 8)$ and interpret your results.

Social Science

47. Ridership on a city's public transportation system averages

$$f(x, y) = \frac{0.1x}{1 + \ln y}$$

thousand people per day, where x is the population of the city in thousands and y is the price of a ticket. The population of the city is currently 100,000 people and the price of a ticket is \$1. Find and interpret $f_x(100, 1)$ and $f_y(100, 1)$.

48. It is estimated that the proportion p of the population who apply for welfare benefits in a community during a year depends on the unemployment rate x and the average years of education y in the community according to the equation

$$p(x, y) = e^{0.08x} - e^{0.01y}.$$

Here $5 \leq x \leq 20$ and $10 \leq y \leq 15$. Find $p_x(8, 12)$ and $p_y(8, 12)$ and interpret your results.

49. Refer to Exercise 48: What happens to the proportion who apply for benefits if the average years of education remains fixed and the unemployment rate increases by 1%?

50. If a politician's campaign spends x thousand dollars on TV ads and the opposition campaign spends y thousand dollars on TV ads, the politician will get $z = 50{,}000 \ln[(x + 1)(y + 2)^{-1}]$ votes. Find $z_x(20, 80)$ and $z_y(20, 80)$ and interpret your results.

8.4 OPTIMIZATION

In this section we discuss **relative maxima, relative minima,** and **saddle points** of functions of two variables and show how to find such points using an analog of the second derivative test for functions of one variable that we introduced in Section 4.2. (Most of the results and techniques presented in this section can be extended to functions of more than two variables, but since the two-variable case illustrates the essential ideas with the fewest possible complications, we will confine ourselves to it.)

We begin by indicating what it means for a function of two variables to have a relative maximum or minimum at a point. If f is such a function, we say that f

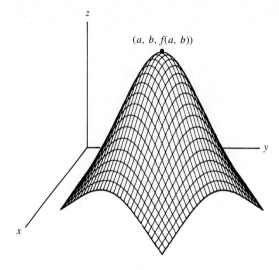

FIGURE 8–11

has a **relative maximum** at the point *(a, b, f(a, b))* if the surface $z = f(x, y)$ has a peak at *(a, b, f(a, b))*, as shown in Figure 8–11; it has a **relative minimum** at *(a, b, f(a, b))* if the surface has a pit at *(a, b, f(a, b))*, as shown in Figure 8–12. Also, Figure 8–13 depicts a function that has a **saddle point** at *(a, b, f(a, b))*. Note that a saddle point is a relative minimum in one direction and a relative maximum in another direction, and hence is itself neither a relative maximum nor a relative minimum.

Suppose that a function of two variables *f* has a relative maximum at *(a, b, f(a, b))* and that the line parallel to the *xz*-plane and tangent to the surface $z = f(x, y)$ at *(a, b, f(a, b))* exists, as in Figure 8–14. Then this line must be

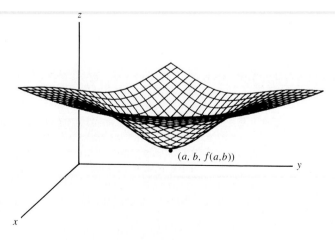

FIGURE 8–12

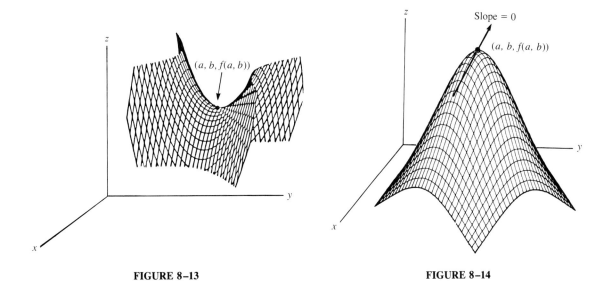

FIGURE 8–13 **FIGURE 8–14**

horizontal, and its slope, which is $f_x(a, b)$, must be zero. Similarly, the line tangent to the surface at $(a, b, f(a, b))$ and parallel to the yz-plane must also have a slope of zero, so $f_y(a, b)$ must be zero. Therefore if f has a relative maximum at $(a, b, f(a, b))$ and $f_x(a, b)$ and $f_y(a, b)$ exist, then

$$f_x(a, b) = 0 \quad \text{and} \quad f_y(a, b) = 0.$$

Similar reasoning holds for relative minima and saddle points.

Relative Maxima, Relative Minima, and Saddle Points

Let f be a function of x and y. If

$$f_x(a, b) = 0 \quad \text{and} \quad f_y(a, b) = 0,$$

then f has either a relative maximum, a relative minimum, or a saddle point at $(a, b, f(a, b))$.

Thus we can find at least some of the points at which such a function has a relative maximum, a relative minimum, or a saddle point by setting $f_x(x, y) = 0$ and $f_y(x, y) = 0$ and then solving these equations simultaneously for x and y. (We say *some* of the points because there may be relative maxima, relative minima, or saddle points at places where f_x or f_y fail to exist.) Unfortunately, knowing that $x = a$, $y = b$ is a solution of these equations is not enough to tell us whether $(a, b, f(a, b))$ yields a relative maximum, a relative minimum, or a saddle point.

EXAMPLE 1   Consider the function f defined by

$$f(x, y) = -x^2 + 2x - y^2 + 4y.$$

Since

$$f_x = -2x + 2 \quad \text{and} \quad f_y = -2y + 4,$$

setting $f_x = 0$ and $f_y = 0$ yields

$$-2x + 2 = 0 \quad \text{and} \quad -2y + 4 = 0,$$

and thus

$$x = 1 \quad \text{and} \quad y = 2.$$

Therefore $(1, 2, f(1, 2)) = (1, 2, 5)$ is either a relative maximum, a relative minimum, or a saddle point, but we cannot tell which of these it is unless we know something else about the function. In this case, the graph of the function is shown in Figure 8–15, so it is clear from the graph that the point $(1, 2, 5)$ is a relative maximum. ☐

Fortunately, there is a method of testing a point (a, b) for which $f_x(a, b) = 0$ and $f_y(a, b) = 0$ to find out whether it yields a relative maximum, a relative minimum, or a saddle point. The method, which we will not prove, is called the **second derivative test.**

The Second Derivative Test

To find relative maxima, relative minima, and saddle points for a function f of x and y, proceed as follows:

1. Find the partial derivatives f_x, f_y, f_{xx}, f_{xy}, and f_{yy}.

2. Set $f_x = 0$ and $f_y = 0$ and solve these equations simultaneously for x and y to obtain all pairs of numbers (a, b) such that $f_x(a, b) = 0$ and $f_y(a, b) = 0$.

3. Let $D(x, y) = f_{xy}^2 - f_{xx}f_{yy}$. Then
   a. If $D(a, b) < 0$ and $f_{xx}(a, b) < 0$, the function has a relative maximum at $(a, b, f(a, b))$.
   b. If $D(a, b) < 0$ and $f_{xx}(a, b) > 0$, the function has a relative minimum at $(a, b, f(a, b))$.
   c. If $D(a, b) > 0$, the function has a saddle point at $(a, b, f(a, b))$.
   d. If $D(a, b) = 0$, the test fails to yield information about the behavior of the function at $(a, b, f(a, b))$.

EXAMPLE 2 Let us apply the second derivative test to the function f defined by

$$f(x, y) = -x^2 + 2xy + 2y^2 - 4x - 14y + 1.$$

1. We have

$$f_x = -2x + 2y - 4, \quad f_y = 2x + 4y - 14,$$
$$f_{xx} = -2, \quad f_{xy} = 2, \quad \text{and} \quad f_{yy} = 4.$$

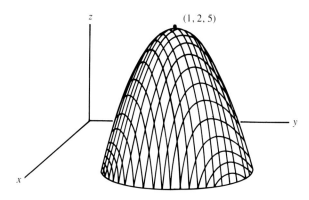

FIGURE 8–15

2. We solve the equations
$$-2x + 2y - 4 = 0 \quad \text{and} \quad 2x + 4y - 14 = 0$$
simultaneously for x and y: from the first equation,
$$2y = 2x + 4,$$
or
$$y = x + 2.$$
Substituting this expression for y into the second equation, we obtain
$$2x + 4(x + 2) - 14 = 0,$$
or
$$6x = 6.$$
Therefore $x = 1$ and $y = 1 + 2 = 3$, and we must test the single pair $(1, 3)$.

3. To test the pair $(1, 3)$, we must evaluate
$$D(x, y) = f_{xy}^2 - f_{xx}f_{yy} = 2^2 - (-2)(4) = 12$$
at $(1, 3)$. But if $D(x, y) = 12$, then
$$D(1, 3) = 12 > 0,$$
so the function has a saddle point at $(1, 3, f(1, 3)) = (1, 3, -22)$. ☐

EXAMPLE 3   Let us apply the second derivative test to the function whose defining equation is
$$f(x, y) = x^4 - 2x^2 + y^2 - 2y + 1.$$

1. We have
$$f_x = 4x^3 - 4x, \quad f_y = 2y - 2,$$
$$f_{xx} = 12x^2 - 4, \quad f_{xy} = 0, \quad \text{and} \quad f_{yy} = 2.$$

2. We solve
$$4x^3 - 4x = 0$$
$$2y - 2 = 0$$
or
$$4x(x - 1)(x + 1) = 0$$
$$2(y - 1) = 0$$
simultaneously for x and y, obtaining
$$x = 0, \quad x = 1, \quad x = -1, \quad \text{and} \quad y = 1.$$
Therefore we must test the pairs $(0, 1)$, $(1, 1)$, and $(-1, 1)$.

3. We test the pairs $(0, 1)$, $(1, 1)$, and $(-1, 1)$ using
$$D(x, y) = 0^2 - (12x^2 - 4)(2) = -8(3x^2 - 1).$$
Since
$$D(0, 1) = -8(3(0)^2 - 1) = 8 > 0,$$
the function has a saddle point at $(0, 1, f(0, 1)) = (0, 1, 0)$.
Since
$$D(1, 1) = -8(3(1)^2 - 1) = -16 < 0,$$
and
$$f_{xx}(1, 1) = 12(1)^2 - 4 = 8 > 0,$$
the function has a relative minimum at $(1, 1, f(1, 1)) = (1, 1, -1)$.
Since
$$D(-1, 1) = -8(3(-1)^2 - 1) = -16 < 0$$
and
$$f_{xx}(-1, 1) = 12(-1)^2 - 4 = 8 > 0,$$
the function has another relative minimum at $(-1, 1, -1)$. □

EXAMPLE 4 Let us apply the second derivative test to the function with defining equation
$$f(x, y) = x^4 + (x - y)^4.$$

1. We have
$$f_x = 4x^3 + 4(x - y)^3, \quad f_y = -4(x - y)^3,$$
$$f_{xx} = 12x^2 + 12(x - y)^2, \quad f_{xy} = -12(x - y)^2,$$
and
$$f_{yy} = 12(x - y)^2.$$

2. Solving

$$4x^3 + 4(x - y)^3 = 0$$
$$-4(x - y)^3 = 0$$

simultaneously for x and y, we obtain

$$x = y = 0$$

as the only solution. (Check this.)

3. Since

$$\begin{aligned}D(x, y) &= (-12(x - y)^2)^2 - (12x^2 + 12(x - y)^2)(12(x - y)^2) \\ &= 144(x - y)^4 - (144x^2(x - y)^2 + 144(x - y)^4) \\ &= -144x^2(x - y)^2,\end{aligned}$$

we have $D(0, 0) = 0$. Therefore the second derivative test fails; it cannot tell us whether $(0, 0, f(0, 0)) = (0, 0, 0)$ is a relative maximum, a relative minimum, or a saddle point. In a case such as this we examine the behavior of the function near the point in question. Since

$$f(0, 0) = 0$$

while

$$f(x, y) = x^4 + (x - y)^4 > 0 \quad \text{if } (x, y) \neq (0, 0),$$

it follows that the point $(0, 0, 0)$ is a relative minimum. ☐

Now we can use the second derivative test to solve optimization problems involving functions of two variables.

EXAMPLE 5   Suppose we wish to construct a rectangular box with a volume of 1000 cubic inches. Let us minimize the cost of the box if the material for it costs 10 cents per square inch.

The box will look like this:

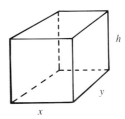

Its volume is $xyh = 1000$ cubic inches, and its cost is

$$C = 2(0.10xy) + 2(0.10xh) + 2(0.10yh)$$

dollars. Since

$$h = \frac{1000}{xy},$$

the cost function may be written as a function of two variables:

$$C(x, y) = 0.2xy + 0.2x\frac{1000}{xy} + 0.2y\frac{1000}{xy},$$

or

$$C(x, y) = 0.2xy + \frac{200}{y} + \frac{200}{x}.$$

The second derivative test shows that $C(x, y)$ has a minimum at $(10, 10, 60)$. (Check this.) Therefore the dimensions of the box of minimum cost are

$$x = 10 \text{ inches}, \quad y = 10 \text{ inches}, \quad h = 10 \text{ inches},$$

and the minimum cost is $60. ◻

EXAMPLE 6 A producer who sells the same product to different classes of customers at different prices is said to practice **price discrimination.** To see how this can occur, suppose that the demand functions for a certain product are different in two regions. For instance, suppose that in region X the demand function for a product is given by

$$p_X = 100 - \frac{x}{5},$$

but in region Y it is given by

$$p_Y = 200 - \frac{y}{15}.$$

Here x is quantity demanded in region X, y is quantity demanded in region Y, and p_X and p_Y are the unit prices in the two regions.

Let us find the producer's total revenue from sales in the two regions. Since x units are sold in region X at price p_X and y are sold in region Y at price p_Y, the revenue function is given by

$$R(x, y) = x\left(100 - \frac{x}{5}\right) + y\left(200 - \frac{y}{15}\right).$$

The second derivative test shows that the revenue function has a maximum at

$$(250, 1500, R(250, 1500)) = (250, 1500, 162{,}500).$$

(Check this.) Therefore to obtain maximum revenue of $162,500, the producer will sell 250 units in region X at a price of $p_X = $50 each and 1500 units in region Y at a price of $p_Y = $100 each, and thus will practice price discrimination. ◻

8.4 EXERCISES

In Exercises 1 through 15, use the second derivative test to find relative maxima, relative minima, and saddle points for the function.

1. $f(x, y) = 2xy$
2. $f(x, y) = x^2 + y^2 - 4$
3. $f(x, y) = 9 - 2x^2 - 3y^2$
4. $f(x, y) = x - (x + y)^2$
5. $f(x, y) = 3x^2 + 2y^2 - 18x + 8y - 2$
6. $f(x, y) = 4x^2 + y^2 - 16x - 6y + 3$
7. $f(x, y) = 4x^2 + 5y^2 + 24x - 100y + 50$
8. $f(x, y) = x^3 + y^3 - 3xy$
9. $f(x, y) = x^3 - y^3 - 3xy$
10. $f(x, y) = 4x^3 + 6x^2 - 72x + 3y^4 - 44y^3 + 180y^2$
11. $f(x, y) = x^3 + 2y^2 - 8xy + 5x + 2$
12. $f(x, y) = 2y^3 + 100y + 10xy - x^2$
13. $f(x, y) = x^4 - 8x^2 + 6y^2 - 12y + 3$
14. $f(x, y) = x^4 - (x + y)^4$
15. $f(x, y) = xy(x + y - 1) - x^2$

16. Find the dimensions of the rectangular box, without a top, of minimum cost if the material for the box costs $2 per square foot and the box must have a volume of 144 cubic feet.

17. Find the dimensions of the rectangular box, without a top, of maximum volume if the surface area of the box must be 12 square meters.

18. Find the dimensions of the rectangular box of maximum volume if the sum of the lengths of its edges must be 96 inches.

19. A carrier for pets is in the shape of a rectangular box that has an opening in its front face. The opening has an area that is two-thirds the area of the front face. Find the dimensions that minimize the surface area of the box if its volume must be 9 cubic feet.

20. The material for a rectangular box with a top and three partitions (see the figure) costs $1 per square foot, and its volume must be 1 cubic foot. Find the dimensions that minimize the cost of building the box.

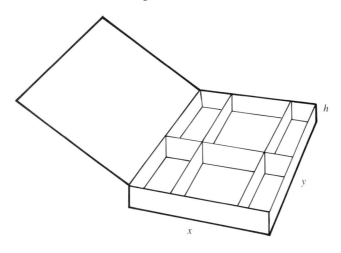

Management

21. If a firm makes and sells x units of one product and y units of another, its profit function is given by
$$P(x, y) = 20x + 30y - 2x^2 - 3y^2.$$
Find the firm's maximum profit.

22. If a company makes x units of one product and y of another, its average cost per unit is given by
$$A(x, y) = 50 + \frac{80}{x} + \frac{216}{y} + 5x + 6y.$$
Find the company's minimum average cost per unit.

23. If a firm makes and sells x thousand units of product X and y thousand units of product Y, its profit is
$$P(x, y) = -x^2 + 40x + xy + 10y - y^2 - 10$$
thousand dollars. Find the firm's maximum profit and the number of units of each product that must be made and sold in order to attain it.

24. The demand functions for a commodity in regions X and Y are given by
$$p_X = 2000 - \frac{x}{20} \quad \text{and} \quad p_Y = 1500 - \frac{y}{15},$$
respectively, with x and y the quantities demanded in the regions, and p_X, p_Y the unit prices in the regions. Find the prices that will maximize revenue. Will the supplier practice price discrimination?

25. Repeat Exercise 24 if
$$p_X = 500 - \frac{x}{4} \quad \text{and} \quad p_Y = 500 - \frac{y}{3}.$$

26. Refer to Exercise 24: if it costs $20 to make each unit sold in region X and $30 to make each unit sold in region Y, find the prices that will maximize profit.

27. The demand functions for two products A and B are as follows:
$$\text{product A:} \quad q_A = 100 - 5p_A - p_B$$
$$\text{product B:} \quad q_B = 96 - p_A - 4p_B.$$
Here p_A, p_B are the unit prices of A and B, respectively. Maximize total revenue from sales of the two products.

28. A retailer sells x units of a product at a unit price of $p_x = 9600 - 0.02x^2$ at one store, and y units at a unit price of $p_y = 6075 - 0.01y^2$ at another store. Find the maximum revenue that the retailer can obtain from the product and the unit prices required in order to attain it. Does the retailer practice price discrimination?

29. A firm can produce
$$f(x, y) = -x^2 - 1.5y^2 + 12x + 9y$$
thousand units of output using x units of labor and y units of capital. Find the firm's maximum output.

*30. A retailer wishes to build a new warehouse at a location that will minimize the sum of the distances from the warehouse to three stores. If the stores are located at the points (10, 20), (−20, 15), and (4, −8), where should the warehouse be located? (*Hint:* The distance from (x, y) to (a, b) is $[(x - a)^2 + (y - b)^2]^{1/2}$.)

31. A firm's sales depend on the amount x it spends on television commercials and the amount y it spends on newspaper advertisements according to the equation
$$S = 100{,}000x + 120{,}000y - x^2 - 4y^2 - 1.6xy.$$
Find the amounts that must be spent on each type of advertising in order to maximize sales.

Life Science

32. If a farmer applies x tons of fertilizer and y tons of pesticide to a crop, the yield will be

$$f(x, y) = 20x + 50y - 10x^2 - 30y^2$$

bushels. How much fertilizer and how much pesticide should be applied if the crop yield is to be maximized?

33. A patient undergoing chemotherapy who receives x units of drug A and y units of drug B per day suffers z units of discomfort, where

$$z = 4x^2 + 5y^2 - 8x - 6y - 5xy + 15$$

and discomfort is measured on a scale ranging from 0 to 15. What drug dosage will minimize discomfort?

8.5 LAGRANGE MULTIPLIERS

In this section we consider the problem of finding maxima and minima for functions of several variables subject to side conditions known as **constraints.** For instance, suppose we wish to find the minimum value of

$$f(x, y) = x^2 + y^2$$

subject to the condition that

$$x + y - 2 = 0.$$

In this case the **constraint equation** $x + y - 2 = 0$ is easily solved for y in terms of x; if we do so, we find that $y = -x + 2$, and this expression may be substituted for y in $f(x, y)$ to obtain

$$f(x, y) = x^2 + (-x + 2)^2 = 2x^2 - 4x + 4.$$

But now f is a function of the single variable x, so it may be written as

$$f(x) = 2x^2 - 4x + 4,$$

and we can then use either the first or second derivative test of Chapter 4 to conclude that the function has a relative minimum of 2 when $x = 1$.

Unfortunately, when we attempt to maximize or minimize a function of several variables subject to a constraint, it is not always possible to solve the constraint equation for one of the variables in terms of the others, as we did above. Therefore we need a method for finding the maximum or minimum values of $z = f(x, y)$ subject to a constraint equation $g(x, y) = 0$ without having to solve $g(x, y) = 0$ for one of the variables. In this section we present such a method: it is known as the **Lagrange multiplier method.**

Suppose we wish to find the maximum or minimum value of $f(x, y)$ subject to the constraint $g(x, y) = 0$. We begin by setting

$$h(x, y) = f(x, y) + \lambda g(x, y).$$

The symbol λ is the Greek letter lambda, which in this context is called the **Lagrange multiplier.** If we consider λ to be a specific number, then we know from our work in Section 8.4 that we can find relative maxima and minima for the function h by solving

$$h_x = 0 \quad \text{and} \quad h_y = 0$$

simultaneously for x and y. If we do this and at the same time ensure that the solutions satisfy $g(x, y) = 0$, then since

$$h(x, y) = f(x, y) \quad \text{when} \quad g(x, y) = 0,$$

it will follow that the solutions that maximize or minimize h and satisfy the constraint equation will also maximize or minimize f and satisfy the constraint equation. But now let us consider λ to be another independent variable, so that h is a function of x, y, and λ. Then

$$h(x, y, \lambda) = f(x, y) + \lambda g(x, y),$$

so h_x and h_y are the same as before. Taking the first partial derivative of h with respect to λ shows that

$$h_\lambda = g(x, y).$$

Therefore we can ensure that the solutions of $h_x = 0$ and $h_y = 0$ satisfy the constraint equation if we take λ to be an independent variable and require that $h_\lambda = 0$.

The Lagrange Multiplier Method

Let f be a function of x and y. To maximize or minimize f subject to the constraint $g(x, y) = 0$:

1. Let

$$h(x, y, \lambda) = f(x, y) + \lambda g(x, y)$$

and find h_x, h_y, and $h_\lambda = g(x, y)$.

2. Set $h_x = 0$, $h_y = 0$, and $h_\lambda = 0$, and solve these equations simultaneously for x and y.

3. The result of step 2 will be pairs of numbers $x = a$, $y = b$ that maximize or minimize f and satisfy the constraint $g(x, y) = 0$. To decide whether a particular pair (a, b) yields a maximum or a minimum for f, compare the value $f(a, b)$ with values of $f(x, y)$ for points (x, y) that are near (a, b) and satisfy the constraint equation $g(x, y) = 0$.

(The simplest way to solve the equations of step 2 simultaneously is usually to first solve $h_x = 0$ and $h_y = 0$ for λ in terms of x and y, equate the two expressions for

FUNCTIONS OF SEVERAL VARIABLES 519

λ, solve for either x or y in terms of the other, and substitute the result into the equation $h_\lambda = 0$.)

EXAMPLE 1 Let us find maxima and minima for
$$f(x, y) = x^2 + y^2$$
subject to the constraint
$$x + y - 2 = 0.$$

1. We form
$$h(x, y, \lambda) = x^2 + y^2 + \lambda(x + y - 2),$$
Then
$$h_x = 2x + \lambda, \qquad h_y = 2y + \lambda,$$
and
$$h_\lambda = x + y - 2.$$

2. We must solve the equations
$$2x + \lambda = 0, \quad 2y + \lambda = 0, \quad \text{and} \quad x + y - 2 = 0$$
simultaneously for x and y. Solving the first two of these equations for λ, we obtain
$$\lambda = -2x \quad \text{and} \quad \lambda = -2y.$$
Thus
$$-2x = -2y,$$
or
$$x = y.$$
Substituting x for y in the equation
$$x + y - 2 = 0$$
now yields
$$2x - 2 = 0,$$
so
$$x = 1 \quad \text{and} \quad y = 1.$$

3. Therefore the function has either a constrained maximum or a constrained minimum when $x = 1$, $y = 1$. To find out whether $f(1, 1) = 2$ is a constrained maximum or a constrained minimum, we compare it with $f(x, y)$ for points (x, y) near $(1, 1)$ that also satisfy $x + y - 2 = 0$. For instance, we have
$$f(1, 1) = 2,$$
$$f(1.05, 0.95) = 2.005,$$
$$f(1.1, 0.9) = 2.02,$$
$$f(0.8, 1.2) = 2.08,$$

and so on. Hence it is clear that $f(1, 1) = 2$ is a constrained minimum for f. ☐

EXAMPLE 2 Let us optimize
$$f(x, y) = 12 - x^2 - y^2$$
subject to the constraint
$$xy = 4.$$

1. We have
$$h(x, y, \lambda) = 12 - x^2 - y^2 + \lambda(xy - 4).$$

Note that the constraint equation was not originally in the form $g(x, y) = 0$, and thus we had to rewrite it in this form before we constructed the function h. Now,
$$h_x = -2x + \lambda y, \quad h_y = -2y + \lambda x, \quad \text{and} \quad h_\lambda = xy - 4.$$

2. We must solve the equations
$$-2x + \lambda y = 0, \quad -2y + \lambda x = 0, \quad \text{and} \quad xy - 4 = 0$$
simultaneously for x and y. From the first two equations,
$$\lambda = \frac{2x}{y} = \frac{2y}{x},$$
or
$$x^2 = y^2.$$
Therefore
$$y = \pm x.$$
If we substitute x for y in
$$xy - 4 = 0,$$
we find that
$$x^2 = 4,$$
so $x = \pm 2$, and we obtain the points $(2, 2)$, $(-2, -2)$. If we substitute $-x$ for y in
$$xy - 4 = 0,$$
we find that
$$x^2 = -4,$$
which has no solutions.

3. Thus $(2, 2)$ and $(-2, -2)$ are the only points that yield constrained maxima or minima for f. Since

$$f(\pm 2, \pm 2) = 4, \quad f\left(\pm 2.1, \frac{4}{\pm 2.1}\right) \approx 3.96, \quad f\left(\pm 2.2, \frac{4}{\pm 2.2}\right) \approx 3.85,$$

and so on, it is clear that $(2, 2)$ and $(-2, -2)$ both yield constrained maxima for f. ☐

Many practical situations require constrained optimization of functions of several variables. We illustrate with an example.

EXAMPLE 3   Consider the production function defined by

$$f(x, y) = 300x^{1/3}y^{2/3}.$$

Here x denotes the number of units of labor and y the number of units of capital required to produce $f(x, y)$ units of product. Suppose that each unit of labor costs \$50 and each unit of capital costs \$100, and assume that the production budget is \$15,000. Then we would like to maximize production $f(x, y)$ subject to the budgetary constraint

$$50x + 100y = 15{,}000.$$

Using the Lagrange multiplier method, we have

$$h(x, y, \lambda) = 300x^{1/3}y^{2/3} + \lambda(50x + 100y - 15{,}000).$$

Thus

$$h_x = 100x^{-2/3}y^{2/3} + 50\lambda = 0,$$

$$h_y = 200x^{1/3}y^{-1/3} + 100\lambda = 0,$$

and

$$h_\lambda = 50x + 100y - 15{,}000 = 0,$$

from which we find that

$$\lambda = -2x^{-2/3}y^{2/3} = -2x^{1/3}y^{-1/3}.$$

Therefore

$$x^{-2/3}y^{2/3} = x^{1/3}y^{-1/3}.$$

Multiplying this last equation by $x^{2/3}y^{1/3}$ yields

$$y = x.$$

Substituting x for y in the equation

$$50x + 100y = 15{,}000,$$

we find that

$$x = y = 100.$$

Therefore the maximum production subject to the budgetary constraint is

$$f(100, 100) = 30,000$$

units.

Note that $x = y = 100$ implies that

$$\lambda = -2(100)^{1/3}(100)^{-1/3} = -2.$$

The absolute value of λ is called the **marginal productivity of money**; it tells us approximately how many additional units could be produced with one additional dollar. In this example, $|\lambda| = |-2| = 2$, so approximately two additional units could be produced if one additional dollar were available for production. ☐

The Lagrange multiplier method can be used with functions of more than two variables: we simply set all first partial derivatives of $h = f + \lambda g$ equal to zero and solve simultaneously.

EXAMPLE 4   Let us find three positive numbers such that the first plus twice the second plus three times the third is 36 and such that their product is a maximum. Thus we wish to maximize

$$f(x, y, z) = xyz$$

subject to the constraint

$$x + 2y + 3z = 36.$$

We have

$$h(x, y, z, \lambda) = xyz + \lambda(x + 2y + 3z - 36),$$

so

$$h_x = yz + \lambda, \quad h_y = xz + 2\lambda, \quad h_z = xy + 3\lambda,$$

and

$$h_\lambda = x + 2y + 3z - 36.$$

Solving

$$h_x = 0, \quad h_y = 0, \quad \text{and} \quad h_z = 0,$$

we find that

$$y = \frac{x}{2} \quad \text{and} \quad z = \frac{x}{3}.$$

Substituting these expressions for y and z in the equation $h_\lambda = 0$, we have $3x = 36$, from which it follows that $x = 12$, $y = 6$, and $z = 4$. Therefore the maximum product is $f(12, 6, 4) = 288$. ☐

8.5 EXERCISES

In Exercises 1 through 12, use the Lagrange multiplier method to find either a maximum or a minimum for the function subject to the given constraint. In each case state whether the value you find is a maximum or a minimum.

1. $f(x, y) = x^2 + y^2 + 2$, subject to $x + y - 3 = 0$.
2. $f(x, y) = 2x^2 + 3y^2$, subject to $x + y - 5 = 0$.
3. $f(x, y) = 9 - x^2 - y^2$, subject to $x + y = 6$.
4. $f(x, y) = 100 - 2x^2 - 4y^2$, subject to $6x + 3y = 12$.
5. $f(x, y) = xy$, subject to $x + 2y = 6$.
6. $f(x, y) = xy$, subject to $x^2 + y^2 = 8$.
7. $f(x, y) = x^3 - y^3$, subject to $x - y = 2$.
8. $f(x, y) = x^3 - y^3$, subject to $xy = 1$.
9. $f(x, y, z) = 2x - 4y + 2z - x^2 - y^2$, subject to $x + y + z = 0$.
10. $f(x, y, z) = x^2 + y^2 + z^2 - x - 2y - 3z$, subject to $10 - x - y - z = 0$.
11. $f(x, y, z) = 2x + 3y - 4z + x^2 - y^2$, subject to $x + 2y + 3z = 0$.
12. $f(x, y, z) = x^2 + y^2 + 2z^2 - x - y + 3z$, subject to $3x + y + 2z = \frac{1}{2}$.
13. Maximize $f(x, y) = x^3 y^2$ subject to $12x + 6y = 24$.
14. Minimize $f(x, y) = x^4 y^4$ subject to $2x - y = 4$.
15. Minimize $f(x, y) = x^2 + y^2 - 3$ subject to $xy = 16$.
16. Minimize $f(x, y) = x^3 + y^3$ subject to $xy = 1$.
17. Minimize $f(x, y) = x^3 + y^3$ subject to $x + y = 1$.
18. Find two positive numbers whose product is 25 and whose sum is a minimum.
19. Find two numbers whose sum is 20 and whose product is a maximum.
20. Find three positive numbers whose sum is 20 and whose product is a maximum.
21. Find two numbers whose product is 64 and such that the square of one plus twice the other is a minimum.
22. Find two numbers such that 3 times one plus $\frac{1}{3}$ of the other equals 24 and the product of their squares is a maximum.
23. Find three numbers whose sum is 33 and such that the square of the first plus twice the square of the second plus 3 times the square of the third is a minimum.
24. Find the minimum distance from the line $x + y = 4$ to the origin $(0, 0)$. (*Hint:* The distance from (x, y) to $(0, 0)$ is $\sqrt{x^2 + y^2}$.)
25. Find the minimum distance from the plane $z = 20 - x - 2y$ to the origin. (*Hint:* The distance from (x, y, z) to $(0, 0, 0)$ is $(x^2 + y^2 + z^2)^{1/2}$.)
26. Find the point on the plane $x + y + z = 1$ that is at a minimum distance from the origin.
27. Repeat Exercise 26 for the plane $2x + 5y + z = 10$.
*28. Find the rectangle of largest area that can be inscribed in a circle of radius r.
*29. Find the rectangular box of largest volume that can be inscribed in a sphere of radius r. (The equation of such a sphere is $x^2 + y^2 + z^2 = r^2$.)

Management

30. A firm can produce
$$f(x, y) = 4000x^{2/5} y^{3/5}$$

units by employing x units of labor and y units of capital. If each unit of labor costs $200, each unit of capital costs $300, and the production budget is $16,000, find the firm's maximum production. Also find and interpret the marginal productivity of money.

31. Repeat Exercise 30 if

$$f(x, y) = x^{1/4} y^{3/4}$$

and each unit of labor costs $20, each unit of capital costs $40, and the production budget is $24,000.

32. A firm can produce $800 x^{1/4} y^{3/4}$ units of output using x units of labor and y units of capital. If each unit of labor costs $50 and each unit of capital costs $80 and the production budget for the firm is $200,000, find the maximum production and the number of units of labor and capital needed to attain it.

33. Suppose the cost of producing x units of product X and y units of product Y is

$$C(x, y) = x^2 + 2y^2 + xy + 200$$

dollars. If it takes 5 worker-hours to make each unit of X, 7 worker-hours to make each unit of Y, and there are 3456 worker-hours available, how many units of each product must be made in order to minimize cost?

34. If a store purchases x radio commercials and y newspaper ads each week, its advertising reaches

$$z = 20x\sqrt{10y + 20}$$

thousand people. If it costs $2000 to run each radio commercial and $1000 to run each newspaper ad, and if the firm budgets $7000 per week for advertising expenses, find the maximum number of people reached by its ads.

35. If a utility spends x million dollars on smokestack scrubbers and y million dollars on equipment that will burn low-sulfur coal, it can remove $z = 600xy - 20x^2 - 30y^2$ particulates per cubic meter from its smoke. If the utility's budget for reducing particulates from its smoke is $65 million, find the values of x and y that will minimize the particulates.

36. Suppose a company's profit function is given by

$$P(x, y, z) = xy + xz + yz + x + y + z - 10,000.$$

If each unit of X requires 2 kilograms (kg) of steel, each unit of Y requires 2 kg, and each unit of Z requires 3 kg, and there are 7999 kg of steel available, find the maximum profit.

Life Science

37. If a farmer applies x tons of fertilizer and y tons of weedkiller to a crop, the yield is

$$z = 10x + 30y - 2x^2 - 5y^2$$

bushels per acre. If each ton of fertilizer costs $500 and each ton of weedkiller costs $800, and if the farmer has exactly $3650 to spend on these products, find the maximum crop yield.

8.6 DOUBLE INTEGRALS

We have seen how functions of several variables may be differentiated by the process of partial differentiation; often they may also be integrated by a process known as **multiple integration.** In this section we illustrate multiple integration for functions of two variables. The integral of a function of two variables is called a **double integral.** Double integrals have many uses, most of which are beyond the scope of this book. However, we will demonstrate how they may be used in finding volumes, areas, and the average value of a function of two variables.

Let f be a function of the two variables x and y whose domain includes a region R in the xy-plane, and such that $f(x, y) \geq 0$ for all (x, y) in R, as in Figure 8–16. Suppose we place a grid of rectangles on R, with each rectangle having width Δx and length Δy. See Figure 8–17. In each rectangle that overlaps R, we choose a point (x_i, y_i) that is also in R, and construct a rectangular box whose base is the rectangle and whose height is $f(x_i, y_i)$. Figure 8–18, page 527, depicts the situation for one rectangle. The rectangular box will touch the surface $z = f(x, y)$ at the point $(x_i, y_i, f(x_i, y_i))$, and its volume will be $f(x_i, y_i) \Delta x \Delta y$. If there are n rectangles that overlap R, then there will be n such rectangular boxes, and the sum of their volumes may be written as the **Riemann sum**

$$\sum_{i=1}^{n} f(x_i, y_i) \Delta x \Delta y = [f(x_1, y_1) + f(x_2, y_2) + \cdots + f(x_n, y_n)] \Delta x \Delta y.$$

This Riemann sum is therefore an approximation to the volume of the solid whose base is R and whose top is the portion of the surface $z = f(x, y)$ that lies over R (Figure 8–19, page 527). If we make the grid on R finer and finer by letting Δx and Δy approach 0, the Riemann sum will approach the volume of this solid; in other words,

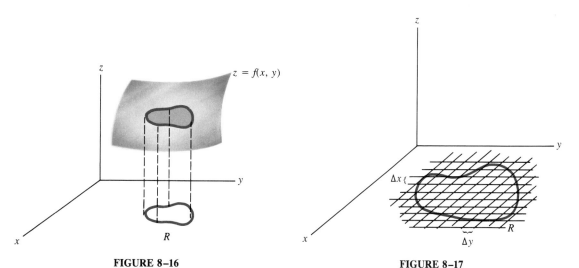

FIGURE 8–16

FIGURE 8–17

$$\text{Lim}_{\substack{\Delta x \to 0 \\ \Delta y \to 0}} \sum_{i=1}^{n} f(x_i, y_i)\, \Delta x\, \Delta y$$

will exist and will be equal to the volume of the solid. We call this limit the **double integral of f over R**. [We have assumed that $f(x, y) \geq 0$ over the region R, but a similar argument shows that if $f(x, y) \leq 0$ over R then the limit of the Riemann sums as Δx and Δy approach 0 is the negative of the volume of the solid bounded by R and the surface $z = f(x, y)$.]

> **The Double Integral**
>
> If f is a function of x and y whose domain includes the region R in the xy-plane, the **double integral of f over R**, denoted by
>
> $$\iint_R f(x, y)\, dA,$$
>
> is defined to be the limit
>
> $$\text{Lim}_{\substack{\Delta x \to 0 \\ \Delta y \to 0}} \sum_{i=1}^{n} f(x_i, y_i)\, \Delta x\, \Delta y,$$
>
> if it exists. If $f(x, y) \geq 0$ for all (x, y) in R, then
>
> $$\iint_R f(x, y)\, dA$$
>
> is the volume of the cylindrical solid bounded by R and the surface $z = f(x, y)$.

Fortunately, we will not need to evaluate double integrals by actually taking limits of Riemann sums; instead, we will evaluate them as **iterated integrals** of the form

$$\int_c^d \int_a^b f(x, y)\, dx\, dy \quad \text{or} \quad \int_a^b \int_c^d f(x, y)\, dy\, dx.$$

Such iterated integrals are integrated from the inside out. Thus, to evaluate

$$\int_c^d \int_a^b f(x, y)\, dx\, dy,$$

we first find the inside integral

$$\int_a^b f(x, y)\, dx$$

by integrating $f(x, y)$ with respect to x from $x = a$ to $x = b$ while treating y as a constant. Since each occurrence of x in the antiderivative will be replaced by a and

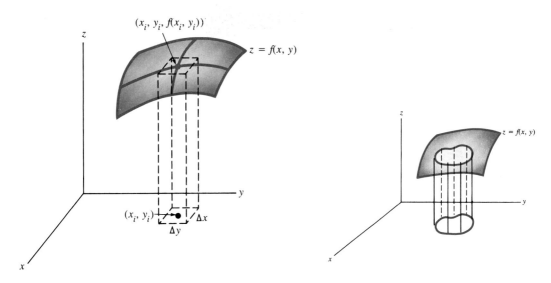

FIGURE 8–18   **FIGURE 8–19**

b, the result of this integration will be a function of y, which we can then integrate from $y = c$ to $y = d$, as the outside integral requires. We illustrate with an example.

EXAMPLE 1   Let us evaluate the iterated integral

$$\int_0^2 \int_0^1 (2x + y) \, dx \, dy.$$

We first evaluate the inside integral by integrating $f(x, y) = 2x + y$ with respect to x while treating y as a constant. Thus

$$\int_0^1 (2x + y) \, dx = x^2 + yx \Big|_{x=0}^{x=1} = [1^2 + y(1)] - [0^2 + y(0)] = 1 + y.$$

Now

$$\int_0^2 \int_0^1 (2x + y) \, dx \, dy = \int_0^2 \left(\int_0^1 (2x + y) \, dx \right) dy$$

$$= \int_0^2 (1 + y) \, dy$$

$$= y + \frac{y^2}{2} \Big|_0^2$$

$$= 4. \quad \square$$

EXAMPLE 2   Let us evaluate the double integral

$$\int_0^1 \int_0^2 (2x + y)\, dy\, dx.$$

This is the integral of Example 1 with the order of integration reversed. We have

$$\int_0^1 \int_0^2 (2x + y)\, dy\, dx = \int_0^1 \left(\int_0^2 (2x + y)\, dy \right) dx$$

$$= \int_0^1 \left(2xy + \frac{y^2}{2} \Big|_{y=0}^{y=2} \right) dx$$

$$= \int_0^1 \left(\left[2x \cdot 2 + \frac{2^2}{2} \right] - \left[2x \cdot 0 + \frac{0^2}{2} \right] \right) dx$$

$$= \int_0^1 (4x + 2)\, dx$$

$$= 2x^2 + 2x \Big|_0^1$$

$$= 4. \quad \square$$

We can interchange the order of integration in an iterated integral without affecting the value of the integral.

Now we return to our consideration of double integrals. It turns out that if the region R over which we are integrating is bounded by lines and curves, then we can evaluate the double integral as an iterated integral.

Evaluating Double Integrals

1. If R is the region bounded by the lines $x = a$ and $x = b$ and the graphs of the equations $y = g_1(x)$ and $y = g_2(x)$, as shown in Figure 8–20, then

$$\iint_R f(x, y)\, dA = \int_a^b \int_{g_1(x)}^{g_2(x)} f(x, y)\, dy\, dx.$$

2. If R is the region bounded by the lines $y = c$ and $y = d$ and the graphs of the equations $x = h_1(y)$ and $x = h_2(y)$, as shown in Figure 8–21, then

$$\iint_R f(x, y)\, dA = \int_c^d \int_{h_1(y)}^{h_2(y)} f(x, y)\, dx\, dy.$$

FUNCTIONS OF SEVERAL VARIABLES

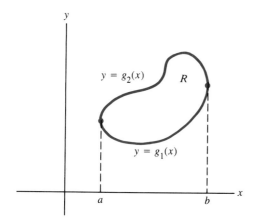

FIGURE 8–20

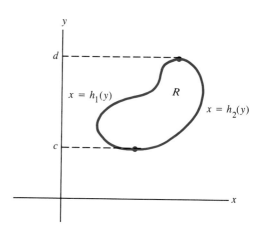

FIGURE 8–21

Note that in the integral

$$\int_a^b \int_{g_1(x)}^{g_2(x)} f(x, y)\, dy\, dx$$

we integrate first with respect to y, and y ranges from the lower boundary $g_1(x)$ to the upper boundary $g_2(x)$ of R. The result of the inner integration will thus be a function of the single variable x, which we then integrate from $x = a$ to $x = b$, as required by the outer integral. Similar remarks apply to the integral

$$\int_c^d \int_{h_1(y)}^{h_2(y)} f(x, y)\, dx\, dy.$$

EXAMPLE 3   Let $f(x, y) = 12 - x - y$, and let R be the rectangle shown in Figure 8–22. Since R is bounded by the lines $x = 1$, $x = 2$, $y = 2$, and $y = 4$, we have

$$\iint_R (12 - x - y)\, dA = \int_1^2 \int_2^4 (12 - x - y)\, dy\, dx$$

$$= \int_1^2 \left(12y - xy - \frac{y^2}{2} \Big|_{y=2}^{y=4} \right) dx$$

$$= \int_1^2 [(48 - 4x - 8) - (24 - 2x - 2)]\, dx$$

$$= \int_1^2 (18 - 2x)\, dx$$

$$= 18x - x^2 \Big|_1^2$$

$$= 15.$$

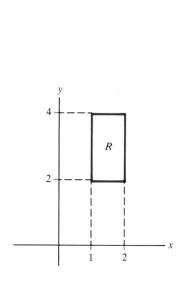

FIGURE 8–22

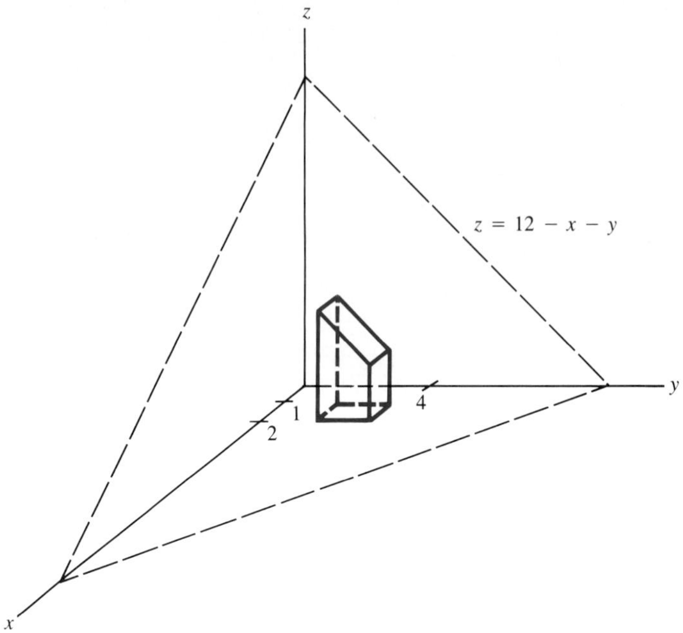

FIGURE 8–23

Thus the volume of the solid pictured in Figure 8–23 is 15. Note that we can also evaluate the double integral of $f(x, y) = 12 - x - y$ over the rectangle R as

$$\iint_R (12 - x - y) \, dA = \int_2^4 \int_1^2 (12 - x - y) \, dx \, dy.$$

This gives the same result. (Check that this is so.) ∎

EXAMPLE 4 Let us find the volume of the solid whose base is the triangle bounded by the x-axis, the y-axis, and the line $x + y = 1$, and whose top is the surface $z = x^2 + y^2$. See Figure 8–24.

The volume of the solid is the double integral

$$\iint_R (x^2 + y^2) \, dA,$$

where R is the given triangle. We seek to evaluate this double integral as an iterated integral of the form

$$\int_a^b \int_{g_1(x)}^{g_2(x)} (x^2 + y^2) \, dy \, dx.$$

Here we are integrating first with respect to y, and hence y must range from the x-axis ($y = 0$) to the line $y = 1 - x$, and x must range from 0 to 1 (Figure 8–25). Thus the volume of the solid is

FUNCTIONS OF SEVERAL VARIABLES

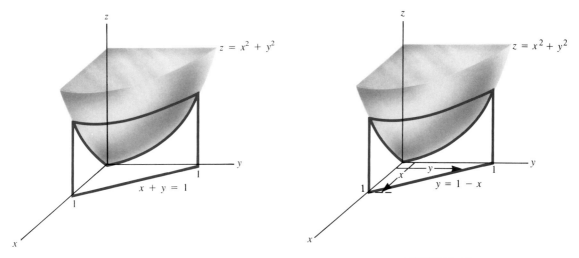

FIGURE 8–24   **FIGURE 8–25**

$$\int_0^1 \int_0^{1-x} (x^2 + y^2)\, dy\, dx = \int_0^1 \left(x^2 y + \frac{1}{3} y^3 \right) \bigg|_{y=0}^{y=1-x} dx$$

$$= \int_0^1 \left[x^2(1-x) + \frac{1}{3}(1-x)^3 \right] dx$$

$$= \int_0^1 \frac{1}{3}(-4x^3 + 6x^2 - 3x + 1)\, dx$$

$$= \frac{1}{6}.$$

(Check the integration.) ☐

If we had wished, we could have set up the iterated integral in the previous example so that we integrated first with respect to x and then with respect to y; in this case the integral would have been

$$\int_0^1 \int_0^{1-y} (x^2 + y^2)\, dx\, dy,$$

since now x must range from the y-axis to the line $x = 1 - y$, and y must range from 0 to 1. See Figure 8–26. Sometimes one way of setting up the iterated integral leads to an easier integration than the other.

EXAMPLE 5   Consider the iterated integral

$$\int_0^1 \int_x^1 e^{y^2}\, dy\, dx.$$

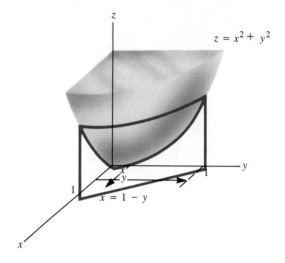

FIGURE 8–26

Here we are integrating first with respect to y, so y ranges from $y = x$ to $y = 1$, and then x ranges from $x = 0$ to $x = 1$. Therefore the region R over which we are integrating is the triangle shown in Figure 8–27. If we attempt to evaluate this integral in the order given, we must first find

$$\int_x^1 e^{y^2}\, dy;$$

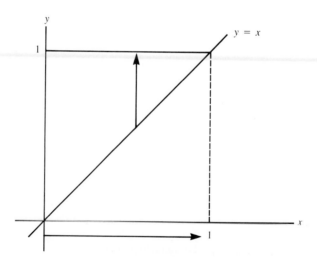

FIGURE 8–27

but since we do not know an antiderivative for e^{y^2}, we cannot carry out the integration. However, if we reverse the order of integration, we have

$$\int_0^1 \int_0^y e^{y^2} \, dx \, dy = \int_0^1 \left(x e^{y^2} \Big|_{x=0}^{x=y} \right) dy$$

$$= \int_0^1 y e^{y^2} \, dy.$$

If we now let $u = y^2$, so that $du = 2y \, dy$, then

$$\int_0^1 y e^{y^2} \, dy = \frac{1}{2} \int_0^1 e^u \, du = \frac{1}{2}(e - 1). \quad \square$$

Suppose we integrate the function f defined by $f(x, y) = 1$ over a region R in the xy-plane. We obtain the volume of the cylinder whose base is the region R and whose top is the plane $z = 1$. See Figure 8–28. But the volume of this cylinder is

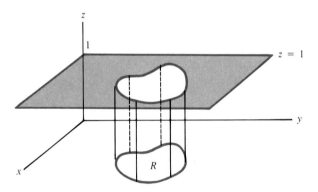

FIGURE 8–28

equal to the area of its base times its height, that is, to the area of R times 1, or the area of R. Thus we can find the area of a region R in the xy-plane by integrating $f(x, y) = 1$ over R:

The Double Integral and Area

If R is a region in the xy-plane, the area of R is given by

$$\iint_R dA.$$

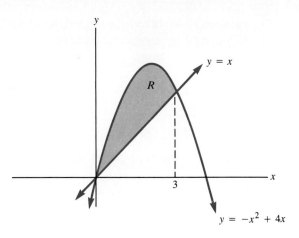

FIGURE 8–29

EXAMPLE 6 Find the area of the region R pictured in Figure 8–29. The area in question is the value of the double integral of 1 over R, and thus is equal to

$$\iint_R dA = \int_0^3 \int_x^{-x^2+4x} dy\, dx$$

$$= \int_0^3 \left(y \Big|_{y=x}^{y=-x^2+4x} \right) dx$$

$$= \int_0^3 (-x^2 + 3x)\, dx$$

$$= -\frac{x^3}{3} + \frac{3}{2}x^2 \Big|_0^3$$

$$= \frac{9}{2}.$$

Note that we could have found this area by the method of Section 6.4, in which case we would have obtained the same result by evaluating the single integral

$$\int_0^3 (-x^2 + 4x - x)\, dx. \quad \square$$

We conclude this section by illustrating how the double integral can be used to find the average value of a function over a region. By means of an argument analogous to that employed in Section 6.5, it is possible to show the following:

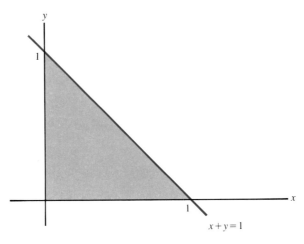

FIGURE 8–30

The Average Value of a Function over a Region

Let f be a function of the two variables x and y, and let R be a region in the domain of f. Then

$$\text{Average value of } f \text{ over } R = \frac{1}{\text{area of } R} \iint_R f(x, y)\, dA.$$

EXAMPLE 7 Let us find the average value of the function f defined by $f(x, y) = 2x + 3y$ over the triangle bounded by the coordinate axes and the line $x + y = 1$. See Figure 8–30. Since the area of this triangle is $\frac{1}{2}$, we have

$$\text{Average value of } f = \frac{1}{\frac{1}{2}} \int_0^1 \int_0^{1-x} (2x + 3y)\, dy\, dx$$

$$= 2\int_0^1 \left(2xy + \frac{3}{2}y^2 \Big|_0^{1-x}\right) dx$$

$$= 2\int_0^1 \left[2x(1 - x) + \frac{3}{2}(1 - x)^2\right] dx$$

$$= 2\int_0^1 \left(-\frac{x^2}{2} - x + \frac{3}{2}\right) dx$$

$$= 2\left(-\frac{x^3}{6} - \frac{x^2}{2} + \frac{3x}{2}\Big|_0^1\right)$$

$$= \frac{5}{3}. \quad \square$$

8.6 EXERCISES

In Exercises 1 through 6, evaluate the iterated integral.

1. $\int_0^1 \int_0^1 x \, dy \, dx$

2. $\int_0^1 \int_0^1 2y \, dx \, dy$

3. $\int_0^2 \int_{-1}^1 (2x + 3y^2) \, dx \, dy$

4. $\int_{-2}^3 \int_{-2}^1 xy \, dy \, dx$

5. $\int_0^1 \int_0^2 xe^{-y} \, dy \, dx$

6. $\int_0^2 \int_0^1 ye^{-x} \, dx \, dy$

In Exercises 7 through 12, sketch the region of integration and evaluate the integral. (In some cases you may find it helpful to reverse the given order of integration.)

7. $\int_0^2 \int_0^1 x^2 \, dy \, dx$

8. $\int_1^2 \int_{-1}^2 (2x - y) \, dx \, dy$

9. $\int_1^2 \int_{-2}^x (x + y) \, dy \, dx$

10. $\int_1^2 \int_{y^2}^y (y^2 + 1) \, dx \, dy$

11. $\int_0^1 \int_y^{\sqrt{y}} y^{1/2} \, dx \, dy$

12. $\int_0^1 \int_{e^{-x}}^{e^x} e^x \, dy \, dx$

In Exercises 13 through 21, evaluate the double integral over the indicated region R in the xy-plane.

13. $\iint_R (x + y) \, dA$, R the rectangle bounded by the x-axis, the y-axis, and the lines $x = 1$, $y = 2$.

14. $\iint_R (2x - 4y) \, dA$, R the square bounded by the lines $x = 1$, $x = 3$, $y = 1$, $y = 3$.

15. $\iint_R (2x - y) \, dA$, R the square with vertices $(1, 1)$, $(3, 1)$, $(3, 3)$, and $(1, 3)$.

16. $\iint_R (x - 2y) \, dA$, R the triangle bounded by the y-axis and the lines $x - y = 0$, $y = 2$.

17. $\iint_R (x - 2y) \, dA$, R the triangle bounded by the x-axis and the lines $x - y = 0$, $x = 2$.

18. $\iint_R (x + y^2) \, dA$, R the triangle bounded by the lines $y = x$, $x = 3$, and $y = 1$.

19. $\iint_R x^2 y \, dA$, R the region bounded by $y = x^2$ and $y = 1 - x^2$.

20. $\iint_R \frac{1}{xy} \, dA$, R the region bounded by $y = e^{-x}$ and the lines $y = 1$, $x = 1$, $x = 2$.

21. $\iint_R xe^y \, dA$, R the region bounded by $y = \ln x$, $y = 0$, and $x = 2$.

FUNCTIONS OF SEVERAL VARIABLES

In Exercises 22 through 31, find the volume of the solid bounded by the given surface and the region R.

22. The plane $z = 2$; the rectangle R bounded by the x-axis, the y-axis, and the lines $x = 2$ and $y = 4$.
23. The plane $z = 8 - x - y$; the rectangle R bounded by the lines $x = 1$, $y = 1$, $x = 2$, and $y = 3$.
24. The plane $z = y$; the square R bounded by $x = 0$, $y = 0$, $x = 2$, and $y = 2$.
25. The plane $z = 20 - 2x - 5y$, R the triangle bounded by the lines $x + y = 1$, $y = 0$, and $x = 0$.
26. The plane $z = 6 - 2x - 3y$; the triangle R bounded by $x = 0$, $y = 0$, and $-2x - 3y = 6$.
27. The plane $z = 4x$; the region R bounded by the nonnegative x-axis, the nonnegative y-axis, and the circle $x^2 + y^2 = 1$.
28. The plane $z = x$, R the region bounded by the nonnegative x-axis, the nonnegative y-axis, and the circle $x^2 + y^2 = 1$.
29. The surface $z = x^2 + y^2$, R the square bounded by the lines $x = 0$, $x = 1$, $y = 0$, and $y = 1$.
30. The surface $z = x^2 + y^2$, R the triangle bounded by the lines $y = x$, $y = 1$, and $x = 0$.
31. The surface $z = e^{-y}$; the region R bounded by the x-axis, $x = e$, and $y = \ln x$.

In Exercises 32 through 35, use a double integral to find the area of the region R.

32. R is bounded by the x-axis and $y = 2x - x^2$.
33. R is bounded by the y-axis, $x = \sqrt{y}$, and $y = 1$.
34. R is bounded by $y = x^2$ and $y = x^3$.
35. R is bounded by $y = x^2 - 4x$ and $y = -x^2 + 4x$.

In Exercises 36 through 39, find the average value of the given function on the given region.

36. $f(x, y) = x + y$, on the rectangle bounded by the lines $x = 1$, $x = 3$, $y = 2$, $y = 5$.
37. $f(x, y) = 2x + y$, on the triangle bounded by the coordinate axes and the line $x + y = 2$.
38. $f(x, y) = x^2 - y^2$, on the region bounded by the x-axis and the parabola $y = 1 - x^2$.
39. $f(x, y) = xe^y$, on the region bounded by the x-axis, the y-axis, the line $y = 1$, and the curve $y = \ln x$.

Management

40. A firm can produce $f(x, y) = 4000x^{0.4}y^{0.6}$ units by employing x thousand worker-hours of labor and y million dollars of capital. If $0 \le x \le 10$ and $0 \le y \le 5$, find the firm's average production.
41. A firm can produce $f(x, y) = 1000x^{1/4}y^{3/4}$ units by employing x units of labor and y units of capital. If the firm's total budget for each of capital and labor must be no more than 500 units, find the firm's average production.

42. If a store runs x newspaper ads and y television commercials during a month, its sales will be $S = 200x + 220y$ thousand dollars. Suppose newspaper ads cost $1000 each and it costs $5000 to run each TV commercial. If the total monthly budget for both types of advertising combined varies from $0 to $100,000, find the store's average monthly sales.

43. Repeat Exercise 42 if in addition to the budget restriction given there, the store always runs at least 20 newspaper ads and at least 5 TV commercials every month.

44. Repeat Exercise 42 if the store's monthly budget for the two types of advertising combined varies from $50,000 to $100,000.

Life Science

45. The time it takes a patient of age t years who receives d units of penicillin per day to recover from pneumonia is

$$f(t, d) = \frac{250t}{d}$$

days, where $21 \leq t \leq 60$ and $500 \leq d \leq 1000$. Find the average amount of time it takes for a patient to recover from pneumonia.

46. The pollution produced by a coal-fired power plant is given by

$$z = \frac{20x}{y + 1}$$

where z is pollution in parts per million, x is the rated capacity of the plant in kilowatt-hours, and y is the amount spent during construction of the plant on pollution-abating equipment. Find the average pollution produced by coal-fired power plants if $100 \leq x \leq 1000$ and $0 \leq y \leq 25$.

47. An animal habitat is rectangular in shape:

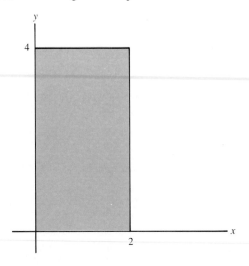

If the density of the animal population at point (x, y) is

$$f(x, y) = 10 - (x - 1)(y - 2)$$

individuals per square kilometer, find the average population density within the habitat.

*48. An animal habitat is circular in shape:

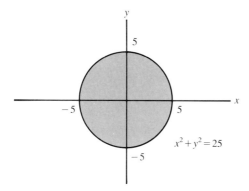

If the density of the animal population at point (x, y) is $f(x, y) = 25 - x - y$ individuals per square kilometer, find the average population density within the habitat.

Social Science

49. Ridership on city mass transit systems is given by

$$r = \frac{100x}{2 + y}.$$

Here r is daily ridership, in thousands, x is the population of the city, in thousands, and y is the price of a ticket, in dollars. Find the average daily ridership for cities whose populations vary from 100,000 to 120,000 and whose ticket prices range from $0.75 to $1.00.

50. The proportion p of the population in a region who will apply for welfare benefits at some time during a month is given by $p = e^{0.08x} - e^{0.01y}$, where x is the monthly unemployment rate and y is the average years of education of the adults in the region. Find the average proportion who will apply for welfare benefits if the unemployment rate ranges between 5 and 20% and the average years of education ranges between 10 and 12 years.

SUMMARY

This summary consists of terms and symbols whose meaning you should know, facts you should know, and techniques and methods you should be able to employ.

Section 8.1

Functions of two variables. The domain of a function of two variables as a set of ordered pairs. Functions of more than two variables. The domain of a function of n variables as a set of ordered n-tuples. Evaluating a function of several variables at a point. The three-dimensional cartesian coordinate system. The origin, the first octant. Plotting points in the three-dimensional cartesian coordinate system. Graphing functions of two variables: the graph as a surface.

Section 8.2

First partial derivatives: the partial derivative with respect to x, the partial derivative with respect to y. Notation: f_x, f_y. Finding first partial derivatives. Evaluating a first partial derivative at a point. Notation: $\partial f/\partial x$, $\partial f/\partial y$. The definition of first partial derivatives. Second partial derivatives: f_{xx}, f_{xy}, f_{yx}, f_{yy}. Finding second partial derivatives. Mixed partials f_{xy}, f_{yx}. Equality of mixed partials. Notation: $\partial^2 f/\partial x^2$, $\partial^2 f/\partial y^2$, $\partial^2 f/\partial x \partial y$, $\partial^2 f/\partial y \partial x$. Third and higher partial derivatives. Finding third and higher partial derivatives.

Section 8.3

The first partial derivatives f_x, f_y as the slopes of lines tangent to the surface $z = f(x, y)$. The significance of the sign of f_x, f_y. The significance of the sign of f_{xx}, f_{yy}. Using partial derivatives to find slopes of tangent lines to surfaces. Using partial derivatives to determine whether the curves generated by slices parallel to the xz-plane or yz-plane are increasing, decreasing, concave upward, or concave downward. The first partial derivative with respect to x as the rate of change of the function when all variables except x are held fixed. Using partial derivatives to find rates of change. Marginality. Competitive and complementary products. Cobb-Douglas production function: the marginal productivity of labor and of capital.

Section 8.4

Relative maximum, relative minimum, saddle point for a function of two variables. Finding relative maxima, relative minima, saddle points by setting the first partial derivatives equal to zero. The second derivative test. Using the second derivative test to find relative maxima, relative minima, and saddle points. The failure of the second derivative test. Price discrimination.

Section 8.5

Optimization subject to constraints. Constraint equation. The Lagrange multiplier method of optimizing a function of two variables subject to a constraint. Using the Lagrange multiplier method to optimize functions of two variables subject to constraints. The marginal productivity of money. Using the Lagrange multiplier method to optimize a function of three variables subject to a constraint.

Section 8.6

Region R in the xy-plane. Grid of rectangles on a region. Riemann sum for a function of two variables over a region R. The double integral as a limit of Riemann sums:

FUNCTIONS OF SEVERAL VARIABLES

$$\iint_R f(x, y)\, dA = \lim_{\substack{\Delta x \to 0 \\ \Delta y \to 0}} \sum_{i=1}^{n} f(x_i, y_i)\, \Delta x\, \Delta y.$$

The double integral is the volume of the cylindrical solid bounded by the region R and the surface $z = f(x, y)$. Iterated integrals of the forms

$$\int_a^b \int_c^d f(x, y)\, dy\, dx, \quad \int_c^d \int_a^b f(x, y)\, dx\, dy$$

Evaluating iterated integrals of the above forms. The double integral as an iterated integral of the form

$$\int_a^b \int_{g_1(x)}^{g_2(x)} f(x, y)\, dy\, dx \quad \text{or} \quad \int_c^d \int_{h_1(y)}^{h_2(y)} f(x, y)\, dx\, dy.$$

Setting up double integrals as iterated integrals. Evaluating double integrals as iterated integrals. Changing the order of integration. Using double integrals to find volumes. Using double integrals to find the area of a region. Using double integrals to find the average value of a function over a region.

REVIEW EXERCISES

1. Find the domain of the function f defined by

$$f(x, y) = x + \frac{2xy}{y - x},$$

and find $f(0, 0)$, $f(1, 0)$, $f(0, -1)$, and $f(3, 4)$, if possible.

2. Find the domain of the function defined by

$$w = e^{-xyz} + \ln 2xy,$$

and, if possible, find w when

(a) $x = y = z = 0$ (b) $x = y = z = 2$
(c) $x = -1, y = 3, z = 2$ (d) $x = 2, y = 1, z = -3$

3. Reddin Toothenclaw's pet store sells both quails and snails. If quails are priced at p_1 dollars each, Reddin can sell

$$q_1 = 200 - 5p_1$$

of them; if snails are priced at p_2 dollars each, Reddin can sell

$$q_2 = 760 - 20p_2$$

of them. Assuming that the prices of quails and snails do not affect one another, express Reddin's revenue as a function of p_1 and p_2 and find the revenue if quails sell for $10 each and snails for $2 each.

4. If a firm makes x units of X and y units of Y its profit is

$$P(x, y) = -8000 + 40x + 30y - 0.05xy$$

dollars. Find the firm's profit if it makes and sells 400 units of X and 300 units of Y. If x is fixed at 400, what can you say about the contribution to profit made by each unit of Y?

5. Find f_x, f_y, f_{xx}, f_{xy}, f_{yx}, and f_{yy} if
$$f(x, y) = 8x^2y - 2xy^3 + 12x + 2y - 5.$$

6. Find $f_x(2, 1)$, $f_y(-1, 0)$, $f_{xx}(2, 2)$, $f_{yy}(-3, 4)$ for the function f of Exercise 5.

7. Let $g(p, q, r) = e^{pq} + e^{-qr} + 2p + r^4$. Find g_{pq}, g_{qqpp}, and g_{qppp}.

8. Find $\dfrac{\partial z}{\partial x}$, $\dfrac{\partial z}{\partial y}$, $\dfrac{\partial z^2}{\partial x \partial y}$, $\dfrac{\partial z^2}{\partial y \partial x}$, $\dfrac{\partial z^2}{\partial x^2}$, and $\dfrac{\partial z^2}{\partial y^2}$, if
$$z = (x^2 + 2xy - y^3)^6.$$

9. Let $f(x, y) = 2x + 3y + 5x^2y^2$. Find the slope of the line tangent to the graph of f at the point $(2, 2, 90)$ and parallel to the
   (a) xz-plane   (b) yz-plane

10. If the number of wolves in a region is given by
$$f(x, y) = x^{1/4}y^{-2},$$
where x is the number of animals the wolves can prey on and y is the number of animals that can prey on the wolves, find $f_x(16, 1)$ and $f_y(16, 1)$ and interpret your results.

11. Suppose a firm's revenue from selling x units of X and y units of Y is
$$R(x, y) = 20x + 25y + 0.1\sqrt{xy}$$
dollars. Find $R_x(25, 100)$ and $R_y(25, 100)$ and interpret your results.

12. Find the relative maxima, relative minima, and saddle points of f if
$$f(x, y) = x^2 + 2y^2 - 1.$$

13. Repeat Exercise 12 if
$$f(x, y) = -2x^4 + 16x^2 + y^2 - 4y + 8.$$

14. Find the dimensions of a rectangular box that is to have a volume of 10.4 cubic meters if the bottom is to be hardwood, which costs $10 per square meter, the sides are to be softwood, which costs $5 per square meter, the top is to be plastic, which costs $3 per square meter, and the cost of the box is to be minimized.

15. The demand functions for a commodity in regions X and Y are given by
$$p_X = 800 - \frac{x}{10} \quad \text{and} \quad p_Y = 500 - \frac{y}{5},$$
respectively. Find the prices that maximize revenue.

16. Find the maximum or minimum value of $f(x, y) = x^3y^3$ subject to the constraint $6x + 2y = 12$, and state whether your answer is a maximum or a minimum.

17. Repeat Exercise 16 for the function defined by
$$f(x, y, z) = x^2y + yz + z^2$$
subject to the constraint $xyz = 64$.

18. Find two positive numbers such that the product of twice the first and three times the second is 384 and their sum is a minimum.

19. A company can produce $f(x, y) = 1000x^{1/5}y^{4/5}$ units of its product, where x is units of labor and y is units of capital. If each unit of labor costs \$20 and each unit of capital costs \$100, and if the company's production budget is \$1000, find the maximum production. Also find and interpret the marginal productivity of money.

20. Evaluate the integral

$$\iint_R x^2 \, dA,$$

where R is the region bounded by the graphs of $y = x$, $y = 5 - x$, $y = 1$, and the x-axis.

21. Find the volume of the cylindrical solid whose base is the triangle in the xy-plane bounded by the lines $y = 2 - x$, $y = x$, and $x = 0$, and whose top is the plane $z = 4 - 2x$.

22. Use double integrals to find the area of the region bounded by $y = e^x$, $y = e^{-x}$, and $x = 1$.

23. Refer to the company of Exercise 19: Suppose the amount of labor available varies from 15 to 30 units and the amount of capital available varies from 4 to 10 units; find the average output.

ADDITIONAL TOPICS

Here are some suggestions for topics not covered in the text that you might want to investigate on your own.

1. Just as there is a chain rule for functions of one variable, so there are chain rules for functions of several variables. For instance, if $z = f(x, y)$, $x = g(u, v)$ and $y = h(u, v)$, then one such chain rule states that

$$\frac{\partial z}{\partial u} = \frac{\partial z}{\partial x}\frac{\partial x}{\partial u} + \frac{\partial z}{\partial y}\frac{\partial y}{\partial u}.$$

Find out about chain rules for partial derivatives and how to use them.

2. In Section 3.3 we developed the formula

$$f(x + \Delta x) \approx f(x) + f'(x) \Delta x$$

for approximation by differentials. There is a corresponding formula for functions of two variables, namely

$$f(x + \Delta x, y + \Delta y) \approx f(x, y) + f_x(x, y) \Delta x + f_y(x, y) \Delta y.$$

The expression $f_x(x, y) \Delta x + f_y(x, y) \Delta y$ is called the **total differential** of f. Find out about total differentials and how they are used to approximate functions and interpret the Lagrange multiplier λ.

3. Calculus in three dimensions is often best studied in the framework of **vectors**. Find out about vectors and how they are used in calculus. In particular, find out about the **directional derivative** and **gradient** of a function, how they are defined and computed, and their meanings.

4. In the Additional Topics at the end of Chapter 1 we suggested that you find out about the **polar coordinate system.** If you have not done so, do so now and then find out how to set up and evaluate double integrals in polar coordinates.

5. Triple integrals are analogous to double integrals: the triple integral of a function f of the variables x, y, and z is an integral over the interior of a three-dimensional solid V that lies within the domain of f, and has the form

$$\iiint_V f(x, y, z)\, dV$$

Triple integrals are evaluated as threefold iterated integrals of the form

$$\int_a^b \int_c^d \int_p^q f(x, y, z)\, dz\, dy\, dx.$$

Find out how to set up and evaluate triple integrals and find some of their uses.

6. In addition to the cartesian or rectangular coordinate system in three dimensions, in which points are located by their distances from the x-, y-, and z-axes, there are two other useful three-dimensional coordinate systems: **cylindrical coordinates** and **spherical coordinates.** Find out how these coordinate systems work and how to set up triple integrals in each of them.

7. The technique of Lagrange multipliers of Section 8.5 is due to the French mathematician Joseph-Louis Lagrange (1736–1813). Lagrange made contributions to many areas of mathematics and physics, and wrote a famous and influential book entitled *Analytical Mechanics*. He was also caught up in the French Revolution, was instrumental in the development of the metric system, and was made a count by Napoleon. Find out about Lagrange's life and his contributions to mathematics and physics.

SUPPLEMENT: LEAST SQUARES

Students sometimes ask how the functions used to model practical situations are "made up." We have seen throughout this book that such functions can often be constructed from an analysis of the situation. For instance, as we saw in Chapter 1, if it costs a firm v dollars to make each unit of its product then its cost function is linear with slope v, and since fixed cost F is the cost of making 0 units, it follows that the cost C of making x units is given by $C = vx + F$.

Often, however, functional models cannot be developed through analysis of a situation, but must be created by fitting curves to experimental data. To illustrate, suppose we want to find a function that models annual family savings as a function of annual family income. To begin with we have no information on the relationship, and thus no reason to prefer any particular type of function (e.g., linear, quadratic, exponential, etc.) as a model. If we collected some data on the relationship between income and savings, perhaps we would be able to decide on the type of function to use as a model. Let us assume we have somehow obtained the following data:

X Annual Family Income (in 1000s of $)	Y Annual Family Savings (in 1000s of $)
42	5.0
48	6.0
37	4.5
52	6.1
60	7.1
54	6.1
44	4.9
48	5.6

Let us plot the data as points (x_i, y_i). The result is the diagram of Figure 8–31, which is called a **scatter diagram** for the data. We now guess the type of curve that would best fit the scatter diagram in the sense that it would come as close as possible to all the points of the diagram. The type of curve that best fits the diagram depends on the pattern formed by the points of the diagram: it might be a line, a parabola, an exponential curve, or some other curve. In this case it looks as if a line would fit the scatter diagram fairly well. See Figure 8–32. Thus our data lead us to choose a linear function to model the relationship between income and savings.

Our next task is to determine the equation $Y = mX + b$ of our linear model by finding the values of m and b that make the line come as close as possible to all the points of the scatter diagram simultaneously. For any scatter diagram and any line, the closeness of the line to the points could be measured by the sum of the deviations of the points from the line: see Figure 8–33, which is a picture of a

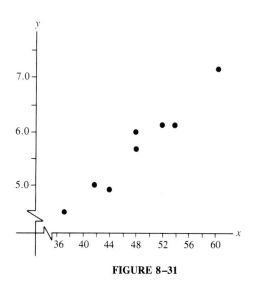

FIGURE 8–31

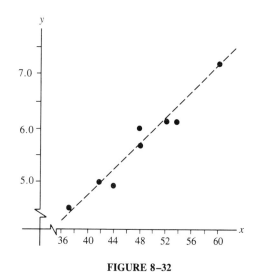

FIGURE 8–32

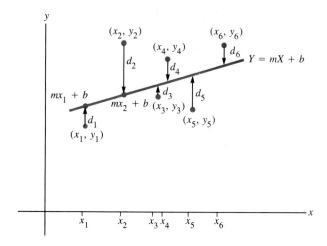

FIGURE 8–33

general scatter diagram and line $Y = mX + b$ with the deviations $d_1, d_2, \ldots, d_n$, where

$$d_i = (mx_i + b) - y_i$$

for each i. However, since some of the d_i's (those corresponding to points that lie below the line) will be positive while others (those corresponding to points that lie above the line) will be negative, there will be some "canceling out" in the sum $d_1 + d_2 + \cdots + d_n$, and this could make the line appear to be closer to the points than it really is. Therefore instead of using the d_i's to measure deviation from the line, we use their squares d_i^2. The line for which $d_1^2 + d_2^2 + \cdots d_n^2$ is a minimum is called the **least squares line** for the data.

The Least Squares Line

Given n data points $(x_1, y_1) \ldots, (x_n, y_n)$, the **least squares line** for the data is the line that best fits the data in the sense that in its equation

$$Y = mX + b,$$

m and b are chosen so as to minimize

$$f(m, b) = \sum_{i=1}^{n} (mx_i + b - y_i)^2.$$

(Note that m and b are the independent variables: the points (x_i, y_i) are known.)

Since $f(m, b)$ is a sum of squares, f has no maxima or saddle points; hence if we set the partial derivatives

$$f_m(m, b) = \sum_{i=1}^{n} 2(mx_i + b - y_i)x_i$$

and

$$f_b(m, b) = \sum_{i=1}^{n} 2(mx_i + b - y_i)$$

equal to zero and solve for m and b, we will find its minimum. Thus we must solve the system of equations

$$\sum_{i=1}^{n} x_i(mx_i + b - y_i) = 0$$

$$\sum_{i=1}^{n} (mx_i + b - y_i) = 0$$

for m and b.

EXAMPLE 1   Let us find the least-squares regression line for the income and savings data with which we began our discussion. Substituting the points (x_i, y_i) into the above equations, we have

$$42(42m + b - 5.0) + 48(48m + b - 6.0) + \cdots$$
$$\cdots + 48(48m + b - 5.6) = 0$$

and

$$(42m + b - 5.0) + (48m + b - 6.0) + \cdots + (48m + b - 5.6) = 0,$$

which reduces to

$$18{,}897m + 385b = 2221.5$$
$$385m + 8b = 45.3.$$

The solution of this system of equations is $m \approx 0.112$ and $b \approx 0.273$. Therefore the equation of the least-squares regression line is

$$Y = 0.112X + 0.273.$$

We can use the least-squares regression line to predict Y from X; when we do this, we are predicting the *average* value of Y for a given value of X. Thus, for instance, if $X = 50$,

$$Y = 0.112(50) + 0.273 = 5.873,$$

so we predict that for families whose annual income is $50,000, the average annual savings is $5873. □

The technique employed above to find m and b for the income and savings data can be used to develop general formulas for m and b. Thus, as we noted previously, to find m and b we must solve the equations

CHAPTER 8

$$\sum_{i=1}^{n} x_i(mx_i + b - y_i) = 0$$

$$\sum_{i=1}^{n} (mx_i + b - y_i) = 0$$

for m and b. If the second of these is solved for b, the result is substituted for b in the first, and the first is then solved for m, the results are as follows:

> **The Least-Squares Line Formulas**
>
> Suppose that $(x_1, y_1), \ldots, (x_n, y_n)$ is a set of data and that $Y = mX + b$ is the least-squares line for the data. Then
>
> $$m = \frac{n \sum_{i=1}^{n} x_i y_i - \left(\sum_{i=1}^{n} x_i\right)\left(\sum_{i=1}^{n} y_i\right)}{n \sum_{i=1}^{n} x_i^2 - \left(\sum_{i=1}^{n} x_i\right)^2}$$
>
> and
>
> $$b = \frac{\sum_{i=1}^{n} y_i - m \sum_{i=1}^{n} x_i}{n}.$$

(See Exercise 1.)

EXAMPLE 2 Consider the following data that relate students' grades on a final examination to their number of absences from class during the semester:

x = Number of Absences	1	2	6	7
y = Grade on Final Exam	91	86	74	73

A scatter diagram (Figure 8–34) shows that the relationship between x and y appears to be linear. Therefore we seek to fit a least-squares line $Y = mX + b$ to the data. But we have

X	Y	XY	X²
1	91	91	1
2	86	172	4
6	74	444	36
7	73	511	49
16	324	1218	90

Therefore

$$m = \frac{4(1218) - (16)(324)}{4(90) - (16)^2} = -3$$

Functions of Several Variables

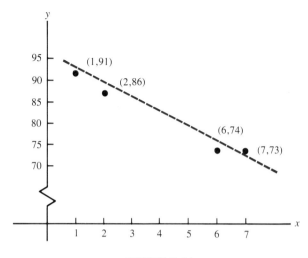

FIGURE 8–34

and

$$b = \frac{324 - (-3)(16)}{4} = 93.$$

Hence the least-squares line is $Y = -3X + 93$. ☐

The technique we have used here to find the least-squares line can also be used to fit other least-squares curves to data sets. We illustrate with an example.

EXAMPLE 3 Consider the following data relating average cost per unit y to the number of units produced x:

x (in thousands of units)	1	2	3	4	5
y (in dollars)	5.90	3.10	2.00	2.90	6.10

A scatter diagram (Figure 8–35) suggests that a parabola will fit the data better than a line. Thus we seek to find the equation of the least-squares parabola

$$Y = aX^2 + bX + c.$$

This will be the parabola that minimizes

$$f(a, b, c) = \sum_{i=1}^{5} (ax_i^2 + bx_i + c - y_i)^2.$$

We minimize f by setting its partial derivatives equal to zero and solving for a, b, and c. We have

$$f_a = \sum_{i=1}^{5} 2x_i^2 (ax_i^2 + bx_i + c - y_i) = 0,$$

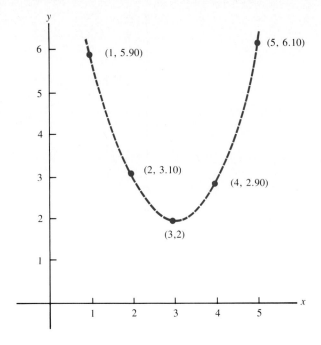

FIGURE 8–35

$$f_b = \sum_{i=1}^{5} 2x_i (ax_i^2 + bx_i + c - y_i) = 0,$$

and

$$f_c = \sum_{i=1}^{5} 2(ax_i^2 + bx_i + c - y_i) = 0.$$

Substituting in the values for x_i and y_i and simplifying, we obtain the system of equations

$$979a + 225b + 55c = 235.2$$
$$225a + 55b + 15c = 60.2$$
$$55a + 15b + 5c = 20.0.$$

This system may be solved for a, b, and c to obtain $a = 1.00$, $b = -5.98$, and $c = 10.94$. Therefore the least-squares quadratic is

$$Y = X^2 - 5.98X + 10.94. \quad \square$$

We can similarly use the method of this supplement to fit cubic, quartic (fourth-power) and other least-squares curves to data sets. We can also use it to fit least-squares equations to three-dimensional data sets. Again, we illustrate with an example.

EXAMPLE 4

Suppose we have the following data that relate savings to income and family size:

$x =$ Annual Income (in 1000s of $)	20	30	40	50
$y =$ Family Size	2	2	4	3
$z =$ Annual Savings (in 1000s of $)	2.10	2.90	3.20	4.25

Note that the data are three-dimensional: the data set is of the form $\{(x_i, y_i, z_i) \mid 1 \le i \le 4\}$.

We wish to express savings z as a function of income x and family size y. Suppose we decide to use a model of the form

$$Z = aX + bY + c.$$

Then the equation of this form that best fits the data will be the one that minimizes

$$f(a, b, c) = \sum_{i=1}^{4} (ax_i + by_i + c - z_i)^2 .$$

Hence we must have

$$f_a = \sum_{i=1}^{4} 2x_i (ax_i + by_i + c - z_i) = 0,$$

$$f_b = \sum_{i=1}^{4} 2y_i (ax_i + by_i + c - z_i) = 0,$$

and

$$f_c = \sum_{i=1}^{4} 2(ax_i + by_i + c - z_i) = 0.$$

Substituting the values of x_i, y_i, and z_i into the equations and simplifying, we obtain the system

$$5400a + 410b + 140c = 469.50$$
$$410a + 33b + 11c = 35.55$$
$$140a + 11b + 4c = 12.45.$$

The solution of this system is $a = 0.08$, $b = -0.25$, and $c = 1$. Therefore the least squares equation is

$$Z = 0.08X - 0.25Y + 1.$$

This equation can be used in the usual manner to predict the average value of z for given values of x and y. For instance, we can predict that the average savings for families of 4 whose income is $25,000 per year will be

$$Z = 0.08(25) - 0.25(4) + 1 = 2$$

thousand dollars per year. □

■ *Computer Manual: Program LESTSQRS*

SUPPLEMENTARY EXERCISES

1. Solve

$$\sum_{i=1}^{n} x_i (mx_i + b - y_i) = 0$$

$$\sum_{i=1}^{n} (mx_i + b - y_i) = 0$$

for b and m.

In Exercises 2 through 5, make a scatter diagram and find the least-squares line $Y = mX + b$.

2.

x	1	2	3	4
y	10	13	19	22

3.

x	2	3	4	5	6
y	8	10	7	11	8

4.

x	12	18	23	24	30	32
y	63	46	38	25	35	16

5.

x	4.1	5.4	6.9	8.2	9.1	9.9
y	42.0	44.0	48.0	61.0	75.0	90.0

6. A cost accountant is attempting to find a linear model that describes the cost of producing a firm's product. She has the following data:

Units Produced (in thousands)	2	3	4	5
Total Cost (in thousands of $)	110	135	155	180

Using the method of Example 1 of this supplement, find the linear model that best describes cost as a function of production and use it to predict the average cost of producing 3500 units.

7. The following table compares weekly sales of gas at a certain gas station with the price per gallon:

Price per Gallon ($)	1.10	1.25	1.35	1.50
Weekly Sales (hundreds of gallons)	16.4	14.3	13.7	12.5

Find the linear model that best describes weekly sales as a function of price, and use it to predict the average weekly sales if gas costs $1.45 per gallon.

8. A researcher has gathered the following data regarding the rate of lung cancer among ex-smokers:

Years Since Stopped Smoking	2	4	5	8	10
Cases of Lung Cancer per 1000 Ex-smokers	42	37	33	24	20

Find the linear model that best describes lung cancer cases as a function of time since smoking was stopped, and then predict the average number of lung cancer cases for those who quit smoking 7 years ago.

9. According to data on page 34 of the June 1988 issue of *Scientific American*, West Germany's birth rate and its population of brooding storks have both been declining in recent years:

Year	1965	1970	1975	1980
Pairs of Brooding Storks	1900	1400	1050	900
Millions of Newborn Babies	1.1	0.88	0.65	0.65

Find a linear least-squares model for births as a function of the number of pairs of brooding storks.

10. According to an article by J. Larry Brown ("Hunger in the U.S.", *Scientific American*, February 1987), participation in the food stamp program as a percentage y of people who are living at or below the poverty level has been as follows:

Year	1980	1981	1982	1983	1984	1985
% y	68	65	60	61	61	59

Using the method of Example 1, find the least-squares line for y as a function of time. Then check your answer by using the formulas for m and b. What are the significances of the slope and y-intercept of the least-squares line?

11. According to the *1989 World Almanac* (Pharos Books, 1989) world gold production over the period 1982–87 was as follows:

Production:	43.1	45.2	46.8	49.2	51.6	52.5
Year t:	0	1	2	3	4	5

The production figures are in millions of troy ounces and the years are coded so that $t = 0$ represents 1982. Make a scatter diagram of production versus time and then find the linear least-squares model for gold production as a function of time. What is the significance of the slope of the model?

12. According to an article in the October 3, 1988, edition of *The Wall Street Journal*, the profits of U.S. oil companies and the price of Mideast light crude oil in 1987 and 1988 were as follows:

	Profits (billions of dollars)	Price of Light Crude (dollars per barrel)
1987 Qtr I	2.936	17.25
II	3.507	17.38
III	3.429	17.65
IV	3.600	17.02
1988 Qtr I	4.395	15.05
II	4.257	14.68

(a) Find the least-squares line for profit as a function of price per barrel of oil. What is the significance of the slope of this line?
(b) Predict the oil companies' profit if the price per barrel is $16.00.

13. An article in the July 3, 1988, edition of *The New York Times* reported that the percentage of eligible voters who voted in the seven presidential elections from 1960 through 1984 was as follows:

Year	1960	1964	1968	1972	1976	1980	1984
% Voting	63.1	61.8	60.7	55.1	53.6	52.6	53.1

The same article also estimated the average hours of television watched per day by U.S. viewers to be as follows:

Year	1960	1964	1968	1972	1976	1980	1984
Avg. Hours per Day	5.0	5.4	5.9	6.5	6.3	6.8	7.2

   (a) Find the linear model that best describes voter participation as a function of time.
   (b) Use the model of part (a) to predict voter participation in the election of 1988. Actual voter participation in 1988 was 50 percent; how does this compare with your prediction?
   (c) Find the linear model that best describes average hours of television watched per day as a function of time.
   (d) Find the linear model that best describes voter participation as a function of hours spent watching television. What is the significance of the slope of this line?

14. An article by Richard Stolarski ("The Antarctic Ozone Hole," *Scientific American*, June 1988) indicates that a NASA satellite collected the following data on ozone levels over Antarctica from 1979 through 1986:

Year	Ozone Level (in Dobson units)
1979	283
1980	237
1981	239
1982	216
1983	196
1984	192
1985	208
1986	168

Find a linear least-squares model for the ozone level as a function of time, and interpret your results.

15. A new drug that shrinks tumors is being tested in various concentrations. The results of the tests are as follows:

X Drug Dosage (units)	Y Time until Tumor Disappears (days)
1	19.0
3	17.5
4	16.6
6	15.4
7	14.4
8	13.7

X Drug Dosage (units)	Y Time until Tumor Disappears (days)
10	12.1
12	10.8
13	9.6
14	9.0

Make a scatter diagram and find the least-squares line for the data. Can you estimate the effect that increasing the dosage by 1 unit has on the tumor?

16. A firm has the following data on its profit from making and selling x units of product:

X Units Made and Sold	Y Profit $
5	−155
10	−114
15	−77
20	−40
25	−2
30	44
35	83
40	120
45	162
50	194

Make a scatter diagram and find the equation of the least-squares line for the data. Can you estimate the firm's fixed cost and the contribution to profit made by each unit?

17. An article by Richard Lester ("Rethinking Nuclear Power," *Scientific American*, March 1986) gives the following data concerning the average construction time, in months, for nuclear power plants in the U.S. during the period 1971–1986:

Year	Average Construction Time (months)
1971	46
1972	73
1973	71
1974	83
1975	78
1976	87
1977	100
1978	112
1979	125
1980	120
1981	140
1982	158
1983	106
1984	134
1985	150
1986	125

Find a linear least-squares model for construction time as a function of time, and interpret your results.

18. Find the quadratic least-squares regression equation for the following data:

x	1	3	4	6	10
y	8.8	5.1	6.0	13.9	53.5

19. A firm has the following data on its average cost per unit A:

Units Made x	10	20	45	50	65	70
Avg Cost/Unit A ($)	120	55	10	20	85	115

Find a quadratic least-squares model for A as a function of x. At what level of production is average cost per unit the lowest?

20. According to the article by Brown cited in Exercise 10, distribution of food to the poor by U.S. foodbanks during the period 1979–1985 was as follows:

Year	'79	'80	'81	'82	'83	'84	'85
Food Distributed (millions of tons)	4	10	18	30	44	70	100

Find a quadratic least-squares model for the amount of food distributed as a function of time, and use it to predict the amount distributed in 1986.

*21. Prove that if $\{(x_i, y_i) \mid 1 \leq i \leq n\}$ is a data set and $Y = aX^2 + bX + c$ is the least-squares quadratic for the data, then a, b, and c can be found by solving the system of equations

$$a \sum_{i=1}^{n} x_i^4 + b \sum_{i=1}^{n} x_i^3 + c \sum_{i=1}^{n} x_i^2 = \sum_{i=1}^{n} x_i^2 y_i$$

$$a \sum_{i=1}^{n} x_i^3 + b \sum_{i=1}^{n} x_i^2 + c \sum_{i=1}^{n} x_i = \sum_{i=1}^{n} x_i y_i$$

$$a \sum_{i=1}^{n} x_i^2 + b \sum_{i=1}^{n} x_i + nc = \sum_{i=1}^{n} y_i.$$

Then use these equations to check your answers for Exercises 18, 19, and 20.

22. Find the least-squares regression equation of the form

$$Y = aX_1 + bX_2 + c$$

for the data

x_1	8	10	14	15	21
x_2	42	36	32	24	15
y	200	170	140	100	30

*23. A firm has the following cost data for the total cost C of making its two products X_1 and X_2:

Units of X_1	50	60	80	100	120	125	140
Units of X_2	20	20	40	50	40	50	70
Cost C ($)	2900	3600	4500	5200	6300	6700	7700

Find a least-squares model of the form $C = aX_1 + bX_2 + c$ for this set of data. What are the significances of the values obtained for a, b, and c?

*24. As in Exercise 21, find a system of equations that when solved will yield the coefficients a, b, and c in the least-squares regression equation $Y = aX_1 + bX_2 + c$. Then use these equations to check your answers for Exercises 22 and 23.

*25. According to an article by Simons, Sherrod, Collopy, and Jenkins ("Restoring the Bald Eagle," *American Scientist*, May–June 1988), the percentage P of eagle colonies founded by x individuals that become extinct is as follows:

x	10	20	30	40
P	47	20	9	4

Find an exponential least-squares equation of the form $P = ae^{bx}$ for this set of data and use it to predict the percentage of colonies founded by 25 individuals that will become extinct. (Hint: $\ln P = \ln a + bx$.)

*26. The following table appeared on page 342 of the July–August 1988 issue of *American Scientist*: it gives the cost of completing the Shoreham, NY, nuclear power plant, as this cost was estimated in the given year.

Year	Estimated Cost C
1966	$ 75 million
1971	$200 million
1976	$600 million
1981	$1.5 billion
1986	$4.5 billion
1988	$5.3 billion

Find an exponential least-squares model of the form $C = ae^{bx}$ and use it to predict the cost of completing the plant in 1990.

9

Differential Equations

A **differential equation** is an equation containing one or more derivatives, such as

$$\frac{dy}{dx} = 8xy \quad \text{or} \quad 2x\frac{d^2y}{dx^2} + 3xy^2\frac{dy}{dx} + y = 0.$$

Since derivatives are rates of change, such equations are really statements about the rate of change of one quantity with respect to another. In fact, that is how they arise: if we do not know the relationship between two quantities but we do know the rate of change of one of them with respect to the other, we can write a differential equation that involves the quantities; solving this differential equation will then yield the relationship between the quantities.

Because rate-of-change problems occur in many different disciplines, differential equations arise in many different ways. For instance, in this chapter we shall use them to solve problems such as the following:

- Find the size of a population when its rate of growth is known.
- Find a firm's cost function when its marginal cost (= rate of change of cost) is known.
- Find the demand function for a commodity when its elasticity of demand is known.
- Find the amount in a savings account when the rate at which it is earning interest is known and the rate at which money is being withdrawn from it is known.

This chapter considers **first-order** differential equations: these are equations in which only first derivatives appear. We begin with a general introduction to first-order equations and their solutions. This is followed by sections devoted to the particular types of first-order equations known as separable, linear, and exact equations. Finally, since not all differential equations can be solved explicitly, we conclude with a consideration of graphical and numerical solutions for first-order equations.

9.1 SOLVING DIFFERENTIAL EQUATIONS

In this section we introduce the terminology of differential equations and show how a few particularly simple first-order equations can be solved and how they can be used to model some practical situations.

Suppose we wish to solve the first-order differential equation

$$\frac{dy}{dx} = 2x + 5$$

for y. This means that we wish to find *all* functions $y = y(x)$ that satisfy the equation. If we consider the derivative to be the ratio of the differentials dy and dx, we may multiply both sides of the differential equation by dx to obtain

$$dy = (2x + 5)\, dx.$$

Then

$$\int dy = \int (2x + 5)\, dx,$$

so

$$y = x^2 + 5x + c.$$

(Strictly speaking, the integration yields the equation

$$y + c_1 = x^2 + 5x + c_2,$$

but the two constants of integration c_1 and c_2 can be combined into a single constant $c = c_2 - c_1$ by subtracting c_1 from both sides of this equation. From now on we will always combine constants of integration in this manner.) We have thus shown that every function $y = y(x)$ that satisfies the differential equation

$$\frac{dy}{dx} = 2x + 5$$

is of the form

$$y = x^2 + 5x + c,$$

and we call y the **general solution** of the differential equation. If we require that the solution of the differential equation satisfy an **initial condition,** such as

$y(0) = 4$, then we can determine the value of the constant c and thereby find the **particular solution** that satisfies the **differential equation with initial condition**

$$\frac{dy}{dx} = 2x + 5, \qquad y(0) = 4.$$

Thus, if we substitute 0 for x in the general solution $y = x^2 + 5x + c$, we obtain

$$4 = y(0) = 0^2 + 5(0) + c = c.$$

Hence $c = 4$, and the particular solution of the differential equation with initial condition is therefore

$$y = x^2 + 5x + 4.$$

EXAMPLE 1 Let us find the general solution of the differential equation

$$\frac{dy}{dx} = x^3 - 3x^2 + 5x + 1.$$

We have

$$dy = (x^3 - 3x^2 + 5x + 1)\, dx,$$

so

$$\int dy = \int (x^3 - 3x^2 + 5x + 1)\, dx,$$

and thus

$$y = \frac{1}{4}x^4 - x^3 + \frac{5}{2}x^2 + x + c. \quad \square$$

EXAMPLE 2 Let us find the particular solution of the differential equation with initial condition

$$\frac{dy}{dx} = x^3 - 3x^2 + 5x + 1, \qquad y(1) = 0.$$

From Example 1, the general solution is

$$y = \frac{1}{4}x^4 - x^3 + \frac{5}{2}x^2 + x + c,$$

so

$$0 = y(1) = \frac{1}{4}(1)^2 - (1)^3 + \frac{5}{2}(1)^2 + 1 + c = \frac{11}{4} + c,$$

and thus $c = -11/4$. Hence the particular solution is

$$y = \frac{1}{4}x^4 - x^3 + \frac{5}{2}x^2 + x - \frac{11}{4}. \quad \square$$

EXAMPLE 3 An object that falls freely in a vacuum under the influence of the earth's gravity accelerates at a constant rate of 32 feet per second per second. Find the velocity of the object 3 seconds after it begins to fall.

Let $v = v(t)$ denote the object's velocity t seconds after it begins to fall. Since acceleration is the derivative of velocity, we have

$$\frac{dv}{dt} = 32.$$

Hence

$$\int dv = \int 32 \, dt,$$

so

$$v = 32t + c.$$

At the instant the object begins its fall, its velocity is 0 feet per second, so $v(0) = 0$. Hence $c = 0$ and

$$v = 32t.$$

Therefore the velocity of the object at the end of 3 seconds is

$$v(3) = 32(3) = 96$$

feet per second. □

Situations that lead to differential equations are often expressed in the language of proportionality. Thus, if we say that the rate of change of y with respect to x is **directly proportional** to u, we mean that

$$\frac{dy}{dx} = ku,$$

where k is some number known as a **constant of proportionality.** Similarly, if the rate of change of y with respect to x is **inversely proportional** to u, then

$$\frac{dy}{dx} = \frac{k}{u},$$

where k is a constant of proportionality. We give three examples to illustrate the construction of differential equations using the language of proportionality.

EXAMPLE 4 Suppose the rate of change of the concentration of pollutants in a lake is directly proportional to the number of summer cottages surrounding the lake. Let $y = y(x)$ denote the concentration of pollutants in the lake, measured in parts per million (ppm), when there are x summer cottages surrounding the lake. Then

$$\frac{dy}{dx} = kx,$$

so

Differential Equations

$$\int dy = \int kx\, dx,$$

and hence

$$y = \frac{k}{2}x^2 + c.$$

To determine the values of k and c, we need to know the value of y for two distinct values of x. For instance, suppose that pollution was 0 ppm when there were no cottages on the lake, and 12 ppm when there were 20 cottages there; that is, suppose that $y(0) = 0$ and $y(20) = 12$. Then

$$0 = y(0) = \frac{k}{2}(0)^2 + c = c,$$

so $c = 0$ and hence

$$y = \frac{k}{2}x^2.$$

But then

$$12 = y(20) = \frac{k}{2}(20)^2 = 200k,$$

and thus $k = 12/200 = 0.06$. Therefore

$$y = 0.03x^2. \quad \square$$

EXAMPLE 5 Suppose that a firm's marginal cost is inversely proportional to the square root of the number of units produced, that its fixed cost is \$1000, and that it costs the firm \$2000 to make four units. If x denotes the number of units the firm produces and C is its cost function, then we have

$$\frac{dC}{dx} = \frac{k}{\sqrt{x}}, \qquad C(0) = 1000, \quad C(4) = 2000.$$

Thus

$$\int dC = \int \frac{k}{\sqrt{x}}\, dx,$$

and

$$C = 2k\sqrt{x} + c.$$

Now $C(0) = 1000$ implies that $c = 1000$, so

$$C = 2k\sqrt{x} + 1000.$$

But then $C(4) = 2000$ implies that $k = 250$ (check this), so

$$C = 500\sqrt{x} + 1000. \quad \square$$

EXAMPLE 6 The rate at which the output of an industry changes is inversely proportional to the quantity $t + 1$, where t is time in years since the founding of the industry. If the output of the industry in its tenth year was $30 million, find an expression that describes output as a function of time.

Suppose we let $y = y(t)$ be the output of the industry t years after its founding, y in millions of dollars. Then we know that

$$\frac{dy}{dt} = \frac{k}{t+1}, \quad y(0) = 0, \quad y(10) = 30, \quad t \geq 0.$$

(Note that $y(0) = 0$; why?) Thus

$$\int dy = k \int \frac{1}{t+1} dt,$$

so

$$y = k \ln(t + 1) + c.$$

Since

$$0 = y(0) = k \ln(0 + 1) + c = k(0) + c = c,$$

we have

$$y = k \ln(t + 1).$$

But then

$$30 = y(10) = k \ln(10 + 1) = k \ln 11,$$

so

$$k = \frac{30}{\ln 11} \approx 12.51.$$

Therefore

$$y = (12.51) \ln(t + 1). \quad \square$$

9.1 EXERCISES

In Exercises 1 through 12, first find the general solution of the differential equation and then find the particular solution that satisfies the initial condition.

1. $\dfrac{dy}{dx} = 3x - 1, \quad y(0) = 0$

2. $\dfrac{dy}{dx} = 4 - 2x, \quad y(0) = 2$

3. $\dfrac{dy}{dx} = x^2 - 2x + 1, \quad y(1) = \dfrac{1}{3}$

4. $\dfrac{dy}{dx} = -2x^{-3}, \quad y(1) = 1$

5. $\dfrac{dy}{dx} = \dfrac{2}{x^2}, \quad y(2) = -1$

6. $\dfrac{dy}{dx} = 3x^{-3} + x, \quad y(1) = 2$

7. $\dfrac{dy}{dt} = \dfrac{4}{t}, \quad y(e) = 1$

8. $\dfrac{dy}{dx} = 2e^x, \quad y(0) = 5$

9. $\dfrac{dy}{dx} = 2e^{-x} + 1, \quad y(0) = -2$

10. $\dfrac{dy}{dt} = \dfrac{1}{t+3}$, $y(-2) = 2$ **11.** $\dfrac{dy}{dx} = x\sqrt{x^2+1}$, $y(0) = 1$ **12.** $\dfrac{dy}{dx} = \dfrac{x}{x^2+1}$, $y(0) = 1$

13. A drag racer's car accelerates at a constant rate of 100 feet per second per second. Find the racer's velocity 2 seconds after the start of a race.

14. A space capsule traveling in orbit at 18,000 mph fired its rockets in order to move to a higher orbit, thus accelerating at a constant rate of 200 mph^2.
    (a) Find the velocity of the capsule 4 seconds after the rockets were fired.
    (b) Find the straight-line distance traveled by the capsule if it fired its rockets for 6 seconds.

15. Repeat Exercise 13 if the car's acceleration is directly proportional to the time since the start of the race and its velocity at the end of 1 second is 100 feet per second.

16. Repeat Exercise 14 if the capsule's acceleration was directly proportional to the time since the rockets began firing and its velocity 1 second after they began firing was 18,200 mph.

17. The acceleration of a particle is inversely proportional to the quantity $(t+1)^2$, where t is time since the particle was set in motion. Find the particle's velocity function if its initial velocity was 10 meters per second (mps) and 5 seconds later it was 9 mps.

Management

18. A firm's marginal profit is directly proportional to the number of units it produces, and its profit from 0 units is $0 while its profit from 10 units is $50. Find the firm's profit function.

19. A firm's marginal cost is directly proportional to the square root of the number of units it produces. If its fixed cost is $10,000 and the cost of making 64 units is $14,096, find the firm's cost function.

20. Repeat Exercise 19 if the marginal cost is inversely proportional to the cube root of the number of units produced.

21. If a firm sells x units of its product, its marginal revenue is inversely proportional to the quantity $x + 1$. Find the firm's revenue function if its revenue from selling 99 units is $2000.

22. A computer has a useful life of five years and depreciates at a rate that is directly proportional to the difference between five years and its present age t, where $0 \le t \le 5$, t in years. If the computer was purchased for $250,000 and one year later its value was $205,000, find an equation that expresses the value of the computer as a function of the time since it was purchased, and then find its value four years after it was purchased.

23. A store's sales change at a rate directly proportional to the square root of the amount it spends on advertising. If the store spends nothing on advertising, its sales will be $50,000; if it spends $16,000 on advertising, its sales will be $100,000. Find an equation that describes the store's sales as a function of the amount spent on advertising.

24. The rate of change of a company's average cost per unit is inversely proportional to the square of the number of units produced. If it costs $50 to make the first unit

and a total of $90 to make the first two units, find a function that describes the company's average cost per unit.

25. The rate at which a company's stock price changes is directly proportional to the dividend the stock pays. Find an equation that describes stock price as a function of the dividend paid if the price was $50 per share when the dividend was $1 per share and $60 per share when the dividend was $2 per share.

26. A firm's marginal cost is directly proportional to the square root of the number of units it produces. Find its cost function C if $C(0) = \$100,000$ and $C(144) = \$215,200$.

27. The rate at which a company's average cost per unit changes is inversely proportional to the cube root of the number of units produced, and was $5 per unit when 1000 units were produced and $4 per unit when 8000 were produced. Find the company's average cost function.

28. The rate at which annual sales of a product change is directly proportional to the difference between the number of years the product has been on the market and 10 years. Suppose that when the product had been on the market for two years, sales were 1 million units per year, and that when it had been on the market for four years they were 0.75 million units per year. Find an expression that describes sales as a function of the time the product has been on the market.

Life Science

29. The concentration of the acid in lake water in a northeastern state has been increasing at a rate of 20 ppm per year and is currently 125 ppm. Find an equation that expresses the concentration of acid as a function of time.

30. The number of bacteria present in a culture changes at a rate inversely proportional to the time since an experiment began. If there were 100,000 bacteria present in the culture when the experiment began and 80,000 present 1 hour later, find an equation that describes the number of bacteria present as a function of time.

31. The concentration of pollutants in the air over a large city is directly proportional to the square root of the number of cars coming into the city each morning. If 250,000 cars cause 40 ppm of pollution and 1,000,000 cars cause 100 ppm, find the pollution that will be caused by 4,000,000 cars.

32. Beginning one day after germination, the rate of growth of a plant is inversely proportional to the time since germination. If the plant grows at a rate of 1 centimeter (cm) per day 1 day after germination and 0.5 cm per day eight days after germination, find an equation that describes the plant's growth as a function of time since germination.

Social Science

33. A subject undergoing psychological testing is allowed varying amounts of time to complete a complex task. The rate at which the subject makes errors is inversely proportional to the amount of time allowed. If the subject makes six errors when 5 minutes are allowed for completion of the task and two errors when 8 minutes are allowed, find an expression that gives the number of errors made as a function of the time allowed.

34. The percentage of eligible voters in a state who do not vote in presidential elections is changing at a rate directly proportional to the fourth root of t, where t is time in years with $t = 0$ representing 1980. If 50% of the eligible voters did not vote in 1980 and 46% did not vote in 1984, predict the percentage who will not vote in 1992.

*35. Let x denote a stimulus and y the resulting response. There is some psychological evidence that the rate of change of the response with respect to the stimulus is inversely proportional to the stimulus. If x_0 is the smallest detectable stimulus (called the **threshold stimulus**) and $y(x_0) = 0$, derive the **Weber-Fechner Law**

$$y = k \ln \frac{x}{x_0}.$$

9.2 SEPARABLE EQUATIONS

In this section we discuss the class of first-order differential equations known as **separable equations.** Separable equations are usually relatively easy to solve, and, as we shall see, the modeling of a practical situation can often result in a separable equation.

Consider the differential equation

$$\frac{dy}{dx} = \frac{x}{2y}.$$

Note that the equation implies that $y \neq 0$. If we multiply both sides of the equation by $2y\, dx$, we obtain

$$2y\, dy = x\, dx,$$

from which it follows that

$$\int 2y\, dy = \int x\, dx,$$

and hence that

$$y^2 = \frac{1}{2}x^2 + c.$$

We may leave the general solution in this form or solve it for y by taking the square root of both sides to get

$$y = \pm\sqrt{\frac{1}{2}x^2 + c}.$$

The technique we have just used to solve the differential equation

$$\frac{dy}{dx} = \frac{x}{2y}$$

is called **separation of variables.** The method of separation of variables requires that we be able to write the differential equation in the form

$$f(y)\,dy = g(x)\,dx.$$

Note that only the dependent variable and its differential appear on the left side of the equation, and only the independent variable and its differential appear on the right side of the equation. Differential equations that can be written in this form are called **separable equations,** and once their variables have been separated they can be solved by integrating both sides of the equation $f(y)\,dy = g(x)\,dx$.

EXAMPLE 1 Let us use separation of variables to solve

$$xy\frac{dy}{dx} = 2, \quad y(e) = 4$$

for y. Separating the variables, we obtain

$$y\,dy = \frac{2}{x}\,dx.$$

Then

$$\int y\,dy = \int \frac{2}{x}\,dx,$$

so

$$\frac{1}{2}y^2 = 2\ln|x| + c.$$

Since $y(e) = 4$, we have

$$\frac{1}{2}4^2 = \frac{1}{2}[y(e)]^2 = 2\ln|e| + c = 2 + c,$$

so $c = 6$. Therefore the solution is

$$\frac{1}{2}y^2 = 2\ln|x| + 6,$$

or

$$y^2 = 4\ln|x| + 12. \quad \square$$

EXAMPLE 2 Let us find the general solution of the differential equation

$$\frac{dy}{dx} = 3(y-1), \quad y \neq 1.$$

Since $y \neq 1$, we can divide both sides of the equation by $y - 1$ to obtain

$$\int \frac{1}{y-1}\,dy = \int 3\,dx.$$

Thus

$$\ln|y-1| = 3x + c.$$

But then
$$|y - 1| = e^{\ln|y-1|} = e^{3x+c} = e^c e^{3x}.$$
Since e^c is a positive constant, we may set $C = e^c$ and write this last equation as
$$|y - 1| = Ce^{3x}, \quad C \text{ a constant}, \quad C > 0.$$
Now, if $y > 1$, then $|y - 1| = y - 1$, and we have
$$y - 1 = Ce^{3x}, \quad C > 0,$$
or
$$y = 1 + Ce^{3x}, \quad C > 0.$$
If $y < 1$, then $|y - 1| = -(y - 1) = -y + 1$, and we have
$$-y + 1 = Ce^{3x}, \quad C > 0,$$
or
$$y = 1 - Ce^{3x}, \quad C > 0.$$
Therefore the solution is in two parts:
$$y = 1 + Ce^{3x}, \quad C > 0, \quad \text{if } y > 1,$$
and
$$y = 1 - Ce^{3x}, \quad C > 0, \quad \text{if } y < 1.$$
If we wish, we can in this case combine the two solution equations into one by introducing a constant K that can be either positive or negative. Thus
$$y = 1 + Ke^{3x}, \quad K \text{ a constant}, K \neq 0. \quad \square$$

Sometimes when we separate variables we must assume that y cannot take on certain values so that we can divide. When this is the case, it is possible to "lose" constant solutions of the differential equation. To illustrate, suppose we wish to solve
$$\frac{dy}{dx} = 3(y - 1)$$
for y. This is the differential equation of Example 2, except that here we do not rule out the possibility that $y = 1$. If we *assume* that $y \neq 1$ and solve the equation, then, as Example 2 shows, the solution is
$$y = 1 + Ke^{3x}, \quad K \text{ a constant}, K \neq 0.$$
But now we must also consider what happens when $y = 1$. If we set $y = 1$ in the differential equation, we have
$$\frac{d}{dx}(1) = 3(1 - 1),$$

or $0 = 0$. Hence $y = 1$ is a **constant solution** of the differential equation. Therefore the general solutions of the differential equation are

$$y = 1 + Ke^{3x}, \quad K \text{ a constant}, K \neq 0,$$

and

$$y = 1.$$

In this case it is possible to combine the two solutions into one by writing

$$y = 1 + Ke^{3x}, \quad K \text{ a constant}.$$

The constant solution $y = 1$ now occurs when $K = 0$.

EXAMPLE 3 Let us solve the differential equation

$$\frac{dy}{dx} = y^2 x.$$

Assuming that $y \neq 0$, separating variables yields

$$\int \frac{1}{y^2} dy = \int x \, dx.$$

Thus

$$-\frac{1}{y} = \frac{x^2}{2} + c,$$

or

$$y = -\frac{2}{x^2 + c}.$$

Setting $y = 0$ in the differential equation shows that $y = 0$ is also a solution. Hence the solution has two parts:

$$y = -\frac{2}{x^2 + c} \quad \text{and} \quad y = 0.$$

In this case the two solutions cannot be combined into a single equation because no value of c in

$$y = -\frac{2}{x^2 + c}$$

can ever yield $y = 0$. ☐

Modeling practical situations with differential equations often results in separable equations. We illustrate with examples.

EXAMPLE 4 Suppose amount P is invested at annual interest rate r, where r is expressed as a decimal, and suppose that interest is compounded continuously. Then the future

DIFFERENTIAL EQUATIONS

value A of the investment is a function of time t, and the rate of change of A at time t is directly proportional to A. Let us find the equation that expresses A as a function of time.

Since the rate of change of A is directly proportional to A, we have

$$\frac{dA}{dt} = kA.$$

Note also that

$$A(0) = P \quad \text{and} \quad \left.\frac{dA}{dt}\right|_{t=0} = rP,$$

because when $t = 0$ the amount of the investment is the principal P and interest is being earned at a rate of rP. Solving the differential equation

$$\frac{dA}{dt} = kA$$

by separation of variables, we have

$$\int \frac{dA}{A} = \int k\, dt,$$

and thus (since A is positive)

$$\ln A = kt + c.$$

Hence

$$e^{\ln A} = e^{kt+c},$$

or

$$A = Ce^{kt}, \quad C > 0.$$

Now $A(0) = P$ implies that

$$P = A(0) = Ce^{k(0)} = C;$$

therefore

$$A = Pe^{kt}.$$

But then

$$\frac{dA}{dt} = kPe^{kt},$$

and the initial condition

$$\left.\frac{dA}{dt}\right|_{t=0} = rP$$

implies that

$$kPe^{k(0)} = rP,$$

or
$$kP = rP,$$
and thus that $k = r$. Therefore the solution is
$$A = Pe^{rt},$$
the formula for the continuous compounding of interest that was introduced in Section 5.1. □

EXAMPLE 5 If $q = f(p)$ is a demand function for a commodity, where p is its unit price and q its quantity demanded, then its elasticity of demand is
$$E(p) = p\frac{dq}{dp} \bigg/ q.$$

(See Section 2.6.) If we solve this equation for dq/dp, we obtain the separable differential equation
$$\frac{dq}{dp} = \frac{qE(p)}{p}.$$

Thus, given the elasticity of demand formula $E(p)$, we can find the demand function $q = f(p)$ by solving this differential equation for q. For instance, if $E(p) = -p$, then
$$\frac{dq}{dp} = \frac{q(-p)}{p},$$
or
$$\frac{dq}{q} = -dp.$$
Thus
$$\ln q = -p + c,$$
which as in Example 4 yields
$$q = Ce^{-p}, \quad C > 0. \quad \square$$

At the beginning of Section 5.5 we said that if
$$\frac{dy}{dt} = ky,$$
then $y = ce^{kt}$ for some c. Now we can prove this by solving the differential equation
$$\frac{dy}{dt} = ky$$
for y. (See Exercise 20 at the end of this section.) Therefore if a situation can be modeled by such a differential equation, and if $y(0) = c$ is positive, then the model

DIFFERENTIAL EQUATIONS

is either one of exponential growth (if $k > 0$) or exponential decline (if $k < 0$). Thus the exponential growth and decline models of Section 5.5 arise as the solutions of differential equations in which the rate of change of the dependent variable is directly proportional to the dependent variable. Similarly, the limited growth and decline models of Section 5.5 arise as the solutions of differential equations in which the rate of change of the dependent variable is directly proportional to the difference between the dependent variable and a constant A. (See Exercise 21.)

EXAMPLE 6   Suppose the size of an animal population is limited by the food available to no more than 10,000 individuals, and that the population at time 0 is 4000 and at time 1 is 4300. If the rate of growth of the population is directly proportional to the difference between the limit of 10,000 and the size of the population, and we let $y = y(t)$ be the size of the population at time t, then we have

$$\frac{dy}{dt} = k(10{,}000 - y), \qquad y(0) = 4000, \quad y(1) = 4300.$$

Separating the variables, we obtain

$$\frac{dy}{10{,}000 - y} = k\,dt.$$

Since $10{,}000 - y$ is positive (why?), integrating this last equation gives

$$-\ln(10{,}000 - y) = kt + c,$$

or

$$\ln(10{,}000 - y) = kt + c.$$

(Here we have multiplied through by -1, absorbing the minus sign into the arbitrary constants k and c.) Thus

$$e^{\ln(10{,}000 - y)} = e^{kt + c},$$

or

$$10{,}000 - y = Ce^{kt}, \qquad C > 0,$$

so

$$y = 10{,}000 - Ce^{kt}, \qquad C > 0.$$

Using the facts that $y(0) = 4000$ and $y(1) = 4300$, we find that $C = 6000$ and $k = -0.051$. (Check this.) Therefore

$$y = 10{,}000 - 6000e^{-0.051t}.$$

This is a limited growth model of the form studied in Section 5.5. Its graph is shown in Figure 9–1. ☐

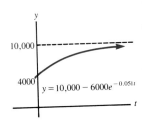

FIGURE 9–1

Another separable differential equation that often occurs in applications is the so-called **logistic equation**

$$\frac{dy}{dt} = ky(A - y).$$

The solution of this equation is a function defined by an equation of the form

$$y = \frac{A}{1 + be^{-rt}}, \quad r > 0.$$

Such a function is known as a **logistic function.** (See the supplement to Chapter 5.) We illustrate with an example.

EXAMPLE 7 In a group of 1000 people, one person starts a rumor, and 1 hour later 10 people have heard it. If the rumor spreads at a rate directly proportional to the product of the number who have heard it and the number who have not, let us find an expression for the number who have heard the rumor as a function of time.

We let $y = y(t)$ be the number of people who have heard the rumor t hours after it was started. We have

$$\frac{dy}{dt} = ky(1000 - y), \quad y(0) = 1, \quad y(1) = 10.$$

Thus

$$\frac{dy}{y(1000 - y)} = k\, dt.$$

Now,

$$\frac{1}{y(1000 - y)} = \frac{1}{1000}\left(\frac{1}{y} + \frac{1}{1000 - y}\right)$$

(check this), so we have

$$\frac{1}{1000}\left(\frac{1}{y} + \frac{1}{1000 - y}\right) dy = k\, dt.$$

Since y and $1000 - y$ are both positive (why?), integration gives

$$\frac{1}{1000}[\ln y - \ln(1000 - y)] = kt + c,$$

or

$$\ln \frac{y}{1000 - y} = kt + c.$$

(Here we multiplied through by 1000, absorbing the 1000 into the arbitrary constants k and c.) But this implies that

$$\frac{y}{1000 - y} = Ce^{kt}, \quad C > 0.$$

DIFFERENTIAL EQUATIONS

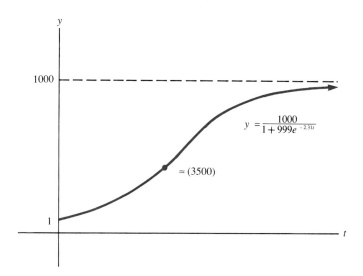

FIGURE 9–2

Substituting 0 for t and using the fact that $y(0) = 1$, we find that $C = 1/999$; then substituting 1 for t and using the fact that $y(1) = 10$, we find that $r \approx 2.31$. (Check this.) Therefore

$$\frac{y}{1000 - y} = \frac{1}{999} e^{2.31t},$$

which can be solved for y to yield

$$y = \frac{1000}{1 + 999e^{-2.31t}}.$$

The graph of this function is shown in Figure 9–2. Notice that the rumor spreads at an increasing rate until $t = (\ln 999)/2.31 \approx 3$ hours and thereafter spreads at a decreasing rate as the number who have heard it approaches 1000. □

9.2 EXERCISES

In Exercises 1 through 14, find the general solution of the differential equation and then use the initial condition to find a particular solution.

1. $y\dfrac{dy}{dx} = 1$, $y(0) = 0$

2. $y\dfrac{dy}{dx} = x$, $y(1) = 3$

3. $x\dfrac{dy}{dx} = 1$, $y(1) = 0$

4. $\dfrac{dy}{dx} = \dfrac{2x}{y}$, $y(0) = 2$

5. $\dfrac{dy}{dx} = \dfrac{x^2}{y}$, $y(1) = 3$

6. $\dfrac{dy}{dx} = -\dfrac{y}{x}$, $y(1) = 1$

7. $\dfrac{dy}{dx} = 3x^2 y$, $y(0) = 2$

8. $\dfrac{dy}{dx} = -\dfrac{y}{x^2}$, $y(1) = e$

9. $\dfrac{dy}{dx} = 2x(y - 1)$, $y(0) = 2$

10. $\dfrac{dy}{dx} = 2x(y - 1)$, $y(0) = 1$

11. $\dfrac{dy}{dx} = (y - 4)(2x + 1)$, $y(0) = 1$

12. $\dfrac{dy}{dx} = (2 - y)(1 - x)$, $y(0) = 4$

13. $x^2 \dfrac{dy}{dx} = y^2$, $y(1) = \dfrac{1}{2}$

14. $2x^2 \dfrac{dy}{dx} = y^3$, $y(1) = 0$

15. A radioactive isotope decays at a rate directly proportional to the amount of it present. If 1000 grams are present at time $t = 0$ and 900 grams are left at time $t = 10$, find an expression that describes the amount of the isotope present at time t.

16. A radioactive isotope decays at a rate directly proportional to the amount of it present. Find the half-life of the isotope if there are 10 grams present to begin with and 20 days later there are 2.6 grams present.

*17. A radioactive isotope decays at a rate directly proportional to the amount of it present. If the half-life of the isotope is t_1 years, find a function that describes the amount of the isotope present as a function of time.

18. Heavy-metal ions are being removed from lake water at a rate directly proportional to the amount of them present. If the concentration at the time removal began was $2.5(10^6)$ ions per liter and two years later was $6.25(10^4)$ ions per liter, find an equation that expresses the concentration as a function of time.

19. A hot steel bar is placed in a large tank of water; according to Newton's Law of Cooling, the rate at which the temperature of the bar changes is directly proportional to the difference between its temperature and that of the water. If the temperature of the water is 10°C and the temperature of the bar when $t = 0$ and $t = 2$ is 600°C and 220°C, respectively, find the temperature of the bar as a function of time.

*20. Solve the differential equation

$$\dfrac{dy}{dt} = ky, \quad y(0) > 0$$

and sketch the graph of the solution when
(a) $k > 0$ (b) $k < 0$

*21. Solve the differential equation

$$\dfrac{dy}{dt} = k(A - y), \quad k > 0, \quad A > 0$$

and sketch the graph of the solution when
(a) $y(0) > A$ (b) $y(0) = A$ (c) $0 \leq y(0) < A$

*22. Solve the logistic equation

$$\dfrac{dy}{dt} = ky(A - y), \quad k > 0, \quad A > 0$$

and sketch the graph of the solution when
(a) $y(0) > A$ (b) $y(0) = A$
(c) $A/2 < y(0) < A$ (d) $0 < y(0) \leq A/2$

Management

23. The rate at which a firm's market share is changing is directly proportional to its market share. If its market share was 25 percent in 1987 and 30 percent in 1989, find and then graph an equation that expresses market share as a function of time.

24. A product has elasticity of demand
$$E(p) = -\frac{5p}{3000 - 5p}.$$
Find its demand function if quantity demanded is 2500 units when $p = \$100$.

25. A commodity has elasticity of demand
$$E(p) = \frac{2p^2}{p^2 - 10{,}000}.$$
Find its demand function if quantity demanded is 0 when $p = \$100$.

*26. Show that if a commodity has elasticity of demand
$$E(p) = \frac{mp}{mp + b}, \quad m < 0, \quad b > 0,$$
then its demand function is of the form $q = mp + b$.

*27. Show that if a product has unit elasticity of demand at every price, then its demand function is of the form $q = cp^{-1}$.

*28. Let r be positive. Show that if a commodity has elasticity of demand equal to r at every price, then its demand function is of the form $q = cp^{-r}$.

29. A worker's production increases at a rate directly proportional to the difference between 400 units per week and the number of units the worker produces per week. If the worker can produce 200 units per week with no experience and 250 units per week with one week's experience, find and graph an equation that describes production as a function of experience.

30. An employee's productivity changes at a rate directly proportional to the difference between the employee's current productivity and 1000 units per day. If the employee's productivity at time $t = 0$ was 1200 units per day and five months later it was 1150 units per day, find an equation that describes the employee's productivity as a function of time.

31. Suppose that the quantity of a commodity supplied to the market is constant at 30,000 units and that its demand function is given by $q = 40{,}000 - 1000p$, where q is quantity demanded. Suppose the price now is $5 per unit and that next month it will be $6 per unit. If the rate of change of price with respect to time is directly proportional to the difference between quantity demanded and quantity supplied, find an equation that describes price as a function of time. What happens to the price in the long run?

*32. Suppose the quantity of a commodity that is supplied to the market is q_0 and its demand function is of the form $q = a - mp$, a and m positive. If the rate of change of price with respect to time is directly proportional to the difference between quantity demanded and quantity supplied, find an equation that describes price as a function of time. Then graph this function and analyze the behavior of price over time.

*33. Repeat Exercise 32 if the quantity demanded of the commodity is constant at q_0 and the quantity supplied is given by $q = -a + mp$, $m > 0$, $a \geq 0$.

34. Suppose the demand function for a commodity is given by $q = 100{,}000 - 2500p$ and the supply function by $q = 3000p$. If the price now is \$5 per unit and in one year it will be \$6 per unit, and if the rate of change of price with respect to time is directly proportional to the difference between quantity demanded and quantity supplied, find an equation that describes price as a function of time. What happens to the price in the long run?

35. Repeat Exercise 34 if the demand function is given by $q = 100{,}000 - 1500p$ and the supply function by $q = -4000 + 2000p$.

*36. Suppose the demand function for a commodity is given by an equation of the form $q = a - mp$ and the supply function by an equation of the form $q = -b + np$, where a, m, and n are positive and b is nonnegative. If the rate of change of price with respect to time is directly proportional to the difference between quantity demanded and quantity supplied, find and analyze an equation that describes price as a function of time.

37. A plant's production rate is directly proportional to the product of its current production and the difference between current production and the maximum designed production of 100,000 units per week. Find an equation that describes the plant's production as a function of time if its production one week after it opened was 10,000 units per week and two weeks after it opened was 20,000 units per week.

38. Sales of a new product introduced in 1986 increase at a rate directly proportional to the product of current sales and the difference between current sales and sales of 4 million units. Find, graph, and analyze sales as a function of time if
   (a) sales were 200,000 units in 1985 and 500,000 units in 1987;
   (b) sales were 2.5 million units in 1985 and 3.2 million in 1987;
   (c) sales were 4.5 million units in 1985 and 4.3 million units in 1987.

Life Science

39. Newton's Law of Cooling states that if a small object is placed in a new environment that is kept at a constant temperature, then the rate at which the temperature of the object changes is directly proportional to the difference between its temperature and that of the environment. Suppose a small cold-blooded animal whose body temperature is initially 20°C is placed in a container whose temperature is kept at 14°C. Find and graph an equation that expresses the animal's temperature as a function of time if its temperature 1 hour after being placed in the container is 18°C.

40. Repeat Exercise 39 if the temperature of the container is 25°C and 1 hour after being placed in the container the animal's temperature is 21°C.

41. An influenza epidemic spreads at a rate directly proportional to the product of the number of people who have already contracted the disease and the number who have not. In a community of 10,000 people, 100 people had influenza on day 0 and 300 had it on day 1. Find an equation that describes the number of people who contracted influenza as a function of time. Then graph this equation and analyze the spread of the disease.

*42. An epidemic spreads at a rate directly proportional to the product of the number of cases of the disease and $e^{-0.2t}$, t in weeks since the start of the epidemic. If the epidemic started with 100 cases of the disease and one week later there were 400 cases, find a function that describes the number of cases as a function of time. Graph this function and analyze the epidemic.

Social Science

43. The percentage of voters who are registered as Independents is changing at a rate directly proportional to the percentage who are not registered as Independents. If 25% of all voters were Independents in 1980 and 30% were Independents in 1984, find an equation that expresses the percentage of voters who are Independents as a function of time.

44. The rate at which a rumor spreads is directly proportional to the product of the proportion of persons who have already heard it and the proportion who have not. Suppose that on day 0 the proportion of people who have heard a certain rumor is 0.01, and on day 4 the proportion is 0.25. Find and graph an equation that describes the proportion who have heard the rumor as a function of time.

*45. A rumor spreads at a rate directly proportional to the product of the number who have heard it and e^{-t}, t in hours since the rumor was started. Find a function that describes the number who have heard the rumor at time t, assuming that it was started by one person and that 1 hour later a total of 20 people (including the person who started it) had heard it. Graph this function and analyze the spread of the rumor.

Transition Rates

Exercises 46 through 49 were suggested by material in Coleman, *Introduction to Mathematical Sociology*, The Free Press of Glencoe, 1964. They are concerned with transition rates as individuals migrate from one group to another. Thus, suppose individuals migrate from Group A to Group B at a rate directly proportional to the size of group A. If $y = y_A(t)$ denotes the size of group A at time t, then

$$\frac{dy_A}{dt} = -k_A y_A, \qquad k_A > 0.$$

The positive constant k_A is called the **transition rate** from A to B.

*46. If members of political party A are switching to political party B at a rate directly proportional to the size of party A, and if $y_A(0) = 10$ million and $y_A(1) = 9.9$ million, find the transition rate from party A to party B.

Now suppose that $y_B = y_B(t)$ is the size of group B at time t and that there is also migration from group B to group A with transition rate k_B. Then group A loses members to group B at a rate of $k_A y_A$ and simultaneously gains members from group B at a rate of $k_B y_B$. Thus

$$\frac{dy_A}{dt} = k_B y_B - k_A y_A.$$

If we assume that the total number of individuals in the two groups is a constant N, so that $y_A + y_B = N$, then we may write

$$\frac{dy_A}{dt} = k_B(N - y_A) - k_A y_A$$

or

$$\frac{dy_A}{dt} = Nk_B - (k_A + k_B)y_A.$$

*47. Show that

$$y_A = N\frac{k_B}{k_A + k_B}(1 - e^{-(k_A + k_B)t}) + ce^{-(k_A + k_B)t}$$

is a solution of the preceding differential equation.

*48. Suppose voters are migrating from political party A to political party B with a transition rate of 0.02, and from B to A with a transition rate of 0.03. Suppose also that at time $t = 0$, party A had 18 million members, party B had 12 million members, and that the total membership in the two parties will remain constant at 30 million.

(a) Find an equation that describes the size of party A as a function of time.
(b) Find the size of party A at time $t = 10$; what will happen to the sizes of the parties in the long run?

*49. If group A is the living and group B is the nonliving, then the transition rate k_A from A to B is the death rate and the transition rate k_B from B to A is the birth rate. Suppose that at time $t = 0$ there are y_0 living individuals. Find the solution of the differential equation

$$\frac{dy_A}{dt} = Nk_B - (k_A + k_B)y_A, \quad y(0) = y_0$$

and express the solution in terms of the **natural rate of increase** $r = k_B/k_A$. Interpret your results.

9.3 LINEAR FIRST-ORDER EQUATIONS

Not all first-order differential equations are separable. For instance, the equation

$$\frac{dy}{dx} = x - \frac{y}{x}$$

is not separable. (Try it and see.) This equation is, however, an example of a **linear differential equation.** Linear first-order differential equations are important, particularly in the sciences, and in this section we show how to solve them.

We begin with the definition of a linear first-order differential equation:

DIFFERENTIAL EQUATIONS

> **Linear First-Order Differential Equations**
> A first-order differential equation is **linear** if it can be written in the form
> $$\frac{dy}{dx} + p(x)y = q(x)$$
> for some expressions $p(x)$ and $q(x)$.

EXAMPLE 1   The differential equation

$$\frac{dy}{dx} = x - \frac{y}{x}$$

with which we began this section is linear because it can be written in the form

$$\frac{dy}{dx} + \frac{1}{x}y = x,$$

and thus it matches the standard form of a linear equation if we take $p(x) = 1/x$ and $q(x) = x$. On the other hand, the differential equation

$$\frac{dy}{dx} + \frac{1}{x}y^2 = x$$

is not linear because of the presence of y^2 in the equation. ∎

How can we solve a nonseparable linear differential equation? It turns out that every first-order linear equation can be solved by multiplying it by an appropriately chosen expression in x known as an **integrating factor.** For the differential equation

$$\frac{dy}{dx} + \frac{1}{x}y = x,$$

the integrating factor is x. (We will see in a moment how such an integrating factor is found.) If we multiply both sides of this equation by x, we have

$$x\frac{dy}{dx} + y = x^2,$$

which we may write as

$$x\frac{dy}{dx} + y\frac{dx}{dx} = x^2,$$

or

$$\frac{d}{dx}(xy) = x^2.$$

But if the derivative of xy is x^2, then

$$xy = \int x^2 \, dx = \frac{x^3}{3} + c.$$

Therefore

$$y = \frac{x^2}{3} + \frac{c}{x}.$$

This method of solution can be repeated for any first-order linear differential equation

$$\frac{dy}{dx} + p(x)y = q(x)$$

using $G(x) = e^{\int p(x)dx}$ as an integrating factor. (Note that above our integrating factor was x, and

$$x = e^{\ln x} = e^{\int (1/x)dx} = e^{\int p(x)dx}.)$$

To see this, note that if

$$G(x) = e^{\int p(x)dx},$$

then

$$\frac{d}{dx}(G(x)) = \frac{d}{dx}(e^{\int p(x)dx})$$

$$= e^{\int p(x)dx} \frac{d}{dx}\left(\int p(x) \, dx\right)$$

$$= e^{\int p(x)dx} p(x)$$

$$= G(x)p(x).$$

Now if we multiply both sides of the linear differential equation by $G(x)$, we obtain

$$G(x)\frac{dy}{dx} + G(x)p(x)y = q(x)G(x);$$

or

$$\frac{d}{dx}(yG(x)) = q(x)G(x).$$

Therefore

$$yG(x) = \int q(x)G(x) \, dx,$$

and hence

$$y = \frac{1}{G(x)}\left[\int q(x)G(x) \, dx\right].$$

> **Solving First-Order Linear Equations**
>
> To solve the first-order linear differential equation
> $$\frac{dy}{dx} + p(x)y = q(x),$$
> proceed as follows:
>
> 1. Find the integrating factor $G(x) = e^{\int p(x)\,dx}$.
> 2. Multiply both sides of the differential equation by $G(x)$; the equation will then reduce to
> $$\frac{d}{dx}(yG(x)) = q(x)G(x).$$
> 3. Thus
> $$yG(x) = \int q(x)G(x)\,dx,$$
> so
> $$y = \frac{1}{G(x)}\left[\int q(x)G(x)\,dx\right].$$

EXAMPLE 2   Let us solve the first-order linear differential equation
$$\frac{dy}{dx} + xy = x.$$
Here $p(x) = x$, so the integrating factor $G(x) = e^{\int x\,dx} = e^{x^2/2}$. Thus
$$e^{x^2/2}\frac{dy}{dx} + xe^{x^2/2}y = xe^{x^2/2},$$
or
$$\frac{d}{dx}(ye^{x^2/2}) = xe^{x^2/2}.$$
Therefore
$$ye^{x^2/2} = \int xe^{x^2/2}\,dx = e^{x^2/2} + c,$$
so
$$y = 1 + ce^{-x^2/2}.$$
(Note that the differential equation of this example is also separable, and thus could have been solved by the method of Section 9.2.) □

We illustrate the uses of linear first-order differential equations with examples.

EXAMPLE 3

Suppose that $50,000 is deposited in an account that pays 10% per year compounded continuously and that money is then withdrawn from the account continuously at the rate of $4000 per year. Let us find a function that describes the amount in the account as a function of time.

We let $A = A(t)$ denote the amount in the account at time t. Then

$$\text{Rate of change of } A = \begin{bmatrix} \text{Rate at which} \\ \text{money is being} \\ \text{added to the} \\ \text{account} \end{bmatrix} - \begin{bmatrix} \text{Rate at which} \\ \text{money is being} \\ \text{subtracted from} \\ \text{the account} \end{bmatrix}$$

$$= 0.10A(t) - 4000.$$

Thus we have

$$\frac{dA}{dt} = 0.10A - 4000, \quad A(0) = \$50,000.$$

Writing this equation in the form

$$\frac{dA}{dt} - 0.10A = -4000,$$

we see that it is linear with integrating factor

$$G(t) = e^{\int -0.10 \, dt} = e^{-0.1t}.$$

Therefore

$$e^{-0.1t} \frac{dA}{dt} - 0.1 e^{-0.1t} A = -4000 e^{-0.1t},$$

or

$$\frac{d}{dt}(e^{-0.1t} A) = -4000 e^{-0.1t}.$$

Hence

$$e^{-0.1t} A = -\int 4000 e^{-0.1t} \, dt = 40,000 e^{-0.1t} + c,$$

so

$$A = 40,000 + c e^{0.1t}.$$

Since $A(0) = \$50,000$, we find that $c = 10,000$, and thus

$$A = 40,000 + 10,000 e^{0.1t}.$$

Figure 9–3 shows the graph of this function. Notice that the account continues to grow and is never depleted: this is so because interest is being earned at a faster rate than money is being withdrawn. □

EXAMPLE 4

During a biology experiment a solution containing 2 milligrams per liter (mg/L) of dissolved lysine flows into the top of a vat at the rate of 0.25 L/min. Simultaneously

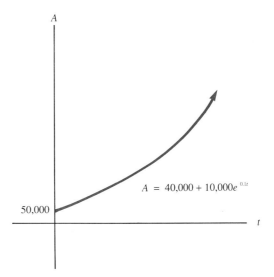

FIGURE 9-3

the solution is drawn out at the bottom of the vat at a rate of 0.10 L/min. If the vat contained 5 mg of lysine dissolved in 1 L of solution when the experiment started, let us find out how much lysine is present in the vat t minutes after the start.

We let $y = y(t)$ be the amount of lysine present at time t. The rate at which the amount of lysine is changing at time t is equal to the rate at which it is entering the vat minus the rate at which it is leaving the vat; that is,

$$\frac{dy}{dt} = \text{(Rate lysine enters at time } t\text{)} - \text{(Rate lysine leaves at time } t\text{)}.$$

Now,

$$\begin{aligned}\text{Rate lysine enters at time } t &= \text{(Incoming concentration)(Incoming flow rate)} \\ &= (2 \text{ mg/L})(0.25 \text{ L/min}) \\ &= 0.5 \text{ mg/min}.\end{aligned}$$

Similarly,

$$\text{Rate lysine leaves at time } t = \text{(Outgoing concentration)(Outgoing flow rate)}.$$

But the outgoing concentration depends on how much solution is in the vat at time t, and this is a function of t, since solution is flowing into the vat faster than it is flowing out. Thus, at time t we have

$$1 + (0.25 - 0.10)t = 1 + 0.15t$$

liters of solution in the vat, and therefore the outgoing concentration of lysine at time t is the amount of lysine present at time t divided by the amount of solution present at time t, or mg/L. Hence

$$\text{Rate lysine leaves at time } t = \left(\frac{y}{1 + 0.15t} \text{ mg/L}\right)(0.10 \text{ L/min})$$

$$= \frac{0.10y}{1 + 0.15t} \text{ mg/min.}$$

Therefore

$$\frac{dy}{dt} = 0.50 - \frac{0.10y}{1 + 0.15t},$$

or

$$\frac{dy}{dt} + \frac{0.10y}{1 + 0.15t} = 0.50,$$

which is a linear differential equation whose general solution is

$$y = 2(1 + 0.15t) + c(1 + 0.15t)^{-2/3}.$$

(Check this.) Since $y(0) = 5$ mg, it follows that $c = 3$, and hence

$$y = 2(1 + 0.15t) + 3(1 + 0.15t)^{-2/3}. \quad \square$$

9.3 EXERCISES

In Exercises 1 through 14, solve the given linear differential equation by the method of this section. If the equation is separable, check your answer by using separation of variables to solve it.

1. $\dfrac{dy}{dx} + y = 1$

2. $\dfrac{dy}{dx} - 3xy = 3x, \quad y(0) = 3$

3. $\dfrac{dy}{dx} = xy + 4x$

4. $\dfrac{dy}{dx} + \dfrac{2}{x}y = 1$

5. $\dfrac{dy}{dx} - \dfrac{y}{x} = x$

6. $x\dfrac{dy}{dx} + y = x$

7. $\dfrac{dy}{dx} + \dfrac{y}{x} = x^{-2}, \quad y(e) = 5$

8. $x^2 \dfrac{dy}{dx} - y = 1$

9. $\dfrac{dy}{dx} = x^{-3}y + x^{-3}$

10. $x\dfrac{dy}{dx} + 2y = 4$

11. $(x + 1)\dfrac{dy}{dx} + y = x$

12. $\dfrac{dy}{dx} + x^{1/2}y = x^{1/2}, \quad y(0) = 2$

13. $e^x \dfrac{dy}{dx} + y = x^2 e^{x+e^{-x}}, \quad y(0) = e$

14. $\dfrac{dy}{dx} + (\ln x)y = x^{-x}$

15. If $100,000 is placed in an account that pays 8% per year compounded continuously and $2000 is continuously withdrawn from the account each year, find an

equation that describes the amount in the account as a function of time. Will the account ever run out of money? If so, when?

16. Repeat Exercise 15 if the amount invested is $20,000, the account pays 12% per year compounded continuously, and $3000 is withdrawn continuously each year.

17. Repeat Exercise 15 if the amount invested is $40,000, the account pays 10% per year compounded continuously, and $3000 is withdrawn continuously each year.

18. Repeat Exercise 15 if the amount invested is $60,000, the account pays 9% per year compounded continuously, and $5400 is withdrawn continuously each year.

*19. If $P is invested in an account that pays $100r$% per year compounded continuously and $W is continuously withdrawn from the account each year, find an equation that describes the amount in the account as a function of time. Then graph this function and analyze the account when

   (a) $P > W/r$ (b) $P = W/r$ (c) $P < W/r$

*20. If $1 million is placed in an account that pays 8% per year compounded continuously, what is the maximum amount that can be continuously withdrawn from the account each year if the $1 million is to be preserved?

21. If $20,000 is placed in an account that pays 10% per year compounded continuously at time $t = 0$, t in years, and if money is then continuously withdrawn from the account at the rate of $1000t$ dollars per year, find an equation that describes the amount in the account as a function of time. Will the account ever run out of money?

22. Repeat Exercise 21 if the initial investment is $100,000; $200,000.

23. A vat contains 1000 gallons of liquid which is 95% water and 5% acetone. At time $t = 0$ the liquid begins to flow out of the vat at a rate of 5 gallons per minute, while another liquid which is 90% water and 10% acetone begins to flow into the vat at the same rate. Find an equation that describes the amount of acetone in the vat as a function of time.

24. Repeat Exercise 23 if the liquid flows out of the vat at a rate of 6 gallons per minute.

25. Repeat Exercise 23 if the liquid flows out of the vat at a rate of 3 gallons per min-

Life Science

26. A vessel contains 5 L of water. Starting at time $t = 0$, a solution containing 2 mg of amino acid per liter flows into the vessel at a rate of 0.3 L/min. Assume the liquid in the vessel is kept well mixed and is drawn out of the vessel at the same rate at which it flows in.
   (a) Find a differential equation that describes the amount of amino acid in the vessel at time t.
   (b) Solve the differential equation of part (a) if there was no amino acid dissolved in the water at the start.
   (c) Solve the differential equation if 10 mg of amino acid were dissolved in the water at the start.

27. Repeat Exercise 26 if the mixture is drawn out of the vessel at the rate of 0.1 L/min.

*28. Refer to Exercise 26: Suppose the solution is flowing into the vessel at a rate of r L/min in a concentration of a mg/L, that the liquid flows out of the vessel at a rate of s L/min, and that initially there are y_0 mg present. Find an equation that describes the amount of amino acid in the vessel as a function of time and analyze it if

(a) $r > s$ (b) $r = s$ (c) $r < s$

9.4 TOTAL DIFFERENTIALS AND EXACT DIFFERENTIAL EQUATIONS

In this section we consider first-order differential equations such as

$$\frac{dy}{dx} = -\frac{y}{x+y}.$$

This equation is neither separable nor linear, but rather is an example of what is known as an **exact** differential equation. Before we discuss exact equations and their solutions, it will be convenient to introduce the notion of the **total differential** of a function of two variables.

The Total Differential

The **total differential** of a function of two variables is analogous to the differential of a function of one variable. (See Section 3.3.) It is defined using the first partial derivatives of the function and the differentials of its independent variables.

> **The Total Differential of f**
>
> Let f be a function of the two variables x and y. Then the **total differential of f** is df, where
>
> $$df = f_x\, dx + f_y\, dy.$$

EXAMPLE 1   If f is defined by

$$f(x, y) = x^2 + 2xy + 2y^2,$$

the first partial derivatives of f are

$$f_x = 2x + 2y = 2(x + y) \quad \text{and} \quad f_y = 2x + 4y = 2(x + 2y).$$

Therefore

$$df = 2(x + y)\, dx + 2(x + 2y)\, dy. \quad \square$$

Now let Δx and Δy be increments in x and y, respectively. Then

$$\Delta f = f(x + \Delta x, y + \Delta y) - f(x, y)$$

is the **change in f** brought about by the simultaneous changes of Δx in x and Δy in y. If we identify dx with Δx and dy with Δy, then, just as was the case for functions of one variable, $\Delta f \approx df$, and therefore we have the following:

Approximation by the Total Differential

If f is a function of the two variables x and y, then

$$\Delta f \approx f_x(x, y)\, \Delta x + f_y(x, y)\, \Delta y.$$

The proportional change in f is

$$\frac{\Delta f}{f} \approx \frac{f_x(x, y)\, \Delta x + f_y(x, y)\, \Delta y}{f(x, y)}.$$

EXAMPLE 2 Suppose $f(x, y) = x^2 + 2xy + 2y^2$, so that

$$df = 2(x + y)\, dx + 2(x + 2y)\, dy,$$

and that x changes from 1 to 1.1 while y changes from 3 to 2.9. Then

$$\Delta x = 0.1 \quad \text{and} \quad \Delta y = -0.1,$$

and thus

$$\Delta f \approx 2(1 + 3)(0.1) + 2(1 + 2(3))(-0.1) = -0.6.$$

The proportional change in f is

$$\frac{\Delta f}{f} \approx \frac{-0.6}{f(1, 3)} = -\frac{0.6}{25} = -0.024.$$

Therefore the percentage change in f is approximately -2.4%. □

EXAMPLE 3 Suppose the quantity demanded q of a commodity depends on its unit price p and the unit price r of another commodity according to the equation

$$q = -2000p^{3/4} + pr + 2000r^{1/2} + 100{,}000.$$

Let us find the approximate change in quantity demanded if both commodities sold for \$16 per unit and the price p then increased by \$1 while the price r decreased by \$1.

We have

$$dq = (-1500p^{-1/4} + r)\, dp + (p + 1000r^{-1/2})\, dr,$$

and hence

$$\Delta q \approx (-1500p^{-1/4} + r)\, \Delta p + (p + 1000r^{-1/2})\, \Delta r.$$

Since $p = r = 16$, $\Delta p = +1$, and $\Delta r = -1$,

$$\Delta q \approx (-1500(16)^{-1/4} + 16)(1) + (16 + 1000(16)^{-1/2})(-1)$$
$$= (-750 + 16) - (16 + 250)$$
$$= -1000.$$

Therefore approximately 1000 fewer units would be demanded. Since $q = 92{,}256$ when $p = r = 16$, it follows that the proportional change in q would be approximately

$$\frac{\Delta q}{q} = \frac{-1000}{92{,}256} \approx -0.0111,$$

and thus the percentage change in q would be approximately -1.11%. □

Exact Differential Equations

Up until now in this section we have been concerned with finding the total differential of a given function. Now we turn this problem around, and ask the following question: if we have an expression of the form $M(x, y)\, dx + N(x, y)\, dy$, when can it be the total differential of a function f? An expression of the form $M(x, y)\, dx + N(x, y)\, dy$ is called a **differential form,** so the question we are asking is: when can a differential form be the total differential of some function?

Suppose that the differential form

$$M(x, y)\, dx + N(x, y)\, dy$$

is the total differential of a function f. Then

$$df = f_x(x, y)\, dx + f_y(x, y)\, dy = M(x, y)\, dx + N(x, y)\, dy,$$

so

$$M(x, y) = f_x(x, y) \quad \text{and} \quad N(x, y) = f_y(x, y).$$

But if this is so, then

$$\frac{\partial M}{\partial y} = \frac{\partial}{\partial y}\left(\frac{\partial f}{\partial x}\right) = f_{xy} \quad \text{and} \quad \frac{\partial N}{\partial x} = \frac{\partial}{\partial x}\left(\frac{\partial f}{\partial y}\right) = f_{yx},$$

and since in this book we consider only functions whose mixed partial derivatives f_{xy} and f_{yx} are equal, it follows that $\partial M/\partial y = \partial N/\partial x$. Therefore if a differential form $M\, dx + N\, dy$ is the total differential of some function, then $\partial M/\partial y = \partial N/\partial x$. On the other hand, it is also true (though we shall not prove this), that if $\partial M/\partial y = \partial N/\partial x$, then there is some function f such that $df = M\, dx + N\, dy$.

Exact Differential Forms

A differential form

$$M(x, y)\, dx + N(x, y)\, dy$$

is said to be **exact** if

$$\frac{\partial M}{\partial y} = \frac{\partial N}{\partial x}.$$

A differential form is exact if and only if it is the total differential of some function.

EXAMPLE 4 Consider the differential form
$$2xy\, dx + (x^2 + 2y)\, dy.$$
Here
$$M(x, y) = 2xy \quad \text{and} \quad N(x, y) = x^2 + 2y.$$
Thus
$$\frac{\partial M}{\partial y} = 2x = \frac{\partial N}{\partial x},$$
so this differential form is exact, and hence there is some function f such that $df = 2xy\, dx + (x^2 + 2y)\, dy$. On the other hand, the differential form
$$x^2 y\, dx + x^2 y^2\, dy$$
is not exact, because
$$\frac{\partial M}{\partial y} = x^2 \quad \text{and} \quad \frac{\partial N}{\partial x} = 2xy^2;$$
therefore there is no function whose total differential is this differential form. ☐

Now that we can tell when a differential form is the total differential of some function f, can we find the defining equation for f? The answer is yes. We illustrate with the exact differential form $2xy\, dx + (x^2 + 2y)\, dy$ of Example 4. Since this differential form is exact, we have
$$df = f_x\, dx + f_y\, dy = 2xy\, dx + (x^2 + 2y)\, dy$$
for some f, and thus
$$f_x = 2xy \quad \text{and} \quad f_y = x^2 + 2y.$$
But f_x is obtained from f by differentiating f with respect to x while treating y as a constant, so we can recover f from f_x by integrating f_x with respect to x while treating y as a constant:
$$f(x, y) = \int 2xy\, dx = x^2 y + g(y),$$
where $g(y)$ is some expression in y. (To see that this is so, find the partial derivative of $x^2 y + g(y)$ with respect to x.) But then $f_y = x^2 + 2y$ implies that
$$f_y = \frac{\partial}{\partial y}(x^2 y + g(y)) = x^2 + g'(y) = x^2 + 2y.$$
Therefore
$$g'(y) = 2y,$$
and hence
$$g(y) = y^2 + c.$$

Thus
$$f(x, y) = x^2y + y^2 + c.$$

The procedure we have just illustrated works for any exact differential form.

Finding f such that $df = M\,dx + N\,dy$

Let $M\,dx + N\,dy$ be an exact differential form. To find a function f such that $df = M\,dx + N\,dy$, proceed as follows:

1. Set
$$f(x, y) = \int M(x, y)\,dx + g(y),$$
where the integral is taken with respect to x while treating y as a constant, and $g(y)$ is an unknown expression in y.

2. Using the expression for f obtained in step 1, set
$$\frac{\partial f}{\partial y} = N(x, y)$$
and solve for $g(y)$.

EXAMPLE 5 The differential form
$$(x^2 + 3xy^2)\,dx + (3x^2y + e^y)\,dy$$
is exact because
$$\frac{\partial}{\partial y}(x^2 + 3xy^2) = 6xy = \frac{\partial}{\partial x}(3x^2y + e^y).$$

Hence it is the total differential of some function f. To find f, we set
$$f(x, y) = \int (x^2 + 3xy^2)\,dx + g(y) = \frac{x^3}{3} + \frac{3x^2y^2}{2} + g(y).$$

Then
$$\frac{\partial f}{\partial y} = N(x, y)$$
implies that
$$3x^2y + g'(y) = 3x^2y + e^y.$$

Therefore
$$g'(y) = e^y,$$

so
$$g(y) = e^y + c,$$
and hence
$$f(x, y) = \frac{x^3}{3} + \frac{3x^2y^2}{2} + e^y + c. \quad \square$$

Now we are ready to discuss exact differential equations. If a first-order differential equation
$$\frac{dy}{dx} = F(x, y)$$
is rewritten in the form
$$M(x, y)\, dx + N(x, y)\, dy = 0$$
and the differential form $M(x, y)\, dx + N(x, y)\, dy$ is exact, we say that the original differential equation is **exact**. But if the differential form is exact, we can find a function f such that $df = M\, dx + N\, dy$. Thus an exact differential equation is equivalent to an equation of the form $df = 0$, and the solution of such an equation is $f(x, y) = c$.

Exact Differential Equations

A first-order differential equation is **exact** if it can be written in the form
$$M(x, y)\, dx + N(x, y)\, dy = 0$$
where $M(x, y)\, dx + N(x, y)\, dy$ is an exact differential form. The general solution of an exact equation is
$$f(x, y) = c,$$
where f is a function such that
$$df = M(x, y)\, dx + N(x, y)\, dy.$$

EXAMPLE 6 The differential equation
$$\frac{dy}{dx} = -\frac{x^2 + 3xy^2}{3x^2y + e^y}$$
is exact because it can be written as
$$(x^2 + 3xy^2)\, dx + (3x^2y + e^y)\, dy = 0,$$

and we saw in Example 5 that this differential form is exact. Therefore its solution is

$$f(x, y) = c,$$

where

$$df = (x^2 + 3xy^2)\, dx + (3x^2y + e^y)\, dy.$$

But in Example 5 we showed that

$$f(x, y) = \frac{x^3}{3} + \frac{3x^2y^2}{2} + e^y + c.$$

Thus the solution of the differential equation is

$$\frac{x^3}{3} + \frac{3x^2y^2}{2} + e^y = c. \quad \square$$

EXAMPLE 7 Let us solve the differential equation

$$\frac{dy}{dx} = -\frac{2xy}{x^2 + y^2}.$$

We rewrite the equation as

$$2xy\, dx + (x^2 + y^2)\, dy = 0.$$

Since

$$\frac{\partial M}{\partial y} = \frac{\partial}{\partial y}(2xy) = 2x = \frac{\partial}{\partial x}(x^2 + y^2) = \frac{\partial N}{\partial x},$$

the equation is exact. We set

$$f(x, y) = \int 2xy\, dx + g(y) = x^2y + g(y).$$

Then

$$\frac{\partial f}{\partial y} = N(x, y)$$

becomes

$$x^2 + g'(y) = x^2 + y^2,$$

which implies that $g'(y) = y^2$. Hence we may take $g(y) = y^3/3 + c$, so

$$f(x, y) = x^2y + \frac{y^3}{3} + c,$$

and the general solution of the differential equation is therefore

$$x^2y + \frac{y^3}{3} = c. \quad \square$$

9.4 EXERCISES

The Total Differential

In Exercises 1 through 4, find the total differential of f.

1. $f(x, y) = 2x^2 - 5xy + 3y^2$
2. $f(x, y) = 4\sqrt{x} + 5y^{3/5}$
3. $f(x, y) = \dfrac{x + y^2}{x + 2y}$
4. $f(x, y) = xye^{xy}$

In Exercises 5 through 8, use the total differential to find the approximate change in f and the approximate proportional change in f brought about by the given changes in x and y.

5. $f(x, y) = x^2 + 2y$, x changes from 1 to 1.1, y from 2 to 2.05.
6. $f(x, y) = xy^{-2}$, x changes from 3 to 3.14, y from 2 to 1.96.
7. $f(x, y) = 2e^{x/y}$, x changes from 0 to 0.6, y from 1 to 0.97.
8. $f(x, y) = \ln(y/x)$, x changes from 1 to 0.8, y from 1 to 1.2.
9. A box has a square base, rectangular sides, and no top. Use the total differential to find the approximate proportional change in its surface area if the width of its base is increased from 10 to 10.1 inches and its height is decreased from 6 to 5.95 inches.
10. Repeat Exercise 9, substituting "volume" for "surface area."
11. A cereal box is a cylinder with a circular base. If the radius of its base is increased from 9 to 9.2 cm and its height is increased from 18 to 18.4 cm, use the total differential to find the approximate proportional change in its volume ($= \pi r^2 h$).
12. Repeat Exercise 11 substituting "surface area" for "volume"; the surface area of a cylinder is $2\pi rh$.
13. The length and the width of a rectangle are measured to be 20 and 10 centimeters with an accuracy of ± 0.02 cm for each. Use the total differential to find the approximate proportional and percentage errors in calculating its area.
14. The length of the base of a triangle is measured to be 450 mm with an accuracy of ± 0.1 mm and its height to be 300 mm with an accuracy of ± 0.2 mm. Use the total differential to find the approximate proportional and percentage errors in calculating its area.
15. A right circular cone has the radius of its base measured to be 100 mm with an accuracy of ± 0.015 mm and its height to be 400 mm with an accuracy of ± 0.003 mm. Use the total differential to find the approximate proportional and percentage errors in calculating its volume ($= \pi r^2 h/3$).

Management

16. A firm's cost function is given by $C(x, y) = 20xy^{1/2} + 10{,}000$, where x is the number of units of product X and y the number of units of product Y produced. If 10 units of X and 25 of Y are currently being produced, use the total differential to find the approximate additional cost of increasing production to 11 units of X and 26 of Y.

17. A firm's profit function is given by
$$P(x, y) = x^2 + 0.5y^2 - xy - 10{,}000,$$
where x is the number of units of product X and y the number of units of product Y made and sold. The firm is currently making and selling 200 units of X and 150 of Y and plans to increase production and sales of X by 10 units and decrease that of Y by 10 units. Use the total differential to find the approximate change in profit that will result.

18. The quantity of a commodity supplied to the market is given by
$$q = 100p_1^{1/2} + 200p_2^{1/3} - 4000,$$
where p_1 is the unit price of the commodity and p_2 is the unit price of a related commodity. Use the total differential to find the approximate change in quantity supplied and the approximate percentage change in quantity supplied if p_1 changes from \$25 to \$26 and p_2 changes from \$27 to \$26.

19. The demand function for a product is given by
$$q = -300p_1^{2/3} + 4000p_2^{1/5} + 50{,}000,$$
where q is quantity demanded, p_1 is the unit price of the product, and p_2 is the unit price of another product. Use the total differential to find the approximate change in quantity demanded and the approximate percentage change in quantity demanded if p_1 changes from \$27 to \$26.50 and p_2 changes from \$32 to \$31.75.

20. A company's production function is given by
$$f(x, y) = 4000x^{1/4}y^{3/4}.$$
Here x is labor in worker-hours and y is capital in dollars. Use the total differential to find the approximate change in production and the approximate percentage change in production if the number of worker-hours available increases from 200 to 208 and the amount of capital available decreases from \$10,000 to \$9950.

21. A firm's production function is given by
$$f(x, y) = 2000x^{2/3}y^{1/3},$$
where x is labor in thousands of worker-hours and y is capital in thousands of dollars. Use the total differential to find the approximate change in production and the approximate percentage change in production if the number of worker-hours available decreases from 1,331,000 to 1,300,000 and the amount of capital available increases from \$13,824,000 to \$14,000,000.

Life Science

22. The pollution in the air of a large city, in parts per million, is given by $p(x, y) = 200x^{4/5} - 40y^{1/2}$, where x is the number of cars on the road, in millions, and y is their average fuel efficiency, in miles per gallon. Use the total differential to find the change in pollution if the number of cars on the road decreases from 1.0 to 0.975 million and their average fuel efficiency increases from 25.0 to 25.5 miles per gallon.

*23. (This exercise was suggested by material in Niebel, *Motion and Time Study*, 8th edition, Irwin, 1988.) The contrast C between an object and its background is given by the equation

$$C = \frac{L - D}{D},$$

where D is the luminance of the darker of the object and its background, and L is the luminance of the lighter of the object and its background, both measured in lamberts. Use the total differential to find the approximate change in contrast and the approximate percentage change in contrast if D decreases by 1% and L increases by 1%.

Social Science

24. Annual housing starts H in a community depend on average weekly income Y and average annual percentage rate I for mortgage loans according to the equation

$$H = 200 + 0.2Y - 0.65I^2.$$

Suppose the average weekly income is currently $300 and is forecast to increase to $310 next year, while the average annual percentage rate for mortgage loans is now 12% and is forecast to increase to 12.5% next year. Use the total differential to forecast the approximate change in housing starts and the approximate percentage change in housing starts from this year to next year.

Exact Differential Equations

In Exercises 25 through 30, state whether the given differential form is exact; if it is exact, find a function f whose total differential is the differential form.

25. $y\, dx + (x + y)\, dy$
26. $(x - 2xy)\, dx - (x^2 + 3y)\, dy$
27. $(x^2 - 3y)\, dx + (3x - y^2)\, dy$
28. $ye^x\, dx + xe^y\, dy$
29. $ye^{xy}\, dx + xe^{xy}\, dy$
30. $xy^{-1}\, dx - 0.5x^2y^{-2}\, dy$

In Exercises 31 through 42, determine if the given differential equation is exact; if it is exact, find its general solution.

31. $\dfrac{dy}{dx} = -\dfrac{y}{x + y}$
32. $\dfrac{dy}{dx} = \dfrac{2x + y}{y - x}$
33. $\dfrac{dy}{dx} = \dfrac{4x + 3y}{2y + 3x}$
34. $\dfrac{dy}{dx} = -\dfrac{x + y}{x + y^2}$
35. $\dfrac{dy}{dx} = -\dfrac{x^2 + 2y^2}{4xy + 1}$
36. $\dfrac{dy}{dx} = \dfrac{2 - y^3}{3xy + 2}$
37. $\dfrac{dy}{dx} = \dfrac{x^2 + 2x + y}{y^2 + 2y + x}$
38. $\dfrac{dy}{dx} = \dfrac{x^{-1/2}(y^{3/2} + 1)}{1 - 3x^{1/2}y^{1/2}}$
39. $\dfrac{dy}{dx} = \dfrac{e^x - 2xe^y}{x^2 + 1}$
40. $\dfrac{dy}{dx} = \dfrac{e^x - 2xe^y}{x^2 e^y + 1}$
41. $\dfrac{dy}{dx} = -\dfrac{y^2 \ln x + 1}{2xy \ln x}$
42. $\dfrac{dy}{dx} = -\dfrac{y^2(\ln x + 1)}{2xy \ln x}$

In Exercises 43 through 46, find the equation of the curve that has the given slope dy/dx and passes through the given point.

43. $\dfrac{dy}{dx} = \dfrac{x + 2y}{y - 2x}$, $(0, 1)$
44. $\dfrac{dy}{dx} = -\dfrac{x^2 + 2y}{2x + y}$, $(0, 2)$
45. $\dfrac{dy}{dx} = -\dfrac{2x(y^3 + 2)}{y^2(3x^2 + 10)}$, $(1, 1)$
46. $\dfrac{dy}{dx} = \dfrac{3y(\sqrt{x} - y)}{6xy - 2x^{3/2}}$, $(4, 1)$

9.5 GRAPHICAL SOLUTIONS

Differential equations cannot always be solved explicitly. In other words, given a differential equation, it is not always possible to find an equation $y = y(x)$ that defines its solutions. Nevertheless, information about the solutions of differential equations can often be obtained by graphical or numerical methods. In this section we consider graphical solutions of differential equations; in the next we discuss numerical solutions.

The solutions of a differential equation form a family of curves. For instance, the differential equation

$$\frac{dy}{dx} = x$$

has general solution

$$y = \frac{1}{2}x^2 + c.$$

(Check this). Each value of c yields a function whose graph is called an **integral curve** of the differential equation. In this case, each integral curve is a parabola with vertex on the y-axis and y-intercept c. Figure 9–4 shows the integral curves for some values of c.

As we shall see in this section, it is often possible to sketch the integral curves of a differential equation without explicitly solving the equation. Such **graphical solutions** of a differential equation can be extremely valuable: not only do they show us how the solutions behave (and this information may be more valuable in a particular instance than an explicit solution is), but they also provide a method of analyzing the equation when no explicit solution can be found.

Consider a differential equation of the form

$$\frac{dy}{dx} = f(x, y).$$

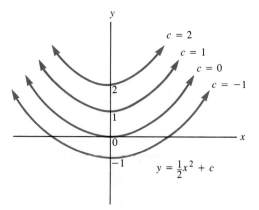

FIGURE 9–4

DIFFERENTIAL EQUATIONS

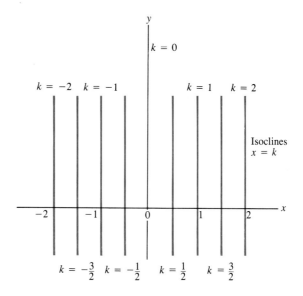

FIGURE 9–5

Since the derivative is the slope of the line tangent to the graph of $y = y(x)$ at the point (x, y), $f(x, y)$ must also be equal to this slope. Thus if k is a constant and we set $f(x, y) = k$, we will obtain a curve along which all solutions $y = y(x)$ of the differential equation have slope k. Such a curve is called an **isocline** of the differential equation. For example, let us again consider the differential equation

$$\frac{dy}{dx} = x.$$

If we set $x = k$, we obtain the family of vertical lines shown in Figure 9–5; these are the isoclines of this differential equation. On an isocline corresponding to a particular value of k, all solutions will have slope k. In Figure 9–6 we have drawn in short line segments of the proper slope on each isocline; thus the line segments on the isocline $x = 0$ have slope 0, those on the isocline $x = 1$ have slope 1, and so on.

Once we have obtained the isoclines and drawn line segments of the proper slope on them, we can sketch the integral curves of the differential equation by making them fit the slopes as they cross the isoclines. For instance, the integral curves of the equation

$$\frac{dy}{dx} = x$$

must fit the slopes depicted in Figure 9–6 and therefore must have the shapes indicated in Figure 9–7. Compare Figures 9–7 and 9–4, and note that we obtained Figure 9–7 *without* explicitly solving the differential equation.

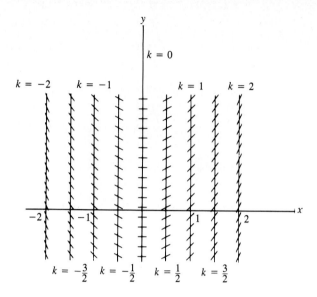

FIGURE 9–6

EXAMPLE 1 Consider the differential equation

$$\frac{dy}{dx} = x + y.$$

We set

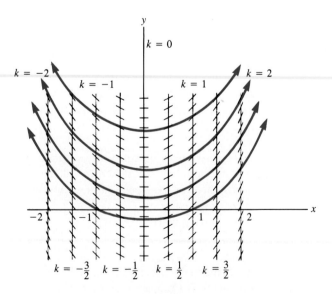

FIGURE 9–7

DIFFERENTIAL EQUATIONS

$$x + y = k,$$

thus obtaining the isoclines

$$y = -x + k.$$

This is a family of lines, and on each line $y = -x + k$ all solutions of the differential equation have constant slope k. See Figure 9–8. We have sketched the integral curves for the differential equation in Figure 9–9. Note that the line $y = -x - 1$ is a solution and that all the other solutions approach this line as an asymptote.

(In this example

$$\frac{dy}{dx} = x + y$$

is linear, so we can solve it explicitly. The general solution is

$$y = -x - 1 + ce^x.$$

Note that the integral curve $y = -x - 1$ is obtained from the general solution by taking $c = 0$.) ☐

EXAMPLE 2   Sketch the integral curves of the differential equation

$$\frac{dy}{dx} = x^2 + y^2.$$

We set

$$x^2 + y^2 = k,$$

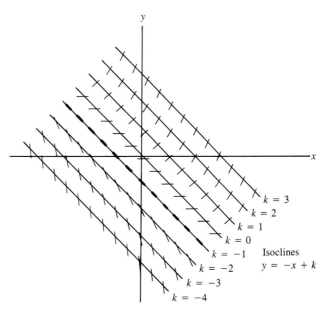

FIGURE 9–8

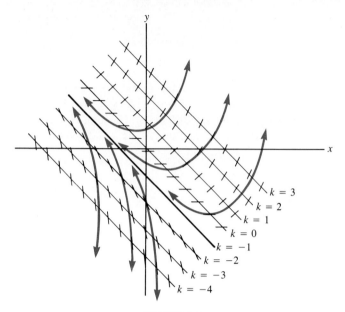

FIGURE 9–9

which when $k > 0$ is the equation of a circle with center at the origin and radius $\sqrt{k}$. Therefore the isoclines are as in Figure 9–10, and the integral curves are as in Figure 9–11. Since this differential equation is neither separable, exact, nor linear, we cannot solve it explicitly. ◻

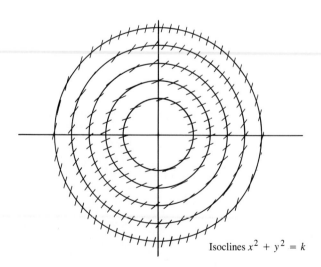

Isoclines $x^2 + y^2 = k$

FIGURE 9–10

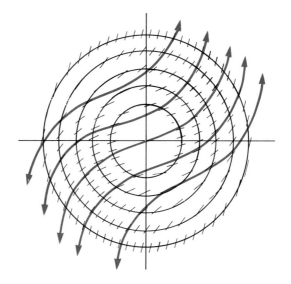

FIGURE 9–11

9.5 EXERCISES

In Exercises 1 through 16, sketch the integral curves of the differential equations.

1. $\dfrac{dy}{dx} = y$
2. $\dfrac{dy}{dx} = x^2$
3. $\dfrac{dy}{dx} = x^3$
4. $\dfrac{dy}{dx} = y^2$
5. $\dfrac{dy}{dx} = \dfrac{y}{x}$
6. $\dfrac{dy}{dx} = \dfrac{x}{y}$
7. $\dfrac{dy}{dx} = y - x$
8. $\dfrac{dy}{dx} = y - 2x$
9. $\dfrac{dy}{dx} = x - y$
10. $\dfrac{dy}{dx} = xy$
11. $\dfrac{dy}{dx} = \dfrac{1}{x}$
12. $\dfrac{dy}{dx} = \dfrac{1}{y}$
13. $\dfrac{dy}{dx} = x^2 + \dfrac{y^2}{4}$
14. $\dfrac{dy}{dx} = x^2 - \dfrac{y^2}{4}$
15. $\dfrac{dy}{dx} = \ln x$
16. $\dfrac{dy}{dx} = \ln y$

Management

17. A firm's marginal cost function is given by

$$\dfrac{dC}{dx} = C + 3x, \qquad x \geq 0, \quad C(0) = 1.$$

Use the method of this section to sketch the firm's cost function. Then check your answer by solving the differential equation for the cost function C and sketching its graph.

18. If a company makes x units, $x \geq 1$, its average cost per unit is $A(x)$ dollars, where

$$\dfrac{dA}{dx} = A(A - 2), \qquad A(1) = 50.$$

Use the method of this section to sketch the graph of A.

Life Science

19. If $y = y(t)$ is the size of a population, in thousands, at time t, $t \geq 0$, then

$$\frac{dy}{dt} = y^3 - 1.$$

Use the method of this section to sketch the graph of y when $y(0) > 1$, $y(0) = 1$, and $0 < y(0) < 1$.

*20. Suppose the rate of change of a population is described by the differential equation

$$\frac{dy}{dt} = (y - 1)(y - 3),$$

where t is time, $t \geq 0$, and y is population in thousands. Sketch the integral curves of this equation and describe the growth of the population if

(a) $0 < y(0) < 1$ (b) $y(0) = 1$ (c) $1 < y(0) \leq 2$
(d) $2 < y(0) < 3$ (e) $y(0) = 3$
(f) $y(0) > 3$

9.6 NUMERICAL METHODS

As we remarked at the beginning of Section 9.5, it is not always possible to solve differential equations explicitly, and therefore it is necessary to develop graphical and numerical methods of solution. In this section we discuss **numerical methods** for initial value problems of the form

$$\frac{dy}{dx} = f(x, y), \quad y(x_0) = y_0.$$

Such methods allow us to find the approximate numerical value $y(x_i)$ of the solution at one or more points x_i. We consider two numerical methods: Euler's method and the Runge-Kutta method.

Euler's Method

Let $y = y(x)$ be the solution of the initial value problem

$$\frac{dy}{dx} = f(x, y), \quad y(x_0) = y_0,$$

let Δx be a small positive number, and set $x_1 = x_0 + \Delta x$. We wish to find $y_1 = y(x_1)$. See Figure 9–12.

From our study of differentials in Section 3.3, we know that when Δx is small,

$$y(x_0 + \Delta x) \approx y(x_0) + y'(x_0) \Delta x.$$

But since

$$y(x_0) = y_0, \quad y(x_0 + \Delta x) = y(x_1) = y_1,$$

Differential Equations

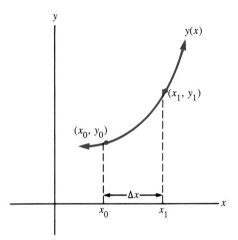

FIGURE 9–12

and

$$y'(x_0) = \frac{dy}{dx}\bigg|_{x=x_0} = f(x_0, y_0),$$

we have

$$y_1 \approx y_0 + f(x_0, y_0)\,\Delta x.$$

Thus we have obtained an approximate value for y_1. If we wish, we may set $x_2 = x_1 + \Delta x$, $y_2 = y(x_2)$, and repeat the procedure to conclude that

$$y_2 \approx y_1 + f(x_1, y_1)\,\Delta x,$$

and so on. Hence:

Euler's Method

To find an approximate numerical solution of the initial value problem

$$\frac{dy}{dx} = f(x, y), \qquad y(x_0) = y_0,$$

let Δx be a small positive number, and for $i = 1, 2, \ldots$, set $x_i = x_{i-1} + \Delta x$. If the numerical value of the solution at $x = x_i$ is y_i, then

$$y_i \approx y_{i-1} + f(x_{i-1}, y_{i-1})\,\Delta x.$$

EXAMPLE 1 Let us use Euler's method to approximate a numerical solution of the initial value problem

$$\frac{dy}{dx} = x + y, \qquad y(0) = 1.$$

We will take $\Delta x = 0.1$ and find the approximate value of y at $x_0 = 0.0$, $x_1 = 0.1$, $x_2 = 0.2$, $x_3 = 0.3$, and $x_4 = 0.4$. We have

$y_0 = 1$,
$y_1 \approx y_0 + f(x_0, y_0) \Delta x = 1 + (0 + 1)(0.1) = 1.1$,
$y_2 \approx y_1 + f(x_1, y_1) \Delta x = 1.1 + (0.1 + 1.1)(0.1) = 1.22$,
$y_3 \approx y_2 + f(x_2, y_2) \Delta x = 1.22 + (0.2 + 1.22)(0.1) = 1.362$,
$y_4 \approx y_3 + f(x_3, y_3) \Delta x = 1.362 + (0.3 + 1.362)(0.1) = 1.5282$. ☐

We noted in Section 9.5 that the general solution of the linear differential equation

$$\frac{dy}{dx} = x + y$$

is

$$y = -x - 1 + ce^x.$$

The initial condition $y(0) = 1$ thus implies that $c = 2$. (Check this.) Hence the solution of the initial value problem of Example 1 is

$$y = -x - 1 + 2e^x.$$

We can find the actual numerical values of $y = y(x)$ at 0.0, 0.1, 0.2, 0.3, and 0.4 by substituting these values for x in the solution. Suppose we do this and then compare the actual numerical values with the approximations we obtained in Example 1 by Euler's method. The result is the following table:

x_i	Actual $y(x_i)$	Euler y_i	Actual − Euler
0.0	1.0000	1.0000	0.0000
0.1	1.1103	1.1000	0.0103
0.2	1.2428	1.2200	0.0228
0.3	1.3997	1.3620	0.0377
0.4	1.5836	1.5282	0.0554

Note that as x_i gets further and further from the initial value x_0, the approximation to y_i in Example 1 becomes less and less accurate. This is a feature of Euler's method: the approximations will become less accurate as we move further away from x_0. Of course, a smaller value of Δx will produce better approximations at every x_i, but in order to obtain reasonable accuracy Euler's method often requires the use of extremely small values of Δx. Therefore Euler's method is not really practical for producing numerical solutions of differential equations. However, it is the foundation for many similar methods that yield better results. One of these, the **Runge-Kutta** method, we now demonstrate.

The Runge-Kutta Method

We will not try to justify the Runge-Kutta method here, but will merely describe it. We begin again with an initial value problem

DIFFERENTIAL EQUATIONS

$$\frac{dy}{dx} = f(x, y), \qquad y(x_0) = y_0$$

and a small, positive Δx. We want to approximate the numerical values of the solution $y = y(x)$ of the problem at $x_1 = x_0 + \Delta x$, $x_2 = x_1 + \Delta x$, $x_3 = x_2 + \Delta x$ . . . . As before, we let $y_i = y(x_i)$ for $i = 0, 1, 2, \ldots$

The Runge-Kutta Method

For each i, $i = 0, 1, 2, \ldots$, set

$$k_1 = f(x_i, y_i)\, \Delta x,$$

$$k_2 = f\left(x_i + \frac{1}{2}\Delta x,\ y_i + \frac{1}{2}k_1\right) \Delta x,$$

$$k_3 = f\left(x_i + \frac{1}{2}\Delta x,\ y_i + \frac{1}{2}k_2\right) \Delta x,$$

$$k_4 = f(x_i + \Delta x,\ y_i + k_3)\, \Delta x,$$

and

$$K = \frac{1}{6}(k_1 + 2k_2 + 2k_3 + k_4).$$

Then

$$y_{i+1} \approx y_i + K.$$

EXAMPLE 2 Let us apply the Runge-Kutta method to the initial value problem

$$\frac{dy}{dx} = x + y, \qquad y(0) = 1$$

of Example 1, again taking $\Delta x = 0.1$. We will round off our calculations to five decimal places. Note that $x_0 = 0$ and $y_0 = 1$. We have

$i = 0$: $\quad k_1 = (0 + 1)(0.1) = 0.1,$

$$k_2 = \left(0 + \frac{0.1}{2} + 1 + \frac{0.1}{2}\right)(0.1) = 0.11,$$

$$k_3 = \left(0 + \frac{0.1}{2} + 1 + \frac{0.11}{2}\right)(0.1) = 0.1105,$$

$$k_4 = (0 + 0.1 + 1 + 0.1105)(0.1) = 0.12105,$$

$$K = \frac{1}{6}(0.1 + 2(0.11) + 2(0.1105) + 0.12105)$$

$$= 0.11034,$$

$$y_1 \approx 1 + 0.11034 = 1.11034.$$

$i = 1$: $\quad k_1 = (0.1 + 1.11034)(0.1) = 0.12103,$

$$k_2 = \left(0.1 + \frac{0.1}{2} + 1.11034 + \frac{0.12103}{2}\right)(0.1) = 0.13209,$$

$$k_3 = \left(0.1 + \frac{0.1}{2} + 1.11034 + \frac{0.13209}{2}\right)(0.1) = 0.13264,$$

$$k_4 = (0.1 + 0.1 + 1.11034 + 0.13264)(0.1) = 0.14430,$$

$$K = \frac{1}{6}(0.12103 + 2(0.13209) + 2(0.13264) + 0.14430) = 0.13247,$$

$$y_2 \approx 1.11034 + 0.13247 = 1.24281.$$

$i = 2$: $\quad k_1 = (0.2 + 1.24281)(0.1) = 0.14428,$

$$k_2 = \left(0.2 + \frac{0.1}{2} + 1.24281 + \frac{0.14428}{2}\right)(0.1) = 0.15650,$$

$$k_3 = \left(0.2 + \frac{0.1}{2} + 1.24281 + \frac{0.15650}{2}\right)(0.1) = 0.15711,$$

$$k_4 = (0.2 + 0.1 + 1.24281 + 0.15711)(0.1) = 0.17000,$$

$$K = \frac{1}{6}(0.14428 + 2(0.15650) + 2(0.15711) + 0.17000) = 0.15692,$$

$$y_3 \approx 1.24281 + 0.15692 = 1.39973.$$

$i = 3$: $\quad k_1 = (0.3 + 1.39973)(0.1) = 0.16997,$

$$k_2 = \left(0.3 + \frac{0.1}{2} + 1.39973 + \frac{0.16997}{2}\right)(0.1) = 0.18347,$$

$$k_3 = \left(0.3 + \frac{0.1}{2} + 1.39973 + \frac{0.18347}{2}\right)(0.1) = 0.18415,$$

$$k_4 = (0.4 + 1.39973 + 0.18415)(0.1) = 0.19839,$$

$$K = \frac{1}{6}(0.16997 + 2(0.18347) + 2(0.18415) + 0.19839) = 0.18393,$$

$$y_4 \approx 1.39973 + 0.18393 = 1.58366.$$

Let us round the approximations y_i to four decimal places and compare them with the true values, as we did following Example 1. We have:

x_i	Actual $y(x_i)$	Runge-Kutta y_i	Actual − Runge-Kutta
0.0	1.0000	1.0000	0.0000
0.1	1.1103	1.1103	0.0000
0.2	1.2428	1.2428	0.0000
0.3	1.3997	1.3997	0.0000
0.4	1.5836	1.5837	−0.0001

Comparing this table with the previous one, we see that the Runge-Kutta method is considerably more accurate than Euler's method. ☐

■ *Computer Manual: Programs EULER, RUNKUT.*

9.6 EXERCISES

Euler's Method

1. Let
$$\frac{dy}{dx} = 2x, \qquad y(0) = 0.$$
   (a) Use Euler's method with $\Delta x = 0.1$ to find approximate values for $y(0.1)$, $y(0.2)$, and $y(0.3)$.
   (b) Repeat part (a) using $\Delta x = 0.05$.
   (c) Compare the results of parts (a) and (b) with each other and with the actual values obtained by solving the differential equation.

2. Let
$$\frac{dy}{dx} = x - y, \qquad y(1) = 1.$$
   Use Euler's method with $\Delta x = 0.1$ to approximate $y(1.1)$, $y(1.2)$, and $y(1.3)$.

3. Let
$$\frac{dy}{dx} = 2 - 3y, \qquad y(1) = 2.$$
   Use Euler's method to approximate $y(1.1)$, $y(1.2)$, $y(1.3)$ when
   (a) $\Delta x = 0.1$
   (b) $\Delta x = 0.05$
   (c) $\Delta x = 0.01$

4. Let
$$\frac{dy}{dx} = 1 - y^2, \qquad y(0) = 0.$$
   Use Euler's method with $\Delta x = 0.2$ to approximate $y(0.2)$ and $y(0.4)$.

5. Let
$$\frac{dy}{dx} = 2xy, \qquad y(0) = 1.$$
   Use Euler's method to approximate $y(0.2)$, $y(0.4)$, $y(0.6)$ when
   (a) $\Delta x = 0.2$
   (b) $\Delta x = 0.1$
   (c) $\Delta x = 0.05$

Management

6. A firm's marginal cost is given by
$$\frac{dC}{dx} = x^2 + C^2, \qquad C(1) = 1,$$

where x is in thousands of units and C is in thousands of dollars. Use Euler's method to estimate the cost of making 1010, 1020, and 1030 units.

The Runge-Kutta Method

7. Repeat Exercise 1 using the Runge-Kutta method, and compare your results with those of Exercise 1.
8. Repeat Exercise 2 using the Runge-Kutta method, and compare your results with those of Exercise 2.
9. Repeat Exercise 3 using the Runge-Kutta method, and compare your results with those of Exercise 3.
10. Repeat Exercise 4 using the Runge-Kutta method, and compare your results with those of Exercise 4.
11. Repeat Exercise 5 using the Runge-Kutta method, and compare your results with those of Exercise 5.

Management

12. Repeat Exercise 6 using the Runge-Kutta method, and compare your results with those of Exercise 6.

SUMMARY

This summary consists of terms and symbols whose meaning you should know, facts you should know, and techniques and methods you should be able to employ.

Section 9.1

Differential equation, first-order differential equation. General solution of a differential equation. Initial condition. Differential equation with initial condition. Particular solution of a differential equation with initial condition. Solving first-order differential equations by integrating. The language of proportionality: directly proportional to, inversely proportional to, constant of proportionality. Setting up differential equations using the language of proportionality. Finding particular solutions of differential equations by using initial conditions to determine the value of the constant of integration and the constant of proportionality.

Section 9.2

Separation of variables. Separable first-order differential equation. Solving separable equations. Constant solution of a separable equation. "Losing" constant solutions. Modeling with separable differential equations: Exponential growth/decline models, Limited growth/decline models, Logistic models.

Section 9.3

Linear first-order differential equation. Integrating factor. Method for solving linear first-order differential equations. Modeling using linear first-order differential equations.

Section 9.4

The total differential $df = f_x\, dx + f_y\, dy$. Finding the total differential. The change in f, Δf. Approximation by total differentials: change in f, proportional change in f. Finding the approximate change, proportional change, percentage change in a function. Differential form $M\, dx + N\, dy$. Exact differential form. A differential form is exact if and only if there is some function f such that the total differential of f is equal to the form. Testing a differential form for exactness. If $M\, dx + N\, dy$ is exact, finding f such that $df = M\, dx + N\, dy$. Exact differential equation. Solution of an exact differential equation is of the form $f = c$, where $df = M\, dx + N\, dy$. Solving exact differential equations.

Section 9.5

The need for graphical solutions of differential equations. Integral curve of a differential equation. Graphical solution of a differential equation. Isocline. The isocline method for sketching integral curves.

Section 9.6

The need for numerical methods. Euler's method. Using Euler's method. The Runge-Kutta method. Using the Runge-Kutta method. Comparison of the accuracy of Euler's method and the Runge-Kutta method.

REVIEW EXERCISES

In Exercises 1 through 4, solve the differential equation.

1. $\dfrac{dy}{dx} = 2x + 2$, $y(0) = 1$

2. $\dfrac{dy}{dx} = x^2 - 3x + 2$, $y(-1) = 4$

3. $\dfrac{dy}{dx} = \dfrac{1}{x}$, $y(e) = 1$

4. $\dfrac{dy}{dx} = \dfrac{2x}{1 - x^2}$, $y(0) = 0$

5. A rocket's acceleration during the first 30 seconds after launch is directly proportional to the time since launch. Find equations that give the rocket's velocity and altitude t seconds after launch, $0 \le t \le 30$, if its altitude 2 seconds after launch is 20 feet.

6. A firm's marginal profit is directly proportional to the cube root of the number of units it makes and sells. Find the firm's profit function if its fixed cost is $25,000 and the profit from making and selling 1000 units is $10,000.

7. A psychologist studying classroom learning estimates that students can absorb new concepts at a rate inversely proportional to $t + 10$, where t is the time in minutes since class started. If students can absorb two new concepts during the first 6 minutes of a class, and three new concepts in the first 10 minutes of the class, find an expression that relates the number of new concepts absorbed to the length of the class.

In Exercises 8 through 11, solve the given differential equation.

8. $x \dfrac{dy}{dx} = \dfrac{1}{y}$

9. $3(y + 2) \dfrac{dy}{dx} = 2x - 1$, $y(0) = 0$

10. $\dfrac{dy}{dx} = \dfrac{y^2}{x^3}$

11. $\dfrac{dy}{dx} = (y + 3)(x - 1)$, $y(0) = 2$

12. Suppose that the rate at which compensation is increasing in a firm is directly proportional to compensation. If the average wage in the firm now is $8 per hour and next year it will be $8.50 per hour, find an equation that relates the average wage to time.

13. The elasticity of demand for a commodity is given by

$$E(p) = \dfrac{-2p}{-2p + 1000}.$$

Find the demand function for the commodity if the quantity demanded is 100,000 units when $p = \$0$.

14. In a community of 10,000 people, 2 people start a rumor at time $t = 0$ and 10 days later 1000 people in the community have heard it. If the rate at which a rumor spreads is inversely proportional to the number of people who have *not* heard it, find an expression for the number who have heard the rumor at time t.

15. The rate at which production of airplanes changes in an aircraft plant is directly proportional to the difference between the total number of planes to be built, which is 200, and the number produced since the start of the production run. If two planes were produced during the first month of the run, and seven during the first two months, find an equation that describes production as a function of time since the production run began.

16. The rate at which the size of a population is changing at time t is directly proportional to the product of the size of the population at time t and the difference between this size and the maximum number of individuals the environment can support. Suppose an environment can support 1 million individuals and the current population size is 100,000 individuals and will be 200,000 individuals 10 years from now. Find an equation that describes population size as a function of time.

In Exercises 17 and 18, solve the given differential equation.

17. $\dfrac{dy}{dx} = -4y + 2x$, $y(0) = 0$

18. $\dfrac{dy}{dx} = -5\dfrac{y}{x} + x^{-5}$

19. A savings account pays 7% per year compounded continuously. If $25,000 is deposited in the account and then each year $3000e^{-0.5t}$ dollars are withdrawn continuously from the account, find an equation that describes the amount in the account as a function of time.

20. If the quantity of a commodity demanded by consumers is

$$q = 4000\frac{p_2}{p_1},$$

where p_1 is the unit price of the commodity and p_2 is the unit price of another commodity, use the total differential to find the approximate change in the quantity demanded and the approximate percentage change in the quantity demanded when p_1 decreases from \$5 to \$4.90 and p_2 increases from \$6 to \$6.20.

In Exercises 21 and 22, solve the given differential equation.

21. $\dfrac{dy}{dx} = \dfrac{5x^2 - 2y}{2(x - y)}$ **22.** $\dfrac{dy}{dx} = -\dfrac{x(x + 3y^2)}{3x^2 y}$

23. Graph the integral curves of

  (a) $\dfrac{dy}{dx} = y + 2x$ **(b)** $\dfrac{dy}{dx} = y^2 + 4$

24. Let

$$\frac{dy}{dx} = x^2 + y, \qquad y(1) = 2.$$

Use **(a)** Euler's method and **(b)** the Runge-Kutta method with $\Delta x = 0.05$ to approximate $y(1.05)$, $y(1.10)$, and $y(1.15)$.

ADDITIONAL TOPICS

Here are some suggestions for topics not covered in the text that you might want to investigate on your own.

1. Sometimes a first-order differential equation that is not separable can be made so by an appropriate substitution. For instance, suppose that

$$\frac{dy}{dx} = f(x, y)$$

   has the property that $f(tx, ty) = f(x, y)$; then the differential equation is said to be **homogeneous,** and the substitution $y = vx$ will convert it to a separable equation in x and v. Show how this is done and how a homogeneous first-order equation can thus be solved.

2. Sometimes a first-order differential equation that is not linear can be made so by an appropriate substitution. For instance, suppose that a differential equation has the form

$$\frac{dy}{dx} + p(x)y = q(x)y^n, \qquad n \neq 1.$$

   Such an equation is called a **Bernoulli equation,** and the substitution $v = y^{1-n}$ will convert it into a linear equation. Show how this is done and how a Bernoulli equation can thus be solved.

3. A first-order differential equation that is not exact can often be converted to an exact equation by multiplying it by an **integrating factor.** Find out how integrating factors are found and how they are used to solve first-order differential equations.

4. An equation of the form $f(x, y) = c$, where c is an arbitrary constant, defines a family of curves. A curve is said to be an **orthogonal trajectory** of the family if it intersects each member of the family in a right angle. Orthogonal trajectories are important in physics and engineering. Find out how first-order differential equations are used to find the orthogonal trajectories of a family of curves.

5. A differential equation of the form

$$a_n \frac{d^n y}{dx^n} + a_{n-1} \frac{d^{n-1} y}{dx^{n-1}} + \cdots + a_1 \frac{dy}{dx} + a_0 y = F(x),$$

where $a_n, a_{n-1}, \ldots, a_1, a_0$ are real numbers with $a_n \neq 0$, is called a **linear differential equation of order n with constant coefficients.** Such equations are very important. Find out how they are solved and some of their uses.

SUPPLEMENT: A PREDATOR–PREY MODEL

A **closed** ecological system is one for which there is no migration of species into or out of the system. Suppose that a closed ecological system contains a predator species and its prey species. (For instance, the predator species might be wolves, and the prey species caribou.) We will develop and analyze a model for such a situation.

Let $X = X(t)$ denote the number of predator individuals and $Y = Y(t)$ the number of prey individuals in the system at time t. We will assume that the rate of growth of the prey species with respect to time is given by

$$\frac{dY}{dt} = (p - qX)Y, \qquad p > 0, \quad q > 0.$$

The rationale for this model is as follows: if there were no predators present (if $X(t)$ were zero for all t), then the prey population would increase without bound, and its rate of increase would be directly proportional to Y according to an equation of the form

$$\frac{dY}{dt} = pY, \qquad p > 0.$$

However, the more predators there are, the more the growth rate of the prey is reduced; thus the growth constant p must be reduced by an amount that is itself directly proportional to the number of predators—that is, by an amount qX, where $q > 0$.

Similarly, assume that the rate of growth of the predator species with respect to time is given by

$$\frac{dX}{dt} = (rY - s)X, \qquad r > 0, \quad s > 0.$$

The rationale is this: if there were no prey, the predators would have nothing to eat, and thus their population would decrease at a rate proportional to X, that is, according to an equation of the form

$$\frac{dX}{dt} = -sX, \quad s > 0.$$

But the more prey there are, the greater the increase in the growth rate of the predator species, so we must increase the growth constant $-s$ by an amount that is directly proportional to the number of prey—that is, by an amount rY, $r > 0$.

Thus we model the predator–prey relationship by the **Lotka-Volterra equations**

$$\frac{dY}{dt} = (p - qX)Y, \quad p > 0, \quad q > 0,$$

$$\frac{dX}{dt} = (rY - s)X, \quad r > 0, \quad s > 0.$$

This is a **system** of differential equations that must be solved simultaneously for $X = X(t)$ and $Y = Y(t)$. Let us analyze the system.

Since $X(0)$ and $Y(0)$ denote the initial number of predators and prey, respectively, we must have $X(0) \geq 0$ and $Y(0) \geq 0$. If $X(0) = 0$ and $Y(0) = 0$, then certainly neither population can ever increase or decrease, so $X(0) = 0$ and $Y(0) = 0$ imply that $X(t) = 0$ and $Y(t) = 0$ for all $t \geq 0$. But $X(t) = 0$, $Y(t) = 0$ are simultaneous constant solutions of the Lotka-Volterra equations. Thus one solution of the system occurs when neither species is present, a result that is neither surprising nor interesting.

Now suppose that $X(0) = 0$ and $Y(0) > 0$. Then as before $X(t) = 0$ for all $t \geq 0$, and the equation

$$\frac{dY}{dt} = (p - qX)Y$$

becomes

$$\frac{dY}{dt} = pY,$$

which has solution $Y(t) = Y(0)e^{pt}$. Therefore when there are no predators present initially, the prey population grows without bound. Similarly, if $Y(0) = 0$ and $X(0) > 0$, we find that $X(t) = X(0)e^{-st}$, and therefore when there is no prey present initially the predator population declines toward zero.

Note also that $X(t) = p/q$, $Y(t) = s/r$ are constant solutions of the Lotka-Volterra equations. Thus if $X(0) = p/q$ and $Y(0) = s/r$, then the populations of the two species will remain at these values indefinitely.

From now on we will assume that $X(0) > 0$ and $Y(0) > 0$. We will not attempt to solve the Lotka-Volterra system under these conditions, but instead will analyze it graphically. Note that

$$X < \frac{p}{q} \quad \text{implies} \quad \frac{dY}{dt} > 0, \quad \text{and hence that } Y \text{ is increasing;}$$

$$X > \frac{p}{q} \quad \text{implies} \quad \frac{dY}{dt} < 0, \quad \text{and hence that } Y \text{ is decreasing;}$$

$$Y < \frac{s}{r} \quad \text{implies} \quad \frac{dX}{dt} < 0, \quad \text{and hence that } X \text{ is decreasing;}$$

$$Y > \frac{s}{r} \quad \text{implies} \quad \frac{dX}{dt} > 0, \quad \text{and hence that } X \text{ is increasing.}$$

It follows that for any configuration of p/q, s/r, $X(0)$, and $Y(0)$, we may sketch the graphs of the solutions $X(t)$, $Y(t)$ on the same set of axes. Figure 9–13 shows how this is done for one such configuration: we begin with $p/q < s/r$, $X(0) < p/q$, and $Y(0) > s/r$. Therefore initially X is increasing and Y is increasing. When X becomes greater than p/q, Y begins to decrease; X continues to increase until Y becomes less than s/r; then X begins to decrease while Y continues to decrease until X becomes less than p/q; and so on. From this we see that the sizes of the two populations oscillate about the constant values p/q and s/r.

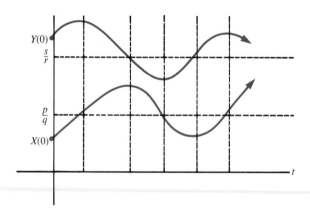

FIGURE 9–13

SUPPLEMENTARY EXERCISES

1. Sketch the graphs of the solutions $X(t)$, $Y(t)$ of the Lotka-Volterra equations on the same set of axes if
   (a) $p/q < s/r$ and $X(0) > p/q$, $Y(0) > s/r$;
   (b) $p/q > s/r$ and $X(0) < p/q$, $Y(0) < s/r$;
   (c) $p/q > s/r$ and $X(0) > p/q$, $Y(0) > s/r$.

*2. It can be shown that there is a time $t = b$ such that
$$X(t + b) = X(t) \quad \text{and} \quad Y(t + b) = Y(t)$$

for all $t \geq 0$. This means that the graphs of $X(t)$ and $Y(t)$ are exactly the same over the interval $[b, 2b]$ as they are over $[0, b]$, the same over $[2b, 3b]$ as they are over $[0, b]$, and so on. In particular, then,
$$X(b) = X(0) \quad \text{and} \quad Y(b) = Y(0).$$

(a) Show that
$$\frac{d}{dt}(\ln Y) = p - qX$$
and then solve this equation for X. Show that
$$\frac{d}{dt}(\ln X) = rY - s$$
and solve this equation for Y.

(b) Use the expressions for X and Y obtained in part (a) to find the average size of both the predator population and the prey population over the interval $[0, b]$. (Don't forget that $X(b) = X(0)$ and $Y(b) = Y(0)$.)

*3. The following system of differential equations may be used to model population size for two competing species X and Y:
$$\frac{dY}{dt} = (p - qX)Y, \quad p > 0, \quad q > 0$$
$$\frac{dX}{dt} = (r - sY)X, \quad r > 0, \quad s > 0.$$

(a) Can you give a rationale for this model?
(b) Analyze this model as we did the predator–prey model. Can one of the species become extinct? Must extinction always occur?

10

Probability

The **probability** of an event is a measure of the likelihood that the event will occur. For instance, a weather forecaster who says "there is a 30 percent chance of rain tomorrow" could just as well have said "the probability is 0.30 that it will rain tomorrow"; both statements express the same estimate of the likelihood of rain tomorrow. Similarly, when asked what the chances are of getting heads if we toss a fair coin once, we might reply "50–50" or "1 chance in 2"; another way of saying the same thing would be to state that the probability of getting heads is 0.50 or $\frac{1}{2}$.

In virtually every sphere of life it is important to consider the likelihood that certain events will occur, so probabilities are used in nearly every field. They are important in particle physics, genetics, quality control, forecasting, marketing, political polling, gambling and gaming, and a host of other areas. In this chapter we introduce the basic notions of probability theory and demonstrate their applications. The first two sections of the chapter are devoted to a general introduction to probabilities, random variables, and probability distributions. The third section discusses two important probability distributions, the binomial and the Poisson. In the fourth section we discuss continuous random variables, probability density functions, and cumulative distribution functions, and we conclude the chapter with an examination of the normal and exponential distributions.

10.1 PROBABILITIES

In this section we begin our study of probability by defining the notions of an **event** associated with an experiment and the **probability** of such an event. We also state the rules of probabilities and show how probabilities are assigned to outcomes.

Suppose we perform the experiment of tossing a fair coin three times and recording whether we get heads or tails on each toss. There are eight distinct ways this experiment can turn out: we can get

$$\text{HHH, HHT, HTH, THH, HTT, THT, TTH, or TTT,}$$

where HTH, for example, stands for "heads on the first toss, tails on the second, and heads on the third," and similarly for the other possibilities. These eight possibilities are called the **outcomes** of the experiment. An **event** for an experiment is a collection of its outcomes. Let us list some events for our coin-tossing experiment and the outcomes that make them up:

Event	Outcomes
Get no heads	TTT
Get 1 head	HTT, THT, TTH
Get 2 or more heads	HHT, HTH, THH, HHH
Get 1 head or 1 tail	HTT, THT, TTH, THH, HTH, HHT
Do not get 1 head	HHH, HHT, HTH, THH, TTT
Get a head on the second toss	HHH, HHT, THH, THT
Get either a head or a tail on the first toss	HHH, HHT, HTH, THH, HTT, THT, TTH, TTT

The descriptions of events are meant to be precise, and thus must be read carefully and taken literally. For instance, with regard to the event "get one head," the phrase "one head" means exactly that; it does not mean "one or more heads," or "at least one head," or anything else except what it says: one head (exactly). Similarly, "get one or more heads" and "get more than one head" are different: the event "get one or more heads" consists of the outcomes HTT, THT, TTH, HHT, HTH, THH, and HHH; the event "get more than one head" consists of the outcomes HHT, HTH, THH, and HHH. Finally, note that there are usually many ways to describe a given event. For example, "get more than one head," "get either two or three heads," "get one or fewer tails," and "do not get more than one tail" all describe the same event for the coin-tossing experiment. (Check this.)

If all the outcomes of the experiment are equally likely to occur, as is the case here, the **probability** of an event is just the ratio of the number of outcomes that make up the event to the total number of outcomes possible. For instance, since the event "get one head" consists of three outcomes and there are eight outcomes possible, there are three chances out of eight of getting one head; therefore the probability of the event "get one head" is $3/8$. If A is any event, we write $P(A)$ for the probability of A. Thus, for the coin-tossing experiment, we have

Event	Number of Outcomes	P(Event)
Get no heads	1	$\frac{1}{8}$
Get 1 head	3	$\frac{3}{8}$
Get 2 or more heads	4	$\frac{4}{8} = \frac{1}{2}$
Get 1 head or 1 tail	6	$\frac{6}{8} = \frac{3}{4}$
Do not get 1 head	5	$\frac{5}{8}$
Get a head on the second toss	4	$\frac{4}{8} = \frac{1}{2}$
Get either a head or a tail on the first toss	8	$\frac{8}{8} = 1$

We summarize what we have done so far:

The Probability of an Event

Suppose an experiment has n distinct outcomes, each of which is equally likely. Let A be an event of the experiment, and let A consist of m distinct outcomes. Then the **probability of A** is

$$P(A) = \frac{m}{n}.$$

Notice that event A is certain to occur if and only if it consists of all possible outcomes of the experiment; but then $m = n$ and

$$P(A) = \frac{m}{n} = \frac{n}{n} = 1.$$

Thus A is certain to occur if and only if $P(A) = 1$. Similarly, A cannot occur if and only if it consists of none of the outcomes of the experiment; but then $m = 0$ and

$$P(A) = \frac{0}{n} = 0.$$

Thus A cannot occur if and only if $P(A) = 0$.

EXAMPLE 1 For the experiment of tossing a fair coin three times, the event "get either a head or a tail on the first toss" is certain to occur and the event "get three heads and two tails" cannot occur. Therefore

$$P(\text{get either a head or a tail on the first toss}) = 1$$

and

$$P(\text{get 3 heads and 2 tails}) = 0. \quad \square$$

Since for any event A we have $P(A) = m/n$, where $0 \leq m \leq n$, it is clear that $P(A)$ is a fraction (or, if divided out, a decimal) between 0 and 1, inclusive. The larger $P(A)$ is, the more likely it is that A will occur; the smaller $P(A)$ is, the less likely it is that A will occur. Therefore $P(A) < P(B)$ if and only if A is less likely to occur than B.

If A and B are two events that have no outcomes in common, then the probability of the event "either A occurs or B occurs" is given by adding the probabilities of A and B. More generally, if $A_1, A_2, \ldots, A_k$ are events no two of which have outcomes in common, then

$$P(A_1 \text{ occurs or } A_2 \text{ occurs or } \ldots \text{ or } A_k \text{ occurs}) = P(A_1) + P(A_2) + \cdots + P(A_k).$$

EXAMPLE 2 In the coin-tossing experiment, the events "get no heads" and "get one head" have no outcomes in common, and hence

$$P(\text{get 1 or fewer heads}) = P(\text{get either no heads or 1 head})$$
$$= P(\text{get no heads}) + P(\text{get 1 head})$$
$$= \frac{1}{8} + \frac{3}{8} = \frac{1}{2}.$$

Similarly,

$$P(\text{get 1 or more heads}) = P(\text{get 1 head}) + P(\text{get 2 heads}) + P(\text{get 3 heads})$$
$$= \frac{3}{8} + \frac{3}{8} + \frac{1}{8} = \frac{7}{8}. \quad \square$$

Suppose that $A_1, A_2, \ldots, A_k$ are events no two of which have outcomes in common. If, in addition, every outcome of the experiment is in one of the events A_i, then the event "either A_1 occurs or A_2 occurs or $\ldots$ or A_k occurs" is certain to occur, and hence

$$1 = P(A_1 \text{ occurs or } A_2 \text{ occurs or } \ldots \text{ or } A_k \text{ occurs})$$
$$= P(A_1) + P(A_2) + \cdots + P(A_k).$$

EXAMPLE 3 In the coin-tossing experiment,

$$P(\text{get no heads}) + P(\text{get 1 head}) + P(\text{get 2 heads}) + P(\text{get 3 heads}) = 1. \quad \square$$

If A is any event, we let $\overline{A}$ denote the event "A does not occur," and call $\overline{A}$ the **complementary event to A**. Since A and $\overline{A}$ are events that have no outcomes in common and since every outcome is in either A or $\overline{A}$, it follows from what we have already done that

$$P(A) + P(\overline{A}) = 1,$$

that is, that

$$P(\overline{A}) = 1 - P(A).$$

PROBABILITY 623

EXAMPLE 4 In the coin-tossing experiment, if A is the event "get three heads," then $\overline{A}$ is the event "do not get three heads"; that is, $\overline{A}$ is the event "get two or fewer heads." Thus

$$P(\text{get 2 or fewer heads}) = 1 - P(\text{get 3 heads}) = 1 - \frac{1}{8} = \frac{7}{8}. \quad \square$$

Let us summarize the basic rules concerning probabilities that we have developed:

The Rules of Probability

If A, B, and $A_1, \ldots, A_k$ are events of an experiment, then

1. $0 \leq P(A) \leq 1$.
2. $P(A) = 1$ if and only if A is certain to occur.
3. $P(A) = 0$ if and only if A cannot occur.
4. $P(A) < P(B)$ if and only if A is less likely to occur than B.
5. If no two of the events $A_1, \ldots, A_k$ have outcomes in common, then

$$P(\text{either } A_1 \text{ occurs or } \ldots \text{ or } A_k \text{ occurs}) = \sum_{i=1}^{n} P(A_i).$$

If, in addition, every outcome is in one of the events A_i, then

$$\sum_{i=1}^{n} P(A_i) = 1.$$

6. If $\overline{A}$ is the complementary event to A, then

$$P(\overline{A}) = 1 - P(A).$$

EXAMPLE 5 A roulette wheel has 38 numbers: 0 and 00, which are colored green, the odd numbers $1, 3, \ldots, 35$, which are colored red, and the even numbers $2, 4, \ldots, 36$, which are colored black. What is the probability that any particular number will come up on one spin of the wheel?

Since there are 38 possible numerical outcomes, all equally likely, it follows that

$$P(\text{any particular number comes up}) = \frac{1}{38} \approx 0.026.$$

What is the probability that a red number will come up on one spin of the wheel? Since the event "a red number comes up" consists of the 18 outcomes $1, 3, \ldots, 35$,

$$P(\text{a red number comes up}) = \frac{18}{38} = \frac{9}{19} \approx 0.474.$$

Similarly,

$$P(\text{a black number comes up}) = \frac{9}{19} \approx 0.474,$$

and

$$P(\text{a green number comes up}) = \frac{2}{38} = \frac{1}{19} \approx 0.053.$$

(Why?) Note that

$$P(\text{any particular number comes up}) < P(\text{a red number comes up}).$$

Hence on any spin of the wheel it is less likely that a given number will come up than that a red one will.

Since it is certain that either a red number, a black number, or a green number will come up,

$$P(\text{either a red, a black, or a green number comes up}) = 1.$$

Also, there are no numbers that are both red and black, so

$$P(\text{a number that is both red and black comes up}) = 0.$$

The events "a black number comes up" and "a green number comes up" have no outcomes in common, so

$P(\text{either a black number or a green number comes up})$

$= P(\text{a black number comes up}) + P(\text{a green number comes up})$

$= \dfrac{9}{19} + \dfrac{1}{19} = \dfrac{10}{19}.$

Alternatively, the event "either a black number or a green number comes up" is the complementary event to "a red number comes up," and therefore

$P(\text{either a black number or a green number comes up})$

$= 1 - P(\text{a red number comes up})$

$= 1 - \dfrac{9}{19} = \dfrac{10}{19}.$ $\square$

The probabilities we have encountered thus far have been inherent in the situations they describe, in the sense that we can calculate them simply by counting outcomes. Sometimes, however, we must resort to assigning probabilities to outcomes on the basis of past experience, or even on the basis of educated guesses or intuition. When we assign probabilities to outcomes by these means, we must make sure that the rules of probability are followed: the probability of each outcome must be a number between 0 and 1, inclusive, and the sum of the probabilities for all the outcomes must be 1. We illustrate with examples.

EXAMPLE 6 A gas station wishes to estimate the probability that it will sell 200 or more gallons of gas during a day. Its records for the past two months show the following:

Gallons Sold per Day	Number of Days This Occurred
Less than 100	12
100 through 199	22
200 through 299	18
300 or more	8
	60

Since the station sold less than 100 gallons $12/60$ of the time, we may assign the outcome "sell less than 100 gallons" a probability of $12/60$ based on past experience. If we proceed similarly for the rest of the outcomes, we have

Gallons Sold per Day	Probability
Less than 100	$12/60$
100 through 199	$22/60$
200 through 299	$18/60$
300 or more	$8/60$
	$60/60 = 1$

Note that the probability for each outcome is a number between 0 and 1 and that the sum of the probabilities for all the outcomes is 1. Now we can see that the probability of the event "sell 200 or more gallons" is equal to the sum of the probabilities of the outcomes "sell 200 through 299" and "sell 300 or more," and hence

$$P(\text{sell 200 or more gallons}) = \frac{18}{60} + \frac{8}{60} = \frac{26}{60} = \frac{13}{30}. \quad \square$$

EXAMPLE 7 A sales manager is asked to give a quick estimate of the probabilities that a new product will sell 100,000 units, 200,000 units, or 300,000 units during its first year on the market. The sales manager's educated guess is that sales of 200,000 units are twice as likely as sales of 100,000 units and three times as likely as sales of 300,000 units. If

$$P(\text{sales will be 100,000 units}) = x,$$

$$P(\text{sales will be 200,000 units}) = y,$$

and

$$P(\text{sales will be 300,000 units}) = z,$$

then the manager's guess implies that

$$y = 2x \quad \text{and} \quad y = 3z.$$

But since the probabilities must add up to 1, it is also the case that

$$x + y + z = 1.$$

Thus

$$x = \frac{y}{2}, \quad z = \frac{y}{3}, \quad \text{and} \quad \frac{y}{2} + y + \frac{y}{3} = 1,$$

which yields $x = 3/11$, $y = 6/11$, and $z = 2/11$. $\square$

10.1 EXERCISES

1. An experiment consists of tossing a fair coin twice.
   (a) List all outcomes of this experiment.
   (b) List the outcomes that make up each of the following events of this experiment:

   get no heads;

   get a head on the second toss;

   get a tail on the first toss;

   get at least one head;

   get at most 1 head;

   get at least 1 tail;

   get fewer than 3 heads;

   get either a head on the first toss or a tail on the second toss;

   get more than 2 heads.

   (c) Find the probability of each of the events of part (b).

2. Repeat Exercise 1 if the experiment consists of tossing the coin four times.

3. A **die** is a cube whose faces are numbered 1 through 6. Suppose an experiment consists of rolling a die and recording the number that comes up. (By "the number that comes up" we mean the number that is on the top face of the die when it has stopped rolling.)
   (a) List all outcomes of this experiment.
   (b) List the outcomes that make up each of the following events of this experiment:

   the number 2 comes up;

   an odd number comes up;

   either an odd number or an even number comes up;

   the number 3 does not come up;

   the number 7 comes up;

   the number 7 does not come up;

   either the number 3 or the number 4 comes up;

PROBABILITY

neither the number 3 nor the number 4 comes up;

a number less than or equal to 4 comes up;

a number greater than 4 comes up.

    (c) Find the probability of each of the events of part (b).

4. An ordinary deck of cards contains two red suits, hearts and diamonds, and two black suits, spades and clubs. Each suit contains 13 cards: those numbered 2 through 10 plus the four face cards jack, queen, king, and ace. If a card is selected at random from an ordinary deck of cards, find

    (a) P(it is an ace);
    (b) P(it is a red king);
    (c) P(it is a 5);
    (d) P(it is a diamond);
    (e) P(it is black);
    (f) P(it is the 3 of spades);
    (g) P(it is either a heart or a spade);
    (h) P(it is either a 2 or a jack);
    (i) P(it is not a heart);
    (j) P(it is not the queen of hearts);
    (k) P(it is not a king);
    (l) P(it is a face card);
    (m) P(it is either a face card or a red 7);
    (n) P(it is numbered 4 through 8).

5. A pair of dice consists of two dies of the form described in Exercise 3. Suppose a pair of dice is rolled and the numbers that come up on each die are added. Find

    (a) P(the numbers total n), where $n = 2, 3, \ldots, 12$.
    (b) P(the numbers total 7 or 11);
    (c) P(the numbers total 2, 3, or 12);
    (d) P(the numbers do not total 2, 3, 7, 11, or 12).

6. A drum contains slips of paper, each of which has written on it one of the 3-digit numbers

$$000, 001, \ldots, 999.$$

Suppose you bet on one of these numbers, then reach into the drum and select a slip of paper at random. Find the probability that the number on the slip will match the number you bet on in

    (a) all 3 of its digits;   (b) exactly 2 of its digits;
    (c) exactly 1 of its digits;   (d) none of its digits.

Management

7. Experts estimate that 68% of new businesses fail within five years. What is the probability that a new business will fail within five years?

8. It is estimated that 7 out of 12 middle managers will be fired at least once in their careers. What is the probability that a middle manager will be fired at least once?

Records show that the number of hot dogs sold by a vendor at a baseball stadium were as follows:

Dozens of Hot Dogs Sold:	1	2	3	4	5	6
Number of Days This Occurred:	2	5	15	13	2	1

Exercises 9 through 12 refer to this situation.

9. Find P(the vendor will sell 1 dozen hot dogs on a given day).
10. Find P(the vendor will sell fewer than 4 dozen hot dogs on a given day).
11. Find P(the vendor will sell more than 4 dozen hot dogs on a given day).
12. Find P(the vendor will sell either 4 or 5 dozen hot dogs on a given day).
13. Records at an auto dealership show the following:

Number of Cars Sold per Day	Number of Days This Occurred
0	15
1	40
2	20
3	8
4	5
5	2

    (a) For $n = 0, \ldots, 5$, assign a value to P(sell n cars during a day).
    (b) Find P(sell at least 1 car during a day).
    (c) Find P(sell more than 2 cars during a day).
    (d) Find P(sell 1, 2, or 3 cars during a day).

14. A research laboratory is about to begin testing of a newly designed computer chip. The director of the laboratory estimates that the new chip will be either 50% faster, 25% faster, or no faster than the standard chip, and that the first of these possibilities is one half as likely as the second and four times as likely as the third. Assign values to

    P(the new chip will be 50% faster than the standard one),
    P(the new chip will be 25% faster than the standard one),

    and

    P(the new chip will be no faster than the standard one).

15. A firm's sales department feels that there are 2 chances out of 10 that next year's sales will increase, 4 chances out of 10 that they will decrease, and 4 chances out of 10 that they will stay the same. Find the probabilities of the events "sales will increase," "sales will decrease," and "sales will stay the same."

Alpha Company is about to introduce a new product. Sanchez, Bernstein, and Johnson are discussing whether the product will be successful. Exercises 16 through 20 refer to this situation.

16. Sanchez is certain the product will be successful. What probability should Sanchez assign to the event "the product will be successful"?
17. Johnson is certain that the product will fail. What probability should Johnson assign to the event "the product will be successful"?
18. Bernstein thinks there is a 40% chance that the product will be successful. What probability should Bernstein assign to the event "the product will be successful"?

PROBABILITY

19. Suppose Chang enters the discussion and states that there are four chances in nine that the product will be successful. What probability should Chang assign to the event "the product will be successful"?

20. Suppose Avanian joins the discussion and states that the chances for the product's success are better than Bernstein thinks but not as good as Chang thinks. What probabilities could Avanian assign to the event "the product will be successful"?

Life Science

21. Medical research shows that 80% of all patients diagnosed with liver cancer will survive for at least three years. Find the probability that a patient diagnosed with liver cancer will survive for at least three years.

22. In a drug-testing experiment, 632 out of 948 patients treated with the drug experienced improvement in their condition. Find the probability that a patient treated with the drug will experience improvement.

23. If it is true that three out of four doctors surveyed recommend Rylenol for pain relief, what is the probability that a doctor selected at random will recommend Rylenol?

Of 200 fledgling robins hatched during a study of bird mortality, 35 lived for fewer than 2 weeks, 44 lived at least 2 weeks but fewer than 4 weeks, 59 lived at least 4 weeks but fewer than 8 weeks, 34 lived at least 8 weeks but fewer than 12 weeks, and the rest lived for 12 or more weeks. Exercises 24 through 26 refer to this study.

24. Find the probability that a fledgling robin will live for at least 4 weeks.

25. Find the probability that a fledgling robin will not live for at least 12 weeks.

26. Find the probability that a fledgling robin will live for fewer than 2 or for 12 or more weeks.

27. One thousand trout were used to stock a river. The following table shows the number still alive at various times after the stocking took place:

Months after Stocking	Number of Trout Still Alive
0	1000
4	450
8	220
12	130
16	70
20	40
24	24
28	14
32	6
36	0

For a single trout placed in the river, estimate
(a) P(it will die in the first 4 months);
(b) P(it will die in the first year);
(c) P(it will not die in the first year);
(d) P(it will live at least 2 years);

28. A team of medical researchers is planning a large-scale test of a new drug. Based on their preliminary studies of the drug, they expect that patients will either be cured, experience some relief, experience no change, or become worse. If the number who will experience some relief is twice the number who will be cured, the number who will experience no change is two-thirds the number who will be cured, and the number who will get worse is one-half the number who will experience no change, assign values to

P(a patient will be cured),
P(a patient will experience some relief),
P(a patient will experience no change),

and

P(a patient will become worse).

Social Science

29. If 44.5% of the adults in a state have a high-school diploma, find the probability that a person chosen at random from the state will have a high-school diploma.
30. If 37.4% of the persons receiving welfare will be off the welfare rolls within two years, what is the probability that a person now on the welfare rolls will be off them within two years?

A poll of voters surveyed before an election revealed their party affiliations to be as follows:

Party:	Democrat	Republican	Conservative	Liberal	None
Number:	247	268	84	44	357

Exercises 31 through 34 refer to this poll.

31. Find P(a voter is a Democrat).
32. Find P(a voter is not a Republican).
33. Find P(a voter is neither a Democrat nor a Republican).
34. Find P(a voter is either a Liberal or a Conservative).

10.2 RANDOM VARIABLES

In this section we introduce the concept of a **random variable,** show how to produce the **probability distribution** of a random variable, and discuss the **expected value** and **variance** of a random variable.

A **random variable** is a function that assigns a number to each outcome of an experiment. For instance, consider again the experiment in which we toss a fair coin three times. For each outcome of this experiment let us set

X(outcome) = Number of heads in the outcome.

Thus, X(HHH) = 3, X(HTH) = 2, and so on. The function X is a random variable.

It is customary to think of a random variable as "taking on" the values it assigns to the outcomes of an experiment. We could have defined the random variable X of the previous paragraph by the statement

Let X = Number of heads obtained.

Then, since the number of heads we obtain every time we perform the experiment must be either 0, 1, 2, or 3, we think of the random variable X as taking on these values. A complete list of the outcomes that correspond to each value of X is as follows:

Outcomes	X = Number of Heads
TTT	0
HTT, THT, TTH	1
HHT, HTH, THH	2
HHH	3

The collection of outcomes that correspond to a particular value that X takes on forms an event, and hence it has a probability: for instance, $X = 3$ corresponds to the event HHH, and therefore we may write

$$P(X = 3) = P(\text{HHH}) = \frac{1}{8}.$$

A complete list of the values that a random variable can take on and their probabilities is called a **probability distribution** for the random variable. The probability distribution for the experiment of tossing a fair coin three times with X = number of heads is as follows:

X	$P(X)$
0	$\frac{1}{8}$
1	$\frac{3}{8}$
2	$\frac{3}{8}$
3	$\frac{1}{8}$
	1

Notice that the sum of the probabilities $P(X)$ is 1: this must always be the case for a probability distribution, because every outcome must correspond to one and only one value of the random variable.

Probability distributions can be given in pictorial form by using **probability histograms,** in which $P(X = x)$ is the height of a rectangle over the value $X = x$. Figure 10–1 shows the probability histogram for the distribution of the preceding paragraph.

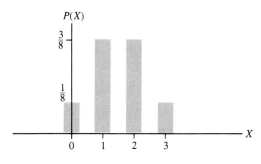

FIGURE 10–1

EXAMPLE 1 Consider again the roulette wheel of Example 5, Section 10.1. If you bet $1 on a particular number and that number comes up, you win $36; if the number you bet on does not come up, you lose your dollar. (Also, as in most gambling games except raffles and lotteries, when you win, the amount which you bet is returned to you.) Define a random variable X by

$$X = \text{Amount won if \$1 is bet on a number.}$$

Then X can take on the two values 36 and -1, representing a win of $36 and a loss of $1, respectively. Since

$$P(X = 36) = P(\text{the particular number bet on comes up}) = \frac{1}{38}$$

and

$$P(X = -1) = P(\text{the particular number bet on does not come up})$$
$$= 1 - P(\text{the particular number bet on does come up})$$
$$= 1 - \frac{1}{38} = \frac{37}{38},$$

the probability distribution for X is

X	$P(X)$
36	$1/38$
-1	$37/38$

The probability histogram for X is shown in Figure 10–2.

If you bet $1 on either red or black in roulette and the color you bet on comes up, you win $1; if it does not come up, you lose your $1. Let

$$Y = \text{Amount won if \$1 is bet on red.}$$

Then the probability distribution for the random variable Y is

Y	$P(Y)$
1	$9/19$
-1	$10/19$

(Check this.) Figure 10–3 shows the histogram for Y. □

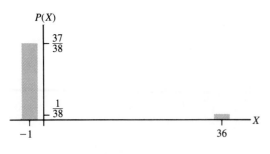

FIGURE 10–2

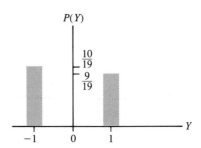

FIGURE 10–3

Let us return to the experiment of tossing a fair coin three times, with

$$X = \text{Number of heads obtained.}$$

If we repeated this experiment over and over, how many heads would we average per toss? In other words, what is the average value of X in the long run? The average value of X is called its **expected value,** and it is found as follows:

Expected Value

Let X be a random variable that takes on the values $x_1, \ldots, x_n$ with

$$P(X = x_1) = p_1, \ldots, P(X = x_n) = p_n.$$

Then the **expected value of X** (also called the **mean** of X) is the sum

$$E(X) = \sum_{i=1}^{n} x_i p_i.$$

The expected value $E(X)$ is the average value of X in the long run.

Thus, to find the expected value of X, we multiply each of its values by the probability that it takes on that value and we add the results. Hence for the experiment of tossing a fair coin three times we have

X	$P(X)$	$X\, P(X)$
0	$1/8$	$0(1/8) = 0$
1	$3/8$	$1(3/8) = 3/8$
2	$3/8$	$2(3/8) = 6/8$
3	$1/8$	$3(1/8) = 3/8$
	1	$E(X) = 12/8 = 1.5$

Therefore in the long run we will average 1.5 heads for every time we perform this experiment.

EXAMPLE 2 For the roulette wheel and the random variables X and Y of Example 1, find the expected values of X and Y. We have

X	$P(X)$	$X\, P(X)$
36	$1/38$	$36/38$
-1	$37/38$	$-37/38$
		$E(X) = -1/38$

and

Y	$P(Y)$	$Y\, P(Y)$
1	$9/19$	$9/19$
-1	$10/19$	$-10/19$
		$E(Y) = -1/19$

Since $E(X) = -1/38 \approx -0.026$, in the long run we will average a loss of approximately $0.026 for every $1 we bet on a number at roulette. Similarly, since $E(Y) = -1/19 \approx -0.053$, in the long run we will average a loss of approximately $0.053 for every $1 we bet on the color red. A game is said to be a **fair game** if the expected value of its winnings is 0, for then in the long run both players will break even. Roulette is therefore not a fair game. ☐

Expected values are often used in decision making. The idea is that if there are two or more possible responses to a situation, the one with the best expected value is the one that should be chosen. We illustrate with an example.

EXAMPLE 3   A film company must choose between two projects, movie A and movie B. Movie A will cost $30 million to make; when it is released, it may become a blockbuster, a modest success, or a flop, and the company estimates that the probabilities of these events are 0.10, 0.30, and 0.60, respectively. (Notice that these probabilities add up to 1; this must be so. Why?) If A is a blockbuster, it will earn $150 million; if it is a modest success, it will earn $50 million; and if it is a flop, it will earn $5 million. Movie B will cost $2 million to make and will earn $18 million if very successful, $12 million if moderately successful, and $3 million if a failure, with the probabilities of these events estimated to be 0.10, 0.80, and 0.10, respectively. The firm has decided to choose the project that has the larger expected profit. Let

$$X = \text{Profit on movie A} \quad \text{and} \quad Y = \text{Profit on movie B},$$

X and Y in millions of dollars. Then

X	$P(X)$	$X\,P(X)$
$150 - 30 = 120$	0.10	12.0
$50 - 30 = 20$	0.30	6.0
$5 - 30 = -25$	0.60	-15.0
		$E(X) = 3.0$

and

Y	$P(Y)$	$Y\,P(Y)$
$18 - 2 = 16$	0.10	1.6
$12 - 2 = 10$	0.80	8.0
$3 - 2 = 1$	0.10	0.1
		$E(Y) = 9.7$

Therefore the film company should choose project B. ☐

The expected value of a random variable tells us its average value. The **variance** and **standard deviation** of a random variable tell us how its values vary from that average.

PROBABILITY

> **Variance and Standard Deviation**
>
> Let X be a random variable that takes on the values $x_1, \ldots, x_n$ with
>
> $$P(X = x_1) = p_1, \ldots, P(X = x_n) = p_n.$$
>
> Then the **variance of X** is the sum
>
> $$V(X) = \sum_{i=1}^{n} (X_i - E(X))^2 p_i,$$
>
> and its **standard deviation** is the square root of its variance.

The variance and standard deviation both measure the extent to which the values of the random variable X deviate from its expected value $E(X)$, the variance in the units of X squared and the standard deviation in the units of X.

EXAMPLE 4   Consider the experiment of tossing a fair coin three times and let

$$X = \text{Number of heads obtained}.$$

Let us find the variance of X. We have previously found that $E(X) = 3/2$. Therefore we have

X	$P(X)$	$X - E(X)$	$(X - E(X))^2$	$(X - E(X))^2 P(X)$
0	$1/8$	$0 - 3/2 = -3/2$	$9/4$	$9/32$
1	$3/8$	$1 - 3/2 = -1/2$	$1/4$	$3/32$
2	$3/8$	$2 - 3/2 = 1/2$	$1/4$	$3/32$
3	$1/8$	$3 - 3/2 = 3/2$	$9/4$	$9/32$

$$V(X) = 24/32 = 3/4$$

Thus the variance of X is $3/4$, and its standard deviation is $\sqrt{3/4} \approx 0.866$. ☐

■ *Computer manual: Programs HISTGRAM, EXPCTVAL.*

10.2 EXERCISES

In Exercises 1 through 6, list the values that the random variable can take on, write the outcomes of the experiment that correspond to each value of the random variable, find the probability distribution for the random variable and draw its probability histogram, find and interpret its expected value, and find its variance.

1. Experiment: toss a fair coin twice.
   Random variable: $X = $ number of heads.

2. Experiment: toss a fair coin four times.
   Random variable: $X = $ number of heads.

3. Experiment: roll a single die.
   Random variable: X = number that comes up.
4. Experiment: draw a card at random from an ordinary deck.
   Random variable: X = number of hearts drawn.
5. Experiment: draw a card at random from an ordinary deck.
   Random variable: X = card value.
   (The value of a numbered card is its number. Face cards are valued as follows: ace = 1, jack = 11, queen = 12, king = 13.)
6. Experiment: roll a pair of dice.
   Random variable: X = total of the numbers that come up.
7. In Europe, roulette wheels have the numbers 0, 1, . . . , 36. (There is no 00.)
   (a) If you bet $1 on a single number and the number comes up, you win $35; otherwise you lose your $1. Find the expected value of your winnings if you bet this way.
   (b) If you bet $1 on red and a red number (1, 3, . . . , 35) comes up, you win $1; otherwise you lose your $1. Find the expected value of your winnings if you bet on red.
   (c) Is European roulette a fair game? If not, is it more or less fair to the individual player than the American roulette analyzed in Examples 1 and 2 of this section?
8. You draw a card from an ordinary deck; if the card is a heart you win, if not, you lose.
   (a) Suppose that if you win, you win $2, but if you lose, you lose $1. Is this a fair game?
   (b) Suppose that if you win, you win $4, but if you lose, you lose $1. Is this a fair game?
   (c) Can you make this game fair by adjusting the payoffs for winning and losing? How?
9. You hold one raffle ticket, which cost you $1. If your ticket is drawn in the raffle, you will win $5000. If 10,000 tickets were sold, find the expected value of your winnings. What should the prize be if the raffle is to be fair?
10. A vase has six red and four white marbles in it. For a $1 bet, you are allowed to reach in and draw one marble from the vase. If it is white, you win, but if it is red, you lose your $1. What amount must you be paid for winning if this is to be a fair game?
11. In the game of numbers, a $1 bet allows you to pick any three-digit number from 000 through 999. If the number you have picked comes up, you win $600. Is this a fair game? What payoff for winning would make it a fair game?
12. You buy a lottery ticket for $1. If the six-digit number on your ticket matches the number drawn by the state, you win $50,000. What are your expected winnings? Is this lottery a fair game?
13. You buy a lottery ticket for $2. The ticket has a five-digit number on it, the first three digits of which are yellow and the last two digits of which are blue. If the entire five-digit number on your ticket matches that drawn by the state, you win $10,000; if only the yellow digits match, you win $1000; if only the blue digits match, you win $100. Find your expected winnings.
14. A survey showed the number of defective light bulbs per carton sampled to be as follows:

PROBABILITY

Defective Bulbs in Carton:	0	1	2	3	4	5
Number of Cartons:	42	36	14	3	1	1

Find the expected number of defective bulbs per carton.

15. You are considering three job offers, A, B, and C. You estimate the salaries each might pay five years from now, and the probabilities that you would be earning those salaries, to be as follows:

Job	Possible Salaries	Probabilities
A	$40,000	0.40
	$50,000	0.60
B	$30,000	0.20
	$50,000	0.50
	$80,000	0.30
C	$20,000	0.35
	$40,000	0.25
	$60,000	0.40

Based on your expected earnings five years from now, which job should you take?

Management

16. A hot dog vendor sells 4 dozen hot dogs per day 40% of the time, 5 dozen per day 30% of the time, 6 dozen per day 20% of the time, and 7 dozen per day 10% of the time. Find the expected number of hot dogs sold per day.

17. A firm can fund one of two new projects: project A will cost $10 million and produce revenues of either $20 million, $10 million, or $4 million, with probabilities of 0.50, 0.20, and 0.30, respectively. Project B will cost $6 million and will produce revenues of $15 million, $7 million, or $3 million, with probabilities of 0.35, 0.55, and 0.10, respectively. On the basis of expected profit, which project should the firm fund?

18. The history of a certain corporation's annual return on investment over the past 30 years has been as follows:

Return on Investment (%)	Number of Years This Occurred
8.2	6
8.8	7
9.6	9
10.3	5
11.0	3

Find the expected value of the corporation's annual return on investment. If the corporation's total investment next year will amount to $250 million dollars, estimate its total return.

19. A restaurant's daily records show the following:

% of capacity	70%	75%	80%	85%	90%	95%
Probability	0.1	0.2	0.2	0.25	0.2	0.05
Revenue ($1000s)	2.0	2.5	3.2	3.8	4.2	4.4

Find the restaurant's expected daily revenue.

20. Ace Company knows that 20% of its salespersons earn commissions of $10,000 per year, 45% earn commissions of $15,000 per year, 30% earn commissions of $20,000 per year, and 5% earn commissions of $30,000 per year. Ace has 200 salespersons, and the company is attempting to estimate the amount that should be budgeted next year for payment of commissions. What should the estimate be?

Life Science

21. Analysis of water samples from 25 wells in a town shows the following concentrations of pollutants, in parts per million:

Pollutants:	2	5	6	8	10	11	12
No. of Wells:	8	4	7	2	2	1	1

    Find the expected concentration of pollutants in the town's wells.

22. The probabilities for the size of a dog's litter are as follows:

Number of pups:	1	2	3	4	5	6
Probability:	0.01	0.11	0.13	0.44	0.22	0.09

    Find the expected size of the litter.

*23. Suppose the probability that two blue-eyed parents will have a blue-eyed child is $\frac{1}{2}$. If two such parents have two children, what is the expected number who will have blue eyes?

*24. Repeat Exercise 23 if the parents have three children.

According to information on page 22 of the May 1986 issue of *Scientific American*, if

$$X = \text{Number of males in a family who will marry,}$$

the probability distribution for X is as follows:

X:	0	1	2	3	4	5	6 or more
P(X):	0.317	0.364	0.209	0.080	0.023	0.005	0.002

Exercises 25 through 31 refer to this random variable.

25. Find P(a family will have at least one male who will marry).
26. Find P(a family will have at most one male who will marry).
27. Find P(a family will have 5 or more males who will marry).
28. Find and interpret $P(X = 3)$.
29. Find and interpret $P(X < 3)$.
30. Find and interpret $P(X > 3)$.
31. Replace the words "6 or more" by just "6," and then find and interpret $E(X)$.

Social Science

32. The length of time it took 21 rats to make their way through a maze on their first and second tries was as follows:

    First Try

Time (seconds):	30	45	60	75	90	105
Number of Rats:	1	5	6	7	1	1

	Second Try					
Time (seconds):	30	45	60	75	90	105
Number of Rats:	4	10	4	2	1	0

What is the expected time for a rat to get through the maze on its first attempt? on its second attempt?

33. Candidates Pusey and Lanimus are running for the same office. Each has just two choices: attack the other or discuss the issues. If Pusey attacks, the probability that Lanimus will also attack is 0.60; if Pusey discusses the issues, the probability that Lanimus will attack is 0.45. If Pusey and Lanimus end up attacking each other, Pusey will receive 48% of the vote; if Pusey attacks and Lanimus discusses the issues, Pusey will get 49% of the vote; if Pusey discusses the issues and Lanimus attacks, Pusey will get 60% of the vote; and if Pusey and Lanimus both discuss the issues, Pusey will get 42% of the vote. Based on expected percentage of the vote, which strategy should Pusey select?

10.3 THE BINOMIAL AND POISSON DISTRIBUTIONS

There are several probability distributions that are used again and again as models for different experiments in different contexts. In this section we examine two of the most important and useful probability distributions, the **binomial distribution** and the **Poisson distribution**, and illustrate their usage.

The Binomial Distribution

The experiment of tossing a fair coin three times, which we have used as a running example in this chapter, is characteristic of an entire class of experiments known as **Bernoulli processes**. A Bernoulli process has the following properties:

- It consists of repeated trials of the same action.
- Each trial has only two possible outcomes, which we can call "success" and "failure."
- No trial is affected by the outcome of any other trial, and the probability of a success on any single trial is always the same.

Thus for the coin-tossing experiment, the trials are the tosses of the coin; each toss has only two possible outcomes, heads or tails, which we can call "success" and "failure," respectively; no toss is affected by the outcome of any other toss; and the probability of a success on any toss is the probability of getting heads on any toss, which is always equal to $\frac{1}{2}$. Thus the coin-tossing experiment is indeed a Bernoulli process.

Bernoulli processes can be modeled by the **binomial probability distribution.** Before we can describe this distribution, we need to define the concept of the **factorial** of a nonnegative integer.

> **Factorials**
>
> Let n be a nonnegative integer. Define n **factorial,** denoted by $n!$, as follows:
> $$0! = 1, \quad 1! = 1,$$
> and if $n \geq 2$,
> $$n! = n(n-1) \cdots 2 \cdot 1.$$

EXAMPLE 1 By definition, $0! = 1$ and $1! = 1$, and then

$$\begin{aligned}
2! &= 2 \cdot 1 = 2, \\
3! &= 3 \cdot 2 \cdot 1 = 6, \\
4! &= 4 \cdot 3 \cdot 2 \cdot 1 = 24, \\
5! &= 5 \cdot 4 \cdot 3 \cdot 2 \cdot 1 = 120, \\
6! &= 6 \cdot 5 \cdot 4 \cdot 3 \cdot 2 \cdot 1 = 720,
\end{aligned}$$

and so on. Also,

$$\frac{10!}{10!0!} = \frac{10 \cdot 9 \cdot 8 \cdot 7 \cdot 6 \cdot 5 \cdot 4 \cdot 3 \cdot 2 \cdot 1}{10 \cdot 9 \cdot 8 \cdot 7 \cdot 6 \cdot 5 \cdot 4 \cdot 3 \cdot 2 \cdot 1 \cdot 1} = 1,$$

$$\begin{aligned}
\frac{10!}{6!4!} &= \frac{10 \cdot 9 \cdot 8 \cdot 7 \cdot 6 \cdot 5 \cdot 4 \cdot 3 \cdot 2 \cdot 1}{6 \cdot 5 \cdot 4 \cdot 3 \cdot 2 \cdot 1 \cdot 4 \cdot 3 \cdot 2 \cdot 1} \\
&= \frac{10 \cdot 9 \cdot 8 \cdot 7}{4 \cdot 3 \cdot 2 \cdot 1} \\
&= \frac{10 \cdot 3 \cdot 3 \cdot 4 \cdot 2 \cdot 7}{4 \cdot 3 \cdot 2 \cdot 1} \\
&= 10 \cdot 3 \cdot 7 \\
&= 210,
\end{aligned}$$

and

$$\begin{aligned}
\frac{10!}{3!(10-3)!} &= \frac{10 \cdot 9 \cdot 8 \cdot 7 \cdot 6 \cdot 5 \cdot 4 \cdot 3 \cdot 2 \cdot 1}{3 \cdot 2 \cdot 1 \cdot 7 \cdot 6 \cdot 5 \cdot 4 \cdot 3 \cdot 2 \cdot 1} \\
&= \frac{10 \cdot 9 \cdot 8}{3 \cdot 2 \cdot 1} \\
&= \frac{5 \cdot 2 \cdot 3 \cdot 3 \cdot 8}{3 \cdot 2 \cdot 1} \\
&= 5 \cdot 3 \cdot 8 \\
&= 120. \quad \square
\end{aligned}$$

Now we return to our discussion of Bernoulli processes. Since any Bernoulli process consists of some number n of trials, each of which has two possible outcomes, "success" and "failure," we can define a random variable X for the process by setting

$$X = \text{Number of successes in the } n \text{ trials.}$$

Then X can take on the values $0, 1, 2, \ldots, n$, and the probability of exactly k successes in the n trials is $P(X = k)$. The probability distribution for X is called the **binomial distribution.** We present it without proof.

The Binomial Distribution

Given a Bernoulli process consisting of n trials, with

$$P(\text{success on any trial}) = p,$$

define a random variable X by

$$X = \text{Number of successes in the } n \text{ trials.}$$

Then for $k = 0, 1, \ldots, n$,

$$P(X = k) = \frac{n!}{k!(n-k)!} p^k (1-p)^{n-k}.$$

Also

$$E(X) = np \quad \text{and} \quad V(X) = np(1-p).$$

EXAMPLE 2 Consider one more time the experiment of tossing a fair coin three times. This is a Bernoulli process with $n = 3$. If we call "heads" a "success," then

$$P(\text{success on any trial}) = P(\text{heads on any trial}) = \frac{1}{2} = p,$$

and

$$P(0 \text{ heads in 3 tosses}) = P(X = 0) = \frac{3!}{0!(3-0)!} \left(\frac{1}{2}\right)^0 \left(1 - \frac{1}{2}\right)^3$$

$$= 1 \left(\frac{1}{2}\right)^0 \left(\frac{1}{2}\right)^3 = \frac{1}{8},$$

$$P(1 \text{ head in 3 tosses}) = P(X = 1) = \frac{3!}{1!(3-1)!} \left(\frac{1}{2}\right)^1 \left(1 - \frac{1}{2}\right)^2$$

$$= 3 \left(\frac{1}{2}\right) \left(\frac{1}{2}\right)^2 = \frac{3}{8},$$

$$P(2 \text{ heads in 3 tosses}) = P(X = 2) = \frac{3!}{2!(3-2)!} \left(\frac{1}{2}\right)^2 \left(1 - \frac{1}{2}\right)^1$$

$$= 3 \left(\frac{1}{2}\right)^2 \left(\frac{1}{2}\right) = \frac{3}{8},$$

$$P(3 \text{ heads in 3 tosses}) = P(X = 3) = \frac{3!}{3!(3-3)!} \left(\frac{1}{2}\right)^3 \left(1 - \frac{1}{2}\right)^0$$

$$= 1 \left(\frac{1}{2}\right)^3 \left(\frac{1}{2}\right)^0 = \frac{1}{8}.$$

Note that

$$E(X) = np = \frac{3}{2} \quad \text{and} \quad V(X) = np(1-p) = \frac{3}{4}.$$

Figure 10–1 shows the histogram for this experiment. □

EXAMPLE 3 Suppose a multiple-choice test consists of six questions, each of which has five choices for its answer. If a student selects the answers at random, the experiment of taking the test is a Bernoulli process with $n = 6$ trials and

$$P(\text{success on any trial}) = P(\text{select the correct answer}) = \frac{1}{5} = 0.2 = p.$$

Therefore

$$P(0 \text{ correct answers}) = \frac{6!}{0!6!} (0.2)^0 (0.8)^6 = (0.8)^6 \approx 0.2621,$$

$$P(1 \text{ correct answer}) = \frac{6!}{1!5!} (0.2)^1 (0.8)^5 = 6(0.2)(0.8)^5 \approx 0.3932,$$

$$P(2 \text{ correct answers}) = \frac{6!}{2!4!} (0.2)^2 (0.8)^4 = 15(0.2)^2 (0.8)^4 \approx 0.2458,$$

$$P(3 \text{ correct answers}) = \frac{6!}{3!3!} (0.2)^3 (0.8)^3 = 20(0.2)^3 (0.8)^3 \approx 0.0819,$$

$$P(4 \text{ correct answers}) = \frac{6!}{4!2!} (0.2)^4 (0.8)^2 = 15(0.2)^4 (0.8)^2 \approx 0.0154,$$

$$P(5 \text{ correct answers}) = \frac{6!}{5!1!} (0.2)^5 (0.8)^1 = 6(0.2)^5 (0.8) \approx 0.0015,$$

$$P(6 \text{ correct answers}) = \frac{6!}{6!0!} (0.2)^6 (0.8)^0 = (0.2)^6 \approx 0.0001,$$

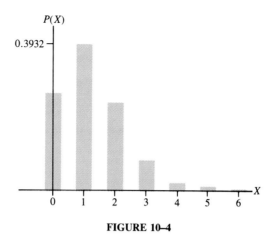

FIGURE 10-4

Figure 10-4 shows the histogram for this distribution. Note that the probabilities of randomly selecting four, five, or six correct answers are very small. Note also that the expected value of X is

$$E(X) = np = 6(0.2) = 1.2;$$

therefore in the long run the strategy of selecting answers at random on five-option multiple choice tests with six questions will produce an average of 1.2 correct answers per test. ☐

The Poisson Distribution

Just as the binomial distribution models Bernoulli processes, so the **Poisson distribution** models **Poisson processes.** A Poisson process is an experiment that involves random arrivals (such as the arrival of cars at a toll booth) and has the following characteristics:

- The probability of an arrival in a small time interval is small, and this probability approaches 0 as the length of the interval approaches 0.
- The probability of two arrivals in a small time interval is virtually 0.
- The number of arrivals in a given time interval is not affected by the number of arrivals in any other nonoverlapping time interval.

For any Poisson process, we may define a random variable X by

$$X = \text{Number of arrivals in a given time interval.}$$

Then X can take on the values 0, 1, 2, . . . , and

$$P(k \text{ arrivals in the given time interval}) = P(X = k).$$

The probability distribution for this random variable is called the **Poisson distribution.** We present it without proof.

The Poisson Distribution

For a Poisson process, let

$$X = \text{Number of arrivals in a given time interval,}$$

and let μ denote the average number of arrivals in the given time interval. Then for $k = 0, 1, 2, \ldots,$

$$P(X = k) = \frac{e^{-\mu}\mu^k}{k!}.$$

Also,

$$E(X) = \mu \quad \text{and} \quad V(X) = \mu.$$

(The symbol μ is the Greek letter mu.)

EXAMPLE 4 Suppose cars arriving at a toll booth constitute a Poisson process. If, on average, three cars per minute arrive at the toll booth, then the given time interval is one minute and $\mu = 3$. Therefore

$$P(0 \text{ cars arrive during a 1-minute period}) = P(X = 0) = \frac{e^{-3}3^0}{0!} \approx 0.0498,$$

$$P(1 \text{ car arrives during a 1-minute period}) = P(X = 1) = \frac{e^{-3}3^1}{1!} \approx 0.1494,$$

$$P(2 \text{ cars arrive during a 1-minute period}) = P(X = 2) = \frac{e^{-3}3^2}{2!} \approx 0.2240,$$

$$P(3 \text{ cars arrive during a 1-minute period}) = P(X = 3) = \frac{e^{-3}3^3}{3!} \approx 0.2240,$$

$$P(4 \text{ cars arrive during a 1-minute period}) = P(X = 4) = \frac{e^{-3}3^4}{4!} \approx 0.1680,$$

and so on. Figure 10–5 shows the histogram for this example. ☐

EXAMPLE 5 Assume that the number of people arriving at a ticket counter forms a Poisson process, with an average of four people arriving per 5-minute period. Find the probability that at least one person will arrive at the counter during any 5-minute period.

Here the given time period is five minutes and $\mu = 4$. Thus

$$P(\text{at least 1 person arrives during a 5-minute period})$$
$$= 1 - P(0 \text{ people arrive during a 5-minute period})$$
$$= 1 - \frac{e^{-4}4^0}{0!}$$
$$\approx 1 - 0.1832$$
$$= 0.8168.$$

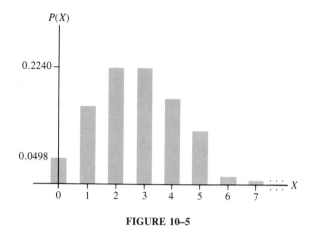

FIGURE 10–5

Now suppose that we wish to find the probability that at least one person will arrive during a 10-minute period. We proceed as we did above, except that we can no longer take $\mu = 4$, because we have changed the basic time period from 5 to 10 minutes. But if an average of four people arrive every 5 minutes, then an average of eight arrive every 10 minutes, and hence

$$P(\text{at least 1 person arrives during a 10-minute period})$$
$$= 1 - P(0 \text{ people arrive during a 10-minute period})$$
$$= 1 - \frac{e^{-8}8^0}{0!}$$
$$\approx 1 - 0.0003$$
$$= 0.9997. \quad \square$$

■ **Computer Manual: Programs BINOMDST, POISSON**

10.3 EXERCISES

The Binomial Distribution

In Exercises 1 through 8, evaluate the expression.

1. $7!$
2. $8!$
3. $9!$
4. $\dfrac{6!}{6!(6-6)!}$
5. $\dfrac{4!}{2!(4-2)!}$
6. $\dfrac{12!}{8!(12-8)!}$
7. $\dfrac{13!}{0!(13-0)!}$
8. $\dfrac{9!}{4!(9-4)!}$

9. A fair coin is tossed four times. Let $X =$ the number of heads obtained. Find the probability distribution for X. Find and interpret $E(X)$ and $V(X)$.

10. A quiz has 10 true-false questions. Let $X =$ number of correct answers obtained by guessing. Find

(a) $P(X = 0)$ (b) $P(X = 1)$ (c) $P(X \geq 1)$ (d) $P(X \geq 3)$ (e) $P(X < 3)$ (f) $P(X \leq 5)$

Find and interpret the expected value of X.

11. You take a multiple-choice quiz having 15 questions, each of which has four choices, and you select your answers by guessing. Find the probability of getting
    (a) exactly three right;
    (b) two or more right;
    (c) fewer than three right;
    (d) four, five, or six right.

    What is the expected number that you will get right?

12. A professional golfer can make a putt of 6 feet or less two-thirds of the time. Find the probability that, of the next three such putts he attempts, he will
    (a) make at least one;
    (b) make all three;
    (c) make none;
    (d) make more than 1.

13. Surveys show that two out of every five people who buy ice cream buy vanilla. Suppose that 10 people buy ice cream at a store. Find the given probability. (You may assume that the trials are independent.)
    (a) $P(4$ of them will buy vanilla$)$
    (b) $P($more than 4 of them will buy vanilla$)$
    (c) $P(8$ or fewer will not buy vanilla$)$

    What is the expected number who will buy vanilla?

Management

14. A machine makes plastic moldings. The probability that any particular molding is defective is 0.10. Find
    (a) $P(0$ defective moldings in the next 20 made$)$;
    (b) $P($at least 3 defective moldings in the next 20 made$)$;
    (c) $P($at least 1 defective molding in the next 50 made$)$;
    (d) $P(48$ or more good moldings in the next 50 made$)$;
    (e) the expected number of defective moldings in the next 100 made.

15. If only 8% of all new businesses succeed, of the next 15 businesses founded, find
    (a) $P(0$ will succeed$)$;
    (b) $P(0$ will fail$)$;
    (c) $P($at least 10 will fail$)$;
    (d) the expected number of failures.

16. A machine stamps out printed circuits. The probability that it will stamp out a defective circuit is 0.05. Find the given probability. (You may assume that the trials are independent.)
    (a) $P(4$ of the next 5 circuits will be good$)$
    (b) $P($at least 14 of the next 15 will be good$)$
    (c) $P(9$ or fewer of the next 10 will be good$)$

17. A computer company gives new employees a training course in computer operation and maintenance. It has been the experience of the company that 8% of those who start the course fail to finish it. Suppose four new employees are hired this week. Find the given probability. (You may assume that the trials are independent.)

(a) P(all of the 4 will finish the course)
(b) P(not all of the 4 will finish the course)
(c) P(at least 3 of the 4 will finish the course)

18. Oil company records show that 13% of wildcat wells strike oil. Find the given probability. (You may assume that the trials are independent.)
    (a) P(2 of the next 3 wildcat wells drilled will strike oil)
    (b) P(at least 1 of the next 5 wildcat wells drilled will strike oil)
    (c) P(1, 2, or 6 of the next 6 wildcat wells drilled will strike oil)

19. (This exercise was suggested by material in Niebel, *Motion and Time Study*, 8th edition, Irwin, 1988.) A machine is said to be ''down'' if it is not working. Suppose a shop has four identical machines and that there is a 5% chance that a machine will be down sometime during the day. Find the given probability. (You may assume that the trials are independent.)
    (a) P(none of the 4 machines will be down during the day)
    (b) P(at least one of the 4 machines will be down during the day)
    (c) P(3 of the 4 machines will be down during the day)
    (d) P(all 4 of the machines will be down during the day)

    What is the expected number of machines that will be down during the day?

Life Science

20. Only one-fifth of all sparrow chicks survive to adulthood. If a nest contains four sparrow chicks, find the probability that
    (a) none will survive to adulthood;   (b) all will survive to adulthood.

    What is the expected number that will survive to adulthood?

21. Suppose that the probability that a child of two brown-eyed parents will have blue eyes is $1/4$. If brown-eyed parents have three children, find the probability that
    (a) none will have blue eyes;   (b) all will have blue eyes;
    (c) at least one will have blue eyes;   (d) at least one will have brown eyes.

22. A new drug improves the condition of 75% of the patients who take it. Suppose 10 patients take the drug. Find the given probability. (You may assume that the trials are independent.)
    (a) P(all 10 will improve)
    (b) P(at least 8 will improve)
    (c) P(fewer than 4 will improve)
    (d) P(5, 6, 7, or 8 will improve)

23. Suppose one-third of all newly hatched robins will live to maturity, and suppose that a nest contains three newly-hatched robins. Find the given probability. (You may assume that the trials are independent.)
    (a) P(none will live to maturity)
    (b) P(at least 2 will live to maturity)
    (c) P(fewer than 3 will live to maturity)
    (d) P(at least 2 will not live to maturity)

24. Suppose the probability that parents who are not red-haired will have a child who has red hair is 0.10. Two parents who are not red-haired have five children. Find the given probability. (You may assume that the trials are independent.)

(a) P(at least 1 of the children will have red hair)
(b) P(none of the children will have red hair)
(c) P(fewer than 2 of the children will have red hair)

What is the expected number who will have red hair?

Sickle-cell anemia is a genetic disease controlled by two genes: the dominant S-gene and the recessive s-gene. A child receives either gene S or gene s from each parent. If the child receives two s-genes (ss), it will develop the disease. If it receives one s and one S (sS), it will not develop the disease, but can pass it on to its offspring and hence is a carrier of the disease. If the child receives two S genes (SS), it cannot develop the disease and cannot be a carrier. A parent can only pass on to a child a gene that the parent has; if the parent is an sS-person, the probability that the child will receive the s-gene is $\frac{1}{2}$. Prospective parents can now be tested to see whether or not they are carriers of the disease and thus be advised as to the chances that their children will develop sickle-cell anemia or be carriers of it. Exercises 25 through 30 refer to this situation.

25. If neither of the parents has sickle-cell anemia, but both are carriers, how many of their four children can be expected to develop the disease?

26. If neither of the parents has sickle-cell anemia, but both are carriers, how many of their four children can be expected to be carriers?

27. If one parent is a carrier and the other does not have sickle-cell anemia and is not a carrier, how many of their four children can be expected to be carriers?

28. If one parent is a carrier and the other has the disease, how many of their four children can be expected to develop the disease?

29. If one parent is a carrier and the other has the disease, how many of their four children can be expected to be carriers?

30. If one parent has the disease and the other does not and is not a carrier, how many of their four children can be expected to be carriers?

Social Science

31. A town contains 2000 voters, 900 of whom are Democrats. If a pollster calls 10 townspeople at random, what is the probability that
    (a) at least two of them will be Democrats;
    (b) at most seven of them will be Democrats.

    What is the expected number of the 10 who will be Democrats?

32. Suppose that 12% of the people in a community are college students and you stop four people at random on the street for a survey. Find the given probability. (You may assume that the trials are independent.)
    (a) P(none of those you stop are college students)
    (b) P(at least 1 is a college student)
    (c) P(not all of them are college students)

The Poisson Distribution

33. Cars arrive at a toll bridge at the rate of one every 20 seconds. Find
    (a) P(no cars arrive during the next 20 seconds);
    (b) P(at least 1 car arrives during the next 20 seconds);

PROBABILITY 649

    (c) $P(3$ or fewer cars arrive during the next 20 seconds);
    (d) $P(5$ or more cars arrive during the next minute).

34. Telephone calls arrive at a switchboard at the rate of three every 4 minutes. Find
    (a) $P(0$ calls arrive during the next 4 minutes);
    (b) $P(4$ or fewer calls arrive during the next 4 minutes);
    (c) $P(5$ calls arrive during the next 10 minutes).

35. Repeat Exercise 33 if the cars arrive at a rate of one every 30 seconds.

36. Repeat Exercise 34 if the calls arrive at a rate of five every 2 minutes.

Management

37. A production line averages three breakdowns per five-day week. If the line must run for 2.5 days to complete a production run, what is the probability that the run will be completed without a breakdown?

38. Repeat Exercise 37 if the line averages six breakdowns per five-day week.

39. A firm that sells computer supplies receives an average of 90 telephone orders per hour. Find the probability that no orders are received during
    (a) any 1-minute period;
    (b) any 2-minute period;
    (c) any 3-minute period.

40. The Poisson distribution can sometimes be used to find the probability that an event will occur a certain number of times in a given area or volume. For instance, suppose a machine produces carpeting that contains 0.2 defects per square yard. Use the Poisson distribution with $\mu = 0.2$ to find
    (a) $P(0$ defects in 1 square yard of carpet);
    (b) $P(1$ defect in 1 square yard of carpet);
    (c) $P(2$ or more defects in 1 square yard of carpet).

10.4 CONTINUOUS RANDOM VARIABLES

A **discrete** random variable is one that takes on discrete numerical values, such as 0, 1, 2, ... All the random variables that we have encountered so far have been discrete. It is possible, however, for a random variable to take on all values in some interval, in which case it is called a **continuous** random variable. For instance, suppose we perform the experiment of selecting at random a number between 0 and 10, inclusive (not necessarily an integer, but *any* number between 0 and 10). If we let X denote the number selected, then X can take on any value in the interval [0, 10], and hence it is a continuous random variable. In this section and the next we study continuous random variables and their probability distributions.

Since a continuous random variable takes on all the values in some interval, it takes on infinitely many values, and hence the probability that it takes on a particular value will be zero. Therefore if X is a continuous random variable, we do not seek to find the probability that X equals some value, since this will be zero, but rather the probability that X falls between two given values, or that it is greater than or less than a given value. To illustrate, suppose that the operating lifetime of a certain type of transistor can be any time between zero and five years, inclusive. If we let

$$X = \text{Operating lifetime of a particular transistor,}$$

X in years, then $0 \leq X \leq 5$, and X is a continuous random variable. Thus

$$P(\text{a transistor's lifetime will be between 2 and 4 years}) = P(2 < X < 4),$$

and

$$P(\text{a transistor's lifetime will be less than 1 year}) = P(X < 1).$$

Note that since the probability that a continuous random variable is equal to a particular value is zero, $P(2 < X < 4)$ will be the same as $P(2 \leq X \leq 4)$; similarly $P(X > 3)$ will equal $P(X \geq 3)$ and $P(X < 1)$ will equal $P(X \leq 1)$. This is so for every continuous random variable, and therefore when working with continuous random variables, we may replace $<$ by $\leq$ and $>$ by $\geq$ in probability statements whenever we wish.

If X is a continuous random variable, the probability distribution of X is usually described by a function called a **probability density function.**

Probability Density Functions

A **probability density function** is a function f such that

$$f(x) \geq 0 \qquad \text{for all } x$$

and

$$\int_{-\infty}^{+\infty} f(x)\, dx = 1.$$

Note that the definition says that f must be defined and nonnegative for all x and that the area under the graph of f must be 1. See Figure 10–6.

EXAMPLE 1 The function defined by

$$f(x) = \begin{cases} -\dfrac{6}{125}(x^2 - 5x), & 0 \leq x \leq 5, \\ 0 & \text{otherwise,} \end{cases}$$

is a probability density function: it is defined and nonnegative for all x (Figure 10–7), and since $f(x) = 0$ when x is outside $[0, 5]$, we have

$$\int_{-\infty}^{+\infty} f(x)\, dx = \int_{0}^{5} -\frac{6}{125}(x^2 - 5x)\, dx = 1.$$

(Check this.) ☐

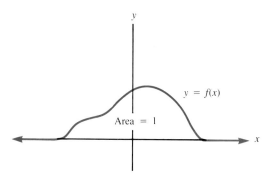
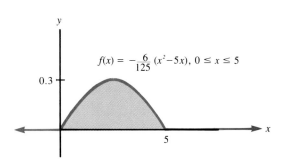

FIGURE 10–6
Probability Density Function

FIGURE 10–7

EXAMPLE 2 Let

$$f(x) = \begin{cases} 0.5e^{-x/2}, & x \geq 0, \\ 0, & x < 0. \end{cases}$$

We claim that f is a probability density function. Certainly f is defined and nonnegative for all x (Figure 10–8), so it suffices to show that the improper integral

$$\int_{-\infty}^{+\infty} f(x)\, dx = \int_{0}^{+\infty} 0.5e^{-x/2}\, dx$$

is equal to 1. We have

$$\int_{0}^{+\infty} 0.5e^{-x/2}\, dx = \lim_{b \to +\infty} \int_{0}^{b} 0.5e^{-x/2}\, dx$$

$$= \lim_{b \to +\infty} -e^{-x/2}\Big|_{0}^{b}$$

$$= \lim_{b \to +\infty} (-e^{-b/2} + e^{0})$$

$$= \lim_{b \to +\infty} (-e^{-b/2}) + 1.$$

But $\lim_{b \to +\infty} e^{-b/2} = 0$, so

$$\int_{0}^{+\infty} 0.5e^{-x/2}\, dx = \lim_{b \to +\infty} (e^{-b/2}) + 1 = 0 + 1 = 1,$$

and we are done. ☐

Given a continuous random variable X, it is usually possible to find a probability density function f such that for any numbers a and b with $a < b$,

$$P(a \leq X \leq b) = \text{Area under } f \text{ from } a \text{ to } b = \int_{a}^{b} f(x)\, dx.$$

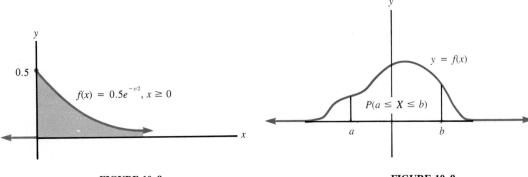

FIGURE 10–8

FIGURE 10–9
Probability as Area

See Figure 10–9. When this is the case, we say that f is a **probability density function for X**. If X is a continuous random variable and f is a probability density function for X, then we can find probabilities involving X by integrating f.

EXAMPLE 3

Let X denote the operating lifetime of a transistor in years, $0 \leq X \leq 5$, and suppose that the probability density function for X is defined by

$$f(x) = \begin{cases} -\dfrac{6}{125}(x^2 - 5x), & 0 \leq x \leq 5, \\ 0, & \text{otherwise.} \end{cases}$$

Then

P(a transistor's lifetime will be between 2 and 4 years)
$= P(2 < X < 4)$

$$= \int_2^4 -\frac{6}{125}(x^2 - 5x)\, dx$$

$$= -\frac{6}{125}\left(\frac{x^3}{3} - \frac{5x^2}{2}\,\bigg|_2^4\right)$$

$$= \frac{68}{125} = 0.544.$$

See Figure 10–10. Similarly,

P(a transistor's lifetime will be greater than 3 years)
$= P(X > 3)$

$$= \int_3^5 -\frac{6}{125}(x^2 - 5x)\, dx$$

$$= \frac{44}{125} = 0.352$$

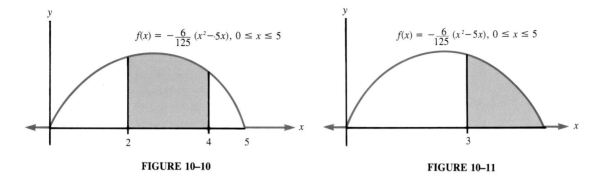

FIGURE 10–10   **FIGURE 10–11**

(Figure 10–11), and

P(a transistor's lifetime will be less than or equal to 1 year)

$$= P(X \leq 1) = P(X < 1)$$

$$= \int_0^1 -\frac{6}{125}(x^2 - 5x)\, dx$$

$$= \frac{13}{125} = 0.104.$$

(Figure 10–12). ☐

EXAMPLE 4   An automaker's quality control department tests engines by running them until they break down. Define

X = Time it takes for an engine to break down, in months.

Since the time it will take for an engine to break down is potentially unlimited, X can take on any value in the interval $[0, +\infty)$. Suppose the probability density function for X is given by

$$f(x) = \begin{cases} 0.5e^{-x/2}, & x \geq 0, \\ 0, & x < 0. \end{cases}$$

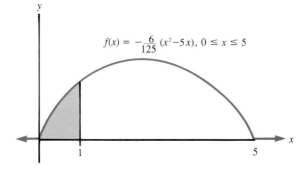

FIGURE 10–12

Then

$$P(1 < X < 3) = \int_1^3 0.5e^{-x/2}\, dx$$

$$= -e^{-x/2}\Big|_1^3 = -e^{-3/2} + e^{-1/2}$$

$$\approx 0.3834.$$

Similarly,

$$P(X \le 2) = \int_0^2 0.5e^{-x/2}\, dx$$

$$= -e^{-x/2}\Big|_0^2 = -e^{-1} + e^0$$

$$= 1 - e^{-1} \approx 0.6321.$$

We could calculate $P(X > 2)$ by evaluating the improper integral

$$\int_2^{+\infty} 0.5e^{-x/2}\, dx;$$

however, it is easier to make use of the work we have already done and the fact that $X > 2$ is the complementary event to $X \le 2$ and write

$$P(X > 2) = 1 - P(X \le 2) = 1 - (1 - e^{-1}) = e^{-1} \approx 0.3679. \quad \square$$

Closely related to the probability density function of a continuous random variable is its **cumulative distribution function.**

The Cumulative Distribution Function

Let X be a continuous random variable. The **cumulative distribution function for X** is the function F defined by

$$F(x) = P(X \le x) \qquad \text{for all } x.$$

Note that if f is a probability density function for X, then

$$F(x) = P(X \le x) = \int_{-\infty}^{x} f(t)\, dt$$

for all x, and thus $F(x)$ is the area under the graph of f from $-\infty$ to x. Hence F is an area function, and, just as in Section 6.4, this means that it is an antiderivative of f. Finally, if $a < b$,

$$P(a \le X \le b) = P(X \le b) - P(X \le a) = F(b) - F(a).$$

PROBABILITY

Let us summarize these properties of the cumulative distribution function:

> **Properties of the Cumulative Distribution Function**
>
> Let X be a continuous random variable. If f is a probability density function for X and F a cumulative distribution function for X, then
>
> 1. $F(x) = \int_{-\infty}^{x} f(t)\, dt \quad$ for all x.
>
> 2. $F'(x) = f(x) \quad$ for all x.
>
> 3. If $a < b$, then
> $$P(a \leq X \leq b) = F(b) - F(a).$$

EXAMPLE 5   Suppose X is the random variable defined on $[0, +\infty)$ whose probability density function f is given by
$$f(x) = \begin{cases} 0.5 e^{-x/2}, & x \geq 0, \\ 0, & x < 0. \end{cases}$$

Let us find the cumulative distribution function F for X.

An antiderivative of f will be of the form
$$F(x) = \begin{cases} -e^{-x/2} + c_1, & x \geq 0, \\ c_2, & x < 0. \end{cases}$$

(Why?) When $x < 0$,
$$0 = P(X \leq x) = F(x) = c_2,$$

so $c_2 = 0$. Also, when $x = 0$, we have
$$0 = P(X \leq 0) = F(0) = -e^0 + c_1 = -1 + c_1,$$

which implies that $c_1 = 1$. Therefore the cumulative distribution function for X is defined by
$$F(x) = \begin{cases} -e^{-x/2} + 1, & x \geq 0, \\ 0, & x < 0. \end{cases}$$

Note that
$$P(1 < X < 3) = F(3) - F(1) = -e^{-3/2} + e^{-1/2}. \quad \square$$

In the previous example we began with a probability density function for the random variable and found its cumulative distribution function. Sometimes it is easier to begin with the cumulative distribution function and find the density function.

EXAMPLE 6   Let us choose a number at random from the interval $[0, 10]$. We let X denote the number chosen. Since each number in $[0, 10]$ is equally likely to be selected, we conclude that there is 1 chance in 10 that the number chosen will be from the interval [0, 1], 2 chances in 10 that it will be from [0, 2], 3.5 chances in 10 that it will be from [0, 3.5], and, in general, x chances in 10 that it will be from [0, x]. Therefore we must have

$$P(0 \leq X \leq x) = \frac{x}{10}, \quad 0 \leq x \leq 10.$$

Of course, if $x < 0$, then $P(X \leq x) = 0$, whereas if $x > 10$, $P(X \leq x) = 1$. Hence the cumulative distribution function F for X is defined by

$$F(x) = P(X \leq x) = \begin{cases} 0, & x < 0, \\ x/10, & 0 \leq x \leq 10, \\ 1, & x > 10, \end{cases}$$

and it follows that the probability density function for X is the function defined by

$$f(x) = F'(x) = \begin{cases} 1/10 & \text{if } 0 \leq x \leq 10, \\ 0 & \text{otherwise.} \end{cases} \quad \square$$

We conclude this section with a brief discussion of the expected value, variance, and standard deviation of a continuous random variable.

Expected Value, Variance, and Standard Deviation

If X is a continuous random variable with probability density function f, then its expected value $E(X)$ is given by

$$E(X) = \int_{-\infty}^{+\infty} x f(x)\, dx$$

and its variance $V(X)$ by

$$V(X) = \int_{-\infty}^{+\infty} x^2 f(x)\, dx - [E(X)]^2.$$

The standard deviation of X is the square root of its variance.

The interpretation of the expected value and variance are the same for continuous random variables as they are for discrete random variables: $E(X)$ is the average value of X in the long run, and $V(X)$ measures the degree to which the values of X vary from its expected value.

EXAMPLE 7 Again let X = number of months a test engine runs before it fails, and suppose that the density function for X is given by

$$f(x) = \begin{cases} 0.5e^{-x/2}, & x \geq 0, \\ 0, & x < 0. \end{cases}$$

PROBABILITY

We have

$$E(X) = \int_{-\infty}^{+\infty} xf(x)\,dx$$

$$= \int_{0}^{+\infty} 0.5xe^{-x/2}\,dx$$

$$= \lim_{b \to +\infty} \int_{0}^{b} 0.5xe^{-x/2}\,dx.$$

Integration by parts gives

$$\int 0.5xe^{-x/2}\,dx = -xe^{-x/2} + \int e^{-x/2}\,dx = -xe^{-x/2} - 2e^{-x/2} + c.$$

(Check this.) Therefore

$$E(X) = \lim_{b \to +\infty} \left(-xe^{-x/2} - 2e^{-x/2} \Big|_{0}^{b} \right)$$

$$= \lim_{b \to +\infty} (-be^{-b/2} - 2e^{-b/2} + 2).$$

Since $\lim_{b \to +\infty} be^{-b/2} = 0$ and $\lim_{b \to +\infty} e^{-b/2} = 0$, we thus have $E(X) = 2$. The interpretation of this result is that on average the engines will run two months before breaking down.

We also have

$$V(X) = \int_{-\infty}^{+\infty} x^2 f(x)\,dx - [E(X)]^2 = \int_{0}^{+\infty} 0.5x^2 e^{-x/2}\,dx - [2]^2$$

$$= \lim_{b \to +\infty} \int_{0}^{b} x^2(0.5)e^{-x/2}\,dx - 4$$

$$= \lim_{b \to +\infty} \left(-x^2 e^{-x/2} - 4xe^{-x/2} - 8e^{-x/2} \Big|_{0}^{b} \right) - 4$$

$$= \lim_{b \to +\infty} (-b^2 e^{-b/2} - 4be^{-b/2} - 8e^{-b/2} + 8) - 4$$

$$= 4.$$

Thus the variance of X is 4, and hence its standard deviation is $\sqrt{4} = 2$. ∎

10.4 EXERCISES

In Exercises 1 through 8, show that f is a probability density function.

1. $f(x) = \begin{cases} 1/3, & 0 \le x \le 3, \\ 0 & \text{otherwise} \end{cases}$

2. $f(x) = \begin{cases} x/2, & 0 \le x \le 2, \\ 0 & \text{otherwise} \end{cases}$

3. $f(x) = \begin{cases} \dfrac{3}{20} x(1-x)^2, & 1 \le x \le 3, \\ 0 & \text{otherwise} \end{cases}$

4. $f(x) = \begin{cases} \dfrac{1}{x^2}, & x \ge 1, \\ 0, & x < 1 \end{cases}$

5. $f(x) = \begin{cases} \dfrac{8}{x^3}, & x \geq 2, \\ 0, & x < 2 \end{cases}$

6. $f(x) = \begin{cases} 2e^{-2x}, & x \geq 0, \\ 0, & x < 0 \end{cases}$

7. $f(x) = \begin{cases} (k-1)x^{-k}, & k > 1, \ x \geq 1, \\ 0, & x < 1 \end{cases}$

8. $f(x) = \begin{cases} ke^{-kx}, & k > 0, \ x \geq 0, \\ 0, & x < 0 \end{cases}$

9. Let X be a continuous random variable whose probability density function is the function of Exercise 1. Find
    (a) $P(0 \leq X \leq 3)$
    (b) $P(2 \leq X \leq 3)$
    (c) $P(2.5 < X < 2.9)$
    (d) $P(X < 1.4)$
    (e) $P(X \geq 0.5)$
    (f) $P(a < X < b), \ 0 \leq a < b \leq 3$

10. Let X be a continuous random variable whose probability density function is the function of Exercise 2. Find
    (a) $P(0 < X < 0.75)$
    (b) $P(0.5 \leq X \leq 1.5)$
    (c) $P(X \leq 1.9)$
    (d) $P(X \geq 1.9)$
    (e) $P(X < b), \ 0 \leq b \leq 2$
    (f) a such that $P(X > a) = 0.10$

11. Let X be a continuous random variable whose probability density function is the function of Exercise 3. Find the cumulative distribution function for X and use it to find
    (a) $P(X \geq 2)$
    (b) $P(X < 2.5)$
    (c) $P(1.4 < X < 2.2)$
    (d) $P(1.5 \leq X \leq 2.4)$
    (e) $P(X > b), \ 1 \leq b \leq 3$
    (f) $P(X < a), \ 1 \leq a \leq 3$

12. Let X be a continuous random variable whose probability density function is the function of Exercise 4. Find
    (a) $P(1 < X < 5)$
    (b) $P(2 < X < 3)$
    (c) $P(X < 10)$
    (d) $P(X > 4)$
    (e) $P(X > b), \ b \geq 1$
    (f) b such that $P(X < b) = 0.25$

13. Let X be a continuous random variable whose probability density function is the function of Exercise 5. Find the cumulative distribution function for X and use it to find
    (a) $P(2 < X < 4)$
    (b) $P(100 < X < 200)$
    (c) $P(X < 10)$
    (d) $P(X > 50)$
    (e) $P(a < X < b), \ 2 \leq a < b$
    (f) b such that $P(X \leq b) = 0.95$

14. Let X be a continuous random variable whose probability density function is the function of Exercise 6. Find the cumulative distribution function for X and use it to find
    (a) $P(1 < X < 2)$
    (b) $P(0.5 \leq X \leq 2.5)$
    (c) $P(X \geq 3)$
    (d) $P(X < \ln 5)$
    (e) $P(X \geq b), \ b \geq 0$
    (f) a such that $P(X > a) = 0.50$

In Exercises 15 through 20, the given function F is a cumulative distribution function for a continuous random variable. Find the probability density function for the random variable.

15. $F(x) = \begin{cases} 0, & x < 0, \\ 3x, & 0 \leq x \leq \frac{1}{3}, \\ 1, & x > \frac{1}{3} \end{cases}$

16. $F(x) = \begin{cases} 0, & x < 0, \\ x^3, & 0 \leq x \leq 1, \\ 1, & x > 1 \end{cases}$

17. $F(x) = \begin{cases} 0, & x < 1, \\ \ln x^{1/3}, & 1 \leq x \leq e^3, \\ 1, & x > e^3 \end{cases}$

18. $F(x) = \begin{cases} 0, & x < 3, \\ \sqrt{x+1} - 2, & 3 \leq x \leq 8, \\ 1, & x > 8 \end{cases}$

19. $F(x) = \begin{cases} 0, & x < 0, \\ 1 - e^{-3x}, & x \geq 0 \end{cases}$

20. $F(x) = \begin{cases} 0, & x < 1, \\ 1 - x^{-5}, & x \geq 1 \end{cases}$

In Exercises 21 through 26, find the cumulative distribution function for the given probability distribution function f, and then use the cumulative distribution function to find the given probabilities.

21. $f(x) = \begin{cases} 1/5, & 0 \leq x \leq 5, \\ 0 & \text{otherwise} \end{cases}$, $P(1 < X < 4), P(X < 2), P(X > 3)$

22. $f(x) = \begin{cases} 1/x, & 1 \leq x \leq e, \\ 0 & \text{otherwise} \end{cases}$, $P(X > 2), P(X < 2), P(1 < X < 2.5)$

23. $f(x) = \begin{cases} x^{-2}, & x \geq 1, \\ 0, & x < 1 \end{cases}$, $P(X < 2), P(2 < X < 3), P(X > 5)$

24. $f(x) = \begin{cases} 8x^{-3}, & x \geq 2, \\ 0, & x < 2 \end{cases}$, $P(4 < X < 6), P(X > 3), P(X < 4)$

25. $f(x) = \begin{cases} 3e^{-3x}, & x \geq 0, \\ 0, & x < 0 \end{cases}$, $P(X > 1), P(X < 2), P(3 < X < 5)$

26. $f(x) = \begin{cases} 4xe^{-2x}, & x \geq 0, \\ 0, & x < 0 \end{cases}$, $P(0 < X < 1), P(X < 3), P(X > \tfrac{1}{2})$

27. Let X denote a number chosen at random from the interval $[0, 4]$. Find the cumulative distribution function and the probability density function for X.

*28. Let a point be selected at random from the circumference or interior of a circle of radius 1, and let X be the distance of the point from the center of the circle. Then X is a continuous random variable on the interval $[0, 1]$. Show that the cumulative distribution function for X is given by

$$F(x) = \begin{cases} 0, & x < 0, \\ x^2, & 0 \leq x \leq 1, \\ 1, & x > 1, \end{cases}$$

and find the probability density function for X. [*Hint:* Show that if $0 \leq x \leq 1$, then $P(0 \leq X \leq x) = x^2$.]

*29. Let a point be selected at random from the perimeter or interior of the square whose vertices are at $(0, 0)$, $(0, 1)$, $(1, 1)$, and $(1, 0)$. If the point selected is (a, b), let X be the maximum of a and b. Then X is a continuous random variable on the interval $[0, 1]$. Show that the cumulative distribution function for X is given by

$$F(x) = \begin{cases} 0, & x < 0, \\ x^2, & 0 \leq x \leq 1, \\ 1, & x > 1. \end{cases}$$

In Exercises 30 through 35, find the expected value and variance of the continuous random variable whose probability density function f is given.

30. $f(x) = \begin{cases} 1/2, & 0 \leq x \leq 2, \\ 0 & \text{otherwise} \end{cases}$

31. $f(x) = \begin{cases} x/3, & 0 \leq x \leq \sqrt{6}, \\ 0 & \text{otherwise} \end{cases}$

32. $f(x) = \begin{cases} 1.2x(x+1), & 0 \leq x \leq 1, \\ 0 & \text{otherwise} \end{cases}$

33. $f(x) = \begin{cases} x^{-2}, & x \geq 1, \\ 0, & x < 1 \end{cases}$

34. $f(x) = \begin{cases} 2e^{-2x}, & x \geq 0, \\ 0, & x < 0 \end{cases}$

35. $f(x) = \begin{cases} xe^{-x}, & x \geq 0, \\ 0, & x < 0 \end{cases}$

*36. Show that if X is a continuous random variable with probability density function f, then

$$V(X) = \int_{-\infty}^{+\infty} (x - E(X))^2 f(x) \, dx.$$

Management

Let X denote the operating lifetime of a certain type of computer disk drive, X in years, $0 \leq X \leq 4$. Suppose the probability density function for X is given by

$$f(x) = \begin{cases} \dfrac{3}{32}(4x - x^2), & 0 \leq x \leq 4, \\ 0 & \text{otherwise.} \end{cases}$$

Exercises 37 through 40 refer to this situation.

37. Find the probability that a disk drive will last less than one year.

38. Find the probability that a disk drive will last more than three years.

39. What percentage of the disk drives will have an operating lifetime of less than six months?

40. Find and interpret $E(X)$.

Let X denote the time, in weeks, between successive inventory reorders. Suppose that the probability density function for X is f, where

$$f(x) = \frac{1}{2}(x + 1)^{-3/2} \quad \text{if } x \geq 0, \qquad f(x) = 0 \quad \text{if } x < 0.$$

Exercises 41 through 44 refer to this situation.

41. Find the probability that the time between successive reorders will be at least one week.

42. If $P(X > b) = 0.05$, find b, and interpret your result.

43. If $P(X < a) = 0.1$, find a and interpret your result.

44. Find the expected time between reorders.

Let X denote the time, in weeks, between breakdowns of a production line, and suppose that X has probability distribution function f, where

$$f(x) = 2(x + 1)^{-3} \quad \text{for } x \geq 0, \qquad f(x) = 0 \quad \text{for } x < 0.$$

Exercises 45 through 48 refer to this situation.

45. Find the probability that the time between successive breakdowns will be less than one week.

46. Find the probability that the time between successive breakdowns will be more than two weeks.

47. Find the probability that the time between successive breakdowns will be more than one week but less than five weeks.

48. Find the expected time between breakdowns.

A manufacturer receives shipments of parts from a supplier at irregular intervals. Let X be the interarrival time, that is, the time between arrivals of successive shipments, with X measured in days. Suppose the probability density function f for X is given by

$$f(x) = 0.1e^{-0.1x} \text{ if } x \geq 0, \qquad f(x) = 0 \text{ if } x < 0.$$

Exercises 49 through 53 refer to this situation.

49. Find the probability that the interarrival time will be between 5 and 10 days.
50. Find the probability that the interarrival time will be more than 14 days.
51. Find the probability that the interarrival time will be less than 30 days.
52. Find the time b such that 50% of the time the interarrival time will be less than or equal to b days.
53. Find the expected interarrival time.

Life Science

Let X denote the time between contraction of a certain disease and death, measured in years. Suppose that the probability distribution function for X is f, where

$$f(x) = \begin{cases} \dfrac{1}{(\ln 11)(x+1)}, & 0 \leq x \leq 10, \\ 0 & \text{otherwise.} \end{cases}$$

Exercises 54 through 58 refer to this situation.

54. What is the probability that a person who contracts the disease will live for more than five years?
55. What percentage of those who contract the disease die within one year?
56. What percentage of those who contract the disease survive at least two years?
57. The 5% who survive the longest after contracting the disease live how many years?
58. What is the expected survival time for those who contract the disease?

Let X denote the recovery time from surgery, measured in days, and suppose the probability density function for X is f, where $f(x) = 0.64xe^{-0.8x}$ for $x \geq 0$, $f(x) = 0$ for $x < 0$. Exercises 59 through 62 refer to this situation.

59. Find the probability that recovery time will be less than three days.
60. Find the percentage of patients whose recovery time will be greater than five days.
*61. The 10% of the patients who recover the quickest recover in approximately how many days? (Hint: try 0.65 and 0.7 days.)
62. Find the expected recovery time.

According to an article by Anthony Nero, Jr. ("Controlling Indoor Pollution," *Scientific American*, May 1988), if X is the concentration of the dangerous radioactive gas radon in the air of a house, then the probability density function for X is approximately given by

$$y = \begin{cases} 0, & x < 0, \\ 13x, & 0 \leq x < 0.2, \\ -12x + 5, & 0.2 \leq x < 0.4, \\ 0.24e^{-0.44x}, & x \geq 0.4. \end{cases}$$

Here x is measured in hundreds of becquerels per cubic meter of air (b/m³). Exercises 63 through 67 refer to this situation.

63. Estimate the percentage of houses that have a radon concentration of less than 10 b/m³.

64. Estimate the percentage of houses that have a radon concentration of between 10 and 30 b/m³.

65. Find the probability that a house will have a radon concentration greater than 300 b/m³.

66. According to the article, people who live for 20 years in a house that has a radon concentration of 1000 b/m³ or more run a 2% to 3% annual risk of developing lung cancer as a result. What percentage of houses contribute to this risk?

67. The 5% of the houses with the greatest concentration of radon have approximately what concentration?

Social Science

A sociologist studying fads among teenagers estimates that if X is the lifetime of a fad, X in months, $0 \leq X \leq 6$, then the probability density function for X is given by

$$f(x) = \begin{cases} -\dfrac{1}{54}x(1-x), & 0 \leq x \leq 6, \\ 0 & \text{otherwise.} \end{cases}$$

Exercises 68 through 70 refer to this situation.

68. Find the probability that a fad will last between two and four months.

*69. The shortest-lived two-thirds of fads will last fewer than how many months?

70. What is the average lifetime of a fad?

10.5 THE NORMAL AND EXPONENTIAL DISTRIBUTIONS

Two of the most useful probability distributions for continuous random variables are the **normal distribution** and the **exponential distribution.** In this section we introduce these distributions and show how to use them.

The Normal Distribution

Continuous random variables that measure naturally occurring phenomena often have probability density functions whose graphs are bell-shaped curves, such as that of Figure 10–13. Examples of such random variables can include those that measure quantities such as the weights or heights of the individuals in a population, scores on standardized tests, the operating lifetimes of electronic components, and many others. A particular class of density functions that give rise to bell-shaped curves are the **normal** density functions.

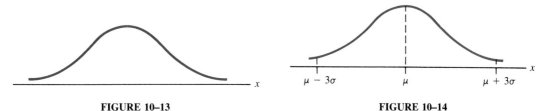

FIGURE 10–13   **FIGURE 10–14**

> **Normal Density Functions**
>
> Let μ and σ be numbers, with $\sigma > 0$. The function f defined by
>
> $$f(x) = \frac{1}{\sigma \sqrt{2\pi}} e^{-(x-\mu)^2/2\sigma^2}$$
>
> for all x is a probability density function called the **normal density function determined by μ and σ.**

(The symbol σ is the Greek letter sigma.)

The graph of the normal density function determined by μ and σ is called the **normal curve determined by μ and σ**; it is the bell-shaped curve shown in Figure 10–14. (See Exercise 42, Section 5.4.) The number μ is the x-coordinate of the top of the bell, and the number σ determines the height and the "flatness" of the bell: the larger σ is, the lower and more "flattened out" the bell is. Figure 10–15 illustrates this with the graphs of two normal density functions that have the same value of μ but different values of σ. Returning to Figure 10–14, note that the normal curve is symmetric about a vertical line through $x = \mu$. Since the total area under the curve is 1, it follows that the area under the curve to the left of $x = \mu$ and the area under the curve to the right of $x = \mu$ are both equal to 0.5. Furthermore, although the curve extends indefinitely to the right and to the left, it turns out that nearly all the area under it lies between $\mu - 3\sigma$ and $\mu + 3\sigma$, and therefore it is customary to draw the curve as we have done in Figure 10–14, extending it just slightly beyond $\mu - 3\sigma$ and $\mu + 3\sigma$.

If X is a random variable whose probability density function is normal, we say that X is a **normal** random variable or that it is **normally distributed**.

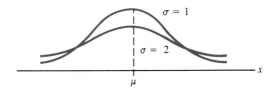

FIGURE 10–15

Normal Random Variables

Let X be a continuous random variable whose probability density function is the normal density function determined by μ and σ. Then

$$E(X) = \mu \quad \text{and} \quad V(X) = \sigma^2,$$

and X is referred to as the **normal random variable with mean μ and standard deviation σ**. The normal random variable for which $\mu = 0$ and $\sigma = 1$ is called the **standard normal** random variable, and is symbolized by Z.

The density function for the standard normal random variable Z is thus given by

$$f(z) = \frac{1}{\sqrt{2\pi}} e^{-z^2/2}.$$

The graph of this density function is shown in Figure 10–16. (See Exercise 41, Section 5.4.) If a and b are any numbers, then

$$P(a < Z < b) = \int_a^b \frac{1}{\sqrt{2\pi}} e^{-z^2/2}\, dz$$

(Figure 10–17). As we noted in Section 7.4, we cannot evaluate integrals of this form by finding an antiderivative for the integrand; however, we can approximate them by methods of numerical integration. This has been done for various values of a and b and the results tabulated in Table 3 at the back of the book. We can use Table 3 and the symmetry of the standard normal curve to find any probability involving the standard normal random variable Z.

EXAMPLE 1 Let us find $P(0 < Z < 1.74)$. See Figure 10–18. We can find this probability directly from Table 3 by looking in the row labeled 1.7 and the column headed by .04. Thus

$$P(0 < Z < 1.74) = 0.4591.$$

Any probability of the form $P(0 < Z < b)$ can similarly be found directly from the table. ☐

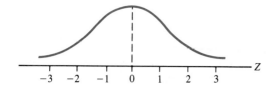

FIGURE 10–16
The Standard Normal Curve

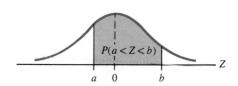

FIGURE 10–17

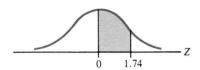 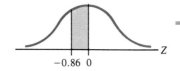 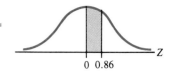

FIGURE 10–18   **FIGURE 10–19**

EXAMPLE 2 Let us find $P(-0.86 < Z < 0)$. By the symmetry of the normal curve,
$$P(-0.86 < Z < 0) = P(0 < Z < 0.86).$$
(See Figure 10–19.) Therefore
$$P(-0.86 < Z < 0) = P(0 < Z < 0.86) = 0.3051,$$
from Table 3. ▢

EXAMPLE 3 Let us find $P(1.22 < Z < 2.03)$ and also $P(-1.80 < Z < -0.25)$. Since
$$P(1.22 < Z < 2.03) = P(0 < Z < 2.03) - P(0 < Z < 1.22)$$
(Figure 10–20), we have
$$P(1.22 < Z < 2.03) = 0.4788 - 0.3888 = 0.0900.$$

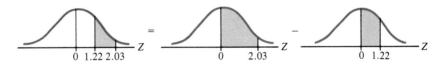

FIGURE 10–20

By the symmetry of the normal curve,
$$P(-1.80 < Z < -0.25) = P(0.25 < Z < 1.80).$$
(See Figure 10–21.) Hence

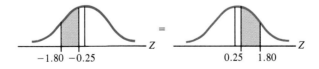

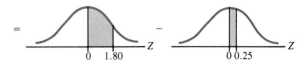

FIGURE 10–21

$$P(-1.80 < Z < -0.25) = P(0 < Z < 1.80) - P(0 < Z < 0.25)$$
$$= 0.4641 - 0.0987$$
$$= 0.3654. \quad \square$$

EXAMPLE 4   Let us find $P(-1.63 < Z < 1.97)$. Since

$$P(-1.63 < Z < 1.97) = P(-1.63 < Z < 0) + P(0 < Z < 1.97)$$

(Figure 10–22), and since

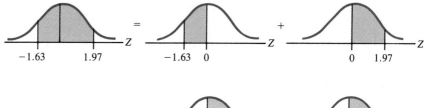

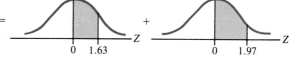

FIGURE 10–22

$$P(-1.63 < Z < 0) = P(0 < Z < 1.63)$$

by the symmetry of the normal curve, we have

$$P(-1.63 < Z < 1.97) = P(0 < Z < 1.63) + P(0 < Z < 1.97)$$
$$= 0.4484 + 0.4756$$
$$= 0.9240. \quad \square$$

EXAMPLE 5   Let us find $P(Z > 1.53)$ and $P(Z > -2.22)$. Since

$$P(Z > 1.53) = P(Z > 0) - P(0 < Z < 1.53)$$

(Figure 10–23), and since $P(Z > 0) = 0.5$, we have

FIGURE 10–23

$$P(Z > 1.53) = 0.5 - 0.4370 = 0.063.$$

Since

$$P(Z > -2.22) = P(-2.22 < Z < 0) + P(Z > 0)$$

(Figure 10–24), and since

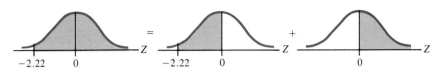

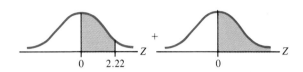

FIGURE 10–24

$$P(-2.22 < Z < 0) = P(0 < Z < 2.22)$$

by the symmetry of the normal curve, we have

$$P(Z > -2.22) = P(0 < Z < 2.22) + P(Z > 0)$$
$$= 0.4868 + 0.5$$
$$= 0.9868. \quad \square$$

EXAMPLE 6 Let us find $P(Z < 0.59)$ and $P(Z < -2.98)$. Since

$$P(Z < 0.59) = P(Z < 0) + P(0 < Z < 0.59)$$

(Figure 10–25), we have

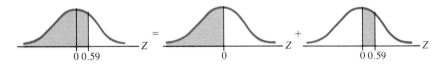

FIGURE 10–25

$$P(Z < 0.59) = 0.5 + 0.2224 = 0.7224.$$

Also

$$P(Z < -2.98) = P(Z < 0) - P(-2.98 < Z < 0)$$
$$= 0.5 - P(0 < Z < 2.98)$$
$$= 0.5 - 0.4986$$
$$= 0.0014. \quad \square$$

The preceding examples demonstrate how we can find probabilities for the standard normal random variable with the aid of Table 3 and the symmetry of the standard normal curve. We can also use the table to find the approximate value of Z that yields a given probability.

EXAMPLE 7 Let us find the approximate value of z such that

$$P(Z < z) = 0.9.$$

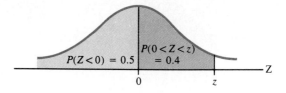

FIGURE 10–26

As Figure 10–26 shows, in order for $P(Z < z) = 0.9$ to hold, we must have

$$P(0 < Z < z) = 0.4.$$

But from the standard normal table,

$$P(0 < Z < 1.28) = 0.3997,$$

and this is the closest we can come to a probability of 0.4. Therefore

$$P(Z < z) = 0.9 \quad \text{implies} \quad z \approx 1.28. \quad \square$$

To find probabilities for a normal random variable that is not standard normal, we must first convert the normal variable to standard normal form.

Converting to Standard Normal

If X is a normal random variable with mean μ and standard deviation σ, then

$$Z = \frac{X - \mu}{\sigma}.$$

We can use the conversion formula to convert normal probabilities $P(a < X < b)$ into equivalent standard normal ones

$$P\left(\frac{a - \mu}{\sigma} < Z < \frac{b - \mu}{\sigma}\right).$$

(Exercise 29 at the end of this section shows why this works.)

We give an example to show how to use the conversion formula.

EXAMPLE 8   Let X denote the height of submariners in the U.S. Navy. Suppose X is normal with a mean height of $\mu = 70$ inches and a standard deviation of $\sigma = 1.2$ inches. Let us find the probability that a submariner selected at random is between 67 and 71 inches tall; that is, let us find $P(67 < X < 71)$. We must convert this normal probability to an equivalent standard normal one by using the conversion formula

$$Z = \frac{X - \mu}{\sigma} = \frac{X - 70}{1.2}.$$

Note that when $X = 67$,
$$Z = \frac{67 - 70}{1.2} = -2.50,$$
and when $X = 71$,
$$Z = \frac{71 - 70}{1.2} \approx 0.83.$$
Therefore
$$\begin{aligned}P(67 < X < 71) &\approx P(-2.50 < Z < 0.83) \\ &= P(-2.50 < Z < 0) + P(0 < Z < 0.83) \\ &= P(0 < Z < 2.50) + P(0 < Z < 0.83) \\ &= 0.4938 + 0.2967 \\ &= 0.7905.\end{aligned}$$

Now let us find the percentage of submariners who are taller than 6 feet (72 inches). We can do this by finding $P(X > 72)$ and multiplying the result by 100. But when $X = 72$,
$$Z = \frac{72 - 70}{1.2} \approx 1.67.$$
Thus
$$\begin{aligned}P(X > 72) &\approx P(Z > 1.67) \\ &= P(Z > 0) - P(0 < Z < 1.67) \\ &= 0.5 - 0.4525 \\ &= 0.0475.\end{aligned}$$

Therefore approximately 4.75% of submariners are taller than 6 feet.

Finally, suppose we ask the following question: the tallest 99% of submariners are taller than what height? In terms of probabilities, we are asking what x must be in order to have
$$P(X > x) = 0.99.$$
See Figure 10–27. If we convert to Z, we find that x must satisfy the equation
$$P\left(Z > \frac{x - 70}{1.2}\right) = 0.99.$$

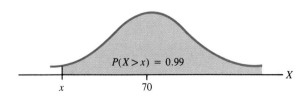

FIGURE 10–27

But as Figure 10–27 indicates, x must be less than $\mu = 70$, and hence

$$\frac{x - 70}{1.2} < 0.$$

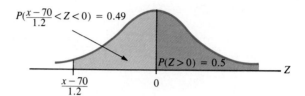

FIGURE 10–28

Figure 10–28 now shows that

$$P\left(\frac{x - 70}{1.2} < Z < 0\right) = 0.49,$$

and thus

$$P\left(\frac{x - 70}{1.2} < Z < 0\right) = P\left(0 < Z < -\frac{x - 70}{1.2}\right) = 0.49.$$

There is no entry in Table 3 that is exactly equal to 0.49; the entry closest to 0.49 is 0.4901, which corresponds to $z = 2.33$. Hence

$$-\frac{x - 70}{1.2} \approx 2.33,$$

and thus

$$x \approx -2.33(1.2) + 70 \approx 67.2.$$

Therefore the tallest 99% of submariners are taller than 67.2 inches, approximately. □

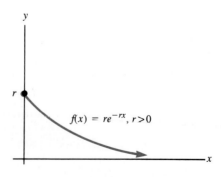

FIGURE 10–29

The Exponential Distribution

Many common situations can be described by the graphs of functions whose defining equations are of the form

$$f(x) = re^{-rx}, \quad r > 0.$$

See Figure 10–29. For instance, such curves can often be used to model the time between arrivals at a tollbooth, service times at airline ticket counters, and (as we did in Example 4 of Section 10.4) time-to-failure of a product.

Exponential Density Functions

If $r > 0$, the function f defined by

$$f(x) = \begin{cases} re^{-rx}, & x \geq 0, \\ 0, & x < 0, \end{cases}$$

is a probability density function, called the **exponential density function determined by r.**

It is easy to show that $f(x) = re^{-rx}$ defines a probability density function. (The proof follows that of Example 2 of Section 10.4; see Exercise 58 at the end of this section.)

If X is a continuous random variable whose probability density function is an exponential density function, we say that X is an **exponential** random variable, or that it is **exponentially distributed.**

Exponential Random Variables

Let X be a continuous random variable whose probability density function is the exponential density function determined by r. Then

$$E(X) = \frac{1}{r} \quad \text{and} \quad V(X) = \frac{1}{r^2},$$

and X is referred to as the **exponential random variable with mean $\frac{1}{r}$.**

We ask for a proof that $E(X) = 1/r$ and $V(X) = 1/r^2$ in Exercises 59 and 60 at the end of this section.

EXAMPLE 9 Let X denote the operating lifetime, in years, of a certain type of computer chip, and suppose X is exponentially distributed with an expected value of five years. Then

$$E(X) = \frac{1}{r} = 5,$$

so

$$r = \frac{1}{5} = 0.2.$$

Thus the density function for X is given by

$$f(x) = \begin{cases} 0.2e^{-0.2x}, & x \geq 0, \\ 0, & x < 0. \end{cases}$$

The probability that a computer chip selected at random will last between three and four years is therefore

$$P(3 < X < 4) = \int_3^4 0.2e^{-0.2x}\, dx$$

$$= -e^{-0.2x} \Big|_3^4 = -e^{-0.8} + e^{-0.6}$$

$$\approx 0.0995.$$

To find the percentage of chips that will last for six or more years, we find $P(X \geq 6) = 1 - P(X < 6)$ and multiply by 100:

$$P(X \geq 6) = 1 - P(X < 6)$$

$$= 1 - \int_0^6 0.2e^{-0.2x}\, dx$$

$$= 1 - \left(-e^{-0.2x} \Big|_0^6 \right) = 1 - (-e^{-1.2} + 1)$$

$$= e^{-1.2} \approx 0.3012.$$

Therefore approximately 30.12% of the chips will last six years or longer.

If a large batch of chips is put into service at one time, how long will it be until 50% of them have failed? To answer this question, we must find b such that

$$P(0 < X < b) = 0.5.$$

But this implies that

$$0.5 = \int_0^b 0.2e^{-0.2x}\, dx = -e^{-0.2b} + 1.$$

Therefore

$$e^{-0.2b} = 0.5,$$

and thus

$$-0.2b = \ln e^{-0.2b} = \ln 0.5 = -0.6931.$$

Hence $b \approx 3.47$ years. (See Exercise 61.) □

■ **Computer Manual: Programs NORMDIST, EXPONTL.**

10.5 EXERCISES

The Normal Distribution

1. Sketch the graphs of the normal density functions determined by μ and σ if
   (a) $\mu = 2, \sigma = 1$ (b) $\mu = 2, \sigma = 3$ (c) $\mu = 25, \sigma = \frac{1}{2}$ (d) $\mu = 25, \sigma = 4$

In Exercises 2 through 8, Z denotes the standard normal random variable.

2. Find $P(0 < Z < 1.40)$, $P(0 \leq Z \leq 2.35)$, $P(-2.03 < Z < 0)$, and $P(-1.67 < Z < 0)$.

3. Find $P(2.14 \leq Z \leq 2.38)$, $P(0.55 < Z < 1.23)$, $P(-1.39 < Z < -0.87)$, and $P(-3.00 < Z \leq -2.79)$.

4. Find $P(-0.45 < Z < 0.45)$, $P(-1.93 < Z < 0.22)$, $P(-2.31 \leq Z < 1.56)$, and $P(-3.00 < Z < 3.00)$.

5. Find $P(Z > 1.99)$, $P(Z > -2.03)$, $P(Z < 2.65)$, and $P(Z < -1.08)$.

6. Find z such that
   $P(0 < Z < z) = 0.25$, $P(0 < Z < z) = 0.49$, $P(z < Z < 0) = 0.4$, $P(z < Z < 0) = 0.351$

7. Find z such that
   $P(Z > z) = 0.1$, $P(Z > z) = 0.33$, $P(Z > z) = 0.95$, $P(Z > z) = 0.556$

8. Find z such that
   $P(Z < z) = 0.25$, $P(Z < z) = 0.02$, $P(Z < z) = 0.99$, $P(Z < z) = 0.846$

Let X be a normally distributed random variable with mean $\mu = 10$ and standard deviation $\sigma = 2$. In Exercises 9 through 14, find

9. $P(10 < X < 14)$
10. $P(9.5 < X < 11.1)$
11. $P(X > 8.8)$
12. $P(X < 10.2)$
13. x such that $P(10 < X < x) = 0.45$
14. x such that $P(X < x) = 0.85$

Let X be a normally distributed random variable with mean $\mu = 25$ and standard deviation $\sigma = 5$. In Exercises 15 through 20, find

15. $P(20 < X < 35)$
16. $P(X > 27)$
17. $P(X < 18)$
18. $P(14 < X < 17)$
19. x such that $P(X > x) = 0.02$
20. x such that $P(X < x) = 0.06$

If X is a normally distributed random variable with mean μ and standard deviation σ, find

21. $P(\mu < X < \mu + \sigma)$
22. $P(\mu + \sigma < X < \mu + 2\sigma)$
23. $P(\mu + 2\sigma < X < \mu + 3\sigma)$
24. $P(\mu - \sigma < X < \mu + \sigma)$
25. $P(\mu - 2\sigma < X < \mu + 2\sigma)$
26. $P(\mu - 3\sigma < X < \mu + 3\sigma)$

27. $P(X > \mu + 3\sigma)$

28. $P(X < \mu + 3\sigma)$

*29. Show why the conversion formula $Z = \dfrac{X - \mu}{\sigma}$ works. $\left[\text{Hint: Express}\right.$

$P(a < X < b)$ as a definite integral and then make the substitution $z = \dfrac{x - \mu}{\sigma}.\Big]$

The scores on an examination are normally distributed with a mean score of 72 and a standard deviation of 8. Exercises 30 through 33 refer to this situation: of the students who took the exam, find the proportion who scored

30. 90 or above **31.** 80 up to 90 **32.** 70 up to 80 **33.** below 70

34. Referring to Exercises 30 through 33, suppose the top 5% of the scores are assigned a grade of A, the next 25% a grade of B, the next 40% a grade of C, the next 25% a grade of D, and the lowest 5% a grade of F. Find the range of scores that corresponds to each grade.

35. The lifetime of a certain brand of light bulb is normally distributed with a mean of 1000 hours and a standard deviation of 250 hours. Find the probability that a light bulb of this brand will last
   (a) more than 1000 hours;
   (b) between 1000 and 1500 hours;
   (c) between 1100 and 1200 hours;
   (d) between 900 and 1100 hours.

36. Refer to Exercise 35: Find the percentage of bulbs that will last
   (a) more than 1350 hours;
   (b) less than 825 hours;
   (c) more than 900 hours;
   (d) less than 1450 hours.

37. Refer to Exercise 35: The company that makes the light bulbs intends to offer a money-back guarantee on any bulb that does not last for a certain length of time. The guarantee will be stated: "If this light bulb does not last for x hours, we will refund your money." What should x be in this statement if the company wants no more than 1% of the bulbs to qualify for the refund?

Management

38. The operating lifetime of a certain type of computer chip is normally distributed with a mean of 5.3 years and a standard deviation of 1.5 years. Find the probability that a chip will last
   (a) between 4 and 6 years;
   (b) between 3 and 5 years;
   (c) more than 7 years;
   (d) fewer than 9.5 years;
   (e) more than 1 year.

39. Refer to Exercise 38:
   (a) The 1% of the chips that last the longest last longer than how many years?
   (b) The 5% of the chips that are shortest-lived last fewer than how many years?

40. The weight of flanges produced by a machine is normally distributed with a mean of 200 grams and a standard deviation of 2 grams. Find the percentage of flanges made by the machine that weigh

(a) more than 200.5 grams;
(b) between 198.8 and 201.5 grams;
(c) more than 202.4 grams;
(d) less than 203.1 grams.

41. Referring to Exercise 40, suppose the heaviest 3% and the lightest 2% of the flanges are unacceptable. What range of weights is acceptable for the flanges?

42. The daily production of a manufacturing plant is normally distributed with a mean of 2000 units per day and a standard deviation of 200 units per day. Find the percentage of days when the plant will produce

(a) more than 2000 units;
(b) between 2000 and 2500 units;
(c) between 2200 and 2300 units;
(d) between 1900 and 1950 units;
(e) between 1630 and 2480 units;
(f) more than 2612 units;
(g) fewer than 1850 units;
(h) more than 1722 units;
(i) fewer than 2700 units.

43. Refer to Exercise 42:

(a) On the 5% of the days when production is greatest, the plant will produce more than how many units?
(b) On the 1% of the days when production is least, the plant will produce fewer than how many units?
(c) Suppose that daily production falls between $2000 - b$ and $2000 + b$ units 75% of the time. What is b?

Life Science

44. Recovery time for patients after a certain surgical procedure is normally distributed with a mean of 5.5 days and a standard deviation of 0.6 days.

(a) Find the probability that a patient's recovery time will be between four and six days.
(b) The 8% of the patients who recover the fastest recover in how many days?
(c) The 12% of the patients who recover the slowest recover in how many days?

45. The growing times for a certain strain of corn are normally distributed with a mean of 112 days and a standard deviation of three days. If you plant this strain of corn, find the probability that your corn crop will be ready for harvesting

(a) between 112 and 115 days after planting;
(b) between 110 and 114 days after planting.

46. Refer to Exercise 45: Find the probability that your corn crop will be ready for harvesting

(a) in 102 or fewer days after planting;
(b) in 108 or more days after planting;
(c) in 116 or more days after planting.

47. Refer to Exercise 45. You intend to sign a contract which states that you will deliver your corn crop to a buyer x days after planting. What should the value of x be if you want to have a 95% chance of honoring the contract and you know that you can deliver the crop to the buyer two days after it is ready to be harvested?

The Normal Approximation to the Binomial

We introduced the binomial distribution in Section 10.3. Finding binomial probabilities can be quite tedious when the number of trials n is large. However, when n is large it is often possible to approximate binomial probabilities by normal probabilities. This is done as follows: suppose we wish to find the binomial probability

$$P(k \text{ successes in } n \text{ trials}),$$

where $p = P(\text{a success on any one trial})$. If we set $\mu = np$ and $\sigma = \sqrt{np(1-p)}$ and let X be normally distributed with mean μ and standard deviation σ, then

$$P(k \text{ successes in } n \text{ trials}) \approx P(k - 0.5 \leq X \leq k + 0.5).$$

This approximation is generally held to be valid if $n \geq 20$ and $np \geq 5$, $n(1-p) \geq 5$. In Exercises 48 through 53 you are to use this approximation technique.

*48. Approximate the binomial probability $P(20 \text{ successes in } 50 \text{ trials})$ if $p = 0.5$.

*49. Approximate the binomial probability $P(15 \text{ or } 16 \text{ successes in } 25 \text{ trials})$ if $p = 0.4$.

*50. If you toss a fair coin 100 times, find the probability that you will obtain at least 45 but not more than 55 heads.

*51. If you toss a fair coin 100 times, find the probability that you will get fewer than 40 heads.

*52. If a basketball player makes 90% of her free throws and shoots 200 free throws in a season, find the probability that she will miss fewer than 10 free throws during the season.

*53. If the probability that a machine turns out a defective part is 0.001 and a lot consists of 10,000 parts, find the probability that there will be at least five defective parts in a lot.

Inventory Control

Exercises 54 through 57 were suggested by material in Stevenson, *Production/Operations Management*, 2nd edition, Irwin, 1986. They are concerned with finding the reorder point for an item. Thus, suppose that when the stock of an item declines to r units, an order for new stock is generated. The number r is called the **reorder point** for the item. Let L denote the time that elapses between the generation of an order for new stock and the arrival of the stock; L is called the **lead time** for an order. **Safety stock** is extra stock that is on hand to lessen the risk of a **stockout**, that is, of running out of stock before the next order arrives. The reorder point r depends on the lead time and the safety stock according to the following equation:

$$r = \text{Expected demand for stock during lead time} + \text{Safety stock}.$$

Now suppose the lead time L for an item is always the same and that daily demand for the item is normally distributed with a mean of μ_d units per day and a standard deviation of σ_d items per day. If the probability of a stockout is p, then

$$\text{Expected demand during lead time} = \mu_d L$$

and

$$\text{Safety stock} = z_p \sigma_d \sqrt{L},$$

where z_p is determined from the following standard normal diagram:

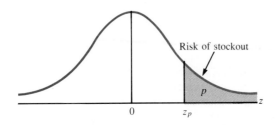

*54. A supermarket's sales of bar soap are normally distributed with average sales of 200 bars per day and a standard deviation of 30 bars per day. The lead time for bar soap is five days. If the store is willing to have $P(\text{stockout of bar soap}) = 0.01$, find the expected demand during lead time, the safety stock, and the reorder point for bar soap.

*55. Repeat Exercise 54 if the store is willing to run a 5% risk of a stockout.

If the lead time L is not always the same for an item, it may itself be normally distributed with a mean time of μ_L and a standard deviation time of σ_L. Then

$$\text{Expected demand during lead time} = \mu_d \mu_L$$

and

$$\text{Safety stock} = z_p \sqrt{\mu_L \sigma_d^2 + \mu_d^2 \sigma_L^2},$$

where z_p is determined as before.

*56. Suppose a service station's gasoline sales are normally distributed with a mean of 400 gallons per day and a standard deviation of 80 gallons per day. Suppose also that the lead time for a delivery of gasoline is normally distributed with a mean of 4.5 days and a standard deviation of 0.4 day. If the station is willing to run a 2% risk of a stockout of gas, find the expected demand during lead time, the safety stock, and the reorder point for gasoline.

*57. Repeat Exercise 56 if the lead time changes so that its standard deviation is 1.1 days.

The Exponential Distribution

*58. Let r be positive. Show that the function f defined by

$$f(x) = \begin{cases} re^{-rx}, & x \geq 0, \\ 0, & x < 0 \end{cases}$$

is a probability density function.

*59. Show that if the probability density function for the random variable X is the exponential density function determined by r, then $E(X) = \dfrac{1}{r}$.

*60. Show that if the probability density function for the random variable X is the exponential density function determined by r, then $V(X) = \dfrac{1}{r^2}$.

*61. Show that if X is an exponentially distributed random variable with expected value $\dfrac{1}{r}$, then $P(X < b) = 0.5$ implies that $b = \dfrac{\ln 2}{r}$.

The operating lifetime of a certain type of TV picture tube is exponentially distributed with an expected lifetime of 6.25 years. Exercises 62 through 66 refer to this situation: if 10,000 such tubes are put into service today, estimate

62. the number still in operation 4 years from now;

63. the number still in operation 10 years from now;

64. the number that will fail within the first year;

65. the number that will fail in years 3 through 6;

66. the number that will fail after 5 years.

Exercises 67 through 69 refer to Exercises 62 through 66.

67. The 15% of the tubes that last the longest will last longer than how many years?

68. The 25% of the tubes that last the shortest time will last fewer than how many years?

69. If 90% of the tubes last fewer than b years, find b.

Management

70. Customers arrive at a ticket counter in such a manner that the time between arrivals is exponentially distributed with an expected value of 0.5 minute.
  - **(a)** Find the probability that the time between the arrival of two successive customers will be between 1 and 2 minutes.
  - **(b)** Find the probability that the time between successive arrivals is either more than 2 minutes or less than 1 minute.

71. The service time for cars in an auto dealer's service department is exponentially distributed with an expected value of 3.2 hours.
  - **(a)** Find the proportion of cars that are serviced in fewer than 2 hours.
  - **(b)** Find the proportion of cars that require more than 5 hours to be serviced.

72. Refer to Exercise 71:
  - **(a)** The 50% of the cars that require the least amount of service time will require fewer than how many hours?
  - **(b)** The 10% of the cars that require the most amount of service time will require more than how many hours?

Life Science

73. In 1980 survival time for patients who had a certain incurable disease was exponentially distributed with an expected value of 2.5 years. Due to improvements in the treatment of the disease, survival time is now exponentially distributed with an expected value of 3.5 years.
  - **(a)** The 5% of patients who had the shortest survival times in 1980 survived fewer than how many years? What is the corresponding time for their survival today?
  - **(b)** The 5% of patients who had the longest survival times in 1980 survived more than how many years? What is the corresponding time for their survival today?

SUMMARY

This summary consists of terms and symbols whose meaning you should know, facts you should know, and techniques and methods you should be able to employ.

Section 10.1

Experiment, outcomes of an experiment. Event. Probability of an event. Using the definition of probability to find probabilities. The event "A or B." The complementary event $\overline{A}$ = "not A." The rules of probability. Finding probabilities using the rules of probability. Finding probabilities using historical data. Assigning probabilities to events.

Section 10.2

Random variable. Probability distribution for a random variable. Probability histogram for a random variable. Constructing the probability distribution and probability histogram for a random variable. The expected value of a random variable. Expected value $E(X)$ as the average value of X in the long run. Finding the expected value of a random variable. Using expected values in decision making. The variance of a random variable. The standard deviation of a random variable. The variance and standard deviation measure the dispersion of the values of X about its expected value. Finding the variance and standard deviation of a random variable.

Section 10.3

Bernoulli process: trials, success, failure, probability of a success on any trial. Factorials. The binomial distribution. The expected value and variance of a binomial random variable X. Constructing the probability distribution for a binomial random variable with given values of n and p. Finding binomial probabilities. Poisson process. Poisson distribution. The expected value and variance of a Poisson random variable X. Constructing the probability distribution for a Poisson random variable with a given value of μ. Finding Poisson probabilities.

Section 10.4

Discrete random variable. Continuous random variable. Probability density function. Showing that a given function is or is not a probability density function. The probability density function of a continuous random variable. Finding probabilities for a continuous random variable by integrating its probability density function. The cumulative distribution function for a continuous random variable X. Properties of the cumulative distribution function. Finding the cumulative distribution function from the probability density function. Using the cumulative distribution function to find probabilities. Finding the probability density function from the cumulative distribution function. The expected value, variance, and standard deviation of a continuous random variable. The expected value $E(X)$ as the long-run average value of X. The variance $V(X)$ and the standard deviation as measures of the degree to which the values of X vary from $E(X)$. Finding $E(X)$, $V(X)$, and the standard deviation of X for a continuous random variable X.

Section 10.5

The normal density function:

$$f(x) = \frac{1}{\sigma\sqrt{2\pi}} e^{-(x-\mu)^2/2\sigma^2}.$$

The normal curve determined by μ and σ. Normally distributed random variable with mean μ and standard deviation σ. The standard normal random variable Z. Finding standard normal probabilities using the standard normal table. Using the standard normal table to find the value of z that yields a given standard normal probability. Converting to standard normal. Using the conversion formula and the standard normal table to find normal probabilities. Using the conversion formula and the standard normal table to find the value of x that yields a given probability. The exponential density function determined by $r > 0$. Exponentially distributed random variable. The expected value and variance of an exponentially distributed random variable. Finding probabilities for exponentially distributed random variables.

REVIEW EXERCISES

1. The integers 0, 1, . . . , 49 are written on slips of paper and placed in a hat. You draw one of the slips from the hat. Find
   (a) P(the number drawn is 18);
   (b) P(the number drawn is positive);
   (c) P(the number drawn is not positive);
   (d) P(the number drawn is less than 40);
   (e) P(the number drawn is greater than 23 and less than 37);
   (f) P(the number drawn is negative);
   (g) P(the number drawn is even);
   (h) P(the number drawn is greater than or equal to 17).

2. A fish market kept track of the length of time required to serve each of its customers during one morning. The results were as follows:

Service Time (minutes)	Number of Customers
Less than 1	2
At least 1 but less than 2	10
At least 2 but less than 3	40
At least 3 but less than 4	32
4 or more	16

   Find
   (a) the probability that it will take less than 1 minute to serve a customer;
   (b) the probability that it will take at least 2 but less than 4 minutes to serve a customer;
   (c) the probability that it will take either less than 2 minutes or 3 or more minutes to serve a customer.

3. A student taking a mathematics course estimates the following:

> it is three times as likely that she will get a B in the course as it is that she will get an A;
>
> it is twice as likely that she will get a B as it is that she will get a C;
>
> it is one half as likely that she will get a D as it is that she will get an A;
>
> there is no chance that she will fail the course.

Find the probability that
   (a) she will get an A in the course;
   (b) she will pass the course with a grade of C or better;
   (c) she will get either a C or a D in the course.

4. For the experiment of Exercise 1, define a random variable X by setting X = sum of the digits in the number drawn. Find $E(X)$ and $V(X)$ and interpret your results.

5. You pay $1 for a lottery ticket. The probability of winning the lottery is 1/1,947,792. Find the expected value of your winnings if a winning ticket will pay
   (a) $1,500,000;
   (b) $2,000,000.

6. You choose a card at random from an ordinary deck. Before selecting the card, you can bet $1 on
   (a) the color of the card (red or black);
   (b) the suit of the card (spades, hearts, diamonds, or clubs);
   (c) whether the card is a face card (jack, queen, king, or ace) or not;
   (d) the number of the card (where ace = 1, jack = 11, queen = 12, and king = 13);
   (e) the exact suit and number of the card (for example, you can bet that the card will be the ace of spades, the 3 of clubs, etc.).

   In each case, if you lose the bet you lose your $1. What should the payoff for winning be in each case if the bet is to be fair?

7. A company that purchases switches by the carton has collected the following information concerning defective switches:

Defectives per Carton	Number of Cartons
0	18
1	42
2	12
3	5
4	10
5	3

Estimate the expected number of defective switches per carton.

8. The probability that an experimental drug will improve the condition of a patient who is treated with it is estimated to be $\frac{2}{3}$.

   (a) Find the probability that of the next 10 patients treated with the drug, 8 will improve.

(b) Find the probability that of the next 10 patients treated with the drug, at least 1 will improve.
(c) Find the probability that of the next 10 patients treated with the drug, fewer than half will improve.
(d) Of the next 100 patients treated with the drug, what is the expected number who will improve?

9. Airplanes enter the landing pattern at a large airport at an average rate of one every 45 seconds. Find the probability that
   (a) no planes will enter the landing pattern during the next 45 seconds;
   (b) at least one plane will enter the landing pattern during the next 45 seconds;
   (c) at least three planes will enter the landing pattern during any 1-minute period;
   (d) fewer than five planes will enter the landing pattern during a 5-minute period.

10. In each case show that the function f is a probability density function on the given interval:

   (a) $f(x) = \begin{cases} 0.25x^{-1/2}, & 0 \le x \le 4, \\ 0 & \text{otherwise} \end{cases}$

   (b) $f(x) = \begin{cases} \dfrac{3}{2}x^{-3/2}, & x \ge 9, \\ 0, & x < 9. \end{cases}$

11. Let X be a random variable whose probability density function is the function of Exercise 10(a). Find
    (a) $P(0 \le X \le 2)$
    (b) $P(X < 1)$
    (c) $P(X > 3.5)$
    (d) b if $P(X > b) = 0.5$
    (e) $E(X)$
    (f) $V(X)$

12. Let X be a random variable whose probability density function is the function of Exercise 10(b). Find the cumulative distribution function for X and use it to find
    (a) $P(16 < X < 25)$
    (b) $P(X < 100)$
    (c) $P(X \ge 400)$
    (d) b if $P(X < b) = 0.9$

13. A company that makes air conditioners tests selected units by running them continuously until they fail. Let X denote the number of years a tested unit runs until it fails. If X has probability density function f given by

$$f(x) = \begin{cases} \dfrac{1}{(x+1)^2}, & x \ge 0, \\ 0, & x < 0, \end{cases}$$

find
(a) the probability that a unit will last at least one year;
(b) the percentage of units that will fail during the first six months of the test.

The 10% of units tested that run the longest will last longer than how many years?

14. If Z denotes the standard normal random variable, find
    (a) $P(0 < Z < 2.35)$;
    (b) $P(-2.11 < Z < 0)$;
    (c) $P(0.55 < Z < 1.62)$;
    (d) $P(-1.79 < Z < -1.25)$;
    (e) $P(Z > 0.87)$;
    (f) $P(Z < -1.42)$;
    (g) $P(Z > -2.40)$;
    (h) $P(Z > 1.32)$;
    (i) z such that $P(Z > z) = 0.23$;
    (j) z such that $P(Z < z) = 0.475$;
    (k) z such that $P(Z < z) = 0.98$;
    (l) z such that $P(Z > z) = 0.645$;
    (m) z such that $P(-z < Z < z) = 0.84$.

15. The time it takes a crop to ripen is normally distributed with a mean of 120 days after planting and a standard deviation of 3.4 days.

(a) What is the probability that the crop will be ready to be harvested within 112 days of planting?
(b) What is the probability that the crop will not be ready to be harvested until at least 126 days have passed since planting?
(c) What is the probability that the crop will ripen sometime between 115 and 125 days after planting?
(d) Ninety-nine percent of the time the crop will ripen before what time?

16. A retailer receives periodic shipments from suppliers. The time between the arrivals of successive shipments is exponentially distributed with an expected value of three days.
    (a) Find the probability that successive orders arrive on the same day.
    (b) Find the probability that the time between the arrival of successive orders is more than four but less than six days.
    (c) If the probability is 0.8 that the time between arrivals of successive orders will be at least b days, what is b?

ADDITIONAL TOPICS

Here are some suggestions for topics not covered in the text that you might want to investigate on your own.

1. The only discrete distributions we have studied are the binomial and the Poisson, but there are others. One that is sometimes discussed in statistics textbooks is the **multinomial** distribution. Find out about the multinomial distribution and some of its applications.

2. The only continuous distributions we have studied are the normal and the exponential, but there are others. The simplest continuous distribution is the **uniform** or **rectangular** distribution. Find out about the uniform distribution and some of its applications.

3. Some other continuous distributions that we have not studied are the student t-distribution, the χ^2 (chi-square) distribution, the β (beta) distribution, and the Γ (gamma) distribution. Find out about these distributions and how they are used.

4. Among the earliest contributors to the theory of probability were the French mathematicians Pierre Fermat (1601?–1665) and Blaise Pascal (1623–1662). Their investigation of probability was instigated by a question about gambling posed to Pascal by a gentleman gambler, and carried out through a series of letters between them in 1654. Find out about this series of letters and the contributions of Fermat and Pascal to probability theory.

SUPPLEMENT: THE FUNDAMENTAL THEOREM OF NATURAL SELECTION

In this supplement we derive a result from the field of population genetics known as the Fundamental Theorem of Natural Selection. Before we begin, we need some preliminaries.

Consider a collection of data that consists of the distinct numbers $x_1, x_2, \ldots, x_n$, with x_i occurring exactly m_i times in the collection, and let

$$N = \text{Number of pieces of data} = m_1 + m_2 + \cdots + m_n.$$

For instance, if the collection of data is

$$90, 90, 90, 80, 80, 80, 80, 80, 70, 70, 70, 50,$$

then

$$\begin{aligned} x_1 &= 90, & m_1 &= 3; \\ x_2 &= 80, & m_2 &= 5; \\ x_3 &= 70, & m_3 &= 3; \\ x_4 &= 50, & m_4 &= 1; \end{aligned}$$

and

$$N = 3 + 5 + 3 + 1 = 12.$$

We define the **arithmetic mean** $\bar{x}$ of the data to be

$$\bar{x} = \frac{\sum_{i=1}^{n} m_i x_i}{N}.$$

Thus the arithmetic mean of a set of data is what we often refer to informally as the "average" of the data.

EXAMPLE 1 Suppose that 12 students take a quiz and that 3 of them get a grade of 90 on the quiz, 5 a grade of 80, 3 a grade of 70, and 1 a grade of 50. Then the grades on the quiz are

$$90, 90, 90, 80, 80, 80, 80, 80, 70, 70, 70, 50,$$

and hence the average grade is

$$\bar{x} = \frac{3(90) + 5(80) + 3(70) + 1(50)}{12} = 77.5. \quad \square$$

(The expected value $E(X)$ of a random variable X is often referred to as the mean of X. The reason for this is that if X takes on the distinct values $x_1, \ldots, x_n$, with x_i taken on m_i times and the probability of each occurrence being equal, then

$$P(X = x_i) = \frac{m_i}{m_1 + \cdots + m_n} = \frac{m_i}{N}$$

and hence

$$E(X) = \sum_{i=1}^{n} x_i P(X = x_i)$$

$$= \sum_{i=1}^{n} x_i \frac{m_i}{N}$$

$$= \frac{\sum_{i=1}^{n} x_i m_i}{N}$$

$$= \bar{x}.)$$

Similarly, if a collection of data consists of the distinct numbers $x_1, x_2, \ldots, x_n$, with x_i occurring exactly m_i times in the collection, and if

$$N = \text{Number of pieces of data} = \sum_{i=1}^{n} m_i,$$

then we define the **variance** V_x of the data to be

$$V_x = \frac{\sum_{i=1}^{n} m_i(x_i - \bar{x})^2}{N}.$$

The variance V_x of a set of data measures the deviation of the data from its mean $\bar{x}$: the larger the variance, the further the data are from $\bar{x}$.

EXAMPLE 2   The variance of the quiz grades of Example 1 is

$$V_x = \frac{3(90 - 77.5)^2 + 5(80 - 77.5)^2 + 3(70 - 77.5)^2 + (50 - 77.5)^2}{12}$$

$$= \frac{1425}{12}$$

$$= 118.75. \quad \square$$

It is easy to show that V_x can also be written in the form

$$V_x = \frac{\sum_{i=1}^{n} m_i x_i^2}{N} - \bar{x}^2. \tag{1}$$

(See Exercise 1 at the end of this supplement). We will use this result later.

The Fundamental Theorem of Natural Selection is one of the most important results of population genetics. It says that the rate at which the fitness of a population is changing at time t is equal to the genetic variance of the population at time t. We will demonstrate the theorem for the special case of a population consisting of subpopulations that do not interbreed and whose fitnesses remain constant.

We let $Y = Y(t)$ denote the size of a population at time t and assume that the rate of growth of the population is proportional to its size. Then we can model population size by the first-order differential equation

$$\frac{dY}{dt} = rY, \tag{2}$$

where the constant of proportionality r is the growth rate of the population. (Of course, as we saw in Chapter 5, such a model implies unlimited growth, and since no population can forever grow without limit, this must be unrealistic in the long run. However, if r is small, the model can be used to approximate population growth over reasonably extensive periods of time.)

Geneticists consider the growth rate r to be a measure of the **fitness** of a population: given two comparable populations, the one with the larger growth rate is reproducing more successfully and hence is presumably more fit in the biological sense. The fact that we are taking r to be constant over time means that we are assuming that fitness is inherited perfectly from one generation to the next.

Now suppose a population S consists of n subpopulations $S_1, S_2, \ldots, S_n$, and assume that the subpopulations do not interbreed. (This would be the case, for instance, if the population consisted of several different species, or if it consisted of one species that was divided into geographically isolated subspecies.) Let $Y_i = Y_i(t)$ be the size of S_i at time t, and apply the model (2) to each S_i to conclude that

$$\frac{dY_i}{dt} = r_i Y_i \tag{3}$$

for each i. As before, the growth rate r_i is a measure of the fitness of the subpopulation S_i and does not change with time.

Since $Y = Y_1 + \cdots + Y_n$, it follows that

$$\frac{dY}{dt} = \sum_{i=1}^{n} \frac{dY_i}{dt} = \sum_{i=1}^{n} r_i Y_i.$$

If we set

$$\bar{r} = \frac{1}{Y} \sum_{i=1}^{n} r_i Y_i, \tag{4}$$

then

$$\frac{dY}{dt} = \bar{r} Y. \tag{5}$$

Equation (5) shows that $\bar{r}$ is the fitness of the entire population. Note that because Y and $Y_1, \ldots, Y_n$ are functions of t, $\bar{r}$ is also a function of t; hence the fitness of the entire population changes with time, even though the fitnesses of the subpopulations do not. (The reason this occurs is that as time goes on the subpopulations that are more fit outbreed those that are less so and become a greater proportion of the whole, thus changing the fitness of the entire population.)

Now we can formally state the Fundamental Theorem of Natural Selection:

The Fundamental Theorem of Natural Selection

The rate of change of the fitness of a population at time t is equal to the genetic variance of the population at time t:

$$\frac{d\bar{r}}{dt} = V_r(t).$$

Let us see why this is so. We have

$$\frac{d\bar{r}}{dt} = \frac{d}{dt}\left(\frac{\sum_{i=1}^{n} r_i Y_i}{Y}\right)$$

$$= \frac{Y \frac{d}{dt}\left(\sum_{i=1}^{n} r_i Y_i\right) - \left(\sum_{i=1}^{n} r_i Y_i\right)\frac{dY}{dt}}{Y^2}$$

$$= \frac{Y\left(\sum_{i=1}^{n} r_i \frac{dY_i}{dt}\right) - \left(\sum_{i=1}^{n} r_i Y_i\right)\frac{dY}{dt}}{Y^2}.$$

By Equation (3),

$$\frac{dY_i}{dt} = r_i Y_i$$

for each i, so

$$\frac{d\bar{r}}{dt} = \frac{Y\left(\sum_{i=1}^{n} r_i^2 Y_i\right) - \left(\sum_{i=1}^{n} r_i Y_i\right)\frac{dY}{dt}}{Y^2}.$$

But also

$$\sum_{i=1}^{n} r_i Y_i = \bar{r} Y$$

by Equation (4) and

$$\frac{dY}{dt} = \bar{r} Y$$

by Equation (5). Thus

$$\frac{d\bar{r}}{dt} = \frac{Y\sum_{i=1}^{n} r_i^2 Y_i - (\bar{r}Y)(\bar{r}Y)}{Y^2},$$

or

$$\frac{d\bar{r}}{dt} = \frac{\sum_{i=1}^{n} r_i^2 Y_i}{Y} - \bar{r}^2.$$

But by Equation (1), the variance $V_r = V_r(t)$ of $r_1, \ldots, r_n$ at time t is

$$V_r = \frac{\sum_{i=1}^{n} Y_i r_i^2}{Y} - \bar{r}^2.$$

Therefore

$$\frac{d\bar{r}}{dt} = V_r(t),$$

and we have demonstrated the Fundamental Theorem of Natural Selection in this special case.

SUPPLEMENTARY EXERCISES

*1. Show that Equation (1) of this supplement is true. (*Hint:* Write
$$(x_i - \bar{x})^2 = x_i^2 - 2x_i\bar{x} + \bar{x}^2$$
and use the facts that
$$\sum_{i=1}^{n} m_i = N \quad \text{and} \quad \sum_{i=1}^{n} m_i x_i = N\bar{x}.)$$

*2. Let a population S be divided into subpopulations $S_1, \ldots, S_n$ that do not interbreed and for each of which fitness is inherited perfectly. Suppose that the subpopulations are all equally fit. Find an expression for the rate of change of the fitness of S, and interpret your result.

11

Trigonometric Functions

In this chapter we introduce a new class of functions, the **trigonometric functions.** Trigonometric functions can be used to model the motion of a satellite in orbit, the movement of a weight at the end of a spring as the spring expands and contracts, the flow of current in an electrical circuit, and many other phenomena. In fact, virtually every situation of a repetitive or cyclical nature requires the use of trigonometric functions in its mathematical description; hence such functions are extremely important in modeling the natural world.

Since trigonometric functions are defined in terms of angles, before we can begin our study of them we must briefly discuss the question of how angles are measured. We next define the sine, cosine, tangent, secant, cosecant, and cotangent of an angle and discuss some important trigonometric identities. In the third section we introduce the trigonometric functions and their graphs. The last two sections of the chapter consider the differentiation and integration of trigonometric functions.

11.1 MEASURING ANGLES

In this section we discuss the measurement of angles. Angles may be measured either by **degrees** or by **radians.** We briefly review degree measure and then introduce radian measure, which we use in the rest of the chapter.

Degrees

Every angle has two **sides** and a **vertex**. Figure 11–1 shows an angle *AOB*: its sides are *OA* and *OB*, and its vertex is the point *O*. An angle is measured from its **initial side** to its **terminal side**, the direction of measurement being indicated by an arrow. Thus the fact that the arrow in Figure 11–1 goes from *OA* to *OB* tells us that *OA* is the initial side of the angle and *OB* is its terminal side. We attach a degree measure to an angle by placing the vertex of the angle at the center of a circle (Figure 11–2); then as we move from the initial side to the terminal side of the angle we sweep

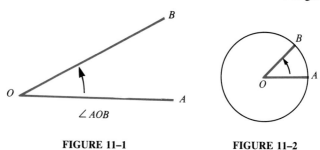

FIGURE 11–1   FIGURE 11–2

out some fraction of the circle. Since by ancient tradition every circle contains 360 degrees (360°), this movement will sweep out some fraction of 360°. This fraction of 360° is the degree measure of the angle. The degree measure of the angle is positive if movement from the initial side to the terminal side is in the counterclockwise direction; it is negative if it is in the clockwise direction.

EXAMPLE 1   Angle *AOB* of Figure 11–3(a) sweeps out one fourth of the circle in a counterclockwise direction; therefore

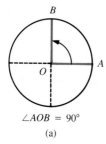

∠*AOB* = 90°
(a)

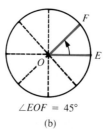

∠*EOF* = 45°
(b)

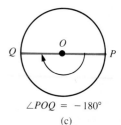

∠*POQ* = −180°
(c)

FIGURE 11–3

$$\text{Degree measure of angle } AOB = \frac{1}{4}(360°) = 90°.$$

Similarly, angle EOF of Figure 11–3(b) sweeps out one eighth of the circle in a counterclockwise direction, so

$$\text{Angle } EOF = \frac{1}{8}(360°) = 45°.$$

On the other hand, angle POQ of Figure 11–3(c) sweeps out one half of the circle in a clockwise direction, so its degree measure is negative, and thus

$$\text{Angle } POQ = -\frac{1}{2}(360°) = -180°. \quad \square$$

EXAMPLE 2 Suppose we are asked to construct an angle of 120°. Since 120° is positive and $120°/360° = \frac{1}{3}$, an angle of 120° must sweep out one third of a circle in a counterclockwise direction. Figure 11–4(a) shows an angle of 120°. Similarly, an angle of 60° must sweep out $60°/360° = \frac{1}{6}$ of a circle in a counterclockwise direction [Figure 11–4(b)]. But since $-270°$ is negative, an angle of $-270°$ must sweep out $270°/360° = \frac{3}{4}$ of a circle in a clockwise direction [Figure 11–4(c)]. $\quad \square$

Figure 11–5 shows some common angles and their degree measures.

Angles can measure more than 360°, in either a positive or negative sense. When this occurs, the angles are equivalent to angles having degree measure between $-360°$ and 360°. For instance, Figure 11–6 shows angles of 480° and $-810°$. Since $480° = 360° + 120°$, to construct an angle of 480°, we must proceed counterclockwise through a full circle (360°) plus another 120°, as the figure indicates. Thus an angle of 480° is equivalent to an angle of 120°. Similarly, since $-810° = 2(-360°) - 90°$, an angle of $-810°$ is equivalent to an angle of $-90°$.

Radians

We can also measure angles in **radians**, and in fact radian measure will be our preferred measure. To obtain the radian measure of an angle AOB, we place its vertex O at the center of a circle of radius 1, as in Figure 11–7. The angle is then measured by the length of the arc PQ which is traced as we move along the circumference of the circle from the initial side to the terminal side of the angle.

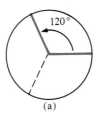

(a)

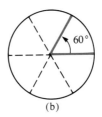

(b)

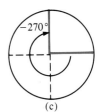

(c)

FIGURE 11–4

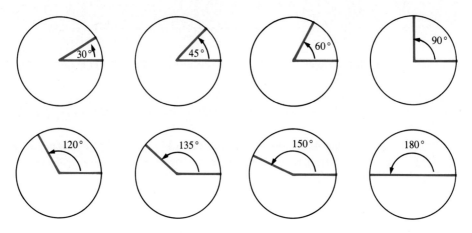

FIGURE 11-5

If the arc is traced in the counterclockwise direction as we move from the initial side to the terminal side, we take the radian measure of the angle to be positive; if it is traced in the clockwise direction, we take the radian measure to be negative. Thus in Figure 11–8 the radian measure of angle *EOF* will be *r* radians while that of angle *GOH* will be −*r* radians. Since the circumference of a circle of radius 1 is 2π, the length of the arc *PQ* will be some fraction of 2π, and hence radian measures are often expressed in terms of the number π.

EXAMPLE 3 Consider the angle *AOB* of Figure 11–9(a): as we move from the initial side to the terminal side of the angle we trace one fourth of the circumference of the circle in a counterclockwise direction. Therefore

$$\text{Radian measure of angle } AOB = \frac{1}{4}(2\pi) = \frac{\pi}{2} \text{ radians}.$$

Similarly, for the angle *EOF* of Figure 11–9(b), as we move from the initial side to the terminal side of the angle we trace one half of the circumference of the circle in a counterclockwise direction, so

$$\text{Angle } EOF = \frac{1}{2}(2\pi) = \pi \text{ radians}.$$

FIGURE 11-6

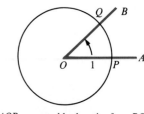

∠ *AOB* measured by length of arc *PQ*

FIGURE 11-7

TRIGONOMETRIC FUNCTIONS   695

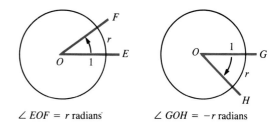

∠ EOF = r radians   ∠ GOH = −r radians

FIGURE 11–8

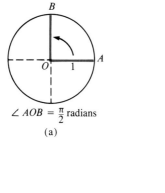

∠ AOB = $\frac{\pi}{2}$ radians
(a)

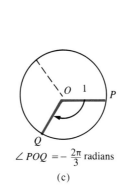

∠ POQ = $-\frac{2\pi}{3}$ radians
(c)

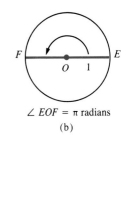

∠ EOF = π radians
(b)

FIGURE 11–9

On the other hand, for the angle POQ of Figure 11–9(c), as we move from the initial to the terminal side of the angle we trace out one third of the circumference of the circle in a clockwise direction, and thus

$$\text{Angle } POQ = -\frac{1}{3}(2\pi) = -\frac{2\pi}{3} \text{ radians.} \quad \square$$

EXAMPLE 4   Suppose we are asked to construct an angle of $\pi/3$ radians. Since the circumference of the circle of radius 1 is equal to 2π, and since

$$\frac{\pi/3}{2\pi} = \frac{1}{6},$$

an angle of $\pi/3$ radians must trace out one sixth of the circumference of the circle. Because $\pi/3$ is positive, the tracing must move in a counterclockwise direction. Figure 11–10(a) shows an angle of $\pi/3$ radians. Similarly, an angle of $-3\pi/2$ radians must trace out

$$\frac{3\pi/2}{2\pi} = \frac{3}{4}$$

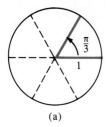 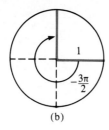 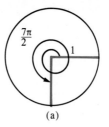 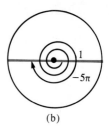

(a)   (b)   (a)   (b)

FIGURE 11–10   **FIGURE 11–11**

of the circumference of the circle. Because $-3\pi/2$ is negative, the tracing must move in a clockwise direction. Figure 11–10(b) depicts an angle of $-3\pi/2$ radians. ▢

Any angle that measures more than 2π radians is equivalent to one that measures between 0 and 2π radians. For instance, since

$$\frac{7\pi}{2} = 2\pi + \frac{3\pi}{2},$$

an angle of $7\pi/2$ radians is equivalent to one of $3\pi/2$ radians. See Figure 11–11(a). Similarly, any angle that measures less than -2π radians is equivalent to one that measures between -2π and 0 radians Figure 11–11(b) shows an angle of -5π radians.

Since we have two ways of measuring angles, we must be able to convert from one to the other. But if angle AOB has degree measure $d°$ and radian measure r radians, then d is the same proportion of 360 as r is of 2π. See Figure 11–12. Thus

$$\frac{d}{360} = \frac{r}{2\pi}.$$

The Conversion Formula for Angle Measurement

If an angle measures $d°$ and r radians, then d and r are related by the equation

$$r = \frac{\pi d}{180}.$$

Note that

$$r = \frac{\pi d}{180} \quad \text{implies that} \quad d = \frac{180r}{\pi}.$$

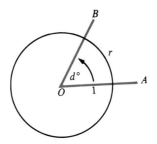

FIGURE 11–12

EXAMPLE 5 Let us convert 1°, 135°, −30°, and 68° to radians.

$$1°: \quad r = \frac{\pi(1)}{180} = \frac{\pi}{180} \text{ radians};$$

$$135°: \quad r = \frac{\pi(135)}{180} = \frac{3\pi}{4} \text{ radians};$$

$$-30°: \quad r = \frac{\pi(-30)}{180} = -\frac{\pi}{6} \text{ radians};$$

$$68°: \quad r = \frac{\pi(68)}{180} = \frac{17\pi}{45} \text{ radians}. \quad \square$$

EXAMPLE 6 Let us convert 1, $\pi/3$, $-5\pi/4$, and 3.8 radians to degrees.

$$1 \text{ radian:} \quad d = \frac{180(1)}{\pi} = \frac{180}{\pi} \approx 57.3°;$$

$$\pi/3 \text{ radians:} \quad d = \frac{180(\pi/3)}{\pi} = 60°;$$

$$-5\pi/4 \text{ radians:} \quad d = \frac{180(-5\pi/4)}{\pi} = -225°;$$

$$3.8 \text{ radians:} \quad d = \frac{180(3.8)}{\pi} = \frac{684}{\pi} \approx 217.7°. \quad \square$$

Figure 11–13 shows some common angles with their radian measures. Note that the angles of Figure 11–13 are exactly the same as the angles of Figure 11–5.

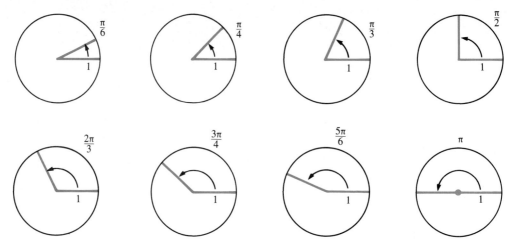

FIGURE 11–13

11.1 EXERCISES

Degrees

In Exercises 1 through 6, state the degree measurement of the angles pictured.

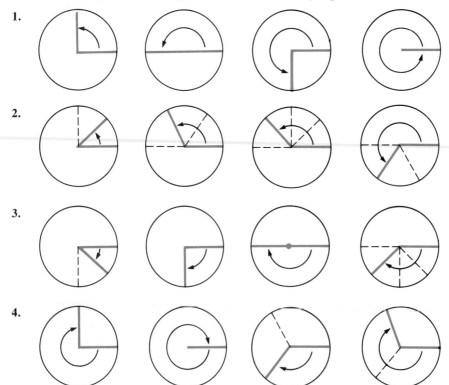

TRIGONOMETRIC FUNCTIONS   699

5.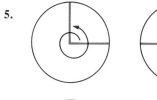

6.

In Exercises 7 through 14, construct the angles whose degree measures are given.

7. 90°, 30°, 180°, 720°
8. 180°, 540°, 315°, −60°
9. 45°, 135°, 60°, 150°
10. 270°, 225°, −120°, 0°, 360°
11. −900°, −30°, 630°, −150°, 15°
12. 495°, 840°, −765°, 22.5°, 40°, 7.5°
13. 36°, 72°, 108°, −252°
14. 12°, 144°, −24°, 18°

Radians

In Exercises 15 through 20, state the radian measure of the angles pictured. Do not use the conversion formula.

15.

16.

17.

18.

19.

20.

In Exercises 21 through 28, construct the angles whose radian measures are given.

21. $\dfrac{3\pi}{2}, \dfrac{5\pi}{2}, -\dfrac{\pi}{2}, \pi$

22. $2\pi, \dfrac{\pi}{3}, -\dfrac{\pi}{6}, \dfrac{2\pi}{3}$

23. $-\dfrac{5\pi}{4}, \dfrac{3\pi}{4}, \dfrac{11\pi}{4}, -\dfrac{3\pi}{2}$

24. $\dfrac{7\pi}{4}, -\dfrac{5\pi}{6}, \dfrac{11\pi}{6}, 4\pi, \dfrac{\pi}{8}$

25. $\dfrac{8\pi}{8}, \dfrac{\pi}{6}, \dfrac{9\pi}{2}, -\dfrac{\pi}{8}, \dfrac{2\pi}{9}$

26. $-3\pi, -\dfrac{2\pi}{3}, \dfrac{11\pi}{8}, \dfrac{\pi}{12}, \dfrac{26\pi}{3}$

27. $\dfrac{\pi}{16}, \dfrac{3\pi}{16}, -\dfrac{5\pi}{16}, \dfrac{23\pi}{16}$

28. $-\dfrac{\pi}{5}, \dfrac{2\pi}{5}, -\dfrac{4\pi}{5}, \dfrac{11\pi}{5}$

29. Convert to radians: 90°, 30°, 125°, −60°

30. Convert to radians: 45°, 12°, 120°, −900°

31. Convert to radians: 135°, −150°, 350°, 624°

32. Convert to radians: 72°, 108°, −50°, 11.25°

33. Convert to degrees: $\dfrac{5\pi}{6}, \dfrac{3\pi}{2}, -\dfrac{8\pi}{3}, \dfrac{9\pi}{2}$ radians.

34. Convert to degrees: $-6\pi, \dfrac{4\pi}{9}, 5\pi, \dfrac{3\pi}{8}$ radians.

35. Convert to degrees: $\dfrac{7\pi}{6}, -\dfrac{5\pi}{12}, \dfrac{22}{7}, 2.5$ radians.

36. Convert to degrees: $\dfrac{7\pi}{10}, \dfrac{7\pi}{5}, \dfrac{4\pi}{11}, -1.32$ radians

11.2 TRIGONOMETRY

In this section we define the sine, cosine, tangent, secant, cosecant, and cotangent of an angle, and discuss some important trigonometric identities. We begin with the definition of the sine and cosine of an angle.

We intend to define the sine and cosine of an angle of x radians, where x may be any number. (From now on, *all angles are measured in radians*.) To accomplish this, we take an angle AOB of x radians and place its vertex at the origin and its initial side along the positive u-axis in the uv-plane, as shown in Figure 11–14. If we now construct a circle with center at the origin and radius 1, the terminal side of the angle will intersect this circle at a point Q. See Figure 11–15. The u-coordinate of the point Q is called the **cosine of x,** written cos x, and its v-coordinate is the **sine of x,** written sin x: thus $Q = (\cos x, \sin x)$. Figure 11–16 shows the location of the point Q for some typical values of x.

Trigonometric Functions

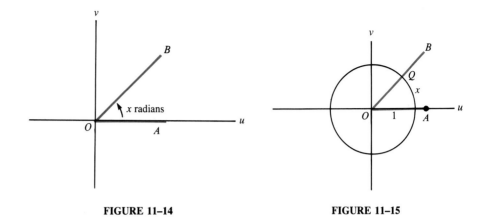

FIGURE 11–14 **FIGURE 11–15**

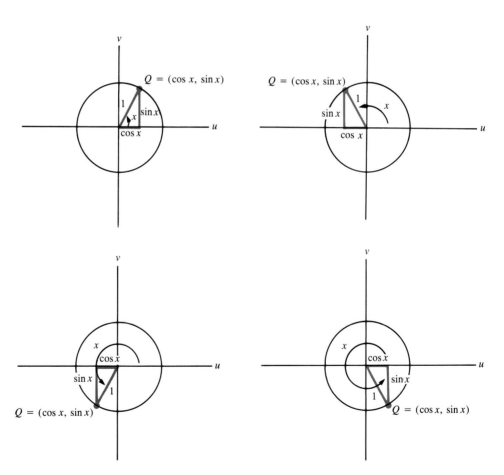

FIGURE 11–16

Note that

cos *x* is positive when $0 < x < \pi/2$ and $3\pi/2 < x < 2\pi$;
cos *x* is negative when $\pi/2 < x < 3\pi/2$;
sin *x* is positive when $0 < x < \pi$;
sin *x* is negative when $\pi < x < 2\pi$.

Note also that

$$-1 \le \cos x \le 1 \quad \text{and} \quad -1 \le \sin x \le 1,$$

since the absolute values of cos *x* and sin *x* can never be larger than the radius of the circle, which is 1.

Let us find cos *x* and sin *x* for some selected values of *x*.

a. If $x = 0$, then $Q = (\cos 0, \sin 0) = (1,0)$. See Figure 11–17(a). Therefore $\cos 0 = 1$ and $\sin 0 = 0$.

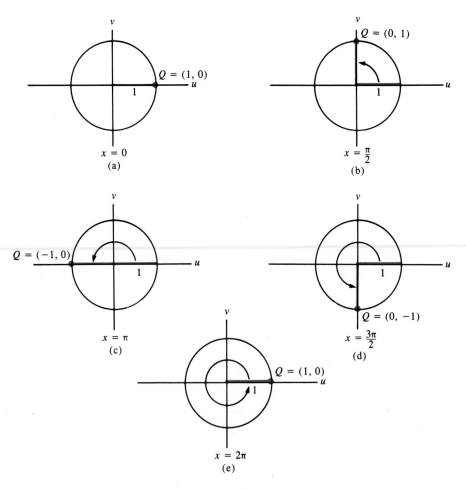

FIGURE 11–17

b. If $x = \pi/2$, then $Q = (\cos \pi/2, \sin \pi/2) = (0,1)$. See Figure 11–17(b). Therefore $\cos \pi/2 = 0$ and $\sin \pi/2 = 1$.

c. If $x = \pi$, then $Q = (\cos \pi, \sin \pi) = (-1,0)$. See Figure 11–17(c). Therefore $\cos \pi = -1$ and $\sin \pi = 0$.

d. If $x = 3\pi/2$, then $Q = (\cos 3\pi/2, \sin 3\pi/2) = (0,-1)$. See Figure 11–17(d). Therefore $\cos 3\pi/2 = 0$ and $\sin 3\pi/2 = -1$.

e. If $x = 2\pi$, then $Q = (\cos 2\pi, \sin 2\pi) = (1,0)$. See Figure 11–17(e). Therefore $\cos 2\pi = 1$ and $\sin 2\pi = 0$.

We summarize in Tables 11–1 and 11–2 the information we have obtained thus far.

TABLE 11–1

	Sign of	
Radians x	$\sin x$	$\cos x$
$0 < x < \pi/2$	+	+
$\pi/2 < x < \pi$	+	−
$\pi < x < 3\pi/2$	−	−
$3\pi/2 < x < 2\pi$	−	+

TABLE 11–2

	Radians				
	0	$\pi/2$	π	$3\pi/2$	2π
sin	0	1	0	−1	0
cos	1	0	−1	0	1

Since the equation of the circle with center at the origin and radius 1 in the uv-plane is

$$u^2 + v^2 = 1,$$

and since $Q = (\cos x, \sin x)$ is a point on this circle, we must have

$$(\cos x)^2 + (\sin x)^2 = 1$$

for all numbers x. An equation such as this that is true for all x is called a **trigonometric identity**. It is common practice to write $\cos^2 x$ for $(\cos x)^2$ and $\sin^2 x$ for $(\sin x)^2$. Hence this trigonometric identity is usually expressed in the form

$$\cos^2 x + \sin^2 x = 1.$$

There are other trigonometric identities that follow from the definitions of the sine and cosine. For instance,

$$\cos(x + 2\pi) = \cos x$$

for all x and

$$\sin(x + 2\pi) = \sin x$$

for all x because adding 2π to x yields the same point Q on the circle, as Figure 11–18 shows. Similarly,

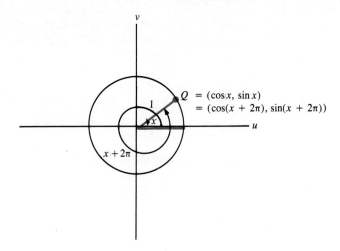

FIGURE 11–18

$$\cos(x - 2\pi) = \cos x \quad \text{and} \quad \sin(x - 2\pi) = \sin x$$

for all x. Furthermore, Figure 11–19 shows that

$$\cos(-x) = \cos x \quad \text{and} \quad \sin(-x) = -\sin x$$

for all x.

The additive identities

$$\sin(x_1 + x_2) = \sin x_1 \cos x_2 + \sin x_2 \cos x_1$$

and

$$\cos(x_1 + x_2) = \cos x_1 \cos x_2 - \sin x_1 \sin x_2$$

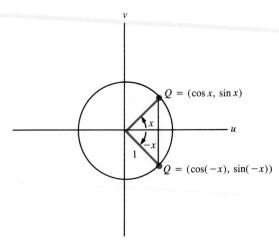

FIGURE 11–19

will be useful to us. Their proofs are outlined in Exercise 44 at the end of this section. As special cases of the additive identities, we have

$$\sin\left(x + \frac{\pi}{2}\right) = \sin x \cos \frac{\pi}{2} + \sin \frac{\pi}{2} \cos x$$
$$= (\sin x)(0) + (1)(\cos x)$$
$$= \cos x$$

and

$$\cos\left(x + \frac{\pi}{2}\right) = \cos x \cos \frac{\pi}{2} - \sin x \sin \frac{\pi}{2}$$
$$= (\cos x)(0) + (\sin x)(1)$$
$$= -\sin x$$

Summary: Trigonometric Identities

The following are true for all x, x_1, and x_2, (x, x_1, x_2 in radians):

1. $\cos^2 x + \sin^2 x = 1$
2. $\cos(x \pm 2\pi) = \cos x$, $\sin(x \pm 2\pi) = \sin x$
3. $\cos(-x) = \cos x$, $\sin(-x) = -\sin x$
4. $\sin(x_1 + x_2) = \sin x_1 \cos x_2 + \sin x_2 \cos x_1$
   $\cos(x_1 + x_2) = \cos x_1 \cos x_2 - \sin x_1 \sin x_2$
5. $\sin\left(x + \frac{\pi}{2}\right) = \cos x$, $\cos\left(x + \frac{\pi}{2}\right) = -\sin x$

These trigonometric identities in combination with the information collected in Tables 11–1 and 11–2 will allow us to evaluate the cosine and sine for additional values of x.

EXAMPLE 1   Let us find $\cos 3\pi$ and $\sin(-5\pi/2)$. Using the identity

$$\cos(x + 2\pi) = \cos x,$$

we have

$$\cos 3\pi = \cos(\pi + 2\pi) = \cos \pi = -1;$$

using the identities

$$\sin(-x) = -\sin x \quad \text{and} \quad \sin(x + 2\pi) = \sin x,$$

we have

$$\sin\left(-\frac{5\pi}{2}\right) = -\sin \frac{5\pi}{2} = -\sin\left(\frac{\pi}{2} + 2\pi\right) = -\sin \frac{\pi}{2} = -1. \quad \square$$

EXAMPLE 2 Let us find $\cos \pi/4$ and $\sin \pi/4$. Since the terminal side of an angle of $\pi/4$ radians bisects the first quadrant (Figure 11–20), the point $(\cos \pi/4, \sin \pi/4)$ lies on the line $v = u$ in the uv-plane, so

$$\cos \frac{\pi}{4} = \sin \frac{\pi}{4}.$$

But then the identity

$$\cos^2 x + \sin^2 x = 1$$

implies that

$$\sin^2 \frac{\pi}{4} + \sin^2 \frac{\pi}{4} = 1,$$

so

$$\sin^2 \frac{\pi}{4} = \frac{1}{2}.$$

Therefore

$$\sin \frac{\pi}{4} = \pm \frac{1}{\sqrt{2}};$$

however, since $\pi/4$ is between 0 and π, its sine is positive. Hence

$$\sin \frac{\pi}{4} = \cos \frac{\pi}{4} = \frac{1}{\sqrt{2}}. \quad \square$$

EXAMPLE 3 Let us use the results of Example 2 and the identities

$$\sin\left(x + \frac{\pi}{2}\right) = \cos x \quad \text{and} \quad \cos\left(x + \frac{\pi}{2}\right) = -\sin x$$

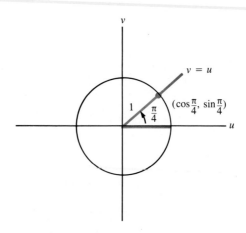

FIGURE 11–20

to find
$$\sin\frac{3\pi}{4}, \quad \cos\frac{3\pi}{4}, \quad \sin\frac{5\pi}{4}, \quad \text{and} \quad \cos\frac{5\pi}{4}.$$

We have
$$\sin\frac{3\pi}{4} = \sin\left(\frac{\pi}{4} + \frac{\pi}{2}\right) = \cos\frac{\pi}{4} = \frac{1}{\sqrt{2}}$$

and
$$\cos\frac{3\pi}{4} = \cos\left(\frac{\pi}{4} + \frac{\pi}{2}\right) = -\sin\frac{\pi}{4} = -\frac{1}{\sqrt{2}}.$$

Then
$$\sin\frac{5\pi}{4} = \sin\left(\frac{3\pi}{4} + \frac{\pi}{2}\right) = \cos\frac{3\pi}{4} = -\frac{1}{\sqrt{2}}$$

and
$$\cos\frac{5\pi}{4} = \cos\left(\frac{3\pi}{4} + \frac{\pi}{2}\right) = -\sin\frac{3\pi}{4} = -\frac{1}{\sqrt{2}}. \quad \square$$

Fortunately we will not often have to evaluate sines and cosines as we did in Examples 2 and 3. It will be helpful, however, to know the values of sin x and cos x when x is between 0 and 2π and is an integral multiple of $\pi/6$, $\pi/4$, or $\pi/3$; these values are included in the following summary:

Selected Values of Sin x and Cos x

radians x:	$\frac{\pi}{6}$	$\frac{\pi}{4}$	$\frac{\pi}{3}$	$\frac{\pi}{2}$	$\frac{2\pi}{3}$	$\frac{3\pi}{4}$	$\frac{5\pi}{6}$	π
sin x:	$\frac{1}{2}$	$\frac{1}{\sqrt{2}}$	$\frac{\sqrt{3}}{2}$	1	$\frac{\sqrt{3}}{2}$	$\frac{1}{\sqrt{2}}$	$\frac{1}{2}$	0
cos x:	$\frac{\sqrt{3}}{2}$	$\frac{1}{\sqrt{2}}$	$\frac{1}{2}$	0	$-\frac{1}{2}$	$-\frac{1}{\sqrt{2}}$	$-\frac{\sqrt{3}}{2}$	-1
radians x:	$\frac{7\pi}{6}$	$\frac{5\pi}{4}$	$\frac{4\pi}{3}$	$\frac{3\pi}{2}$	$\frac{5\pi}{3}$	$\frac{7\pi}{4}$	$\frac{11\pi}{6}$	2π
sin x:	$-\frac{1}{2}$	$-\frac{1}{\sqrt{2}}$	$-\frac{\sqrt{3}}{2}$	-1	$-\frac{\sqrt{3}}{2}$	$-\frac{1}{\sqrt{2}}$	$-\frac{1}{2}$	0
cos x:	$-\frac{\sqrt{3}}{2}$	$-\frac{1}{\sqrt{2}}$	$-\frac{1}{2}$	0	$\frac{1}{2}$	$\frac{1}{\sqrt{2}}$	$\frac{\sqrt{3}}{2}$	-1

(See Exercises 9 through 13.)

For our purposes it will not be necessary to find sines or cosines of any angles other than those we have already considered. In practice, sines and cosines are usually found with the aid of an electronic calculator that has sin and cos keys. Mathematical handbooks also have extensive tables of sines and cosines.

Now that we have defined the sine and cosine of an angle of x radians, we can use them to define other trigonometric functions of angles.

Definition of Tangent, Cotangent, Secant, and Cosecant

Let x be an angle measured in radians.

1. If x is not an odd multiple of $\pi/2$, define the **tangent of x**, written $\tan x$, by

$$\tan x = \frac{\sin x}{\cos x}.$$

2. If x is not an integral multiple of π, define the **cotangent of x**, written $\cot x$, by

$$\cot x = \frac{\cos x}{\sin x}.$$

3. If x is not an odd multiple of $\pi/2$, define the **secant of x**, written $\sec x$, by

$$\sec x = \frac{1}{\cos x}.$$

4. If x is not an integral multiple of π, define the **cosecant of x**, written $\csc x$, by

$$\csc x = \frac{1}{\sin x}.$$

Note that, because $\cos x = 0$ if and only if x is an odd multiple of $\pi/2$, $\tan x$ and $\sec x$ are defined if and only if x is *not* an odd multiple of $\pi/2$. Similarly, because $\sin x = 0$ if and only if x is an integral multiple of π, $\cot x$ and $\csc x$ are defined if and only if x is *not* an integral multiple of π.

EXAMPLE 4 We have

$$\tan \frac{\pi}{4} = \frac{\sin \pi/4}{\cos \pi/4} = \frac{1/\sqrt{2}}{1/\sqrt{2}} = 1, \qquad \cot \frac{\pi}{4} = \frac{\cos \pi/4}{\sin \pi/4} = \frac{1/\sqrt{2}}{1/\sqrt{2}} = 1,$$

$$\sec \frac{\pi}{4} = \frac{1}{\cos \pi/4} = \frac{1}{1/\sqrt{2}} = \sqrt{2}, \qquad \csc \frac{\pi}{4} = \frac{1}{\sin \pi/4} = \frac{1}{1/\sqrt{2}} = \sqrt{2},$$

$$\tan \frac{2\pi}{3} = \frac{\sqrt{3}/2}{-1/2} = -\sqrt{3}, \qquad \cot \frac{2\pi}{3} = \frac{-1/2}{\sqrt{3}/2} = -\frac{1}{\sqrt{3}},$$

$$\sec \frac{2\pi}{3} = \frac{1}{-1/2} = -2 \qquad \csc \frac{2\pi}{3} = \frac{1}{\sqrt{3}/2} = 2/\sqrt{3}.$$

Also, $\tan 0 = 0$, $\sec 0 = 1$, $\cot 0$ and $\csc 0$ are undefined, $\cot \pi/2 = 0$, $\csc \pi/2 = 1$, and $\tan \pi/2$ and $\sec \pi/2$ are undefined. ∎

We conclude this section with a brief discussion of the relationship between the trigonometric functions and sides of right triangles. Consider the right triangle ABC of Figure 11-21: if the angle at vertex A of the triangle measures x radians as indicated in the figure, then

$$\sin x = \frac{\text{Length of side } BC}{\text{Length of side } AC},$$

which we write less formally as

$$\sin x = \frac{\text{Opposite side}}{\text{Hypotenuse}}.$$

This is so because if we place the vertex A of the triangle at the origin, as in Figure 11-22, then triangle APQ is similar to triangle ABC, and hence

$$\frac{\text{Length of } PQ}{\text{Length of } AQ} = \frac{\text{Length of } BC}{\text{Length of } AC},$$

or

$$\frac{\sin x}{1} = \frac{\text{Length of } BC}{\text{Length of } AC}.$$

Similar reasoning shows that

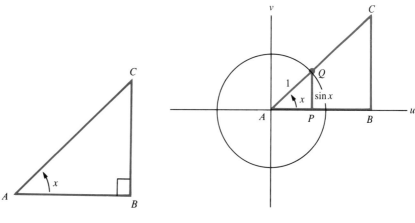

FIGURE 11-21   FIGURE 11-22

$$\cos x = \frac{\text{Length of } AB}{\text{Length of } AC} = \frac{\text{Adjacent side}}{\text{Hypotenuse}},$$

$$\tan x = \frac{\text{Opposite side}}{\text{Adjacent side}}, \quad \cot x = \frac{\text{Adjacent side}}{\text{Opposite side}},$$

$$\sec x = \frac{\text{Hypotenuse}}{\text{Adjacent side}}, \quad \csc x = \frac{\text{Hypotenuse}}{\text{Opposite side}}.$$

EXAMPLE 5 Let us find sin x, cos x, tan x, sec x, csc x, and cot x for the angle of Figure 11–23. The length of the hypotenuse of this triangle is $\sqrt{2^2 + 3^2} = \sqrt{13}$. Thus

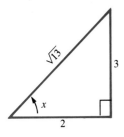

FIGURE 11–23

$$\sin x = \frac{\text{Opposite side}}{\text{Hypotenuse}} = \frac{3}{\sqrt{13}},$$

$$\cos x = \frac{\text{Adjacent side}}{\text{Hypotenuse}} = \frac{2}{\sqrt{13}},$$

$$\tan x = \frac{\text{Opposite side}}{\text{Adjacent side}} = \frac{3}{2},$$

$$\sec x = \frac{\text{Hypotenuse}}{\text{Adjacent side}} = \frac{\sqrt{13}}{2},$$

$$\csc x = \frac{\text{Hypotenuse}}{\text{Opposite side}} = \frac{\sqrt{13}}{3},$$

and

$$\cot x = \frac{\text{Adjacent side}}{\text{Opposite side}} = \frac{2}{3}. \quad \square$$

EXAMPLE 6 Let us find the radian measure x of the angle of Figure 11–24(a) and the lengths r and s of Figure 11–24(b). From Figure 11–24(a),

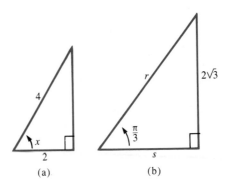

FIGURE 11–24

$$\cos x = \frac{2}{4} = \frac{1}{2}.$$

The angle between 0 and $\pi/2$ whose cosine is $\frac{1}{2}$ is $\pi/3$. Thus $x = \pi/3$. From Figure 11–24(b),

$$\sin \frac{\pi}{3} = \frac{2\sqrt{3}}{r},$$

or

$$\frac{\sqrt{3}}{2} = \frac{2\sqrt{3}}{r},$$

which implies that $r = 4$, and hence that

$$s = \sqrt{r^2 - (2\sqrt{3})^2} = \sqrt{16 - 12} = 2. \quad \square$$

■ *Computer Manual: Program TRIGFNS*

11.2 EXERCISES

1. Find $\cos 3\pi$, $\sin 3\pi$, $\cos\left(-\frac{3\pi}{2}\right)$, and $\sin\left(-\frac{3\pi}{2}\right)$.

2. Find $\cos(-\pi)$, $\sin(-\pi)$, $\cos\frac{5\pi}{2}$, and $\sin\frac{5\pi}{2}$.

3. Find $\sin\frac{7\pi}{2}$, $\cos\frac{7\pi}{2}$, $\sin(-2\pi)$, and $\cos(-2\pi)$.

4. Find $\sin\left(-\frac{9\pi}{2}\right)$, $\cos\left(-\frac{9\pi}{2}\right)$, $\sin 6\pi$, and $\cos 6\pi$.

5. (a) For what values of x is $\sin x = 0$?
   (b) For what values of x is $\cos x = 0$?

6. (a) For what values of x is $\sin x = 1$?
   (b) For what values of x is $\sin x = -1$?

7. (a) For what values of x is $\cos x = 1$?
   (b) For what values of x is $\cos x = -1$?

8. Use the results of Examples 2 and 3 and trigonometric identities to show that
$$\cos\frac{7\pi}{4} = \frac{1}{\sqrt{2}} \text{ and } \sin\frac{7\pi}{4} = -\frac{1}{\sqrt{2}}.$$

9. Let ABC be an equilateral triangle each of whose sides has length 2. Since the angles of a triangle add up to 180° and the angles of an equilateral triangle are equal, it follows that each angle of triangle ABC measures 60° or $\pi/3$ radians:

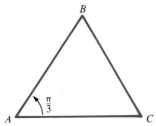

If we construct a perpendicular from vertex B to side AC, it will bisect AC and also bisect the angle at B. Thus we obtain the triangle ABP:

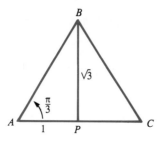

(The length of side BP is equal to $\sqrt{3}$ because ABP is a right triangle; hence $2^2 = 1^2 + (\text{length of } BP)^2$.) Use triangle ABP to show that $\sin \dfrac{\pi}{3} = \dfrac{\sqrt{3}}{2}$ and $\cos \dfrac{\pi}{3} = \dfrac{1}{2}$.

10. Use the triangle ABP of Exercise 9 to show that $\sin \dfrac{\pi}{6} = \dfrac{1}{2}$ and $\cos \dfrac{\pi}{6} = \dfrac{\sqrt{3}}{2}$.

11. Use trigonometric identities and the results of Exercise 9 to show that
$$\sin \dfrac{2\pi}{3} = \dfrac{\sqrt{3}}{2} \text{ and } \cos \dfrac{2\pi}{3} = -\dfrac{1}{2}.$$

12. Use trigonometric identities and the results of Exercises 10 and 11 to show that
$$\sin \dfrac{5\pi}{6} = \dfrac{1}{2} \text{ and } \cos \dfrac{5\pi}{6} = -\dfrac{\sqrt{3}}{2}.$$

13. Use trigonometric identities and the results of Exercise 11 to show that
$$\sin \dfrac{4\pi}{3} = -\dfrac{\sqrt{3}}{2} \text{ and } \cos \dfrac{4\pi}{3} = -\dfrac{1}{2}.$$

14. Use trigonometric identities and the results of Exercises 9 and 13 to show that
$$\sin \dfrac{5\pi}{3} = -\dfrac{\sqrt{3}}{2} \text{ and } \cos \dfrac{5\pi}{3} = \dfrac{1}{2}.$$

15. Use trigonometric identities and the results of previous exercises to show that
$$\sin \dfrac{7\pi}{6} = -\dfrac{1}{2}, \cos \dfrac{7\pi}{6} = -\dfrac{\sqrt{3}}{2}, \sin \dfrac{11\pi}{6} = -\dfrac{1}{2}, \text{ and } \cos \dfrac{11\pi}{6} = \dfrac{\sqrt{3}}{2}.$$

*16. Prove the double-angle identities
  (a) $\sin 2x = 2 \sin x \cos x$
  (b) $\cos 2x = \cos^2 x - \sin^2 x$

  (*Hint:* Use the sum formulas.)

*17. Use the results of Exercise 16 to prove the following identities:
  (a) $\sin^2 x = (1 - \cos 2x)/2$
  (b) $\cos^2 x = (1 + \cos 2x)/2$

18. Use the identities of Exercise 17 to find $\sin \dfrac{\pi}{8}$ and $\cos \dfrac{\pi}{8}$.

19. Use the identities of Exercise 17 to find $\sin \dfrac{\pi}{12}$ and $\cos \dfrac{\pi}{12}$.

20. Use the identities of Exercise 17 and the results of Exercise 18 to find $\sin \dfrac{\pi}{16}$ and $\cos \dfrac{\pi}{16}$.

21. Use the identities of Exercise 17 and the results of Exercise 19 to find $\sin \dfrac{\pi}{24}$ and $\cos \dfrac{\pi}{24}$.

22. If possible, find
   (a) $\tan 2\pi$, $\cot 2\pi$, $\sec 2\pi$, and $\csc 2\pi$
   (b) $\tan \dfrac{\pi}{6}$, $\cot \dfrac{\pi}{6}$, $\sec \dfrac{\pi}{6}$, and $\csc \dfrac{\pi}{6}$

23. If possible, find
   (a) $\tan\left(-\dfrac{\pi}{2}\right)$, $\cot\left(-\dfrac{\pi}{2}\right)$, $\sec\left(-\dfrac{\pi}{2}\right)$, and $\csc\left(-\dfrac{\pi}{2}\right)$
   (b) $\tan \dfrac{\pi}{3}$, $\cot \dfrac{\pi}{3}$, $\sec \dfrac{\pi}{3}$, and $\csc \dfrac{\pi}{3}$

24. If possible, find
   (a) $\tan \pi$, $\cot \pi$, $\sec \pi$, $\csc \pi$
   (b) $\tan \dfrac{5\pi}{3}$, $\cot \dfrac{5\pi}{3}$, $\sec \dfrac{5\pi}{3}$, and $\csc \dfrac{5\pi}{3}$

25. Find
   (a) $\tan \dfrac{5\pi}{6}$, $\cot \dfrac{5\pi}{6}$, $\sec \dfrac{5\pi}{6}$, and $\csc \dfrac{5\pi}{6}$
   (b) $\tan\left(-\dfrac{3\pi}{4}\right)$, $\cot\left(-\dfrac{3\pi}{4}\right)$, $\sec\left(-\dfrac{3\pi}{4}\right)$, and $\csc\left(-\dfrac{3\pi}{4}\right)$

26. Find
   (a) $\tan \dfrac{5\pi}{4}$, $\cot \dfrac{5\pi}{4}$, $\sec \dfrac{5\pi}{4}$, and $\csc \dfrac{5\pi}{4}$
   (b) $\tan\left(-\dfrac{7\pi}{6}\right)$, $\cot\left(-\dfrac{7\pi}{6}\right)$, $\sec\left(-\dfrac{7\pi}{6}\right)$, and $\csc\left(-\dfrac{7\pi}{6}\right)$

27. Find $\tan \dfrac{\pi}{8}$, $\cot \dfrac{\pi}{8}$, $\sec \dfrac{\pi}{8}$, and $\csc \dfrac{\pi}{8}$.

28. Find $\tan \dfrac{\pi}{12}$, $\cot \dfrac{\pi}{12}$, $\sec \dfrac{\pi}{12}$, and $\csc \dfrac{\pi}{12}$.

*29. Prove the following identity:
$$\tan(x + y) = \frac{\tan x + \tan y}{1 - \tan x \tan y}.$$

30. Use the result of Exercise 29 to find $\tan \dfrac{5\pi}{12}$ and $\tan \dfrac{11\pi}{12}$.

*31. Use the result of Exercise 29 to prove the identity

$$\tan 2x = \frac{2 \tan x}{1 - \tan^2 x}$$

32. Use the result of Exercise 31 and the results of Exercises 27 and 28 to find $\tan \frac{\pi}{16}$ and $\tan \frac{\pi}{24}$.

*33. Prove the identity $\sec^2 x = 1 + \tan^2 x$.

34. Use the result of Exercise 33 and the results of Exercise 32 to find $\sec \frac{\pi}{16}$ and $\sec \frac{\pi}{24}$.

In Exercises 35 through 37, find sin x, cos x, tan x, sec x, csc x, and cot x.

35.   **36.**   **37.**

In Exercises 38 through 40, find the lengths r and s.

38.   **39.**   **40.**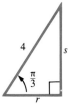

In Exercises 41 through 43, find the radian measure x of the angle.

41.   **42.**   **43.**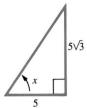

*44. In this exercise we outline a proof of the additive trigonometric identities

$$\sin(x + y) = \sin x \cos y + \sin y \cos x$$

and

$$\cos(x + y) = \cos x \cos y - \sin x \sin y.$$

Let angle *POQ* measure x radians, angle *POR* measure y radians, and angle *POS* measure $x + y$ radians:

Trigonometric Functions

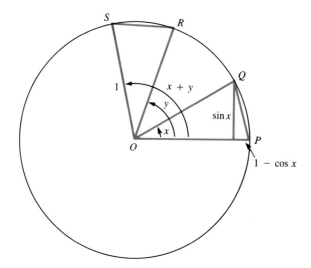

By definition,

$$Q = (\cos x, \sin x), \quad R = (\cos y, \sin y),$$

and

$$S = (\cos(x + y), \sin(x + y)).$$

Also, since angle *ROS* measures $x + y - y = x$ radians and angle *POQ* also measures x radians, it follows that

$$\text{Length of segment } SR = \text{Length of segment } QP.$$

(a) Show that the square of the length of *SR* is equal to

$$(\cos y - \cos(x + y))^2 + (\sin y - \sin(x + y))^2$$

and that the square of the length of *QP* is equal to

$$\sin^2 x + (1 - \cos x)^2.$$

(b) Simplify the expressions of part(a), set them equal to each other, and solve for $\cos x$ to obtain

$$\cos x = \cos y \cos(x + y) + \sin y \sin(x + y).$$

(c) Since the result of part (b) holds for all x and all y, we may substitute $x + y$ for x and $-x$ for y in it. Do this and show that the result is the trigonometric identity

$$\cos(x + y) = \cos x \cos y - \sin x \sin y. \tag{1}$$

(d) Now substitute $-\dfrac{\pi}{2}$ for x in identity (1) and show that the result is the identity

$$\cos\left(y - \dfrac{\pi}{2}\right) = \sin y. \tag{2}$$

(e) Now substitute $y - \dfrac{\pi}{2}$ for y in (2) and expand the result using (1) to obtain the identity

$$\sin\left(y - \dfrac{\pi}{2}\right) = -\cos y. \tag{3}$$

(f) Now substitute $x + y$ for y in (2) to get

$$\sin(x + y) = \cos\left(x + y - \frac{\pi}{2}\right) = \cos\left(x + \left(y - \frac{\pi}{2}\right)\right);$$

expand $\cos\left(x + \left(y - \frac{\pi}{2}\right)\right)$ using (1) and simplify using (2) and (3) to obtain the identity

$$\sin(x + y) = \sin x \cos y + \sin y \cos x.$$

11.3 THE TRIGONOMETRIC FUNCTIONS

Now that we have defined the sine, cosine, tangent, cosecant, secant, and cotangent of an angle of x radians, we can define the **trigonometric functions** $y = \sin x$, $y = \cos x$, $y = \tan x$, $y = \csc x$, $y = \sec x$, and $y = \cot x$, x in radians. In this section we discuss these functions, their properties, and their graphs. We begin with the sine function.

The **sine function** is the function defined by $y = \sin x$ for all real numbers x. (We emphasize here that by $\sin x$ we mean the sine of an angle of x *radians*.) The domain of the sine function is the set **R** of real numbers. The information of Section 11.2 allows us to sketch the graph of the sine function over the interval $[0, 2\pi]$. See Figure 11–25. However, the identity

$$\sin(x + 2\pi) = \sin x$$

implies that the graph of $y = \sin x$ looks the same over the interval $[2\pi, 4\pi]$ as it does over the interval $[0, 2\pi]$, the same over $[4\pi, 6\pi]$ as over $[0, 2\pi]$, the same over $[-2\pi, 0]$ as over $[0, 2\pi]$, and so on. Hence the complete graph of $y = \sin x$ is as shown in Figure 11–26. Note that the sine function is continuous for all x.

As we have seen, the graph of the sine function repeats itself every 2π units. A function whose graph repeats itself is said to be **periodic**; its **period** is the length of the interval of repetition. Thus the sine function is periodic with period 2π. Note also that the "waves" of the sine function go from a value of $y = -1$ to a value of $y = 1$, that is, that the range of the function is the interval $[-1, 1]$; we express this by saying that the sine function has **amplitude** 1.

We can create functions whose graphs have the same shape as $y = \sin x$, but with different periods and amplitudes, by multiplying x by a constant b and $\sin x$

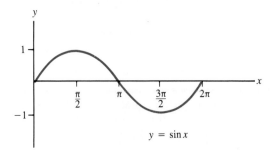

FIGURE 11–25

Trigonometric Functions

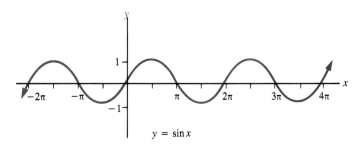

FIGURE 11–26

by a constant a: multiplying x by b changes the period of the sine curve, and multiplying $\sin x$ by a changes its amplitude.

EXAMPLE 1 Let us graph the function $y = 3 \sin 2x$. The graph will have the same shape as that of $y = \sin x$, but its period will be π rather than 2π. The reason for this is that as x ranges from 0 to π, $2x$ ranges from 0 to 2π; hence $\sin 2x$ will begin to repeat as soon as x reaches π. Furthermore, since

$$-1 \leq \sin 2x \leq 1,$$

we must have

$$-3 \leq 3 \sin 2x \leq 3,$$

or

$$-3 \leq y \leq 3.$$

Hence the amplitude of this function will be 3. A table of values for $y = 3 \sin 2x$ over the interval $[0, \pi]$ is as follows:

x:	0	$\dfrac{\pi}{6}$	$\dfrac{\pi}{4}$	$\dfrac{\pi}{3}$	$\dfrac{\pi}{2}$	$\dfrac{2\pi}{3}$	$\dfrac{3\pi}{4}$	$\dfrac{5\pi}{6}$	π
$2x$:	0	$\dfrac{\pi}{3}$	$\dfrac{\pi}{2}$	$\dfrac{2\pi}{3}$	π	$\dfrac{4\pi}{3}$	$\dfrac{3\pi}{2}$	$\dfrac{5\pi}{3}$	2π
$\sin 2x$:	0	$\dfrac{\sqrt{3}}{2}$	1	$\dfrac{\sqrt{3}}{2}$	0	$-\dfrac{\sqrt{3}}{2}$	-1	$-\dfrac{\sqrt{3}}{2}$	0
$3 \sin 2x$:	0	$\dfrac{3\sqrt{3}}{2}$	3	$\dfrac{3\sqrt{3}}{2}$	0	$-\dfrac{3\sqrt{3}}{2}$	-3	$-\dfrac{3\sqrt{3}}{2}$	0

The graph of $y = 3 \sin 2x$ is shown in Figure 11–27. ☐

In general, a function whose defining equation is of the form $y = a \sin bx$, $a > 0$, $b > 0$, will have as its graph a sine curve with period $2\pi/b$ and amplitude a, and hence the graph will be as in Figure 11–28. See Exercise 22.

Now let us consider what will happen to the graph of a sine function if we add a constant to the independent variable before we take its sine.

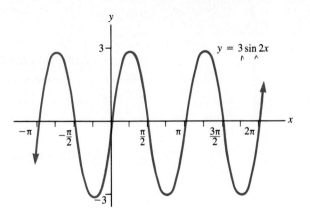

FIGURE 11–27

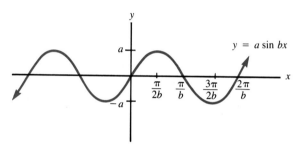

FIGURE 11–28

EXAMPLE 2 The graph of the function defined by

$$y = \sin\left(x + \frac{\pi}{4}\right)$$

is the same as that of $y = \sin x$ except that it is shifted $\pi/4$ units to the left. See Figure 11–29. (Check this.) Similarly, the graph of

$$y = 3 \sin\left(2x - \frac{\pi}{3}\right)$$

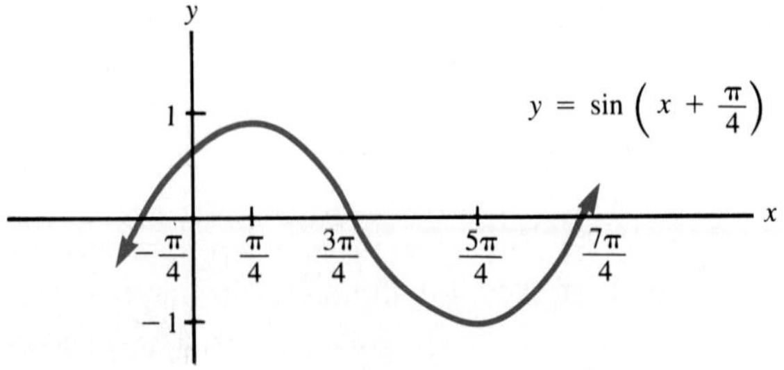

FIGURE 11–29

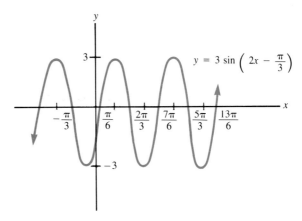

FIGURE 11-30

is shown in Figure 11–30; it is the graph of $y = 3 \sin 2x$ shifted $\pi/6$ units to the right. ☐

In general, the graph of $y = a \sin(bx + c)$, $a > 0$, $b > 0$, will be the same as that of $y = a \sin bx$, but shifted c/b units to the left if c is positive and c/b units to the right if c is negative. See Exercise 23.

Because the sine function is periodic, it is often used to model repetitive situations in nature. For instance, if sales of a product depend on the season in a repetitive manner, it may be possible to model sales using a sine function.

EXAMPLE 3 Sam's Ski Shop sells y thousand dollars of ski equipment per month t, where

$$y = 25 \sin\left(\frac{\pi}{6}t + \frac{2\pi}{3}\right) + 32.$$

Here $0 \le t \le 12$, with $t = 0$ representing January 1. The graph of this function is shown in Figure 11–31. Notice how sales are high during autumn and winter,

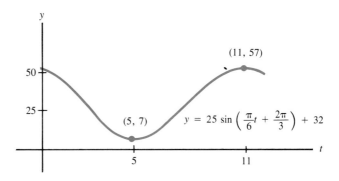

FIGURE 11-31

reaching a maximum of $57,000 when $t = 11$ (that is, in November). They are low during spring and summer, with a minimum of $7000 in May. □

Now let us consider the **cosine function** defined by $y = \cos x$ for all x. Since

$$\cos x = \sin\left(x + \frac{\pi}{2}\right),$$

by what we have already done the graph of the cosine function is the same as that of $y = \sin x$, except that it is shifted $\pi/2$ units to the left. See Figure 11–32. Thus

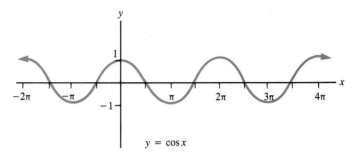

FIGURE 11–32

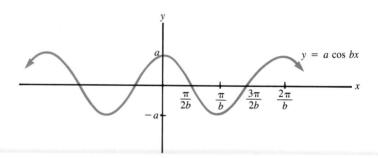

FIGURE 11–33

the cosine function is continuous, periodic with period 2π, and has amplitude 1. Just as for sine functions, the graph of $y = a \cos bx$, $a > 0$, $b > 0$, will be a cosine curve with period $2\pi/b$ and amplitude a, as in Figure 11–33. (See Exercise 24.) The graph of $y = a \cos(bx + c)$, $a > 0$, $b > 0$, will be the same as that of $y = a \cos bx$, except that it will be shifted c/b units to the left if c is positive and c/b units to the right if c is negative. (See Exercise 25.)

EXAMPLE 4   The graph of

$$y = 2 \cos 3x$$

is shown in Figure 11–34. The graph of

TRIGONOMETRIC FUNCTIONS

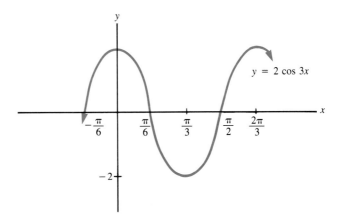

FIGURE 11-34

$$y = 2 \cos\left(3x - \frac{\pi}{6}\right)$$

is shown in Figure 11-35. ☐

The **tangent function** is defined by $y = \tan x$ for all x not an odd multiple of $\pi/2$. The graph of the tangent function is shown in Figure 11-36. Notice the vertical asymptotes at odd multiples of $\pi/2$: if n is an odd integer, then

$$\lim_{x \to n\pi/2^-} \tan x = +\infty \quad \text{and} \quad \lim_{x \to n\pi/2^+} \tan x = -\infty.$$

The tangent function is continuous where it is defined, and its graph repeats every π units, so it is periodic with period π. A function whose defining equation is of the form $y = a \tan bx$, $a > 0$, $b > 0$, will have a graph like that of $y = \tan x$, except that its asymptotes will be located at odd multiples of $\pi/2b$, and

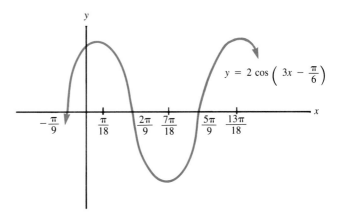

FIGURE 11-35

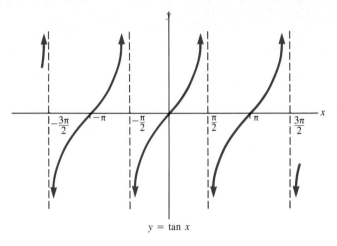

FIGURE 11–36

thus its period will be π/b. A function whose defining equation is of the form $y = a \tan(bx + c)$, $a > 0$, $b > 0$, will have a graph like that of $y = a \tan bx$, but shifted to the left or the right by c/b units, depending on whether c is positive or negative. See Exercises 36 and 37.

EXAMPLE 5 The graph of $y = 2 \tan 3x$ is shown in Figure 11–37. Its asymptotes are located at odd multiples of $\pi/2b = \pi/6$. The graph of $y = 2 \tan(3x - \pi/3)$ is shown in Figure 11–38. ☐

EXAMPLE 6 The cost of removing x percent of the pollutants from the air over a large city would be $\$y$ billion, where

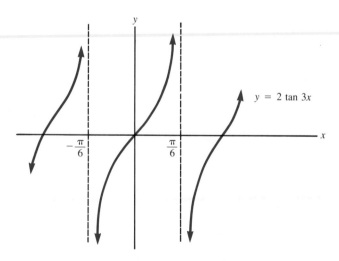

FIGURE 11–37

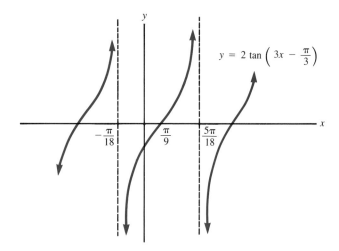

FIGURE 11–38

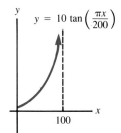

FIGURE 11–39

$y = 10 \tan(\pi x/200), \quad 0 \le x < 100.$

This function has a vertical asymptote at $\pi/[2(\pi/200)] = 100$. See Figure 11–39. Note that the cost of removing the last few vestiges of the pollution approaches $+\infty$. ☐

There are three more trigonometric functions:

the **cosecant function,** defined by $y = \csc x$, for all x not an integral multiple of π;

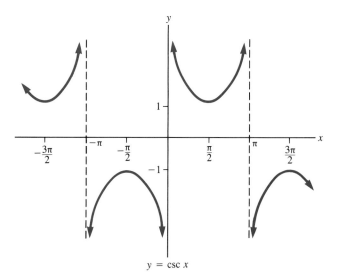

FIGURE 11–40

the **secant function,** defined by $y = \sec x$, for all x not an odd multiple of $\pi/2$;

the **cotangent function,** defined by $y = \cot x$, for all x not an integral multiple of π.

These are not as common as the sine, cosine, and tangent functions, so we merely present their graphs in Figures 11–40, 11–41, and 11–42. Note that the cosecant and secant functions have period 2π, while the cotangent function has period π.

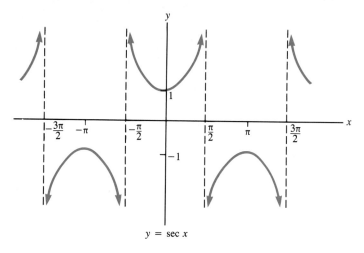

$y = \sec x$

FIGURE 11–41

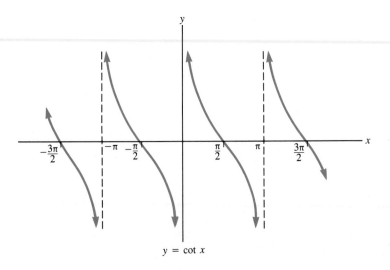

$y = \cot x$

FIGURE 11–42

11.3 EXERCISES

1. Does $\lim_{x \to +\infty} \sin x$ exist, either as a finite number or as $\pm\infty$? What about $\lim_{x \to +\infty} \cos x$ and $\lim_{x \to +\infty} \tan x$?

In Exercises 2 through 21, graph the function defined by the given equation and state its period and amplitude.

2. $y = \sin 3x$
3. $y = \sin \dfrac{x}{2}$
4. $y = \cos 2x$
5. $y = \cos \dfrac{x}{3}$
6. $y = \sin \dfrac{2x}{3}$
7. $y = \cos \dfrac{3x}{2}$
8. $y = 2 + \cos 2x$
9. $y = 3 + \sin 3x$
10. $y = 2 \sin 3x$
11. $y = \dfrac{1}{2} \sin 2x$
12. $y = 3 \sin \dfrac{x}{4}$
13. $y = 4 \cos 3x$
14. $y = 2 \cos 2\pi x$
15. $y = \dfrac{3}{2} \cos \dfrac{\pi x}{4}$
16. $y = 5 \sin \left(x - \dfrac{\pi}{2}\right)$
17. $y = 2 \sin \left(2x + \dfrac{\pi}{3}\right)$
18. $y = \dfrac{3}{4} \sin \left(\dfrac{x}{2} + \dfrac{\pi}{4}\right)$
19. $y = 4 + 6 \sin \left(8x - \dfrac{\pi}{6}\right)$
20. $y = 3 + 4 \cos \left(x + \dfrac{\pi}{4}\right)$
21. $y = 5 - 3 \cos \left(\pi x - \dfrac{4\pi}{3}\right)$

*22. Sketch the graph of $y = a \sin bx$, $a > 0$, $b > 0$, and thus show that the period of the function defined by this equation is $2\pi/b$ and its amplitude is a.

*23. Sketch the graph of $y = a \sin(bx + c)$, $a > 0$, $b > 0$, when
   (a) $c > 0$ (b) $c < 0$
   and thus show that it is the graph of $y = a \sin bx$ shifted c/b units to the left or the right, respectively.

*24. Sketch the graph of $y = a \cos bx$, $a > 0$, $b > 0$, and thus show that the period of the function defined by this equation is $2\pi/b$ and its amplitude is a.

*25. Sketch the graph of $y = a \cos(bx + c)$, $a > 0$, $b > 0$, when
   (a) $c > 0$ (b) $c < 0$
   and thus show that it is the graph of $y = a \cos bx$ shifted c/b units to the left or the right, respectively.

In Exercises 26 through 35, graph the function defined by the given equation and state its period.

26. $y = \tan 2x$
27. $y = \tan \dfrac{x}{2}$
28. $y = 3 \tan 2x$
29. $y = 4 + \tan \dfrac{x}{3}$
30. $y = \dfrac{2}{3} \tan 4x$
31. $y = \dfrac{5}{4} \tan \dfrac{3x}{4}$
32. $y = 6 + \tan \left(2x - \dfrac{\pi}{2}\right)$
33. $y = 3 \tan \left(3x + \dfrac{\pi}{3}\right)$
34. $y = 2 \tan \left(\dfrac{x}{2} + \dfrac{2\pi}{3}\right)$
35. $y = \dfrac{1}{2} \tan \left(\dfrac{3x}{2} - \dfrac{\pi}{6}\right)$

*36. Graph the function defined by $y = a \tan bx$, $a > 0$, $b > 0$, and thus show that its period is π/b. How does the size of the constant a affect the graph?

*37. Sketch the graph of $y = a \tan(bx + c)$, $a > 0$, $b > 0$, when
  (a) $c > 0$ (b) $c < 0$

and thus show that it is the graph of $y = a \tan bx$ shifted c/b units to the left or the right, respectively.

In Exercises 38 through 47, graph the function defined by the given equation and state its period.

*38. $y = \csc 2x$

*39. $y = \sec 3x$

*40. $y = \cot \dfrac{x}{2}$

*41. $y = 2 \cot 3x$

*42. $y = 2 \csc 4x$

*43. $y = 3 \sec \dfrac{x}{3}$

*44. $y = \dfrac{3}{4} \sec\left(2x + \dfrac{\pi}{3}\right)$

*45. $y = 3 \csc\left(2x - \dfrac{\pi}{4}\right)$

*46. $y = 5 \sec\left(\dfrac{2x}{3} - \dfrac{\pi}{6}\right)$

*47. $y = 4 \cot\left(\dfrac{3x}{4} + \dfrac{\pi}{2}\right)$

Management

48. The monthly sales of a store's swimsuit department follow a seasonal pattern: in month t, $0 \le t \le 12$, they are $\$y$ thousand, where

$$y = 20 \sin\left(\dfrac{\pi}{6}t - \dfrac{\pi}{2}\right) + 25.$$

Here $t = 0$ represents January 1 of each year. Graph this function and analyze the sales of the swimsuit department.

49. Repeat Exercise 48 if $y = 20 \sin\left(\dfrac{\pi}{6}t - \dfrac{2\pi}{3}\right) + 25.$

*50. Refer to Exercises 48 and 49: Can you find an equation that describes sales as a function of time if sales follow a sine curve with a single maximum of $40,000 in August and a single minimum of $4000 in February?

51. A firm's weekly revenue is $\$y$ thousand, where

$$y = 2 + 2 \cos \dfrac{\pi t}{26}, \qquad 0 \le t \le 52.$$

Here t is in weeks, with $t = 0$ representing January 1 of each year. Graph this function and analyze the firm's revenue over the course of a year.

52. Repeat Exercise 51 if $y = 2 + 2 \cos\left(\dfrac{\pi t}{26} + \dfrac{3\pi}{4}\right).$

53. Repeat Exercise 51 if $y = 4 + 2 \cos \dfrac{\pi t}{13}.$

54. During the year beef prices fluctuated according to the equation

$$p = 0.3 \sin \dfrac{\pi t}{3}.$$

Here p is the price per pound and t is time in months, $0 \leq t \leq 12$, with $t = 0$ representing January 1. Graph this function and analyze the fluctuations in beef prices during the year.

*55. Suppose that during the course of a year the price of pork follows a cosine curve, ranging from $2.25 per pound to $2.75 per pound and peaking at $2.75 per pound at the end of April, August, and December. Can you find an equation that describes the price of pork over the course of the year?

56. During one year the price of a company's stock varied according to the equation

$$p = 6.50 \cos\left(\frac{\pi t}{13} + \frac{\pi}{4}\right).$$

Here p is the price per share and t is time in weeks, with $0 \leq t \leq 52$ and $t = 0$ representing January 1. Graph this function and analyze the changes in the price of the company's stock over the course of the year.

Life Science

57. An individual's blood pressure reading at time t is

$$P = 110 + 25 \cos 140\pi t,$$

where $0 \leq t \leq 1$, t in minutes. What are the individual's systolic and diastolic blood pressure readings? What is the individual's pulse rate?

58. A patient's temperature varies during the day according to the equation

$$y = 98.6 + 1.1 \sin\left(\frac{\pi t}{12} + \frac{\pi}{12}\right), \qquad 0 \leq t \leq 24.$$

Here temperature y is in degrees Fahrenheit and time t is in hours, with $t = 0$ representing noon. Graph this function and analyze the patient's temperature over the course of the day.

59. If x per cent of the acid in a lake that has been stressed by acid rain is to be neutralized, it will cost

$$y = 12 + \tan \frac{\pi x}{200}$$

million dollars. Here $0 \leq x < 100$. Graph this function and analyze the cost of deacidifying the lake.

60. Repeat Exercise 59 if $y = 8 \tan\left(\frac{\pi x}{220} + \frac{\pi}{6}\right)$.

11.4 DIFFERENTIATION OF TRIGONOMETRIC FUNCTIONS

In this section we develop formulas for differentiating the trigonometric functions and illustrate some of their uses. We begin with the derivative of the sine function.

As we will soon see, in order to develop the formula for the derivative of $y = \sin x$ we will need to evaluate the limits

$$\lim_{h \to 0} \frac{\sin h}{h} \quad \text{and} \quad \lim_{h \to 0} \frac{\cos h - 1}{h}.$$

Tables 11–3 and 11–4, which were obtained with the aid of a calculator, suggest that the first of these limits is 1 and the second is 0.

TABLE 11–3

h	1	0.5	0.25	0.1
$\sin h$	0.841471	0.479426	0.247404	0.099833
$\dfrac{\sin h}{h}$	0.841471	0.958852	0.989616	0.99833
h	-1	-0.5	-0.25	-0.1
$\sin h$	-0.841471	-0.479426	-0.247404	-0.099833
$\dfrac{\sin h}{h}$	0.841471	0.958852	0.989616	0.99833

TABLE 11–4

h	1	0.5	0.25	0.1
$\cos h$	0.540302	0.877583	0.968912	0.995004
$\dfrac{\cos h - 1}{h}$	-0.459698	-0.244834	-0.124352	-0.049960
h	-1	-0.5	-0.25	-0.1
$\cos h$	0.540302	0.877583	0.968912	0.995004
$\dfrac{\cos h - 1}{h}$	0.459698	0.244834	0.124352	0.049960

We can verify these facts with the aid of Figure 11–43, which depicts an angle POQ of h radians. From the figure we see that

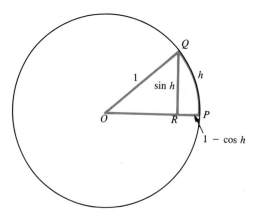

FIGURE 11–43

TRIGONOMETRIC FUNCTIONS

$$\frac{\sin h}{h} = \frac{\text{Length of segment } RQ}{\text{Length of arc } PQ}.$$

As h approaches 0, that is, as the angle POQ becomes smaller and smaller, the point Q slides down the circle toward the point P; hence the segment RQ and the arc PQ come closer and closer to coinciding, and therefore the ratio of their lengths tends toward 1. Thus

$$\lim_{h \to 0} \frac{\sin h}{h} = 1.$$

Figure 11-43 also shows that the length of the segment RP is very small compared to the length of the arc PQ, and the ratio

$$\frac{1 - \cos h}{h} = \frac{\text{Length of segment } RP}{\text{Length of arc } PQ}$$

becomes smaller and smaller as h approaches 0. Therefore

$$\lim_{h \to 0} \frac{1 - \cos h}{h} = 0,$$

and hence

$$\lim_{h \to 0} \frac{\cos h - 1}{h} = \lim_{h \to 0} \frac{-(1 - \cos h)}{h} = -\lim_{h \to 0} \frac{1 - \cos h}{h} = 0.$$

Now we are ready to find the derivative of $\sin x$. To do so, we must evaluate

$$\lim_{h \to 0} \frac{\sin(x + h) - \sin x}{h}.$$

But

$$\lim_{h \to 0} \frac{\sin(x + h) - \sin x}{h} = \lim_{h \to 0} \frac{\sin x \cos h + \sin h \cos x - \sin x}{h}$$

$$= \lim_{h \to 0} \frac{\sin x (\cos h - 1) + \sin h \cos x}{h}$$

$$= \lim_{h \to 0} \sin x \frac{\cos h - 1}{h} + \lim_{h \to 0} \cos x \frac{\sin h}{h}$$

$$= (\sin x) \lim_{h \to 0} \frac{\cos h - 1}{h} + (\cos x) \lim_{h \to 0} \frac{\sin h}{h}$$

$$= (\sin x)(0) + (\cos x)(1)$$

$$= \cos x.$$

Hence the derivative of the sine is the cosine.

EXAMPLE 1   If $y = 3 \sin x$, then

$$\frac{dy}{dx} = \frac{d}{dx}(3 \sin x) = 3 \frac{d}{dx}(\sin x) = 3 \cos x. \quad \square$$

Now consider a composite function of the form

$$y = \sin u,$$

where u is a function of x. By the chain rule,

$$\frac{dy}{dx} = \frac{d}{dx}(\sin u) = \frac{d}{du}(\sin u)\frac{du}{dx} = (\cos u)\frac{du}{dx}.$$

Derivative of the Sine

$$\frac{d}{dx}(\sin x) = \cos x$$

$$\frac{d}{dx}(\sin u) = (\cos u)\frac{du}{dx}$$

EXAMPLE 2 If $y = \sin 2x$, then

$$\frac{dy}{dx} = \frac{d}{dx}(\sin 2x) = (\cos 2x)\frac{d}{dx}(2x) = (\cos 2x)(2) = 2\cos 2x. \quad \square$$

EXAMPLE 3 If $y = \sin(x^2 - 3x + 5)$, then

$$\frac{dy}{dx} = \frac{d}{dx}(\sin(x^2 - 3x + 5))$$

$$= [\cos(x^2 - 3x + 5)]\frac{d}{dx}(x^2 - 3x + 5)$$

$$= [\cos(x^2 - 3x + 5)](2x - 3)$$

$$= (2x - 3)\cos(x^2 - 3x + 5). \quad \square$$

Now let us find the derivative of the cosine function. Since

$$\cos x = \sin\left(x + \frac{\pi}{2}\right) \quad \text{and} \quad \cos\left(x + \frac{\pi}{2}\right) = -\sin x$$

for all x, we have

$$\frac{d}{dx}(\cos x) = \frac{d}{dx}\left(\sin\left(x + \frac{\pi}{2}\right)\right)$$

$$= \left[\cos\left(x + \frac{\pi}{2}\right)\right]\frac{d}{dx}\left(x + \frac{\pi}{2}\right)$$

$$= \left[\cos\left(x + \frac{\pi}{2}\right)\right](1)$$

$$= \cos\left(x + \frac{\pi}{2}\right)$$

$$= -\sin x.$$

TRIGONOMETRIC FUNCTIONS

The chain rule now shows that if

$$y = \cos u,$$

then

$$\frac{dy}{dx} = \frac{d}{dx}(\cos u) = \frac{d}{du}(\cos u)\frac{du}{dx} = -(\sin u)\frac{du}{dx}.$$

Derivative of the Cosine

$$\frac{d}{dx}(\cos x) = -\sin x$$

$$\frac{d}{dx}(\cos u) = -(\sin u)\frac{du}{dx}$$

EXAMPLE 4 If $y = 5 \cos x$, then

$$\frac{dy}{dx} = \frac{d}{dx}(5 \cos x) = 5\frac{d}{dx}(\cos x) = 5(-\sin x) = -5 \sin x. \quad \square$$

EXAMPLE 5 If $y = \cos(x^2 + 1)$, then

$$\frac{dy}{dx} = \frac{d}{dx}(\cos(x^2 + 1))$$

$$= -[\sin(x^2 + 1)]\frac{d}{dx}(x^2 + 1)$$

$$= -[\sin(x^2 + 1)](2x)$$

$$= -2x \sin(x^2 + 1). \quad \square$$

EXAMPLE 6 If $y = \sin 2x \cos 3x$, then by the product rule

$$\frac{dy}{dx} = (\sin 2x)\frac{d}{dx}(\cos 3x) + \left[\frac{d}{dx}(\sin 2x)\right](\cos 3x)$$

$$= (\sin 2x)(-\sin 3x)\frac{d}{dx}(3x) + (\cos 2x)\left[\frac{d}{dx}(2x)\right](\cos 3x)$$

$$= -3 \sin 2x \sin 3x + 2 \cos 2x \cos 3x. \quad \square$$

EXAMPLE 7 If $y = \sin^2 x$, then [since $\sin^2 x$ means $(\sin x)^2$] application of the general power rule shows that

$$\frac{dy}{dx} = \frac{d}{dx}(\sin^2 x)$$

$$= \frac{d}{dx}[(\sin x)^2]$$

$$= 2(\sin x)\frac{d}{dx}(\sin x)$$

$$= 2 \sin x \cos x. \quad \square$$

EXAMPLE 8   If $y = e^{\cos^2 x}$, then

$$\frac{dy}{dx} = \frac{d}{dx}(e^{\cos^2 x})$$

$$= e^{\cos^2 x}\left[\frac{d}{dx}(\cos^2 x)\right]$$

$$= e^{\cos^2 x}(2 \cos x)\frac{d}{dx}(\cos x)$$

$$= e^{\cos^2 x}(2 \cos x)(-\sin x)$$

$$= -2(\sin x \cos x)e^{\cos^2 x}. \quad \square$$

EXAMPLE 9   Let us use the curve-sketching procedure to graph the function f defined by

$$f(x) = \sin x + \cos x$$

on the interval $[0, 2\pi]$.

Note that

$$f'(x) = \cos x - \sin x.$$

Setting $f'(x) = 0$ yields

$$\cos x - \sin x = 0$$

or

$$\cos x = \sin x.$$

The only solutions of this equation in the interval $[0, 2\pi]$ are $x = \pi/4$ and $x = 5\pi/4$. Thus we have the diagram

Choosing 0 from the interval $[0, \pi/4)$, π from the interval $(\pi/4, 5\pi/4)$, and 2π from the interval $(5\pi/4, 2\pi]$, we have

$$f'(0) = \cos 0 - \sin 0 = 1 - 0 = 1 > 0,$$
$$f'(\pi) = \cos \pi - \sin \pi = -1 - 0 = -1 < 0,$$

and
$$f'(2\pi) = \cos 2\pi - \sin 2\pi = 1 - 0 = 1 > 0.$$
Therefore the diagram becomes

and f has a relative maximum at
$$\left(\frac{\pi}{4}, f\left(\frac{\pi}{4}\right)\right) = \left(\frac{\pi}{4}, \frac{2}{\sqrt{2}}\right) = \left(\frac{\pi}{4}, \sqrt{2}\right)$$
and a relative minimum at
$$\left(\frac{5\pi}{4}, f\left(\frac{5\pi}{4}\right)\right) = \left(\frac{5\pi}{4}, -\frac{2}{\sqrt{2}}\right) = \left(\frac{5\pi}{4}, -\sqrt{2}\right).$$

Since
$$f''(x) = -\sin x - \cos x,$$
setting $f''(x) = 0$ yields
$$\sin x = -\cos x,$$
the solutions of which on $[0, 2\pi]$ are $x = 3\pi/4$ and $x = 7\pi/4$. Thus we have

```
0           3π/4         7π/4         2π
```

Again selecting 0, π, and 2π as our test numbers, we have
$$f''(0) = -\sin 0 - \cos 0 = 0 - 1 = -1 < 0,$$
$$f''(\pi) = -\sin \pi - \cos \pi = 0 - (-1) = 1 > 0,$$
and
$$f''(2\pi) = -\sin 2\pi - \cos 2\pi = 0 - 1 = -1 < 0.$$
Therefore the diagram for the second derivative becomes

```
       ∩         ∪         ∩
0           3π/4         7π/4         2π
```

so there are points of inflection at
$$\left(\frac{3\pi}{4}, f\left(\frac{3\pi}{4}\right)\right) = \left(\frac{3\pi}{4}, 0\right) \quad \text{and} \quad \left(\frac{7\pi}{4}, f\left(\frac{7\pi}{4}\right)\right) = \left(\frac{7\pi}{4}, 0\right).$$
The graph of f is shown in Figure 11–44. ∎

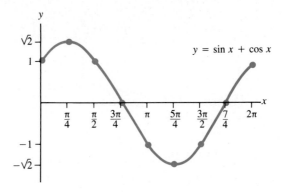

FIGURE 11–44

Now that we know the derivatives of the sine and cosine functions, we can easily find the derivatives of the rest of the trigonometric functions. For instance,

$$\frac{d}{dx}(\tan x) = \frac{d}{dx}\left(\frac{\sin x}{\cos x}\right)$$

$$= \frac{(\cos x)\frac{d}{dx}(\sin x) - (\sin x)\frac{d}{dx}(\cos x)}{(\cos x)^2}$$

$$= \frac{(\cos x)(\cos x) - (\sin x)(-\sin x)}{\cos^2 x}$$

$$= \frac{\cos^2 x + \sin^2 x}{\cos^2 x}$$

$$= \frac{1}{\cos^2 x}$$

$$= \sec^2 x.$$

Similarly,

$$\frac{d}{dx}(\sec x) = \sec x \tan x,$$

$$\frac{d}{dx}(\csc x) = -\csc x \cot x,$$

and

$$\frac{d}{dx}(\cot x) = -\csc^2 x.$$

(See Exercises 29 through 31 at the end of this section.)

TRIGONOMETRIC FUNCTIONS

Derivatives of the Trigonometric Functions

$$\frac{d}{dx}(\sin u) = (\cos u)\frac{du}{dx} \qquad \frac{d}{dx}(\cos u) = -(\sin u)\frac{du}{dx}$$

$$\frac{d}{dx}(\tan u) = (\sec^2 u)\frac{du}{dx} \qquad \frac{d}{dx}(\cot u) = -(\csc^2 u)\frac{du}{dx}$$

$$\frac{d}{dx}(\sec u) = (\sec u \tan u)\frac{du}{dx} \qquad \frac{d}{dx}(\csc u) = -(\csc u \cot u)\frac{du}{dx}$$

EXAMPLE 10
$$\frac{d}{dx}(\tan^3 x) = (3\tan^2 x)\frac{d}{dx}(\tan x) = 3\tan^2 x \sec^2 x. \quad \square$$

EXAMPLE 11
$$\frac{d}{dx}(\sec e^{2x}) = (\sec e^{2x} \tan e^{2x})\frac{d}{dx}(e^{2x})$$

$$= (\sec e^{2x} \tan e^{2x})e^{2x}\frac{d}{dx}(2x)$$

$$= 2e^{2x} \sec e^{2x} \tan e^{2x}. \quad \square$$

Since we can now take derivatives of trigonometric functions, we can analyze the behavior of angles as they change and the behavior of quantities that are described by trigonometric functions. We illustrate with examples.

EXAMPLE 12 A helicopter takes off and rises straight up. If an observer 100 feet away watches the helicopter, and if its altitude t seconds after takeoff is

$$y = 50t$$

feet, find the rate at which the observer's angle of observation is changing 2 seconds after takeoff.

Let x be the radian measure of the angle of observation. See Figure 11–45. The rate at which the angle of observation is changing at time t is the derivative of x with respect to t. From Figure 11–45 and the relationship between trigonometric functions and the sides of right triangles discussed at the end of Section 11.2, we see that

$$\tan x = \frac{\text{Opposite side}}{\text{Adjacent side}} = \frac{y}{100} = \frac{50t}{100} = \frac{t}{2}.$$

Therefore

$$\frac{d}{dt}(\tan x) = \frac{d}{dt}\left(\frac{t}{2}\right),$$

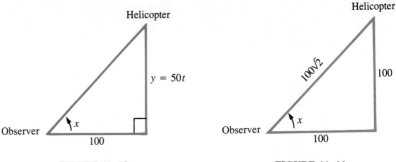

FIGURE 11-45 **FIGURE 11-46**

or

$$(\sec^2 x)\frac{dx}{dt} = \frac{1}{2}.$$

Hence

$$\frac{dx}{dt} = \frac{1}{2\sec^2 x} = \frac{\cos^2 x}{2}.$$

But when $t = 2$, $y = 50(2) = 100$, and Figure 11-45 becomes Figure 11-46, which shows that

$$\cos x = \frac{100}{100\sqrt{2}} = \frac{1}{\sqrt{2}}.$$

Therefore

$$\left.\frac{dx}{dt}\right|_{t=2} = \frac{(1/\sqrt{2})^2}{2} = \frac{1}{4},$$

so the observer's angle of observation is increasing at a rate of $\frac{1}{4}$ radian per second when $t = 2$. ◻

EXAMPLE 13 Suppose a store's sales in week t of the year are y thousand dollars, where

$$y = 10 + 4\sin\left(\frac{\pi t}{26} + \frac{\pi}{4}\right), \qquad 0 \le t \le 52.$$

Let us find the rate at which sales are changing when $t = 12$ and also the maximum and minimum weekly sales during the year. The rate of change of sales with respect to time is

$$\frac{dy}{dt} = \frac{4\pi}{26}\cos\left(\frac{\pi t}{26} + \frac{\pi}{4}\right).$$

Therefore when $t = 12$, sales are changing at the rate of

$$\left.\frac{dy}{dt}\right|_{t=12} = \frac{4\pi}{26}\cos\left(\frac{12\pi}{26} + \frac{\pi}{4}\right) \approx \frac{4\pi}{26}\cos(2.2354)$$

thousand dollars per week. Using a calculator with a cosine key, we find that $\cos(2.2354) \approx -0.6167$. Hence when $t = 12$, sales are changing at a rate of approximately

$$\frac{4\pi}{26}(-0.6167) = -0.2981$$

thousand dollars per week, and thus are decreasing at a rate of approximately \$298 per week.

Since sales follow a sine curve, their absolute maxima will occur at peaks of the sine curve, and hence will be relative maxima. Similarly, absolute minima will be relative minima. But relative maxima and minima will occur at the values of t for which $dy/dt = 0$, that is, at the values of t for which

$$\cos\left(\frac{\pi t}{26} + \frac{\pi}{4}\right) = 0.$$

But this equation implies that

$$\frac{\pi t}{26} + \frac{\pi}{4} = \frac{\pi}{2} \quad \text{or} \quad \frac{\pi t}{26} + \frac{\pi}{4} = \frac{3\pi}{2},$$

and thus that $t = 6.5$ or $t = 32.5$. Since

$$\frac{\pi t}{26} + \frac{\pi}{4} = \frac{\pi}{2}$$

when $t = 6.5$, it follows that when $t = 6.5$,

$$y = 10 + 4 \sin \frac{\pi}{2} = 10 + 4 = 14.$$

Similarly, when $t = 32.5$,

$$\frac{\pi t}{26} + \frac{\pi}{4} = \frac{3\pi}{2},$$

so

$$y = 10 + 4 \sin \frac{3\pi}{2} = 10 - 4 = 6.$$

Therefore there is a relative maximum at $(6.5, 14)$ and a relative minimum at $(32.5, 6)$. Hence the maximum weekly sales are \$14,000 when $t = 6.5$ and the minimum weekly sales are \$6000 when $t = 32.5$. ▫

EXAMPLE 14 At time $t = 0$ an electromotive force (generator, battery, etc.) is turned on and begins to supply current to an electrical circuit. At time t the current I in the circuit is given by

$$I = \sin t - \sqrt{3} \cos t + \sqrt{3} e^{-10t},$$

where t is in seconds and I is in amperes. Let us analyze the flow of current in the circuit.

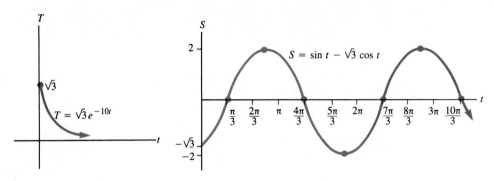

FIGURE 11-47 **FIGURE 11-48**

Note that $I(0) = 0$, so initially there is no current in the circuit. Note also that I is the sum of the **steady-state** current

$$S = \sin t - \sqrt{3} \cos t$$

and the **transient** current

$$T = \sqrt{3} e^{-10t}.$$

The transient current quickly becomes negligible (Figure 11–47). Thus long-term current flow in the circuit is described by the steady-state current. Applying the curve-sketching procedure to

$$S = \sin t - \sqrt{3} \cos t$$

yields the graph of Figure 11–48. (Check this.) From the graph we see that the steady-state current alternates, building up to a peak of 2 amperes in one direction then switching and building to a peak of 2 amperes in the other. The time between successive peaks in the same direction is $t = 2\pi$ seconds. ☐

■ *Computer Manual: Programs LIMIT, DFQUOTNT*

11.4 EXERCISES

In Exercises 1 through 18, find $\dfrac{dy}{dx}$.

1. $y = 3 \cos x$
2. $y = -5 \sin x$
3. $y = 5 \sin 3x$
4. $y = 6 \cos \pi x$
5. $y = \sin(x^2 + 8x)$
6. $y = -4 \sin \sqrt{2x + 1}$
7. $y = 2 \cos x \sin 2x$
8. $y = -3 \cos 4x \sin(-6x)$
9. $y = \cos^2 7x$
10. $y = \sin^3 x^2$
11. $y = \sin^2 x \cos^3 x$
12. $y = \dfrac{\sin x}{\sqrt{\cos x}}$

13. $y = \left(\dfrac{\cos 4x}{\sin 3x}\right)^4$ **14.** $y = e^{2\sin x}$ **15.** $y = e^{-x} \sin 2x$

16. $y = \ln(\cos x)$ **17.** $y = \sin(\cos x)$ **18.** $y = \cos(\sin(\tan(2x)))$

In Exercises 19 through 28, sketch the graph of the given function over the given interval.

19. $y = \sin x - \cos x$, $[0, 2\pi]$ **20.** $y = 2 \sin x + 3 \cos x$, $[0, 2\pi]$

21. $f(x) = -2 \sin 2x + 2 \cos 2x$, $[0, \pi]$ **22.** $f(x) = 3 \sin x/3 - \cos x/3$, $(-\infty, +\infty)$

23. $y = \sqrt{3} \sin 2x + \cos 2x$, $[0, \pi]$ **24.** $y = 2 \cos \dfrac{1}{2}x + 2\sqrt{3} \sin \dfrac{1}{2}x$, $(-\infty, +\infty)$

25. $y = x + \sin x$, $[0, +\infty)$ **26.** $y = x - \cos x$, $[0, +\infty)$

27. $f(x) = \sin^2 x$, $(-\infty, +\infty)$ **28.** $f(x) = \cos^2 x$, $(-\infty, +\infty)$

*29. Using only the chain rule and the fact that $\dfrac{d}{dx}(\cos x) = -\sin x$, show that
$\dfrac{d}{dx}(\sec x) = \sec x \tan x.$

*30. Using only the chain rule and the fact that $\dfrac{d}{dx}(\sin x) = \cos x$, show that
$\dfrac{d}{dx}(\csc x) = -\csc x \cot x.$

*31. Using only the chain rule and the fact that $\dfrac{d}{dx}(\tan x) = \sec^2 x$, show that
$\dfrac{d}{dx}(\cot x) = -\csc^2 x.$

In Exercises 32 through 45, find $\dfrac{dy}{dx}$.

32. $y = \tan 3x$ **33.** $y = \sec x^2$ **34.** $y = 3 \csc(x^3 + x^2 + x + 1)$

35. $y = 2 \tan^2 x$ **36.** $y = 2 \sec(\sqrt{2x + 1})$ **37.** $y = \cot e^x$

38. $y = e^{\cot x}$ **39.** $y = \sec x \tan x$ **40.** $y = \csc 3x \sec^2 2x$

41. $y = 1 + \tan^2 x$ **42.** $y = \ln(\csc x)$ **43.** $y = \ln(\sec x + \tan x)$

44. $y = \ln(\cot^2 x)$ **45.** $y = \tan^2 x^2$

46. A rocket rises straight up from its launch pad and t seconds after liftoff is at an altitude of $y = 100t^2$ meters. An observer watches the rocket from a point 2000 meters from the launch pad. Find the rate at which the observer's angle of observation is changing 3 seconds after liftoff.

47. At time $t = 0$ an airplane passes directly over an observer on the ground. If the plane flies at an altitude of 6000 feet and its ground speed is 600 feet per second, find the rate at which the observer's angle of observation is changing when $t = 10$.

48. Suppose the current I flowing in a circuit at time t is given by
$$I = \sin 2t - \cos 2t + e^{-20t},$$
I in amperes and t in seconds. Analyze current flow in the circuit as we did in Example 14.

49. Repeat Exercise 48 if $I = 2\sqrt{3} \sin 4t - 2 \cos 4t + 2e^{-5t}$.

50. If a weight is hung on the end of a spring, allowed to come to rest, then pulled down from its rest position and released, it will begin to oscillate about its rest position as the spring expands and contracts. Let

$$y = \sin 3t + \cos 3t$$

be the displacement of the weight from its rest position at time t. Here t is in seconds since the weight was released and y is in inches; a positive value of y denotes displacement below the rest position and a negative value of y displacement above the rest position.

(a) Find the weight's position, relative to its rest position, when $t = 0$, $t = \frac{\pi}{6}$, and $t = \frac{\pi}{4}$.

(b) Find the velocity and acceleration of the weight when $t = 0$, $t = \frac{\pi}{6}$, and $t = \frac{\pi}{4}$.

(c) Sketch the graph of y. What is the maximum displacement of the weight? How long does it take the weight to make one complete oscillation?

*51. The model of Exercise 50 is unrealistic because it assumes that the weight will never stop oscillating about its rest position. A more realistic description of the movement of the weight is given by the equation

$$y = e^{-t}(2 \sin 3t + \cos 3t).$$

Answer the questions of Exercise 50 using this model.

Management

52. The monthly sales of a store's shoe department follow a seasonal pattern: in month t, $0 \le t \le 12$, they are $\$y$ thousand, where

$$y = 105 + 40 \sin \frac{\pi t}{6}.$$

Here $t = 0$ represents January 1 of each year.
(a) Find the rate at which sales are changing in April and in August.
(b) Without drawing the graph of the function, find the store's maximum and minimum sales and when during the year they occur.

53. Repeat Exercise 52 if $y = 105 - 40 \sin \left(\frac{\pi t}{6} - \frac{\pi}{3} \right)$.

54. A firm's weekly profit is $\$y$ thousand, where

$$y = 8 - 3.5 \cos \frac{\pi t}{26}, \qquad 0 \le t \le 52.$$

Here t is in weeks, with $t = 0$ representing January 1 of each year.
(a) Find the rate at which the firm's profit is changing during the twentieth and fiftieth weeks of the year.
(b) Without drawing the graph of the function, find the firm's maximum and minimum weekly profit and when these occur.

55. Repeat Exercise 54 if $y = 10 + 2 \cos \left(\frac{\pi t}{26} + \frac{\pi}{4} \right)$.

56. During the year cotton prices fluctuated according to the equation

$$p = 47 - 8.5 \sin\left(\frac{t}{3} + \frac{\pi}{3}\right).$$

Here p is the price per bale and t is time in months, $0 \leq t \leq 12$, with $t = 0$ representing January 1.

(a) Find the rate of change of cotton prices when $t = 2$ and when $t = 7$.
(b) Without drawing the graph of the function, find the maximum and minimum price for cotton and when during the year these occurred.

57. During one year the price of a company's stock varied according to the equation

$$p = 24 - 6 \cos \frac{\pi t}{13} - 2 \sin \frac{\pi t}{13}.$$

Here p is the price per share and t is time in weeks, with $0 \leq t \leq 52$ and $t = 0$ representing January 1.

(a) Find the rate at which the price of the stock was changing when $t = 24$ and when $t = 48$.
*(b) Without drawing the graph of the function, find the maximum and minimum prices of the stock during the year and the times when these occurred.

Life Science

58. A patient's temperature varies during the day according to the equation

$$y = 98.6 - 2.8 \sin\left(\frac{t}{12} + \frac{\pi}{12}\right), \quad 0 \leq t \leq 24.$$

Here temperature y is in degrees Fahrenheit and time t is in hours, with $t = 0$ representing noon.

(a) Find the rate at which the patient's temperature is changing at noon, 6 P.M., midnight, and 6 A.M.
(b) Without drawing the graph of the function, find the patient's maximum and minimum temperatures and when during the day these occurred.

11.5 INTEGRATION OF TRIGONOMETRIC FUNCTIONS

In this section we discuss the integration of trigonometric functions. Since every differentiation formula yields an integration formula, we immediately have the following:

Trigonometric Integration Formulas

$$\int \sin u \, du = -\cos u + c, \qquad \int \cos u \, du = \sin u + c,$$

$$\int \sec^2 u \, du = \tan u + c, \qquad \int \csc^2 u \, du = -\cot u + c,$$

$$\int \sec u \tan u \, du = \sec u + c, \qquad \int \csc u \cot u \, du = -\csc u + c.$$

EXAMPLE 1   Let us find
$$\int \cos 2x \, dx.$$
Since we know from the formulas preceding the example that $\sin u$ is an antiderivative of $\cos u$, we can guess that an antiderivative of $\cos 2x$ must involve $\sin 2x$. However, since
$$\frac{d}{dx}(\sin 2x) = 2 \cos 2x,$$
it is clear that in fact $\frac{1}{2} \sin 2x$ is an antiderivative of $\cos 2x$. Thus
$$\int \cos 2x \, dx = \frac{1}{2} \sin 2x + c. \quad \square$$

EXAMPLE 2   Let us find
$$\int_0^{\pi/4} 2 \sec^2 3x \, dx.$$
We have
$$\int 2 \sec^2 3x \, dx = 2 \int \sec^2 3x \, dx = 2 \left(\frac{1}{3} \tan 3x \right) + c = \frac{2}{3} \tan 3x + c.$$
Therefore
$$\int_0^{\pi/4} 2 \sec^2 3x \, dx = \frac{2}{3} \tan 3x \Big|_0^{\pi/4}$$
$$= \frac{2}{3} \left(\tan \frac{3\pi}{4} - \tan 0 \right)$$
$$= \frac{2}{3}(-1 - 0)$$
$$= -\frac{2}{3}. \quad \square$$

EXAMPLE 3   We will use integration by substitution to find
$$\int x \sin(x^2 + 1) \, dx.$$
We recognize that, except for a constant factor of 2, x is the derivative of $x^2 + 1$. Therefore the substitution $u = x^2 + 1$ will simplify the integral. If we set
$$u = x^2 + 1,$$
we have
$$du = 2x \, dx,$$

or

$$\frac{1}{2} du = x \, dx.$$

Thus

$$\int x \sin(x^2 + 1) \, dx = \frac{1}{2} \int \sin u \, du$$

$$= -\frac{1}{2} \cos u + c$$

$$= -\frac{1}{2} \cos(x^2 + 1) + c. \quad \square$$

Integration by substitution is often useful when it is necessary to integrate products and quotients of trigonometric functions. We illustrate with examples.

EXAMPLE 4   Find

$$\int \sin^2 x \cos x \, dx.$$

Since $\cos x$ is the derivative of $\sin x$, the substitution $u = \sin x$ will simplify the integral. We have

$$u = \sin x, \qquad du = \cos x \, dx,$$

and thus

$$\int \sin^2 x \cos x \, dx = \int u^2 \, du = \frac{u^3}{3} + c = \frac{\sin^3 x}{3} + c. \quad \square$$

EXAMPLE 5   Find

$$\int \frac{\sin 2x}{1 + \cos 2x} \, dx.$$

Note that $\sin 2x$ is the derivative of $1 + \cos 2x$ (except for a constant factor), and thus the substitution $u = 1 + \cos 2x$ will simplify the integral. We have

$$u = 1 + \cos 2x, \qquad du = (-2 \sin 2x) \, dx,$$

and therefore

$$\int \frac{\sin 2x}{1 + \cos 2x} \, dx = -\frac{1}{2} \int \frac{du}{u}$$

$$= -\frac{1}{2} \ln |u| + c$$

$$= -\frac{1}{2} \ln |1 + \cos 2x| + c. \quad \square$$

EXAMPLE 6 Consider the shaded region in Figure 11–49. The area of this region is

$$\int_0^{\pi/4} (\cos x - \sin x)\, dx = \sin x + \cos x \Big|_0^{\pi/4}$$

$$= \left(\sin \frac{\pi}{4} + \cos \frac{\pi}{4}\right) - (\sin 0 + \cos 0)$$

$$= \left(\frac{1}{\sqrt{2}} + \frac{1}{\sqrt{2}}\right) - (0 + 1)$$

$$= \frac{2}{\sqrt{2}} - 1. \quad \square$$

EXAMPLE 7 Let us find the average value of the function $y = \sin x$ over the interval $[0, \pi/2]$. We have

$$\text{Average value} = \frac{1}{\pi/2 - 0} \int_0^{\pi/2} \sin x\, dx$$

$$= \frac{2}{\pi}\left(-\cos x \Big|_0^{\pi/2}\right)$$

$$= \frac{2}{\pi}\left(-\cos \frac{\pi}{2} + \cos 0\right)$$

$$= \frac{2}{\pi}(0 + 1)$$

$$= \frac{2}{\pi}. \quad \square$$

See Figure 11–50.

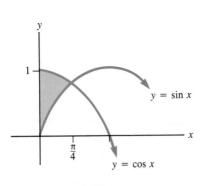

FIGURE 11–49

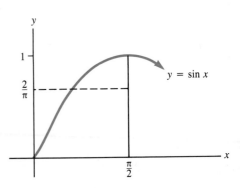

FIGURE 11–50

EXAMPLE 8 A firm's marginal revenue function is given by

$$r(x) = 2x + 5 \sin \pi x.$$

If the firm is currently selling two units, let us find the additional revenue from selling one more unit. It is

$$\int_2^3 (2x + 5 \sin \pi x)\, dx = x^2 - \frac{5}{\pi} \cos \pi x \Big|_2^3$$

$$= \left(3^2 - \frac{5}{\pi} \cos 3\pi\right) - \left(2^2 - \frac{5}{\pi} \cos 2\pi\right)$$

$$= \left(9 - \frac{5}{\pi}(-1)\right) - \left(4 - \frac{5}{\pi}(1)\right)$$

$$\approx 8.18.$$

Therefore the additional revenue will be approximately $8.18. □

We know that

$$\int \sin u\, du = \cos u + c$$

and

$$\int \cos u\, du = -\sin u + c,$$

but we do not yet possess integration formulas for tan u, cot u, sec u, and csc u. We will now derive the integration formulas for these functions.

To find

$$\int \tan x\, dx,$$

we write

$$\tan x = \frac{\sin x}{\cos x}$$

and make the substitution $u = \cos x$, since then $du = -\sin x\, dx$ and we obtain

$$\int \tan x\, dx = \int \frac{\sin x}{\cos x}\, dx$$

$$= -\int \frac{du}{u}$$

$$= -\ln|u| + c$$

$$= -\ln|\cos x| + c.$$

Similarly,

$$\int \cot x\, dx = \ln|\sin x| + c.$$

Finding

$$\int \sec x\, dx$$

is not so easy. The derivation of the integration formula for sec x rests upon the observation that

$$\frac{d}{dx}(\sec x + \tan x) = \sec x \tan x + \sec^2 x$$
$$= \sec x (\sec x + \tan x).$$

Thus if we write

$$\int \sec x \, dx = \int \frac{\sec x (\sec x + \tan x)}{\sec x + \tan x} \, dx,$$

the numerator of the integrand on the right side of the equation is the derivative of its denominator. Hence if

$$u = \sec x + \tan x,$$

then

$$du = \sec x (\sec x + \tan x) \, dx,$$

and

$$\int \sec x \, dx = \int \frac{\sec x (\sec x + \tan x)}{\sec x + \tan x} \, dx,$$
$$= \int \frac{du}{u}$$
$$= \ln |u| + c$$
$$= \ln |\sec x + \tan x| + c.$$

Similarly,

$$\int \csc x \, dx = -\ln |\csc x + \cot x| + c.$$

Integrals of the Trigonometric Functions

$$\int \sin x \, dx = -\cos x + c,$$

$$\int \cos x \, dx = \sin x + c,$$

$$\int \tan x \, dx = -\ln |\cos x| + c,$$

$$\int \cot x \, dx = \ln |\sin x| + c,$$

$$\int \sec x \, dx = \ln |\sec x + \tan x| + c,$$

$$\int \csc x \, dx = -\ln |\csc x + \cot x| + c.$$

Trigonometric Functions

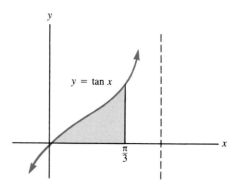

FIGURE 11-51

EXAMPLE 9 Let us find the area of the region bounded by the x-axis and the graph of $y = \tan x$ from $x = 0$ to $x = \pi/3$. See Figure 11-51. We have

$$\int_0^{\pi/3} \tan x \, dx = -\ln |\cos x| \Big|_0^{\pi/3}$$

$$= -\ln \left|\cos \frac{\pi}{3}\right| + \ln |\cos 0|$$

$$= -\ln \left|\frac{1}{2}\right| + \ln |1|$$

$$= -\ln \frac{1}{2}$$

$$= \ln 2. \quad \square$$

Finally we consider integrals that "mix" trigonometric functions with other types of functions, such as

$$\int x \sin x \, dx \quad \text{and} \quad \int e^x \cos x \, dx.$$

Integrals such as these can often be handled with the aid of the integration by parts formula

$$\int u \, dv = uv - \int v \, du$$

which we introduced in Section 7.2. We illustrate with examples.

EXAMPLE 10 Find

$$\int x \sin x \, dx.$$

If we let

$$u = x \quad \text{and} \quad dv = \sin x \, dx,$$

then

$$du = dx \quad \text{and} \quad v = \int \sin x \, dx = -\cos x.$$

Therefore

$$\int x \sin x \, dx = \int u \, dv$$
$$= uv - \int v \, du$$
$$= x(-\cos x) - \int -\cos x \, dx$$
$$= -x \cos x + \int \cos x \, dx$$
$$= -x \cos x + \sin x + c. \quad \square$$

EXAMPLE 11   Let us find

$$\int e^x \cos x \, dx.$$

We will use integration by parts with

$$u = e^x \quad \text{and} \quad dv = \cos x \, dx,$$

so that

$$du = e^x \, dx \quad \text{and} \quad v = \int \cos x \, dx = \sin x.$$

Then

$$\int u \, dv = uv - \int v \, du$$

becomes

$$\int e^x \cos x \, dx = e^x \sin x - \int e^x \sin x \, dx.$$

Now we try integration by parts on $\int e^x \sin x \, dx$. If we let

$$u = e^x \quad \text{and} \quad dv = \sin x \, dx,$$

then

$$du = e^x \, dx \quad \text{and} \quad v = \int \sin x \, dx = -\cos x,$$

so

$$\int e^x \sin x \, dx = e^x(-\cos x) - \int -e^x \cos x \, dx.$$

Therefore

$$\int e^x \cos x \, dx = e^x \sin x - \left[-e^x \cos x + \int e^x \cos x \, dx \right],$$

or
$$\int e^x \cos x \, dx = e^x \sin x + e^x \cos x - \int e^x \cos x \, dx.$$

Notice that the integral
$$\int e^x \cos x \, dx$$
appears on the right side of the last equation preceded by a minus sign. Therefore if we add it to both sides of the equation we obtain
$$2 \int e^x \cos x \, dx = e^x (\sin x + \cos x).$$

Hence
$$\int e^x \cos x \, dx = \frac{1}{2} e^x (\sin x + \cos x) + c. \quad \square$$

■ *Computer Manual: Program DEFNTGRL*

11.5 EXERCISES

In Exercises 1 through 18, find the integral.

1. $\int \sin 3x \, dx$
2. $\int 2 \cos(-4x) \, dx$
3. $\int \csc^2(-\pi x) \, dx$
4. $\int_{\pi/6}^{\pi/3} \sec x \tan x \, dx$
5. $\int_0^{\pi} x \cos x^2 \, dx$
6. $\int x^2 \sec^2 x^3 \, dx$
7. $\int \frac{\sin \sqrt{x}}{\sqrt{x}} \, dx$
8. $\int \frac{\cos x}{2 + \sin x} \, dx$
9. $\int_{-\pi/4}^{\pi/4} \sin 2x \cos 2x \, dx$
10. $\int_\pi^{2\pi} \cos^2 x \sin x \, dx$
11. $\int \sin^3 2x \cos 2x \, dx$
12. $\int \tan x \sec^2 x \, dx$
13. $\int_0^{\pi/2} e^{-\cos x} \sin x \, dx$
14. $\int \frac{1}{x^2} \sec \frac{1}{x} \tan \frac{1}{x} \, dx$
15. $\int \sqrt{1 - \cos^2 x} \, dx$
16. $\int_0^{\pi/6} \frac{\csc x \cot x}{2 + \cot x} \, dx$
17. $\int \sin^2 x \, dx$ [*Hint:* Use the identity $\sin^2 x = (1 - \cos 2x)/2$.]
18. $\int \sin^3 x \, dx$ (*Hint:* Use the identity $\sin^2 x + \cos^2 x = 1$.)
19. Sketch and find the area of
    (a) the region bounded by the x-axis and the graph of $y = \sin x$ from $x = 0$ to $x = \pi$;
    (b) the region bounded by the x-axis and the graph of $y = \cos x$ from $x = 0$ to $x = \frac{\pi}{2}$.

20. Sketch and find the area of

   (a) the region bounded by $y = \sin x$ and $y = \cos x$ from $x = \dfrac{\pi}{4}$ to $x = \dfrac{5\pi}{4}$;

   (b) the region bounded by $y = \sin x$ and $y = \cos x$ from $x = \dfrac{5\pi}{4}$ to $x = 2\pi$.

21. Sketch and find the area of the region bounded by the graphs of

   (a) $y = \sin x$ and $y = \dfrac{2}{\pi} x$, $x \geq 0$;

   (b) $y = \cos x$, $y = \dfrac{3}{2\pi} x$, and the y-axis, $x \geq 0$.

22. Over the interval $[0, \pi]$, find the average value of
   (a) $y = \sin x$
   (b) $y = \cos x$

23. Over the interval $[0, \pi]$, find the average value of
   (a) $y = \sin 2x$
   (b) $y = \cos 2x$

24. Over the interval $[0, \pi]$, find the average value of
   (a) $y = \sin (x/3)$
   (b) $y = \cos (x/3)$

In Exercises 25 through 30, find the integral.

25. $\displaystyle\int \tan 5x \, dx$

26. $\displaystyle\int_{\pi/4}^{\pi/3} \cot 2x \, dx$

27. $\displaystyle\int x \csc x^2 \, dx$

28. $\displaystyle\int e^{2x} \sec e^{2x} \, dx$

29. $\displaystyle\int_0^{\pi/4} (1 + \tan^2 x) \, dx$ (Hint: Use the identity $1 + \tan^2 x = \sec^2 x$.)

30. $\displaystyle\int \tan^3 x \, dx$.

31. Sketch and find the area of the region bounded by $y = \sec x$ and $y = 1$ from $x = -\dfrac{\pi}{3}$ to $x = \dfrac{\pi}{3}$.

32. Sketch and find the area of the region bounded by $y = \tan x$, the y-axis, and $y = \sqrt{3}$.

33. Find the average value of $y = \sec x$ over the interval $\left[-\dfrac{\pi}{4}, \dfrac{\pi}{4}\right]$.

34. Find the average value of $y = \tan x$ over the interval $\left[\dfrac{\pi}{6}, \dfrac{\pi}{4}\right]$.

In Exercises 35 through 40, use integration by parts to find the integral.

35. $\displaystyle\int x \cos x \, dx$

36. $\displaystyle\int x \sec^2 x \, dx$

37. $\displaystyle\int_0^{\pi/2} x^2 \sin x \, dx$

38. $\displaystyle\int_0^{\pi} e^{-x} \sin x \, dx$

39. $\displaystyle\int e^{2x} \cos 2x \, dx$

40. $\displaystyle\int \sin^2 x \, dx$

Let $I = \sin 2t - \cos 2t$ be the steady-state current in an electrical circuit at time $t \geq 0$, where t is in seconds and I is in amperes. In Exercises 41 through 43, find the average current in the circuit over the given interval.

41. $\left[0, \dfrac{\pi}{4}\right]$

42. $\left[0, \dfrac{\pi}{2}\right]$

43. $[0, \pi]$

Management

44. A firm's marginal revenue function is given by

$$r(x) = 0.6x + 1.5 \sin \frac{\pi}{2} x.$$

If the firm is currently selling five units, find the additional revenue if it sells one more unit.

45. A company's income from selling a fad item accrues at the rate of y million dollars per month, where

$$y = 6 \cos \frac{\pi t}{16}, \quad 0 \le t \le 8.$$

Here t is time in months since the introduction of the item. Find the total income the company will derive from the item.

46. A company's marginal cost function is given by

$$c(x) = 10,000 + 2x + 12 \cos \frac{\pi}{3} x.$$

If the company is currently making 10 units, find the additional cost of making 2 more units.

47. The savings due to new equipment accrue at a rate of

$$y = 40 \sin \frac{\pi t}{24}, \quad 0 \le t \le 24,$$

where y is savings in thousands of dollars and t is in months since the installation of the new equipment. Find the total savings due to the new equipment.

48. A retailer's inventory at time t is y units, where

$$y = 20 + 10 \sin \frac{\pi t}{6}, \quad 0 \le t \le 12.$$

Here t is in months. Find the retailer's average inventory over
(a) the first six months of the year;
(b) the entire year.

49. A retailer's inventory at time t is y units, where

$$y = 100\left(1 - \cos \frac{\pi t}{26}\right), \quad 0 \le t \le 52.$$

Here t is in weeks. Find the retailer's average inventory over
(a) the first four weeks of the year;
(b) the entire year.

Life Science

50. A patient undergoing chemotherapy has a powerful drug in his blood in a concentration of y milligrams per liter, where

$$y = 8.2 - 3.4 \cos \frac{2\pi t}{15}.$$

Here t is time in days since the last treatment. Find the average concentration of the drug in the patient's blood during the first five days after a treatment.

SUMMARY

This summary consists of terms and symbols whose meaning you should know, facts you should know, and techniques and methods you should be able to employ.

Section 11.1

Angle, sides of an angle, vertex of an angle. Initial and terminal sides of an angle. Degree measure of an angle. Positive and negative degree measure. Finding the degree measure of an angle. Constructing an angle of a given degree measure. Radians. Radian measure of an angle. Positive and negative radian measure. Finding the radian measure of an angle. Constructing an angle of a given radian measure. The conversion formula. Converting from degree to radians and from radians to degrees.

Section 11.2

Definition of sin x, cos x, x in radians, as coordinates of a point on the unit circle with center at the origin. The sign of sin x, cos x. Finding sin x, cos x for x an integral multiple of $\pi/2$. Trigonometric identities. Using trigonometric identities to find sin x, cos x. Values of sin x, cos x for x integral multiples of $\pi/6$, $\pi/4$. Definition of tangent, cotangent, secant, and cosecant of an angle of x radians. Finding tan x, cot x, sec x, csc x. Relationship between trigonometric functions and the sides of a right triangle. Finding an angle of a right triangle given its sides. Finding the sides of a right triangle given its angles.

Section 11.3

The trigonometric functions. The graph of the sine function $y = \sin x$. Periodic function, period. The sine function is periodic of period 2π. Amplitude. Amplitude of sine function is 1. Graphs of $y = a \sin bx$, $y = a \sin (bx + c)$, $a > 0$, $b > 0$. The graph of the cosine function $y = \cos x$. The cosine function is periodic of period 2π and has amplitude 1. Graphs of $y = a \cos bx$, $y = a \cos (bx + c)$, $a > 0$, $b > 0$. The graph of the tangent function $y = \tan x$. The tangent function is periodic of period π, has vertical asymptotes at odd multiples of $\pi/2$. Graphs of $y = a \tan bx$, $y = a \tan (bx + c)$, $a > 0$, $b > 0$. Graphs of $y = \csc x$, $y = \sec x$, $y = \cot x$. The cosecant and secant functions have period 2π, the cotangent function has period π.

Section 11.4

Differentiation formulas for sin x. Differentiation formulas for cos x. Finding derivatives of functions that involve sines and cosines. Using the curve-sketching procedure to sketch the graphs of functions that involve sines and cosines. Differentiation formulas for tangent, secant, cosecant, and cotangent. Finding derivatives of functions that involve tangents, secants, cosecants, and

TRIGONOMETRIC FUNCTIONS

cotangents. Finding rates of change, relative maxima and minima of functions that involve sines and cosines.

Section 11.5

Integral formulas for trigonometric functions. Integrating trigonometric functions by guessing and by substitution. Finding areas, average values of functions that involve trigonometric functions. Integration formulas for tangent, cotangent, secant, cosecant. Integration by parts using trigonometric functions.

REVIEW EXERCISES

In Exercises 1 through 4, state the degree measure and the radian measure of the given angles. Do not use the conversion formula.

1.   2.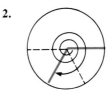

3.   4.

In Exercises 5 through 8, construct the angles whose measures are given.

5. $-60°, 225°, 240°, -300°$

6. $20°, 50°, 25°, 15°$

7. $-5\pi, \dfrac{5\pi}{3}, \dfrac{11\pi}{4}, -\dfrac{7\pi}{6}$ radians

8. $\dfrac{3\pi}{8}, \dfrac{3\pi}{16}, \dfrac{5\pi}{12}, \dfrac{\pi}{10}$ radians

9. Convert to radians: $135°, -240°, 22.5°, -4°$

10. Convert to degrees: $\dfrac{4\pi}{3}, -\dfrac{5\pi}{4}, \dfrac{\pi}{15}, 3.25$ radians

In Exercises 11 through 16, find $\sin x$, $\cos x$, $\tan x$, $\csc x$, $\sec x$, and $\cot x$ for the given value of x, if possible.

11. $x = \dfrac{3\pi}{2}$ 12. $x = \dfrac{5\pi}{4}$ 13. $x = -4\pi$ 14. $x = \dfrac{7\pi}{3}$ 15. $x = \dfrac{17\pi}{6}$ 16. $x = -\dfrac{11\pi}{6}$

17. Graph the function $y = \dfrac{1}{2} \cos(-2x)$ and state its period.

18. Graph the function $y = 2 \sin \frac{3x}{4}$ and state its period.

19. Graph the function $y = 4 + 3 \sin \left(\frac{2x}{3} + \frac{\pi}{4}\right)$.

20. Graph the function $y = 5 + 2 \cos \left(\frac{\pi x}{2} - \frac{2\pi}{3}\right)$.

21. A retailer's daily profit is P dollars, where

$$P = 620 + 225 \cos \left(\frac{\pi t}{6} - 2.4\right),$$

and t is in months, $0 \le t \le 12$, with $t = 0$ representing January 1. Graph this equation and analyze the retailer's profits over the course of the year.

22. Graph the functions

   (a) $y = \tan \frac{x}{2}$ (b) $y = 2 \csc (x + \pi)$

23. Find

   (a) $\sin x$, $\cos x$, $\tan x$, $\csc x$, $\sec x$, and $\tan x$ if

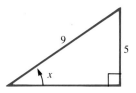

   (b) a and b if

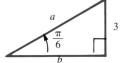

   (c) the radian measure of x if

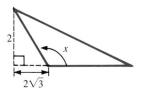

In Exercises 24 through 33, find $\frac{dy}{dx}$.

24. $y = \sin 3x$

25. $y = \cos 2\sqrt{x}$

26. $y = \frac{\sin x}{x}$

27. $y = \frac{1}{6} \sin 3x \cos 2x$

28. $y = e^{\tan x}$

29. $y = 2 \cot 3x^2$

30. $y = \frac{\tan 4x}{\csc 2x}$

31. $y = \sin^3(-4x)$

32. $y = \sec^2 (\tan x)$

33. $y = \ln(\sin x + \csc x)$

34. Sketch the graph of $y = \sin 2x + \cos 2x$ over the interval $[0, \pi]$.

35. Sketch the graph of $y = \sin 3x - \sqrt{3} \cos 3x$.

36. Sketch the graph of $y = x - \tan x$ over the interval $(-\pi/2, \pi/2)$.

37. A company's sales are forecast to be

$$y = 80 + 14 \sin \left(\frac{\pi t}{5} + 0.7\right)$$

million dollars t years from now.

   (a) Find the rate at which the company's sales will be changing one and four years from now.

(b) Without drawing the graph of the function, determine the company's maximum and minimum sales and when these will occur.

In Exercises 38 through 50, find the integral.

38. $\int_{-\pi/6}^{\pi/2} \cos 5x \, dx$

39. $\int 2x \sin x^2 \, dx$

40. $\int 2 \csc 5x \cot 5x \, dx$

41. $\int \cos^4 x \sin x \, dx$

42. $\int_{\pi/8}^{\pi/6} 3 \tan 2x \, dx$

43. $\int \sqrt{1 - \sin^2 2x} \, dx$

44. $\int \cos x \ln(\sin x) \, dx$

45. $\int \sec 2x \tan 2x \, e^{\sec 2x} \, dx$

46. $\int \dfrac{\cos \sqrt{x}}{\sqrt{x}} \, dx$

47. $\int_0^{\pi/4} 3x \sec(x^2 + 2) \, dx$

48. $\int x \sin 2x \, dx$

49. $\int_{2\pi/3}^{\pi} \dfrac{1 - \sin x}{\cos x + x} \, dx$

50. $\int e^{3x} \cos 2x \, dx$

51. Find the average value of

   (a) $y = \sin x \cos x$ over $[0, \pi]$

   (b) $y = \sec^2 x$ over $\left[-\dfrac{\pi}{3}, \dfrac{\pi}{3}\right]$

52. Sketch and find the area of the region bounded by $y = \cos x$, $y = x$, the y-axis, and $x = \dfrac{\pi}{6}$.

53. A factory orders flanges at the beginning of each quarter. If the factory's inventory of flanges t days after the arrival of an order is y units, where

$$y = 1200 - 1150 \cos \dfrac{\pi t}{180}, \quad 0 \le t \le 90,$$

find the average inventory of flanges.

54. A firm's marginal cost function is given by

$$y = 5000 + 4x - 20 \sin \dfrac{\pi x}{20}.$$

Find the increase in the cost if production goes from four to five units.

ADDITIONAL TOPICS

Here are some suggestions for topics not covered in the text that you might want to investigate on your own.

1. Suppose C_1 and C_2 are curves that intersect at a point P, and assume that each of them has a unique tangent line at P. The **angle between the curves** is then defined to be the angle θ formed by the tangent lines. If m_1 is the slope of the tangent line to C_1 at P and m_2 the slope of the tangent line to C_2 at P, show that

(a) if $m_1 m_2 = -1$, then $\theta = \pi/2$;
(b) if $m_1 m_2 \neq -1$, then

$$\tan \theta = \frac{|m_2 - m_1|}{|1 + m_1 m_2|}.$$

2. If the domain of the sine function $y = \sin x$ is restricted to the interval $[-\pi/2, \pi/2]$, the result is a one-to-one function. Hence over $[-\pi/2, \pi/2]$ the function $y = \sin x$ has an inverse function $y = \sin^{-1} x$, called the **inverse sine function**. (The inverse sine function is sometimes denoted by $y = \arcsin x$, and called the **arcsine of x**.) Similarly, the other trigonometric functions also have inverse functions. Find out how these inverse trigonometric functions are defined and what their graphs look like.

3. Find the derivatives of the inverse trigonometric functions referred to above. Show how the inverse trigonometric functions appear as indefinite integrals of expressions such as

$$\frac{1}{\sqrt{1-x^2}}, \quad \frac{1}{x\sqrt{x^2-1}}, \quad \text{and} \quad \frac{1}{1+x^2}.$$

4. Find the indefinite integrals of the inverse trigonometric functions.

5. Many integrals can be handled by means of **trigonometric substitutions**. For instance,

$$\int \sqrt{a^2 - x^2} \, dx$$

can be simplified by means of the trigonometric substitution $x = a \sin u$. Find out how this and other trigonometric substitutions are used to integrate expressions such as $\sqrt{a^2 - x^2}$, $\sqrt{x^2 - a^2}$, and $\sqrt{a^2 + x^2}$.

SUPPLEMENT: PERIODICITY AND THE DIFFERENTIAL EQUATION $y'' + ky = 0$

We noted at the beginning of this chapter that many natural phenomena are periodic in the sense that they repeat themselves over and over. Since the trigonometric functions are periodic, it would seem plausible that they could serve to model periodic phenomena, and this is indeed the case. In particular, periodic models often arise as solutions to differential equations of the form $y'' + ky = 0$, $k > 0$. In this supplement we examine this differential equation and its solutions.

We assume that $y = y(t)$ is a continuous function of t and that all its derivatives exist and are continuous also. The differential equation

$$y'' + ky = 0, \quad k > 0$$

may be written as

$$y'' = -ky, \quad k > 0,$$

and hence any solution must have the property that, up to a constant, its second derivative is its negative. But the sine and cosine functions have this property:

$$\frac{d^2}{dt}(\sin bt) = -b^2 \sin bt \tag{1}$$

$$\frac{d^2}{dt}(\cos bt) = -b^2 \cos bt. \tag{2}$$

(Check this.) In fact, Equations (1) and (2) with $b = \sqrt{k}$ show that

$$y = \sin \sqrt{k}\, t \quad \text{and} \quad y = \cos \sqrt{k}\, t$$

are solutions of $y'' = -ky$, $k > 0$. But if $y = \sin \sqrt{k}\, t$ and $y = \cos \sqrt{k}\, t$ are solutions of this differential equation, so is

$$y = c_1 \sin \sqrt{k}\, t + c_2 \cos \sqrt{k}\, t \tag{3}$$

for any choice of constants c_1, c_2. (See Exercise 1.) In fact, it is shown in differential equations courses that every solution of $y'' = -ky$, $k > 0$ is of the form given in (3).

Note that the function defined by Equation (3) is periodic of period $2\pi/\sqrt{k}$. (See Exercise 2.) Also, its points of inflection occur at those values of t, and only those values of t, such that

$$y'' = -ky = 0,$$

that is, at those values of t, and only those values of t, such that $y = 0$. Hence the points of inflection of the solution y are the same as its t-intercepts.

Now consider a periodic natural phenomenon having the property that its points of inflection are the same as its t-intercepts. For example, consider the motion of a pendulum, as in Figure 11–52. We let $y = y(t)$ denote the angle the pendulum makes with the vertical at time t, with angles to the right of vertical positive and those to the left negative. In the absence of friction or air resistance, the pendulum once put into motion will remain in motion, swinging back and forth and repeating its path forever. Furthermore, as the pendulum approaches the point E in Figure 11–52 from the left, it speeds up, and thus the rate of change of y with respect to t increases; however, as soon as the pendulum passes the point E, it begins to slow down, and thus the rate of change of y with respect to t decreases. But a point where the rate of change of a function changes from increasing to decreasing (or vice versa) is a point of inflection for the function. Therefore y has a point of inflection every time it passes through E, that is, at every t-intercept.

It can be shown that a function such as the pendulum function that is periodic and has its t-intercepts and points of inflection the same must be a solution of a differential equation of the form $y'' + ky = 0$, for some $k > 0$, and therefore must be of the form

$$y = c_1 \sin \sqrt{k}\, t + c_2 \cos \sqrt{k}\, t$$

for some c_1, c_2.

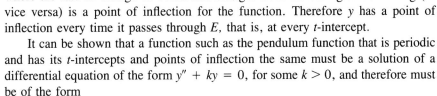

FIGURE 11–52

EXAMPLE 1 Suppose the pendulum function y must satisfy the differential equation

$$y'' + 4y = 0, \qquad y(0) = -\pi/4, \quad y'(0) = 1.$$

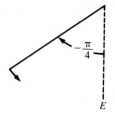

FIGURE 11-53

Thus $y(0) = -\pi/4$ is the initial angle formed by the pendulum; see Figure 11-53. Also, $y'(0) = 1$ is the initial angular velocity of the pendulum.

Since $k = 4$, the solution must be of the form

$$y = c_1 \sin \sqrt{4}\, t + c_2 \cos \sqrt{4}\, t,$$

or

$$y = c_1 \sin 2t + c_2 \cos 2t.$$

But then

$$-\pi/4 = y(0) = c_1 \sin 0 + c_2 \cos 0 = c_2,$$

so that

$$y = c_1 \sin 2t - \frac{\pi}{4} \cos 2t.$$

But now

$$y' = 2c_1 \cos 2t + \frac{\pi}{2} \sin 2t,$$

and hence

$$1 = y'(0) = 2c_1 \cos 0 + \frac{\pi}{2} \sin 0 = 2c_1.$$

Therefore $c_1 = 1/2$, and

$$y = \frac{1}{2} \sin 2t - \frac{\pi}{4} \cos 2t. \quad \square$$

Animal populations sometimes exhibit periodic behavior. What typically happens is that the population grows until it exhausts its food supply, then declines

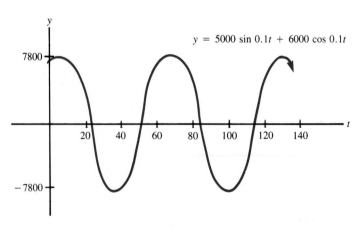

FIGURE 11-54

Trigonometric Functions

until the food supply is regenerated, then grows again until it exhausts its food supply, and so on. In such a case the size of the population can oscillate in a periodic manner about its average size. If we let $y = y(t)$ denote the difference between the size of the population at time t and its average size (so that $y(1) = 200$ means that at time 1 the population is 200 above the average and $y(2) = -300$ means that at time 2 the population is 300 below the average, for instance), then an analysis of the situation may show that y must satisfy a differential equation of the form $y'' + ky = 0$ for some $k > 0$.

EXAMPLE 2 Suppose $y = y(t)$ is the difference between the size of a population at time t and its average population, and suppose that y satisfies the differential equation

$$y'' + 0.01y = 0, \quad y(0) = 6000, \quad y'(0) = 500.$$

Thus initially the size of the population is 6,000 and it is increasing at a rate of 500 individuals per time period.

Since $\sqrt{0.01} = 0.1$, the solution y must have the form

$$y = c_1 \sin 0.1t + + c_2 \cos 0.1t.$$

Then $y(0) = 6000$ implies that $c_2 = 6000$, and $y'(0) = 500$ implies that $c_1 = 5000$. Therefore

$$y = 5000 \sin 0.1t + 6000 \cos 0.1t.$$

See Figure 11–54. ☐

FIGURE 11–55

Sometimes a measure of business activity, such as sales, will follow a linear long-term trend with cyclical behavior superimposed on the trend, as in Figure 11–55. In such a case, if we let $y = y(t)$ be the amount that sales are above or below the long-term trend at time t, it may be possible to find y as a solution to a differential equation of the form $y'' + ky = 0$, for some $k > 0$.

EXAMPLE 3 Suppose the sales of a firm follow a long-term trend given by

$$S = 500 + 12t.$$

Here S is in millions of dollars and t is time in quarter-years, with $t = 0$ representing quarter I of 1989. Let $y = y(t)$ be the amount sales are above or below the trend line at time t, in millions of dollars, and suppose that y satisfies the differential equation

$$y'' + 9y = 0, \quad y(0) = 10, \quad y'(0) = -6.$$

Thus in quarter I of 1989 ($t = 0$), we had $S = 500$ and $y(0) = 10$, so that sales were \$510 million.

The solution of the differential equation has the form

$$y = c_1 \sin 3t + c_2 \cos 3t.$$

The initial conditions $y(0) = 10$ and $y'(0) = -6$ imply that $c_1 = -2$ and $c_2 = 10$. Therefore

$$y = -2 \sin 3t + 10 \cos 3t.$$

Let us use this result to forecast sales in quarter I of 1991 ($t = 8$). When $t = 8$, the long-term trend will contribute

$$S = 500 + 12(8) = 596$$

million dollars to sales, and the cyclical component will contribute

$$y = -2 \sin 24 + 10 \cos 24 \approx 6.1$$

million dollars. Therefore our forecast for sales during the first quarter of 1991 is $596 + 6.1 = 602.1$ million dollars. □

SUPPLEMENTARY EXERCISES

*1. Show that $y = c_1 \sin \sqrt{k}\, t + c_2 \cos \sqrt{k}\, t$ is a solution of $y'' + ky = 0$, $k > 0$, for any choice of c_1, c_2.

*2. Graph $y = c_1 \sin \sqrt{k}\, t + c_2 \cos \sqrt{k}\, t$ for $c_1 > 0$, $c_2 > 0$, and thus show that the function defined by this equation is periodic with period $2\pi/\sqrt{k}$ and has amplitude $\sqrt{c_1^2 + c_2^2}$.

3. Suppose the motion of a pendulum satisfies the differential equation

$$y'' + 5y = 0, \quad y(0) = -\pi/2, \quad y'(0) = 1.5.$$

Find and graph the solution of this differential equation.

In the differential equation $y'' + ky = 0$ that describes the motion of a pendulum, the constant k depends on the length s of the pendulum: if we measure length in feet and time in seconds, then $k = 32/s$. Exercises 4 through 6 refer to this situation.

4. Suppose one pendulum has a length of 1 foot and another a length of 2 feet, and that both of them have $y(0) = -\pi/4$ and $y'(0) = 1$. Find and graph the equations of motion of these pendulums and compare your results.

5. Suppose two pendulums both have a length of 1 foot and have $y'(0) = 1$, but that for one of them $y(0) = -\pi/6$ while for the other $y(0) = -\pi/3$. Find and graph their equations of motion and compare your results.

6. Suppose two pendulums both have a length of 1 foot and $y(0) = -\pi/4$, but that the initial velocity for one is $y'(0) = 1$ while for the other it is $y'(0) = 2$. Find and graph their equations of motion and compare your results.

7. Suppose y denotes the difference between the size of an animal population at time t and its average population size. If y satisfies the differential equation

$$y'' + 0.0225y = 0, \quad y(0) = 20{,}000, \quad y'(0) = -1000,$$

find and graph the solution of this differential equation.

*8. Let y denote the difference between the size of an animal population and its average size. If y satisfies the differential equation

$$y'' + ky = 0, \quad k > 0, \quad y(0) = y_0, \quad y'(0) = y_1,$$

find y. What happens to y if k is increased? if y_0 is increased? if y_1 is increased?

9. The sales of a business are described by a long-term trend line
$$S = 20 + 1.5t,$$
where S is in thousands of dollars and t is in months with $t = 0$ representing January 1, 1990. The cyclical component of sales is described by $y = y(t)$, where $y(t)$ is the amount sales are above or below the trend line at time t, in thousands of dollars. Suppose that y satisfies the differential equation
$$y'' + 3y = 0, \quad y(0) = 0.5, \quad y'(0) = 0.2.$$
Forecast the firm's monthly sales as of January 1991 ($t = 13$).

10. Repeat Exercise 9 if
$$S = 430 - 22t$$
and
$$y'' + 25y = 0, \quad y(0) = -20, \quad y'(0) = -5.$$

If a weight is suspended from the end of a spring and the spring is allowed to come to rest, it will hang as in the figure:

Now suppose the weight is pulled down below E and released with a certain initial velocity in the upward direction. Then the weight, pulled by the compression of the spring, will travel upward at an increasing rate until it passes through E; after it passes through E the compression of the spring will slow it down. Let $y = y(t)$ denote the position of the weight at time t, with position below E positive and position above E negative. Exercises 11 through 16 refer to this situation.

*11. The function y must satisfy a differential equation of the form
$$y'' + ky = 0, \quad k > 0, \quad y(0) = y_0, \quad y'(0) = v_0,$$
for some k, some initial position y_0 and some initial velocity v_0. Why?

*12. Suppose the differential equation is
$$y'' + 2y = 0, \quad y(0) = 1, \quad y'(0) = -2.$$
Find and graph y.

*13. The constant k depends on the spring and on the mass of the weight. For a given spring, $k = c/m$, where c is a positive constant and m is the mass of the weight. Find the equation of motion of the weight.

*14. Referring to Exercise 13: suppose the initial position and initial velocity remain the same but that the mass of the weight is increased. What effect does this have on the motion of the weight?

*15. Answer the question of Exercise 14 if the mass and the initial velocity remain the same but the initial position is moved further down from E.

*16. Answer the question of Exercise 14 if the mass and the initial position remain the same but the initial velocity upward is increased.

12

Sequences and Series

Polynomial, rational, exponential, logarithmic, and trigonometric functions, and functions that are combinations of these, are examples of **elementary functions**. Hence all the functions we have used so far in this book have been elementary functions. Many functions are not elementary, however, and these can arise in quite natural ways. For instance, at the beginning of Section 7.4, we stated that it is not possible to find a simple expression for any antiderivative of

$$f(x) = e^{-x^2}.$$

A more precise way of saying the same thing is this: No antiderivative of $f(x) = e^{-x^2}$ is an elementary function. Similarly, no antiderivative of

$$g(x) = \sin x^2$$

is an elementary function. Thus it is clear that if we wish to study calculus in its full generality, we will have to study nonelementary functions. One way to do this is through the use of **power series**, which are introduced in this chapter.

We begin the chapter with a discussion of sequences of numbers. This will prepare us for a brief study of infinite series of numbers, which in turn will provide us with the necessary background for the introduction of power series in Section 12.3. Section 12.4 is devoted to the special type of power series known as a Taylor series.

12.1 SEQUENCES

A **sequence** of numbers is an ordered set of numbers generated from the positive integers 1, 2, 3, . . . by means of some rule. For instance, the set of numbers $\{1, \frac{1}{2}, \frac{1}{3}, \frac{1}{4}, \ldots\}$ is a sequence: It is generated from the positive integers by the rule that the nth number in the sequence, counting from the left, is equal to $1/n$. The numbers in the sequence are called its **terms**. In this example, 1 is the **first term** of the sequence, $\frac{1}{2}$ is its **second term**, $\frac{1}{3}$ its **third term**, and so on; the n**th term** of the sequence is thus $1/n$. We can write this sequence more compactly as $\{1/n\}$. Here the notation tells us that we are dealing with a sequence that is to be generated by substituting first 1 for n, then 2 for n, then 3 for n, and so on, in the expression $1/n$ that is enclosed within the braces. In a similar manner, $\{a_n\}$ denotes the sequence to be generated by substituting 1, 2, 3, . . . for n in the expression a_n.

EXAMPLE 1 Consider the sequences $\{n^2\}$, $\left\{\dfrac{n}{n+1}\right\}$, and $\{(-1)^n\}$.

We have

$$\{n^2\} = \{1^2, 2^2, 3^2, 4^2, \ldots\}$$
$$= \{1, 4, 9, 16, \ldots\},$$

$$\left\{\frac{n}{n+1}\right\} = \left\{\frac{1}{1+1}, \frac{2}{2+1}, \frac{3}{3+1}, \frac{4}{4+1}, \ldots\right\}$$
$$= \left\{\frac{1}{2}, \frac{2}{3}, \frac{3}{4}, \frac{4}{5}, \ldots\right\},$$

and

$$\{(-1)^n\} = \{(-1)^1, (-1)^2, (-1)^3, (-1)^4, \ldots\}$$
$$= \{-1, 1, -1, 1, \ldots\}. \quad \square$$

A sequence is really a function whose domain is the set of positive integers. For instance, the sequence $\{1/n\}$ is really the function f defined by

$$f(n) = \frac{1}{n}, \quad \text{for all positive integers } n.$$

Thus when we write $\{1/n\}$ for this sequence, we are simply saving time by not bothering to give the function a name and by stating only the right side of its defining equation.

Since a sequence is a function, it has a graph. Figure 12–1 shows the graph of the sequence $\{1/n\}$, and Figure 12–2 the graph of the sequence $\{n^2\}$. Notice in Figure 12–1 that as n increases without bound, the terms of the sequence come arbitrarily close to 0 and stay there. Therefore we say that the **limit of the sequence is 0**, or that the **sequence converges to 0**, and we write

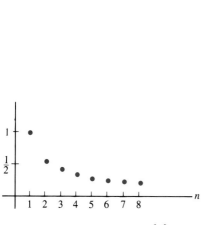

FIGURE 12–1 Graph of $\left\{\dfrac{1}{n}\right\}$

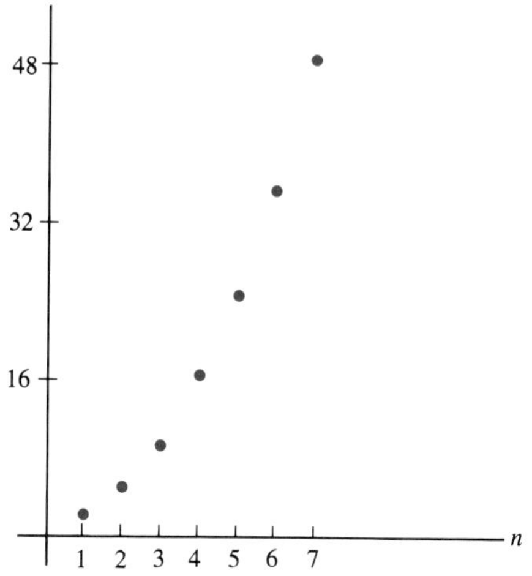

FIGURE 12–2 Graph of $\{n^2\}$

$$\operatorname*{Lim}_{n\to\infty}\frac{1}{n}=0.$$

On the other hand, as Figure 12–2 shows, the terms of the sequence $\{n^2\}$ do not come arbitrarily close to a single number as n increases, and hence we say that the sequence **has no limit** or that it **diverges**.

The Limit of a Sequence

Let $\{a_n\}$ be a sequence. If, as n increases, there is a single number L such that the terms of $\{a_n\}$ come arbitrarily close to L and remain arbitrarily close to L, then L is called the **limit** of the sequence, and we say that $\{a_n\}$ **converges to L**. We write this as

$$\operatorname*{Lim}_{n\to\infty} a_n = L.$$

If there is no such number L, we say that the sequence $\{a_n\}$ **diverges**.

EXAMPLE 2   Figure 12–3 shows the graph of the sequence

$$\left\{\frac{n}{n+1}\right\} = \left\{\frac{1}{2},\frac{2}{3},\frac{3}{4},\frac{4}{5},\ldots\right\}.$$

From the figure we see that as n increases, the terms of the sequence come arbitrarily close to, and remain arbitrarily close to, the number 1. Hence the sequence converges to 1 and

$$\operatorname*{Lim}_{n\to\infty}\frac{n}{n+1}=1. \quad \square$$

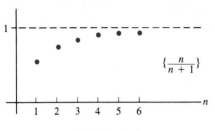

FIGURE 12–3

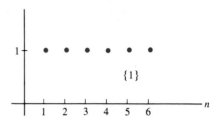

FIGURE 12–4

EXAMPLE 3 Consider the sequence $\{1\}$, that is, the sequence $\{a_n\}$ where $a_n = 1$ for every positive integer n. Figure 12–4 shows the graph of this sequence. Certainly all terms of this sequence come arbitrarily close to, and remain arbitrarily close to, the number 1. Therefore

$$\operatorname*{Lim}_{n \to \infty} a_n = \operatorname*{Lim}_{n \to \infty} 1 = 1. \quad \square$$

EXAMPLE 4 Figure 12–5 depicts the graph of the sequence

$$\left\{2 + \frac{(-1)^n}{n}\right\} = \left\{1, \frac{5}{2}, \frac{5}{3}, \frac{9}{4}, \frac{9}{5}, \frac{13}{6}, \ldots\right\}.$$

Since the terms of the sequence eventually come arbitrarily close to 2 and remain there, we have

$$\operatorname*{Lim}_{n \to \infty} \left(2 + \frac{(-1)^n}{n}\right) = 2. \quad \square$$

Recall that if n is a nonnegative integer then the number $n!$ (**n factorial**) is defined by

$$0! = 1, \qquad 1! = 1, \qquad \text{and} \qquad n! = n(n-1) \cdots 2 \cdot 1 \qquad \text{for } n \geq 2.$$

EXAMPLE 5 The graphs of $\{n!\} = \{1, 2, 6, 24, \ldots\}$ and $\{1/n!\} = \{1, \frac{1}{2}, \frac{1}{6}, \frac{1}{24}, \ldots\}$ (Figures 12–6 and 12–7, respectively) show that $\{n!\}$ diverges and that $\{1/n!\}$ converges to

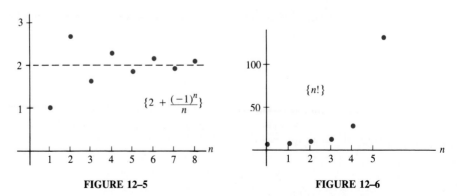

FIGURE 12–5

FIGURE 12–6

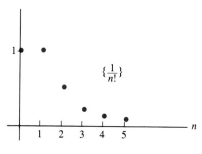

FIGURE 12–7 **FIGURE 12–8**

0. Since the terms of $\{n!\}$ grow large without bound, we say that it **diverges to $+\infty$**. ☐

EXAMPLE 6 Figure 12–8 shows the graph of the sequence $\{(-1)^n\}$. Since, as n increases, there is no single number L such that the terms of the sequence come arbitrarily close to L and remain arbitrarily close to L, the sequence diverges. ☐

EXAMPLE 7 Let

$$a_n = \begin{cases} \dfrac{1}{n} & \text{if 4 does not divide } n \text{ evenly,} \\ 1 & \text{if 4 does divide } n \text{ evenly.} \end{cases}$$

Then

$$\{a_n\} = \left\{1, \frac{1}{2}, \frac{1}{3}, 1, \frac{1}{5}, \frac{1}{6}, \frac{1}{7}, 1, \frac{1}{9}, \frac{1}{10}, \frac{1}{11}, 1, \ldots\right\}.$$

See Figure 12–9. Note that as n increases the terms of this sequence come arbitrarily close to 0, but that they do not remain arbitrarily close to 0 because every fourth term is equal to 1. Hence the sequence diverges. ☐

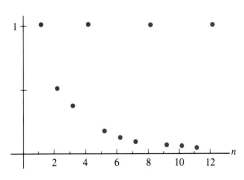

FIGURE 12–9

EXAMPLE 8

Suppose an investment of $5000 earns 8% simple interest per year. (Thus each year it earns interest of 0.08($5000) = $400.) Let us write a sequence that gives the value of the investment at the end of each year.

We let a_n denote the value of the investment at the end of year n, $n \geq 0$. Then

$$a_0 = 5000,$$
$$a_1 = 5000 + 400,$$
$$a_2 = 5000 + 2(400),$$

and so on. Hence

$$a_n = 5000 + n(400),$$

so the sequence is

$$\{a_n\} = \{5000 + 400n\}. \quad \square$$

A **geometric sequence** is a sequence of the form $\{r^n\}$, where r is some number. For instance, $\{(-1)^n\}$ is a geometric sequence with $r = -1$, and

$$\left\{\left(\frac{1}{2}\right)^n\right\} = \left\{\frac{1}{2^n}\right\} = \left\{\frac{1}{2}, \frac{1}{4}, \frac{1}{8}, \frac{1}{16}, \ldots\right\}$$

is a geometric sequence with $r = \frac{1}{2}$.

Let us analyze geometric sequences. First of all, if $r = 1$ we have

$$\{r^n\} = \{1^n\} = \{1, 1, 1, \ldots\},$$

which converges to 1 by Example 3. Similarly, if $r = 0$, then

$$\{r^n\} = \{0^n\} = \{0, 0, 0, \ldots\},$$

which converges to 0. (Why?) Also, if $r = -1$, then $\{r^n\} = \{(-1)^n\}$, which diverges by Example 6.

Now suppose that $0 < r < 1$. Since the value of r^n then becomes smaller and smaller as n increases (Figure 12–10), $\{r^n\}$ converges to 0 when $0 < r < 1$. Similarly, $\{r^n\}$ converges to 0 when $-1 < r < 0$ (Figure 12–11). Finally, if $r > 1$ or $r < -1$, then the absolute value of r^n grows without bound as n increases (Figures 12–12, 12–13), and hence $\{r^n\}$ diverges in both cases. Let us summarize these results.

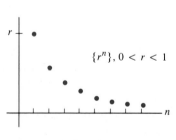
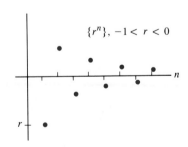

FIGURE 12–10   **FIGURE 12–11**

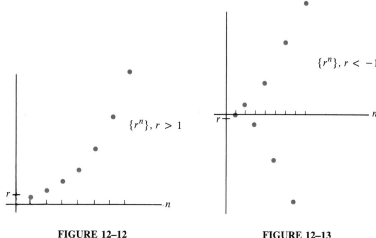

FIGURE 12-12   FIGURE 12-13

Geometric Sequences

Let r be a number. The geometric sequence $\{r^n\}$ converges to 1 if $r = 1$, converges to 0 if $-1 < r < 1$, and diverges otherwise.

EXAMPLE 9 The geometric sequence $\left\{\dfrac{1}{2^n}\right\}$ converges to 0, and the geometric sequence $\{2^n\}$ diverges. □

EXAMPLE 10 A company's monthly sales are increasing by 1.5% per month and are currently $50,000. Let us write a sequence that will give the company's monthly sales n months from now.

If we let a_n denote the company's sales n months from now, $n \geq 0$, then

$$a_0 = 50{,}000,$$

$$a_1 = 50{,}000 + 0.015(50{,}000) = 1.015(50{,}000),$$

$$a_2 = 1.015(50{,}000) + 0.015[1.015(50{,}000)]$$

$$= (1.015)^2 (50{,}000),$$

and so on. Therefore

$$\{a_n\} = \{(1.015)^n (50{,}000)\}.$$

This is not a geometric sequence (why not?), but each of its terms is 50,000 times the corresponding term of the geometric sequence $\{(1.015)^n\}$. Note also that if b_n is the total sales over the next n months, then

$$b_0 = 50{,}000,$$
$$b_1 = 50{,}000 + 1.015(50{,}000)$$
$$b_2 = 50{,}000 + (1.015)(50{,}000) + (1.015)^2(50{,}000),$$

and so on. Thus the sequence

$$\{b_n\} = \left\{ \sum_{k=0}^{n} 50{,}000(1.015)^k \right\}$$

gives the company's total sales over the next n months, $n \geq 0$. ◻

We can state properties of limits of sequences that are analogous to the properties of limits of functions presented in Section 2.1.

Properties of Limits of Sequences

Let $\{a_n\}$ and $\{b_n\}$ be convergent sequences, and let c be a number. Then

1. $\lim_{n \to \infty} c = c.$

2. $\lim_{n \to \infty} ca_n = c \cdot \lim_{n \to \infty} a_n.$

3. $\lim_{n \to \infty} (a_n \pm b_n) = \lim_{n \to \infty} a_n \pm \lim_{n \to \infty} b_n.$

4. $\lim_{n \to \infty} (a_n b_n) = \left(\lim_{n \to \infty} a_n \right) \left(\lim_{n \to \infty} b_n \right).$

5. If $b_n \neq 0$ for all n and $\lim_{n \to \infty} b_n \neq 0$, then

$$\lim_{n \to \infty} \frac{a_n}{b_n} = \frac{\lim_{n \to \infty} a_n}{\lim_{n \to \infty} b_n}.$$

6. If $b_n \neq 0$ for all n, $\lim_{n \to \infty} a_n \neq 0$, and $\lim_{n \to \infty} b_n = 0$, then $\lim_{n \to \infty} \frac{a_n}{b_n}$ does not exist.

We can use these properties of limits of sequences and the results we have already obtained to find limits of sequences without resorting to their graphs.

EXAMPLE 11 Consider the constant sequence $\{a_n\} = \{3\}$. By Property 1,

$$\lim_{n \to \infty} a_n = \lim_{n \to \infty} 3 = 3. \quad \Box$$

SEQUENCES AND SERIES

EXAMPLE 12   Let $\{a_n\} = \{4/n\}$. Using Property 2 and the previously established fact that

$$\lim_{n \to \infty} \frac{1}{n} = 0,$$

we have

$$\lim_{n \to \infty} a_n = \lim_{n \to \infty} \frac{4}{n} = \lim_{n \to \infty} 4 \cdot \frac{1}{n} = 4 \cdot \lim_{n \to \infty} \frac{1}{n} = 4 \cdot 0 = 0. \quad \square$$

EXAMPLE 13   Using Properties 1, 2, and 3, we have

$$\lim_{n \to \infty} \left(\frac{5}{n} + 2\right) = \lim_{n \to \infty} \frac{5}{n} + \lim_{n \to \infty} 2 = 5 \cdot \lim_{n \to \infty} \frac{1}{n} + 2 = 5 \cdot 0 + 2 = 2. \quad \square$$

EXAMPLE 14   Using Property 4, we have

$$\lim_{n \to \infty} \frac{1}{n^2} = \lim_{n \to \infty} \frac{1}{n} \cdot \frac{1}{n} = \left(\lim_{n \to \infty} \frac{1}{n}\right)\left(\lim_{n \to \infty} \frac{1}{n}\right) = 0 \cdot 0 = 0. \quad \square$$

EXAMPLE 15   Consider

$$\lim_{n \to \infty} \frac{2n^2 + 3n + 5}{5n^2 - 6n + 1}.$$

To evaluate the limit of such a quotient, we divide each term in the quotient by the highest power of n present and use Properties 1, 2, 3, and 5:

$$\lim_{n \to \infty} \frac{2n^2 + 3n + 5}{5n^2 - 6n + 1} = \lim_{n \to \infty} \frac{2 + 3/n + 5/n^2}{5 - 6/n + 1/n^2}$$

$$= \frac{\lim_{n \to \infty}(2 + 3/n + 5/n^2)}{\lim_{n \to \infty}(5 - 6/n + 1/n^2)}$$

$$= \frac{\lim_{n \to \infty} 2 + 3 \cdot \lim_{n \to \infty}(1/n) + 5 \cdot \lim_{n \to \infty}(1/n^2)}{\lim_{n \to \infty} 5 - 6 \cdot \lim_{n \to \infty}(1/n) + \lim_{n \to \infty}(1/n^2)}$$

$$= \frac{2 + 3 \cdot 0 + 5 \cdot 0}{5 - 6 \cdot 0 + 0}$$

$$= \frac{2}{5}. \quad \square$$

EXAMPLE 16   Since

$$\lim_{n \to \infty} \frac{n^2 + 1}{n} = \lim_{n \to \infty} \frac{1 + 1/n^2}{1/n}$$

and $\text{Lim}_{n\to\infty} \left(1 + \dfrac{1}{n^2}\right) = 1 + 0 = 1$ while $\text{Lim}_{n\to\infty} \dfrac{1}{n} = 0$, the sequence $\left\{\dfrac{n^2 + 1}{n}\right\}$ diverges by Property 6. □

Sequences occur in nature in many ways. The following example shows how a sequence known as the Fibonacci sequence occurs in the study of population growth.

EXAMPLE 17 Let us suppose that a pair of rabbits always consists of one male and one female. Let us also suppose that, starting exactly two months after they are born, every pair will produce another pair at one-month intervals. If we begin with one newly born pair of rabbits, how many pairs will we have n months later if no rabbits die?

A newly born pair will not reproduce until two months from their day of birth, so during both the first and the second month we will have just one pair of rabbits. At the beginning of the third month, the original pair will produce another pair; thus during the third month we will have two pairs of rabbits. At the beginning of the fourth month the original pair will again produce another pair, so during the fourth month we will have three pairs, and at the beginning of the fifth month the original pair will produce still another pair while the second pair will now also produce a pair. Thus we will have five pairs during the fifth month, and so on.

Let a_n denote the number of pairs of rabbits we have during month n. Then, as we have seen, $a_1 = a_2 = 1$, and, for $n \geq 3$,

a_n = Number of pairs we have during month $n - 1$
$\quad$ + Number of pairs born at the beginning of month n.

But

Number of pairs we have during month $n - 1 = a_{n-1}$

and

Number of pairs born at the beginning of month n
= Number of pairs we have during month $n - 2$
= a_{n-2}.

Therefore

$a_1 = a_2 = 1,$ $\quad$ and $\quad$ $a_n = a_{n-1} + a_{n-2}$ $\quad$ for $n \geq 3$.

The sequence $\{a_n\}$ defined by the preceding equations is known as the **Fibonacci sequence**. Thus

$$\{a_n\} = \{1, 1, 2, 3, 5, 8, 13, 21, 34, 55, 89, 144, \ldots\}.$$

Note that (except for its first two terms), each term of the Fibonacci sequence is the sum of the two preceding terms.

The Fibonacci sequence is a divergent sequence (why?), but it has many in-

SEQUENCES AND SERIES

teresting properties. For instance, suppose we consider the ratio of successive terms of the sequence. Let

$$b_n = \frac{a_{n+1}}{a_n} \quad \text{for } n \geq 1.$$

The graph of the sequence

$$\{b_n\} = \left\{1, 2, \frac{3}{2}, \frac{5}{3}, \frac{8}{5}, \frac{13}{8}, \ldots\right\}$$

is shown in Figure 12–14. It appears from the graph that the sequence converges to a number L between 1 and 2.

Note that the Fibonacci relation

$$a_n = a_{n-1} + a_{n-2}, \quad n \geq 3,$$

implies that

$$\frac{a_n}{a_{n-1}} = \frac{a_{n-1}}{a_{n-1}} + \frac{a_{n-2}}{a_{n-1}}, \quad n \geq 3,$$

or

$$\frac{a_n}{a_{n-1}} = 1 + \frac{1}{a_{n-1}/a_{n-2}}, \quad n \geq 3,$$

or

$$b_{n-1} = 1 + \frac{1}{b_{n-2}}, \quad n \geq 3.$$

But if $n \geq 3$, the sequences

$$\{b_{n-1}\} = \{b_2, b_3, b_4, \ldots\}$$

and

$$\{b_{n-2}\} = \{b_1, b_2, b_3, b_4, \ldots\}$$

both have limit L, and thus

$$L = \lim_{n \to \infty} b_{n-1} = \lim_{n \to \infty} \left(1 + \frac{1}{b_{n-2}}\right) = 1 + \frac{1}{\lim_{n \to \infty} b_{n-2}} = 1 + \frac{1}{L}.$$

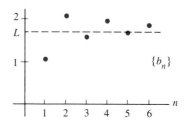

FIGURE 12–14

Hence

$$L = 1 + \frac{1}{L}, \quad \text{or} \quad L^2 - L - 1 = 0,$$

and the positive root of this quadratic equation is

$$L = \frac{1 + \sqrt{5}}{2} \approx 1.618.$$

When n is large, b_n is close to L, so

$$b_n \approx 1.618, \quad \text{or} \quad a_{n+1} \approx (1.618)a_n.$$

This shows that in the long run, the number of rabbits we have during each month will be approximately 1.618 times the number we had during the previous month. □

■ **Computer Manual: Program SEQUENCE**

12.1 EXERCISES

In Exercises 1 through 8, graph the sequence and state whether it converges or diverges. If it converges, find its limit.

1. $\left\{\dfrac{1}{2n}\right\}$

2. $\{2, -3, 2, -3, 2, -3, \ldots\}$

3. $\{n\}$

4. $\left\{\dfrac{n+1}{n}\right\}$

5. $\left\{\dfrac{1}{2}, -1, \dfrac{1}{3}, -1, \dfrac{1}{4}, -1, \ldots\right\}$

6. $\left\{3 - \dfrac{1}{n}\right\}$

7. $\left\{\dfrac{n!}{n^2}\right\}$

8. $\left\{2, \dfrac{1}{2}, 3, \dfrac{1}{3}, \dfrac{1}{9}, 4, \dfrac{1}{4}, \dfrac{1}{16}, \dfrac{1}{64}, \ldots\right\}$

In Exercises 9 through 32, state whether the sequence converges or diverges. If it converges, find its limit.

9. $\{n + 1\}$

10. $\{3n\}$

11. $\left\{\dfrac{1}{3^n}\right\}$

12. $\{3^{-n}\}$

13. $\left\{\dfrac{(-1)^n}{5^n}\right\}$

14. $\left\{\dfrac{2^n}{3^{n+1}}\right\}$

15. $\left\{\dfrac{7}{(2/5)^n}\right\}$

16. $\{e^n\}$

17. $\{e^{-2n} + 1\}$

18. $\{-2\}$

19. $\{\pi\sqrt{5}\}$

20. $\left\{\dfrac{2}{n} + \dfrac{3}{n^2}\right\}$

21. $\left\{\dfrac{1}{n} - n\right\}$

22. $\left\{\dfrac{2}{n!}\right\}$

23. $\left\{\dfrac{(n+1)!}{2n}\right\}$

24. $\left\{\dfrac{2n}{(n+1)!}\right\}$

25. $\left\{\dfrac{3n+2}{n+3}\right\}$

26. $\left\{\dfrac{5n+1}{7-2n}\right\}$

27. $\left\{\dfrac{n+1}{n^2+3}\right\}$

28. $\left\{\dfrac{n^2+5}{n+1}\right\}$

29. $\left\{\dfrac{2n^3 + 3n + 1}{n^2 + 2}\right\}$

30. $\left\{\dfrac{4n^3 + 3n + 1}{2n^4 + 3n^2}\right\}$

31. $\left\{\sin\dfrac{n\pi}{2}\right\}$

32. $\left\{\dfrac{n+1}{n}\cos 2n\pi\right\}$

*33. If p is a positive integer, show that $\{n^{-p}\}$ converges to 0.

*34. If p is a positive integer, show that $\{n^p\}$ diverges.

If a sequence $\{a_n\}$ converges to L, f is a continuous function, and $f(a_n)$ is defined for all n, then the sequence $\{f(a_n)\}$ converges to $f(L)$. In Exercises 35 through 42, use this fact to find the limit of the given sequence.

*35. $\{\sqrt{n/(n+1)}\}$ ***36.** $\{\sqrt{(2n+1)/n}\}$ ***37.** $\{e^{1/n}\}$

*38. $\{e^{(n+1)/n}\}$ ***39.** $\{\sin(\pi/n)\}$ ***40.** $\{\cos(\pi/n^2)\}$

*41. $\{\ln[(n+1)/2n]\}$ ***42.** $\{\ln[(3n^2+2)/(5n^2+1)]\}$

43. An investment of \$10,000 earns 10% simple interest per year. (Thus it earns $0.1(\$10,000) = \$1000$ per year.) Write a sequence $\{a_n\}$, where a_n is the value of the investment at the end of n years, $n \geq 0$.

44. Repeat Exercise 43 if the investment earns 8% simple interest per year.

45. Suppose an investment of \$$P$ earns $100r\%$ simple interest per year, and let A_n denote the value of the investment at the end of n years, $n \geq 0$. Write the sequence $\{A_n\}$.

46. An investment of \$10,000 earns 10% interest per year compounded annually. Let a_n be the value of the investment at the end of n years, $n \geq 0$. Write the sequence $\{a_n\}$.

47. Repeat Exercise 46 if the investment earns 8% interest per year compounded annually.

48. Suppose an investment of \$$P$ earns $100r\%$ interest per year, compounded annually. Let A_n denote the value of the investment at the end of n years, $n \geq 0$. Write the sequence $\{A_n\}$.

49. If you save \$1 today, \$2 tomorrow, \$4 the next day, and so on, write a sequence $\{a_n\}$, where a_n is the amount saved on day n.

Management

50. A firm's payroll is currently \$1 million per month and is increasing at a rate of 5% per month.
  (a) Let a_n be the amount of the firm's payroll n months from now. Write the sequence $\{a_n\}$.
  (b) Let b_n be the total amount the firm's payroll will cost over the next n months. Write the sequence $\{b_n\}$.

51. A firm's profits are currently \$500,000 and are decreasing by 6% per year. Let a_n be the firm's profits n years from now and b_n the total of the firm's profits over the next n years.
  (a) Write the sequence $\{a_n\}$.
  (b) Write the sequence $\{b_n\}$.

Life Science

52. A certain strain of bacteria increases its population by 12% every hour. If a colony of this strain is started with 100 bacteria, how many will there be in the colony 24 hours later?

53. Suppose that a pair of rabbits always consists of a male and a female and that each pair will produce another pair every month, beginning three months after their birth. Let a_n be the number of pairs of rabbits alive in month n. Assuming no rabbits die, write the sequence $\{a_n\}$ and analyze it in the manner of Example 17.

12.2 INFINITE SERIES OF CONSTANTS

Nearly 2500 years ago, the Greek mathematician Zeno (fifth century B.C.) presented an argument that seems to show that it is impossible to traverse any finite distance while traveling at a constant speed. The argument may be stated as follows: Suppose we wish to traverse a distance D at a constant speed. It will take some finite time T to cover the first half of D; then (because we are moving at a constant speed), it will take time $T/2$ to cover one-half of the remaining distance, time $(1/2)(T/2) = T/4$ to cover one half of the distance that still remains, and so on. See Figure 12–15. The total time required to traverse the distance D will thus be given by the unending sum

$$T + \frac{1}{2}T + \frac{1}{4}T + \frac{1}{8}T + \cdots.$$

Zeno argued that since this sum adds up infinitely many positive numbers, its total must be larger than any positive number, or, in other words, its total must be infinite. But if the time required to traverse the distance D is infinite, it must actually be impossible to traverse D.

Since we know that it is indeed possible to traverse a finite distance in a finite time while traveling at a constant speed, there must be some fallacy in Zeno's argument. Where is it? The fallacy lies in the statement that the sum

$$T + \frac{1}{2}T + \frac{1}{4}T + \frac{1}{8}T + \cdots$$

is infinite. As we shall see later on, this sum is actually finite and equal to $2T$. Thus it follows that the distance D can be traversed at a constant speed in exactly twice the time it takes to cover the first half of D, a result that agrees with our experience.

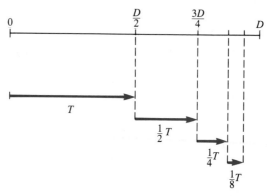

FIGURE 12–15

The sum

$$T + \frac{1}{2}T + \frac{1}{4}T + \frac{1}{8}T + \cdots$$

is an example of an **infinite series**. In this section we study infinite series.

Informally, an **infinite series of constants** is an infinite sum

$$\sum_{k=1}^{\infty} a_k = a_1 + a_2 + a_3 + \cdots,$$

where $a_1, a_2, a_3, \ldots$ are numbers. This is not an adequate definition, however, for it is not clear what the infinite sum on the right side of the equation means: How are we supposed to add infinitely many numbers? The proper definition of an infinite series identifies it as a particular kind of sequence.

Infinite Series

An **infinite series**

$$\sum_{k=1}^{\infty} a_k,$$

where $a_1, a_2, \ldots$ are numbers, is defined as follows: For every positive integer n, let

$$S_n = \sum_{k=1}^{n} a_k = a_1 + a_2 + \cdots + a_n.$$

Then

$$\sum_{k=1}^{\infty} a_k$$

is the sequence $\{S_n\}$. The number S_n is called the ***n*th partial sum** of the infinite series.

EXAMPLE 1   By definition, the infinite series

$$\sum_{k=1}^{\infty} \frac{1}{k} = 1 + \frac{1}{2} + \frac{1}{3} + \frac{1}{4} + \cdots$$

is the sequence $\{S_n\}$, where

$$S_1 = \sum_{k=1}^{1} \frac{1}{k} = \frac{1}{1} = 1,$$

$$S_2 = \sum_{k=1}^{2} \frac{1}{k} = \frac{1}{1} + \frac{1}{2} = \frac{3}{2},$$

$$S_3 = \sum_{k=1}^{3} \frac{1}{k} = \frac{1}{1} + \frac{1}{2} + \frac{1}{3} = \frac{11}{6},$$

$$S_4 = \sum_{k=1}^{4} \frac{1}{k} = \frac{1}{1} + \frac{1}{2} + \frac{1}{3} + \frac{1}{4} = \frac{25}{12},$$

and so on. ☐

EXAMPLE 2 The infinite series

$$\sum_{k=1}^{\infty} \frac{(-1)^k}{k!} = \frac{(-1)^1}{1!} + \frac{(-1)^2}{2!} + \frac{(-1)^3}{3!} + \frac{(-1)^4}{4!} + \cdots$$

$$= -1 + \frac{1}{2} - \frac{1}{6} + \frac{1}{24} - \cdots$$

is the sequence

$$\{S_n\} = \left\{ \sum_{k=1}^{n} \frac{(-1)^k}{k!} \right\}$$

$$= \left\{ \sum_{k=1}^{1} \frac{(-1)^k}{k!}, \sum_{k=1}^{2} \frac{(-1)^k}{k!}, \sum_{k=1}^{3} \frac{(-1)^k}{k!}, \cdots \right\}$$

$$= \left\{ -1, -1 + \frac{1}{2}, -1 + \frac{1}{2} - \frac{1}{6}, -1 + \frac{1}{2} - \frac{1}{6} + \frac{1}{24}, \cdots \right\}$$

$$= \left\{ -1, -\frac{1}{2}, -\frac{2}{3}, -\frac{15}{24}, \cdots \right\}. \quad ☐$$

The summation of an infinite series need not begin with $k = 1$; this was just a convenience in our definition.

EXAMPLE 3 The infinite series

$$\sum_{k=0}^{\infty} \frac{1}{2^k} = 1 + \frac{1}{2} + \frac{1}{4} + \frac{1}{8} + \frac{1}{16} + \cdots$$

is the sequence

$$\left\{ \sum_{k=0}^{0} \frac{1}{2^k}, \sum_{k=0}^{1} \frac{1}{2^k}, \sum_{k=0}^{2} \frac{1}{2^k}, \sum_{k=0}^{3} \frac{1}{2^k}, \cdots \right\}$$

$$= \left\{ 1, 1 + \frac{1}{2}, 1 + \frac{1}{2} + \frac{1}{4}, 1 + \frac{1}{2} + \frac{1}{4} + \frac{1}{8}, \cdots \right\}$$

$$= \left\{ 1, \frac{3}{2}, \frac{7}{4}, \frac{15}{8}, \cdots \right\}$$

$$= \left\{ 2 - \frac{1}{2^{n-1}} \right\}. \quad ☐$$

Since an infinite series is really a sequence, it must be either convergent or divergent.

> **Convergence of Infinite Series**
>
> The infinite series
> $$\sum_{k=1}^{\infty} a_k$$
> **converges to L** if its sequence of partial sums
> $$\left\{ \sum_{k=1}^{n} a_k \right\}$$
> converges to L; it **diverges** if its sequence of partial sums diverges. If
> $$\sum_{k=1}^{\infty} a_k$$
> converges to L, we say that L is the **sum** of the series, and we write
> $$\sum_{k=1}^{\infty} a_k = L.$$

EXAMPLE 4 Consider the infinite series
$$\sum_{k=0}^{\infty} \frac{1}{2^k} = 1 + \frac{1}{2} + \frac{1}{4} + \frac{1}{8} + \cdots.$$
From Example 3, the sequence of partial sums for this series is
$$\left\{ 1, \frac{3}{2}, \frac{7}{4}, \frac{15}{8}, \ldots \right\} = \left\{ 2 - \frac{1}{2^{n-1}} \right\},$$
and
$$\operatorname*{Lim}_{n \to \infty} \left(2 - \frac{1}{2^{n-1}} \right) = 2 - \operatorname*{Lim}_{n \to \infty} \frac{1}{2^{n-1}} = 2 - 0 = 2.$$
Therefore the series converges to 2, and we write
$$\sum_{k=0}^{\infty} \frac{1}{2^k} = 2.$$
See Figure 12–16. ☐

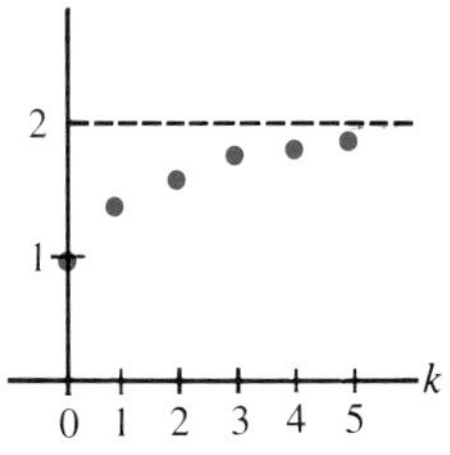

FIGURE 12–16

EXAMPLE 5 The infinite series
$$\sum_{k=1}^{\infty} k = 1 + 2 + 3 + 4 + \cdots$$

diverges: Its sequence of partial sums is $\{1, 3, 6, 10, \ldots\}$, which diverges to $+\infty$. ☐

Generally speaking, it is not an easy task to determine whether a particular infinite series converges or diverges, or, if it does converge, to find its sum. However, there is one type of infinite series, the **geometric series**, for which questions of convergence and sum are easily answered.

Geometric Series

A **geometric series** is one that can be written in the form

$$\sum_{k=0}^{\infty} ar^k = a + ar + ar^2 + ar^3 + \cdots$$

for some numbers a and r. The number a is called the **first term** of the series, and the number r is its **ratio**.

EXAMPLE 6 The series

$$\sum_{k=0}^{\infty} \frac{1}{2^k} = 1 + \frac{1}{2} + \frac{1}{4} + \frac{1}{8} + \cdots$$

$$= 1 + 1\left(\frac{1}{2}\right) + 1\left(\frac{1}{2}\right)^2 + 1\left(\frac{1}{2}\right)^3 + \cdots$$

is a geometric series with first term $a = 1$ and ratio $r = \frac{1}{2}$. Note that by Example 4 this series converges to 2. ☐

EXAMPLE 7 The series

$$\sum_{k=1}^{\infty} 1 = 1 + 1 + 1 + 1 + \cdots = 1 + 1 \cdot 1 + 1 \cdot 1^2 + 1 \cdot 1^3 + \cdots$$

is geometric with $a = 1$ and $r = 1$. This series diverges to $+\infty$ because its sequence of partial sums is $\{1, 2, 3, 4, \ldots\}$. In fact, any geometric series having $a \neq 0$ and $r = 1$ must diverge, for such a series has the form

$$\sum_{k=0}^{\infty} a \cdot 1^k = a + a + a + \cdots,$$

and hence its sequence of partial sums is the divergent sequence

$$\{a, 2a, 3a, \ldots\}. \quad \square$$

Example 7 shows that a geometric series with first term $a \neq 0$ and ratio $r = 1$ must be divergent. Now let us consider convergence for geometric series with $a \neq 0$ and $r \neq 1$. If

$$\sum_{k=0}^{\infty} ar^k = a + ar + ar^2 + \cdots$$

SEQUENCES AND SERIES

is such a series, its nth partial sum is
$$S_n = a + ar + ar^2 + \cdots + ar^{n-1}.$$
Then
$$rS_n = ar + ar^2 + ar^3 + \cdots + ar^n,$$
and
$$S_n - rS_n = a - ar^n,$$
or
$$(1 - r)S_n = a(1 - r^n).$$
Since $r \neq 1$, we may divide by $1 - r$ to obtain
$$S_n = \frac{a}{1-r}(1 - r^n).$$
Thus
$$\mathrm{Lim}_{n \to \infty} S_n = \mathrm{Lim}_{n \to \infty} \frac{a}{1-r}(1 - r^n) = \frac{a}{1-r} \cdot \mathrm{Lim}_{n \to \infty}(1 - r^n)$$
$$= \frac{a}{1-r}\left(1 - \mathrm{Lim}_{n \to \infty} r^n\right).$$

But $\{r^n\}$ is a geometric sequence, and we know from our work in Section 12.1 that
$$\mathrm{Lim}_{n \to \infty} r^n = 0 \quad \text{if } -1 < r < 1,$$
and
$$\mathrm{Lim}_{n \to \infty} r^n \quad \text{does not exist if } r > 1 \text{ or } r \leq -1.$$

Since $\{S_n\}$ is the sequence of partial sums of the geometric series, this shows that the series converges to $a/(1 - r)$ if $-1 < r < 1$ and diverges if $r > 1$ or $r \leq -1$. Combining this with the result of Example 7, we have the following characterization of geometric series.

Limit of a Geometric Series

If $a \neq 0$, the geometric series
$$\sum_{k=0}^{\infty} ar^k = a + ar + ar^2 + \cdots$$
converges to $\dfrac{a}{1-r}$ if $-1 < r < 1$ and diverges if $r \geq 1$ or $r \leq -1$.

(Of course, if $a = 0$, the series becomes
$$\sum_{k=0}^{\infty} 0 r^k = 0 + 0 + \cdots,$$
which converges to 0.)

EXAMPLE 8 The geometric series

$$\sum_{k=0}^{\infty} \frac{1}{2^k} = 1 + \frac{1}{2} + \frac{1}{4} + \frac{1}{8} + \cdots$$

has $a = 1$ and $r = \frac{1}{2}$. Therefore (as we saw in Example 4) it converges to

$$\frac{a}{1-r} = \frac{1}{1 - \frac{1}{2}} = \frac{1}{\frac{1}{2}} = 2. \quad \square$$

EXAMPLE 9 We can now justify the claim made at the beginning of this section that Zeno's series

$$T + \frac{1}{2}T + \frac{1}{4}T + \frac{1}{8}T + \cdots$$

sums to $2T$. The series is clearly geometric with $a = T \neq 0$ and $r = \frac{1}{2}$, so it converges to

$$\frac{T}{1 - \frac{1}{2}} = 2T. \quad \square$$

EXAMPLE 10 The geometric series

$$\sum_{k=0}^{\infty} 2\left(\frac{3}{2}\right)^k$$

has $a = 2$ and $r = \frac{3}{2}$ and hence diverges. $\quad \square$

EXAMPLE 11 The series

$$\frac{4}{3} - 1 + \frac{3}{4} - \frac{9}{16} + \frac{27}{64} - \cdots$$

may be written in the form

$$\frac{4}{3} + \frac{4}{3}\left(-\frac{3}{4}\right) + \frac{4}{3}\left(-\frac{3}{4}\right)^2 + \frac{4}{3}\left(-\frac{3}{4}\right)^3 + \frac{4}{3}\left(-\frac{3}{4}\right)^4 + \cdots$$

and hence is geometric with $a = \frac{4}{3}$ and $r = -\frac{3}{4}$. Therefore it converges to

$$\frac{4/3}{1 - (-3/4)} = \frac{4/3}{7/4} = \frac{16}{21}. \quad \square$$

Geometric series are important in many applications. The following example shows how they can arise in probability problems.

EXAMPLE 12 An assembly line produces microchips, most of which are good but some of which are defective. Define a random variable X by

$X = $ Number of consecutive good chips produced.

Suppose that

$$P(k \text{ consecutive good chips are produced}) = P(X = k) = (0.01)(0.99)^k$$

for $k = 0, 1, 2, \ldots$. (This is so if the probability that any particular chip is a good one is 0.99.)

First of all, is this a legitimate assignment of probabilities for the random variable X? To show that it is, we must show that the sum of the probabilities equals 1, that is, that

$$\sum_{k=0}^{\infty} P(X = k) = 1.$$

But

$$\sum_{k=0}^{\infty} P(X = k) = \sum_{k=0}^{\infty} (0.01)(0.99)^k$$
$$= 0.01 + (0.01)(0.99) + (0.01)(0.99)^2 + \cdots,$$

which is a geometric series with $a = 0.01$ and $r = 0.99$ and thus converges to

$$\frac{0.01}{1 - 0.99} = 1.$$

Now suppose we wish to know the probability of producing 40 or more good microchips consecutively. We have

$$P(X \geq 40) = \sum_{k=40}^{\infty} P(X = k)$$
$$= \sum_{k=40}^{\infty} (0.01)(0.99)^k$$
$$= (0.01)(0.99)^{40} + (0.01)(0.99)^{41} + \cdots$$
$$= \frac{(0.01)(0.99)^{40}}{1 - 0.99}$$
$$\approx 0.6690. \quad \square$$

For a series that is not geometric, convergence or divergence can often be established by a **convergence test**. There are many such tests; each can be applied to some series but not to others. As examples of convergence tests, we will present two of the more widely applicable ones: the **integral test** and the **ratio test**.

Suppose we have an infinite series

$$\sum_{k=1}^{\infty} a_k, \quad a_k \geq 0 \text{ for } k = 1, 2, \ldots,$$

and we can find a decreasing function f defined on $[1, +\infty)$ such that $f(k) = a_k$ for $k = 1, 2, 3, \ldots$. In Figure 12–17, the area of the shaded rectangles, which is

$$\sum_{k=2}^{\infty} a_k,$$

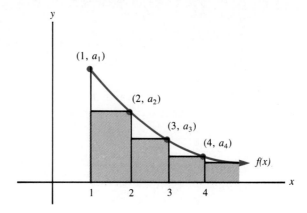

FIGURE 12-17

is certainly no larger than the area under f from 1 to $+\infty$, so

$$\sum_{k=2}^{\infty} a_k \leq \int_1^{+\infty} f(x)\,dx.$$

But then

$$\sum_{k=1}^{\infty} a_k \leq a_1 + \int_1^{+\infty} f(x)\,dx,$$

so the sum on the left-hand side of this last inequality will be finite if the integral on the right-hand side is finite, that is, if the improper integral converges. On the other hand, as Figure 12–18 shows,

$$\sum_{k=1}^{\infty} a_k \geq \int_1^{+\infty} f(x)\,dx,$$

so if the improper integral diverges, the sum cannot be finite. Hence we have the **integral test**.

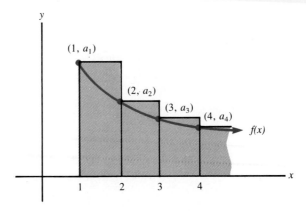

FIGURE 12-18

The Integral Test

Suppose

$$\sum_{k=1}^{\infty} a_k$$

is an infinite series with the property that

$$a_k \geq 0 \quad \text{for } k = 1, 2, 3, \ldots,$$

and let f be a decreasing function defined on $[1, +\infty)$ such that

$$f(k) = a_k \quad \text{for } k = 1, 2, 3, \ldots.$$

Then the series $\sum_{k=1}^{\infty} a_k$ converges if the improper integral $\int_{1}^{+\infty} f(x)\,dx$ converges, and diverges if the integral diverges.

EXAMPLE 13 The series

$$\sum_{k=1}^{\infty} \frac{1}{k^2}$$

is not geometric, so let us try to apply the integral test to it. If we let $f(x) = x^{-2}$, then f is defined on $[1, +\infty)$, f is decreasing on $[1, +\infty)$, (why?) and

$$f(k) = k^{-2} = \frac{1}{k^2} = a_k \quad \text{for } k = 1, 2, 3, \ldots.$$

Therefore

$$\sum_{k=1}^{\infty} \frac{1}{k^2}$$

converges if the improper integral

$$\int_{1}^{+\infty} x^{-2}\,dx$$

converges, and diverges if the integral diverges. But

$$\int_{1}^{+\infty} x^{-2}\,dx = \lim_{t \to +\infty} \int_{1}^{t} x^{-2}\,dx$$

$$= \lim_{t \to +\infty} \left[-\frac{1}{x} \bigg|_{1}^{t} \right]$$

$$= \lim_{t \to +\infty} \left[-\frac{1}{t} + 1 \right]$$

$$= 0 + 1$$

$$= 1.$$

Therefore the series $\sum_{k=1}^{\infty} 1/k^2$ converges. □

EXAMPLE 14 The series

$$\sum_{k=1}^{\infty} \frac{1}{k}$$

is known as the **harmonic series**. We can use the integral test to show that the harmonic series diverges. Let f be defined by $f(x) = 1/x$; then f is defined on $[1, +\infty)$, decreasing on $[1, +\infty)$ (why?), and

$$f(k) = \frac{1}{k} = a_k \quad \text{for } k = 1, 2, 3, \ldots.$$

Therefore $\sum_{k=1}^{\infty} 1/k$ converges if and only if $\int_{1}^{+\infty} (1/x)\,dx$ converges. But by Example 3 of Section 7.3, this improper integral diverges. Therefore the harmonic series diverges. □

We present the ratio test without proof.

The Ratio Test

Let

$$\sum_{k=1}^{\infty} a_k$$

be an infinite series and form the sequence

$$\left\{ \left| \frac{a_{k+1}}{a_k} \right| \right\}.$$

Then

1. The series converges if

$$\lim_{k \to \infty} \left| \frac{a_{k+1}}{a_k} \right| < 1.$$

2. The series diverges if

$$\lim_{k \to \infty} \left| \frac{a_{k+1}}{a_k} \right| > 1$$

or if the sequence

$$\left\{ \left| \frac{a_{k+1}}{a_k} \right| \right\}$$

diverges to $+\infty$.

SEQUENCES AND SERIES

EXAMPLE 15   The infinite series

$$\sum_{k=0}^{\infty} \frac{1}{k!}$$

is not geometric, nor can we apply the integral test to it. (Why not?) Let us try to use the ratio test. We have

$$\frac{a_{k+1}}{a_k} = \frac{1/(k+1)!}{1/k!} = \frac{1}{(k+1)!} \cdot \frac{k!}{1} = \frac{1}{(k+1)k!} \cdot \frac{k!}{1} = \frac{1}{k+1}.$$

Therefore

$$\lim_{k \to \infty} \left| \frac{a_{k+1}}{a_k} \right| = \lim_{k \to \infty} \left| \frac{1}{k+1} \right| = \lim_{k \to \infty} \frac{1}{k+1}$$

$$= \lim_{k \to \infty} \frac{1/k}{1 + 1/k}$$

$$= \frac{0}{1 + 0}$$

$$= 0.$$

Hence the series $\sum_{k=0}^{\infty} 1/k!$ converges. □

EXAMPLE 16   Let us apply the ratio test to the series

$$\sum_{k=1}^{\infty} (-1)^k \frac{2^k}{3k}.$$

We have

$$\lim_{k \to \infty} \left| \frac{a_{k+1}}{a_k} \right| = \lim_{k \to \infty} \left| \frac{(-1)^{k+1} 2^{k+1}/3(k+1)}{(-1)^k 2^k/3k} \right|$$

$$= \lim_{k \to \infty} \left| \frac{(-1)^{k+1}}{(-1)^k} \cdot \frac{2^{k+1}}{2^k} \cdot \frac{3k}{3(k+1)} \right|$$

$$= \lim_{k \to \infty} \left| -2 \frac{k}{k+1} \right|$$

$$= 2 \cdot \lim_{k \to \infty} \frac{k}{k+1}$$

$$= 2 \cdot \lim_{k \to \infty} \frac{1}{1 + 1/k}$$

$$= 2 \cdot 1$$

$$= 2.$$

Thus the series

$$\sum_{k=1}^{\infty} (-1)^k \frac{2^k}{3k}$$

diverges. ☐

EXAMPLE 17 The ratio test shows that the series

$$\sum_{k=0}^{\infty} k!$$

diverges, because the sequence

$$\left\{ \left| \frac{a_{k+1}}{a_k} \right| \right\} = \left\{ \frac{(k+1)!}{k!} \right\} = \{k+1\} = \{1, 2, 3, 4, \ldots\}$$

diverges to $+\infty$. ☐

The ratio test is easy to apply, but it is sometimes inconclusive. For instance, if

$$\operatorname*{Lim}_{k \to \infty} \left| \frac{a_{k+1}}{a_k} \right| = 1,$$

the ratio test does not tell us whether the series converges or diverges, and in fact either alternative is possible.

EXAMPLE 18 Applying the ratio test to the harmonic series

$$\sum_{k=1}^{\infty} \frac{1}{k}$$

yields

$$\operatorname*{Lim}_{k \to \infty} \left| \frac{1/(k+1)}{1/k} \right| = \operatorname*{Lim}_{k \to \infty} \frac{k}{k+1} = 1.$$

Therefore the ratio test gives us no information about the convergence or divergence of the harmonic series. Recall that in Example 14 we used the integral test to show that the harmonic series diverges.

Now consider the series

$$\sum_{k=1}^{\infty} \frac{1}{k^2}.$$

Applying the ratio test to this series, we have

$$\operatorname*{Lim}_{k \to \infty} \left| \frac{1/(k+1)^2}{1/k^2} \right| = \operatorname*{Lim}_{k \to \infty} \frac{k^2}{(k+1)^2} = \operatorname*{Lim}_{k \to \infty} \frac{k^2}{k^2 + 2k + 1} = 1.$$

Once again the ratio test gives us no information regarding the convergence or divergence of the series. However, in this case, the series converges. (See Example 13.) ☐

■ *Computer Manual: Program SERIES*

12.2 EXERCISES

In Exercises 1 through 4, write out the first five terms of the sequence of partial sums of the series.

1. $\sum_{k=0}^{\infty} \frac{1}{3^k}$
2. $\sum_{k=1}^{\infty} k^2$
3. $\sum_{k=0}^{\infty} k!$
4. $\sum_{k=1}^{\infty} \frac{(-1)^k}{k}$

In Exercises 5 through 14, determine whether the series converges or diverges. If it converges find its sum.

5. $\sum_{k=0}^{\infty} \frac{1}{3^k}$
6. $\sum_{k=0}^{\infty} (2.5)^k$
7. $\sum_{k=0}^{\infty} (-1)^k$
8. $\sum_{k=2}^{\infty} (-1)^{k+1} 2^{-2k}$
9. $\sum_{k=1}^{\infty} (-1)^k 3 \left(\frac{2}{5}\right)^k$
10. $\sum_{k=0}^{\infty} 2(1.01)^k$
11. $\sum_{k=0}^{\infty} 3(0.99)^k$
12. $-4 + 2 - 1 + \frac{1}{2} - \frac{1}{4} + \frac{1}{8} - \cdots$
13. $5 + \frac{10}{3} + \frac{20}{9} + \frac{40}{27} + \cdots$
14. $\sum_{k=1}^{\infty} \frac{3}{10^k}$

15. The repeating decimal $d = 0.2222 \ldots$ may be written in series form as

$$d = \frac{2}{10} + \frac{2}{10^2} + \frac{2}{10^3} + \cdots.$$

Show that $d = \frac{2}{9}$.

16. Show that
    (a) $0.232323 \ldots = \frac{23}{99}$.
    (b) $\frac{1}{2} = 0.49999 \ldots$.

17. A rubber ball is thrown from a height of 10 feet onto a concrete floor. On its first bounce it reaches a height of 6 feet; thereafter each bounce attains one third the height of the previous bounce. How far does the ball travel?

In Exercises 18 through 26, use the integral test to establish convergence or divergence of the series.

18. $\sum_{k=1}^{\infty} \frac{1}{k^3}$
19. $\sum_{k=1}^{\infty} \frac{1}{\sqrt{k}}$
20. $\sum_{k=1}^{\infty} \frac{1}{k\sqrt{k}}$
21. $\sum_{k=1}^{\infty} \frac{2}{4k + 3}$
22. $\sum_{k=1}^{\infty} ke^{-k}$
23. $\sum_{k=0}^{\infty} ke^{-k^2}$
24. $\sum_{k=1}^{\infty} \frac{\ln(k + 2)}{k + 2}$
25. $\sum_{k=1}^{\infty} \frac{1}{(k + 1)\ln(k + 1)}$
26. $\sum_{k=1}^{\infty} \frac{1}{(k + 1)[\ln(k + 1)]^2}$

*27. A series of the form $\sum_{k=1}^{\infty} 1/k^p$, where p is a positive number, is called a ***p*-series**.
Show that a *p*-series
(a) converges if $p > 1$
(b) diverges if $0 < p \leq 1$

In Exercises 28 through 41, apply the ratio test. If the ratio test is inconclusive, establish convergence or divergence by other means.

28. $\sum_{k=0}^{\infty} \frac{2^k}{k!}$

29. $\sum_{k=1}^{\infty} (-1)^k \frac{2^k}{3^{k+1}}$

30. $\sum_{k=0}^{\infty} \frac{3^{k+4}}{(k+1)!}$

31. $\sum_{k=1}^{\infty} (-1)^{k+1} \frac{5^{k+1}}{3^{k-1}}$

32. $\sum_{k=1}^{\infty} \frac{1}{2k+1}$

33. $\sum_{k=1}^{\infty} \frac{(-1)^k 10^k}{k^2}$

34. $\sum_{k=2}^{\infty} \frac{k(k+1)}{k!}$

35. $\sum_{k=1}^{\infty} \frac{1}{(3k-1)^2}$

36. $\sum_{k=0}^{\infty} \frac{k!}{(2k+1)!}$

37. $\sum_{k=0}^{\infty} k! e^{-k}$

38. $\sum_{k=1}^{\infty} \frac{(k+1)(k+2)}{10^{k-1}}$

39. $\sum_{k=1}^{\infty} (-1)^k \frac{2^k}{k \cdot 3^k}$

40. $\sum_{k=1}^{\infty} \frac{(k!)^2}{(2k)!}$

41. $\sum_{k=0}^{\infty} (-1)^k \frac{1 + 3^{-k}}{(4/3)^k}$

42. Show that the series $\sum_{k=1}^{\infty} kx^{k-1}$ converges if $-1 < x < 1$ and diverges if $x \geq 1$ or $x \leq -1$.

Management

43. Let X denote the number of consecutive good parts produced by a stamping machine.
    (a) Show that $P(X = k) = \frac{1}{20}(0.95)^k$ is a legitimate assignment of probabilities to X.
    (b) Find the probability that the machine will produce 20 or more consecutive good parts.
    (c) What is the probability that the machine will produce 50 or fewer consecutive good parts?

44. Repeat Exercise 43 if $P(X = k) = (0.1)(0.9)^k$

*45. Let X denote the number of consecutive successes produced by some process which always produces either a success or a failure. Let $0 < p < 1$.
    (a) Show that $P(X = k) = (1 - p)p^k$ is a legitimate assignment of probabilities to X.
    (b) If $m \geq 0$, find the probability that the process will produce m or more consecutive successes.

46. Suppose that the people in a country spend 95% of their disposable income. Thus, of each $10,000 in newly acquired disposable income, 95% of $10,000, or $9500, is spent; then 95% of the $9500, or $9025, is spent; then 95% of the $9025 is spent, and so on. (This is called the **multiplier effect**.) If a tax cut increases disposable income in the country by $20 billion, how much stimulation is applied to the economy?

47. Repeat Exercise 46 if the people spend 80% of their disposable income.

Life Science

48. A patient undergoing prolonged treatment takes a daily dose of a certain drug. As the concentration of the drug in the patient's blood increases, more of each dose is

eliminated by the body. Suppose that the concentration of the drug as of day n of the treatment is

$$\sum_{k=1}^{n} \frac{100}{2^k}$$

micrograms per cubic centimeter (μg/cc). What concentration will the drug approach as the treatment is continued?

12.3 POWER SERIES

Now we are ready to begin our study of power series. A **power series in x** is an infinite series of the form

$$\sum_{k=0}^{\infty} a_k x^k = a_0 + a_1 x + a_2 x^2 + a_3 x^3 + \cdots,$$

where x is a variable and $a_0, a_1, a_2, a_3, \ldots$ are numbers. For instance,

$$\sum_{k=0}^{\infty} x^k = x^0 + x^1 + x^2 + x^3 + \cdots = 1 + x + x^2 + x^3 + \cdots$$

and

$$\sum_{k=0}^{\infty} \frac{(-1)^k x^k}{2k+1} = \frac{(-1)^0 x^0}{2 \cdot 0 + 1} + \frac{(-1)^1 x^1}{2 \cdot 1 + 1} + \frac{(-1)^2 x^2}{2 \cdot 2 + 1} + \cdots$$

$$= 1 - \frac{x}{3} + \frac{x^2}{5} - \frac{x^3}{7} + \cdots$$

are power series in x. In this section we show how power series can be used to define functions, and we examine the differentiation and integration of such functions.

Consider the power series

$$\sum_{k=0}^{\infty} x^k = 1 + x + x^2 + x^3 + \cdots.$$

If we substitute a number for x in this power series, we obtain an infinite series of constants, which may or may not converge. Thus, if we substitute $-\frac{1}{2}$ for x, we obtain the infinite series

$$\sum_{k=0}^{\infty} \left(-\frac{1}{2}\right)^k = 1 - \frac{1}{2} + \frac{1}{4} - \frac{1}{8} + \cdots,$$

which is geometric with first term $a = 1$ and ratio $r = -\frac{1}{2}$, and hence converges to

$$\frac{1}{1-(-\frac{1}{2})} = \frac{2}{3}.$$

On the other hand, if we substitute 3 for x, we obtain the infinite series

$$\sum_{k=0}^{\infty} 3^k = 1 + 3 + 9 + 27 + \cdots,$$

which is geometric with $a = 1$ and $r = 3$ and therefore diverges.

Note that, no matter what number is substituted for x, the series

$$\sum_{k=0}^{\infty} x^k = 1 + x + x^2 + x^3 + \cdots$$

is geometric with $a = 1$ and $r = x$. Therefore the power series converges if $-1 < x < 1$ and diverges if $x \geq 1$ or $x \leq -1$. Since the power series converges for every number x in the interval $(-1, 1)$, we may define a function g on this interval by setting

$$g(x) = \sum_{k=0}^{\infty} x^k, \qquad -1 < x < 1.$$

Of course, when $-1 < x < 1$, the geometric series

$$\sum_{k=0}^{\infty} x^k = \frac{1}{1-x}.$$

Thus we have two ways to write the function g: in **closed form** as

$$g(x) = \frac{1}{1-x}, \qquad -1 < x < 1,$$

and in **power series form** as

$$g(x) = \sum_{k=0}^{\infty} x^k, \qquad -1 < x < 1.$$

The power series

$$\sum_{k=0}^{\infty} x^k$$

is called the **power series expansion** of the function g.

Any power series can be used to define a function in the way we have just done: Given a series

$$\sum_{k=0}^{\infty} a_k x^k,$$

we define a function f by setting

$$f(x) = \sum_{k=0}^{\infty} a_k x^k$$

for all x for which the series converges. The function f may turn out to be an elementary function that we can also write in closed form (as was the case for the function g above), or it may be a nonelementary function that cannot be written in closed form, but only in power series form.

In order to use a power series to define a function, we obviously must know the values of x for which the series converges. The following result, which we shall not prove, shows that there are only three possibilities for the convergence of any power series.

Convergence of Power Series

A power series

$$\sum_{k=0}^{\infty} a_k x^k$$

must satisfy exactly one of the following:

1. It converges for all numbers x.
2. There is some positive number R such that it converges for $-R < x < R$ and diverges for $x > R$ and $x < -R$.
3. It converges only for $x = 0$.

The positive number R of alternative 2 is called the **radius of convergence** of the power series. By analogy, we say that the radius of convergence of a series is $+\infty$ if it converges for all numbers x, and that its radius of convergence is 0 if it converges only for $x = 0$. Thus every power series has a radius of convergence R, $0 \leq R \leq +\infty$. (Note that alternative 2 does not say whether the power series converges at $x = R$ and $x = -R$. In fact, it is possible for a power series to converge at both of these values, at neither of them, or at one and not the other. We will not need to determine which of these cases occurs for any of our series.)

EXAMPLE 1 We have already seen that the power series

$$\sum_{k=0}^{\infty} x^k$$

converges for $-1 < x < 1$ and diverges for $x > 1$ and for $x < -1$. Therefore its radius of convergence is $R = 1$. ☐

EXAMPLE 2 Let us find the radius of convergence for the power series

$$\sum_{k=0}^{\infty} \frac{x^k}{k!}.$$

Applying the ratio test to this series, we have

$$\text{Lim}_{k \to \infty} \left| \frac{x^{k+1}/(k+1)!}{x^k/k!} \right| = \text{Lim}_{k \to \infty} \left| x \cdot \frac{1}{k+1} \right|$$

$$= |x| \cdot \text{Lim}_{k \to \infty} \frac{1}{k+1}$$

$$= |x| \cdot 0$$

$$= 0.$$

Since the limit is always less than 1, the ratio test tells us that the power series converges no matter what x is, and thus it converges for all numbers x. Hence its radius of convergence is $R = +\infty$. □

EXAMPLE 3 Consider the power series

$$\sum_{k=1}^{\infty} \frac{x^k}{k 2^k}.$$

If we apply the ratio test to this series, we have

$$\text{Lim}_{k \to \infty} \left| \frac{x^{k+1}/(k+1)2^{k+1}}{x^k/k 2^k} \right| = \text{Lim}_{k \to \infty} \left| \frac{x}{2} \cdot \frac{k}{k+1} \right|$$

$$= \frac{|x|}{2} \text{Lim}_{k \to \infty} \frac{k}{k+1}$$

$$= \frac{|x|}{2} \cdot 1$$

$$= \frac{|x|}{2}.$$

But $|x|/2 < 1$ if $-2 < x < 2$ and $|x|/2 > 1$ if $x > 2$ or $x < -2$. Hence the series converges if $-2 < x < 2$ and diverges if $x > 2$ or $x < -2$, so its radius of convergence is $R = 2$. □

EXAMPLE 4 The power series

$$\sum_{k=0}^{\infty} k! x^k$$

converges only for $x = 0$ and thus has radius of convergence $R = 0$. To see this, apply the ratio test and note that the terms of the sequence

$$\left\{ \left| \frac{a_{k+1}}{a_k} \right| \right\} = \left\{ \left| \frac{(k+1)! x^{k+1}}{k! x^k} \right| \right\} = \{|(k+1)x|\} = \{|x|, 2|x|, 3|x|, \ldots\}$$

grow larger without bound unless $x = 0$. □

Now suppose that a power series

$$\sum_{k=0}^{\infty} a_k x^k$$

has radius of convergence $R \neq 0$, so that it converges for $-R < x < R$. (If $R = +\infty$, it converges for $-\infty < x < +\infty$, that is, for all x.) As we explained previously, we may use the power series to define a function f by setting

$$f(x) = \sum_{k=0}^{\infty} a_k x^k, \qquad \text{for } -R < x < R.$$

The function f can then be differentiated by differentiating the power series term by term.

Differentiation of Power Series

If f is defined by

$$f(x) = \sum_{k=0}^{\infty} a_k x^k$$

for $-R < x < R$, then

$$\frac{df}{dx} = \sum_{k=0}^{\infty} \frac{d}{dx}(a_k x^k) = \sum_{k=1}^{\infty} k a_k x^{k-1}$$

for $-R < x < R$.

EXAMPLE 5 Suppose g is the function defined by

$$g(x) = \sum_{k=0}^{\infty} x^k = 1 + x + x^2 + x^3 + \cdots, \qquad -1 < x < 1.$$

Then

$$\frac{dg}{dx} = \frac{d}{dx}(1) + \frac{d}{dx}(x) + \frac{d}{dx}(x^2) + \frac{d}{dx}(x^3) + \cdots, \qquad -1 < x < 1,$$

so

$$\frac{dg}{dx} = 0 + 1 + 2x + 3x^2 + \cdots, \qquad -1 < x < 1,$$

or

$$\frac{dg}{dx} = \sum_{k=1}^{\infty} k x^{k-1}, \qquad -1 < x < 1. \quad \square$$

Differentiation can often be used to find power series expansions for functions.

EXAMPLE 6 Suppose we wish to find a power series expansion for the function h defined by

$$h(x) = \frac{1}{(1-x)^2}, \qquad -1 < x < 1.$$

We have seen previously that

$$\frac{1}{1-x} = \sum_{k=0}^{\infty} x^k, \qquad -1 < x < 1.$$

Differentiating both sides of this last equation yields

$$\frac{1}{(1-x)^2} = \sum_{k=1}^{\infty} kx^{k-1}, \qquad -1 < x < 1,$$

and we have found a power series expansion for h. ☐

EXAMPLE 7 In Example 12 of Section 12.2 we defined a random variable X to be the number of consecutive good microchips produced by an assembly line. If

$$P(X = k) = (0.01)(0.99)^k,$$

what is the expected value of X?

We have

$$E(X) = \sum_{k=0}^{\infty} kP(X = k)$$

$$= \sum_{k=0}^{\infty} k(0.01)(0.99)^k$$

$$= 0 + (0.01)(0.99) + 2(0.01)(0.99)^2 + 3(0.01)(0.99)^3 + \cdots$$

$$= (0.01)(0.99)[1 + 2(0.99) + 3(0.99)^2 + \cdots].$$

But from Example 6,

$$\frac{1}{(1-x)^2} = \sum_{k=1}^{\infty} kx^{k-1} = 1 + 2x + 3x^2 + \cdots, \qquad -1 < x < 1.$$

Since $-1 < 0.99 < 1$, it follows that

$$\frac{1}{(1-0.99)^2} = 1 + 2(0.99) + 3(0.99)^2 + \cdots,$$

or

$$10{,}000 = 1 + 2(0.99) + 3(0.99)^2 + \cdots.$$

Therefore

$$E(X) = (0.01)(0.99)(10{,}000) = 99.$$

Hence on average the assembly line will produce 99 good microchips between any two defective ones. ☐

In Example 2 we showed that the power series

$$\sum_{k=0}^{\infty} \frac{x^k}{k!}$$

converges for all x. Let us define a function f by setting

$$f(x) = \sum_{k=0}^{\infty} \frac{x^k}{k!} = 1 + x + \frac{x^2}{2!} + \frac{x^3}{3!} + \frac{x^4}{4!} + \cdots$$

for all x. Then

$$f'(x) = 0 + 1 + \frac{2x}{2!} + \frac{3x^2}{3!} + \frac{4x^3}{4!} + \cdots$$

$$= 1 + x + \frac{x^2}{2!} + \frac{x^3}{3!} + \cdots$$

for all x. Therefore $f'(x) = f(x)$ for all x. Since also

$$f(0) = 1 + 0 + \frac{0^2}{2!} + \frac{0^3}{3!} + \cdots = 1,$$

this shows that the function f is a solution of the differential equation with initial condition

$$\frac{dy}{dx} = y, \quad y(0) = 1.$$

But this is a separable differential equation whose unique solution is

$$y = e^x.$$

(See Section 9.2.) Thus $f(x) = e^x$, and we have the following result:

Power Series Expansion for e^x

$$e^x = \sum_{k=0}^{\infty} \frac{x^k}{k!} \quad \text{for all } x.$$

The power series expansion for e^x is quite important. For one thing, it allows us to find the value of any power of e to any desired degree of accuracy. For instance, if $x = 1$, then

$$e = \sum_{k=0}^{\infty} \frac{1^k}{k!} = 1 + 1 + \frac{1}{2!} + \frac{1}{3!} + \frac{1}{4!} + \cdots.$$

Summing the terms of this series will provide us with an approximate value for e; the more terms we sum, the better the approximation will be. For instance, if we sum the first four terms of the series, we obtain

$$e \approx 1 + 1 + \frac{1}{2!} + \frac{1}{3!} = 2.6667,$$

rounded to four decimal places, whereas if we sum the first seven terms of the series we obtain

$$e \approx 1 + 1 + \frac{1}{2!} + \frac{1}{3!} + \frac{1}{4!} + \frac{1}{5!} + \frac{1}{6!} = 2.7181.$$

Similar calculations can be carried out for any power of e.

Another reason the power series expansion for e^x is important is that it often turns up in mathematical models. We illustrate with an example.

EXAMPLE 8   In Section 10.3 we discussed the Poisson distribution. Recall that the Poisson distribution is used to model situations in which an event occurs an average of μ times in a given time interval. If X is a Poisson random variable, then

$$P(X = k) = P(\text{the event occurs } k \text{ times in the given interval})$$
$$= \frac{e^{-\mu} \mu^k}{k!}.$$

In Section 10.3 we stated without proof that $E(X) = \mu$. Now we can prove this:

$$E(X) = \sum_{k=0}^{\infty} k P(X = k) = \sum_{k=0}^{\infty} \frac{k e^{-\mu} \mu^k}{k!}$$
$$= \frac{e^{-\mu} \mu^1}{1!} + \frac{2 e^{-\mu} \mu^2}{2!} + \frac{3 e^{-\mu} \mu^3}{3!} + \cdots$$
$$= \mu e^{-\mu} \left(1 + \mu + \frac{\mu^2}{2!} + \frac{\mu^3}{3!} + \cdots \right).$$

But the power series expansion for e^x with $x = \mu$ shows that

$$e^{\mu} = 1 + \mu + \frac{\mu^2}{2!} + \frac{\mu^3}{3!} + \cdots.$$

Therefore

$$E(X) = \mu e^{-\mu} e^{\mu} = \mu. \quad \square$$

Just as a power series can be differentiated term by term inside its interval of convergence, so also can it be integrated term by term there.

Integration of Power Series

If f is defined by

$$f(x) = \sum_{k=0}^{\infty} a_k x^k$$

for $-R < x < R$, then

$$\int f(x)\, dx = \sum_{k=0}^{\infty} \int a_k x^k\, dx + C$$
$$= \sum_{k=0}^{\infty} a_k \frac{x^{k+1}}{k+1} + C$$

for $-R < x < R$.

SEQUENCES AND SERIES

EXAMPLE 9 If
$$g(x) = \sum_{k=0}^{\infty} x^k = 1 + x + x^2 + \cdots, \qquad -1 < x < 1,$$
then
$$\int g(x)\, dx = \int 1\, dx + \int x\, dx + \int x^2\, dx + \cdots + C$$
$$= x + \frac{x^2}{2} + \frac{x^3}{3} + \cdots + C$$
$$= \sum_{k=1}^{\infty} \frac{x^k}{k} + C, \qquad -1 < x < 1. \quad \square$$

EXAMPLE 10 Let g be as in Example 9. Since the power series
$$\sum_{k=1}^{\infty} \frac{x^k}{k}$$
is an antiderivative for g, we can use it to evaluate the definite integral
$$\int_0^{1/2} g(x)\, dx.$$
Thus
$$\int_0^{1/2} g(x)\, dx = \sum_{k=1}^{\infty} \frac{x^k}{k}\bigg|_0^{1/2}$$
$$= \left(\sum_{k=1}^{\infty} \frac{(\tfrac{1}{2})^k}{k}\right) - \left(\sum_{k=1}^{\infty} \frac{0^k}{k}\right)$$
$$= \sum_{k=1}^{\infty} \frac{1}{k 2^k}.$$
Again, in closed form
$$g(x) = \frac{1}{1-x},$$
so
$$\int_0^{1/2} g(x)\, dx = \int_0^{1/2} \frac{1}{1-x}\, dx = -\ln(1-x)\bigg|_0^{1/2}$$
$$= -\ln\frac{1}{2}$$
$$= \ln 2.$$
Therefore
$$\ln 2 = \sum_{k=1}^{\infty} \frac{1}{k 2^k}. \quad \square$$

Term-by-term integration can also be used to find power series expansions for functions.

EXAMPLE 11 If we integrate both sides of the expansion

$$\frac{1}{1-x} = \sum_{k=0}^{\infty} x^k, \quad -1 < x < 1,$$

we find that

$$-\ln(1-x) = \sum_{k=1}^{\infty} \frac{x^k}{k}, \quad -1 < x < 1,$$

or

$$\ln(1-x) = -\sum_{k=1}^{\infty} \frac{x^k}{k} = \sum_{k=1}^{\infty} -\frac{x^k}{k}, \quad -1 < x < 1.$$

Therefore we have obtained the power series expansion

$$\ln(1-x) = \sum_{k=1}^{\infty} -\frac{x^k}{k}, \quad -1 < x < 1$$

for $\ln(1-x)$. ◻

Once a power series expansion for a function has been found, expansions for related functions can often be created by substitution. To illustrate, suppose we want to find an expansion for e^{2x}. Since

$$e^x = \sum_{k=0}^{\infty} \frac{x^k}{k!}$$

for all numbers x, and since $2x$ is a number, we may substitute $2x$ for x to obtain

$$e^{2x} = \sum_{k=0}^{\infty} \frac{(2x)^k}{k!} = \sum_{k=0}^{\infty} \frac{2^k x^k}{k!}$$

for all x.

EXAMPLE 12 Let us find a power series expansion for $\ln(1+x)$. We begin with the expansion

$$\ln(1-x) = \sum_{k=1}^{\infty} -\frac{x^k}{k}, \quad -1 < x < 1$$

of Example 11. Now, if $-1 < x < 1$, then also $-1 < -x < 1$, so we may substitute $-x$ for x in this expansion to obtain

$$\ln(1-(-x)) = \sum_{k=1}^{\infty} -\frac{(-x)^k}{k}, \quad -1 < x < 1,$$

or

$$\ln(1+x) = \sum_{k=1}^{\infty} -\frac{(-1)^k x^k}{k}, \quad -1 < x < 1,$$

or
$$\ln(1+x) = \sum_{k=1}^{\infty} \frac{(-1)^{k+1}x^k}{k}, \qquad -1 < x < 1. \quad \square$$

As an application of these ideas, let us use power series to evaluate the definite integral
$$\int_0^1 e^{-x^2}\, dx.$$

We begin by finding a power series expansion for e^{-x^2}.

Since
$$e^x = \sum_{k=0}^{\infty} \frac{x^k}{k!}$$
for all numbers x, and since $-x^2$ is a number, we may substitute $-x^2$ for x in the power series to obtain
$$e^{-x^2} = \sum_{k=0}^{\infty} \frac{(-x^2)^k}{k!} = \sum_{k=0}^{\infty} \frac{(-1)^k x^{2k}}{k!}$$
for all x. Then
$$\int e^{-x^2}\, dx = \sum_{k=0}^{\infty} \frac{(-1)^k}{k!} \int x^{2k}\, dx + C$$
$$= \sum_{k=0}^{\infty} \frac{(-1)^k x^{2k+1}}{k!(2k+1)} + C,$$
for all x. In other words,
$$\int e^{-x^2}\, dx = f(x) + C,$$
where f is the function defined by
$$f(x) = \sum_{k=0}^{\infty} \frac{(-1)^k x^{2k+1}}{k!(2k+1)}$$
for all x. As stated at the beginning of this chapter, f is not an elementary function, so we cannot write it in closed form. Nevertheless, it is an antiderivative of e^{-x^2}, and thus
$$\int_0^1 e^{-x^2}\, dx = \sum_{k=0}^{\infty} \frac{(-1)^k x^{2k+1}}{k!(2k+1)} \Big|_0^1$$
$$= \sum_{k=0}^{\infty} \frac{(-1)^k}{k!(2k+1)}$$
$$= 1 - \frac{1}{3} + \frac{1}{5 \cdot 2!} - \frac{1}{7 \cdot 3!} + \frac{1}{9 \cdot 4!} - \cdots.$$

The sum of the first five terms of this series is

$$1 - \frac{1}{3} + \frac{1}{5 \cdot 2!} - \frac{1}{7 \cdot 3!} + \frac{1}{9 \cdot 4!} = 0.7475$$

to four decimal places, so

$$\int_0^1 e^{-x^2} \, dx \approx 0.7475.$$

We could obtain a better approximation for the value of the integral by summing more terms of the series

$$\sum_{k=0}^{\infty} \frac{(-1)^k}{k!(2k+1)}.$$

We conclude this section by remarking that power series can be added to yield new power series:

$$\sum_{k=0}^{\infty} a_k x^k + \sum_{k=0}^{\infty} b_k x^k = \sum_{k=0}^{\infty} (a_k + b_k) x^k.$$

When this is done, the radius of convergence of the sum is the smaller of the radii of convergence of the summands. Also, adding a power series and a polynomial, or multiplying a power series by a polynomial, will give a power series that has the same radius of convergence as the original series. These operations give us still another method for creating new power series expansions from old ones. We illustrate with an example.

EXAMPLE 13 Since

$$\frac{1}{1-x} = 1 + x + x^2 + x^3 + \cdots, \qquad -1 < x < 1,$$

and

$$\frac{1}{1+x} = 1 - x + x^2 - x^3 + \cdots, \qquad -1 < x < 1,$$

we have

$$\begin{aligned}\frac{2}{1-x^2} &= \frac{1}{1-x} + \frac{1}{1+x} \\ &= (1 + x + x^2 + x^3 + \cdots) \\ &\quad + (1 - x + x^2 - x^3 + \cdots) \\ &= 2 + 2x^2 + 2x^4 + \cdots \\ &= 2(1 + x^2 + x^4 + \cdots), \qquad -1 < x < 1.\end{aligned}$$

Note that the radius of convergence of the expansion for $2/(1-x^2)$ is the smaller of the radii of convergence of its components, and hence is also equal to 1.

Now we can find power series expansions for

$$\frac{1}{1-x^2}, \quad \frac{x}{1-x^2} \quad \text{and} \quad \frac{1}{1-x^2} - (x+1):$$

$$\frac{1}{1-x^2} = \frac{1}{2}\frac{2}{1-x^2}$$

$$= \frac{1}{2}\cdot 2(1 + x^2 + x^4 + \cdots)$$

$$= 1 + x^2 + x^4 + \cdots, \quad -1 < x < 1;$$

$$\frac{x}{1-x^2} = x(1 + x^2 + x^4 + \cdots)$$

$$= x + x^3 + x^5 + \cdots, \quad -1 < x < 1;$$

and

$$\frac{1}{1-x^2} - (x+1) = (1 + x^2 + x^4 + \cdots) - (x+1)$$

$$= -x + x^2 + x^4 + \cdots, \quad -1 < x < 1. \quad \square$$

■ *Computer Manual: Program SERIES.*

12.3 EXERCISES

In Exercises 1 through 12, find the radius of convergence of the power series.

1. $\sum_{k=0}^{\infty} 2^k x^k$
2. $\sum_{k=0}^{\infty} \frac{x^k}{2^k}$
3. $\sum_{k=0}^{\infty} \frac{(-1)^{k+1} x^{k+2}}{2k+1}$
4. $\sum_{k=1}^{\infty} \frac{kx^k}{100^k}$
5. $\sum_{k=0}^{\infty} \frac{x^{2k}}{3^k}$
6. $\sum_{k=0}^{\infty} \frac{2^k x^k}{k!}$
7. $\sum_{k=0}^{\infty} \frac{k! x^k}{(2k)!}$
8. $\sum_{k=0}^{\infty} \frac{(2k)! x^k}{k!}$
9. $\sum_{k=1}^{\infty} \frac{x^{2k+1}}{r^k(2k+1)}, \quad r \neq 0$
10. $\sum_{k=1}^{\infty} \frac{x^{mk}}{k!}, \quad m > 0$
11. $\sum_{k=1}^{\infty} \frac{kx^{10k+40}}{(k-1)!}$
12. $\sum_{k=1}^{\infty} \frac{k^k x^k}{k!}$

13. Show that

$$e^{-0.2} = \sum_{k=0}^{\infty} \frac{(-1)^k (0.2)^k}{k!}.$$

Estimate $e^{-0.2}$ by summing the first five terms of this series. How does your result compare with the value of $e^{-0.2}$ obtained from a calculator or from Table 1 at the back of the book?

14. Show that

$$\ln 1.5 = \sum_{k=1}^{\infty} \frac{(-1)^{k+1}}{k 2^k}.$$

Estimate $\ln 1.5$ by summing the first four terms of this series. Compare your result with that obtained from a calculator or from Table 2 at the back of the book.

15. Use power series to find an approximation of ln 0.95 that is accurate to at least four decimal places.

16. Use power series to find an approximation of $e^{0.1}$ that is accurate to at least five decimal places.

17. Define a function f by setting $f(x) = \sum_{k=1}^{\infty} \dfrac{kx^k}{2^k}$ for x in the interval of convergence of the series. Find

    (a) $f'(x)$ (b) $\int f(x)\, dx$ (c) $\int_0^1 f(x)\, dx$

18. Define a function f by setting $f(x) = \sum_{k=0}^{\infty} \dfrac{x^{2k+1}}{2k+1}$ for x in the interval of convergence of the series. Find

    (a) $f'(x)$ (b) $\int f(x)\, dx$ (c) $\int_{-1/2}^{1/2} f(x)\, dx$

19. Show that the power series
$$\sum_{k=0}^{\infty} \dfrac{x^k}{3^k}$$
converges for $-3 < x < 3$. If f is defined by
$$f(x) = \sum_{k=0}^{\infty} \dfrac{x^k}{3^k}, \qquad -3 < x < 3,$$
find

    (a) $f'(x)$ (b) $\int f(x)\, dx$ (c) $\int_1^2 f(x)\, dx$

20. Show that the power series
$$\sum_{k=0}^{\infty} \dfrac{x^{2k}}{(2k)!}$$
converges for all x. If f is defined by
$$f(x) = \sum_{k=0}^{\infty} \dfrac{x^{2k}}{(2k)!},$$
for all x, find

    (a) $f'(x)$ (b) $\int f(x)\, dx$ (c) $\int_0^1 f(x)\, dx$

21. Repeat Exercise 20 for the series
$$\sum_{k=0}^{\infty} \dfrac{x^{2k+1}}{(2k+1)!}$$

22. Show that the radius of convergence of the power series
$$\sum_{k=0}^{\infty} \dfrac{(-1)^k x^{2k+1}}{(2k+1)}$$
is $R = 1$. If f is defined by
$$f(x) = \sum_{k=0}^{\infty} \dfrac{(-1)^k x^{2k+1}}{(2k+1)}, \qquad -1 < x < 1,$$

find

(a) $f'(x)$ (b) $\int f(x)\,dx$ (c) $\int_0^{1/2} f(x)\,dx$

23. The series

$$\sum_{k=0}^{\infty} \frac{(-1)^k x^{2k+1}}{(2k+1)!} \quad \text{and} \quad \sum_{k=0}^{\infty} \frac{(-1)^k x^{2k}}{(2k)!},$$

both have radius of convergence $R = +\infty$. (See Exercises 20 and 21.) If

$$s(x) = \sum_{k=0}^{\infty} \frac{(-1)^k x^{2k+1}}{(2k+1)!} \quad \text{and} \quad c(x) = \sum_{k=0}^{\infty} \frac{(-1)^k x^{2k}}{(2k)!}$$

show that $s'(x) = c(x)$ and $c'(x) = -s(x)$. (We shall see in the next section that $s(x) = \sin x$ and $c(x) = \cos x$.)

In Exercises 24 through 37, find a power series expansion for the given expression.

24. e^{-x}
25. $e^x - 1$
26. xe^x
27. e^{4x^2}
28. $\dfrac{e^{2x} - 1}{2}$
29. $\ln(1 + 2x)$
30. $\ln(1 + x^2)$
31. $\dfrac{\ln(1 + x)}{5}$
32. $\ln\left(\dfrac{1 + x}{1 - x}\right)$
33. $\dfrac{x + 1}{1 + x^2}$
34. $\dfrac{1}{(1 - x)^3}$
35. $\dfrac{1}{(1 + x)^2}$
36. $\dfrac{x}{(1 - x)^3}$
37. $\dfrac{x^2}{(1 + x)^2} + 1 - x^2$

*38. If X is a random variable and $P(X = k) = p^k(1 - p)$, $k = 0, 1, 2, \ldots$, find a closed-form formula for $E(X)$.

*39. Let X be a Poisson random variable with expected value μ. Show that the variance of X is also μ.

Management

40. If X denotes the number of consecutive good parts produced by a stamping machine and $P(X = k) = \dfrac{1}{20}(0.95)^k$, find and interpret $E(X)$. (See Exercise 38.)

12.4 TAYLOR SERIES

In the previous section we found power series expansions of various functions by using several techniques, including differentiation, integration, and substitution. It would be convenient to have a general method for producing power series expansions of functions. Such a method does indeed exist: It produces what is known as a **Taylor series.** In this section we introduce Taylor series and show how they can be used to find functional values.

To start with, suppose a function f has a power series expansion on an interval, say

$$f(x) = \sum_{k=0}^{\infty} a_k x^k = a_0 + a_1 x + a_2 x^2 + a_3 x^3 + a_4 x^4 + \cdots$$

for $-R < x < R$. Then f possesses derivatives of all orders on the interval $(-R, R)$:

$$f'(x) = \sum_{k=1}^{\infty} k a_k x^{k-1} = a_1 + 2a_2 x + 3a_3 x^2 + 4a_4 x^3 + \cdots,$$

$$f''(x) = \sum_{k=2}^{\infty} k(k-1) a_k x^{k-2} = 2a_2 + 2 \cdot 3a_3 x + 3 \cdot 4a_4 x^2 + \cdots,$$

$$f'''(x) = \sum_{k=3}^{\infty} k(k-1)(k-2) a_k x^{k-3} = 2 \cdot 3a_3 + 2 \cdot 3 \cdot 4a_4 x + \cdots,$$

and so on. Notice that

$$f(0) = a_0, \quad \text{so} \quad a_0 = \frac{f(0)}{0!},$$

$$f'(0) = a_1, \quad \text{so} \quad a_1 = \frac{f'(0)}{1!},$$

$$f''(0) = 2a_2, \quad \text{so} \quad a_2 = \frac{f''(0)}{2!},$$

$$f'''(0) = 2 \cdot 3 a_3, \quad \text{so} \quad a_3 = \frac{f'''(0)}{3!},$$

and so on. For convenience, let $f^{(0)}$ denote f and let $f^{(k)}$ denote the kth derivative of f, for $k = 1, 2, \ldots$. Then we have

$$a_0 = \frac{f^{(0)}(0)}{0!}, \quad a_1 = \frac{f^{(1)}(0)}{1!}, \quad a_2 = \frac{f^{(2)}(0)}{2!}, \quad a_3 = \frac{f^{(3)}(0)}{3!},$$

and so on. Hence if

$$f(x) = \sum_{k=0}^{\infty} a_k x^k, \quad -R < x < R,$$

then

$$a_k = \frac{f^{(k)}(0)}{k!}$$

for $k = 0, 1, 2, \ldots$.

EXAMPLE 1   Define f by

$$f(x) = \sum_{k=0}^{\infty} (-1)^k x^k = 1 - x + x^2 - x^3 + x^4 - \cdots, \quad -1 < x < 1.$$

Then

SEQUENCES AND SERIES

$$f^{(0)}(x) = 1 - x + x^2 - x^3 + x^4 - \cdots, \quad \text{so} \quad f^{(0)}(0) = 1,$$
$$f^{(1)}(x) = -1 + 2x - 3x^2 + 4x^3 - \cdots, \quad \text{so} \quad f^{(1)}(0) = -1,$$
$$f^{(2)}(x) = 2 - 6x + 12x^2 - \cdots, \quad \text{so} \quad f^{(2)}(0) = 2,$$
$$f^{(3)}(x) = -6 + 24x - \cdots, \quad \text{so} \quad f^{(3)}(0) = -6,$$
$$f^{(4)}(x) = 24 - \cdots, \quad \text{so} \quad f^{(4)}(0) = 24.$$

Therefore

$$\frac{f^{(0)}(0)}{0!} = \frac{1}{1} = 1 = a_0,$$

$$\frac{f^{(1)}(0)}{1!} = \frac{-1}{1} = -1 = a_1,$$

$$\frac{f^{(2)}(0)}{2!} = \frac{2}{2} = 1 = a_2,$$

$$\frac{f^{(3)}(0)}{3!} = \frac{-6}{6} = -1 = a_3,$$

$$\frac{f^{(4)}(0)}{4!} = \frac{24}{24} = 1 = a_4. \quad \square$$

We have shown that if

$$f(x) = \sum_{k=0}^{\infty} a_k x^k, \quad -R < x < R,$$

then f has derivatives of all orders on $(-R, R)$ and

$$f(x) = \sum_{k=0}^{\infty} \frac{f^{(k)}(0)}{k!} x^k, \quad -R < x < R.$$

Now suppose we turn this around: We start with a function f that has derivatives of all orders at $x = 0$ and construct the power series

$$\sum_{k=0}^{\infty} \frac{f^{(k)}(0)}{k!} x^k \quad \text{for} \quad k = 0, 1, 2, \ldots.$$

Taylor Series

If f is a function that has derivatives of all orders at $x = 0$, then the power series

$$\sum_{k=0}^{\infty} \frac{f^{(k)}(0)}{k!} x^k$$

is called the **Taylor series for f about $x = 0$**, or the **Maclaurin series for f**.

EXAMPLE 2 Let $f(x) = e^x$. Then $f^{(k)}(x) = e^x$ for $k = 0, 1, 2, \ldots$, so $f^{(k)}(0) = e^0 = 1$ for $k = 0, 1, 2, \ldots$. Therefore the Taylor series for $f(x) = e^x$ about $x = 0$ is

$$\sum_{k=0}^{\infty} \frac{f^{(k)}(0)}{k!} x^k = \sum_{k=0}^{\infty} \frac{1}{k!} x^k.$$

But this is the power series expansion for e^x;

$$e^x = \sum_{k=0}^{\infty} \frac{1}{k!} x^k.$$

Thus the Taylor series about $x = 0$ for f converges to f for all x. ☐

In the interest of brevity, let us agree from now on that "Taylor series" will mean "Taylor series about $x = 0$."

EXAMPLE 3 If $f(x) = 1/(1 - x)$, we have

$$f^{(1)}(x) = \frac{1}{(1-x)^2},$$

$$f^{(2)}(x) = \frac{2}{(1-x)^3},$$

$$f^{(3)}(x) = \frac{2 \cdot 3}{(1-x)^4},$$

and, in general,

$$f^{(k)}(x) = \frac{k!}{(1-x)^k}.$$

Thus

$$f^{(k)}(0) = k!$$

for $k = 0, 1, 2, \ldots$. Hence the Taylor series for f is the series

$$\sum_{k=0}^{\infty} \frac{f^{(k)}(0)}{k!} x^k = \sum_{k=0}^{\infty} \frac{k!}{k!} x^k = \sum_{k=0}^{\infty} x^k.$$

In Section 12.2 we showed that the series $\sum_{k=0}^{\infty} x^k$ converges to $1/(1 - x)$ for $-1 < x < 1$. Therefore, once again the Taylor series for the function converges to the function within its interval of convergence. ☐

EXAMPLE 4 Let $g(x) = \sin x$. Then

$$g^{(0)}(x) = \sin x, \quad g^{(1)}(x) = \cos x, \quad g^{(2)}(x) = -\sin x, \quad g^{(3)}(x) = -\cos x,$$

$$g^{(4)}(x) = \sin x, \quad g^{(5)}(x) = \cos x, \quad g^{(6)}(x) = -\sin x, \quad g^{(7)}(x) = -\cos x,$$

and the derivatives continue to repeat in this pattern. Thus

$$g^{(k)}(0) = \sin 0 = 0 \quad \text{if} \quad k = 0, 4, 8, 12, \ldots,$$

$$g^{(k)}(0) = \cos 0 = 1 \quad \text{if} \quad k = 1, 5, 9, 13, \ldots,$$

$$g^{(k)}(0) = -\sin 0 = 0 \qquad \text{if} \qquad k = 2, 6, 10, 14, \ldots,$$
$$g^{(k)}(0) = -\cos 0 = -1 \qquad \text{if} \qquad k = 3, 7, 11, 15, \ldots.$$

Therefore the Taylor series for $\sin x$ is

$$\frac{0}{0!}x^0 + \frac{1}{1!}x^1 + \frac{0}{2!}x^2 + \frac{-1}{3!}x^3 + \frac{0}{4!}x^4 + \frac{1}{5!}x^5 + \frac{0}{6!}x^6 + \frac{-1}{7!}x^7 + \cdots$$
$$= x - \frac{x^3}{3!} + \frac{x^5}{5!} - \frac{x^7}{7!} + \cdots,$$

which we may write as

$$\sum_{k=0}^{\infty} \frac{(-1)^k x^{2k+1}}{(2k+1)!}.$$

The ratio test shows that this series converges for all x. (Check this.) In fact, though we shall not prove it, this Taylor series for $\sin x$ converges to $\sin x$ for all x. ☐

Every Taylor series is a power series and therefore has some radius of convergence R, $0 \leq R \leq +\infty$. If $R \neq 0$, the Taylor series converges for $-R < x < R$. What does it converge to? Examples 2, 3, and 4 suggest that it converges to the function from which it was constructed. This is not always so, but for many functions it is true that, within its interval of convergence, the Taylor series for the function converges to the function.

The power series expansions we obtained by various means in Section 12.2 are all Taylor series. In fact, any power series expansion for any function f, no matter how it is obtained, is the Taylor series for the function because, as we demonstrated at the beginning of this section, if

$$f(x) = \sum_{k=0}^{\infty} a_k x^k, \qquad \text{then} \qquad a_k = \frac{f^{(k)}(0)}{k!}.$$

EXAMPLE 5 From Example 4,

$$\sin x = \sum_{k=0}^{\infty} \frac{(-1)^k x^{2k+1}}{(2k+1)!}, \qquad \text{for all } x.$$

Differentiating both sides of this equation yields the power series expansion

$$\cos x = \sum_{k=0}^{\infty} \frac{(-1)^k x^{2k}}{(2k)!} \qquad \text{for all } x.$$

Therefore this power series expansion for $\cos x$ is the Taylor series for $\cos x$. ☐

We mentioned in the previous section that power series expansions are useful for calculating functional values. For instance, electronic calculators and computers find the values of exponential, logarithmic, and trigonometric functions with the aid of power series, and all tables of such functions ultimately depend on power series for their construction. We will now show how to use power series to find approximate values of functions. The key here is that we not only wish to approximate functional values by adding the first few terms of a series, but we also want to have some idea of how good the resulting approximation is.

Suppose f is a function whose Taylor series converges to it at x, so that

$$f(x) = \sum_{k=0}^{\infty} \frac{f^{(k)}(0)}{k!} x^k.$$

If we add the first $n + 1$ terms of the series, that is, all the terms up to and including the one that contains x^n, we obtain an approximation for $f(x)$:

$$f(x) \approx \sum_{k=0}^{n} \frac{f^{(k)}(0)}{k!} x^k = \frac{f^{(0)}(0)}{0!} + \frac{f^{(1)}(0)}{1!} x + \cdots + \frac{f^{(n)}(0)}{n!} x^n.$$

The sum of the terms we have not used is called the **nth remainder** of the Taylor series, and is denoted by $r_n(x)$. Thus

$$r_n(x) = \frac{f^{(n+1)}(0)}{(n+1)!} x^{n+1} + \cdots = \sum_{k=n+1}^{\infty} \frac{f^{(k)}(0)}{k!} x^k.$$

Since

$$f(x) = \sum_{k=0}^{n} \frac{f^{(k)}(0)}{k!} x^k + r_n(x),$$

we have

$$r_n(x) = f(x) - \left(\sum_{k=0}^{n} \frac{f^{(k)}(0)}{k!} x^k \right);$$

thus $r_n(x)$ is the error of the approximation

$$f(x) \approx \sum_{k=0}^{n} \frac{f^{(k)}(0)}{k!} x^k.$$

If we could find the value of $r_n(x)$, we would know how good (or bad) the approximation is. What we need is a closed-form expression for $r_n(x)$. The following result, which we will not prove, gives us such an expression.

Taylor Series Remainder Formula

Given a Taylor series

$$\sum_{k=0}^{\infty} \frac{f^{(k)}(0)}{k!} x^k,$$

let

$$r_n(x) = \sum_{k=n+1}^{\infty} \frac{f^{(k)}(0)}{k!} x^k.$$

Then there is some number c between 0 and x such that

$$r_n(x) = \frac{f^{(n+1)}(c)}{(n+1)!} x^{n+1}.$$

EXAMPLE 6 Let us use the Taylor series for e^x to find an approximate value for e. We have

$$e^x = \sum_{k=0}^{\infty} \frac{x^k}{k!},$$

so

$$e = \sum_{k=0}^{\infty} \frac{1}{k!}.$$

If we approximate e by summing the first four terms of this series, we obtain

$$e \approx 1 + 1 + \frac{1}{2!} + \frac{1}{3!} = 2.6667,$$

rounded to four decimal places. The error in this approximation is

$$r_3(1) = \frac{1}{4!} + \frac{1}{5!} + \cdots.$$

But there is some number c between 0 and $x = 1$ such that

$$r_3(1) = \frac{f^{(4)}(c)}{4!} 1^4.$$

Since $f(x) = e^x$, $f^{(4)}(x) = e^x$ also, and hence $f^{(4)}(c) = e^c$. Therefore

$$r_3(1) = \frac{e^c}{4!}.$$

Clearly, $r_3(1)$ is positive, and because c is between 0 and 1,

$$e^c < e^1 = e.$$

Thus

$$0 < r_3(1) < \frac{e}{4!}.$$

We cannot use the value of e here (that is what we are trying to determine), but surely $e < 3$, so

$$r_3(1) < \frac{e}{4!} < \frac{3}{4!} = 0.125.$$

Hence

$$0 < r_3(1) < 0.125,$$

or

$$0 < e - 2.6667 < 0.125,$$

which shows that

$$2.6667 < e < 2.6667 + 0.125 = 2.7917.$$

Therefore the approximation $e \approx 2.6667$ is too small and the true value of e lies between 2.6667 and 2.7917. ☐

EXAMPLE 7

Suppose we wish to estimate the value of e with at least three-decimal-place accuracy. In order to achieve this accuracy, the error of our approximation will have to be less than

$$\frac{1}{2}(10^{-3}) = 0.0005.$$

In other words, the absolute value of the difference between the true value of e and our approximation must be less than 0.0005:

$$|e - \text{approximation to } e| < 0.0005.$$

If we approximate e with the sum of the first $n + 1$ terms of its Taylor series, then

$$e - \text{approximation to } e = r_n(1).$$

Therefore we must have

$$|r_n(1)| < 0.0005.$$

However, as in Example 6,

$$|r_n(1)| = \left|\frac{f^{(n+1)}(0)}{(n+1)!} 1^{n+1}\right| = \frac{e^c}{(n+1)!} < \frac{e^1}{(n+1)!} < \frac{3}{(n+1)!}.$$

Thus if we choose n so that

$$\frac{3}{(n+1)!} < 0.0005,$$

we will ensure at least three-decimal-place accuracy. The smallest value of n for which

$$\frac{3}{(n+1)!} < 0.0005$$

is $n = 7$, so if we sum the Taylor series through the term corresponding to x^7 our approximation will be of the desired accuracy. Therefore the approximation

$$e \approx 1 + 1 + \frac{1}{2!} + \cdots + \frac{1}{7!} = 2.7182788$$

is accurate to at least three decimal places. (In fact, it is accurate to four decimal places.) ☐

EXAMPLE 8

Let us estimate $\sin 0.2$ with a minimum of four-decimal-place accuracy. We use the Taylor series for $\sin x$,

$$\sin x = \sum_{k=0}^{\infty} \frac{(-1)^k x^{2k+1}}{(2k+1)!}.$$

Thus

$$\sin 0.2 = 0.2 - \frac{(0.2)^3}{3!} + \frac{(0.2)^5}{5!} - \frac{(0.2)^7}{7!} + \cdots.$$

For four-place accuracy, we must have

$$|r_n(0.2)| < \frac{1}{2}(10^{-4}) = 0.00005.$$

Now
$$r_n(0.2) = \frac{f^{(n+1)}(c)}{(n+1)!}(0.2)^{n+1},$$

where $f(x) = \sin x$ and c is between 0 and 0.2. Since $f(x) = \sin x$, $f^{(n+1)}(c) = \pm \sin c$ or $f^{(n+1)}(c) = \pm \cos c$; in either case,

$$|f^{(n+1)}(c)| \le 1.$$

Therefore

$$|r_n(0.2)| = \frac{|f^{(n+1)}(c)|}{(n+1)!}(0.2)^{n+1} \le \frac{1}{(n+1)!}(0.2)^{n+1}.$$

Hence if we choose n so that

$$\frac{(0.2)^{n+1}}{(n+1)!} < 0.00005,$$

we will attain the desired accuracy. The smallest value of n for which this inequality holds is $n = 4$. Therefore we must add the terms of the series through the one that contains x^4. Since this term is missing in the series for $\sin 0.2$, we need only add the terms through the one that contains x^3. Hence the approximation

$$\sin 0.2 \approx 0.2 - \frac{(0.2)^3}{3!} = 0.1986667$$

is accurate to at least four decimal places. (It is actually accurate to five decimal places.) ☐

We conclude this section with a reference table of Taylor series expansions.

Taylor Series Expansions

1. $\dfrac{1}{1-x} = \sum\limits_{k=0}^{\infty} x^k,$ $-1 < x < 1.$

2. $\dfrac{1}{1+x} = \sum\limits_{k=0}^{\infty} (-1)^k x^k,$ $-1 < x < 1.$

3. $e^x = \sum\limits_{k=0}^{\infty} \dfrac{x^k}{k!},$ all $x.$

4. $\ln(1+x) = \sum\limits_{k=1}^{\infty} \dfrac{(-1)^{k+1} x^k}{k},$ $-1 < x \le 1.$

5. $\sin x = \sum\limits_{k=0}^{\infty} \dfrac{(-1)^k x^{2k+1}}{(2k+1)!},$ all $x.$

6. $\cos x = \sum\limits_{k=0}^{\infty} \dfrac{(-1)^k x^{2k}}{(2k)!},$ all $x.$

Note that the Taylor series expansion for $\ln(1 + x)$ also converges at the endpoint $x = 1$ of its interval of convergence.

12.4 EXERCISES

In Exercises 1 through 14, find the Taylor series for the function.
1. $f(x) = (x + 1)^2$
2. $f(x) = x^3 - 3x^2 + 4x - 2$
3. $f(x) = \sqrt{x + 1}$
4. $f(x) = \sqrt{x + 2}$
5. $f(x) = \dfrac{1}{1 + x^4}$
6. $f(x) = \dfrac{1}{1 - x^4}$
7. $f(x) = e^{3x}$
8. $f(x) = \dfrac{1}{2}(e^x + e^{-x})$
9. $f(x) = \dfrac{1}{2}(e^x - e^{-x})$
10. $f(x) = \ln(x^2 + 1)$
11. $f(x) = \ln(x + 2)$
12. $f(x) = \cos 2x$
13. $f(x) = \sin(2x + \pi)$
14. $f(x) = \sin x^2$

15. Show that if p is a polynomial function, then $p(x)$ is identical to its Taylor series.
16. (a) Find the Taylor series for the function
$$f(x) = \dfrac{4}{1 + x^2}.$$
   (b) Find the Taylor series for the antiderivative F of f such that $F(0) = 0$.
   (c) It is known that $F(1) = \pi$. Use the first six terms of the series for $F(1)$ to estimate π.
17. The result of Exercise 16(c) suggests that the series for $F(1)$ is not a good one for estimating π. However, it is also known that
$$\pi = F(1/2) + F(1/3).$$
Estimate π by adding the first six terms of the series for $F(1/2)$ to the first six terms of the series for $F(1/3)$.
18. (a) Find an approximate value for $e^{0.2}$ by summing the first four terms of its Taylor series.
   (b) Using the approximation of part (a) and the method of Example 6, find an interval in which the true value of $e^{0.2}$ lies.
19. (a) Find an approximate value for $\ln 1.1$ by summing the first three nonzero terms of its Taylor series.
   (b) Use the approximation of part (a) to find an interval in which the true value of $\ln 1.1$ lies.
20. (a) Find an approximate value for $\sin 0.5$ by summing the first three nonzero terms of its Taylor series.
   (b) Use the approximation of part (a) to find an interval in which the true value of $\sin 0.5$ lies.
21. Repeat Exercise 20 for $\cos 0.2$.
22. The Taylor series for the function $f(x) = (1 + x)^r$ is referred to as the **binomial series**. It converges for $-1 < x < 1$.

(a) Find the binomial series.
(b) Estimate the value of $(1.1)^{0.4}$ by summing the first three terms of the binomial series.
(c) Use your result of part (b) to find an interval in which the true value of $(1.1)^{0.4}$ lies.

23. How many terms of the Taylor series for e would you have to sum in order to estimate e with an accuracy of at least
    (a) five decimal places?
    (b) six decimal places?
    Assume that all you know about the value of e is that it lies between 2 and 3.

24. How many terms of the Taylor series for $\sin 0.5$ would you have to sum in order to estimate $\sin 0.5$ with an accuracy of at least
    (a) three decimal places?
    (b) five decimal places?
    Assume that all you know about the value of $\sin x$ and $\cos x$ is that they lie between -1 and 1 for all x.

25. Referring to Exercise 22:
    (a) How many terms of the binomial series would you have to sum in order to estimate $(1.1)^{0.4}$ with at least six-decimal-place accuracy?
    (b) How many terms would you have to sum in order to obtain at least 12-decimal-place accuracy?

26. Estimate $\cos 0.4$ with at least five-decimal-place accuracy.

27. Estimate $\ln 1.2$ with at least four-decimal-place accuracy.

28. Find $\int \sin x^2 \, dx$.

29. Use the first four nonzero terms of the series of Exercise 28 to approximate the value of
$$\int_0^1 \sin x^2 \, dx.$$

30. If Z is the standard normal random variable, then
$$P(0 < Z < 0.5) = \frac{1}{\sqrt{2\pi}} \int_0^{0.5} e^{-z^2/2} \, dz.$$
Use the power series for $e^{-z^2/2}$ to find the approximate value of this probability, and compare your result with the probability given in Table 3 at the back of the book.

31. (a) Find the Taylor series for $x^2 e^x$.
    (b) Obtain an infinite series that sums to e by integrating the series of part (a) from 0 to 1.

32. Approximate the value of e by summing the first six terms of the series of Exercise 31(b). Compare your result with the approximation obtained by summing the first six terms of the Taylor series for e. If you had to estimate the value of e to 20 decimal places, which series would you rather use?

SUMMARY

This summary consists of terms and symbols whose meaning you should know, facts you should know, and techniques and methods you should be able to employ.

Section 12.1

Sequence, terms of a sequence. nth term a_n of a sequence. Notation $\{a_n\}$ for sequences. Sequences as functions from the positive integers to R. The graph of a sequence. The limit of a sequence as n approaches infinity. Convergent sequence, divergent sequence. Determining whether a sequence converges or diverges from its graph. Geometric sequence $\{r^n\}$, r a real number. Determining when a geometric sequence converges and when it diverges. Properties of limits of sequences. Determining convergence or divergence of a sequence using the properties of limits of sequences. Fibonacci sequence.

Section 12.2

Zeno's paradox. Infinite series. nth partial sum of an infinite series. Writing infinite series as sequences of partial sums. Convergent, divergent infinite series. Sum of a convergent infinite series. Determining convergence or divergence of an infinite series by examining its sequence of partial sums. Geometric series $\sum_{k=0}^{\infty} ar^k$, r a real number. First term a, ratio r of a geometric series. Determining convergence and divergence for geometric series. Finding the sum of a convergent geometric series. Resolution of Zeno's paradox. The integral test. Using the integral test to establish convergence or divergence. The harmonic series. The harmonic series diverges. The ratio test. Using the ratio test to establish convergence or divergence. If the ratio test fails, then the series may converge or it may diverge.

Section 12.3

Power series in x. Convergence of power series. Using a power series to define a function. The closed form of a function, the power series form of a function. The power series expansion of a function. The radius of convergence and interval of convergence of a power series. Finding the radius of convergence and interval of convergence of a power series. Using a power series to define a function on its interval of convergence. Differentiation of power series. Finding power series expansions by differentiation. Power series for e^x. Integration of power series. Evaluating definite integrals using power series. Finding power series expansions by integration. Finding power series expansions by substitution into known power series. Finding power series expansions by adding and subtracting power series. Finding power series expansions by multiplying power series by polynomials.

Section 12.4

If a function f has a power series expansion on an interval, it possesses derivatives of all orders there. If $f(x) = \sum_{k=0}^{\infty} a_k x^k$, then

SEQUENCES AND SERIES

$$a_k = \frac{f^{(k)}(0)}{k!} \quad \text{for } k = 0, 1, 2, \ldots$$

Taylor series for f about 0. Finding the Taylor series for a function. The Taylor series for $(1 - x)^{-1}$, $(1 + x)^{-1}$, e^x, $\ln(1 + x)$, $\sin x$, $\cos x$. For many functions, including those above, the Taylor series of f converges to f on its interval of convergence. Approximating $f(x)$ using the first $n + 1$ terms of the Taylor series for f. The nth remainder of the approximation, $r_n(x)$. The Taylor series remainder formula. Using Taylor series to approximate functional values. Using the remainder formula with an approximation to find an interval in which the true value lies. Using the remainder formula to determine how many terms of the Taylor series must be added in order to obtain an approximation of the desired accuracy.

REVIEW EXERCISES

In Exercises 1 through 14, state whether the sequence converges or diverges. If it converges, find its limit.

1. $\left\{\dfrac{n}{2}\right\}$
2. $\left\{\dfrac{n}{2^n}\right\}$
3. $\left\{\dfrac{5^n}{3^{n+2}}\right\}$
4. $\left\{\dfrac{3^{n+2}}{5^n}\right\}$
5. $\{-3\}$
6. $\left\{\dfrac{3n}{n+1}\right\}$
7. $\left\{(-1)^n \left(\dfrac{2}{7}\right)^n\right\}$
8. $\left\{\dfrac{n^2 + 1}{2n^2 - 15}\right\}$
9. $\left\{\dfrac{n^2 + 2n}{3n + 2}\right\}$
10. $\left\{\dfrac{n^3 + 1}{n^4 + n}\right\}$
11. $\left\{-1, 2, -\dfrac{1}{2}, 1, -\dfrac{1}{4}, \dfrac{1}{2}, -\dfrac{1}{8}, \dfrac{1}{4}, \ldots\right\}$
12. $\left\{1, \dfrac{1}{2}, \dfrac{1}{3}, 0, \dfrac{1}{5}, \dfrac{1}{6}, \dfrac{1}{7}, 0, \dfrac{1}{9}, \dfrac{1}{10}, \dfrac{1}{11}, 0, \ldots\right\}$
13. $\left\{1, \dfrac{1}{2}, \dfrac{1}{3}, 1, \dfrac{1}{5}, \dfrac{1}{6}, \dfrac{1}{7}, 1, \dfrac{1}{9}, \dfrac{1}{10}, \dfrac{1}{11}, 1, \ldots\right\}$
14. $\left\{4 + \dfrac{(-1)^{n+1}}{n+1} \sin(2n - 1)\dfrac{\pi}{2}\right\}$
15. A person deposits $100 in a savings account at the beginning of every month. If the account pays 12% per year compounded monthly and a_n is the amount in the account at the end of n months, write the sequence $\{a_n\}$.

In Exercises 16 through 21, state whether the infinite series converges or diverges. If it converges, find its sum.

16. $\displaystyle\sum_{k=0}^{\infty} (-1)^{k+1} \dfrac{2^k}{3^k}$
17. $\displaystyle\sum_{k=0}^{\infty} \dfrac{3}{4^k}$
18. $\displaystyle\sum_{k=1}^{\infty} (-1)^{k+1}(1.2)^k$
19. $\displaystyle\sum_{k=2}^{\infty} \dfrac{73}{10^{2k}}$
20. $\dfrac{8}{3} - 3 + \dfrac{27}{8} - \dfrac{243}{64} + \cdots$
21. $210 - 205.8 + 201.684 - \cdots$

In Exercises 22 through 29, use the ratio test or the integral test to establish convergence or divergence.

22. $\sum_{k=1}^{\infty} (-1)^k \dfrac{2^k}{11^{k+3}}$

23. $\sum_{k=1}^{\infty} \dfrac{1}{(k+2)^2}$

24. $\sum_{k=1}^{\infty} k^2 e^{-k}$

25. $\sum_{k=1}^{\infty} \dfrac{2^k}{(k+2)!}$

26. $\sum_{k=1}^{\infty} \dfrac{(-1)^k k!}{2^{2k}}$

27. $\sum_{k=1}^{\infty} \dfrac{2^{2k}}{k!}$

28. $\sum_{k=1}^{\infty} \dfrac{1 \cdot 3 \cdot 5 \cdots (2k+1)}{(2k)!}$

29. $\sum_{k=1}^{\infty} \dfrac{1 \cdot 3 \cdot 5 \cdots (2k+1)}{k!}$

30. The probability that a salesperson will make k consecutive calls on customers without generating a sale is $(0.4)(0.6)^k$.
    (a) Find the probability that the salesperson will make 10 or more consecutive calls without a sale.
    (b) Find the salesperson's expected number of consecutive calls without a sale.

In Exercises 31 through 36, find the radius of convergence of the power series.

31. $\sum_{k=0}^{\infty} \dfrac{x^{2k+1}}{5^{k+1}}$

32. $\sum_{k=1}^{\infty} \dfrac{(k+1)x^k}{k^2}$

33. $\sum_{k=1}^{\infty} \dfrac{2^k x^k}{k 3^{k+1}}$

34. $\sum_{k=0}^{\infty} \dfrac{(-1)^{k+1} 1 \cdot 3 \cdots (2k+1) x^{2k}}{(2k)!}$

35. $\sum_{k=0}^{\infty} \dfrac{1 \cdot 3 \cdots (2k+1)}{2 \cdot 4 \cdots (2k+2)} x^k$

36. $\sum_{k=0}^{\infty} \dfrac{1^2 \cdot 3^2 \cdots (2k+1)^2}{2 \cdot 4 \cdots (2k+2)} x^k$

37. Let f be defined by
$$f(x) = \sum_{k=0}^{\infty} \dfrac{(-1)^{k+1} 1 \cdot 3 \cdots (2k+1)}{(2k)!} x^k,$$
for $-R < x < R$, where R is the radius of convergence of the series. Find
    (a) $f'(x)$
    (b) $\int f(x)\,dx$
    (c) $\int_0^1 f(x)\,dx$

38. Find $\int \dfrac{1}{1+x^2}\,dx$.

39. Find $\int \dfrac{\sin x}{x}\,dx$. (Hint: Multiply the series for $\sin x$ by $\dfrac{1}{x}$.)

In Exercises 40 through 45, find a Taylor series expansion for the function.

40. $f(x) = \cos\left(\dfrac{x+\pi}{2}\right)$

41. $f(x) = \ln(1 - x^3)$

42. $f(x) = (1+x)^{1/3}$

43. $f(x) = -1 + \cos x$

44. $f(x) = \dfrac{1}{2}(1 + \cos 2x)$

45. $f(x) = \dfrac{2x}{(1+x^2)^2}$

46. Find an approximate value for $\tan 0.3$, accurate to four decimal places, by using the Taylor series for $\sin x$ and $\cos x$.

47. Use the first six terms of the Taylor series for e^x to estimate e^{-1}. Use your estimate to find an interval in which the true value of e^{-1} lies, assuming that $2 < e < 3$.

48. How many terms of the Taylor series for e^{-1} used in Exercise 47 would you need to sum in order to obtain at least six-decimal-place accuracy in your estimate?

SEQUENCES AND SERIES

ADDITIONAL TOPICS

Here are some suggestions for topics not covered in the text that you might want to investigate on your own.

1. The definition of the limit of a sequence that we gave in Section 12.1 is an informal one. A formal definition of the limit of a sequence $\{a_n\}$ is as follows:

   $$\lim_{n \to +\infty} a_n = L \text{ if and only if for every positive real number}$$

   ϵ there is some positive integer m such that $n \geq m$ implies that $|a_n - L| < \epsilon$.

   Show how this definition is equivalent to the one we gave.

2. Use the formal definition of the limit of a sequence to prove the properties of limits of sequences.

3. There are many tests for convergence and/or divergence of infinite series besides the integral test and the ratio test. One such is the **divergence test:**

   $$\sum_{k=1}^{\infty} a_k \text{ diverges if } \lim_{k \to +\infty} a_k \neq 0.$$

   Prove the divergence test and find out how it is used.

4. Another test for convergence and divergence of infinite series is the **root test:**

   if $a_k \geq 0$ for $k = 1, 2, \ldots$, then

   $$\sum_{k=1}^{\infty} a_k \text{ converges if } \lim_{k \to +\infty} \sqrt[k]{a_k} < 1,$$

   $$\sum_{k=1}^{\infty} a_k \text{ diverges if } \lim_{k \to +\infty} \sqrt[k]{a_k} > 1 \text{ or if the sequence } \{\sqrt[k]{a_k}\} \text{ diverges to } +\infty.$$

   Prove the root test and show how it is used.

5. The most useful test for the convergence and divergence of infinite series is the **comparison test:** Suppose that $a_k \geq 0$ for $k = 1, 2, \ldots$.

   (1) If $\sum_{k=1}^{\infty} b_k$ converges and $a_k \leq b_k$ for $k = 1, 2, \ldots$, then $\sum_{k=1}^{\infty} a_k$ converges;

   (2) If $\sum_{k=1}^{\infty} b_k$ diverges and $a_k \geq b_k$ for $k = 1, 2, \ldots$, then $\sum_{k=1}^{\infty} a_k$ diverges.

   Prove the comparison test and show how it is used.

6. If f is a function and a is any real number, the **Taylor series for f about $x = a$** is the series

   $$\sum_{k=0}^{\infty} \frac{f^{(k)}(a)}{k!} (x - a)^k.$$

   Note that the Taylor series about $x = 0$ that we studied in Section 12.4 is the special case of this obtained by taking $a = 0$. Most of the work we did in Section

12.4 using Taylor series about $x = 0$ can be repeated using Taylor series about $x = a$, for any a. Find out how this is done and why it is useful.

7. Taylor series about $x = a$ can sometimes be useful in evaluating limits. Thus, suppose we wish to evaluate

$$\lim_{x \to a} \frac{f(x)}{g(x)},$$

where $f(a) = g(a) = 0$. If we can write f and g in terms of their Taylor series about $x = a$ and then substitute these series for $f(x)$ and $g(x)$ in the quotient $f(x)/g(x)$, some simplification may occur that will make it possible to determine the limit. Find out how this is done, and use the technique to find

$$\lim_{x \to 0} \frac{\sin x}{x}, \quad \lim_{x \to 1} \frac{\ln(x-1)}{x-1}, \quad \text{and} \quad \lim_{x \to 0} \frac{e^x - 1 - x}{x^2}.$$

SUPPLEMENT: DISCRETE INCOME STREAMS

In Section 7.1 and in the Chapter 7 Supplement we discussed continuous streams of income. Sometimes, however, it is not appropriate to consider income as arriving continuously; for instance, if an investment pays $1000 at the end of each year, the resulting stream of income is not continuous, but discrete. In this supplement we consider discrete streams of income that extend into the indefinite future.

It is often necessary to analyze a discrete income stream that will continue indefinitely far into the future. In such cases it is useful to assume that the stream will last forever, and if this is done, the total income received may be expressed as an infinite series. For example, if the investment previously referred to, which pays $1000 at the end of each year, continues indefinitely into the future, we may consider its value to be

$$\$1000 + \$1000 + \cdots = \sum_{k=1}^{\infty} \$1000.$$

Of course, this infinite series does not converge, and its "sum" is infinite. What we need is a method of attaching a *finite* number to the total income series that will allow us to compare it with other such series. How can we do this? One way that sometimes works is to consider the **average income per period**. For example, the investment that pays $1000 per year has an average value of $1000 per year, as the following table shows:

Year	Total Income Received	Average Income per Year
1	$1000	$\frac{\$1000}{1} = \$1000$
2	$1000 + $1000 = $2000	$\frac{\$2000}{2} = \$1000$
3	$2000 + $1000 = $3000	$\frac{\$3000}{3} = \$1000$
⋮	⋮	⋮

Sequences and Series

Let us apply the concept of the average income period to a more complex stream of income.

EXAMPLE 1 Suppose a firm's future income stream is expected to be as follows:

Year:	1	2	3	4	5	6	7	8	9	...
Income (millions of $):	2	5	3	2	5	3	2	5	3	...

Note that the income stream is repetitive over a three-year cycle.

Suppose we focus on years when the income is the same, for instance, the years 1, 4, 7, 10, ... Since each three-year cycle increases total income by $2 + 5 + 3 = 10$ million dollars, we have

Year	Total Income Received	Average Income per Year
1	2	$\dfrac{2 + 0(10)}{1 + 0(3)} = 2$
4	$2 + 1(10) = 12$	$\dfrac{2 + 1(10)}{1 + 1(3)} = 3$
7	$2 + 2(10) = 22$	$\dfrac{2 + 2(10)}{1 + 2(3)} = \dfrac{22}{7}$
10	$2 + 3(10) = 32$	$\dfrac{2 + 3(10)}{1 + 3(3)} = \dfrac{32}{10}$
⋮	⋮	⋮

Thus we can write a sequence that gives the average income per year of the firm in the years 1, 4, 7, 10, ... : the sequence is

$$\left\{ \frac{2 + n(10)}{1 + n(3)} \right\},$$

where $n = 0, 1, 2, \ldots$. The limit of this sequence as $n \to \infty$ is $10/3$, so the firm's average annual income for the years 1, 4, 7, 10, ... approaches $10/3$.

Now suppose we perform a similar analysis for the years 2, 5, 8, 11, ... If we do so, we obtain the sequence

$$\left\{ \frac{7 + n(10)}{2 + n(3)} \right\}$$

(check this), and the limit of this sequence is also $10/3$. Finally, the sequence for the years 3, 6, 9, 12, ... is

$$\left\{ \frac{10 + n(10)}{3 + n(3)} \right\},$$

which also has a limit of $10/3$. Therefore the firm's average income per year for all years approaches $10/3$. □

The procedure utilized in the preceding example is known as the **average income per period method** of analyzing discrete income streams. It can be carried out for any strictly repetitive stream. If

$$a_1, a_2, \ldots, a_p, a_1, a_2, \ldots, a_p, \ldots$$

is such a stream, and if m is a positive integer, $1 \leq m \leq p$, then the average income per period for the periods $m, m + p, m + 2p, \ldots$ is given by

$$\lim_{n \to \infty} \left[\frac{\sum_{k=1}^{m} a_k + n \sum_{k=1}^{p} a_k}{m + np} \right] = \frac{\sum_{k=1}^{p} a_k}{p}.$$

> **Average Income per Period**
>
> Let $a_1, a_2, \ldots, a_p, a_1, a_2, \ldots, a_p, \ldots$ be a repetitive discrete stream of income. Then the average income per period approaches
>
> $$\frac{1}{p} \sum_{k=1}^{p} a_k.$$

EXAMPLE 2 Two investments generate discrete income streams that continue indefinitely into the future. The annual incomes from the investments are as follows:

Investment A: 2, 5, 4, 1, 3, 2, 5, 4, 1, 3, . . . ,
Investment B: 6, 0, 2, 6, 0, 2, 6, 0, 2, ...

The average income per year from investment A approaches

$$\frac{2 + 5 + 4 + 1 + 3}{5} = 3,$$

and that of B approaches

$$\frac{6 + 0 + 2}{3} = \frac{8}{3}.$$

Therefore if we use the criterion of average income per year to evaluate these investments, we can say that in the long run A is preferable to B. ☐

It is not always possible to attach a limiting value to the average income per period of a discrete income stream. For instance, the stream 1, 4, 9, 16, 25, . . . has average income per period 1, 2, 3, 4, 5, . . . ; since this sequence diverges, we cannot say that the average income per period approaches any finite number.

Even when the average income per period does approach a finite number, it may not be appropriate to use this number as a measure of the value of the income stream. This is so because the average income per period method ignores the time

SEQUENCES AND SERIES   823

value of money: it treats all income as equal, no matter how far in the future it may be received. If we wish to take into account the time value of money, we must consider the **present value** of discrete income streams. We illustrate with an example.

EXAMPLE 3   Suppose a firm's stream of income is expected to be constant at $10 million per year for the indefinite future so that the stream is 10, 10, 10, ... If the firm discounts future income at 10% per year compounded annually, then (since it receives the first $10 million in the current year but has to wait one year for the second $10 million, two years for the third $10 million, and so on) the present value of its income stream is

$$10 + 10(1.1)^{-1} + 10(1.1)^{-2} + \cdots = \sum_{k=0}^{\infty} 10(1.1)^{-k}.$$

This is a geometric series with first term 10 and ratio $(1.1)^{-1}$, so it converges to

$$\frac{10}{1 - (1.1)^{-1}} = 110.$$

Hence the present value of the firm's future stream of income is $110 million.   □

If a discrete stream of income is constant at $P per period and is discounted at a compound rate of $100r\%$ per period, then its present value is

$$V = \sum_{k=0}^{\infty} P(1+r)^{-k} = \frac{P}{1-(1+r)^{-1}} = \frac{P(1+r)}{r}.$$

Present Value of a Constant Stream of Income

If a discrete stream of income is constant at $P per period for the indefinite future and is discounted at a compounded rate of $100r\%$ per period, then its present value is

$$V = \frac{P(1+r)}{r}$$

dollars.

EXAMPLE 4   The present value of the monthly income stream 100, 100, 100, . . . discounted at 12% per year compounded monthly is

$$V = \frac{100(1.01)}{0.01} = 10{,}100. \quad □$$

The concept of present value can be extended to repetitive income streams.

EXAMPLE 5   Consider the following streams of annual income:

Stream X: 3, 4, 3, 4, . . .
Stream Y: 6, 1, 6, 1, ...

Let us find the present value of each of these streams using a discount rate of 10% per year compounded annually. (Note that average income per year approaches $7/2$ for both of these streams.)

For income stream X we have

$$V_X = 3 + 4(1.1)^{-1} + 3(1.1)^{-2} + 4(1.1)^{-3} + \cdots$$

$$= 3(1 + (1.1)^{-2} + (1.1)^{-4} + \cdots)$$

$$\quad + 4(1.1)^{-1}(1 + (1.1)^{-2} + (1.1)^{-4} + \cdots)$$

$$= (3 + 4(1.1)^{-1}) \frac{1}{1 - (1.1)^{-2}}$$

$$= 38.238095.$$

Similarly, for income stream Y,

$$V_Y = 6 + 1(1.1)^{-1} + 6(1.1)^{-2} + 1(1.1)^{-3} + \cdots$$

$$= 6(1 + (1.1)^{-2} + (1.1)^{-4} + \cdots)$$

$$\quad + 1(1.1)^{-1}(1 + (1.1)^{-2} + (1.1)^{-4} + \cdots)$$

$$= (6 + (1.1)^{-1}) \frac{1}{1 - (1.1)^{-2}}$$

$$= 39.809524.$$

Thus stream Y has a larger present value than stream X when both are discounted at 10% per year compounded annually; if we use present value with this discount rate as our criterion, we will therefore prefer stream Y to stream X. □

The present value of a stream of income depends on the discount rate r as well as on the stream itself. It may be that one income stream has a larger present value than another no matter what r is; it may also be that one stream's present value is larger than that of another for only some values of r.

EXAMPLE 6 Let the income streams X and Y be as in Example 5. If r is the discount rate in decimal form, then a calculation similar to that of Example 5 shows that

$$V_X = (3 + 4(1 + r)^{-1}) \left(\frac{1}{1 - (1 + r)^{-2}} \right)$$

$$= \left(3 + \frac{4}{1 + r} \right) \left(\frac{(1 + r)^2}{(1 + r)^2 - 1} \right)$$

$$= \left(\frac{3(1 + r) + 4}{1 + r} \right) \left(\frac{(1 + r)^2}{r(r + 2)} \right)$$

$$= \frac{(7 + 3r)(1 + r)}{r(r + 2)}.$$

Similarly,

$$V_Y = (6 + (1+r)^{-1})\left(\frac{1}{1-(1+r)^{-2}}\right)$$

$$= \left(6 + \frac{1}{1+r}\right)\left(\frac{(1+r)^2}{(1+r)^2 - 1}\right)$$

$$= \left(\frac{6(1+r) + 1}{1+r}\right)\left(\frac{(1+r)^2}{r(r+2)}\right)$$

$$= \frac{(7 + 6r)(1+r)}{r(r+2)}.$$

Thus

$$V_Y - V_X = \frac{(7+6r)(1+r) - (7+3r)(1+r)}{r(r+2)}$$

$$= \frac{3(1+r)}{r+2}.$$

Since r is a discount rate, it is positive. But if $r > 0$, then also

$$\frac{3(1+r)}{r+2} > 0.$$

Hence $V_Y - V_X > 0$ for $r > 0$, and thus stream Y has a larger present value than stream X no matter what discount rate we use. □

EXAMPLE 7 Let income stream Y be as in Examples 5 and 6, and let income stream Z be as follows:

Stream Z: 9, 1, 1, 1, ...

Then from Example 6,

$$V_Y = \frac{(7+6r)(1+r)}{r(r+2)},$$

and

$$V_Z = 9 + 1(1+r)^{-1} + 1(1+r)^{-2} + \cdots$$

$$= 9 + \frac{1}{1-(1+r)^{-1}}$$

$$= 9 + \frac{1+r}{r}$$

$$= \frac{10r+1}{r}.$$

CHAPTER 12

Thus

$$V_Y - V_Z = \frac{(7 + 6r)(1 + r)}{r(r + 2)} - \frac{10r + 1}{r}$$

$$= \frac{(6r^2 + 13r + 7) - (10r^2 + 21r + 2)}{r(r + 2)}$$

$$= \frac{-4r^2 - 8r + 5}{r(r + 2)}$$

$$= \frac{-4(r + 5/2)(r - 1/2)}{r(r + 2)}.$$

Since $r > 0$, it follows that

$$V_Y - V_Z > 0 \quad \text{if } 0 < r < \frac{1}{2}$$

and

$$V_Y - V_Z < 0 \quad \text{if } r > \frac{1}{2}.$$

Therefore stream Y has a larger present value than stream Z if the discount rate is between 0 and 0.5, whereas Z has the larger present value if the discount rate is greater than 0.5. When $r = 0.5$, the two income streams have the same present value. $\square$

SUPPLEMENTARY EXERCISES

1. Apply the average income per period method to rank the annual income streams

   A: 2, 3, 4, 2, 3, 4, . . .

   B: 2, 2, 0, 5, 2, 2, 0, 5, . . .

   and

   C: 4, 4, 4, 4, . . .

   in order of preference.

2. Show that the average income per period method cannot be applied to the income stream 1, 2, 3, 2, 4, 6, 3, 6, 9, ...

3. Find the present value of the income stream C of Exercise 1 at a discount rate of
   (a) 10%   (b) 80%

   per year compounded annually.

4. Compare the income streams A and C of Exercise 1 using a discount rate of
   (a) 10% per year compounded annually
   (b) 80% per year compounded annually

5. Compare the income streams A and B of Exercise 1 using a discount rate of
   (a) 10% per year compounded annually
   (b) 80% per year compounded annually
6. Using an arbitrary discount rate, compare the income stream C of Exercise 1 and the stream
$$D: \quad 7, 0, 7, 0, 7, 0, \ldots .$$
7. Using an arbitrary discount rate, compare the income streams
$$X: \quad 5.1, 2.2, 5.1, 2.2, \ldots$$
and
$$Y: \quad 5.0, 2.35, 5.0, 2.35, \ldots .$$
8. Using an arbitrary discount rate, compare the income streams A and C of Exercise 1.

Tables

Table 1: Powers of *e*
Table 2: Natural Logarithms
Table 3: Areas under the Standard Normal Curve

TABLE 1 Powers of e

x	e^x	e^{-x}	x	e^x	e^{-x}	x	e^x	e^{-x}	x	e^x	e^{-x}	x	e^x	e^{-x}
0.00	1.00000	1.00000	0.42	1.52196	0.65705	0.84	2.31637	0.43171	1.26	3.52542	0.28365	2.40	11.0232	0.09072
0.01	1.01005	0.99005	0.43	1.53726	0.65051	0.85	2.33965	0.42741	1.27	3.56085	0.28083	2.45	11.5883	0.08629
0.02	1.02020	0.98020	0.44	1.55271	0.64404	0.86	2.36316	0.42316	1.28	3.59664	0.27804	2.50	12.1825	0.08209
0.03	1.03045	0.97045	0.45	1.56831	0.63763	0.87	2.38691	0.41895	1.29	3.63279	0.27527	2.55	12.8017	0.07808
0.04	1.04081	0.96079	0.46	1.58407	0.63128	0.88	2.41090	0.41478	1.30	3.66930	0.27253	2.60	13.4637	0.07427
0.05	1.05127	0.95123	0.47	1.59999	0.62500	0.89	2.43513	0.41066	1.31	3.70617	0.26982	2.65	14.1540	0.07065
0.06	1.06184	0.94176	0.48	1.61607	0.61878	0.90	2.45960	0.40657	1.32	3.74342	0.26714	2.70	14.8797	0.06721
0.07	1.07251	0.93239	0.49	1.63232	0.61263	0.91	2.48432	0.40252	1.33	3.78104	0.26448	2.75	15.6426	0.06393
0.08	1.08329	0.92312	0.50	1.64872	0.60653	0.92	2.50929	0.39852	1.34	3.81904	0.26185	2.80	16.4447	0.06081
0.09	1.09417	0.91393	0.51	1.66529	0.60050	0.93	2.53451	0.39455	1.35	3.85743	0.25924	2.85	17.2878	0.05784
0.10	1.10517	0.90484	0.52	1.68203	0.59452	0.94	2.55998	0.39063	1.36	3.89619	0.25666	2.90	18.1741	0.05502
0.11	1.11628	0.89583	0.53	1.69893	0.58861	0.95	2.58571	0.38674	1.37	3.93535	0.25411	2.95	19.1060	0.05234
0.12	1.12750	0.88692	0.54	1.71601	0.58275	0.96	2.61170	0.38289	1.38	3.97490	0.25158	3.00	20.0855	0.04979
0.13	1.13883	0.87810	0.55	1.73325	0.57695	0.97	2.63794	0.37908	1.39	4.01485	0.24908	3.05	21.1153	0.04736
0.14	1.15027	0.86936	0.56	1.75067	0.57121	0.98	2.66446	0.37531	1.40	4.05520	0.24660	3.10	22.1980	0.04505
0.15	1.16183	0.86071	0.57	1.76827	0.56553	0.99	2.69123	0.37158	1.41	4.09596	0.24414	3.15	23.3361	0.04285
0.16	1.17351	0.85214	0.58	1.78604	0.55990	1.00	2.71828	0.36788	1.42	4.13712	0.24171	3.20	24.533	0.04076
0.17	1.18531	0.84366	0.59	1.80399	0.55433	1.01	2.74560	0.36422	1.43	4.17870	0.23931	3.30	27.113	0.03688
0.18	1.19722	0.83527	0.60	1.82212	0.54881	1.02	2.77320	0.36059	1.44	4.22070	0.23693	3.40	29.964	0.03337
0.19	1.20925	0.82696	0.61	1.84043	0.54335	1.03	2.80107	0.35701	1.45	4.26312	0.23457	3.50	33.115	0.03020
0.20	1.22140	0.81873	0.62	1.85893	0.53794	1.04	2.82922	0.35345	1.46	4.30596	0.23224	3.60	36.598	0.02732
0.21	1.23368	0.81058	0.63	1.87761	0.53259	1.05	2.85765	0.34994	1.47	4.34924	0.22993	3.70	40.447	0.02472
0.22	1.24608	0.80252	0.64	1.89648	0.52729	1.06	2.88637	0.34646	1.48	4.39295	0.22764	3.80	44.701	0.02237
0.23	1.25860	0.79453	0.65	1.91554	0.52205	1.07	2.91538	0.34301	1.49	4.43710	0.22537	3.90	49.402	0.02024
0.24	1.27125	0.78663	0.66	1.93479	0.51685	1.08	2.94468	0.33960	1.50	4.4817	0.22313	4.00	54.598	0.01832
0.25	1.28403	0.77880	0.67	1.95424	0.51171	1.09	2.97427	0.33622	1.55	4.7115	0.21225	4.10	60.340	0.01657
0.26	1.29693	0.77105	0.68	1.97388	0.50662	1.10	3.00417	0.33287	1.60	4.9530	0.20190	4.20	66.686	0.01500
0.27	1.30996	0.76338	0.69	1.99372	0.50158	1.11	3.03436	0.32956	1.65	5.2070	0.19205	4.30	73.700	0.01357
0.28	1.32313	0.75578	0.70	2.01375	0.49659	1.12	3.06485	0.32628	1.70	5.4739	0.18268	4.40	81.451	0.01228
0.29	1.33643	0.74826	0.71	2.03399	0.49164	1.13	3.09566	0.32303	1.75	5.7546	0.17377	4.50	90.017	0.01111
0.30	1.34986	0.74082	0.72	2.05443	0.48675	1.14	3.12677	0.31982	1.80	6.0496	0.16530	4.60	99.484	0.01005
0.31	1.36343	0.73345	0.73	2.07508	0.48191	1.15	3.15819	0.31664	1.85	6.3598	0.15724	4.70	109.947	0.00910
0.32	1.37713	0.72615	0.74	2.09594	0.47711	1.16	3.18993	0.31349	1.90	6.6859	0.14957	4.80	121.510	0.00823
0.33	1.39097	0.71892	0.75	2.11700	0.47237	1.17	3.22199	0.31037	1.95	7.0287	0.14227	4.90	134.290	0.00745
0.34	1.40495	0.71177	0.76	2.13828	0.46767	1.18	3.25437	0.30728	2.00	7.3891	0.13534	5.00	148.413	0.00674
0.35	1.41907	0.70469	0.77	2.15977	0.46301	1.19	3.28708	0.30422	2.05	7.7679	0.12873	5.10	164.022	0.00610
0.36	1.43333	0.69768	0.78	2.18147	0.45841	1.20	3.32012	0.30119	2.10	8.1662	0.12246	5.20	181.272	0.00552
0.37	1.44773	0.69073	0.79	2.20340	0.45384	1.21	3.35348	0.29820	2.15	8.5849	0.11648	5.30	200.337	0.00499
0.38	1.46228	0.68386	0.80	2.22554	0.44933	1.22	3.38719	0.29523	2.20	9.0250	0.11080	5.40	221.406	0.00452
0.39	1.47698	0.67706	0.81	2.24791	0.44486	1.23	3.42123	0.29229	2.25	9.4877	0.10540	5.50	244.692	0.00409
0.40	1.49182	0.67032	0.82	2.27050	0.44043	1.24	3.45561	0.28938	2.30	9.9742	0.10026	5.60	270.427	0.00370
0.41	1.50682	0.66365	0.83	2.29332	0.43605	1.25	3.49034	0.28650	2.35	10.4856	0.09537	5.70	298.867	0.00335
												5.80	330.300	0.00303
												5.90	365.038	0.00274
												6.00	403.429	0.00248
												6.10	445.858	0.00224
												6.20	492.749	0.00203
												6.30	544.572	0.00184
												6.40	601.845	0.00166
												6.50	665.142	0.00150
												6.60	735.09	0.00136
												6.70	812.41	0.00123
												6.80	897.85	0.00111
												6.90	992.27	0.00101
												7.00	1096.63	0.00091
												7.10	1211.97	0.00083
												7.20	1339.43	0.00075
												7.30	1480.30	0.00068
												7.40	1635.98	0.00061
												7.50	1808.04	0.00055
												7.60	1998.20	0.00050
												7.70	2208.35	0.00045
												7.80	2440.60	0.00041
												7.90	2697.28	0.00037
												8.00	2980.96	0.00034
												8.10	3294.47	0.00030
												8.20	3640.95	0.00027
												8.30	4023.87	0.00025
												8.40	4447.07	0.00022
												8.50	4914.77	0.00020
												8.60	5431.66	0.00018
												8.70	6002.91	0.00017
												8.80	6634.25	0.00015
												8.90	7331.97	0.00014
												9.00	8103.08	0.00012
												9.10	8955.29	0.00011
												9.20	9897.13	0.00010
												9.30	10938.0	0.00009
												9.40	12088.4	0.00008
												9.50	13359.7	0.00007
												9.60	14764.8	0.00007
												9.70	16317.6	0.00006
												9.80	18033.7	0.00006
												9.90	19930.4	0.00005

TABLE 2 Natural Logarithms

x	$\ln x$	x	$\ln x$	x	$\ln x$	x	$\ln x$	x	$\ln x$	x	$\ln x$	x	$\ln x$	x	$\ln x$
0.01	−4.6052	0.51	−0.6733	1.01	0.0100	1.51	0.4121	2.01	0.6981	2.55	0.9361	5.05	1.6194	7.55	2.0215	10.05	2.3076
0.02	−3.9120	0.52	−0.6539	1.02	0.0198	1.52	0.4187	2.02	0.7031	2.60	0.9555	5.10	1.6292	7.60	2.0281	10.10	2.3125
0.03	−3.5066	0.53	−0.6349	1.03	0.0296	1.53	0.4253	2.03	0.7080	2.65	0.9746	5.15	1.6390	7.65	2.0347	10.15	2.3175
0.04	−3.2189	0.54	−0.6162	1.04	0.0392	1.54	0.4318	2.04	0.7130	2.70	0.9933	5.20	1.6487	7.70	2.0412	10.20	2.3224
0.05	−2.9957	0.55	−0.5978	1.05	0.0488	1.55	0.4383	2.05	0.7178	2.75	1.0116	5.25	1.6582	7.75	2.0477	10.25	2.3273
0.06	−2.8134	0.56	−0.5798	1.06	0.0583	1.56	0.4447	2.06	0.7227	2.80	1.0296	5.30	1.6677	7.80	2.0541	10.30	2.3321
0.07	−2.6593	0.57	−0.5621	1.07	0.0677	1.57	0.4511	2.07	0.7275	2.85	1.0473	5.35	1.6771	7.85	2.0605	10.35	2.3370
0.08	−2.5257	0.58	−0.5447	1.08	0.0770	1.58	0.4574	2.08	0.7324	2.90	1.0647	5.40	1.6864	7.90	2.0669	10.40	2.3418
0.09	−2.4079	0.59	−0.5276	1.09	0.0862	1.59	0.4637	2.09	0.7372	2.95	1.0818	5.45	1.6956	7.95	2.0732	10.45	2.3466
0.10	−2.3026	0.60	−0.5108	1.10	0.0953	1.60	0.4700	2.10	0.7419	3.00	1.0986	5.50	1.7047	8.00	2.0794	10.50	2.3514
0.11	−2.2073	0.61	−0.4943	1.11	0.1044	1.61	0.4762	2.11	0.7467	3.05	1.1151	5.55	1.7138	8.05	2.0857	10.55	2.3561
0.12	−2.1203	0.62	−0.4780	1.12	0.1133	1.62	0.4824	2.12	0.7514	3.10	1.1314	5.60	1.7228	8.10	2.0919	10.60	2.3609
0.13	−2.0402	0.63	−0.4620	1.13	0.1222	1.63	0.4886	2.13	0.7561	3.15	1.1474	5.65	1.7317	8.15	2.0980	10.65	2.3656
0.14	−1.9661	0.64	−0.4463	1.14	0.1310	1.64	0.4947	2.14	0.7608	3.20	1.1632	5.70	1.7405	8.20	2.1041	10.70	2.3702
0.15	−1.8971	0.65	−0.4308	1.15	0.1398	1.65	0.5008	2.15	0.7655	3.25	1.1787	5.75	1.7492	8.25	2.1102	10.75	2.3749
0.16	−1.8326	0.66	−0.4155	1.16	0.1484	1.66	0.5068	2.16	0.7701	3.30	1.1939	5.80	1.7579	8.30	2.1163	10.80	2.3795
0.17	−1.7720	0.67	−0.4005	1.17	0.1570	1.67	0.5128	2.17	0.7747	3.35	1.2090	5.85	1.7664	8.35	2.1223	10.85	2.3842
0.18	−1.7148	0.68	−0.3857	1.18	0.1655	1.68	0.5188	2.18	0.7793	3.40	1.2238	5.90	1.7750	8.40	2.1282	10.90	2.3888
0.19	−1.6607	0.69	−0.3711	1.19	0.1740	1.69	0.5247	2.19	0.7839	3.45	1.2384	5.95	1.7834	8.45	2.1342	10.95	2.3933
0.20	−1.6094	0.70	−0.3567	1.20	0.1823	1.70	0.5306	2.20	0.7885	3.50	1.2528	6.00	1.7918	8.50	2.1401	11.00	2.3979
0.21	−1.5606	0.71	−0.3425	1.21	0.1906	1.71	0.5365	2.21	0.7930	3.55	1.2669	6.05	1.8001	8.55	2.1459	11.05	2.4024
0.22	−1.5141	0.72	−0.3285	1.22	0.1989	1.72	0.5423	2.22	0.7975	3.60	1.2809	6.10	1.8083	8.60	2.1518	11.10	2.4069
0.23	−1.4697	0.73	−0.3147	1.23	0.2070	1.73	0.5481	2.23	0.8020	3.65	1.2947	6.15	1.8165	8.65	2.1576	11.15	2.4114
0.24	−1.4271	0.74	−0.3011	1.24	0.2151	1.74	0.5539	2.24	0.8065	3.70	1.3083	6.20	1.8245	8.70	2.1633	11.20	2.4159
0.25	−1.3863	0.75	−0.2877	1.25	0.2231	1.75	0.5596	2.25	0.8109	3.75	1.3218	6.25	1.8326	8.75	2.1691	11.25	2.4204
0.26	−1.3471	0.76	−0.2744	1.26	0.2311	1.76	0.5653	2.26	0.8154	3.80	1.3350	6.30	1.8405	8.80	2.1748	11.30	2.4248
0.27	−1.3039	0.77	−0.2614	1.27	0.2390	1.77	0.5710	2.27	0.8198	3.85	1.3481	6.35	1.8485	8.85	2.1804	11.35	2.4292
0.28	−1.2730	0.78	−0.2485	1.28	0.2469	1.78	0.5766	2.28	0.8242	3.90	1.3610	6.40	1.8563	8.90	2.1861	11.40	2.4336
0.29	−1.2379	0.79	−0.2357	1.29	0.2546	1.79	0.5822	2.29	0.8286	3.95	1.3737	6.45	1.8641	8.95	2.1917	11.45	2.4380
0.30	−1.2040	0.80	−0.2231	1.30	0.2624	1.80	0.5878	2.30	0.8329	4.00	1.3863	6.50	1.8718	9.00	2.1972	11.50	2.4423
0.31	−1.1712	0.81	−0.2107	1.31	0.2700	1.81	0.5933	2.31	0.8372	4.05	1.3987	6.55	1.8795	9.05	2.2028	13.0	2.5649
0.32	−1.1394	0.82	−0.1985	1.32	0.2776	1.82	0.5988	2.32	0.8416	4.10	1.4110	6.60	1.8871	9.10	2.2083	17.0	2.8332
0.33	−1.1087	0.83	−0.1863	1.33	0.2852	1.83	0.6043	2.33	0.8459	4.15	1.4231	6.65	1.8946	9.15	2.2138	19.0	2.9444
0.34	−1.0788	0.84	−0.1744	1.34	0.2927	1.84	0.6098	2.34	0.8502	4.20	1.4351	6.70	1.9021	9.20	2.2192	23.0	3.1355
0.35	−1.0498	0.85	−0.1625	1.35	0.3001	1.85	0.6152	2.35	0.8544	4.25	1.4469	6.75	1.9095	9.25	2.2246	29.0	3.3673
0.36	−1.0217	0.86	−0.1508	1.36	0.3075	1.86	0.6206	2.36	0.8587	4.30	1.4586	6.80	1.9169	9.30	2.2300	31.0	3.4340
0.37	−0.9943	0.87	−0.1393	1.37	0.3148	1.87	0.6259	2.37	0.8629	4.35	1.4702	6.85	1.9242	9.35	2.2354	37.0	3.6109
0.38	−0.9676	0.88	−0.1278	1.38	0.3221	1.88	0.6313	2.38	0.8671	4.40	1.4816	6.90	1.9315	9.40	2.2407	41.0	3.7136
0.39	−0.9416	0.89	−0.1165	1.39	0.3293	1.89	0.6366	2.39	0.8713	4.45	1.4929	6.95	1.9387	9.45	2.2460	43.0	3.7612
0.40	−0.9163	0.90	−0.1054	1.40	0.3365	1.90	0.6419	2.40	0.8755	4.50	1.5041	7.00	1.9459	9.50	2.2513	47.0	3.8501
0.41	−0.8916	0.91	−0.0943	1.41	0.3436	1.91	0.6471	2.41	0.8796	4.55	1.5151	7.05	1.9530	9.55	2.2565	53.0	3.9703
0.42	−0.8675	0.92	−0.0834	1.42	0.3507	1.92	0.6523	2.42	0.8838	4.60	1.5261	7.10	1.9601	9.60	2.2618	59.0	4.0775
0.43	−0.8440	0.93	−0.0726	1.43	0.3577	1.93	0.6575	2.43	0.8879	4.65	1.5369	7.15	1.9671	9.65	2.2670	61.0	4.1109
0.44	−0.8210	0.94	−0.0619	1.44	0.3646	1.94	0.6627	2.44	0.8920	4.70	1.5476	7.20	1.9741	9.70	2.2721	67.0	4.2047
0.45	−0.7985	0.95	−0.0513	1.45	0.3716	1.95	0.6678	2.45	0.8961	4.75	1.5581	7.25	1.9810	9.75	2.2773	71.0	4.2627
0.46	−0.7765	0.96	−0.0408	1.46	0.3784	1.96	0.6729	2.46	0.9002	4.80	1.5686	7.30	1.9879	9.80	2.2824	73.0	4.2905
0.47	−0.7550	0.97	−0.0305	1.47	0.3853	1.97	0.6780	2.47	0.9042	4.85	1.5790	7.35	1.9947	9.85	2.2875	79.0	4.3694
0.48	−0.7340	0.98	−0.0202	1.48	0.3920	1.98	0.6831	2.48	0.9083	4.90	1.5892	7.40	2.0015	9.90	2.2925	83.0	4.4188
0.49	−0.7134	0.99	−0.0101	1.49	0.3988	1.99	0.6881	2.49	0.9123	4.95	1.5994	7.45	2.0082	9.95	2.2976	89.0	4.4886
0.50	−0.6931	1.00	0.0000	1.50	0.4055	2.00	0.6931	2.50	0.9163	5.00	1.6094	7.50	2.0149	10.00	2.3026	97.0	4.5747

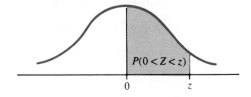

TABLE 3 Areas under the Standard Normal Curve

Normal Deviate z	.00	.01	.02	.03	.04	.05	.06	.07	.08	.09
0.0	.0000	.0040	.0080	.0120	.0160	.0199	.0239	.0279	.0319	.0359
0.1	.0398	.0438	.0478	.0517	.0557	.0596	.0636	.0675	.0714	.0753
0.2	.0793	.0832	.0871	.0910	.0948	.0987	.1026	.1064	.1103	.1141
0.3	.1179	.1217	.1255	.1293	.1331	.1368	.1406	.1443	.1480	.1517
0.4	.1554	.1591	.1628	.1664	.1700	.1736	.1772	.1808	.1844	.1879
0.5	.1915	.1950	.1985	.2019	.2054	.2088	.2123	.2157	.2190	.2224
0.6	.2257	.2291	.2324	.2357	.2389	.2422	.2454	.2486	.2517	.2549
0.7	.2580	.2611	.2642	.2673	.2704	.2734	.2764	.2794	.2823	.2852
0.8	.2881	.2910	.2939	.2967	.2995	.3023	.3051	.3078	.3106	.3133
0.9	.3159	.3186	.3212	.3238	.3264	.3289	.3315	.3340	.3365	.3389
1.0	.3413	.3438	.3461	.3485	.3508	.3531	.3554	.3577	.3599	.3621
1.1	.3643	.3665	.3686	.3708	.3729	.3749	.3770	.3790	.3810	.3830
1.2	.3849	.3869	.3888	.3907	.3925	.3944	.3962	.3980	.3997	.4015
1.3	.4032	.4049	.4066	.4082	.4099	.4115	.4131	.4147	.4162	.4177
1.4	.4192	.4207	.4222	.4236	.4251	.4265	.4279	.4292	.4306	.4319
1.5	.4332	.4345	.4357	.4370	.4382	.4394	.4406	.4418	.4429	.4441
1.6	.4452	.4463	.4474	.4484	.4495	.4505	.4515	.4525	.4535	.4545
1.7	.4554	.4564	.4573	.4582	.4591	.4599	.4608	.4616	.4625	.4633
1.8	.4641	.4649	.4656	.4664	.4671	.4678	.4686	.4693	.4699	.4706
1.9	.4713	.4719	.4726	.4732	.4738	.4744	.4750	.4756	.4761	.4767
2.0	.4772	.4778	.4783	.4788	.4793	.4798	.4803	.4808	.4812	.4817
2.1	.4821	.4826	.4830	.4834	.4838	.4842	.4846	.4850	.4854	.4857
2.2	.4861	.4864	.4868	.4871	.4875	.4878	.4881	.4884	.4887	.4890
2.3	.4893	.4896	.4898	.4901	.4904	.4906	.4909	.4911	.4913	.4916
2.4	.4918	.4920	.4922	.4925	.4927	.4929	.4931	.4932	.4934	.4936
2.5	.4938	.4940	.4941	.4943	.4945	.4946	.4948	.4949	.4951	.4952
2.6	.4953	.4955	.4956	.4957	.4959	.4960	.4961	.4962	.4963	.4964
2.7	.4965	.4966	.4967	.4968	.4969	.4970	.4971	.4972	.4973	.4974
2.8	.4974	.4975	.4976	.4977	.4977	.4978	.4979	.4979	.4980	.4981
2.9	.4981	.4982	.4982	.4983	.4984	.4984	.4985	.4985	.4986	.4986
3.0	.4987	.4987	.4987	.4988	.4988	.4989	.4989	.4989	.4990	.4990

Adapted by permission from Ernest Kurnow, Gerald J. Glasser, and Frederick R. Ottman. *Statistics for Business Decisions* (Homewood, Ill.: Richard D. Irwin, 1959), p. 501. © 1959 by Richard D. Irwin, Inc.

Solutions to Exercises

1.1 EXERCISES

1.
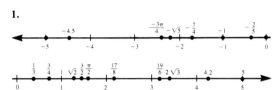

3.
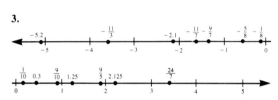

5. $0, 1/10, 3/20, 3/10, 7/20, 1/2, 11/20, 6/10, 3/4, 9/10, 19/20, 1$ **7.** True
9. True **11.** True **13.** False **15.** True **17.** True
19. True **21.** $S \geq \$1{,}500{,}000$ **23.** $S > \$1{,}500{,}000$
25. $\$86.50 \leq P \leq \$92.50$ **27.** $(-2, 7]$ **29.** $(-3, +\infty)$
31. $[1/2, 3/2)$ **33.** $(-\infty, -5/3]$ **35.** $-4 < x < 8$ **37.** $x \geq -5$ **39.** $x < -4$ **41.** $-2, -1.99, -3/4, -1, 2/3, \pi/2, 9/5$
43. d = dividend per share, t = year
$$d = \begin{cases} \$1.40 & \text{if } t \in [1980, 1986] \\ \$1.60 & \text{if } t \in [1986, 1989] \end{cases}$$
45. c = cost, r = rating
$$c = \begin{cases} \$275{,}000 & \text{if } r \in [0, 21.0) \\ \$300{,}000 & \text{if } r \in [21.0, +\infty) \end{cases}$$
47. d = dosage in mg, a = age in years
$$d = \begin{cases} 50 & \text{if } a \in (0, 5) \\ 100 & \text{if } a \in [5, 12] \\ 200 & \text{if } a \in (12, +\infty) \end{cases}$$
49. 4 **51.** -8 **53.** $4/5$ **55.** 5 **57.** $2 - \sqrt{2}$ **59.** p = pressure in lb/in^2, in safe range if $|p - 20| \leq 2.5$ **61.** 64
63. 1 **65.** $9/16$ **67.** 4 **69.** undefined **71.** $1/8$
73. $25z^{10}/x^2$ **75.** $8x^6z^6/y^5$ **77.** $1/x^7y^4z^{10}$ **79.** $|x|y^{3/2}$
81. $3x^{2/3}y/z^{2/3}$ **83.** \$200,000 **85.** \$40.048
87. $\approx (0.469, 0.531)$

1.2 EXERCISES

1. $\mathbb{R}$ **3.** $-3, 14/3, -11.2$ **5.** $\{x \mid x \neq 0\}$ **7.** $-1/3, -1/9, 2, 20$ **9.** $\mathbb{R}$ **11.** $3/4, -0.24, 4 - 3\sqrt{2}, \pi^2 - 3\pi + 2$

13. $\{x \mid x \neq 3\}$ **15.** $-1, 5/14, -5, 7$ **17.** $[-1, +\infty)$
19. $2\sqrt{2}, 2\sqrt{3}, 2/\sqrt{2}, 1$ **21.** $\{q \mid q \neq 0\}$ **23.** $2/\sqrt{2}, 1/\sqrt{2}, 1/5, 1/50$ **25.** $y = 2, 2x + y = 6, x + 2y = 6, y = x^2, y = |x|$ **27.(a)** $-\$200{,}000$ **(b)** \$0 **(c)** \$25,000
(d) $-\$75{,}000$ **29.(a)** \$432 **(b)** \$36 **(c)** \$32.04
31.(a) 480 **(b)** 445 **(c).** ≈ 444 **(d)** 460 **33.(a)** $\{x \mid x \in \mathbb{Q}, x > 0\}$ **(b)** $2 - \sqrt{2} \approx 0.59, 1.5 - \sqrt{1.25} \approx 0.38$
(c) $f(2/3) - f(2/4) \approx 0.083$ **(d)** $f(2/10) - f(2/11) \approx 0.015$
(e) $f(3/5) - f(2/5) \approx 0.111$

35.
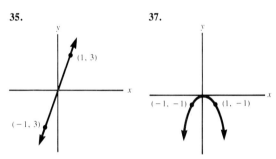

intercepts: (0, 0)
range: $\mathbb{R}$

37.

intercepts: (0, 0)
range: $(-\infty, 0]$

39.
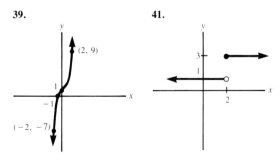

intercepts: (0, 1), $(-1, 0)$
range: $\mathbb{R}$

41.

intercepts: (0, 1)
range: $\{1, 3\}$

833

43.

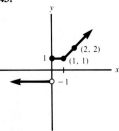

intercepts: (0, 1)
range: $\{y \mid y \geq 1 \text{ or } y = -1\}$

45. not a function **47.** is a function **49.** not a function
51. is a function **53.** not a function

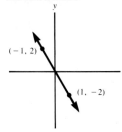   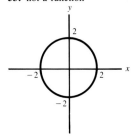

55. not a function

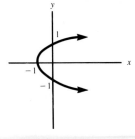

57.(a) 0, 2, 8, 1000, −3, −7, −99 **(b)** 0, 1, −1, −4, 1, 0, −2, −1 **(c)** 0, 6, −5, 18, −1, −13, −4 **(d)**

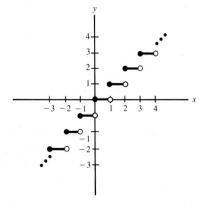

59.(a) $29 **(b)** 22 **(c)** 20, $20 **(d)** 10, 40
 (e) decreases to $20 then increases without bound
61.(a) $40,000 **(b)** $60,000 **(c)** 3 or more **(d)** increases toward a bound of $100,000
63.(a)

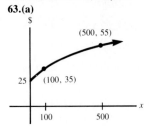

(b) $B(x) = \begin{cases} 25 + 0.1x, & 0 < x < 100 \\ 30 + 0.05\,x, & 100 \leq x < 500 \\ 45 + 0.02\,x, & x \geq 500 \end{cases}$

65.(a) 300 ppm **(b)** 250 ppm **(c)** 6 years **(d)** declines toward 0

67.(a)

(b) $d = \begin{cases} 15x, & 0 < x < 10 \\ 200, & 10 \leq x < 15 \\ 400, & 15 \leq x < 25 \\ 800, & 25 \leq x < 50 \\ 1200, & 50 \leq x < 100 \\ 1600, & x \geq 100 \end{cases}$

1.3 EXERCISES

1. $5x + 4$, $-x - 14$, $6x^2 + 3x - 45$, $\dfrac{2x - 5}{3x + 9}$; 14, −14, 18, $-\dfrac{7}{6}$; all x, $\{x \mid x \neq -3\}$ **3.** $-3x^4 - 16x^3 + 29x^2 - 98x + 130$, $\dfrac{x^2 + 6x - 10}{-3x^2 + 2x - 13}$; 42, $\dfrac{10}{13}$ **5.** $\dfrac{3x^2 + 9x + 10}{x^2 + 4x + 3}$, $\dfrac{-x^2 - 11x - 14}{x^2 + 4x + 3}$, $\dfrac{2x^2 - 8}{x^2 + 4x + 3}$, $\dfrac{x^2 - x - 2}{2x^2 + 10x + 12}$, $\dfrac{11}{4}$, $-\dfrac{1}{6}$; $\{x \mid x \neq -3, -1\}$, $\{x \mid x \neq -3, -1\}$

7. $\dfrac{x^3 - 6x^2 + 15x - 26}{-2x^2 + 2x + 4}, \dfrac{3x^3 - 8x^2 - 13x + 30}{-2x^2 + 2x + 4},$
$\dfrac{-6x^2 + 22x - 20}{x^3 - 7x - 6}$; $\{x|x \neq -1, 2\}, \{x|x \neq -1, 2\}, \{x|x \neq -2, -1, 3\}$ **9.(a)** $P = 30x - 10{,}000$ **(b)** $-\$7000$; $\$30$ **11.(a)** $P = -x^2 + 50x - 400, -\$175, \$0, \$200,$ $\$0, -\$1000$ **(b)** $C(b+1) - C(b) = 4b + 2, R(b+1) - R(b) = 2b + 51, P(b+1) - P(b) = -2b + 49$
13.

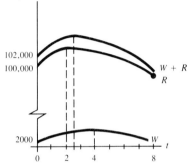

$t = 4, t = 2, t = 2.5$ **15.** $3x^2 + 18, 9x^2 + 36x + 40$
17. $x^4 + 4hx^3 + (6h^2 + 5)x^2 + (4h^3 + 10)x + h^4 + 5h^2 + 1, x^4 + 5x^2 + h + 1$ **19.** $\dfrac{1}{x+h}, \dfrac{1}{x} + h$
21. $\dfrac{-3x^2 + 10x + 5}{7x^2 + 10x - 5}, \dfrac{24x^2 + 68x - 56}{-40(x+1)}$ **23.** $y = 4t$
25. $t = y/4$ **27.** $f(g(x)), g(x) = 5 - 7x, f(x) = x^{-4}$
29. $f(g(x)), g(x) = 4x - 1, f(x) = \sqrt{x}$ **31.** $f(g(x)), g(x) = x^2 + 1, f(x) = \sqrt[3]{x}$ **33.** $f(g(x)), g(x) = 1 - 2x^{-1},$ $f(x) = x^{-3/2}$ **35.** $q = \dfrac{10{,}000}{0.02t + 5}, \approx 1953$ units **37.** $f(t) = \dfrac{100}{1.92t + 4.52}$
39. is a one-to-one function   **41.** is a one-to-one function

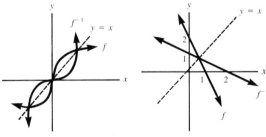

43. not a function **45.** one-to-one, $f^{-1}(x) = \dfrac{x+1}{3}$
47. one-to-one, $f^{-1}(x) = -x$ **49.** one-to-one, $f^{-1}(x) = \sqrt[3]{x}$ **51.** one-to-one, $f^{-1}(x) = x^2, x \geq 0$ **53.** $x = -3, 4$ **55.** $x = -4, 4$ **57.** $x = -2$ **59.** $x = -1, 4$
61. $x = -2, 0, 1$ **63.** $x = -\tfrac{3}{2}, 1$ **65.** $\tfrac{45}{8}$ sec **67.** 5

in **69.** $x = 30$ units **71.** $x = 85$ units **73.** \$88 or \$112
75. 2002 ($t = 12$)

1.4 EXERCISES

1.

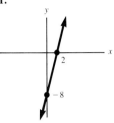

3.

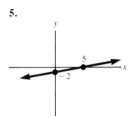

5.

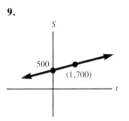

7.

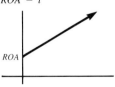

9.

\$500,000; increasing; by \$200,000

11.(a) 0.42 **(b)** $A = \dfrac{E(1-t)}{A}$
(c) ROE increases by
$ROA - i$ **(d)** $ROE = ROA$

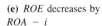

(e) ROE decreases by
$ROA - i$

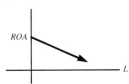

(f) *ROA* > *i*: increase *L* (by taking on more debt and/or reducing *OE*); increase *ROA* by decreasing asset value as a proportion of earnings; *ROA* = *i*: increase *ROA*; *ROA* < *i*: decrease *L* (by retiring debt and/or increasing *OE*); increase *ROA*
13. 100 dB; 1000 dB; increases by 2 dB

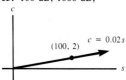

15.(a) 840 msec;

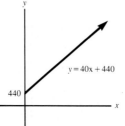

(b) 630 msec; **(c)** yes

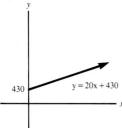

17. $y = -3x + 2$ **19.** $y = 6x - 16$ **21.** $y = 2x + 2$
23. $y = 2$ **25.** $C = \frac{5}{9}F - \frac{160}{9}$ **27.** $y = 0.6t + 59.5$
(a) in 25 years **(b)** in 75 years **29.** $y = 10t + 150$
(a) 220 **(b)** 1994 **31.** $y = 108t + 3320$ **(a)** 3,644,000 tons **(b)** 2nd quarter of 1994 ($t \approx 6.3$) **33.** $y = 200 - 4t$
(a) 112 **(b)** 39.75 days after arrival of the last order
35. $25x + 35y = 105,000$

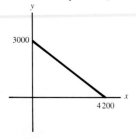

2450 small and 1250 large; 4200 small, 3000 large
37. $80x + 45y = 18,000$; no; yes; 225; 400 **39.(a)** $y = 2.1x + 53$ **(b)** 62.45 cm **(c)** 5.2 mo **41.** $80x + 125y = 2000$; 25; 12.5 **43.** $y = 2000x + 15,000$; sales increase by $2000 for every increase of 1 in test score; a score of 0 on the test corresponds to sales of $15,000 **45.** $y = 160t + 2240$ ($t = 0$ is last year) **(a)** 3200 **(b)** 4.75

years from now **47.** $(-4, -13)$ **49.** $(36, 20)$ **51.** none, lines are parallel **53.** $2500 in *A*, $7500 in *B*
55.(a) single processor: $C = 12x + 250,000$
multiprocessor: $C = 22x + 200,000$

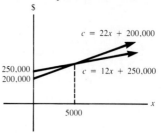

(b) $x > 5000$

57. Location Production Range Where Cheapest
  A $70,000 < x < 100,000$
  B $0 \le x < 20,000$
  C $20,000 < x < 70,000$
  D $x > 100,000$
59. 20.2 years **61.** 1995 **63.** $C(x) = 6x + 100,000$, $R(x) = 10x$, (25,000, $250,000)
65.

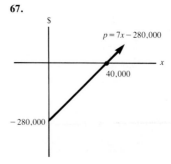

$\approx (2984.5, \$165,341.09)$

67.

$p = 7x - 280,000$

40,000 units

69.

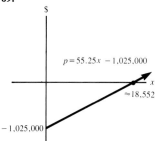

≈18,552 units

71.

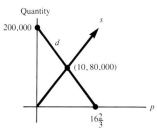

($10, 80,000); each $1 rise in price brings 8000 more units onto the market; no units are supplied to the market when the price is $0; each $1 rise in price causes quantity demanded to decrease by 12,000 units; when price is $0, 200,000 units are demanded; no units are demanded when price is ≈ $16.67.
73. ($8, 62,000) **75.** $S = 2500p - 10,000$; $d = -3000p + 120,000$; ≈ ($23.64, 49,091) **77.** supply: $q = 220p - 11,250$, demand: $q = -510p + 80,000$

1.5 EXERCISES

1.

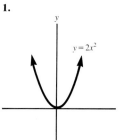

3.

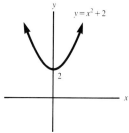

5.

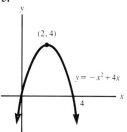

7.

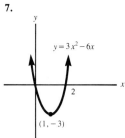

9.

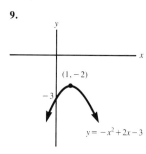

11.

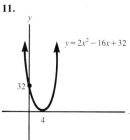

13.

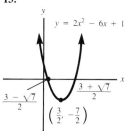

15.

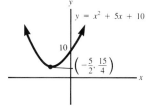

19.

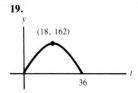

(a) 162 mi; 18 min after launch **(b)** 36 min **21.(a)** $R(x) = 150x$

(b)

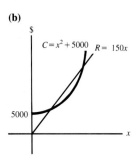

(c) $x = 50, 100$ **(d)** $P(x) = -x^2 + 150x - 5000$; max profit is $625 when $x = 75$ units

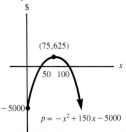

(e) $625 when $x = 75$

23.(a)

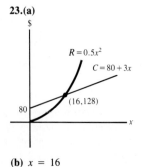

(b) $x = 16$

(c) $P(x) = 0.5x^2 - 3x - 80$

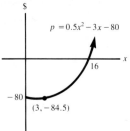

(d) no max profit

25.

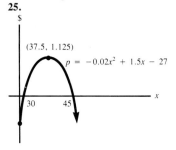

(a) $30 < x < 45$ **(b)** $1.125 million when $x = 37{,}500$

27.

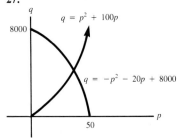

$40 **29.** $p = \$1.25$, $q = 5$ million **31.** $R(x) = 0.2x^2 + 2x$ **33.(a)** $9.6°$, $t = 30$ min **(b)** 70 min **35.** 32.4 mg; 36 hrs later; 72 hrs later **37.** $y = -x^2 - 12x + 100$; ≈ 5.7 days **39.** 4; $480,000 **41.** $y = -43.2t^2 + 57.6t + 60$, $t = $ yrs in office; 79.2 when $t = \tfrac{2}{3}$

1 REVIEW EXERCISES

1. true **2.** false **3.** false **4.** false **5.** true **6.** false **7.** false **8.** true **9.** true **10.** true **11.** true **12.** false
13. ○——○ x
     0   1
14. ●——● x
     0   1
15. ←——● x
         1

Solutions to Exercises

16. ———○———→ x
           0

17.(a) $p \le 25$ **(b)** $T \ge 20°$ **(c)** $P < \$2$ **(d)** $\$2 \le P < \$2.25$ **18.(a)** $243/4$ **(b)** $3/64$ **(c)** $2/125$ **(d)** 1331
(e) $\dfrac{1}{3600}$ **19.(a)** $\dfrac{x^2 z}{y^2}$ **(b)** $\dfrac{x}{y^{6/5} z^2}$ **20.** 5 **21.** -1 **22.** 8
23. $5 + 3a$ **24.** -14 **25.** $\dfrac{7}{4}$ **26.** 0 **27.** $2 - b$ **28.** 4
29. $-\dfrac{1}{2}$ **30.** 0 **31.** undefined **32.** \$155, \$59, \$47

33.(a)

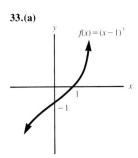

domain **R**, range **R**

(b)

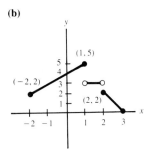

domain $[-2, 3]$, range $[0, 5]$

34. $2x^3 + 5x^2 - 3x - 2$, $-6x^3 + 5x^2 + 9x - 10$, $-8x^6 + 20x^5 + 24x^4 - 62x^3 + 2x^2 + 48x - 24$, $\dfrac{-2x^3 + 5x^2 + 3x - 6}{4x^3 - 6x + 4}$ **35.** $\dfrac{-x^3 + 5x^2 + 6x - 36}{-2x^2 + 12x - 18}$, $\dfrac{-x^3 + x^2 - 2x + 24}{-2x^2 + 12x - 18}$, $\dfrac{x^3 + 5x^2 - 2x - 10}{-2x^2 + 12x - 18}$, $\dfrac{-x^3 + 3x^2 + 2x - 6}{2x^2 + 4x - 30}$ **36.** $10x + 7$, $10x + 17$, $\dfrac{3x + 1}{x - 1}$, $\dfrac{2}{5x + 1}$ **37.** $C(w) = 4.5w + 6\sqrt{w + 1} + 106.5$
38. $f(g(x))$, $g(x) = 3x^2 + 5x + 1$, $f(x) = \sqrt[3]{x}$

39.(a) one-to-one, $f^{-1}(x) = 1/\sqrt{x}$;

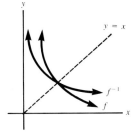

(b) not one-to-one **(c)** not one-to-one **(d)** one-to-one, $f^{-1}(x) = x/a - b/a$; graph of f^{-1} is a line that depends on signs of a and b **40.** $-8, 8$ **41.** $-8, 4$ **42.** 4 **43.** $-1, 0, 1$ **44.** $-3, 0, 4$ **45.** $-\dfrac{3}{2}, \dfrac{1}{2}$ **46.** \$23 or \$34

47.

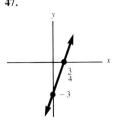

48.

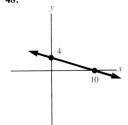

49.

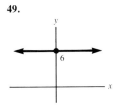

50.

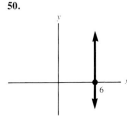

no slope

51. $y = \dfrac{1}{3}x + \dfrac{16}{3}$ **52.** $y = -3x - 1$ **53.** $y = 12$
54. $x = 5$ **55.** $y = -6000t + 120{,}000$; 20 hrs **56.** $S = 124{,}000t + 368{,}000$; \$1,360,000; 1992 **57.(a)** $(9, 33)$
(b) $\left(\dfrac{63}{26}, \dfrac{3}{13}\right)$ **58.** 2.5 lb and 1.4 lb/week **59.** $C = 46x + 92{,}000$; $R = 62x$; $P = 16x - 92{,}000$; $(5750, \$356{,}500)$
60. each \$1 rise in price brings 6250 units onto the market; when $p = \$2.50$ (or less), no units are supplied to the market; each \$1 rise in price causes quantity demanded to decrease by 8500 units; at a price of \$0, 183,500 units would be demanded; when $p \approx \$21.59$ (or more), no units will be demanded; \$13.50 **61.(a)** $x = 30, 70$

(b) $P(x) = -0.5x^2 + 50x - 1050$

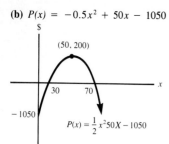

(c) $200 when $x = 50$ **62.(a)** $d = -1000p + 50,000$
(b) $P = -1000p^2 + 60,000p - 500,000$
(c) $400,000; $30

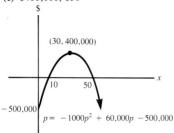

63. $q = 800p^2 - 4000p - 2000$, $p \approx \$5.458$

2.1 EXERCISES

1.

x	2.8	2.9	2.99	2.999	2.9999
$f(x)$	-1.16	-0.59	-0.0599	-0.005999	-0.0005999
x	3.2	3.1	3.01	3.001	3.0001
$f(x)$	1.24	0.61	0.0601	0.006001	0.0006000

$\lim\limits_{x \to 3} (x^2 - 9) = 0$

3.

x	-0.1	-0.01	-0.001	-0.0001	-0.00001
$f(x)$	-10	-100	-1000	$-10,000$	$-100,000$
x	0.1	0.01	0.001	0.0001	0.00001
$f(x)$	10	100	1000	10,000	100,000

$\lim\limits_{x \to 0} \dfrac{1}{x}$ does not exist.

5. does not exist **7.** 2 **9.** 2 **11.** does not exist **13.** 2
15. -1 **17.** does not exist **19.** -2 **21.** 2 **23.** 5
25. $\sqrt{3}$ **27.** 8 **29.** $\frac{1}{3}$ **31.** 8 **33.** 243 **35.** does not exist **37.** 14 **39.** 0 **41.** 18 **43.** -17 **45.** 25
47. $\frac{321}{16}$ **49.** -8 **51.** -36 **53.** does not exist **55.** $\frac{11}{3}$
57. -8 **59.** $\frac{1}{7}$ **61.** does not exist **63.** $\frac{16}{15}$

65.

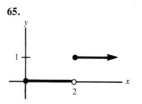

(a) does not exist
(b) 0 **(c)** 1

67.

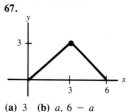

(a) 3 **(b)** $a, 6 - a$

69.

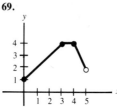

(a) 4 **(b)** 4 **(c)** does not exist **(d)** $1 + a$, 4, $12 - 2a$, 0

71.

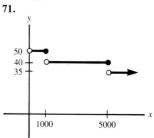

(a) does not exist, does not exist **(b)** 50, 40, 35

73.

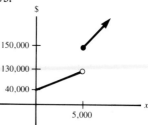

$40,000 + 18a$, does not exist, $50,000 + 20a$

75.

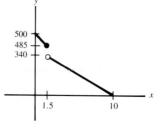

$500 - 10a$, does not exist, $400 - 40a$

2.2 EXERCISES

1. 1, −1, does not exist **3.** 2, 2, 2 **5.** 1, 2, does not exist, −1, 1, does not exist **7.** 2 **9.** 19 **11.** 0 **13.** 0 **15.** 0 **17.** 0, −1, does not exist **19.** 0, 0, 0 **21.** 1, −1, does not exist **23.** 20,000, 20,000, 31,000, 35,000 **25.** 10.50, 10.50, 46.50, 46.50 **27.** $+\infty, +\infty, x = 1$ **29.** $-\infty, -\infty, x = 3$ **31.** $+\infty, -\infty$ **33.** $-\infty, +\infty$ **35.** ½, ½ **37.** $-\infty, +\infty$ **39.** $+\infty, -\infty$ **41.** $+\infty, -\infty$ **43.** $300 billion, all can be removed at a cost of $300 billion **45.** $+\infty$, cannot remove as much as 50% **47.** $+\infty$, all cannot be obtained **49.** $\lim_{x \to +\infty} f(x) = -1$, $\lim_{x \to -\infty} f(x) = 2$ **51.** $\lim_{x \to +\infty} f(x) = 3$, $\lim_{x \to -\infty} f(x) = 3$ **53.** $\lim_{x \to +\infty} f(x) = +\infty$, $\lim_{x \to -\infty} f(x) = 0$ **55.** $+\infty$ **57.** 0 **59.** 0 **61.** 1 **63.** $-\frac{2}{3}$ **65.** $+\infty$ **67.** 0 **69.** $\frac{5}{11}$ **71.** $+\infty$ **73.** vertical: $x = \frac{2}{3}$; horizontal: $y = 2$; oblique: none **75.** vertical: $x = 3$; horizontal: $y = \frac{5}{2}$; oblique: none **77.** vertical: $x = 0$; horizontal: $y = 3$; oblique: none **79.** vertical: $x = \pm 2$; horizontal: $y = 0$; oblique: none **81.** vertical: $x = 0$; horizontal: none; oblique: $y = 2x$ **83.** 5; as production increases avg cost/unit approaches $5 **85.** 300; as production increases worker-hrs required to build a unit approache 300 **87.** 2; as time passes, numbers recalled approache 2

2.3 EXERCISES

1. 1 **3.** 3 **5.** $-2x$ **7.** $3x^2$ **9.** $-3/x^2$ **11.** $1 - x^{-2}$ **13.** $2(x + 1)^{-2}$ **15.** $0.5(x + 1)^{-1/2}$ **17.** $y = x$ **19.** $y = 3x + 2$ **21.** $y = -6x + 10$ **23.** $y = 3x + 2$ **25.** $y = -3x/4 + 3$ **27.** $y = 2$ **29.** does not exist **31.** does not exist **33.** $x^{-2/3}/3$ **35.** all are differentiable everywhere **37.(a)** $(-\infty, -1), (-1, +\infty)$ **(b)** $x = -1$ **(c)** $x = -1, x = 2$ **39.(a)** $(-\infty, +\infty)$ **(b)** none **(c)** $x = \pm 2$ **41.(a)** $(-\infty, +\infty)$ **(b)** none **(c)** none **43.(a)** $(-\infty, 1), (1, +\infty)$ **(b)** $x = 1$ **(c)** $x = \pm 1$ **45.(a)** $(-\infty, 2), (2, +\infty)$ **(b)** $x = 2$ **(c)** $x = \pm 2, x = 0$ **47.** continuous for all x **49.** continuous on $(-\infty, 2)$, $(2, +\infty)$, continuous where defined **51.** continuous on $(-\infty, -2), (-2, +\infty)$, continuous where defined **53.** continuous on $[0, 1), (1, 2), (2, 3]$, discontinuous at $x = 1$, $x = 2$ **55.** continuous on $[0, 3), (3, +\infty)$, discontinuous at $x = 3$ **57.** continuous on $(-\infty, 0), (0, +\infty)$, discontinuous at $x = 0$

59. $f(x) = \begin{cases} 0, & x = 0 \\ 12, & 0 < x \le 1 \\ 24, & 1 < x \le 2 \\ 36, & 2 < x \le 3 \\ 48, & 3 < x \le 4 \end{cases}$

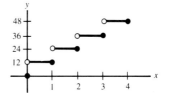

continuous on $(0, 1), (1, 2), (2, 3), (3, 4]$; discontinuous at $x = 0, 1, 2, 3$; nondifferentiable at $x = 0, 1, 2, 3, 4$ **61.** continuous on $(0, 1000), (1000, 5000), (5000, +\infty)$; discontinuous at $x = 1000$, $x = 5000$; nondifferentiable at $x = 1000$, $x = 5000$ **63.** continuous on $[0, +\infty)$; discontinuous nowhere; nondifferentiable at $p = \$1.50$ **65.**

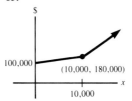

continuous on $[0, +\infty)$; discontinuous nowhere; nondifferentiable at $x = 10,000$

67. $C(x) = \begin{cases} 9.00x, & 0 < x < 10 \\ 8.60x, & 10 \le x \le 24 \\ 8.20x, & x > 24 \end{cases}$

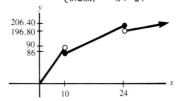

continuous on $(0, 10), (10, 24), (24, +\infty)$; discontinuous at $x = 10, x = 24$; nondifferentiable at $x = 10, x = 24$ **69.**

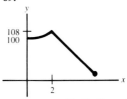

continuous on $[0, 6]$; discontinuous nowhere; nondifferentiable at $x = 2$

2.4 EXERCISES

1. 0 **3.** 0 **5.** $7x^6$ **7.** $-2x^{-3}$ **9.** $-5x^{-6}$ **11.** $7x^{2/5}/5$ **13.** $-2.7x^{-3.7}$ **15.** $x^{-4/5}/5$ **17.** 3 **19.** $-6x^2$ **21.** $3x^{1/2}$ **23.** $-2x^{-1/2}$ **25.** $6x^{-3}$ **27.** $-x^{-4/3}$ **29.** -8 **31.** $2x - 5$ **33.** $16x - 12$ **35.** $3x^2 - 14x + 9$ **37.** $4x^3 -$

$9x^2 - 6x + 5$ **39.** $-200x^{99} + 297x^{98}$ **41.** $1.5x^{-1/2} - 9x^{-5/2}$ **43.** $-4x^{-2} - 2x^{7/5}/5$ **45.** $2x^{-2/3} + 5x^{-1/2}/2 + 2x^{-3/2}$ **47.** $12x - 3$ **49.** $-x^{-2}$ **51.** $y = 5x - 3$ **53.** $y = -7x + 10$ **55.** $y = 12x + 15$ **57.** $y = -13x/3 + 6$ **59.** $y = x/2 - 1$

61.(a)

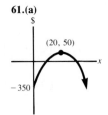

63.(a)

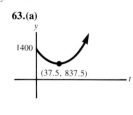

(b) $50,000 when $x = 20$ units **(b)** 837.5 units/day, $t = 37.5$ months **65.** $x = 10$, avg cost = $200/unit **67.** $t = 4$, $t = 8$ months (April, August)

69.(a)

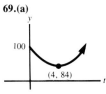

(b) 84 cubs in 1984 **71.** $(5, 56\frac{2}{3})$, $(2, 269\frac{1}{15})$ **73.** $(9, 6.7991)$, $(40, 3.82)$ **75.** ≈ -9.42, ≈ -11.11, ≈ -26.19; the approximate percentage changes brought about by a 1% rise in interest rates

2.5 EXERCISES

1. $v(t) = 40$ mph, $a(t) = 0$ mph² **3.(a)** 0.2, 6.4, 48.6 m **(b)** 0.5, 4, 13.5 mps **(c)** 0.75, 1.5, 2.25 mps² **5.(a)** 134, 206 fps **(b)** 24 fps² for both **7.** 50 ft, 0 fps, neither (at peak), -4 fps² **9.** 0 ft, -20 fps, is hitting the ground, -4 fps² **11.** 233.75 ft, 3 fps, up, -2 fps² **13.** 233.75 ft, -3 fps, down, -2 fps² **15.** 236 ft when $t = 4$ sec **17.** $s(t) = 16t^2$ **19.** 64 ft, 64 fps, 32 fps² **21.** 1.6 mps² **23.** 0.968 m, 1.76 mps, 1.6 mps² **25.** 15.488 m, 7.04 mps, 1.6 mps² **27.** 0.5, 5, 50 $/unit **29.(a)** $104,000, $-$1000/month, decreasing **(b)** $94,000, $-$5000/month, decreasing **31.** 12,000 units/$, 2,002,000 units/$ **33.(a)** 11 million units/$ **(b)** 11.5 million units/$ **(c)** $11\frac{2}{3}$ million units/$ **(d)** 11.75 million units/$ **(e)** 11.8 million units/$ **35.(a)** $216/unit **(b)** $72/unit **(c)** $0/unit **(d)** $-$8/unit **(e)** $-$8/unit **(f)** $0/unit **(g)** $27/unit **(h)** $72/unit **37.** $t = 1$: -63 units/wk², decreasing; $t = 4$: 0 units/wk², neither; $t = 6$: 12 units/wk², increasing; $t = 8$: 0 units/wk², neither; $t = 9$: -15 units/wk², decreasing **39.** $t = 0.5$: 0.24 gps; $t = 1$: 0.12 gps; $t = 1.5$: 0.08 gps; $t = 2$: 0.04 gps **41.(a)** 0.6 ppm/million cars **(b)** 2.4 ppm/million cars **(c)** 5.4 ppm/million cars **(d)** 6.3375 ppm/million cars **(e)** 7.35 ppm/million cars **43.** $t = 1$: 72,000/day, increasing; $t = 2$: 0/day, neither; $t = 3$: $-80,000$/day, decreasing; $t = 4$: $-108,000$/day, decreasing; $t = 5$: 0/day, neither; $t = 6$: 352,000/day, increasing **45.** $t = 5$: forgets at rate of 0.2/min; $t = 10$: forgets at rate of 0.4/min; $t = 15$: forgets at rate of 0.6/min; $t = 25$: forgets at rate of 1/min **47.** $t = 5$: 0.219/month, increasing; $t = 9$: 0/month, neither; $t = 30$: -0.126/month, decreasing; $t = 40$: 0, neither **49.** $t = 10$: -0.001/day; $t = 5$: -0.004/day; $t = 1$: -0.1/day

2.6 EXERCISES

1. $145,000, $180,000, $35,000 **3.** $600 **5.** $80,000, $x = 600$ **7.** $532,500, $4400 **9.** $40,000, $x = 60$ **11.** $100 **13.** $-$300 **15.** $120,000, $120,000, $0 **17.** $400 **19.** $22,500, $x = 350$, $p = 525 **21.** $2,850,000, $45,500 **23.** $3,795,500, $x = 110$, $p = $39,600$ **25.** $300,000, $500 **27.** $-$212,500, $x = 50$, $p = 6000 **29.** $3,750,000, $x = 1250$, $p = 9875 **31.** $180,000, $x = 100$, $p = 2500 **33.** marg avg cost: $-$2/unit; marg avg rev: $0/unit; marg avg profit: $2/unit **35.** marg avg cost: $-$195/unit; marg avg rev: $-$230/unit; marg avg profit: $-$35/unit **37.** -1.0; $p = 20$: 60,000, $1,200,000; $p = 19: 63,000, $1,197,000; $p = 21: 57,000, $1,197,000 **39.** -1.5; $p = 150: 10,000, $1,500,000; $p = 153: 9700, $1,484,100; $p = 147: 10,300, $1,514,100 **41.** $-\frac{2}{3}$; $p = 100: 15,000, $1,500,000; $p = 102: 14,800, $1,509,600; $p = 98: 15,200, $1,489,600 **43.** -1; $p = 20: 40,000, $800,000; $p = 20.02: 39,598, $799,880; $p = 19.80: 40,398, $799,880 **45.** -0.2, -5.0 **47.** $p = 300; $p = 315: 1425, $448,875; $p = 285: 1575, $448,875 **49.** elastic: $p > 14.50; inelastic: $0 < p < 14.50; unit elasticity: $p = 14.50 **51.** $E(p) = -p(900 - p)^{-1}$ **53.** elastic, so increase in price will decrease revenue, decrease in price will increase revenue **55.** $E(p) = -2p^2(7500 - p^2)^{-1}$ **57.** inelastic, so increase in price will increase revenue, decrease in price will decrease revenue **59.** elastic: $p > 2; inelastic: $0 < p < 2; unit elasticity: $p = 2; revenue will increase **61.(a)** accounting **(b)** neither **(c)** marketing **63.** $E(q) = (q - 160,000)/q$ **65.** $-\frac{1}{3}$ **67.** -7, $-\frac{1}{3}$ **69.** -0.3

2.7 EXERCISES

1. $2x - 5$ **3.** $3x^2 - 10x + 6$ **5.** $2.5x^{-1/2} - 4x^{-1/3}/3$ **7.** $21t^6 - 2t^{-3}$ **9.** $-2x^{-2} + 12x^{-4}$ **11.** 9 **13.** $\frac{19}{2}$ **15.** $\frac{265}{6144}$ **17.** $3a^2 + 4a$ **19.** $-\frac{3}{2}$, $-\frac{2}{3}$ **21.** $-0.25t^{-2}$, $-0.25y^{-2}$ **23.** $6x + 2$, 6 **25.** $60x^3 - 72x$ **27.** 720 **29.** $12x^2 - 24x$, $24x - 24$ **31.** $12x^{-4} - 2x^{-3}$, $240x^{-6} - 24x^{-5}$ **33.** $12x^{-5}$, $-60x^{-6}$ **35.** $-2520x^{-8}$ **37.** 0 **39.** 0 **41.** 8, -18 **43.** $1680a^4 + 24$

2 REVIEW EXERCISES

1. does not exist 2. $+\infty$ 3. $-\infty$ 4. 2 5. 2 6. 2
7. -1 8. 1 9. does not exist 10. 0 11. 0 12. 0
13. -1 14. -1 15. -1 16. 0 17. 0 18. 0
19. $+\infty$ 20. 1 21. 0 22. 4 23. 24 24. 13 25. $^{106}/_9$
26. -1 27. 1 28. does not exist 29. $^5/_2$ 30. $+\infty$
31. $+\infty$. 32. does not exist 33. $^5/_2$ 34. $+\infty$ 35. $^1/_5$
36. 0 37. $^7/_2$ 38. $+\infty$ 39. $^5/_7$ 40. $+\infty$ 41. $^1/_4$ 42. 0
43. $x = 2$, $y = -^3/_2$ 44. $x = 4$ 45. $x = 0$, $y = x - 3$
46. $15x^2$, $y = 135x + 270$ 47. $(3x)^{-2/3}$, $x = 0$
48.(a) $x = -1, 0, 3, 5$ (b) $x = -1, 3$ 49. continuous on $(-\infty, +\infty)$, no discontinuities, differentiable everywhere
50. continuous on $(-\infty, -5)$, $(-5, -4)$, $(-4, +\infty)$, no discontinuities, differentiable everywhere 51. continuous on $(-\infty, 0)$, $(0, 2)$, $(2, +\infty)$, discontinuous at $x = 0$ and $x = 2$, nondifferentiable at $x = -1, 0, 1, 2$ 52. 0
53. $\sqrt{3}$ 54. $6x + 6$ 55. $-6x^5 + 20x^4 - 36x^2 + 32x - 8$ 56. $-6x^{-4} + x^{-2}$ 57. $-8x^{-3} - 6x^{-4}$ 58. $2x^{-1/2}$
59. $4x^{-1/5} + x^{-3/2} - 2x^{-5/3}$ 60. $y = 2\sqrt{3}$ 61. $y = 18x - 34$ 62. $y = 3x + 12$ 63. $y = -x/2 + ^7/_2$
64. $y = 24x/5 - ^{128}/_5$ 65. $y = -x/729 + ^{34}/_{27}$
66. $t = 3$: vel $= 120$ mps, acc $= -10$ mps^2; $t = 6$: vel $= 90$ mps, acc $= -10$ mps^2; max altitude $= 1125$ m at 30 sec
67. $-\$96,000$/yr, $-\$48,000$/yr, $-\$48/\sqrt{2}$ thousand/yr $\approx -\$33,941.13$/yr 68. \$72, $x = 14$, $p = \$42$ 69.(a) $-^1/_3$, $-1, -3$ (b) \$1,800,000, \$2,400,000, \$1,800,000
(c) \$1,914,000, \$2,376,000, \$1,386,000 70. elastic: $p > \$20$; inelastic: $0 < p < \$20$; unit elasticity: $p = \$20$; decrease, increase 71. $8x^3 - 10x$, $24x^2 - 10$, 96
72. $x^{-1/2} - x^{-3/2}$, $-0.5x^{-3/2} + 1.5x^{-5/2}$ 73. $9x^2 - 16x^{-1/5}/5$, $18x + 16x^{-6/5}/25$, $\dfrac{2154}{125}$ 74. $4u + 3b + u^{-1/2}$, $4 - u^{-3/2}/2$ 75. $^8/_{11}$, $^{11}/_8$

3.1 EXERCISES

1. $12x + 7$ 3. $15x^2 + 2x - 10$ 5. $4x^3 + 4x$
7. $-4/(x + 2)^2$ 9. $-3/2\sqrt{x}$ $(\sqrt{x} - 2)^2$ 11. $-1/(2x - 1)^2$ 13. $-22/(5x - 1)^2$ 15. $(x^2 + 4x + 6)/(x + 2)^2$
17. $(7 + 6x - x^2)/(3 - x)^2$ 19. $4x(x^2 + 6x + 8)$
21. $-24/x^4$ 23. $2(5x^{-2} + 8x^{-3})$ 25. $-(3x^2 + 4x + 6)/(x - 2)^2$ 27. $\dfrac{2x^2 + 6x + 2}{x^4 - 2x^2 + 1}$ 29. $6x^5 + 5x^4 + 4x^3 + 6x^2 + 2x + 1$ 31. $-(10x^2 + 24x + 45)/3(2x - 1)^2$ $(x + 3)^2$ 33. $(-30x^3 - 79x^2 + 88x + 159)/(5x + 11)^2$
35. $(-3x^6 + 8x^5 - x^4 + 4x^3 + 4x^2 + 4x - 4)/(x^2 + 1)^2$ $(x^3 + 2)^2$ 37. $6x^5 + 5x^4 + 4x^3 + 6x^2 + 1$
39. $(-311x^2 - 162x - 31)/(7x + 2)^2$ $(x - 1)^2$
41. $t = 1$: $\approx \$3.306$ million/wk; $t = 2$: $\approx \$2.778$ million/wk; $t = 5$: $\approx \$1.778$ million/wk; total approaches \$40 million 43. $x = 1$: $\approx \$0.494$/unit; $x = 10$: $\approx \$0.004$/unit;

$x = 100$: $\approx \$0.00004$/unit; avg cost per unit approaches \$4
45.(a) $-\$11.944$/unit, $-\$4.625$/unit, $\approx -\$0.227$/unit
(b) cost per unit approaches $+\infty$ 47. $q = 5$: $= \$1.04$/million units; $q = 25$: $\approx \$0.295$/million units; $q = 100$: $\approx \$0.253$/million units; price approaches \$0.25 49. $t = 0$: ≈ 77.1 symbols/month; $t = 6$: ≈ 10.5 symbols/month; $t = 12$: ≈ 3.9 symbols/month; total number of symbols approaches 270 51. $x = 0$: 70%/repetition; $x = 2$: ≈ 7.78%/repetition; $x = 5$: ≈ 1.94%/repetition

3.2 EXERCISES

1. $30(3x + 5)^9$ 3. $28(5 - 7x)^{-5}$ 5. $10(x - 1)(x^2 - 2x)^4$
7. $100(10x - 3)(5x^2 - 3x + 1)^{99}$ 9. $-3/2\sqrt{1 - 3x}$
11. $2x/3(x^2 + 1)^{2/3}$ 13. $-3/x^2(1 - 2x^{-1})^{5/2}$ 15. $x^2(5x^2 - 1)(35x^2 - 3)$ 17. $x(2x^2 + 1)^{10}$ $(x^3 - 1)^8 (98x^3 + 27x - 44)$ 19. $x^2(4x + 3)^{-2/3} (40x/3 + 9)$ 21. $x(2x^3 - x - 1)/(x^2 - 1)(x^2 - 1)^{1/2} (x^3 - 1)^{2/3}$ 23. $-3(4\sqrt{x} - 3x^{-2})^{-4} (4x^{4/3} + 10)^{-4} (88x^{5/6}/3 + 8x^{-5/3} + 20x^{-1/2} + 60x^{-3})$ 25. $2x(x^2 - 1)/\sqrt{(x^2 + 1)(x^2 - 3)}$
27. $-16(x + 2)^3/(x - 2)^5$ 29. $(9 - x)/(x^2 + 3)^{3/2}$
31. $\dfrac{9x^{1/2}(1 + x^{3/2})^2}{(1 - x^{3/2})^4}$ 33. $8t - 16$ 35. $100t + 5$
37. $-9(v^2 - 1)^2/2\sqrt{1 - v}$ 39.(a) $-4/(4t - 5)^2$
(b) $750(15,625t^6 + 75t^2 - 1)(625t^4 + 1)$ 41. $24t + 142$
43. $-320/r^2\sqrt{320 + 2r^2}$ 45. $R'(1) = \$0.9405 = $ approximate revenue from selling 2nd unit, $R'(10) = \$0 =$ approximate revenue from selling 11th unit, $R'(20) = \$57 =$ approximate revenue from selling 21st unit; $P'(1) = \$0.1905 = $ approximate profit from making and selling 2nd unit, $P'(10) \approx -\$0.99 =$ approximate profit from making and selling 11th unit, $P'(20) \approx \$56 =$ approximate profit from making and selling 21st unit 47. \$4.5 million when $x = 2000$ units 49. $p = \$20$: $50/\sqrt{1000} \approx 1.58$ units/\$; $p = \$30$: $50/\sqrt{2000} \approx 1.19$ units/\$; $p = \$50$: $50/\sqrt{4000} \approx 0.79$ units/\$ 51. $\dfrac{34}{13^{3/2}18^{1/2}} \approx 171$ and $\dfrac{34}{50^{1/2}21^{3/2}} \approx 50$ individuals per yr; $\lim_{t \to +\infty} y = 2$, so population size will approach 2000 53. $t = 1$: $300,000/\sqrt{10,300} \approx 2956$ bacteria/hr; $t = 2$: $600,000/\sqrt{11,200} \approx 5669$ bacteria/hr; $t = 10$: $3,000,000/\sqrt{40,000} \approx 15,000$ bacteria/hr 55. ≈ -0.486% per year 57. $t = 8$: ≈ -0.017/election; $t = 9$: ≈ -0.015/election 59.(a) \$771.84/employee
(b) $-\$304.64$/employee

3.3 EXERCISES

1. $df = (6x - 2)dx$ 3. $dy = 8(x^3 - 5x + 2)^7 (3x^2 - 5)dx$ 5. $dy = (5x^4 + 8x^3 + 8x + 8)dx$ 7. $dy = [(-x^4 + 5x^2 + 2)/(x^3 + 2x)^2]dx$ 9. $\Delta f \approx 3x^2 \Delta x$, $\Delta f \approx 0.6$

11. $\Delta f \approx -x^{-2}\,\Delta x$, $\Delta f \approx -0.005$ **13.** $\Delta f \approx \Delta x/2\sqrt{x}$, $\Delta f \approx -0.025$ **15.** $\Delta f \approx x^{-2/3}\,\Delta x/3$, $\Delta f \approx -1.6$ **19.** $\Delta f/f \approx \Delta x/x$, ≈ 0.05, $\approx 5\%$ **21.** $\Delta f/f \approx 3\,\Delta x/x$, ≈ -0.071, $\approx -7.1\%$ **23.** $\Delta f/f \approx \Delta x/2x$, ≈ 0.0125, $\approx 1.25\%$ **25.(a)** $\Delta x/x$ **(b)** ± 0.01, $\pm 1\%$ **27.(a)** $2\,\Delta r/r$ **(b)** ± 0.02, $\pm 2\%$ **29.(a)** ± 0.000333, $\pm 0.0333\%$ **(b)** ± 0.0005, $\pm 0.05\%$ **31.** 0.03, 3% **33.** -0.03, -3% **35.(a)** within $\pm 1\%$ of the true value **(b)** within $\pm 2\%$ of the true value **37.(a)** within $\pm 0.05\%$ of the true value **(b)** within $\pm 0.033\%$ of the true value **39.(a)** $\Delta p \approx -2500q^{-3/2}\,\Delta q$ **(b)** $\Delta p \approx \$0.231$ **41.** $\approx \$750$ **43.(a)** $\Delta P/P \approx (-2x+80)\,\Delta x/(-x^2+80x-200)$ **(b)** 0.04, 4% **45.** ≈ -0.078, -7.8% **47.(a)** $\Delta y \approx -1200(400-120t)^{-1/2}$ **(b)** ≈ -120 **49.** ≈ -10 mg/cm^3 **51.(a)** $\Delta y \approx 100\Delta t/\sqrt{t}$ **(b)** ≈ 33 **57.** ≈ 0.235 or 23.5% **59.** by increasing debt; yes, if try to grow too fast can take on too much debt **61.** ≈ 0.075 or 7.5%

3.4 EXERCISES

1. $dy/dx = 1/2y$ **3.** $dy/dx = 2x/3y^2$ **5.** $dy/dx = -6x^2/(y-1)$ **7.** $dy/dx = 2x/3y^2$ **9.** $dy/dx = y/2x$ **11.** $dy/dx = y(x^2+y^3)/x(x^2+2y^3)$ **13.** $dy/dx = -x$ **15.** $dy/dx = \sqrt{y}\,(2y\sqrt{x}+x^2\sqrt{y}-4x^{5/2})/x^{3/2}(2\sqrt{y}-x^{3/2}-6x\sqrt{y})$ **17.** $y = 0.4x - 2$ **19.** $y = x$ **21.** $y = 2x - 4$ **23.** $x = 1$ **25.** $dy/dx = -25x/48 + 97/48$ **27.** $dy/dt = -(x/y)(dx/dt)$ **29.** $dy/dt = (2x/y)(dx/dt)$ **31.** $dy/dt = [(2x - y^2 + 2)/(2xy + 3)](dx/dt)$ **33.** $d^2y/dx^2 = -y^{-3}$ **35.** $d^2y/dx^2 = (4y^6 - 6y^3 - 18)/9x^2y^5$ **37.** $d^2y/dx^2 = (16x^3y^2 - 16x^6 - y^4)/4x^2y^3$ **39.(a)** -20 **(b)** -20 **41.(a)** -0.2 units/\$ **(b)** ≈ -0.105 units/\$

3.5 EXERCISES

1. 128π cm^3/sec **3.** 2400 cm^3/sec **5.** $2/14.7 \approx 0.136$ in^3/min **7.** 1.25 fps **9.** $-40\sqrt{2} + 30 \approx -26.6$ mph **11.** $20{,}000/\sqrt{38{,}400} \approx 102.1$ mph **13.** $-1/4\pi$ fps **15.** $\pm 5\sqrt{63}/2 \approx \pm 19.84$ mph ($-$ if ship is approaching observer, $+$ if it is receding from observer) **17.** $-\$103.50$/day **19.** -79.2/month **21.** \$29,963.40/year **23.** \$25/7 $\approx$ \$3.57 thousand/ad **25.** $3{,}840{,}000\pi$ ft^2/hr **27.** $dy/dt = 500/\sqrt{x}$ **29.** $0.4k$ mm/min

3 REVIEW EXERCISES

1. $30x^4 - 8x^3 + 48x - 8$ **2.** $-\dfrac{44}{(5x-3)^2}$ **3.** $\dfrac{4x^2 + 12x - 3}{2(2x+3)^2}$ **4.** $9x^2 + 7x^{5/2} - x^{-1/2}$

5. $\dfrac{5(2x^4 - 3x^2 - 4x + 3)}{(2x^2+1)^2}$ **6.** $(-5x^4 + 65x^2 + 20x + 20)/(5x+2)^2(x^2+1)^2$ **7.** $(6x^6 + 2x^5 + 10x^4 + 4x^3 + x^2 - 4x - 1)/(x^2+1)^2$ **8.** $(4x^7 - 8x^5 + 2x^4 + 4x^3 - 6x^2 + 2x)/(x+1)^2(x-1)^2$ **9.** $70(7x-9)^9$ **10.** $12(6x+5)(3x^2+5x+1)^{11}$ **11.** $(9x^2-1)/\sqrt{6x^3-2x+3}$ **12.** $-(5x^2+4x+10)/2(5x+2)^{1/2}(x^2-3)^{3/2}$ **13.** $4(2x+3)^5(x^2-5x)^7(11x^2-23x-30)$ **14.** $10(6x^2-2x+4)(2x-2)^{19}(x^2+1)^4$ **15.** $\dfrac{7(3x-1)^6(-3x^2+2x+6)}{(x^2+2)^8}$

16. $-\dfrac{5(3x+4)^4(9x^2+32x+3)}{(5x^2-1)^5}$ **17.** $384t^3 - 880t$

18. $\dfrac{dy}{dt} = 128t - 88$ **19.** $\dfrac{du}{dw} = 11\left(7 - \dfrac{3}{2\sqrt{w}}\right)(4v^3 - 2)(v^4 - 2v + 11)^{10}$ **20.** $\dfrac{118125(120)^{-1/4}}{2} \approx 17{,}845$ units/dollar **21.(a)** $\Delta p \approx -400{,}000(10q)^{-3/2}\,\Delta q$ **(b)** $-\$2$ **22.(a)** $\Delta p/p \approx -5(10q)^{-1}\,\Delta q$ **(b)** 0.004, 0.4% **23.** ± 0.005, $\pm 0.5\%$ **24.** within $\pm 0.5\%$ of the true value **25.** $dy/dx = 1/8y$ **26.** $dy/dx = (2xy - y^3)/(3xy^2 - x^2)$ **27.** $dy/dx = -(2xy^2 + y)/(2y^3 - x^2)$ **28.** $dy/dx = -(2x^3 + y)/(2xy - 1)$ **29.** $y = -x/2 + 3$ **30.** $y = 17x/2 + 57/2$ **31.** $d^2y/dx^2 = 2y/x^2$ **32.** $d^2y/dx^2 = (2y - 10xy^{3/2} + 6x^2y^2)x^{-2}$ **33.** $\dfrac{dy}{du} = \dfrac{y(2-3x^2y)}{2x(3x^2y+1)}\dfrac{dx}{du}$

34. -50 units/month **35.** $\dfrac{2815}{\sqrt{1489}}$ mph

3 SUPPLEMENTARY EXERCISES

1. in both cases, 0.6; \$0.60 of each additional \$1 of income is spent **3.(a)** 0.5; when $Y = \$400$ billion, \$0.50 of each additional \$1 of income is spent **(b)** 1/3; when $Y = \$900$ billion, 1/3 of each additional \$1 of income is spent **5.(a)** 1; when $Y = \$125$ million, each additional \$1 of income is spent **(b)** 0.25; when $Y = \$1000$ million, \$0.25 of each additional \$1 of income is spent **(c)** 0.0625; when $Y = \$8000$ million, \$0.0625 of each additional \$1 of income is spent **7.** ≈ 1.47; each additional \$1 of savings generates $\approx \$1.47$ additional income **9.(a)** 4/3; each additional \$1 of savings generates $\approx \$1.33$ additional income **(b)** 10/9; each additional \$1 of savings generates $\approx \$1.11$ additional income **11.** 0.4; of each additional \$1 of income, \$0.40 is saved **13.** $Y = \$400$ billion: 0.5; of each additional \$1 of income, \$0.50 is saved; $Y = \$900$ billion: 2/3; of each additional \$1 of income, $\approx \$0.67$ is saved **15.** $Y = \$125$ million: 0; of each additional \$1 of income, none is saved; $Y = \$1000$ million: 0.75; of each additional \$1 of income, \$0.75 is saved; $Y = \$8000$ million: 0.9375; of each additional \$1 of income, \$0.9375 is saved

4.1 EXERCISES

1. crit values: $x = -1, 1, 2, 3$; rel max: $(-1, 2), (2, 1)$; rel min: $(1, -1), (3, -2)$ **3.** crit values: $x = -2, 0, 2, 3, 4$; rel max: $(-2, 2), (2, 2), (4, 2)$; rel min: $(0, 0), (3, 1)$ **5.** crit values: $x = 2$; rel max: $(2, 2)$; no rel min **7.** decreasing: $(-\infty, 5)$; increasing: $(5, +\infty)$ **9.** increasing: $(-\infty, 0), (0, +\infty)$ **11.** decreasing: $(-\infty, -5), (2, +\infty)$; increasing: $(-5, 2)$ **13.** increasing: $(-\infty, 1/2), 5/2, +\infty)$; decreasing: $(1/2, 5/2)$ **15.** decreasing: $(-\infty, -2), (0, 2)$; increasing: $(-2, 0), (2, +\infty)$ **17.** increasing: $(0, +\infty)$ **19.** decreasing: $(-\infty, 0), (0, +\infty)$ **21.** decreasing: $(-\infty, 2), (2, +\infty)$ **23.** rel min: $(2, -1)$ **25.** rel min: $(-6, -18)$ **27.** rel max: $(0, 0)$; rel min: $(4/3, -32/27)$ **29.** rel min: $(-2, -16)$; rel max: $(2, 16)$ **31.** rel max: $(-3, 27)$; rel min $(1, -5)$ **33.** no rel max or min **35.** rel max: $(-6, 560)$; rel min: $(3, -169)$ **37.** rel min: $(-1, -1), (1, -1)$; rel max: $(0, 0)$ **39.** rel min: $(-2, -32), (1, -5)$; rel max: $(0, 0)$ **41.** rel max: $(-2, 74)$; rel min: $(2, -54)$ **43.** rel min: $(-3, -729), (3, -729)$; rel max: $(0, 0)$ **45.** no rel max or min **47.** rel min: $(-3, -1/6)$; rel max: $(-1, -1/2)$ **49.** rel max: $(-2, -4)$; rel min: $(0, 0)$ **51.** rel max: $(-1, -2)$; rel min: $(1, 2)$ **53.** rel max: $(-2, -32/3)$; rel min: $(2, 32/3)$ **55.** increasing; $[0, 18)$; decreasing: $(18, +\infty)$; max of 166.2 units when $t = 18$ yrs **57.** increasing: $[0, 20)$; decreasing: $(20, 40]$; max of $152,000 when spend $20,000 **59.** increasing: $[1, 6)$; decreasing: $(6, 12]$; max of $23/60 \approx 38.3\%$ when $t = 6$ **61.(a)** increasing: $[0, 5)$; decreasing: $(5, 10]$ **(b)** $25,000 when $p = $5 **(c)** inelastic: $[0, 5)$; elastic: $(5, 10]$; unit elasticity: 5 **65.** increasing: $[0, 2)$; decreasing: $(2, +\infty)$; max population of 2.5 million 2 months from now **67.** increasing: $[0, 3), (15, 20]$; decreasing: $(3, 15)$; max of 6.6281 million in 1993, min of 1.9625 million in 2005

4.2 EXERCISES

1. concave upward: $(-\infty, +\infty)$ **3.** concave upward: $(0, +\infty)$; concave downward: $(-\infty, 0)$ **5.** concave upward: $(2, +\infty)$; concave downward: $(-\infty, 2)$ **7.** concave upward: $(-\infty, +\infty)$ **9.** concave upward: $(-2, 3)$; concave downward: $(-\infty, -2), (3, +\infty)$ **11.** concave upward: $(-\infty, 2)$; concave downward: $(2, +\infty)$ **13.** rel max: $(3, 14)$ **15.** rel max: $(-6, 108)$; rel min: $(0, 0)$ **17.** rel max: $(-4, 521)$; rel min: $(9, -1676)$ **19.** rel min: $(-3, -108)$; rel max: $(7, 392)$ **21.** rel min: $(\sqrt{2}, 6), (-\sqrt{2}, 6)$; rel max: $(0, 10)$ **23.** rel max: $(1, 9), (3, 9)$; rel min: $(2, 8)$ **25.** no rel max or min **27.** rel min: $(0, -50), (2, -42)$; rel max: $(1, -31), (3, -23)$ **29.** rel min: $(0, 0)$ **31.** no rel max or min **33.** rel min: $(-3, -1/6)$; rel max: $(1, 1/2)$ **35.** $72/unit when $x = 30$ **37.** $280,000 at $x = 60$ **41.** 8000 when $t = 3$ **43.** 75/thousand in 1995

4.3 EXERCISES

1.(a) abs max: 3 at $x = 0$; abs min: -2 at $x = 3$ **(b)** abs max: 4 at $x = 6$; abs min: -2 at $x = 3$ **(c)** abs max: 3 at $x = 0$; abs min: none **(d)** abs max: none; abs min: -2 at $x = 3$ **3.** on $[0, 5]$: abs max: -5 at $x = 0$; abs min: -32 at $x = 3$; on $[6, 8]$: abs max: 43 at $x = 8$; abs min: -5 at $x = 6$; on $[0, +\infty)$: abs max: none; abs min: -32 at $x = 3$; on $(-\infty, +\infty)$: abs max: none; abs min: -32 at $x = 3$ **5.** on $[0, 6]$: abs max: 144 at $x = 6$; abs min: -16 at $x = 2$; on $[-1, 1]$: abs max: 11 at $x = -1$; abs min: -11 at $x = 1$; on $[-6, 6]$: abs max: 144 at $x = 6$; abs min: -144 at $x = -6$; on $(-\infty, +\infty)$: abs max: none; abs min: none **7.** on $[0, 12]$: abs max: 200 at $x = 0$; abs min: -8 at $x = 4$; on $[0, 5]$: abs max: 200 at $x = 0$; abs min: -8 at $x = 4$; on $[5, 12]$: abs max: 100 at $x = 10$; abs min: 0 at $x = 5$; on $(-\infty, +\infty)$: abs max: none; abs min: none **9.** on $(-\infty, 0]$: abs max: none; abs min: 0 at $x = 0$; on $[-3, 3]$: abs max: 177 at $x = -3$; abs min: -48 at $x = 2$; on $[0, 4]$: abs max: 128 at $x = 4$; abs min: -48 at $x = 2$; on $[0, +\infty)$: abs max: none; abs min: -48 at $x = 2$ **11.** on $(-\infty, 1)$: abs max: none; abs min: none; on $(1, +\infty)$: abs max: none; abs min: none; on $[0, 2]$: abs max: none; abs min: none; on $[2, 4]$: abs max: $1/3$ at $x = 4$; abs min: -1 at $x = 2$ **13.** on $(-3, 3)$: abs max: 0 at $x = 0$; abs min: none; on $(3, +\infty)$: abs max: none; abs min: none; on $[-2, 2]$: abs max: 0 at $x = 0$; abs min: $-4/5$ at $x = 2, -2$; on $(-\infty, +\infty)$: abs max: none; abs min: none **15.** 30 mph **21.** 160 ft by 80 ft **23.** 120 m **25.** no cut, all into square **27.** $5 - 3^{-1/2} \approx 4.42$ miles from A **29.** base 2 ft by 4 ft, height 40/9 ft **31.** radius $= 3/\sqrt[3]{\pi}$ in, height $= 6/\sqrt[3]{\pi}$ in **33.(a)** $66,400 **(b)** $54,000 **35.** $50,000,000, $500/unit; $48,000,000, $600/unit **37.** 200 boxes, 1 order every other week, $52 **41.** $605,000 when sell 11,000 tickets at $55 each **43.** make 60 regular and 70 deluxe for max profit of $1310 **45.** max revenue of $1920 when price is $16/bouquet **47.** at 10-yr mark **49.** 34,225 eggs, 185 chickens **51.** max of ≈ 1333 in 1985, min of ≈ 771 in 1970 **53.** velocity is greatest on the central axis

4.4 EXERCISES

1.

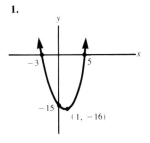

3.

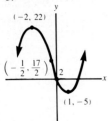

5.

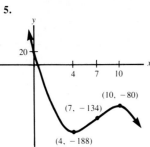

7.

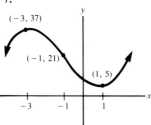

9.

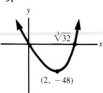

11.

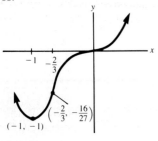

13.

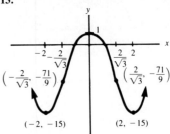

15.

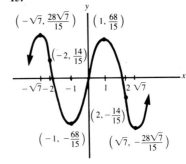

17.

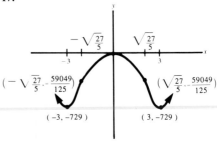

19.

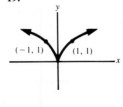

21.

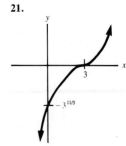

23.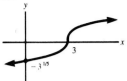

25.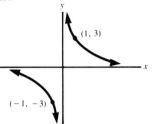

27.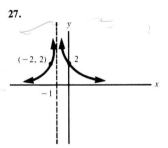

29.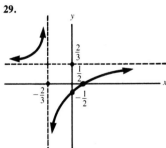

31.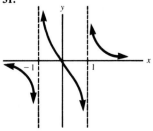

33.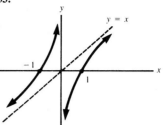

35. **37.**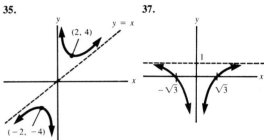

41.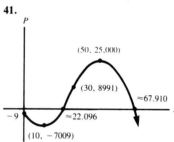

43.

45.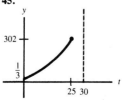

47.
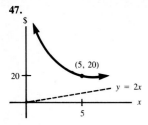

49.
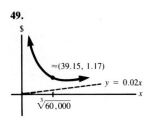

51.

53.
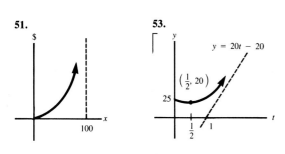

55.
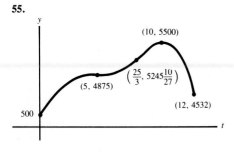

57.

59.(a)
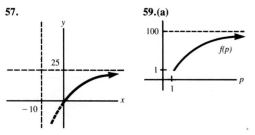

59.(b)   **(c)**
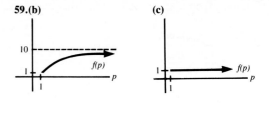

4 REVIEW EXERCISES

1. rel max: (1, 5), (5, 3), (7, 5); rel min: (0, 4), (4, 2) **2.** $x = -1, 0, 1, 4, 5, 6, 7, 8$ **3.** increasing: (0, 1), (4, 5), (6, 7); decreasing: $(-\infty, -1)$, $(-1, 0)$, (1, 3), (3, 4), (5, 6), (7, 8), $(8, +\infty)$ **4.** concave upward: $(-\infty, -1)$, (3, 4.5), (5.5, 6), (6, 7), (7, 8); concave downward: $(-1, 0)$, (0, 3), (4.5, 5.5), $(8, +\infty)$ **5.** pts of inflection: $(-1, 5)$, (4.5, 2.5), (5.5, 2.5), (8, 4) **6.** rel max: (7/2, 33/4) **7.** rel max: $(-10, 200)$; rel min: $(6, -1848)$ **8.** rel min: $(-3, -27)$ **9.** rel max: $(-\sqrt{8}, 200 + 448\sqrt{8})$; rel min: $(\sqrt{8}, 200 - 448\sqrt{8})$ **10.** no rel max or min **11.** rel min: $((-1 - \sqrt{33})/2, -2\sqrt{33}/(33 + \sqrt{33}))$; rel max: $((-1 + \sqrt{33})/2, 2\sqrt{33}/(33 + \sqrt{33}))$ **12.** $78.50/unit **13.** 7/12 of the trees **14.** rel max: (9/4, 57/8) **15.** rel max: $(-5, 100)$; rel min: $(1, -8)$ **16.** rel max: (0, 1); rel min: $(3/2, -65/16)$, $(-3/2, -65/16)$ **17.** rel max: (0, 1); rel min: $(2, -43/5)$ **18.** rel min: (0, 0) **19.** rel min: (0, 0) **20.** $8,000,000,000 **21.** abs max: 4 at $x = 5$; abs min: 1 at $x = 1$ **22.** abs max: 3 at $x = 3$; abs min: 1 at $x = 1$ **23.** abs max: 4 at $x = 5$; abs min: -2 at $x = -2$ **24.** abs max: 3 at $x = 3$; abs min: -2 at $x = -2$ **25.** abs max: 4 at $x = 5$; abs min: none **26.** abs max: 4 at $x = 5$; abs min: none. **27.** abs max: 0 at $x = -3$; abs min: none **28.** abs max: 2 at $x = 0$; abs min: none **29.(a)** abs max: none; abs min: -256 at $x = \pm 4$ **(b)** abs max: 0 at $x = 0$; abs min: -31 at $x = \pm 1$ **(c)** abs max: 144 at $x = 6$; abs min: -256 at $x = \pm 4$ **(d)** abs max: 144 at $x = 6$; abs min: -256 at $x = 4$ **30.** $10,000, $64,000 **31.** 20 cars, $4,840,000 **32.** radius = $1/\sqrt{\pi}$ ft, height = $2/\sqrt{\pi}$ ft

33.
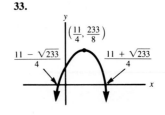

34.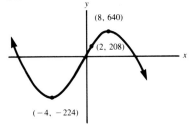

35.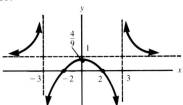

36.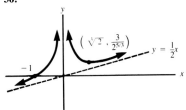

37.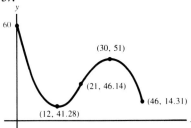

38.

39.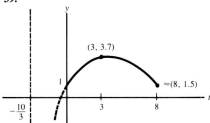

4 SUPPLEMENTARY EXERCISES

1.

3.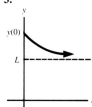

5.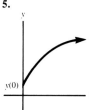

7.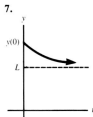

5.1 EXERCISES

1.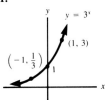

3.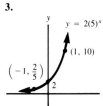

5.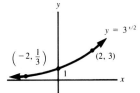

7.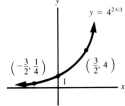

9.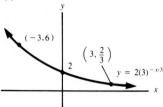

11.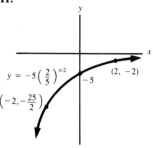

13.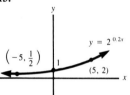

15.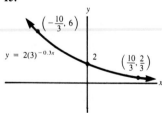

17.(a)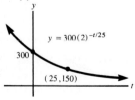

(b) $300/\sqrt{2} \approx 212.132$ g (c) 25 yrs (d) 0; in the long run the isotope disappears **19.** $16,200 **21.** $9004.07 **23.** $12,661.08 **25.** $9505.12 **27.** $23,654.80 **29.** $34,946.25 **31.** $A = 22.5(2)^{t/8}$, $t = 0$ represents 1985; $45.00 **33.** $D = \$10,000(0.6)^{t-1}$, $3600, $1296 **35.** $D = \$80,000(14/15)^{t-1}$, $80,000, $42,995.30, $15,274.70 **37.** present value $-$ cost $\approx \$0.412$ million for A, $0.063 million for B, $-\$0.394$ million for C, so invest in A **39.** 1.061837 **41.** 2.117 **43.** 0.367879 **45.** 0.030197

47. **49.**

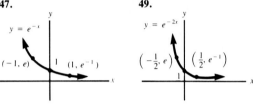

51. **53.**

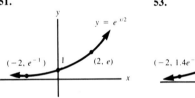

55.

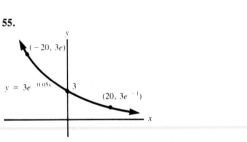

57. $y = 3e^{0.2772t}$ **59.** $y = e^{-2.0793x}$ **61.** $y = 2e^{-8.7888x}$

63.(a)

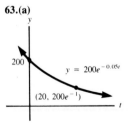

(b) ≈ 121.306 yr (c) ≈ 13.862 yr (d) 0; in the long run y approaches 0 **65.** $10,832.87 **67.** $20,247.88

69.

Compoundings/yr	Future Value of $10,000 at 6% for 1 yr
1	$10,600.00
2	$10,609.00
4	$10,613.64
12	$10,616.79
continuous	$10,618.37

71. $7408.18 **73.** $67,032.01

75.

Compoundings/yr	Present Value of $10,000 at 9% for 1 yr
1	$9174.31
2	$9157.30
4	$9148.43
12	$9142.38
continuous	$9139.31

77.(a)

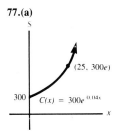

$300,000 **(b)** $447,547.41 **79.** $S = 5e^{0.0692t}$
81. present value $-$ cost $\approx$ $39,500 for X, $55,800 for Y, so buy Y **83.** present value $-$ cost $\approx -$683 for X, $-$12,000 for Y, so buy neither
85.(a)

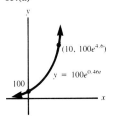

(b) $\approx$99,227, $\approx$989,713

5.2 EXERCISES

1. 4, 7, $-$2, 2, $-$3 **3.** 0, 1, 0, 1, $-$1 **5.** 16 **7.** 6
9. 13 **11.** $-$3 **13.** 5.1 **15.** 4.25 **17.** 125 **19.** e^3
21. 4/9 **23.** $-$3

25.

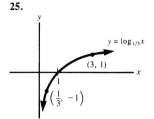

27.

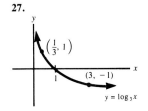

29. $y = e^{2.30259t}$ **31.** $y = 2e^{7.78364x}$ **33.** $y = 1.4427(\ln x)$
35. $y = -6.21335(\ln x)$ **37.(a)** 2 **(b)** 4 **(c)** 7
39. between 10^8 and 10^9 times
41.

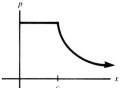

when income is $\leq c$, 100% do not purchase the good within 1 yr; as income grows past c, the percentage who do not purchase it within 1 yr declines toward 0 **43.** $15,000: 100%; $30,000: $100e^{-0.5}$% $\approx$ 60.7%; $50,000: $100e^{-1.5}$% $\approx$ 22.3% **45.** For a given income greater than c, the proportion who do not purchase the good within 1 yr becomes smaller as a increases **47.** $x = -0.7993$
49. $x = 1.73287$ **51.** $x = 1.58496$ **53.** $x = 4.09564$
55. $x = 5.89097$ **57.** $y = 2^{-6.64386x}$ **59.** $y = 1.09861 (\log_3 x)$ **61.** 11.91917 $(\log_{11} x)$ **63.** 34.6574 days, 115.1293 days **65.** 7.7016 yrs **67.** 18.4472 yrs
69. 18.3258% **71.** 7.7494% **73.** 13.5155% **75.** $t = 32.189$ (A.D. 2017) **77.** 13.288 hrs **79.** 198.97 yrs after 1988 (A.D. 2187); 352.48 yrs after 1988 (A.D. 2340)
81. IRR $= (A/P)^{1/t} - 1$ **83.** $i \leq 10.24948$
85. 0.0570086% **87.** 0.0302778%

5.3 EXERCISES

1. $3/x$ **3.** $7/(7x + 2)$ **5.** $(2x + 5)/(x^2 + 5x)$ **7.** $16x(x^2 + 1)/(x^4 + 2x^2 + 200)$ **9.** $2(\ln 3x + 1)$ **11.** $2x^2(3 \ln x + 1)$ **13.** $2/(2x + 1) - 3/(3x + 2)$ **15.** $\ln(6/5)/x(\ln 6x)^2$

17. $3(\ln x)^2/x$ **19.** $(3x^2 + 6x + 4)[\ln(x^3 + 3x^2 + 4x + 3)]^{-1/2}/2(x^3 + 3x^2 + 4x + 3)$ **21.** $1/x(\ln x)(\ln(\ln x))$

23.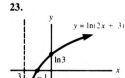

25.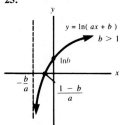

27.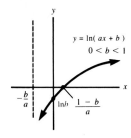

29. 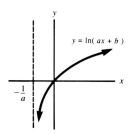

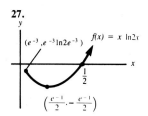

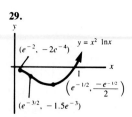

31.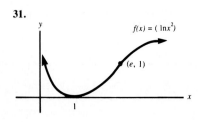

33.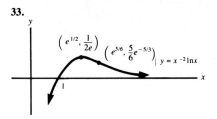

35. $1/(\ln 2)x$ **37.** $3/(\ln 7)x$ **39.** $8(x + 2)/(\ln 3)x(x + 4)$ **41.** $2x[\log_{10}(2x - 1) + x/(\ln 10)(2x - 1)]$ **43.** $5/24 \approx 0.20833$ million \$/yr; $5/114 \approx 0.04386$ million \$/yr **45.** $48/\sqrt{15}\,(\sqrt{15} + 3) \approx 1.80323$ million tons/yr, $48/\sqrt{25}\,(\sqrt{25} + 3) = 1.2$ million tons/yr, $48/\sqrt{35}\,(\sqrt{35} + 3) \approx 0.90998$ million tons/yr

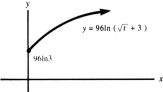

47. $100e^{-1} + 20 \approx \$56.78794$ million, when $t = 10(e - 0.1) \approx 26.182$ (A.D. 1976) **49.** $2/7$ pH/yr, $1/6$ pH/yr

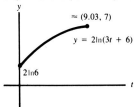

51. $36(10^{12/5})/(20 + 60(10^{2/5})) \approx 53$ persons/yr

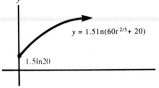

53. $dr/dx = 1/(\ln 10)x$ **55.** $f(x)[1/x + 9x^2/(x^3 + 1) + 25x^4/(x^5 + 1)]$ **57.** $f(x)[6x^2/(x^3 + 1) + 6x/(x^2 + 1) - 8(x + 2)/(x^2 + 4x + 1)]$ **59.** $2x^{2x}(\ln x + 1)$ **61.** $-x^{-x}(\ln x + 1)$ **63.** $2x^{2(x+1)}[(x + 1)/x + \ln x]$ **65.** $(\ln x)^x[\ln(\ln x) + 1/\ln x]$

5.4 EXERCISES

1. $2e^{2x}$ **3.** $-2e^{-x}$ **5.** $(e^x + e^{-x})/2$ **7.** $5(2x + 3)e^{x^2+3x+4}$ **9.** $6x^2\, e^{x^3}$ **11.** $3x^2$ **13.** $2e^{-4x}(1 - 4x)$ **15.** $e^x(x - 1)/x^2$

17. $e^x(\ln x + 1/x)$ **19.** $e^{2x}(2\ln 3x + 1/x)$ **21.** e^{1+ex}
23. ex^{e-1}

25.
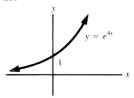

27.
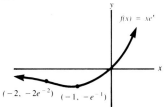

29.
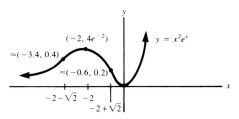

31.
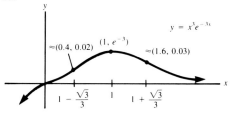

33.
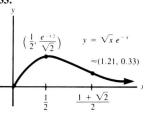

35.
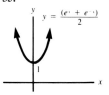

37.
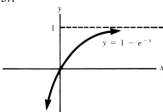

39.
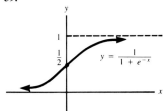

41.
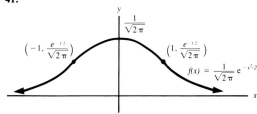

43. $(\ln 2)2^x$ **45.** $-6(\ln 5)5^{-2x}$ **47.** $x5^x(2 + x \ln 5)$
49. $3^{2x}[(2\ln 3)/x - 1/x^2]$ **51.** \$1193.46/yr, \$1780.43/yr
53. rPe^{rt}, $r = i/100$ **55.** -20 g/min, -7.358 g/min
57. $-\$7357.59$/yr, $-\$2706.71$/yr
59. $-\$0.110$/unit, \$0.027/unit, \$0.137/unit

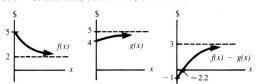

61. ≈ 21.75 days **63.** $10/9 \approx 1.111$ μg/L, $-(4/9)\ln 3 \approx -0.488$ μg/L **65.** $1000(\ln 5) \approx 1609$, $1000(\ln 5)5^{-1.2} \approx 233$, $1000(\ln 5)5^{-2.4} \approx 34$

67.

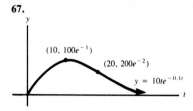

max is $100e^{-1} \approx 36.8$ mg/mL when $t = 10$ **69.** ≈ 1.099 students/hr, ≈ 213.761 students/hr, ≈ 68.781 students/hr   **71.** yes; 9 yrs   **73.** the value of t for which $dI/dt = rI$

5.5   EXERCISES

1. $y = 10,000e^{0.08t}$ **3.** $y = P_0e^{rt}$, $r = \dfrac{\ln(P_n/P_0)}{n}$ **5.(a)** $y = ce^{-0.043t}$ **(b)** $\approx 65.1\%$ **(c)** ≈ 53.5 days   **7.** $y = ce^{-rt}$, $r = 0.6931/T$ **9.(a)** $y = ce^{-0.0001209t}$ **(b)** ≈ 2379.5 yrs old   **(c)** $\approx 13,312$ yrs old   **11.(a)** $y = 100,000e^{0.4t}$ **(b)** \$1.644 million   **(c)** mid-1992 ($t \approx 12.5$)   **13.(a)** $y = 2.5e^{-0.193t}$ **(b)** \$0.363 million   **(c)** end of 1992 ($t \approx 11.9$)   **15.** 2008 ($t \approx 28$)   **17.** $q = 20,000e^{0.04p}$, $p \approx \$17.88$ **19.** $\approx 211,458$ **21.** $V = 0.0267e^{-0.1438t}$, $\approx 0.634\%$

23.

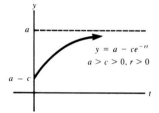

25. $y = 200 - 140e^{-0.113t}$; ≈ 7.5 weeks

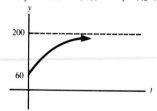

27. $y = 100,000 - 80,000e^{-0.144t}$ **29.** $y = 120 - 62.5e^{-0.223t}$, y in thousands; 120,000 units/yr   **31.** $y = 4000 + 1828.57e^{-0.1335x}$

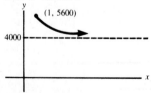

$\approx 21,296.5$ hrs

33. $y = 2750 + 2082e^{-0.51x}$, 2750 hrs   **35.** $y = 1200 - 1140e^{-0.082t}$, ≈ 698 **37.** $y = 100(1 + e^{-0.255t})$, $t = 0$ represents 1988   **39.(a)** $y_c = 100(1 - e^{-0.05x})$ **(b)** $y_n = 100(1 - e^{-0.05(x-30)})$
(c)

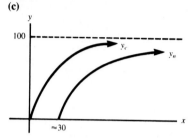

No

41. $y = 100 - 78e^{-0.0094t}$, $t = 0$ represents 1970; $t \approx 47.3$ yrs (A.D. mid-2017)   **43.(a)** $y_L = 100(1 - e^{-0.086x})$ **(b)** $y_U = 100(1 - e^{-0.136x})$ **(c)**

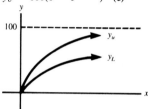

between 15.8% and 23.8%

45. 8500, 7225, 6141.25, 5270.0625 hrs   **47.** $y = 10,000x^{-0.152}$ **49.** $y = 5000x^{-0.184}$, ≈ 3273 worker-hrs   **51.** $y = 5000x^{-0.3221}$, 80%   **53.**

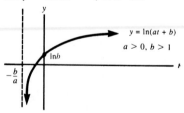

55. $y = \ln(9.43x + 2.6)$, x, y in millions of \$ **57.** $y = \ln(20x + 3)$, x in thousands of units, y in millions of \$; $\approx \$5,313,000$; $\approx 1,101,000$ units   **59.** $y = \ln(1.64t + 7.39)$, $t = 0$ represents 1988   **61.** $y = \ln(23.231x - 7.134)$, x in millions of students, y in billions of \$; $\approx \$1.92$ billion   **63.(a)** $y = \ln(0.52x + 23.58)$ **(b)** ≈ 3.90 **(c)** ≈ 1.8 kilotons

5 REVIEW EXERCISES

1.

2.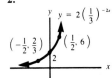

3.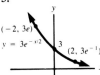

4.

6. $y = 12(3)^{t/20}$ **7.** $6259.23; $6304.69 **8.** $14,684.73; $14,425.31 **9.** 1 **10.** 0 **11.** 6 **12.** 200 **13.** -6 **14.** -35 **15.** $\log_7 22$ **16.** 5.2 **17.** 4.2 **18.** 1/25 **19.** -5 **20.** 13 **21.** 1.7481 **22.** -0.531

23.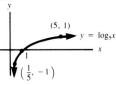

24.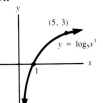

25.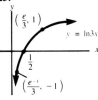

26.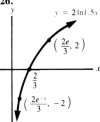

27. $f(x) = e^{3.8918x}$ **28.** $y = 7^{-3.55x}$ **29.** $y = 0.7797(\ln 3x)$ **30.** $g(x) = 0.774(\log_5 x)$ **31.**(a) 1 (b) 3 (c) 6 (d) ≈ 6.4 (e) 10^{14} **32.** $x = -0.536479$ **33.** $x = 0.623950$ **34.** $x = -0.764010$ **35.** $x = -225.98998$

36.(a)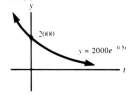

(b) ≈ 1213 (c) ≈ 6.9 days **37.**(a) 87.5% (b) 16 months **38.** 5.36479% **39.** 5.3768% **40.** 10.9861 yrs **41.** $3(x^2 + 1)/(x^3 + 3x - 2)$ **42.** $4x/(x^2 + 1)$ **43.** $2/x - 2/(2x - 1)$ **44.** $[(2/x)\ln(2x - 1) - (4/(2x - 1))\ln x]/(\ln(2x - 1))^2$ **45.** $10e^{5x}$ **46.** $4xe^{-x}(2 - x)$ **47.** $2e^{x^2}(2x \ln x + 1/x)$ **48.** $2e^x/(e^x + 1)^2$ **49.** $-2e^{-2x}e^{e^{-2x}}$ **50.** $-x^{1-x^2}(2 \ln x + 1)$ **51.** $4/(\ln 2)(4x - 3)$ **52.** $e^x(\dfrac{-2}{\ln 10} + \log_{10} e^{-2x})$ **53.** $-20(\ln 2)2^{-5x}$ **54.** $x(10^{-3x})(2 - 3x \ln 10)$

55.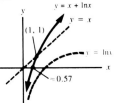

56.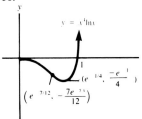

57.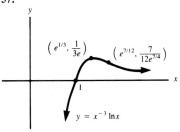

58.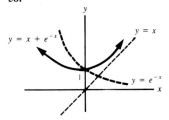

59.

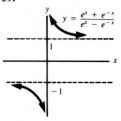

60.

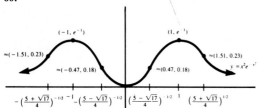

61. $\frac{3}{16}(\ln 8)^{-1/2} \approx \$0.13$ million/yr **62.** $-40e^{-1} \approx -14.7$ ppm/yr **63.** ≈ -3678.6/hr **64.** $y = 59.5e^{0.00263t}$ (a) $\approx 63.5°F$ (b) ≈ 266.7 yrs **65.** $y = 2000e^{-0.1t}$

66.(a)

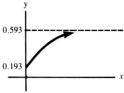

(b) will increase toward 0.593 (c) will decrease toward 0.193

67. $y = 2400 - 800e^{-0.1335t}$

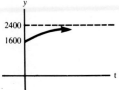

in ≈ 5.2 months

68. $y = 250 + 173e^{-0.1431x} \approx 1490$ hrs, $\approx 26/3$ units

69. $y = 200 - 87e^{-0.0297t}$, midday Aug 19th ($t \approx 18.6$)
70. $y = 237.5 + 129.8e^{-0.1431t}$ **71.** $y = \ln(0.1763t + e)$, y in hundreds of thousands of \$, $t = 0$ representing 1980; A.D. mid-2033 ($t \approx 53.7$)

5 SUPPLEMENTARY EXERCISES

3.(a)

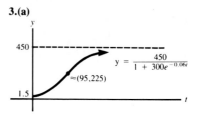

(b) $\approx \$1.5$ billion, $\approx \$449$ billion (c) $\approx$ A.D. 1895

5.(a)

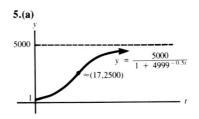

(b) ≈ 17 days after it started (c) $5/12 \approx 41.7\%$

9.(a)

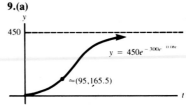

(b) ≈ 0; $\approx \$449$ billion (c) $\approx$ A.D. 1895

13.(a)

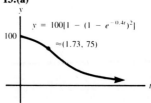

(b) 100; ≈ 25.2, ≈ 3.6 (c) $t \approx 7.42$ mins

15.

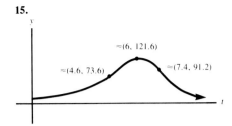

17.

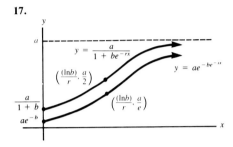

6.1 EXERCISES

1. $3x + c$ **3.** c **5.** $x^{11}/11 + c$ **7.** $-t^{-3}/3 + c$
9. $2x^{5/2}/5 + c$ **11.** $x^4/2 + c$ **13.** $-3x^{-5}/5 + c$
15. $4t^{1/2} + c$ **17.** $2\ln|p| + c$ **19.** $3\ln(-t) + c$
21. $2e^x + c$ **23.** $x^2/2 + 2x + c$ **25.** $2v^3/3 - 3v^2/2 + v + c$ **27.** $x^4/2 - 2x^3 + 9x^2 - 3x + c$ **29.** $-x^4 + x^3 - x^2 + 7x + c$ **31.** $6e^x + x^2/2 + c$ **33.** $7x^3/3 - 7x^2 + 2\ln|x| + 5x^{-2}/2 + c$ **35.** $x^{100} + x^{99} + x^{98} + \cdots + x^2 + x + c$ **37.** $2x^{3/3} + 7x^2/2 - 15x + c$ **39.** $-(4 - x)^4/4 + c$ **41.** 600 km **43.(a)** 1000 fps **(b)** 5000 ft
45. $C(x) = x^2 + x + 5000$, $R(x) = 15x$, $P(x) = -x^2 + 14x - 5000$ **47.** $P(x) = x^3/3 - 50x^2 + 1600x - 18$
49. $q = -25p^2 + 275,000$ **51.** $F(t) = 2t^{3/2}/3 + 4t + 2$
53. $F(t) = 100e^{-t} + 300$ **55.** $F(t) = 2\ln|t + 1| + 80$

6.2 EXERCISES

1. $2(x + 1)^{3/2}/3 + c$ **3.** $(4x - 3)^{3/2}/6 + c$ **5.** $(6x - 2)^{7/3}/14 + c$ **7.** $(7x - 3)^{13}/91 + c$ **9.** $-5(2x + 3)^{-1/2} + c$ **11.** $-3(4 - 7x)^4/14 + c$ **13.** $-\ln|1 - 2x| + c$ **15.** $-4\ln|8 - 3x|/3 + c$ **17.** $2e^{3x}/3 + c$
19. $-3e^{-7x}/7 + c$ **21.** $(x^2 + 1)^5/10 + c$ **23.** $(2x^5 + 10)^2/20 + c$ **25.** $-5(2 - x^2)^{3/2}/3 + c$ **27.** $(x^2 + 3x + 1)^5/5 + c$ **29.** $\ln(x^2 + 1) + c$ **31.** $-7\ln|t^3 + 2|/3 + c$
33. $3e^{x^2}/2 + c$ **35.** $e^{2x^2+2x}/2 + c$ **37.** $e^{e^x} + c$ **39.** $(\ln x)^3/3 + c$ **41.** $-(\ln x)^{-1} + c$ **43.** $C(x) = (x^2 + 1)^{3/2} + 999$, $R(x) = (x^2 + 1)^{7/4} - 1$ **45.** $P(x) = (2x^2 + 9)^{1/2}/2 + 48$ **47.** $F(t) = -100e^{-2t} + 200$ **49.** $F(t) = -10(2t^2 + 1)^{-1/2} + 260$ **51.** $F(t) = 2e^{-0.25t} + 1.56$
53. $F(t) = -10\ln(2t + 5) + 56.09$

6.3 EXERCISES

1. 8 **3.** 0 **5.** 6 **7.** -2 **9.** $^{14}/_3$ **11.** $^5/_4$ **13.** $^1/_2$
15. $3(e^3 - 1)$ **17.** 15 **19.** $-4 \cdot \ln 2$ **21.** $2a^{n+1}/(n + 1)$
23. cost: \$226,500; rev: \$543,000 **25.** \$500
27. \$5833.33 **29.** \$100,000 **31.** 24,827 people
33. 4160 people **35.** $^{19}/_3$ **37.** $1 - e^{-4}$ **39.** $2(e^3 - 1)$
41. $\ln 3$ **43.** -1 **45.** $4(10^{3/2} - 1)/9$ **47.** $(0.5) \ln (^{20}/_3)$
49. $54\sqrt{3}$ **51.** $\approx 512{,}711$ units; ≈ 18.2 yrs **53.** 469.1 units, 586.6 units **55.** \$20,239.89 **57.** $5 \ln 6 \approx 8.96$ million ft^3; $5 \ln(^{46}/_{21}) \approx 3.92$ million ft^3 **59.** $200(1 + \ln 11)$ ≈ 680 animals **61.** $(1200) \ln \left(\dfrac{850}{840.5}\right) \approx 13$; $(1200) \ln \left(\dfrac{2050}{2000.5}\right) \approx 29$ **63.** $(0.2) \ln 11 \approx 47.96\%$

6.4 EXERCISES

1. $^{92}/_3$ **3.** $^{14}/_3$ **5.** $^{52}/_3$ **7.** $^{863}/_6$ **9.** $^{81}/_4$ **11.** $1 + \ln 2$
13. $(1 - e^{-3})/3$ **15.** $^5/_2$
17.(a)

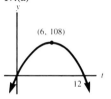

(b) 640 **19.** \$125,000; total savings **21.** 9719.489 units; total production for year **23.** $50 \ln(^5/_2) + 10 \approx \$55{,}814.54$
25. 304; total population gain during period **27.** $^7/_3$ **29.** $e^2 - e - ^3/_2$ **31.** $^1/_2$ **33.** $^9/_2$ **35.** $^1/_3$ **37.** $^{128}/_{15}$
39. $2\ln(^5/_3)$ **41.** 6 **43.** 21 **45.** $2(1 - e^{-1})$ **47.** $^{320}/_3$
49. \$45,000; total net savings **51.** \$1000 **53.(a)** $\approx 191{,}606$ people **(b)** $\approx -536{,}711$ people

6.5 EXERCISES

1. 28 **3.** 19 **5.** 15 **7.** 168 **9.** 34 **11.** 12 **13.** $^3/_2$
15. 6 **17.** 20 **19.** $^1/_3$ **21.** $^{10}/_3$ **23.** $^2/_3$ **25.** 9 **27.** $^{14}/_9$
29. $^5/_4$ **31.** $e^2 - 1$ **33.** $1/e(e - 1)$ **35.** 0 **37.** 64 fps
39. \$412,000 **41.** 407.8 units **43.** 350 units **45.** $\dfrac{1600}{3}$ units **47.** 70 units **49.** $\approx \$1.04$ per unit
51. $\approx \$85{,}882.82$ **53.** 1.75 g **55.** $(187.5) \ln 25 \approx 604$ animals

6 REVIEW EXERCISES

1. $-x^6/6 + x^4/4 + 3x^3 - x + c$ **2.** $-2x^6/3 + x^3 - 4x^2 + 14x + c$ **3.** $x + x^2/2 + x^3/6 + x^4/24 + x^5/120 + x^6/720 + c$ **4.** $2u^{1/2} + 3u^{2/3}/2 + 4u^{3/4}/3 + 5u^{4/5}/4 + c$
5. $2x^{-1} + 15x^{9/5}/9 + 4x + c$ **6.** $t^{-2} - 2t^{-1} + 8t^{3/2}/3 + t + c$ **7.** $2e^x + c$ **8.** $-3\ln|x| + c$ **9.** $2\ln|x| - $

$3e^x + c$ **10.** $(25/8)x^{8/5} + x^6 - 2\ln|x| - 9x^{-1/3} + 2x^{3/2}/3 - e^{2x}/2 + c$ **11.** $(4x + 1)^{3/2}/6 + c$ **12.** $3(x^2 + 4)^4/8 + c$ **13.** $(x^2 + 2x - 8)^6/12 + c$ **14.** $25(x^3 + 6x - 2)^{9/5}/27 + c$ **15.** $0.25\ln(4x^2 + 1) + c$ **16.** $-3(3x^2 + 2x + 1)^{-1/3} + c$ **17.** $3(x^2 - 4x + 1)^{1/3}/2 + c$ **18.** $e^{x^3} + c$ **19.** $-5e^{-x^2}/2 + c$ **20.** $2(\ln x)^{1/2} + c$ **21.** $C(x) = 3x^2 + 2x + 5000$, $R(x) = 10x^2$ **22.** $F(t) = 200t^{3/2} + 400$ **23.** $F(t) = 1000e^{0.12t}$ **24.** $F(t) = 25\ln(t + 1) + 1000$ **25.** -28 **26.** 26 **27.** $-4/3$ **28.** $-49/3$ **29.** $7/3$ **30.** $(11\sqrt{11} - 3\sqrt{3})/3$ **31.** $\dfrac{461}{36}$ **32.** 2 **33.** $1 - e^{-1/2}$ **34.** $\dfrac{1}{3}\ln 9$ **35.** $\ln 2$ **36.** $2/3$ **37.** 112 **38.** $\approx \$177.16$ **39.** $\dfrac{15}{8}$ **40.** $\dfrac{1}{2}\ln 5$ **41.** $\dfrac{125}{6}$ **42.** $253/12$ **43.** $\dfrac{863}{6}$ **44.** $\dfrac{125}{6}$ **45.** $-4 + 2e + 2e^{-1}$ **46.(a)** $50\ln\dfrac{b + 2}{2}$; total increase in junk bonds over period from 1984 to $1984 + b$ **(b)** last half of 1996 ($t \approx 12.8$) **49.** $(1 - e^{-4})/4$ **50.(a)** $\$68,908$ **(b)** $\$47,730$ **51.** 582.3 units **52.** $51,600$

7.1 EXERCISES

1. PS = $\$1452$, CS = $\$22,748$ **3.** PS = $\$41.67$, CS = $\$625$ **5.** PS = $\$42.67$, CS = $\$64$ **7.** PS = $\$1.45$, CS = $\$3.87$ **9.** PS = $\$250$, CS = $\$25$ **11.** $\$58,309.80$ **13.(a)** $\approx \$5.535$ million **(b)** ≈ 5.88 yrs **15.** $\approx \$6.925$ million **17.** $\approx \$849,702$ **19.(a)** $\approx \$314,681,000$ **(b)** $\approx \$350,385,000$ **21.** ≈ 8.66 yrs **23.** $\approx \$245,488$ **25.** $\$50,000$ **29.** $\$8960$ **31.** ≈ 2353 caribou **35.** $A(5) \approx 65,085$, $A(10) \approx 92,365$, $A(20) \approx 114,580$ individuals **39.** 4605 individuals

7.2 EXERCISES

1. $e^x(x - 1) + c$ **3.** $x^2 e^x - 2xe^x + 2e^x + c$ **5.** $6 - 16e^{-1}$ **7.** $0.5x^2(\ln x - 0.5) + c$ **9.** $0.25x^4(\ln x - 0.25) + c$ **11.** $x(\ln x)^2 - 2x(\ln x) + 2x + c$ **13.** $x(\ln x)^3 - 3x(\ln x)^2 + 6x(\ln x) - 6x + c$ **15.** $\dfrac{19,022}{21}$ **17.** 110.1 **19.** $3x^2(x + 2)^{4/3}/4 - 9x(x + 2)^{7/3}/14 + 63(x + 2)^{10/3}/140 + c$ **21.** $116/15$ **23.** $C(x) = 2(x + 1)^{3/2}(3x - 2)/15 + 4/15$ **25.** PS = $2e^{10} - 22 \approx \$44,031$, CS = $3e^{10} - 33 \approx \$66,046$ **27.** $(1 + \ln 3)/15 - (1 + \ln 13)/65 \approx 0.085$ mg/mL **29.** $-\dfrac{1}{9}\ln\left|\dfrac{9 + \sqrt{81 - x^2}}{x}\right| + c$ **31.** $\left(\dfrac{3}{4}\right)[\ln(4 + \sqrt{17}) - \ln(2 + \sqrt{5})] \approx 0.4883$ **33.** $-\dfrac{5^{-3x}}{3(\ln 5)} + c$ **35.** $(x/3) - 2\ln|3x + 2|/9 + c$ **37.** $-\dfrac{1}{6}\ln\left|\dfrac{2 + \sqrt{4 + 9x^2}}{x}\right| + c$ **39.** $\dfrac{1}{3}\ln\left|\dfrac{x}{x + 3}\right| + c$ **41.** $\dfrac{1}{12}[\ln|6x + 5| + 5(6x + 5)^{-1}] + c$ **43.** $\dfrac{1}{2}\ln\left|\dfrac{x}{x + 3}\right| + c$ **45.** $\dfrac{1}{2\sqrt{2}}[\sqrt{2}\,x\sqrt{2x^2 - 1} - \ln|\sqrt{2}\,x + \sqrt{2x^2 - 1}|] + c$ **47.** $-\dfrac{1}{3\sqrt{2}}\ln\left|\dfrac{\sqrt{2} + \sqrt{2 - 4x^2}}{2x}\right| + c$ **49.** $x(\log_2 2x - 1/\ln 2) + c$ **51.** $R(x) = 10\ln(2x + \sqrt{4x^2 - 1}) + 63.76$ **53.** $32(\ln 2 - 0.5) \approx 6.2$ concepts

7.3 EXERCISES

1. divergent **3.** convergent; 1 **5.** convergent; $1/8$ **7.** divergent **9.** divergent **11.** divergent **13.** divergent **15.** divergent **17.** convergent; $e^2/2$ **19.** convergent; $1/2$ **21.** convergent; $1/3$ **23.** divergent **25.** divergent **27.** divergent **29.** convergent; $(\ln 2)^{1-r}/(r - 1)$ **31.** $\$40,000$ **33.** $\$2$ million **35.** $250,000$ bbls **37.** $\$200,000$ **39.** 100 million tons

7.4 EXERCISES

1. 24 **3.** 2.33102 **5.** 0.69191 **7.** 0.45193 **9.** 0.52352 **11.** 0.78670 **13.** ≈ 7000 ft^2 (using $n = 3$) **15.** 2.33796 **17.** 2.33449 **19.** 0.69377 **21.** 1.14871 **23.** 11.37189 **25.** 0.74298 **27.** $\approx \$2070$ **29.** ≈ 680; worker's total production **31.** ≈ 19.25 mg/mL **33.** ≈ 185 concepts **35.** 2.33333 **37.** 0.25 **39.** 0.69315 **41.** 1.14779 **43.** 11.37058 **45.** 0.74686 **47.** $\approx 36,950$ worker-hrs **49.** $\approx -\$123.33$ **51.** $\approx 35,067$ ft^2

7 REVIEW EXERCISES

1. PS = $\$250$, CS = $\$100$ **2.** PS = $\$32,546.85$, CS = $\$14,552.91$ **3.** PS = $\$187.50$, CS = $\$208.33$ **4.** $\approx \$1,160,563,000$ **5.** $\approx \$361,770,000$ **6.** $\approx \$1,249,045,000$ **8.** $31\pi/5$ **9.** $\pi(1 - e^{-4})/2$ **10.** $-e^{-4x}(x + 0.25) + c$ **11.** $3(7x - 5)(x + 5)^7/56 + c$ **12.** $2^{7/2}(\ln 2 - 0.4)/5 + 4/25$ **13.** $3x^3(x + 1)^{7/3}/7 - 27x^2(x + 1)^{10/3}/70 + 81x(x + 1)^{13/3}/455 - 243(x + 1)^{16/3}/7280$ **14.** $\dfrac{1}{2}\ln\left|\dfrac{x}{x + 2}\right| + c$ **15.** $-\dfrac{1}{16}\ln\left|\dfrac{x - 2}{x + 2}\right| + c$ **16.** $-\dfrac{1}{4\sqrt{2}}\ln\left(\dfrac{1 + \sqrt{2}}{2\sqrt{2} + \sqrt{7}}\right)$ **17.** $\dfrac{1}{6}[\ln|x + 2| + 2(x + 2)^{-1}] + c$ **18.** convergent; $1/k$ **19.** divergent **20.** convergent; $1/24$ **21.** divergent **22.** convergent; $1/(r + 1)$ **23.** convergent; $5/6$ **24.** 3.80461 **25.** 3.79077 **26.** 3.79990 **27.** ≈ 3554 crimes **28.** ≈ 3081 crimes **29.** ≈ 3910 ft^2 **30.** ≈ 3953.33 ft^2

7 SUPPLEMENTARY EXERCISES

1. $\approx \$9.024$ million **3.(a)** $\approx \$40.240$ million **(b)** $\approx \$46.180$ million **5.(a)** $\approx \$15.523$ million **(b)** $\approx \$38.167$ million **(c)** $\approx \$45.833$ million

7.(a) ≈$14.777 million **(b)** $500 million **9.(a)** ≈$82,420 **(b)** ≈$199,526 **(c)** $250,000 **11.(a)** ≈$119,208 **(b)** ≥$183,701 **13.(a)** for $100: yes; for $50: no **(b)** ≥$38.52 per share

8.1 EXERCISES

1. $\{(x, y) | x \in \mathbf{R}, y \in \mathbf{R}\}$, 0, 16 **3.** $\{(p, q) | p \in \mathbf{R}, q \in \mathbf{R}, p \geq 0, q \geq 0\}$, 4, 11 **5.** $\{(u, v) | u \in \mathbf{R}, v \in \mathbf{R}, uv \geq 0\}$, 1, undefined **7.** $\{(x, y, z) | x \in \mathbf{R}, y \in \mathbf{R}, z \in \mathbf{R}\}$, 6, 29 **9.** $\{(x, y, z) | x \in \mathbf{R}, y \in \mathbf{R}, z \in \mathbf{R}\}$, $x > 0, y > 0, z > 0\}$, 0, 14 **11.** -12; 17; $\{(x, y) | x \in \mathbf{R}, y \in \mathbf{R}\}$ **13.** $-\frac{1}{2}$; $-\frac{1}{2}$; $1/a^2$; $\{(x, y) | x \in \mathbf{R}, y \in \mathbf{R}, x \neq 0, y \neq 0\}$ **15.** undefined; $\sqrt{5}$; $b|a|$; $\{(x, y, z) | x \in \mathbf{R}, y \in \mathbf{R}, z \in \mathbf{R}, |x| \geq |y|\}$ **17.** $50; $205 **19.** $P = 60A + 75B - 30,000$; $P = $25,500$ **21.** 156 units **23.** it is $200 - 8p_A + 3b$, so will decrease by 8 units for every $1 increase in p_A **25.** $2000 **27.** each additional unit of Y produced costs $20; each additional unit of X produced costs $40 **29.** avg cost per unit if make 150 units of X and y units of Y is $(8000 + 20y)/(150 + y)$ dollars; avg cost per unit if make 100 units of Y and x units of X is $(4000 + 40x)/(100 + x)$ dollars **31.** $A(x, y) = (12xy^{1/2} - 40x - 20y - 2000)/(x + y)$; $32 **33.** $A(150, y) = (1800y^{1/2} - 20y - 8000)/(150 + y)$; $A(x, 100) = (80x - 4000)/(x + 100)$ **35.** $C = 32z + 43,050$ **37.** 7 million units **39.** if p_1, p_2 increase, p_3 decreases **41.** $\{(s, m, t) | s \in \mathbf{R}, m \in \mathbf{R}, t \in \mathbf{R}, s \geq 0, m \geq 0, t \geq 0, s + t \neq 0\}$ **43.** 5 **45.** $26.4875 **47.** 12 days **49.** $f(t, b) = 1.2t/b$; recovery time depends directly on age **51.** $\frac{1}{2}$ **53.** $C = 1 - D_0/L$; increases toward 1 **55.** decrease D, increase L **57.** $H = 106.4 + 0.2Y$; increases by 0.2 for every $1 increase in avg income/wk

8.2 EXERCISES

1. $f_x = 5, f_y = -7$ **3.** $f_x = 8y, f_y = 8x - 4y$ **5.** $\frac{\partial f}{\partial x} = 3 + \frac{4x}{y^2}, \frac{\partial f}{\partial y} = -\frac{4x^2}{y^3}$ **7.** $g_x = \sqrt{\frac{y}{2x}}, g_y = \sqrt{\frac{x}{2y}}$ **9.** $\frac{\partial h}{\partial u} = 4(2u + v)(u^2 + uv + v^3)^3, \frac{\partial h}{\partial v} = 4(u + 3v^2)(u^2 + uv + v^3)^3$ **11.** $\frac{\partial z}{\partial x} = 2 \ln y, \frac{\partial z}{\partial y} = \frac{2x}{y}$ **13.** $\frac{\partial w}{\partial x} = \frac{3y}{(x + y)^2}, \frac{\partial w}{\partial y} = -\frac{3x}{(x + y)^2}$ **15.** $f_x = z + 2xz^2$, $f_y = z^3 - 5, f_z = x + 2x^2z + 3yz^2$ **17.** $f_x(1, 3) = 6$, $f_y(-2, 1) = -4$ **19.** $f_x(1, 0) = 2e, f_y(0, 1) = 3e^{-1}$ **21.** $\frac{\partial z}{\partial x}\bigg|_{(0,e)} = -1, \frac{\partial z}{\partial y}\bigg|_{(0,e)} = e^{-1}$ **23.** $f_x(1, -1, 2) = 24.5, f_y(1, -1, 2) = 2.5, f_z(1, -1, 2) = 11.25$ **25.** $f_{xx} = 0, f_{xy} = f_{yx} = 0, f_{yy} = 6$ **27.** $g_{xx} = 2, g_{xy} = g_{yx} = -6y^2$, $g_{yy} = 2 - 12xy$ **29.** $f_{xx} = 0, f_{xy} = f_{yx} = -\frac{4}{3y^3}, f_{yy} = \frac{4x}{y^4}$ **31.** $\frac{\partial^2 z}{\partial x^2} = \frac{48x^2}{y} + \frac{12y^2}{x^4}, \frac{\partial^2 z}{\partial y \partial x} = \frac{\partial^2 z}{\partial x \partial y} = -\frac{16x^3}{y^2} - \frac{8y}{x^3}$,
$\frac{\partial^2 z}{\partial y^2} = \frac{8x^4}{y^3} + \frac{4}{x^2}$ **33.** $f_{xx} = ye^x - \frac{2}{x^2}, f_{xy} = f_{yx} = e^x, f_{xz} = f_{zx} = \frac{1}{z}, f_{yy} = 0, f_{yz} = f_{zy} = 0, f_{zz} = -\frac{x}{z^2}$ **35.** $f_{xyxy} = e^{xy}(2 + 4xy + x^2y^2), f_{xyyy} = x^2e^{xy}(3 + xy)$ **37.** $g_{pqq} = -6qp^{-2}$ **39.** $f_{xxyz} = 72xyz^2$ **41.** $f_{xx}(1, 1) = 10$, $f_{xy}(2, 2) = -3, f_{xx}(3, -5) = -3, f_{yy}(-2, 7) = 0$ **43.** $f_{xyx}(1, 1) = 6, f_{yyx}(-1, -1) = 12$

8.3 EXERCISES

1. slope of line parallel to xz-plane and tangent to surface at $(2, 1, 11)$ is -2; curve on surface parallel to xz-plane is decreasing and is neither concave upward nor concave downward at $(2, 1, 11)$ (it is a straight line); slope of line parallel to yz-plane and tangent to surface at $(2, 1, 11)$ is -5; curve on surface parallel to yz-plane is decreasing and is neither concave upward nor concave downward at $(2, 1, 11)$ (it is a straight line) **3.** slope of line parallel to xz-plane and tangent to surface at $(1, -1, 3)$ is -4; curve on surface parallel to xz-plane is decreasing and concave downward at $(1, -1, 3)$; slope of line parallel to yz-plane and tangent to surface at $(1, -1, 3)$ is 3; curve on surface parallel to yz-plane is increasing and neither concave upward nor concave downward at $(1, -1, 3)$ **5.** slope of line parallel to xz-plane and tangent to surface at $(1, -1, 4)$ is 0; curve on surface parallel to xz-plane is neither increasing nor decreasing and is concave downward at $(1, -1, 4)$; slope of line parallel to yz-plane and tangent to surface at $(1, -1, 4)$ is 0; curve on surface parallel to yz-plane is neither increasing nor decreasing and is concave downward at $(1, -1, 4)$ **7.** slope of line parallel to xz-plane and tangent to surface at $(1, b, 1)$ is 2; curve on surface parallel to xz-plane is increasing and concave upward at $(1, b, 1)$; slope of line parallel to yz-plane and tangent to surface at $(1, b, 1)$ is 0; curve on surface parallel to yz-plane is neither increasing nor decreasing and is neither concave upward nor concave downward at $(1, b, 1)$ **9.** $C_x = 20, C_y = 14$, if y is held fixed, each additional unit of A made will cost $20; if x is held fixed, each additional unit of B made will cost $14 **11.** $P_x = 5, P_y = 6$: if y is held fixed, each additional unit of A made and sold will contribute $5 to profit; if x is held fixed, each additional unit of B made and sold will contribute $6 to profit **13.** $C_x = $230, C_y = 142 **15.** increases cost by ≈$142 **17.** $R_x = 35 + 2y, R_y = 50 + 2x$ **19.** increases revenue by ≈$435 **21.** increases by $35; increases by $50 **23.** cost increases by ≈$5/(10x + 25)^{1/2}$, revenue by $2, profit by ≈$2 - $5/(10x + 25)^{1/2}$; cost increases by ≈$10/(20y + 25)^{1/2}$, revenue by $1, and profit by ≈$1 - $10/(20y + 25)^{1/2}$ **25.** cost increases by ≈$y/2(xy + 25)^{1/2}$, revenue by $2, profit by ≈$2 - $y/2(xy + 25)^{1/2}$ **27.** $V_V = -500, V_W = 300, W_V = 800, W_W = -400$; increase of $1 in price of V while price of W is held fixed results in quantity of V demanded decreasing by 500 units and quantity of W demanded increasing by 800 units; increase of $1 in price of W while price of V is held fixed results in quantity of V demanded increasing by 300 units and quantity of W demanded decreasing by 400 units; products are competitive   **29.** $f_x(10, 10) = -3500, f_y(10, 10) = 300, g_x(10,10) = 400, g_y(10,10) = -7000$; when y is fixed at $10, an increase of $1 in the price of X will result in quantity of X demanded decreasing by ≈ 3500 units and quantity demanded of Y increasing by ≈ 400 units; when x is fixed at $10, an increase of $1 in the price of Y will result in quantity of X demanded increasing by ≈ 300 units and quantity of Y demanded decreasing by ≈ 7000 units; the products are competitive at this combination of prices   **31.** $f_x = 200, f_y = -300, g_x = -400, g_y = 200$; if y is held fixed, an increase of $1 in the price of X will result in quantity of X supplied increasing by 200 units and quantity of Y supplied decreasing by 400 units; if x is held fixed, an increase of $1 in the price of Y will result in quantity of X supplied decreasing by 300 units and quantity of Y supplied increasing by 200 units   **33.** $A_x(7, 3) = 58, A_y(7, 3) = 52, B_x(7, 3) = -10, B_y(7, 3) = 188$; if y is held fixed at $3, an increase of $1 in the price of X will result in quantity of X supplied increasing by ≈ 58 units and quantity of Y supplied decreasing by ≈ 10 units; if x is held fixed at $7, an increase of $1 in the price of Y will result in quantity of X supplied increasing by ≈ 52 units and quantity of Y supplied increasing by ≈ 188 units   **35.** increases by $\approx 6x^{2/5}y^{-2/5}$ dollars; increases by $\approx 4x^{-3/5}y^{3/5}$ dollars   **37.** $f(81,625) = 4500, f_x(81,625) = {}^{125}\!/_9, f_y(81,625) = {}^{27}\!/_5$; when 81 worker-hrs of labor and $625 of capital are available each week, production is 4500 units per week; if capital is held fixed at $625 and an additional 1 worker-hr per week is made available, production will increase by $\approx {}^{125}\!/_9$ units per week; if labor is held fixed at 81 worker-hrs per week and an additional $1 of capital is made available each week, production will increase by $\approx {}^{27}\!/_5$ units per week   **39.** increases by $\approx 0.01xe^{0.01y}$ thousand units/wk; increases by $\approx e^{0.01y}$ thousand units/wk   **41.** $\partial C/\partial D = -1.25A^2(A + 1.25D)^{-2}$; if total audience A is fixed, increasing the sum of the duplicated audiences by 1 will result in a change in the net (unduplicated) audience of $\approx -1.25A^2(A + 1.25D)^{-2}$ **43.** $P_x(2, 20, 3) = 20e^{0.4} + 40 \approx 69.8, P_T(2, 20, 3) = 404, P_t(2, 20, 3) = 10,000$; if T is fixed at 20°, an additional 1 gram of nutrient present at time $t = 3$ will result in an increase of ≈ 70 bacteria; if x is held fixed at 2 grams, an increase of 1° in the temperature at time $t = 3$ will result in an increase of ≈ 404 bacteria; if x is held fixed at 2 grams and T at 20°, running the experiment 1 additional hour will result in an increase of 10,000 bacteria   **45.** decreases by $\approx -4\sqrt{x}/y^2$; increases by $\approx 2/y\sqrt{x}$ **47.** $f_x(100, 1) = 0.1, f_y(100, 1) = -10$; if the price of a ticket is held fixed at $1, an increase of 1000 in the city's population will result in an increase in ridership of ≈ 100 persons per day; if the population of the city remains fixed at 100,000, an increase of $1 in the price of a ticket will result in a decrease in ridership of $\approx 10,000$ per day   **49.** increases by $\approx 0.08e^{0.08x}$

8.4 EXERCISES

1. saddle pt: (0, 0, 0)   **3.** rel max: (0, 0, 9)   **5.** rel min: (3, −2, −37)   **7.** rel min: (−3, 10, −486)   **9.** saddle pt: (0, 0, 0), rel max: (−1, 1, 1)   **11.** rel min: (5, 10, −48), saddle pt: $\left(\dfrac{1}{3}, \dfrac{2}{3}, \dfrac{76}{27}\right)$ **13.** saddle pt: (0, 1, −6), rel min: (±2, 1, −19)   **15.** saddle pt: (0, 1, 0), (−1, 1, 0), (0, 0, 0), rel max: $\left(-\dfrac{1}{3}, \dfrac{2}{3}, \dfrac{1}{27}\right)$ **17.** base 2 m by 2 m, height 1 m   **19.** width = $3/\sqrt[3]{2}$ ft, length = $\sqrt[3]{4}$ ft, height = $3/\sqrt[3]{2}$ ft   **21.** $125   **23.** $690,000, $x = 30,000$ and $y = 20,000$ units   **25.** $p_A = p_B = \$250$; no   **27.** $p_A = \$8, p_B = \$10$, max revenue = $880   **29.** 49,500 units   **31.** $x = \$45,238.10, y = \$5952.38$ **33.** 2 units of A, 1.6 units of B

8.5 EXERCISES

1. min: $f\left(\dfrac{3}{2}, \dfrac{3}{2}\right) = \dfrac{13}{2}$ **3.** max: $f(3, 3) = -9$ **5.** max: $f\left(3, \dfrac{3}{2}\right) = \dfrac{9}{2}$ **7.** min: $f(1, -1) = 2$ **9.** max: $f(0, -3, 3) = 9$ **11.** min: 5.25 when $x = -{}^5\!/_3, y = {}^{17}\!/_6, z = -{}^4\!/_3$ **13.** $f\left(\dfrac{6}{5}, \dfrac{8}{5}\right) = \dfrac{13,824}{3125}$ **15.** $f(\pm 4, \pm 4) = 29$ **17.** $f\left(\dfrac{1}{2}, \dfrac{1}{2}\right) = \dfrac{1}{4}$ **19.** 10, 10   **21.** 4, 16   **23.** 18, 9, 6   **25.** $10\sqrt{6}/3$ **27.** (⅔, ⅔, ⅓)   **29.** cube of side $2r/\sqrt{3}$ **31.** max production is $f(300, 450) = 300^{1/4} 450^{3/4} \approx 406.6$ units; marginal productivity of money = $\dfrac{3}{160}\left(\dfrac{300}{450}\right)^{1/4} \approx$ 0.0169 units; 1 additional dollar could be used to produce approximately 0.0169 additional units   **33.** 351 of X, 243 of Y **35.** $x = \$33$ million, $y = \$32$ million   **37.** 57.5 bushels/acre

8.6 EXERCISES

1. $\dfrac{1}{2}$ **3.** 16   **5.** $\dfrac{1}{2}(1 - e^{-2})$

7.

9.

11.

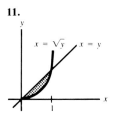

$\frac{1}{10}$

13. 3 **15.** 8 **17.** 0 **19.** $\sqrt{2}/30$ **21.** $\frac{5}{6}$ **23.** 9
25. $5\frac{3}{6}$ **27.** $\frac{4}{3}$ **29.** $\frac{2}{3}$ **31.** $e - 2$ **33.** $\frac{2}{3}$ **35.** $\frac{64}{3}$
37. 2 **39.** $(e^2 + e + 1)/6$ **41.** $\approx 228{,}571$ units
43. $\approx \$9{,}573{,}300$ **45.** ≈ 14.04 days **47.** 10 animals/km^2
49. $\approx 3{,}828{,}500$ riders

8 REVIEW EXERCISES

1. $\{(x, y)\,|\,x \neq y\}$; undefined, 1, 0, 27 **2.** $\{(x, y, z)\,|\,xy > 0\}$
(a) undefined **(b)** $e^{-8} + \ln 8$ **(c)** undefined **(d)** $e^6 +$
$\ln 4$ **3.** $R(p_1, p_2) = 200p_1 - 5p_1^2 + 750p_2 - 20p_2^2$; \$2920
4. \$11,000; contributes \$10 to profit **5.** $f_x = 16xy - 2y^3$
$+ 12, f_y = 8x^2 - 6xy^2 + 2, f_{xx} = 16y, f_{xy} = f_{yx} = 16x$
$- 6y^2, f_{yy} = -12xy$ **6.** $f_x(2, 1) = 42, f_y(-1, 0) = 10,$
$f_{xx}(2, 2) = 32, f_{yy}(-3, 4) = 144$ **7.** $g_{pq} = e^{pq}(pq + 1),$
$g_{qqpp} = e^{pq}(p^2q^2 + 4pq + 2), g_{qppp} = q^2e^{pq}(pq + 3)$
8. $\dfrac{\partial z}{\partial x} = 12(x + y)(x^2 + 2xy - y^3)^5, \dfrac{\partial z}{\partial y} = 6(2x - 3y^2)$
$(x^2 + 2xy - y^3)^5, \dfrac{\partial^2 z}{\partial x\,\partial y} = 12(x^2 + 2xy - y^3)^4(11x^2 +$
$12xy - 15xy^2 - 16y^3), \dfrac{\partial^2 z}{\partial y\,\partial x} = \dfrac{\partial^2 z}{\partial x\,\partial y}, \dfrac{\partial^2 z}{\partial x^2} = 12(x^2 +$
$2xy - y^3)^4(11x^2 + 22xy + 10y^2 - y^3), \dfrac{\partial^2 z}{\partial y^2} = 6(x^2 +$
$2xy - y^3)^4(-6x^2y - 72xy^2 + 51y^4 + 20x^2)$ **9.(a)** 82
(b) 83 **10.** $f_x(16, 1) = \dfrac{1}{32}, f_y(16, 1) = -4$; if predators
remain fixed at 1, an additional prey animal (the 17th one)
would cause the number of wolves to increase by
approximately $\dfrac{1}{32}$; if prey remains fixed at 16, an additional
predator (the 2nd one) would cause the number of wolves to
decrease by approximately 4 **11.** $R_x(25, 100) = \$20.10,$
$R_y(25,100) = \$25.025$; if y is held fixed at 100 units, an
additional unit of x (the 26th) could be sold for
approximately \$20.10; if x is held fixed at 25 units, an
additional unit of y (the 101st) could be sold for
approximately \$25.025 **12.** rel min: $(0, 0, -1)$ **13.** rel
max: $(0, 2, 4)$, saddle pt: $(\pm 2, 2, 36)$ **14.** base is 2 m by 2
m, height is 2.6 m **15.** $p_X = \$400, p_Y = \$250$ **16.** max:
$f(1, 3) = 27$ **17.** min: $f(2, 8, 4) = 80$ **18.** 8, 8
19. max productivity is $f(10, 8) = 1000\,(10^{1/5}8^{4/5}) \approx 8365;$
marginal productivity of money $= 2(10^{4/5}\,8^{-4/5}) \approx 8.3625;$
an additional \$1 could be used to increase production by
approximately 8.3625 units **20.** $\dfrac{92}{3}$ **21.** $\dfrac{10}{3}$ **22.** $e - e^{-1}$
$- 2$ **23.** ≈ 8770 units

8 SUPPLEMENTARY EXERCISES

3.

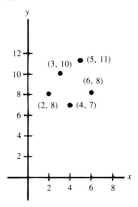

$Y = 0.1X + 8.4$

5.

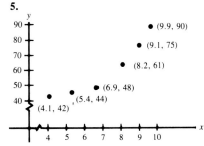

$Y = 8X + 1.81$ **7.** $Y = -9.53X + 26.61, \approx 1279$
gallons **9.** $Y = 0.048X + 0.189$, X in hundreds of pairs,
Y in millions of babies

11.

$Y = 1.96t + 43.17$; production is increasing at an avg rate of 1.96 million troy oz/yr **13.(a)** $Y = -1.98t + 63.09$ **(b)** 49.23% **(c)** $Y = 0.35t + 5.11$ **(d)** $Y = -5.44X + 90.64$; for every increase of 1 hr in avg TV viewing, avg voter participation decreases by 5.44%
15.

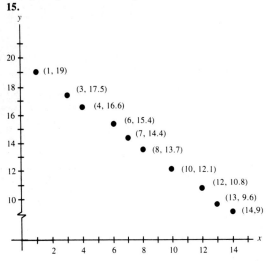

$Y = -0.77X + 19.81$; for every increase of 1 unit in dosage, time until tumor will disappear decreases by ≈ 0.77 day **17.** $Y = 5.86t + 62.81$; avg yearly increase in construction time is 5.86 months **19.** $Y = 0.1228X^2 - 9.853X + 204.94$; ≈ 40.1 units **23.** $Y = 434.54 + 47.92X_1 + 5.22X_2$, Y in hundreds of $; fixed cost is $\approx \$434.54$, variable cost per unit of X_1 when X_2 is held fixed is $\approx \$47.92$, variable cost per unit of X_2 when X_1 is held fixed is $\approx \$5.22$ **25.** $P = 105.1e^{-0.0819x}$; $\approx 13.6\%$

$\frac{16}{3}$ **29.** $y = 20t + 15$ **31.** $\frac{60}{1000^{3/2} - 250^{3/2}}(4000^{3/2} - 250^{3/2}) + 40 \approx 580$ ppm **33.** $y = -8.51 \ln t + 19.70$

9.2 EXERCISES

1. $y^2 = 2x + c$, $y^2 = 2x$ **3.** $y = \ln|x| + c$, $y = \ln|x|$ **5.** $y^2 = \frac{2}{3}x^3 + c$, $y^2 = \frac{2}{3}x^3 + \frac{25}{3}$ **7.** $y = ce^{x^3}$, $y = 2e^{x^3}$ **9.** $y = 1 + ce^{x^2}$, $y = 1 + e^{x^2}$ **11.** $y = 4 + ce^{x^2+x}$, $y = 4 - 3e^{x^2+x}$ **13.** $y = \frac{x}{cx+1}$ and $y = 0$, $y = \frac{x}{x+1}$ **15.** $y = 1000e^{-0.01t}$ **17.** $y = ce^{-rt}$, $r = (\ln 2)/t_1$ **19.** $y = 10 + 590e^{-0.5165t}$ **21.** $y = A + Ce^{-kt}$
(a) **(b)** **(c)** **(d)** **23.** $y = 25e^{0.091t}$

25. $q = c(p^2 - 10{,}000)$, $c > 0$ **29.** $y = 400 - 200e^{-0.2877t}$

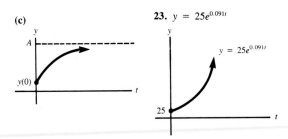

31. $p = 10 - 5e^{-0.2231t}$, approaches $10/unit

9.1 EXERCISES

1. $y = 3x^2/2 - x + c$, $y = 3x^2/2 - x$ **3.** $y = x^3/3 - x^2 + x + c$, $y = x^3/3 - x^2 + x$ **5.** $y = -2/x + c$, $y = -2/x$ **7.** $y = 4\ln|t| + c$, $y = 4\ln|t| - 3$ **9.** $y = -2e^{-x} + x + c$, $y = -2e^{-x} + x$ **11.** $y = \frac{1}{3}(x^2 + 1)^{3/2} + c$, $y = \frac{1}{3}(x^2 + 1)^{3/2} + \frac{2}{3}$ **13.** 200 ft/sec **15.** 400 ft/sec **17.** $v = \frac{6}{5(t+1)} + \frac{44}{5}$ **19.** $C = 8x^{3/2} + 10{,}000$ **21.** $R = \frac{2000}{\ln 100}\ln(x+1)$ **23.** $S = \frac{25}{32}x^{3/2} + 50$, S in $1000s **25.** $p = \frac{10}{3}d^2 + \frac{140}{3}$ **27.** $A = -\frac{1}{300}x^{2/3} + $

33. $p = \dfrac{a + q_0}{m} + ce^{-mkt}$

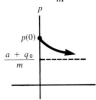
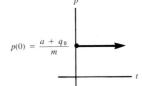
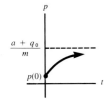

35. $p = 29.71 - 24.71e^{-0.041t}$, approaches $29.71/unit

37. $y = \dfrac{100{,}000}{1 + 20.25e^{-0.811t}}$

39. $y = 14 + 6e^{-0.405t}$

41. $y = \dfrac{10{,}000}{1 + 99e^{-1.119t}}$,

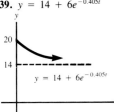
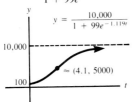

43. $y = 100 - 75e^{-0.017t}$
45. $y = 114.3e^{-4.739e^{-t}}$;

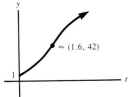

49. $y = \dfrac{Nr}{1 + r} - \left(\dfrac{Nr}{1 + r} - y_0\right)e^{-kA(1+r)t}$; if $y_0 > \dfrac{Nr}{1 + r}$, population declines toward $\dfrac{Nr}{1 + r}$; if $y_0 = \dfrac{Nr}{1 + r}$, population is stable at $\dfrac{Nr}{1 + r}$; if $y_0 < \dfrac{Nr}{1 + r}$, population increases toward $\dfrac{Nr}{1 + r}$

9.3 EXERCISES

1. $y = ce^{-x} + 1$ **3.** $y = ce^{x^2/2} - 4$ **5.** $y = x^2 + cx$
7. $y = x^{-1}[\ln x + 5e - 1]$ **9.** $y = ce^{-x-2/2} - 1$

11. $y = (x + 1)^{-1}(0.5x^2 + c)$ **13.** $y = e^{e^{-x}}\left(\dfrac{1}{3}x^3 + 1\right)$

15. $y = 25{,}000 + 75{,}000e^{0.08t}$, no **17.** $y = 30{,}000 + 10{,}000e^{0.1t}$, no

19. $y = \dfrac{W}{r} + \left(P - \dfrac{W}{r}\right)e^{rt}$

(a)

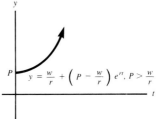

(b)

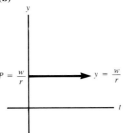

(c)

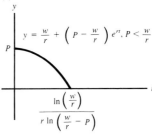

21. $y = 100{,}000 + 10{,}000t - 80{,}000e^{0.1t}$, yes **23.** $y = 100 - 50e^{-0.005t}$ **25.** $y = 100 + 0.2t - 50(10)^{9/2}(1000 + 2t)^{-3/2}$ **27.(a)** $y' = 0.6 - \dfrac{0.1y}{5 + 0.2t}$, $y(0) = $ initial concentration **(b)** $y = 10 + 0.4t - 10\sqrt{5}\,(5 + 0.2t)^{-1/2}$
(c) $y = 10 + 0.4t$

9.4 EXERCISES

1. $df = (4x - 5y)dx + (6y - 5x)dy$ **3.** $df = \left(\dfrac{y(2 - y)}{(x + 2y)^2}\right)dx + \left(\dfrac{2(xy - x + y^2)}{(x + 2y)^2}\right)dy$ **5.** 0.3, 0.06
7. 1.2, 0.6 **9.** ≈ -0.0071 **11.** ≈ 0.067 **13.** $\approx \pm 0.003$, $\approx \pm 0.3\%$ **15.** $\approx \pm 0.0003$, $\approx \pm 0.03\%$ **17.** $\approx \$3000$
19. ≈ 20.83 units, $\approx 0.038\%$ **21.** $\approx 65{,}534$ units, $\approx 1.13\%$

23. $\approx \dfrac{0.02L}{D}, \approx \dfrac{2L}{L-D}$ **25.** exact; $f = xy + \dfrac{y^2}{2} + c$
27. not exact **29.** exact; $f = e^{xy} + c$ **31.** exact; $xy + \dfrac{y^2}{2} = c$ **33.** not exact **35.** exact; $\dfrac{x^3}{3} + 2xy^2 + y = c$
37. exact; $\dfrac{x^3}{3} + x^2 + xy + y^2 + \dfrac{y^3}{3} = c$ **39.** not exact
41. not exact **43.** $-x^2 - 4xy + y^2 = 1$ **45.** $3x^2y^3 + 6x^2 + 10y^3 = 19$

9.5 EXERCISES

1.
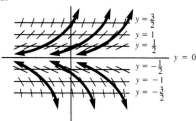

3.
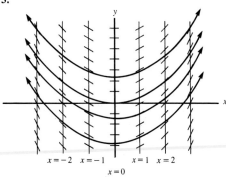

5.
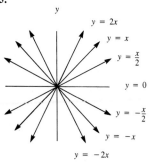

7.
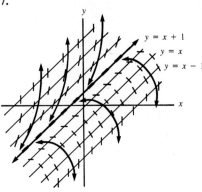

9.
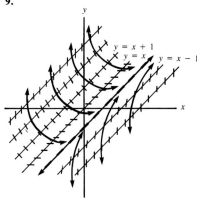

11.
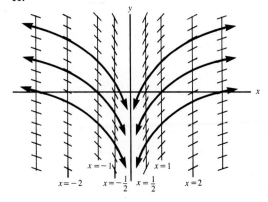

13.

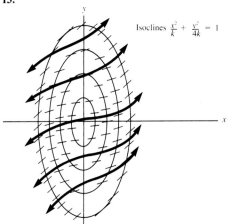

Isoclines $\frac{x^2}{k} + \frac{y^2}{4k} = 1$

15.

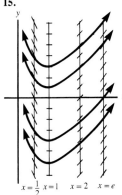

$x = \frac{1}{2}$ $x = 1$ $x = 2$ $x = e$

17. $C = -3 - 3x + 4e^x$,

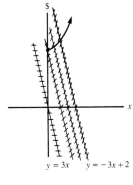

$y = 3x$ $y = -3x + 2$

19.

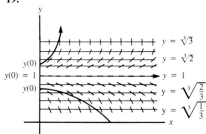

$y(0)$
$y(0) = 1$
$y(0)$

$y = \sqrt[3]{3}$
$y = \sqrt[3]{2}$
$y = 1$
$y = \sqrt[3]{\frac{2}{3}}$
$y = \sqrt[3]{\frac{1}{3}}$

9.6 EXERCISES

1.(a) $y(0.1) \approx 0.00$, $y(0.2) \approx 0.02$, $y(0.3) \approx 0.06$
(b) $y(0.1) \approx 0.005$, $y(0.2) \approx 0.030$, $y(0.3) \approx 0.075$

(c) x	Euler y, $\Delta x = 0.1$	Euler y, $\Delta x = 0.05$	Actual y
0.1	0.00	0.005	0.01
0.2	0.02	0.030	0.04
0.3	0.06	0.075	0.09

3.(a) $y(1.1) \approx 1.60000$, $y(1.2) \approx 1.32000$, $y(1.3) \approx 1.12400$ **(b)** $y(1.1) \approx 1.63000$, $y(1.2) \approx 1.36268$, $y(1.3) \approx 1.16953$ **(c)** $y(1.1) \approx 1.64990$, $y(1.2) \approx 1.39173$, $y(1.3) \approx 1.20134$ **5.(a)** $y(0.2) \approx 1.00000$, $y(0.4) \approx 1.08000$, $y(0.6) \approx 1.25280$ **(b)** $y(0.2) \approx 1.02000$, $y(0.4) \approx 1.12445$, $y(0.6) \approx 1.33584$ **(c)** $y(0.2) \approx 1.03028$, $y(0.4) \approx 1.14830$, $y(0.6) \approx 1.38244$ **7.(a)** $y(0.1) \approx 0.01$, $y(0.2) \approx 0.04$, $y(0.3) \approx 0.09$ **(b)** $y(0.1) \approx 0.01$, $y(0.2) \approx 0.04$, $y(0.3) \approx 0.09$

(c) x	Runge-Kutta y $\Delta x = 0.1$	Runge-Kutta y $\Delta x = 0.05$	Actual y
0.1	0.01	0.01	0.01
0.2	0.04	0.04	0.04
0.3	0.09	0.09	0.09

9.(a) $y(1.1) \approx 1.65445$, $y(1.2) \approx 1.39845$, $y(1.3) \approx 1.20880$ **(b)** $y(1.1) \approx 1.65443$, $y(1.2) \approx 1.39842$, $y(1.3) \approx 1.20876$ **(c)** $y(1.1) \approx 1.65442$, $y(1.2) \approx 1.39842$, $y(1.3) \approx 1.20876$ **11.(a)** $y(0.2) \approx 1.04081$, $y(0.4) \approx 1.17351$, $y(0.6) \approx 1.43332$ **(b)** $y(0.2) \approx 1.04081$, $y(0.4) \approx 1.17351$, $y(0.6) \approx 1.43333$ **(c)** same as (b)

9 REVIEW EXERCISES

1. $y = x^2 + 2x + 1$ **2.** $y = \frac{x^3}{3} - \frac{3}{2}x^2 + 2x + \frac{47}{6}$
3. $y = \ln|x|$ **4.** $y = -\ln|1 - x^2|$ **5.** $a = 15t$, $v = 7.5t^2$, $h = 2.5t^3$ **6.** $P = 3.5x^{4/3} - 25{,}000$ **7.** $y = 4.48 \ln(t + 10) - 10.43$ **8.** $y^2 = 2\ln|x| + c$ **9.** $y^2 + 4y = \frac{2}{3}(x^2 - x)$ **10.** $y = \frac{2x^2}{1 + cx^2}$ and $y = 0$ **11.** $y = -3 + 5e^{-x + x^2/2}$ **12.** $y = 8e^{0.061t}$ **13.** $q = 100{,}000 - 200p$
14. $10{,}000y - \frac{1}{2}y^2 = (948{,}000.2)t + 19{,}998$
15. $y = 200 - 203.13e^{-0.0256t}$ **16.** $y = \frac{1}{1 + 9e^{-0.0811t}}$
17. $y = \frac{1}{8}(e^{-4x} + 4x - 1)$ **18.** $y = x^{-4} + cx^{-5}$
19. $y = 25{,}000e^{-0.05t}$ **20.** ≈ 640 units, $\approx 19.2\%$
21. $2xy - \frac{5x^3}{3} - y^2 = c$ **22.** $y^2 = \frac{2(c - x^3)}{9x^2}$
23.(a)

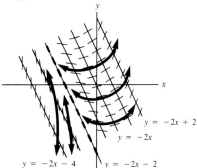

$y = -2x + 2$
$y = -2x$
$y = -2x - 4$ $y = -2x - 2$

(b)

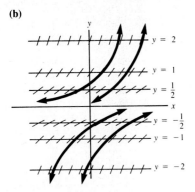

24.(a) $y(1.05) \approx 2.1500$, $y(1.10) \approx 2.3126$, $y(1.15) \approx 2.4888$ **(b)** $y(1.05) \approx 2.1564$, $y(1.10) \approx 2.3262$, $y(1.15) \approx 2.5103$

9 SUPPLEMENTARY EXERCISES

1.(a)

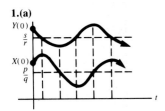

(b)

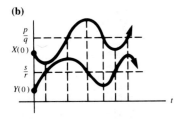

(c)

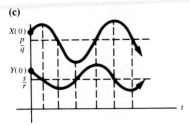

10.1 EXERCISES

1.(a) HH, HT, TH, TT **(b)** TT; HH, TH; TH, TT; HH, HT, TH; HT, TH, TT; HT, TH, TT; HH, HT, TH, TT; HH, HT, TT; none **(c)** $\frac{1}{4}; \frac{1}{2}; \frac{1}{2}; \frac{3}{4}; \frac{3}{4}; \frac{3}{4}; 1; \frac{3}{4}; 0$ **3.(a)** 1, 2, 3, 4, 5, 6 **(b)** 2; 1, 3, 5; 1, 2, 3, 4, 5, 6; 1, 2, 4, 5, 6; none; 1, 2, 3, 4, 5, 6; 3, 4; 1, 2, 5, 6; 1, 2, 3, 4; 5, 6 **(c)** $\frac{1}{6}; \frac{1}{2};$ 1; $\frac{5}{6}$; 0; 1; $\frac{1}{3}; \frac{2}{3}; \frac{2}{3}; \frac{1}{3}$ **5.(a)** $n = 2$, $n = 12$: $\frac{1}{36}$; $n = 3$, $n = 11$: $\frac{1}{18}$; $n = 4$, $n = 10$: $\frac{1}{12}$; $n = 5$, $n = 9$: $\frac{1}{9}$; $n = 6$, $n = 8$: $\frac{5}{36}$; $n = 7$: $\frac{1}{6}$ **(b)** $\frac{2}{9}$ **(c)** $\frac{1}{9}$ **(d)** $\frac{2}{3}$ **7.** 0.68 **9.** $\frac{1}{19}$ **11.** $\frac{3}{38}$ **13.(a)** $\frac{1}{6}$, $\frac{4}{9}$, $\frac{2}{9}$, $\frac{4}{45}$, $\frac{1}{18}$, $\frac{1}{45}$ **(b)** $\frac{5}{6}$ **(c)** $\frac{1}{6}$ **(d)** $\frac{34}{45}$ **15.** 0.2, 0.4, 0.4 **17.** 0 **19.** $\frac{4}{9}$ **21.** 0.8 **23.** 0.75 **25.** $\frac{43}{50}$ **27.(a)** 0.55 **(b)** 0.87 **(c)** 0.13 **(d)** 0.044 **29.** 0.445 **31.** 0.247 **33.** 0.485

10.2 EXERCISES

1.

X	Outcomes	P(X)
0	TT	1/4
1	HT, TH	1/2
2	HH	1/4

$E(X) = 1$: if the experiment is repeated many times, will average 1 head per repetition; $V(X) = \dfrac{1}{2}$

3.

X	Outcomes	P(X)
1	1	1/6
2	2	1/6
3	3	1/6
4	4	1/6
5	5	1/6
6	6	1/6

$E(X) = 3.5$: if the experiment is repeated many times, the number obtained will average 3.5; $V(X) = \dfrac{35}{12}$

Solutions to Exercises

5.

X	Outcomes	P(X)
1	Ace	1/13
2	2	1/13
3	3	1/13
4	4	1/13
5	5	1/13
6	6	1/13
7	7	1/13
8	8	1/13
9	9	1/13
10	10	1/13
11	jack	1/13
12	queen	1/13
13	king	1/13

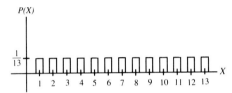

$E(X) = 7$: if the experiment is repeated many times, will average a card value of 7; $V(X) = 14$ **7.(a)** $-\$1/37 \approx -\$0.027$ **(b)** $-\$1/37 \approx -\$0.027$ **(c)** not fair, fairer overall than American version **9.** $-\$0.50; \$10,000$ **11.** no; \$999 **13.** 1 **15.** B **17.** B **19.** \$3350 **21.** 5.48 ppm **23.** 1 **25.** 0.683 **27.** 0.007 **29.** 0.89; the probability that a family will have fewer than three males who marry **31.** 1.151; the average number of males in a family who will marry **33.** discuss issues

10.3 EXERCISES

1. 5040 **3.** 362,880 **5.** 6 **7.** 1

9.

X	P(X)
0	1/16
1	1/4
2	3/8
3	1/4
4	1/16

$E(X) = 2$, so in the long run will average 2 heads for each time the experiment is performed; $V(X) = 1$ **11.(a)** 0.2252 **(b)** 0.9198 **(c)** 0.2361 **(d)** 0.4821; $E(X) = 3.75$ **13.(a)** 0.2508 **(b)** 0.3669 **(c)** 0.9536; $E(X) = 4$ **15.(a)** 0.2863 **(b)** $\approx 3.5(10^{-17})$ **(c)** 0.9994 **(d)** 13.8 **17.(a)** 0.7164 **(b)** 0.2836 **(c)** 0.9656 **19.(a)** 0.8145 **(b)** 0.1855 **(c)** 0.000475 **(d)** 0.0000062; $E(X) = 0.2$ **21.(a)** 0.4219 **(b)** 0.0156 **(c)** 0.5781 **(d)** 0.9844 **23.(a)** $8/27$ **(b)** $7/27$ **(c)** $26/27$ **(d)** $12/27$ **25.** 1 **27.** 2 **29.** 2 **31.(a)** $1 - [(0.55)^{10} + 10(0.45)(0.55)^9] \approx 0.9767$ **(b)** $1 - [45(0.45)^8(0.55)^2 + 10(0.45)^9(0.55) + (0.45)^{10}] \approx 0.9726$; $E(X) = 4.5$ **33.(a)** 0.3679 **(b)** 0.6321

(c) 0.9810 **(d)** 0.0037 **35.(a)** 0.5134 **(b)** 0.4866 **(c)** 0.9951 **(d)** 0.0527 **37.** 0.2231 **39.(a)** 0.8187 **(b)** 0.1637 **(c)** 0.0176

10.4 EXERCISES

9.(a) 1 **(b)** $\frac{1}{3}$ **(c)** $\frac{2}{15}$ **(d)** $\frac{7}{15}$ **(e)** $\frac{5}{6}$ **(f)** $(b-a)/3$

11. $F(x) = \frac{3}{20}\left(\frac{1}{2}x^2 - \frac{2}{3}x^3 + \frac{1}{4}x^4\right) - \frac{1}{80}$ **(a)** 73/80

(b) 459/1280 **(c)** 0.16 **(d)** 0.27 **(e)** $1 - \frac{3}{20}\left(\frac{1}{2}b^2 - \frac{2}{3}b^3 + \frac{1}{4}b^4\right)$ **(f)** $\frac{3}{20}\left(\frac{1}{2}a^2 - \frac{2}{3}a^3 + \frac{1}{4}a^4\right)$ **13.** $F(x) = 1 - \frac{4}{x^2}$ **(a)** $\frac{3}{4}$ **(b)** 3/10,000 **(c)** $\frac{24}{25}$ **(d)** $\frac{1}{625}$ **(e)** $4\left(\frac{1}{a^2} - \frac{1}{b^2}\right)$ **(f)** $4\sqrt{5}$

15. $f(x) = \begin{cases} 3, & 0 \le x \le 1/3 \\ 0, & \text{otherwise} \end{cases}$

17. $f(x) = \begin{cases} 1/3x, & 1 \le x \le e^3 \\ 0, & \text{otherwise} \end{cases}$

19. $f(x) = \begin{cases} 3e^{-3x}, & x \ge 0 \\ 0, & x < 0 \end{cases}$

21. $F(x) = \begin{cases} 0, & x < 0 \\ x/5, & 0 \le x \le 5 \\ 1, & x > 5 \end{cases}$

3/5, 2/5, 2/5

23. $F(x) = \begin{cases} 0, & x < 1 \\ 1 - 1/x, & x \ge 1 \end{cases}$

1/2, 1/6, 1/5

25. $F(x) = \begin{cases} 0, & x < 0 \\ 1 - e^{-3x}, & x \ge 0 \end{cases}$

$e^{-3}, 1 - e^{-6}, e^{-9} - e^{-15}$

27. $F(x) = \begin{cases} 0, & x < 0 \\ x/4, & 0 \le x \le 4 \\ 1, & x > 4 \end{cases}$

$f(x) = \begin{cases} 0, & x < 0 \\ 1/4, & 0 \le x \le 4 \\ 0, & x > 4 \end{cases}$

31. $E(x) = 6\sqrt{6}/9, V(x) = 1/3$ **33.** $E(x) = V(x) = +\infty$ **35.** $E(x) = 2, V(x) = 2$ **37.** 5/32 **39.** $\frac{11}{256} 100\% \approx 4.3\%$ **41.** $1/\sqrt{2} \approx 0.7071$ **43.** ≈ 0.235 **45.** 3/4 **47.** 2/9 **49.** $e^{-0.5} - e^{-1} \approx 0.2387$ **51.** $1 - e^{-3} \approx 0.9502$ **53.** 10 days **55.** $(\ln 2/\ln 11) 100\% \approx 28.9\%$ **57.** $11^{0.95} - 1 \approx 8.76$ yrs **59.** 0.6916 **61.** ≈ 0.66 day **63.** 6.5% **65.** $\approx 14.6\%$ **67.** $\approx 543 \, b/m^3$ **69.** $\approx 5\frac{1}{3}$ months

10.5 EXERCISES

1.(a), (b)

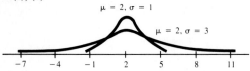

(c), (d)

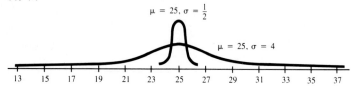

3. 0.0075, 0.1819, 0.1099, 0.0013 **5.** 0.0233, 0.9788, 0.9960, 0.1401 **7.** ≈1.28, 0.44, −1.645, −0.14
9. 0.4772 **11.** 0.7257 **13.** $x \approx 13.29$ **15.** 0.8185
17. 0.0808 **19.** $x \approx 35.25$ **21.** 0.3413 **23.** 0.0215
25. 0.9544 **27.** 0.0013 **31.** 0.1465 **33.** 0.4013
35.(a) 0.5 (b) 0.4772 (c) 0.1327 (d) 0.3108
37. $x \approx 1582.5$ **39.**(a) ≈8.795 yrs (b) ≈2.8325 yrs
41. ≈195.9 to 203.76 g **43.**(a) $x \approx 2329$ (b) $x \approx 1534$
(c) 230 **45.**(a) 0.3413 (b) 0.6568 **47.** $x \approx 119$ days
49. 0.0289 **51.** 0.0179 **53.** 0.9591 **55.** exp. dem. during lead time = 1000 bars; safety stock ≈ 110 bars; reorder point ≈ 1110 bars **57.** exp. dem. during lead time = 1800 gal; safety stock ≈ 967 gal; reorder point ≈ 2767 gal **63.** 2019 **65.** 2359 **67.** ≈11.9 yrs
69. ≈14.4 yrs **71.**(a) $1 - e^{-0.625} \approx 0.4647$ (b) $e^{-1.5625} \approx 0.2096$ **73.**(a) in 1980: ≈ 0.128 yrs; now: ≈ 0.180 yrs (b) in 1980: ≈ 7.49 yrs; now: ≈ 10.49 yrs

11.(a) $\dfrac{\sqrt{2}}{2}$ (b) $\dfrac{1}{2}$ (c) $1 - \dfrac{1}{2}\sqrt{3.5}$ (d) 1 (e) $\dfrac{4}{3}$
(f) $\dfrac{16}{5}$ **12.** $F(x) = -\dfrac{3}{\sqrt{x}} + 1$ (a) $\dfrac{3}{20}$ (b) $\dfrac{7}{10}$ (c) $\dfrac{3}{20}$
(d) 900 **13.**(a) $\dfrac{1}{2}$ (b) $\dfrac{1}{3}$; 9 yrs **14.**(a) 0.4906
(b) 0.4826 (c) 0.2386 (d) 0.0689 (e) 0.1922
(f) 0.0778 (g) 0.9918 (h) 0.0934 (i) $z \approx 0.74$ (j) $z \approx -0.065$ (k) $z \approx 2.05$ (l) $z \approx -0.37$ (m) $z \approx 1.405$
15.(a) ≈0.0094 (b) ≈0.0392 (c) ≈0.8584
(d) ≈127.92 days after planting **16.**(a) $1 - e^{-1/3} \approx 0.2835$ (b) $e^{-4/3} - e^{-2} \approx 0.1283$ (c) ≈ 0.669 days

10 REVIEW EXERCISES

1.(a) $\dfrac{1}{50}$ (b) $\dfrac{49}{50}$ (c) $\dfrac{1}{50}$ (d) $\dfrac{4}{5}$ (e) $\dfrac{13}{50}$ (f) 0 (g) $\dfrac{1}{2}$
(h) $\dfrac{33}{50}$ **2.**(a) 0.02 (b) 0.72 (c) 0.60 **3.**(a) $\dfrac{2}{12}$ (b) $\dfrac{11}{12}$
(c) $\dfrac{1}{3}$ **4.** $E(X) = 6.5$; in the long run, the sum of the digits will average 6.5; $V(X) = 10.25$ **5.**(a) ≈ −$0.230
(b) ≈$0.027 **6.**(a) $1 (b) $3 (c) $2.25 (d) $12
(c) $51 **7.** 68/45 ≈ 1.51 per carton **8.**(a) 11,520/59,049 ≈ 0.1951 (b) 59,048/59,049 ≈ 0.99998 (c) 4521/59,049 ≈ 0.0766 (d) $66\dfrac{2}{3}$ **9.**(a) $e^{-1} \approx 0.3679$ (b) $1 - e^{-1} \approx 0.6321$ (c) $1 - e^{-4/3}\left(1 + \dfrac{4}{3} + \dfrac{8}{9}\right) \approx 0.1506$
(d) $e^{-20/3}\left(1 + \dfrac{20}{3} + \dfrac{200}{9} + \dfrac{4000}{81} + \dfrac{20{,}000}{243}\right) \approx 0.2056$

11.1 EXERCISES

1. 90°, 180°, 270°, 360° **3.** −45°, −90°, −180°, −135°
5. 450°, 540°, 30°, 150°
7.

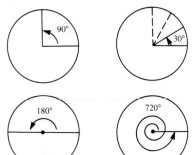

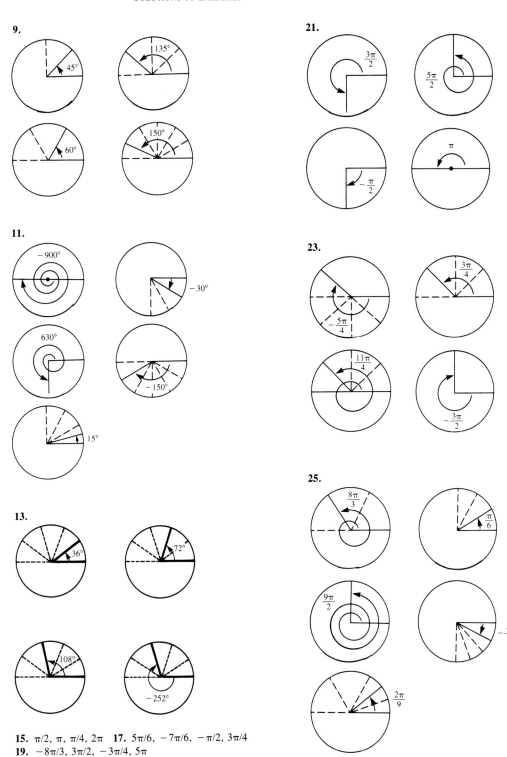

15. $\pi/2$, π, $\pi/4$, 2π **17.** $5\pi/6$, $-7\pi/6$, $-\pi/2$, $3\pi/4$
19. $-8\pi/3$, $3\pi/2$, $-3\pi/4$, 5π

27.

29. $\pi/2, \pi/6, 25\pi/36, -\pi/3$ **31.** $3\pi/4, -5\pi/6, 35\pi/18,$
$52\pi/15$ **33.** $150°, 270°, -480°, 810°$ **35.** $210°, -75°,$
$3960/7\pi \approx 180.07°, 450/\pi \approx 143.24°$

11.2 EXERCISES

1. $-1, 0, 0, 1$ **3.** $-1, 0, 0, 1$ **5.(a)** $n\pi, n = 0, \pm 1,$
$\pm 2, \ldots$ **(b)** $n\pi/2, n = \pm 1, \pm 3, \pm 5, \ldots$
7.(a) $2n\pi, n = 0, \pm 1, \pm 2, \ldots$ **(b)** $n\pi, n = \pm 1, \pm 3,$
$\pm 5 \ldots$ **19.** $\sin\dfrac{\pi}{12} = \dfrac{1}{2\sqrt{2+\sqrt{3}}}, \cos\dfrac{\pi}{12} =$
$\dfrac{\sqrt{2+\sqrt{3}}}{2}$ **21.** $\sin\dfrac{\pi}{24} = \dfrac{\sqrt{2-\sqrt{2+\sqrt{3}}}}{2}, \cos\dfrac{\pi}{24} =$
$\dfrac{\sqrt{2+\sqrt{2+\sqrt{3}}}}{2}$ **23.(a)** undefined, 0, undefined, -1
(b) $\sqrt{3}, 1/\sqrt{3}, 2, 2/\sqrt{3}$ **25.(a)** $-1/\sqrt{3}, -\sqrt{3},$
$-2/\sqrt{3}, 2$ **(b)** $1, 1, -\sqrt{2}, -\sqrt{2}$ **27.** $\sqrt{\dfrac{\sqrt{2}-1}{\sqrt{2}+1}},$
$\sqrt{\dfrac{\sqrt{2}+1}{\sqrt{2}-1}}, \sqrt{\dfrac{2\sqrt{2}}{\sqrt{2}+1}}, \sqrt{\dfrac{2\sqrt{2}}{\sqrt{2}-1}}$ **35.** $4/5, 3/5, 4/3, 5/3, 5/4,$
$3/4$ **37.** $5/8, \sqrt{39}/8, 5/\sqrt{39}, 8/\sqrt{39}, 8/5, \sqrt{39}/5$
39. $r = 2, s = 2\sqrt{2}$ **41.** $x = \pi/6$ **43.** $x = \pi/3$

11.3 EXERCISES

1. the limits do not exist, either as finite numbers or as $\pm\infty$.

3. period $= 4\pi$, amplitude $= 1$

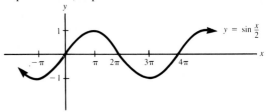

5. period $= 6\pi$, amplitude $= 1$

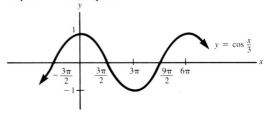

7. period $= 4\pi/3$, amplitude $= 1$

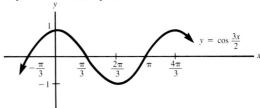

9. period $= 2\pi/3$, amplitude $= 1$

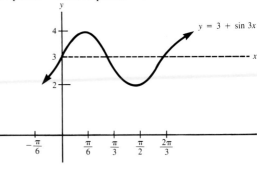

11. period $= \pi$, amplitude $= 1/2$

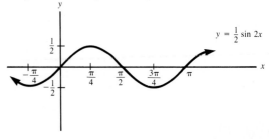

13. period = $2\pi/3$, amplitude = 4

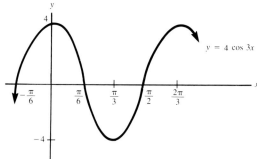

15. period = 8, amplitude = $3/2$

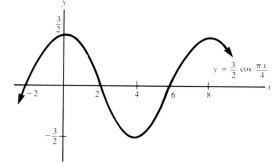

17. period = π, amplitude = 2

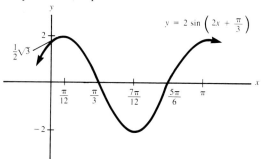

19. period = $\pi/4$, amplitude = 6

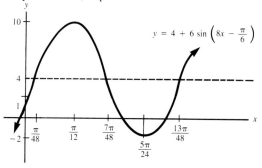

21. period = 2, amplitude = 3

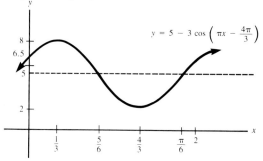

23.(a)

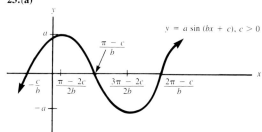

(b)

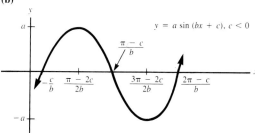

25.(a)

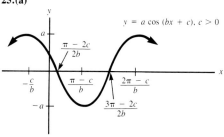

(b)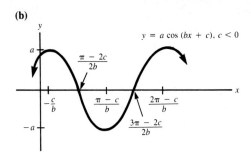

33. period = $\pi/3$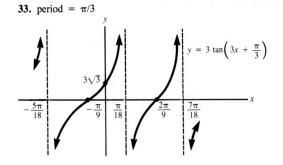

27. period = 2π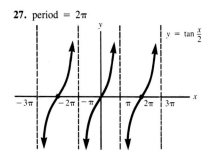

35. period = $2\pi/3$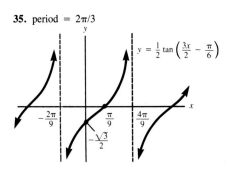

29. period = 3π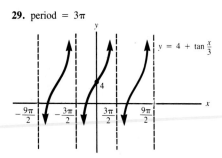

37.(a)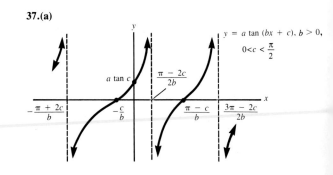

31. period = $4\pi/3$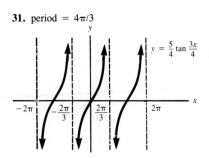

(b)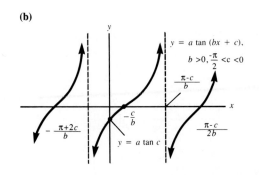

39. period = $2\pi/3$

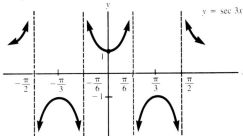

47. period = $4\pi/3$

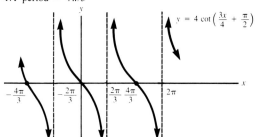

41. period = $\pi/3$

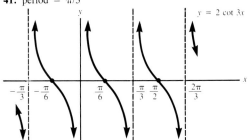

49.

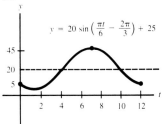

43. period = 6π

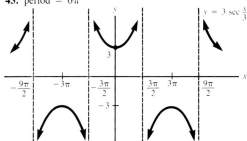

51.

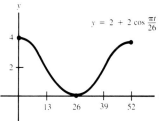

53.

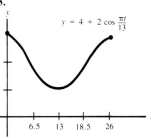

45. period = π

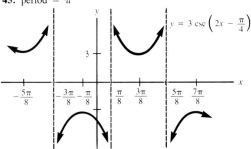

55. $p = 2.5 + 0.25 \cos \dfrac{\pi t}{2}$ **57.** 135, 85, 70

59.

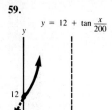

11.4 EXERCISES

1. $\dfrac{dy}{dx} = -3\sin x$ **3.** $\dfrac{dy}{dx} = 15 \cos 3x$ **5.** $\dfrac{dy}{dx} = 2(x+4)\cos(x^2+8x)$ **7.** $\dfrac{dy}{dx} = 2(2\cos x \cos 2x - \sin x \sin 2x)$

9. $\dfrac{dy}{dx} = -14 \cos 7x \sin 7x$ **11.** $\dfrac{dy}{dx} = 2 \sin x - 7 \sin^3 x + 5 \sin^5 x$ **13.** $\dfrac{dy}{dx} = -\dfrac{4\cos^3 4x(4\sin 3x \sin 4x + 3 \cos 4x \cos 3x)}{\sin^5 3x}$ **15.** $\dfrac{dy}{dx} = e^{-x}(2\cos 2x - \sin 2x)$ **17.** $\dfrac{dy}{dx} = -\sin x \cos(\cos x)$

19.

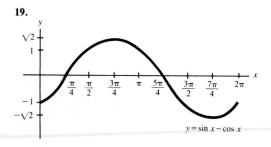

21.

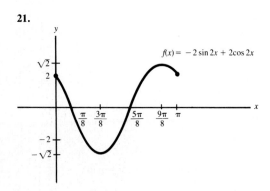

23.

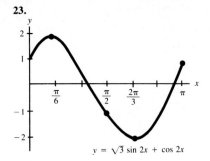

25.

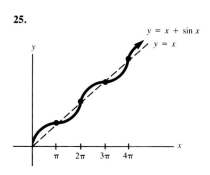

27.

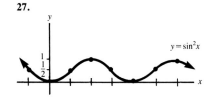

33. $\dfrac{dy}{dx} = 2x \sec x^2 \tan x^2$ **35.** $\dfrac{dy}{dx} = 4 \tan x \sec^2 x$ **37.** $\dfrac{dy}{dx} = -e^x \csc^2 e^x$ **39.** $\dfrac{dy}{dx} = \sec x (2 \tan^2 x + 1)$ **41.** $\dfrac{dy}{dx} = 2 \tan x \sec^3 x$ **43.** $\dfrac{dy}{dx} = \sec x$ **45.** $\dfrac{dy}{dx} = 4x \tan x^2 \sec^2 x^2$ **47.** $-\tfrac{1}{20}$ rad/sec
49. transient current $T = 2e^{-5t}$;

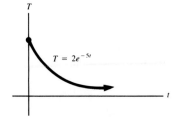

steady-state current $S = 2\sqrt{3} \sin 4t - 2\cos 4t$

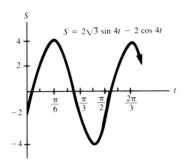

51.(a) $t = 0$; 1 in below rest position; $t = \pi/6$: $2e^{-\pi/6} \approx 1.18$ in below rest position; $t = \pi/4$: $e^{-\pi/4}/\sqrt{2} \approx 0.32$ in below rest position **(b)** $t = 0$: velocity = 5 in/sec, acceleration = -20 in/sec²; $t = \pi/6$: velocity = $-5e^{-\pi/6} \approx -2.96$ in/sec, acceleration = $-10e^{-\pi/6} \approx -5.92$ in/sec²; $t = \pi/4$: velocity = $-5\sqrt{2}\, e^{-\pi/4} \approx -3.22$ in/sec, acceleration = $5\sqrt{2}\, e^{-\pi/4} \approx 3.22$ in sec²
(c)

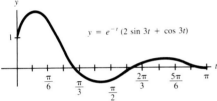

$e^{-\pi/12}\left(2\sin\dfrac{\pi}{4} + \cos\dfrac{\pi}{4}\right) \approx 1.63$ in below rest position; $2\pi/3$ **53.(a)** $-\dfrac{10\pi}{3} \approx -10.472$ thousand dollars per month, $\dfrac{20\pi}{3} \approx 20.944$ thousand dollars per month **(b)** max of $145,000 when $t = 11$, min of $65,000 when $t = 5$
55.(a) $-\dfrac{\pi}{3}\sin\dfrac{53\pi}{52} \approx 0.063$ thousand dollars per week, $-\dfrac{\pi}{3}\sin\dfrac{113\pi}{52} \approx -0.542$ thousand dollars per week
(b) max of $12,000 when $t = 45.5$, min of $8000 when $t = 19.5$ **57(a)** $\dfrac{2\pi}{13}\left(3\sin\dfrac{24\pi}{13} - \cos\dfrac{24\pi}{13}\right) \approx -$1.10 per week, $\dfrac{2\pi}{13}\left(3\sin\dfrac{48\pi}{13} - \cos\dfrac{48\pi}{13}\right) \approx -$1.47 per week **(b)** max of $\approx$ $30.32 when $t \approx 14.33, 40.33$; min of $\approx$ $17.68 when $t \approx 1.33, 27.33$

11.5 EXERCISES

1. $-\dfrac{1}{3}\cos 3x + c$ **3.** $\dfrac{1}{\pi}\cot(-\pi x) + c$ **5.** $\dfrac{1}{2}\sin \pi^2$
7. $-2\cos\sqrt{x} + c$ **9.** 0 **11.** $\dfrac{1}{8}\sin^4 2x + c$

13. $1 - e^{-1}$ **15.** $-\cos x + c$ **17.** $\dfrac{1}{2}x - \dfrac{1}{4}\sin 2x + c$
19.(a)

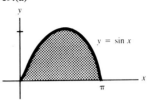

2

(b)

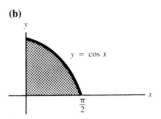

1

21.(a)

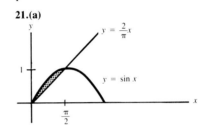

$1 - \pi/4$
(b)

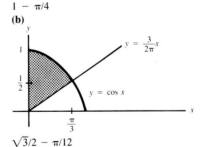

$\sqrt{3}/2 - \pi/12$

23.(a) 0 **(b)** 0 **25.** $-\dfrac{1}{5}\ln|\cos 5x| + c$ **27.** $-\dfrac{1}{2}\ln|\csc x^2 + \cot x^2| + c$ **29.** 1 **31.** $-\dfrac{2\pi}{3} + \ln\left(\dfrac{2+\sqrt{3}}{2-\sqrt{3}}\right)$ **33.** $\dfrac{2}{\pi}\ln\left(\dfrac{\sqrt{2}+1}{\sqrt{2}-1}\right)$ **35.** $x \sin x + \cos x + c$ **37.** $\pi - 2$ **39.** $\dfrac{1}{4}e^{2x}(\sin 2x + \cos 2x)$ **41.** 0
43. 0 **45.** $\dfrac{96}{\pi} \approx 30.558$ million dollars **47.** $\dfrac{1920}{\pi} \approx 611.155$ thousand dollars **49.(a)** ≈ 3.85 units **(b)** 100 units

11 REVIEW EXERCISES

1. 22.5°, π/8; −150°, −5π/6 **2.** −840°, −14π/3; 720°, 4π **3.** 210°, 7π/6; 405°, 9π/4 **4.** −75°, −5π/12; 100°, 5π/9

5.

6.

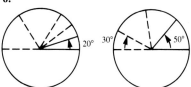

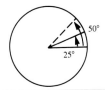 

7.

 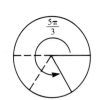

8.

 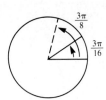

9. 3π/4, −4π/3, π/8, −π/45 **10.** 240°, −225°, 12°, 585°/π ≈ 186.2° **11.** sin x = −1, cos x = 0, tan x is undefined, csc x = −1, sec x is undefined, cot x = 0 **12.** sin x = −1/√2, cos x = −1/√2, tan x = 1, csc x = −√2, sec x = −√2, cot x = 1 **13.** sin x = 0, cos x = 1, tan x = 0, csc x is undefined, sec x = 1, cot x is undefined **14.** sin x = √3/2, cos x = ½, tan x = √3, csc x = 2/√3, sec x = 2, cot x = 1/√3 **15.** sin x = ½, cos x = −√3/2, tan x = −1/√3, csc x = 2, sec x = −2/√3, cot x = −√3 **16.** sin x = ½, cos x = √3/2, tan x = 1/√3, csc x = 2, sec x = 2/√3, cot x = √3

17.

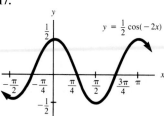

π

18.

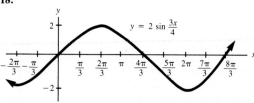

8π/3

19.

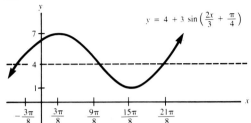

20.

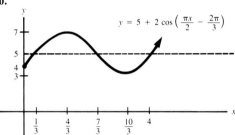

21.

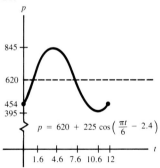

22.(a)

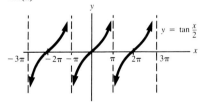

(b)

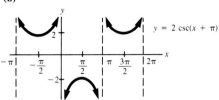

23.(a) $\sin x = 5/9$, $\cos x = 2\sqrt{14}/9$, $\tan x = 5/2\sqrt{14}$, $\csc x = 9/5$, $\sec x = 9/2\sqrt{14}$, $\cot x = 2\sqrt{14}/5$ **(b)** $a = 6$, $b = 3\sqrt{3}$ **(c)** $x = 5\pi/6$ radians **24.** $\dfrac{dy}{dx} = 3\cos 3x$ **25.** $\dfrac{dy}{dx} = -\dfrac{\sin 2\sqrt{x}}{\sqrt{x}}$ **26.** $\dfrac{dy}{dx} = \dfrac{x\cos x - \sin x}{x^2}$ **27.** $\dfrac{dy}{dx} = \dfrac{1}{6}(3\cos 2x \cos 3x - 2\sin 2x \sin 3x)$ **28.** $\dfrac{dy}{dx} = e^{\tan x}\sec^2 x$ **29.** $\dfrac{dy}{dx} = -12x(\csc^2 3x^2)$ **30.** $\dfrac{dy}{dx} = \dfrac{2(2\sec^2 4x + \cot 2x \tan 4x)}{\csc 2x}$ **31.** $\dfrac{dy}{dx} = -12\cos 4x \sin^2 4x$ **32.** $\dfrac{dy}{dx} = 2\sec^2(\tan x)\tan(\tan x)\sec^2 x$ **33.** $\dfrac{dy}{dx} = \dfrac{\cos x - \csc x \cot x}{\sin x + \csc x}$

34.

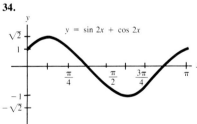

35.

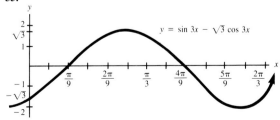

36.

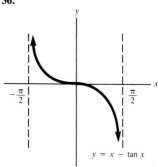

5. $y = \frac{1}{\sqrt{32}} \sin \sqrt{32}\, t - \frac{\pi}{6} \cos \sqrt{32}\, t$, $y = \frac{1}{\sqrt{32}} \sin \sqrt{32}\, t - \frac{\pi}{3} \cos \sqrt{32}\, t$

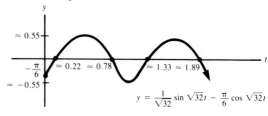

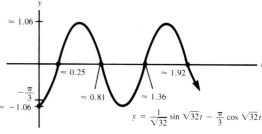

37. (a) $\frac{14\pi}{5} \cos\left(\frac{\pi}{5} + 0.7\right) \approx 2.112$ million dollars per yr,

$\frac{14\pi}{5} \cos\left(\frac{4\pi}{5} + 0.7\right) \approx -8.774$ million dollars per yr

(b) max of $94 million when $t = 2.5 - \frac{3.5}{\pi} + 10n \approx$ 1.39 + 10n, n a nonnegative integer; min of $66 million when $t = 7.5 - \frac{3.5}{\pi} + 10n \approx 6.39 + 10n$, n a nonnegative integer 38. $\frac{3}{10}$ 39. $-\cos x^2 + c$ 40. $-\frac{2}{5}$ csc $5x + c$ 41. $-\frac{\cos^5 x}{5} + c$ 42. $\frac{3}{2} \ln \sqrt{2}$ 43. $\frac{1}{2} \sin 2x + c$ 44. $\sin x [\ln(\sin x) - 1] + c$ 45. $\frac{1}{2} e^{\sec 2x} + c$

46. $2 \sin \sqrt{x} + c$ 47. $\frac{3}{2}\left[\ln\left|\sec\left(\frac{\pi^2}{16} + 2\right) + \tan\left(\frac{\pi^2}{16} + 2\right)\right| - \ln|\sec 2 + \tan 2|\right]$ 48. $-\frac{x \cos 2x}{2} + \frac{\sin 2x}{4} + c$

49. $\ln\left(\frac{6(\pi - 1)}{4\pi - 3}\right)$ 50. $\frac{e^{3x}}{13}(2 \sin 2x + 3 \cos 2x) + c$

51. (a) 0 (b) $3\sqrt{3}/\pi$ 52. $\frac{1}{2} - \frac{\pi^2}{72}$ 53. $\approx$468 units

54. $\approx$$5005

7. $y = -\frac{20{,}000}{3} \sin 0.15t + 20{,}000 \cos 0.15\, t$

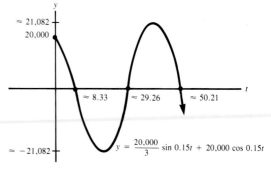

11 SUPPLEMENTARY EXERCISES

3. $y = \frac{1.5}{\sqrt{5}} \sin \sqrt{5}\, t - \frac{\pi}{2} \cos \sqrt{5}\, t$

9. $\approx$$39,000 11. the equation of motion of the weight is continuous, and both its points of inflection and its intercepts occur as the weight passes through E 13. $y = v_0 \sqrt{\frac{m}{c}}$ $\sin \sqrt{\frac{c}{m}}\, t + y_0 \cos \sqrt{\frac{c}{m}}\, t$ 15. period remains the same, but amplitude is increased and intercepts are shifted to the left; i.e., the weight makes larger swings and passes through E sooner

12.1 EXERCISES

1.

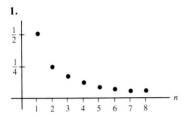

converges to 0

3.

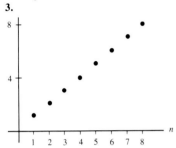

diverges

5.

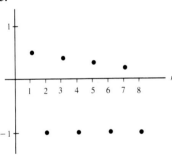

diverges

7.

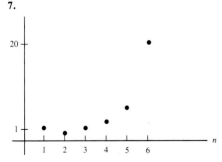

diverges **9.** diverges **11.** converges to 0 **13.** converges to 0 **15.** diverges **17.** converges to 1 **19.** converges to $\pi\sqrt{5}$ **21.** diverges **23.** diverges **25.** converges to 3 **27.** converges to 0 **29.** diverges **31.** diverges **35.** 1 **37.** 1 **39.** 0 **41.** $-\ln 2$ **43.** $\{10{,}000 + 1000n\}$, $n \geq 0$

45. $\{P(1 + rn)\}$, $n \geq 0$ **47.** $\{10{,}000(1.08)^n\}$, $n \geq 0$
49. $\{2^{n-1}\}$, $n \geq 1$ **51.(a)** $\{500{,}000(0.94)^n\}$, $n \geq 0$
(b) $\left\{\sum_{k=0}^{n} 500{,}000(0.94)^k\right\}$, $n \geq 0$ **53.** $a_1 = a_2 = a_3 = 1$,
$a_n = a_{n-1} + a_{n-3}$, $n \geq 4$

12.2 EXERCISES

1. $\left\{1, \dfrac{4}{3}, \dfrac{13}{9}, \dfrac{40}{27}, \dfrac{121}{81}, \ldots\right\}$ **3.** $\{1, 2, 4, 10, 34, \ldots\}$
5. converges to $\dfrac{3}{2}$ **7.** diverges **9.** converges to $-\dfrac{6}{7}$
11. converges to 300 **13.** converges to 15 **17.** 28 feet
19. diverges **21.** diverges **23.** converges **25.** diverges
29. converges **31.** diverges **33.** diverges **35.** converges
37. diverges **39.** converges **41.** converges
43.(b) 0.3585 **(c)** 0.9269 **45.(b)** p^m **47.** $100 billion

12.3 EXERCISES

1. $R = \dfrac{1}{2}$ **3.** $R = 1$ **5.** $R = \sqrt{3}$ **7.** $R = +\infty$
9. $|r|^{1/2}$ **11.** $+\infty$ **13.** $e^{-0.2} \approx 0.8187333$ **15.** -0.0513
17.(a) $\sum_{k=2}^{\infty} \dfrac{k^2 x^{k-1}}{2^k}$ **(b)** $\sum_{k=1}^{\infty} \dfrac{kx^{k+1}}{(k+1)2^k} + c$ **(c)**
$\sum_{k=1}^{\infty} \dfrac{k}{(k+1)2^k}$ **19.(a)** $\sum_{k=1}^{\infty} \dfrac{kx^{k-1}}{3^k}$, $-3 < x < 3$ **(b)** $\sum_{k=0}^{\infty} \dfrac{x^{k+1}}{(k+1)3^k} + C$ **(c)** $\sum_{k=0}^{\infty} \dfrac{2^{k+1}-1}{(k+1)3^k}$ **21.(a)** $\sum_{k=0}^{\infty} \dfrac{x^{2k}}{(2k)!}$ for all x **(b)** $\sum_{k=1}^{\infty} \dfrac{x^{2k}}{(2k)!} + C$ **(c)** $\sum_{k=1}^{\infty} \dfrac{1}{(2k)!}$ **25.** $\sum_{k=1}^{\infty} \dfrac{x^k}{k!}$ for all x
27. $\sum_{k=0}^{\infty} \dfrac{4^k x^{2k}}{k!}$ for all x **29.** $\sum_{k=1}^{\infty} \dfrac{(-1)^{k+1} 2^k x^k}{k}$, $-\dfrac{1}{2} < x < \dfrac{1}{2}$
31. $\sum_{k=1}^{\infty} \dfrac{(-1)^{k+1} x^k}{5k}$, $-1 < x < 1$ **33.** $\sum_{k=0}^{\infty} (-1)^k x^{2k}(x+1)$,
$-1 < x < 1$ **35.** $\sum_{k=0}^{\infty} (-1)^k (k+1) x^k$, $-1 < x < 1$
37. $1 + \sum_{k=1}^{\infty} (-1)^k (k+1) x^{k+2}$, $-1 < x < 1$

12.4 EXERCISES

1. $1 + 2x + x^2$ **3.** $1 + \dfrac{x}{2} - \dfrac{x^2}{8} + \dfrac{x^3}{16} - \dfrac{5x^4}{128} + \cdots$
5. $\sum_{k=0}^{\infty} (-1)^k x^{4k}$ **7.** $\sum_{k=0}^{\infty} \dfrac{3^k x^k}{k!}$ **9.** $\sum_{k=0}^{\infty} \dfrac{x^{2k+1}}{(2k+1)!}$ **11.** $\ln 2$
$+ \sum_{k=1}^{\infty} \dfrac{(-1)^{k+1} x^k}{k 2^k}$ **13.** $\sum_{k=1}^{\infty} \dfrac{(-1)^k 2^{2k-1} x^{2k-1}}{(2k-1)!}$
17. ≈ 3.141562 **19.(a)** ≈ 0.0953333 **(b)** $0.0953083 <$
$\ln 1.1 < 0.0953163$ **21.(a)** ≈ 0.9800666 **(b)** 0.9800661
$< \cos 0.2 < 0.9800666$ **23.(a)** 1st 10 terms (through x^9)

(b) 1st 11 terms (through x^{10}) **25.**(a) 4 (b) 10 **27.** $\sum_{k=1}^{5} \frac{(0.2)^k}{k} = 0.1823306$ **29.** 0.3102681 **31.**(a) $\sum_{k=0}^{\infty} \frac{x^{k+2}}{k!}$

(b) $2 + \sum_{k=0}^{\infty} \frac{1}{k!(k+3)}$

35. $R = 1$ **36.** $R = 0$ **37.**(a) $\sum_{k=1}^{\infty} \frac{(-1)^{k+1} 1 \cdot 3 \cdot 5 \cdots (2k+1)}{(2k)!} kx^{k-1}$

(b) $\sum_{k=0}^{\infty} \frac{(-1)^{k+1} 1 \cdot 3 \cdot 5 \cdots (2k+1)}{(2k)!(k+1)} x^{k+1}$

(c) $\sum_{k=0}^{\infty} \frac{(-1)^{k+1} 1 \cdot 3 \cdot 5 \cdots (2k+1)}{(2k)!(k+1)}$ **38.** $\sum_{k=0}^{\infty} \frac{(-1)^k x^{2k+1}}{2k+1}$

39. $\sum_{k=0}^{\infty} \frac{(-1)^k x^{2k+1}}{(2k+1)!(2k+1)}$ **40.** $\sum_{k=0}^{\infty} \frac{(-1)^{k+1} x^{2k+1}}{2^{2k+1}(2k+1)!}$

41. $\sum_{k=1}^{\infty} \frac{(-1)^{2k+1} x^{3k}}{k}$

42. $1 + \frac{1}{3}x - \frac{1}{9}x^2 + \frac{5}{81}x^3 - \frac{10}{243}x^4 + \cdots$

43. $\sum_{k=1}^{\infty} \frac{(-1)^k x^{2k}}{(2k)!}$ **44.** $1 + \sum_{k=1}^{\infty} \frac{(-1)^k (2x)^{2k}}{2(2k)!}$

45. $\sum_{k=1}^{\infty} (-1)^{k+1} 2kx^{2k-1}$ **46.** 0.3093 **47.** $e^{-1} \approx$ 0.3666666, $0.3666666 < e^{-1} < 0.3680555$

48. first 10 terms

12 REVIEW EXERCISES

1. diverges **2.** converges to 0 **3.** diverges **4.** converges to 0 **5.** converges to -3 **6.** converges to 3

7. converges to 0 **8.** converges to $\frac{1}{2}$ **9.** diverges

10. converges to 0 **11.** converges to 0 **12.** converges to 0

13. diverges **14.** converges to 4 **15.** $\left\{ \sum_{k=1}^{n} 100(1.01)^k \right\}$

16. converges to $-\frac{3}{5}$ **17.** converges to 4 **18.** diverges

19. converges to $\frac{73}{9900}$ **20.** diverges **21.** converges to 10,500 **22.** converges **23.** converges **24.** converges **25.** converges **26.** diverges **27.** converges **28.** converges **29.** diverges **30.**(a) 0.006 (b) 1.5

31. $R = \sqrt{5}$ **32.** $R = 1$ **33.** $R = \frac{3}{2}$ **34.** $R = +\infty$

12 SUPPLEMENTARY EXERCISES

1. Average income per year: $C = 4$, $A = 3$, $B = \frac{9}{4}$

3.(a) 44 (b) 9 **5.**(a) $V_A = 32.302115$, $V_B = 23.896143$
(b) $V_A = 5.9155629$, $V_B = 4.3862871$ **7.** $V_Y > V_X$ when $0 < r < 0.5$, $V_X > V_Y$ when $r > 0.5$, $V_Y = V_X$ when $r = 0.5$

Applications Index

Management

Advertising, 14, 62–63, 217, 241, 248, 260, 328–29, 341, 343, 347, 506, 516, 524, 538, 565
Approximation by differentials, 199, 201–3, 219
Audience share, 242
Auto emissions, 116
Average
  cost, 7, 15, 26, 29, 83, 118, 139, 159, 184, 241–42, 248, 259, 273, 277, 300, 302, 328, 341, 350, 486–87, 516, 549–50, 556, 565–66, 603
  inventory, 412, 414–15, 419
  production, 537, 543
  profit, 159, 486
  revenue, 159
  sales, 30, 538

Break-even analysis, 48, 55–56, 67, 73–74, 77–78, 84
Budgeting, 62–63
Business failures, 627, 646

Carrying cost, 437–38
Cash flow, 402
Commissions, 62
Competitive, complementary products, 501–2, 504–5
Compound interest, 313
Computer chip speed, 628
Consumer purchases and income, 311–12
Consumers' surplus, 430–31, 436, 446–47, 467
Consumption function, 221–23
Cost, 7, 10, 17, 26, 28–31, 34–35, 44–45, 55–56, 65–67, 73–74, 77–78, 83–84, 116, 132, 157–59, 199, 202–3, 216, 260, 273, 278, 382, 385, 400, 418, 453, 458–59, 462, 477, 480, 485–88, 524, 552, 556, 566, 595, 609–10, 751, 755

Cost—*Cont.*
  advertising, 14
  average, 15, 26, 29, 83, 118, 139, 159, 184, 241–42, 248, 259, 273, 277, 300, 302, 328, 341, 350, 486–87, 516, 549–50, 556
  functions, 17, 26, 30–31, 34–35, 65–67, 73–74, 77–78, 84, 114, 116, 132, 147, 150, 170, 302, 365, 370–71, 378, 418, 446, 485, 552, 556, 563, 565, 603
  of goods sold, 63
  inventory, 255–56, 259–60, 278
  marginal, 151–52, 157–59, 184, 193, 199, 242, 248, 503–4
  production, 10, 132
  replacement, 260–61
  shipping, 114
  variable, 103, 131
Curve-sketching problems, 270–71, 273, 278, 320–21, 328, 359–60, 726–27, 754

Decision-making using expected value, 634, 637
Defective production, 646, 649, 676, 681, 782–83, 790, 805
Demand, 26, 46, 56–58, 68, 74, 78, 84–85, 103, 131–32, 161, 328, 341, 372, 415, 480–81, 486, 501–2, 504–5, 578, 589–90, 596, 612–13
  elasticity of, 154–57, 159–62, 170–71, 242
  functions, 26, 103, 110, 131–32, 147–48, 153, 159–62, 170, 184, 191, 193, 202, 209–10, 212, 216, 219–20, 328, 572, 577, 612
Depreciation, 48, 56–58, 68, 74, 78, 84, 170, 565
  double-declining balance, 300
Dividends, 14

Earnings, 436
Elasticity of demand, 154–57, 159–62, 170–71, 242, 572, 577, 612
Employee
  firings, 627
  productivity, 139, 144, 148, 241, 335–37, 342–43, 352, 463, 577
  training, 30, 343, 646–47
  wages, 14
Fees, 31
Gross profit, 63

Income streams
  average income per period, 820–22, 826
  discrete, 820–27
  future value, 432–37, 453, 467–68
  present value, 470–74, 823–27
Industrial output, 564
Information transmittal, 378
Interarrival times
  customers, 678
  orders, 661, 683
Internal rate of return, 313–14
Inventory, 62
  average, 412, 414–15, 419, 751, 755
  carrying cost for, 437–38
  control, 676–77
  optimal reorder quantity, 255–56, 259–60, 278
  reorder times, 660
  safety stock, 660
Investment, 301, 303, 313

Learning curves, 328, 335–38, 342–43, 345–46, 352, 378, 387, 400, 415, 577, 612

Machine
  downtime, 647
  tending time, 487

Marginal
  average cost, 159
  cost, 151–52, 157–59, 242, 248, 503–4
  productivity of capital/labor, 502, 505–6
  productivity of money, 522, 523–24, 543
  profit, 151–52, 157–59, 193, 501, 504
  propensity to consume, 222–23
  propensity to save, 224
  revenue, 151–52, 157–59, 183–84, 193, 242, 320, 503–4, 542
  revenue product, 194–95
Market equilibrium, 56–58, 68, 74, 78, 84, 328, 341
Market share, 62, 577
Maximization problems
  audience, 242, 260, 524
  employee productivity, 144, 241
  plant productivity, 139, 343
  prices, 741
  production, 62, 320, 516, 521–24, 543
  profit, 74, 77–79, 84, 138, 152–53, 157–59, 170, 193, 241, 248, 252, 259, 278, 320, 328–29, 515–16, 524, 740
  revenue, 242, 248, 259–60, 277, 516, 542
  sales, 63, 241, 516, 737, 740, 755
Minimization problems
  average cost, 139, 241–42, 248, 259, 277, 516
  cost, 65–66, 260, 278, 524
  distance, 516
  inventory, 255–56, 259–60, 278
  plant productivity, 138
  pollution by utility, 524
  prices, 741
  production time, 343
  profit, 22, 740
  replacement cost, 260–61
  sales, 737, 740, 755
Movies, 634
Multiplier effect, 222–24, 790

Net
  audience, 506
  income, 15
Nuclear power plant, 116, 273, 555, 557

Oil
  drilling, 103, 116, 647
  production, 302, 446, 453
Operating lifetime of product, 650, 652–54, 660, 671–74

Payroll, 755
Plant
  location, 65–66
  productivity, 138, 343
Platinum use, 62
Present value, 301, 303, 313–14, 823–27
Price
  discrimination, 514, 516
  fluctuation, 726–27, 741
Pricing, 48, 83, 103, 131, 260
Producers' surplus, 428–30, 436, 446–47, 467
Product
  success of, 628–29
  testing, 653–54, 656–57, 682
Production, 53, 59, 62, 65, 378, 387, 399, 414, 516, 521–23, 543, 578, 596, 675
  average, 537, 543
  breakdowns, 649, 660
  function, 202, 217, 502, 505–6, 521–24, 537, 543, 596
  process, 14
  time, 328, 337–38, 343, 352, 387, 465, 612
Product use, 6–7
Profit, 3, 17, 22, 26, 29–30, 35, 44–46, 48, 63, 67, 73–74, 77–79, 82, 84–85, 152–53, 157–59, 170, 242, 248, 252, 320, 328–29, 341, 347, 353, 385, 387, 414, 436–37, 453, 463, 465, 472–73, 485–86, 515–16, 524, 541–42, 553, 555, 596, 755
  average, 159, 486
  function, 26, 35, 44–45, 55–56, 67, 74–74, 77–79, 84, 105, 114, 138, 147, 150, 157–59, 193, 202, 217, 241, 252, 259, 270–72, 278, 371, 378, 485–86, 553, 555, 565, 611
  marginal, 151–52, 157–58, 193, 500–501, 504
  per share, 82
  seasonal, 754
Proportional/percentage change, 202–3, 219, 596

Proportionality problems, 565–66, 577–78, 589–90, 611–13
Purchasing equipment, 301, 303

Related rates problems, 212, 216–17
Reorder point, 676–77
Return
  on assets, 59–60
  on equity, 59–60
  on investment, 13, 301, 303, 313, 419, 637
Revenue, 18, 34–35, 44–45, 48, 63, 67, 74, 77–79, 83–84, 159–62, 170–71, 183, 194–95, 242, 248, 259–60, 277, 341, 347, 356–57, 385, 471–73, 479, 485–86, 516, 541–42, 637, 745, 751
  average, 159
  function, 34–35, 55–56, 67, 73–74, 77–79, 84, 148, 150, 157–59, 203, 216–17, 320, 371, 378, 418, 447, 485–86, 552
  marginal, 151–52, 157–58, 183–84, 193, 320, 503–4, 542
  seasonal, 726

Safety stock, 676–77
Sales, 13, 30, 51, 58–59, 62–63, 84, 139, 147–48, 204–5, 217, 241, 273, 285, 300, 302, 313, 320, 325, 341, 347, 355–56, 359, 361, 414, 516, 552, 565–66, 578, 625–28, 637
  average, 538
  calls, 818
  commissions, 630
  seasonal, 719–20, 726, 736–37, 740, 754–55, 759, 761, 769–70
Savings, 382–83, 385–86, 400–402, 437, 751
Service times, 678, 680
Shipping, 114
Software, 103, 131
Steel mills, 67
Stock prices, 13, 566, 727, 741
Stream of income, 432–37, 453, 467–68, 470–74
  discrete, 820–27
Supply, 46, 56–58, 68, 74, 78, 84–85, 328, 341, 372, 415, 487, 505, 596
  function, 56–58, 68, 74, 78, 84–85, 116, 147–48, 199, 202, 216, 219, 577–78

Sustainable growth rate, 204–5
Synchronous service function, 487

Telephone orders, 649
Value of industry's production, 320

Wages, 14, 612
Weight of parts, 674–75
Worker-hours, 118
Yield on government securities, 273

Life Science
AIDS, 303, 438–39
Amino acids, 584–88
Anesthesia, 342
Animal activity, 15
Antibody production, 379
Antivenin shot, 118
Approximation by differentials, 203
Artery, volume of, 148, 203

Bactericide, effects of, 84, 139, 149, 415, 500, 507
Birch trees, 277
Bird mortality, 629, 647
Births, 139
Blood
 pressure, 217
 velocity of, 217–18, 439, 502–3

Cell growth, 64, 360
Chemical reactions, 148
Chemotherapy, 517, 751
Competing species, 617
Concentration of solute in a cell, 372–73
Cost, 107, 116, 274, 727
Crop yield, 261, 344, 477, 506, 517, 524
Curve-sketching problems, 273–74, 280–83, 321, 329, 360–61, 727, 760

Deacidification of lake, 321, 348, 727
Desert, encroachment by, 448
Diet supplements, 84
Disease, cases of, 342, 388, 438–39
Doctor's recommendation, 629

Drug
 concentration, 79, 182, 203, 313, 329, 344, 378, 415, 447, 463, 751, 790–91
 dosage, 8–9, 14, 33, 64, 185
 effectiveness, 358–59, 361, 554–55, 647
 effects of, 148, 681–82
 testing, 554, 629–30
 toxicity, 344

Egg production, 261
Endemic disease, 283
Enzyme production, 464, 466
Epidemic, 303, 329, 360, 386, 418, 438–39, 578–79
Extinction of eagles, 303, 557

Fick's law, 372–73
Fish eggs, 60

Genetics
 eye color, 638, 647
 fitness, 683–88
 hair color, 647–48
 sickle-cell anemia, 648
Greenhouse effect, 61, 352
Growing time, 675, 682–83

Heart disease, 3
HIV, 438–39

Immunization, 274
Infant growth, 63
Intravenous solution, 388

Litter size, 638
Lung cancer, 552–53
Lysine solution, 584–86

Males in family who marry, 638
Maximization problems
 air pollution, 79
 crop yield, 261, 517, 524
 drug concentration, 79, 344
 egg production, 261
 inoculations, 243
 population size, 139, 242, 261

Maximization problems—*Cont.*
 proportion of trees affected by disease, 277
 temperature, 76, 79, 248, 741
 velocity of blood, 261
Mental imagery, 60
Minimization problems
 discomfort, 517
 population size, 79, 139, 248, 261
 temperature, 741
Mixing problems, 584–88

Ozone depletion, 46, 61, 194, 554

Ph.D.'s in life sciences, 66
Plant growth, 386, 566
Poiseuille's law, 217–18, 261, 439, 502–3
Pollen count, 352
Pollution
 air, 46, 48, 79, 107, 116, 148, 194, 217, 250, 329, 344, 348, 379, 447, 453, 538, 566, 596, 722–23
 water, 31–32, 60–61, 203, 217, 352, 386, 466, 566, 638
Population
 density, 538–39
 gain, 388, 401
Population size, 26, 66, 280–83, 365, 573, 604, 612, 758–60
 bacteria, 19, 31, 84, 132, 139, 148, 184, 193–94, 203, 248, 313, 329, 342–44, 352, 370, 372, 378, 415, 506, 566, 775
 birds, 79, 242, 274, 303, 352, 372, 553, 557
 fish, 203, 261, 342, 344, 351, 629
 furbish lousewort, 79
 gypsy moths, 242
 mammals, 43, 63–64, 139, 193, 273–74, 290–91, 331–32, 335, 372, 378, 415, 438, 772–76
Predator-prey relationships, 39, 45, 488, 506–7, 542, 614–17
Proportional/percentage change, 203, 597
Proportionality problems, 566, 578–79, 612
Public health, 217, 243, 274
 bilharzia, 194
 influenza, 329, 360, 386, 418

Radioactive iodine, 320–21, 347
Radon concentration, 661–62
Recovery time, 448, 538, 661, 675
Related rates problems, 217–18

Scar, visibility of, 149
Signing by chimpanzee, 185
Snake bite, 118
Stimulus, smallest, 60
Survival-renewal functions, 438
Survival time, 629, 661, 678

Temperature
  animal, 15, 578
  human, 32, 75–76, 79, 248, 358–59, 727, 741

Vision, 60–61
Visual
  contrast, 488, 596–97
  processing, 60–61
Vitamins, 55, 66

Weber's law, 60
Weed growth, 149

Social Science
Approximation by differentials, 203
Aptitude test, 64

Birth rate, 553, 580
Bureaucracy, size of, 289–90, 321, 329, 348

Campaign
  spending, 507
  strategy, 639
Charity contributions, 80
Cost, 274
Course evaluation, 26–27
Crime rate, 249, 469
Curve-sketching problems, 274, 278, 321, 330, 359–61

Death rate, 580
Debt service, 185
Decision-making using expected value, 639

Education, 330, 348, 630
Emigration, 402

Fads, 33, 361, 662
Foodbanks, 556
Food stamp program, 553

High school
  dropout rate, 64
  enrollment, 227
Housing starts, 274, 488–89, 597

Immigration, 402
Income distribution, 402–3

Language
  comprehension, 149
  retention, 203, 389
Learning, 448, 464, 612
Likert scales, 26–27
Literacy rate, 64
Lorenz curves, 402–3

Maximization problems
  charity contributions, 80
  cohort of 18–25 year olds, 249
  school enrollment, 227, 243
  voter approval, 80
  voter interest, 80
  votes, 243
Memory, 118–19, 149, 203
Minimization problems
  cohort of 18–25 year olds, 249
  crime rate, 249
  school enrollment, 227, 243
  voter approval, 73, 140
Municipal budget, 185

Political polls, 15–16, 648
Population
  changes, 381–82, 402, 418
  density, 440
  size, 3, 249
Prison overcrowding, 440
Proportional/percentage change, 203, 597
Proportionality problems, 566–67, 579, 612
Psychological testing, 566

Psychology experiments, 185, 342, 353, 638–39

Related rates problem, 218
Religious cult, 439–40
Rumors, 203, 329, 360, 574–75, 579, 612

School enrollment, 227, 243, 274
Social
  services, 321, 351–52
  workers, 218
Stimulus-response, 567
Survey, 648

Tax burden, 372
Television viewing, 554
Tests, 26, 64, 66
Transition rates, 579–80
Transit system ridership, 48, 419, 507, 539

Unemployment, 321, 386
  rate, 140, 149, 274–75

Voter
  affiliation, 61, 66, 344, 372, 579–80, 630
  approval ratings, 73, 80, 139–40, 278
  interest, 80
  participation, 64, 194, 554, 567
  preference, 15–16
  support for candidates, 150, 243, 389

Weber-Fechner law, 567
Welfare, 149, 507, 539

Other
Acceleration problems, 144–47, 170
Altitude, 47, 77, 142–44, 146–47, 170, 611
Angle of observation, 735–36, 739
Area, 391–99, 401, 418–19, 461, 465, 469, 534, 537, 543, 744, 749–50, 755

Applications Index

Arrivals
  airplanes, 682
  autos, 644, 648–49
  people, 644–45
  telephone calls, 649
Automatic pilot, 9
Average daily yield, 314

Baseball, 215
Basketball, 676
Bonds, 140, 314
Bouncing ball, 789
Boyle's law, 215

Chemical reaction, 82
Class absences, 548–49
Coin-tossing, 620–23, 626, 630–31, 633, 635, 639, 641–42, 645, 676
Cold fusion, 278
Compound interest, 292, 298–99, 302, 309–10, 312–13, 328, 350–51, 570–72, 584, 586–87, 612, 775, 817
Computer science, 27, 119, 275, 344–45
Cost, 128, 131
Cruise control, 14
Curve-sketching problems, 272, 278, 283, 321, 359–61, 740, 760–61

Decibel scale, 350–51
Decision-making using expected value, 637
Defective light bulbs, 636–37
Dice, 626–27, 636
Dimension problems, 47, 515
Distance problems, 523, 565, 789
Drawing
  cards from a deck, 627, 636, 681
  marbles from a vase, 636
  numbers, 627, 680–81

Earthquakes, 311, 321
Electric current, 737–39, 750
Energy efficiency, 360
Exam
  answers, 642–43, 645–46
  grading, 674

Family income and savings, 545, 547
Family size, income, and savings, 551

Gold production, 553
Golf, 646
Grade in course, 681
Grades and absences, 548–49
Gross national product, 418

Height of submariners, 668–70
Hurricanes, 386

Inflation, 272, 312, 340
Investments, 65, 292, 298–99, 301–3, 309–10, 312–13, 328, 340, 350–51, 418, 584, 586–87, 612, 775, 817
  optimum holding period for, 330

Job offers, 636–37
Junk bonds, 419

Lottery, 636

Manufacturing capacity, 338–39, 347, 359–61
Maximization problems
  altitude, 77, 143, 146, 170
  area, 257–58
  displacement of spring, 740
  electric current, 738
  return on investment, 330
  volume, 253–55, 258, 278, 515, 523
Measurement errors
  area, 200–202
  volume, 201–2, 219–20
Metal ion removal, 576
Minimization problems
  cost, 257–58, 513–15, 542
  distance, 523
  electric current, 738
  perimeter, 252–53, 258
  sunspots, 77
  surface area, 259, 515
Mixing problems, 587

Newton's Law of Cooling, 576
Nuclear
  strategy, 345
  testing, 321, 348
Numbers, 630

Operating lifetimes, 674, 677–78

Patent applications, 283
Pendulum, 757–58, 760
Position problems, 365
Postal function, 128
Present value, 293, 298–303, 313–14, 350
Pressure, 14, 82
Proportional/percentage change, 200–202, 219–20, 595
Proportionality problems, 565, 576, 611

Radioactivity, 294–96, 299, 301–2, 309, 312, 328, 332–33, 340–41, 351, 576
Radiocarbon dating, 340–41
Raffle, 636
Related rates problems, 210–16, 220, 735–36, 739
Rental income, 473–74
Richter scale, 311, 321
Roulette, 623–24, 631, 633–34, 636

Savings, 360, 545, 547, 551
Simple interest, 768, 775
Springs, 739–40, 761
Stefan's law, 147, 371
Stockbroker's commissions, 62, 485
Stocks, 474
Storks and births, 553
Sunspots, 77
Superconductivity, 241

Telephone calls, 131
Temperature, 61, 82, 241

VCR rental, 131
Velocity, 140–47, 170, 371, 565
Volume, 530–31, 537, 543
  of solid of revolution, 468

Windmills, 352

Subject Index

Absolute maximum, minimum, 249–56
Absolute value, 9
  function, 21
Acceleration, 144
Addition
  of polynomial functions, 34–36
  of power series, 802
  of rational functions, 36–37
Amplitude, 716
Angle(s), 692–98, 700–11
  between curves, 755
  conversion formula, 696
  degree measure of, 692–93, 696–97
  radian measure of, 693–98
  sides of, 692
  trigonometric functions of, 700–11
  vertex of, 692
Antiderivative, 365–67
Antidifferentiation, 365–66
Approximation
  of definite integrals, 453–60, 470
  by differentials, 196–200, 589–90
  using Taylor series, 810–13
Archimedes, 421
Arc length, 470
Area
  bounded by two curves, 394–99
  and double integrals, 533–34
  function, 389
  under a curve, 388–93
Arithmetic of functions, 34–37
Arithmetic mean, 684
Asymptote
  horizontal, 108
  oblique, 111
  vertical, 106–7
Average daily yield, 314
Average value of a function
  of one variable, 410–13
  of two variables, 534–35
Axes, 19

Barrow, Isaac, 424
Base, 9

Bernoulli
  equation, 613
  process, 639
Beta distribution, 683
Binomial distribution, 639–43
  normal approximation to, 675–76
Bond duration, 140
Break-even
  analysis, 55–56
  point, 55
  quantity, 55

Calculus
  development of, 172–75
  fundamental theorem of, 410, 424–25
Capital
  marginal productivity of, 502
  value, 471–72
Cardano, Girolamo, 86
Cartesian
  coordinate system, 19, 481
  product, 85
Cauchy's mean value theorem, 279
Chain rule, 185–91, 543
  first formulation, 186
  second formulation, 190
Change in f, 196, 589
Change-of-base formulas, 307–8
Chi-square distribution, 683
Closed ecological system, 614
Cobb-Douglas production function, 502
Coefficient
  of inequality, 403
  of a polynomial, 36
Common
  factor, 42
  logarithm, 307
Competitive products, 501
Complementary
  event, 622
  products, 501
Complex numbers, 86
Composite function, 38

Compound interest, 291–93, 296–98
  continuous, 296–98
  future value formulas, 292, 297
  and present value, 293, 298
Concavity, 244
Constant
  function, 53
  of integration, 367
  multiple rule, 135
  of proportionality, 562
  rule, 133
  solution of differential equation, 570
Constraint equation, 517
Consumers' surplus, 430–31
Consumption, 221–23
  function, 221
Continuity, 125–29
  and differentiability, 128–29
  limit definitions of, 127
Continuous
  compounding, 297
  function, 125
  random variable, 649
  stream of income, 432–35, 470–73
Convergence tests, 783–88, 819
  comparison, 819
  integral, 785
  ratio, 786
  root, 819
Convergent
  improper integral, 450
  infinite series, 779
  power series, 793
  sequence, 765
Converting
  normal to standard normal, 668
  degrees to radians, 696
Coordinate planes, 482
Coordinates
  cartesian, 19, 481
  cylindrical, 544
  polar, 85, 544
  spherical, 544
Cosecant, 708

Cosecant function, 723
   derivative of, 735
   integral of, 746
Cosine, 700, 702–3, 707
Cosine function, 720
   derivative of, 731
   integral of, 741
Cost
   fixed, 55
   function, 34, 55, 150
   of goods sold, 63
   marginal, 151
   variable, 55
Cotangent, 708
Cotangent function, 724
   derivative of, 735
   integral of, 746
Critical value, 230
Cubic formula, 86
Cumulative distribution function, 655
Curve sketching, 261–71
   procedure, 263

Decibel scale, 350
Decimals, 45, 85
Decline, exponential, 330–33
Decreasing function, 231
Definite integral, 365
   as area, 389–99
   development of, 420–25
   as limit of Riemann sums, 407–11
   as total change in antiderivative, 379–83
Degree measure, 692–93
Demand, 55–68, 74, 154–57
   elastic, 155
   function, 56
   inelastic, 155
Dependent variable, 18
Derivative, 89, 119–24
   definition of, 121
   higher, 164–65
   notation for, 163–64
   partial, 489–94
   as rate of change, 140–45
   rules, 132–37, 178–82, 185–91, 315–18, 322, 326, 730–31, 735
   second, 164
   sign of, 231
   and three-step rule, 123
Descartes, Rene, 85, 423
Difference quotient, 122
Differentiability, 124–25
   and continuity, 128–29

Differentiable function, 124
Differential equation(s), 280–82, 560–64, 756–61
   Bernoulli, 613
   exact, 593–94
   first order, 560
   graphical solution of, 598–603
   homogeneous, 613
   with initial condition, 561
   linear, 580–86, 614
   logistic, 573
   numerical solution of, 604–9
   qualitative solution of, 281
   separable, 567–75
   solution of, 280, 560–61
   system of, 616
Differential form, 590–93
   exact, 590
Differential notation, 163–64
Differentials, 173, 195–200, 588–90
   approximation by, 196–200, 589–90
   total, 588
Differentiation
   of exponential functions, 322–27
   implicit, 205–8
   logarithmic, 318–19
   of logarithmic functions, 314–18
   of power series, 795
   rules of, 132–37, 178–82, 185–91, 315–18, 322, 326, 730–31, 735
   of trigonometric functions, 727–38
Directional derivative, 543
Directly proportional, 562
Discontinuous function, 125
Discounting, 293
Discrete income stream, 820–26
Discriminant, 70
Disk method, 470
Distance function, 141
Divergence test, 819
Divergent
   improper integral, 450
   sequence, 765
Domain, 17
Double
   inequality, 7
   integrals, 525–35

e, 293
Elasticity of demand, 154–57
   formula, 155
   and revenue, 156
Electric current, 738

Element, 4
Elementary function, 763
Equation(s)
   constraint, 517
   defining, 16
   exponential, 308
   of a line, 51
   parametric, 221
   polynomial, 43
   quadratic, 70
Euler, Leonhard, 353
Euler's method, 604–6
Event, 620
   complementary, 622
Exact differential
   equation, 593
   form, 590
Exhaustion, method of, 421–22
Expected value, 633–34, 656–57
Explicit function, 205
Exponential
   distribution, 671–72
   equations, 308–10
   function, 285–87
   growth, decline models, 330–38, 354–59
Exponential functions, 285–98
   with base b, 287
   with base e, 294
   differentiation of, 322–27
   properties of, 289
Exponentials, 286–93
   change-of-base formula, 308
Exponents, 9–12
   integral, 9
   rational, 10
   rules of, 11
Extended mean value theorem, 279

Factorials, 639–40, 766
Factoring polynomials, 42–43
Fair game, 634
Fermat, Pierre, 683
Fibonacci sequence, 772
First
   derivative test, 234
   partial derivatives, 489–92
First order differential equation, 560
Fixed cost, 55
Fluents, 172
Fluxions, 172
Fontana, Nicolo, 86

Subject Index

Formula
  cubic, 86
  quadratic, 70
  quartic, 86
Function(s), 16–24, 34–40
  absolute value, 21
  antiderivative of, 365–66
  area, 389
  arithmetic of, 34–35
  average value of, 411, 535
  and cartesian products, 85
  closed form of, 792
  Cobb-Douglas production, 502
  composite, 38
  composition of, 38–40
  concavity of, 244
  constant, 53
  consumption, 221
  continuous, 125
  cosecant, 723
  cosine, 720
  cost, 34, 150
  cotangent, 724
  cumulative distribution, 654
  decreasing, 231
  defining equation for, 16
  demand, 56
  dependent variable of, 18
  derivative of, 121
  differentiable, 124
  discontinuous, 125
  distance, 141
  domain of, 17
  elementary, 763
  explicit, 205
  exponential, 285, 287, 294
  Gompertz, 356
  graph of, 19
  greatest integer, 28
  implicit, 205
  increasing, 231
  independent variable of, 16
  intercepts of, 20
  inverse, 41, 303
  linear, 48
  linear approximation to, 279
  logarithmic, 285, 304, 307
  logistic, 354, 574
  marginal, 151, 159
  natural logarithm, 307
  nondifferentiable, 124
  one-to-one, 40
  polynomial, 36
  power, 133

Function(s)—Cont.
  power series expansion of, 792
  probability density, 650
  profit, 34, 150
  quadratic, 68
  range of, 18
  rational, 36
  renewal, 438
  revenue, 34, 150
  of several variables, 477
  secant, 724
  sigmoid, 357
  sine, 716
  supply, 56
  survival, 438
  synchronous service, 487
  tangent, 721
  vertical line test for, 24
Functional
  notation, 16
  value, 17
Fundamental theorem
  of calculus, 410, 424–25
  of natural selection, 687
Future value, 292
  of a stream of income, 432–35

Gamma distribution, 683
General
  form of equation of a line, 53
  power rule, 189
  solution of a differential equation, 560
Geometric
  sequence, 768–70
  series, 780–83
Gompertz function, 356
Gradient, 543
Graph, 19
Graphical solution of differential equations, 598–603
Graphing, 18–24, 49–51, 69–72, 261–71
  of functions, 18–24, 261–71
  of linear functions, 49–51
  procedure, 20
  of quadratic functions, 69–72
  of relations, 23–24
Greatest integer function, 28
Gross profit, 63
Growth
  exponential, 330–33
  logarithmic, 338–39
Growth rate of functions, 353

Half-life, 295
Harmonic series, 786
Histogram, 631
Homogeneous differential equation, 613

Implicit
  differentiation, 205–8
  function, 205
Income, stream of, 432–35, 470–73, 820–26
Increasing function, 231
Increment, 195
Inequalities, 6–7
Infinite limits, 106–7
Infinite series, 776–88
  convergence of, 779
  geometric, 780–83
  sum of, 779
Infinity symbols, 8, 106
Inflection point, 262
Initial condition, 560
Insanity, 345
Integers, 4
Integral
  as accumulator, 410, 428, 435
  and consumers' surplus, 430–31
  curve, 598
  definite, 365, 379–84
  double, 525–35
  improper, 448–52
  indefinite, 365, 367–70, 373–77
  iterated, 526–28
  and producers' surplus, 428–30
  sign, 367
  and streams of income, 432–35, 470–73
  test, 785
Integral exponents, 9
Integrand, 367
Integrating factor, 581, 614
Integration
  constant of, 367
  by guessing, 373–74
  limits of, 380
  numerical, 453–60
  by parts, 440–44
  of power series, 798
  and Riemann sums, 405–10
  rules of, 368
  by substitution, 374–77, 383–84
  using tables of integrals, 444–46
  of trigonometric functions, 741–49
Intercepts, 20, 49, 70–71

Interest
  compound, 291–93, 297–98
  simple, 768
Internal rate of return, 313–14
Intersection of lines, 54–58
Intervals, 7–9
Inventory
  control, 676–77
  problems, 355–56
Inverse function, 40–42, 303
  rule, 220
Inversely proportional, 562
Irrational numbers, 4
Isocline, 599
Iterated integral, 526–28

Keys to
  integration by parts, 442
  integration by substitution, 376
  related rates problems, 211

Labor, marginal productivity of, 502
Lagrange, Joseph-Louis, 544
Lagrange multiplier, 518
  method, 517–22
Lead time, 676
Learning curves, 336–38
Least-squares line, 546
Leibniz, Gottfried Wilhelm, 171–74, 424
Leverage, 59
L'Hopital's rule, 279–80
Life cycle curve, 354
Likert scale, 26–27
Limit(s)
  and continuity, 127
  of a function, 89–94, 171
  infinite, 106–7
  at infinity, 108–12
  of integration, 380
  one-sided, 104–6
  of polynomial functions, 96
  properties of, 94–98, 171
  of rational functions, 96–98
  of Riemann sums, 407–8
  of a sequence, 764–65, 819
Limited growth, decline, 333–38
Line(s)
  finding equation of, 51–54
  graphing, 49–51
  intersection of, 54–58
  parallel, 54, 86
  perpendicular, 86

Line(s)—Cont.
  secant, 119
  slope of, 50, 52
  tangent, 119, 497
  vertical, 50–51
Linear
  approximation, 279
  differential equation, 580, 614
  regression, 544–49
Linear functions, 48–58
  cost, revenue, profit models, 55–56
  finding equations of, 51–53
  graphing, 49–50
  intersections of, 54–55
  slope of, 50, 52
  supply, demand models, 56–58
Line diagram, 7
Logarithmic
  differentiation, 318–19
  function, 285, 304–5, 314–18
  growth, 338–39
Logarithms, 303–10
  change-of-base formula, 308
  common, 307
  and exponential equations, 308–10
  natural, 307
  properties of, 306
Logistic
  equation, 573–74
  function, 354–56, 574
Lorenz curve, 402–3
Lotka-Volterra equations, 615

Maclaurin series, 807
Marginal
  average cost, 159
  cost, 151
  productivity of capital, labor, 502
  productivity of money, 522
  profit, 151
  propensity to consume, 222
  propensity to save, 224
  revenue, 151
  revenue product, 194–95
Market equilibrium, 57–58, 74
Maximum
  absolute, 249
  relative, 228, 507–8
Mean
  of a data set, 684
  of a random variable, 633
Mean value theorem
  for derivatives, 279
  for integrals, 420

Midpoint rule, 454–56
Minimum
  absolute, 249
  relative, 228, 507–8
Mixed partials, 493
Model(s)
  cost, revenue, profit, 55–56, 73–74
  linear, 55–58
  quadratic, 72–76
  predator-prey, 614–16
  supply and demand, 56–58, 74
Multinomial distribution, 683
Multiplication
  of polynomials, 35
  of rational functions, 36–37
Multiplier, 222
  effect, 221, 790
  Lagrange, 518

Napier, John, 353
Natural
  logarithm, 307
  numbers, 4
  rate of increase, 580
Negative exponents, 11
Net present value, 314
Newton, Isaac, 171–74, 424
Newton's method, 171
Nominal interest rate, 291
Nondifferentiable function, 124
Normal
  approximation to the binomial, 675–76
  curve, 663
  density function, 663
  distribution, 662–73
  random variable, 663
Notation
  functional, 16
  interval, 8
  set-builder, 4
nth root, 10
Number systems, 4–5, 86
Numerical integration, 453–60
  errors of, 470
Numerical solution of differential equations, 604–9
  Euler's method, 605
  Runge-Kutta method, 607

Octant, 481
One-sided limit, 104
One-to-one function, 40

Optimal reorder quantity, 255
Origin, 19, 481
Orthogonal trajectory, 614
Outcome, 620

Parabola, 68
Parallel lines, 54
Partial
  fractions, 470
  sum, 777
Partial derivatives, 489–94, 496–503
  definition of, 492
  first, 489–92
  geometric significance of, 497–99
  higher, 492–94
  mixed, 493
  as rates of change, 500–503
  second, 492
Particular solution of a differential equation, 561
Parts, integration by, 440–44
Pascal, Blaise, 683
Pendulum, 757–58
Percentage-of-sales, 63
Perfect elasticity, inelasticity, 162
Period, 716
Perpendicular lines, 86
Plotting points, 19, 481
Point of inflection, 262
Poisson
  distribution, 643–45
  process, 643
Polar coordinates, 85, 544
Polynomial(s), 34–36, 42–43, 85–86
  arithmetic of, 34–36, 85–86
  coefficients of, 36
  factoring of, 42–43
  function, 36, 107, 110
Positive integers, 4
Power, 9
  function, 133
  rule, 134
Power series, 763, 791–803
  algebra of, 802–3
  convergence of, 793–94
  differentiation of, 974–96
  expansion, 792
  integration of, 798–800
  and substitution, 800–801
Predator-prey model, 614–16
Present value, 293, 298, 470–73, 823–26
  and continuous compounding, 298

Present value—Cont.
  formulas, 293, 298, 471, 472
  net, 314
  of streams of income, 470–73, 823–26
Price discrimination, 514
Prime notation, 163
Principal, 291
Probability, 619, 620–26
  and cumulative distribution function, 655–56
  and density function, 652–54
  histogram, 631
  rules of, 623
Probability density function, 650
  exponential, 671
  normal, 663
Probability distribution, 631
  beta, 683
  binomial, 641–43
  chi-square, 683
  exponential, 671–73
  gamma, 683
  normal, 662–70
  poisson, 643–45
  student t, 683
  uniform, 683
Producers' surplus, 428–30
Product rule, 178–79
Profit
  function, 34, 150
  marginal, 151
  maximization, 152–54
Properties
  of antiderivatives, 367
  of the cumulative distribution function, 655
  of the definite integral, 381
  of exponential functions, 289
  of exponents, 11
  of limits, 94
  of limits of sequences, 770
  of logarithms, 306
Proportional change in f, 199–200, 589–90
Proportionality, 562

Quadratic
  formula, 70
  functions, 68–72
  models, 72–76
Quartic formula, 86
Quotient rule, 179–82

Radian measure, 693–98
Radiocarbon dating, 340
Radius of convergence, 793
Random variable, 630–35
  binomial, 641
  continuous, 649–57
  and cumulative distribution function, 654
  discrete, 649
  expected value of, 630, 633–34, 656–57
  exponential, 671
  normal, 663
  Poisson, 644
  and probability density function, 651–52
  standard deviation of, 634–35, 651–52
  standard normal, 664
  variance of, 634–35, 651–52
Range, 18
Rates of change, 140–45
  and partial derivatives, 500–503
  and related rates, 210–14
Ratio
  of geometric series, 780
  test, 786
Rational
  exponents, 10
  functions, 36–37, 107, 110–11
  numbers, 4
Real
  line, 5–6
  numbers, 4–5
Regression, 544–51
  linear, 544–49
  multiple, 550–51
  quadratic, 549–50
Related rates, 210–14
Relative maxima, minima, 228–39, 243–47, 507–14
Remainder, Taylor series, 810
Reorder point, 676
Return
  on assets, 59
  on equity, 59
Revenue
  and elasticity of demand, 156
  function, 34, 55, 150
  marginal, 151
Riemann, Bernhard, 420
Riemann sum, 403, 405, 525
Rolle's theorem, 279
Root test, 819

Rule(s)
    of differentiation, 133–35, 178, 180, 186, 189–90, 220, 315–18, 322, 326
    of exponents, 11
    of integration, 368
    L'Hopital's, 279–80
    midpoint, 455
    of probability, 623
    of rational arithmetic, 37
    Simpson's, 460
    three-step, 123
    trapezoidal, 457
Runge-Kutta method, 606–9

Saddle point, 507–8
Safety stock, 676
Scatter diagram, 545
S-curves, 354–59
Secant, 708
Secant function, 724
    derivative of, 734–35
    integral of, 746
Secant line, 119
Second derivative, 164
    test, 245, 510
Second partials, 492–94
Separable differential equation, 568
Separation of variables, 567–75
Sequence(s), 764–74
    convergent, 764–65
    divergent, 765, 767
    Fibonacci, 772–73
    geometric, 768–70
    limit of, 764–65, 819
    properties of limits of, 770–72
    term of, 764
Series
    harmonic, 786
    infinite, 777
    p-, 789
    power, 791
    Taylor, 807
Set(s), 4–5
    cartesian product of, 85
Shell method, 470
Sigmoid function, 357–58
Simple interest, 768
Simpson's rule, 459–60

Sine, 700, 702–3, 707
Sine function, 716
    derivative of, 730
    integral of, 741
Slope of a line, 50, 52
Solid of revolution, 468, 470
Solution of a differential equation, 280, 560–61
Square root, 10
Standard deviation, 634–35, 656–57
Standard normal distribution, 664–68
Steady-state current, 738
Stockout, 676
Stream of income, 432–35, 470–73
    capital value of, 471
Student t distribution, 683
Substitution, integration by, 374–77, 383–84
Subtraction
    of polynomials, 34–35
    of rational functions, 37
Sum
    rule, 135–36
    of a series, 779
Summation notation, 404–5
Supply, 56–58, 74
    function, 56
Surface of revolution, 470
Sustainable growth rate, 204–5
Synchronous service function, 487

Tangent, 708
Tangent function, 721
    derivative of, 734–35
    integral of, 745–46
Tangent line, 119, 497
Taylor series, 805–14, 819–20
    and approximation, 810–13
    convergence of, 807–9
    remainder formula, 810
Techniques of integration
    by guessing, 373–74
    by parts, 440–44
    by substitution, 374–77
    using tables, 444–46
Theory of the firm, 150–57
Three-step rule, 123
Total differential, 543, 588–90
    approximation by, 589–90

Trajectory, orthogonal, 614
Transient current, 738
Transition rates, 579–80
Trapezoidal rule, 456–59
Trigonometric functions, 716–24
    differentiation of, 727–38
    integration of, 741–49
    and right triangles, 709–11
Trigonometric identities, 703–5
Triple integral, 544

Uniform distribution, 683
Variable
    cost per unit, 55
    dependent, 18
    independent, 16
Variance, 634–35, 656–57
    of a data set, 685
Vector, 543
Velocity, 141–42
Vertex
    of an angle, 692
    of a parabola, 68–69, 71
Vertical line, 24, 50–51
    test, 24
Volume, 468, 470, 525–26, 529–33

Weber-Fechner law, 567
Work, 470

x-axis, 19
x-coordinate, 19
x-intercept, 20
xy-plane, xz-plane, 482

y-axis, 19
y-coordinate, 19
y-intercept, 20
Yield
    curve, 273
    to maturity, 140
yz-plane, 482

Zeno's paradox, 776
Zero product law, 43

Table of Integrals

1. $\int u^n \, du = \dfrac{u^{n+1}}{n+1} + c, \quad n \neq -1$

2. $\int \dfrac{1}{u} \, du = \ln |u| + c$

3. $\int e^{au} \, du = \dfrac{1}{a} e^{au} + c$

4. $\int b^{au} \, du = \dfrac{1}{a \ln b} b^{au} + c$

5. $\int \ln au \, du = u(\ln au - 1) + c$

6. $\int \log_b au \, du = u \left(\log_b au - \dfrac{1}{\ln b} \right) + c$

7. $\int \sqrt{u^2 + a^2} \, du = \dfrac{1}{2} [u \sqrt{u^2 + a^2} + a^2 \ln |u + \sqrt{u^2 + a^2}|] + c$

8. $\int \sqrt{u^2 - a^2} \, du = \dfrac{1}{2} [u \sqrt{u^2 - a^2} - a^2 \ln |u + \sqrt{u^2 - a^2}|] + c$

9. $\int \dfrac{1}{\sqrt{u^2 + a^2}} \, du = \ln |u + \sqrt{u^2 + a^2}| + c$

10. $\int \dfrac{1}{\sqrt{u^2 - a^2}} \, du = \ln |u + \sqrt{u^2 - a^2}| + c$

11. $\int \dfrac{1}{u^2 - a^2} \, du = \dfrac{1}{2a} \ln \left| \dfrac{u - a}{u + a} \right| + c$

12. $\int \dfrac{1}{u \sqrt{a^2 + u^2}} \, du = -\dfrac{1}{a} \ln \left| \dfrac{a + \sqrt{a^2 + u^2}}{u} \right| + c$

13. $\int \dfrac{1}{u \sqrt{a^2 - u^2}} \, du = -\dfrac{1}{a} \ln \left| \dfrac{a + \sqrt{a^2 - u^2}}{u} \right| + c$